We wish to thank the members of our families for their continuing support, suggestions, tolerance, and their humoring of our fluctuating moods as this multi-faceted text, which combines theoretical concepts, computer usage, and experimental verification, evolved from an idea to reality.

The Enns Family	The McGuire Family
Karen	Lynda
Russell	Colleen
Jennifer	Sheelo
Heather	Michael
Nicole	Serag
Robert	Kevin
Justine	Ruth
Gabrielle	

Richard H. Enns
George C. McGuire

Nonlinear Physics
with *Mathematica*
for Scientists and Engineers

Springer Science+Business Media, LLC

Richard H. Enns
Department of Physics
Simon Fraser University
Burnaby, BC V5A 1S6
Canada

George C. McGuire
Department of Physics
University College of the Fraser Valley
Abbotsford, BC V2S 7M9
Canada

Library of Congress Cataloging-in-Publication Data

Enns, Richard H.
   Nonlinear physics with Mathematica for scientists and engineers / Richard H. Enns and
George C. McGuire.
      p. cm.
   Includes bibliographical references and index.
   ISBN 978-0-8176-4223-5   ISBN 978-1-4612-0211-0 (eBook)
   DOI 10.1007/978-1-4612-0211-0
      1. Nonlinear theories–Data processing. 2. Mathematical physics–Data processing. 3.
Mathematica (Computer file) I. McGuire, George, 1940- II. Title.

QC20.7.N6.E57 2001
530.15–dc21   2001035590
                                                                   CIP

AMS Subject Classifications: 00A79, 70Kxx

Printed on acid-free paper
© 2001 Springer Science+Business Media New York
Originally published by Birkhäuser Boston in 2001

Additional material to this book can be downloaded from http://extras.springer.com.

ISBN 978-0-8176-4223-5   SPIN 10794415

Typeset by the authors
Cover design by Jeff Cosloy, Newton, MA

9 8 7 6 5 4 3 2 1

Contents

Preface

Philosophy of the Text

This text presents an introductory survey of the basic concepts and applied mathematical methods of nonlinear science as well as an introduction to some simple related nonlinear experimental activities. Students in engineering, physics, chemistry, mathematics, computing science, and biology should be able to successfully use this book. In an effort to provide the reader with a cutting edge approach to one of the most dynamic, often subtle, complex, and still rapidly evolving, areas of modern research—nonlinear physics—we have made extensive use of the symbolic, numeric, and plotting capabilities of a powerful computer algebra software system applied to examples from these disciplines.

Currently, the two dominant computer algebra or symbolic computation software systems are Mathematica and Maple. In an effort to introduce nonlinear physics to as wide an audience as possible, we have created two different versions of this text, an earlier edition making use of Maple having already been published[1]. This edition is based on Mathematica 4.1, the current Mathematica release at the time of writing. Since the two software systems have different strengths and the subject of nonlinear physics and the interests of the authors continues to evolve, this Mathematica version is not simply a verbatim translation of the earlier Maple text into Mathematica. For example, in this text we have introduced new nonlinear computer algebra files which make use of Mathematica's extensive programming, graphics, and sound production capabilities and have included three additional experimental activities.

No prior knowledge of Mathematica or programming is assumed, the reader being gently introduced to Mathematica as an auxiliary tool as the concepts of nonlinear science are developed. Just as the Maple version was not intended to teach you everything you would like to know about programming in Maple, this text will not begin to cover the vast number of commands and options that are available in Mathematica. The CD-ROM provided with this book gives a wide variety of illustrative nonlinear examples solved with Mathematica, the command structures being introduced on a need to know basis. In addition, numerous annotated examples are sprinkled throughout the text and also placed on the CD. An accompanying set of experimental activities keyed to the theory developed in Part I of the book is given in Part II. These activities allow the student the option of "hands on" experience in exploring nonlinear phenomena in the REAL world. Although the experiments are easy to perform, they give

---

[1] *Nonlinear Physics with Maple for Scientists and Engineers*, Birkhäuser, Boston

rise to experimental and theoretical complexities which are not to be underestimated.

The Level of the Text

The essential prerequisites for the first nine chapters of the theory portion of this text would normally be one semester of ordinary differential equations and an intermediate course in classical mechanics. The last three chapters of Part I are mathematically more sophisticated and assume that the student has some familiarity with partial derivatives, has encountered the wave, diffusion, and Schrödinger equations, and knows something about their solutions.

Most of the experimental activities in Part II may be approached on three levels:

- simplest—for non-physicists—investigate the features of the nonlinear phenomena with the minimum of data gathering and analysis,

- moderate—for physics majors and engineers—more emphasis on data gathering and analysis,

- complex—for experimentalists—deeper and more profound analysis required with modifications suggested to stimulate ideas for research projects.

The material in this text has been successfully used to introduce nonlinear physics to students ranging from the junior year to first year graduate level. The book is designed to permit the instructor to pick and choose topics according to the level and background of the students as well as to their inclination towards theory or experiment.

Suggestions to the Student

We suggest that you do not just passively read the material. The book is dynamic in that it asks you to actively participate. This is obvious if you choose to do some of the Experimental Activities in Part II, but it is also true in progressing through Part I. In the theory part, this means carefully studying the worked-out Mathematica examples appearing in the text, running the additional Mathematica files that appear on the CD-ROM, and doing the associated problems as they are encountered. If this technique is followed, it will provide you with a more profound and broader understanding of the material. The Mathematica code in the files and in the text can be used to produce all the text's plots. The code can provide you with help when you do the problems, and more importantly it allows you to explore and investigate the frontiers of nonlinear science. Since we do not presuppose any knowledge of Mathematica or computer programming skills, it is essential that the text examples be studied and the Mathematica files used. In this way you will acquire the Mathematica programming skills and the confidence to use Mathematica as an auxiliary tool to help you understand the concepts and to do the problems.

Suggestions to the Instructor

This book, with its three-pronged approach of developing the necessary nonlinear theory, introducing the reader to the Mathematica computer algebra system, and providing detailed writeups of associated experimental activities, can be effectively used in a variety of ways. Here are several possibilities:

1. A one-or two-semester course where the instructor explains the underlying concepts in short "talk and chalk" sessions that are interspersed with longer interludes during which each student at his or her own computer terminal runs the given Mathematica files, creates new files, solves problems, and explores nonlinear systems. During this time the instructor is free to visit each station and provide individual help and/or answer questions. The sage has abandoned the stage! In this approach, if desired, many of the Experimental Activities of Part II are short and simple enough to be done as demonstrations or assigned as take-home projects.

2. A mainly experimental approach that requires the students to concentrate on doing the Experimental Activities referenced in the theory portion of the text and covered in detail in Part II. In this approach, Part I is used to deepen the understanding of the theoretical concepts underlying the experiments. Engineers might find this method of learning more satisfying and appealing. It is essential that even in this approach the students have easy access to computer terminals containing Mathematica so that they can check their experimental results and run their files.

3. A combination of the two approaches listed above for universities that require that all physics courses have a lab component.

4. Finally, the text could be effectively used in a "distance learning" or in a self-taught mode where financial and personal constraints make one or more of these modes mandatory.

Whatever approach to covering the material in this text is used, we hope that you will enjoy this introduction to one of the most exciting topics of contemporary science. Best wishes from the authors on your excursion into the wonderful world of nonlinear physics.

Richard H. Enns and George C. McGuire

Nonlinear Physics
with *Mathematica*
for Scientists and Engineers

Part I

THEORY

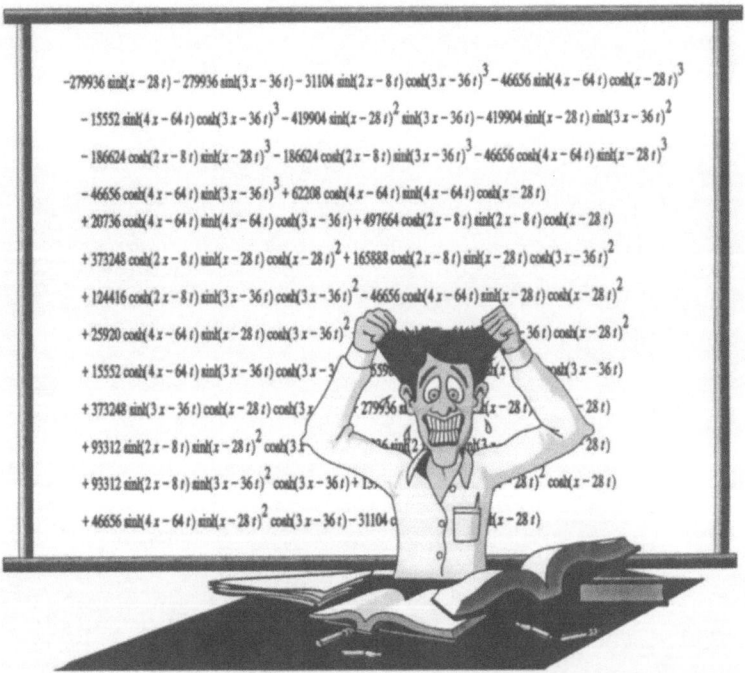

I would be less frustrated if my Prof would allow me to use Mathematica to solve this problem.

Chapter 1

Introduction

1.1 It's a Nonlinear World

In this text on nonlinear physics, we are primarily interested in the problem of
how to deal with physical phenomena described by nonlinear ordinary or partial
differential equations (ODEs or PDEs), i.e., by equations which are nonlinear
functions of the dependent variables. For the familiar simple pendulum (Fig-

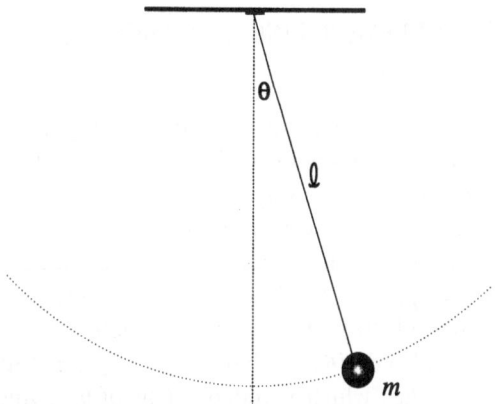

Figure 1.1: Simple plane pendulum.

ure 1.1) of classical mechanics, a mass m attached to a rigid massless rod with
a length ℓ, the relevant equation of motion is

$$\ddot{\theta} + \omega_0^2 \sin \theta = 0 \tag{1.1}$$

with $\omega_0 = \sqrt{g/\ell}$, g being the acceleration due to gravity, and dots denoting derivatives with respect to time. The term $\sin\theta$ is a nonlinear function of θ. In elementary physics courses, one limits the angle θ to sufficiently small values, so that $\sin\theta \simeq \theta$, and Equation (1.1) reduces to the linear simple harmonic oscillator equation,

$$\ddot{\theta} + \omega_0^2\theta = 0. \tag{1.2}$$

This same linearization approach is usually taken in most other undergraduate physics problems. One restricts the transverse displacement of a plucked string and the density or pressure variation of longitudinal sound waves in a fluid to small values, in each case deriving a linear wave equation.

First assume all animals are horses. This eliminates all superfluous feathers and scales. Next the horse is assumed to be spherical or better yet a point mass. Further, to make all the oscillations isochronous, the amplitudes should be kept as small as possible. Are there any questions?

SOLVING PROBLEMS IN LINEAR PHYSICS

However, if the range of the dependent variable in Equation (1.1) is increased sufficiently, the linearization process is no longer valid, and one must work with the full nonlinearity. How the physical behavior is affected depends on the particular nonlinear equations considered. In the case of the pendulum, the motion doesn't appear to be much changed if the pendulum swings between, say $\pm\theta_{max} = 150°$ as opposed to $\pm\theta_{max} = 10°$. Yet, accurate and precise measurement of the period (T_0), the time to complete one cycle, reveals that it is dependent on the amplitude (θ_{max}). Indeed[1], for $\theta_{max} = 180°$, $T_0 = \infty$! From Equation (1.2), $T_0 = 2\pi/\omega_0$ which is independent of θ_{max} and is clearly wrong for a sufficiently large angular amplitude.

For other nonlinear equations, we shall see qualitatively new behavior as the amplitude of the dependent variable is increased. It is these latter types of nonlinear systems that are of extreme interest to modern researchers in physics,

---

[1]Think about how long it would take the above pendulum to fall if it were initially oriented with the mass m vertically upward, i.e., at $\theta = 180°$. Remember that our pendulum makes use of a (light) supporting rod rather than a string.

chemistry, biology, and other scientific disciplines. A wide variety of such examples will be surveyed in the next two chapters.

To study nonlinear systems, we shall have to learn new mathematical techniques. Ideas that are ingrained through our prolonged undergraduate exposure to linear differential equations are no longer valid or useful. For example, the

First assume all point masses are animals that have feathers or scales. Describe the animal with at least three equations using a minimum of three state variables. Further ensure that the oscillations have large amplitudes. Are there any questions?

SOLVING PROBLEMS IN NONLINEAR PHYSICS

familiar linear superposition principle does not hold for nonlinear systems. Let's consider the linear simple harmonic oscillator Equation (1.2). If θ_1 and θ_2 are solutions then $\theta = A\theta_1 + B\theta_2$, where A, B are arbitrary constants, is also clearly a solution. On the other hand, for the simple pendulum Equation (1.1), if θ_1 and θ_2 are separate solutions, then

$$\ddot{\theta}_1 + \ddot{\theta}_2 + \sin\theta_1 + \sin\theta_2 = 0. \tag{1.3}$$

But, substituting the linear combination $\theta = \theta_1 + \theta_2$ (with $A = B = 1$ for convenience) into (1.1) yields

$$\ddot{\theta}_1 + \ddot{\theta}_2 + \sin(\theta_1 + \theta_2) = \ddot{\theta}_1 + \ddot{\theta}_2 + \sin\theta_1 \cos\theta_2 + \sin\theta_2 \cos\theta_1 \tag{1.4}$$

which is nonzero because it differs from (1.3) by the inclusion of the cosine terms. So, although θ_1 and θ_2 separately satisfy (1.1), their sum does not.

As a consequence, the "bread and butter" mathematical technique of Fourier analysis, which relies on linear superposition, is not very useful in dealing with truly nonlinear systems. We have to learn new mathematical approaches, many of which apply to only certain classes of equations. The study of physically interesting nonlinear differential equations is not yet developed much beyond the infancy stage as compared to the situation for the linear case. Further, the present knowledge base for nonlinear PDEs is not as advanced as for ODEs, so we shall concentrate more on physical systems described by the latter.

In tackling nonlinear systems of differential equations, we shall use three broad avenues of approach, often in combination, namely:

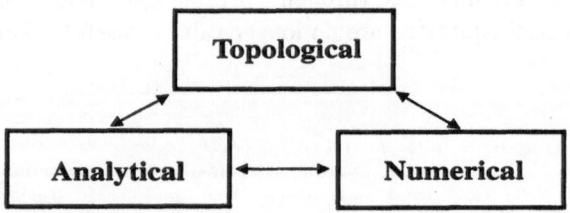

In the topological approach our goal is to obtain at least some qualitative knowledge of *all* the possible solutions that can occur for a given nonlinear system described by ODEs. Although only a few nonlinear ODEs or PDEs of physical interest have easily derived exact analytical solutions, the analytic approach is still important and consists mainly of finding approximate analytic solutions or finding some special physically important solutions. In the numerical approach the differential equations (ordinary or partial) are replaced by finite difference approximations, which are then solved for specific initial conditions using a computer. These various approaches will be examined in the ensuing chapters and applied to the examples of Chapters 2 and 3 as well as to many others.

1.2   Symbolic Computation

In the last several years, a number of powerful computer software packages have been developed that allow one to perform complicated symbolic manipulations, one of these being the Mathematica software system created by Stephen Wolfram. The Mathematica package allows one to perform a wealth of functions, including

- analytical differentiation

- analytical and numerical integration

- Taylor series expansions of functions to arbitrary order

- algebraic manipulation of equations including expanding, simplifying, collecting of terms, factoring, solving, etc.

- producing simple 2-dimensional plots as well as more complicated vector field and 3-dimensional plots

- numerical solution of ordinary and partial differential equations

- summation of series

- animation of analytic and numerical solutions

- production of sound

Since the Mathematica system is undergoing continual evolution, each new release adds to the types of mathematical operations that may be carried out. This edition of Nonlinear Physics is based on Mathematica Version 4.1.

In the following subsection, we present a series of simple examples illustrating how Mathematica may be used to perform some of the operations listed above. The purpose of these examples is to give you the flavor of what can be done with a computer algebra system. It is not intended as a systematic introduction to Mathematica, but rather as a way of providing the reader with the first stepping stones in the exhilarating climb into the stratospheric world of nonlinear physics.

Whether you are experienced with the Mathematica system or a beginner, we recommend that you carefully study these examples, execute the command steps, and test your understanding by trying the associated problems at the end of this section. For your convenience, all examples have been placed on the accompanying CD-ROM. The recommended way of accessing the associated notebook files (bearing the suffix .nb) is to open the master file **00master.nb** on the CD and follow the instructions contained therein. All the notebook files are hyperlinked to this master file.

If you wish to learn more about programming with Mathematica than is provided in this text, we recommend that you consult the two associated reference books that are available, namely the voluminous *Mathematica Book* by Stephen Wolfram [Wol99a] and the companion publication *Wolfram Research, Mathematica 4.0 Standard Add-on Packages* [Wol99b]. We have also found that two very useful texts which illustrate scientific and engineering applications of Mathematica are *Mathematica Navigator* by Heikki Ruskeepää [Rus99] and *Mathematica for Scientists and Engineers* by Richard Gass [Gas98].

1.2.1 Examples of Mathematica Operations

In presenting the following examples and those in ensuing chapters, it should be stressed that there are often several different ways of achieving the same result, and the reader will learn about some of these optional approaches in progressing through the text. Our first example illustrates the ease with which analytic differentiation and integration of complicated functions may be performed.

Example 1-1: Differentiation and Integration

Analytically differentiate or integrate the given functions as indicated:

a. $\dfrac{d}{dx}\left(\dfrac{x^3 \tan(2\pi x)}{\sqrt{1 + \log(x)\,e^x}}\right),$

b. $\dfrac{\partial^2}{\partial y^2}\left(\dfrac{x^3 y^2 \cosh(2x/y)}{1 + x^4 y^2}\right),$

c. $\displaystyle\int x \tanh(3x)\,dx,$

d. $\displaystyle\int_0^{\pi/2} x \tanh(3x)\,dx.$

`Solution:a`. The expression $x^3 \tan(2\pi x)/\sqrt{1 + \log(x)\, e^x}$ is differentiated once
with respect to the variable x in the following input command line[2], the basic
syntax being of the form `D[expression,variable]`:

```
D[x^3*Tan[2*Pi*x]/Sqrt[1+Log[x]*E^x],x]
```

In entering the form of the expression, it should be noted that Mathematica uses `*`, `/`, `+`, `-`, `^` for multiplication, division, addition, subtraction, and
powers, respectively. For multiplication, one may alternatively use a space between numbers, symbols, or functions, instead of inserting the asterisk `*`. That
is to say, e.g., `x y` is equivalent to the multiplication `x*y`, whereas `xy` is not equal
to `x` times `y`, but instead is a new symbol. However, it should be noted that
both `2x` and `2 x`, i.e., a number times a symbol, both give 2 times `x`. Except
where confusion may possibly occur, we shall generally use a space instead of
an asterisk when carrying out multiplications.

It should also be noted that the names of built-in Mathematica functions and
constants, e.g., the square root function, `Sqrt`, and the exponential constant, `E`,
always begin with capital letters.[3] Further, the arguments of built-in functions
are always enclosed in square brackets, e.g., the natural logarithm of `x`, `Log[x]`.
Round brackets, `()`, are used *only* to group terms, while curly brackets, `{ }`,
are used for *lists*, and double square brackets, `[[]]`, for extracting elements
from lists. The use of these other types of brackets will be illustrated in ensuing
examples.

On an "ordinary" keyboard, such as found on most laptop computers, the
input command line will be executed by simultaneously striking the Shift and
Enter keys. Using the Enter key alone will simply move the cursor down for
inputting another command line within the same Mathematica "cell". On an
"extended" keyboard, such as those associated with most desktop computers,
use the Enter key on the far right of the keyboard to execute the input command
line and the other Enter key to add more input commands within a given cell.

On executing the input command line, the following output[4] results.

$$\frac{2\pi x^3 \sec(2\pi x)^2}{\sqrt{1+e^x\,\log(x)}} + \frac{3x^2 \tan(2\pi x)}{\sqrt{1+e^x\,\log(x)}} - \frac{x^3\left(\dfrac{e^x}{x} + e^x\,\log(x)\right)\tan(2\pi x)}{2\left(1+e^x\,\log(x)\right)^{\frac{3}{2}}}$$

By making use of the Cell option on the Mathematica tool bar at the top of
the computer screen, we have chosen to show the output in TraditionalForm,
the form that resembles traditional mathematical notation. By opening Default
Output Format Type, other output forms may be selected. For the input, we
have chosen StandardForm. TraditionalForm was not used for the input because
it lacks the precision that is needed to provide reliable input to Mathematica. It

---

[2] A bolder type style shall be used for all Mathematica input.

[3] Mathematica is case-sensitive.

[4] As a consequence of exporting the Mathematica output into the text as *LATEX*, the
ordering and grouping of output terms in the text may occasionally differ slightly from what
appears on the computer screen. It should be further noted that Mathematica has its own
unique symbols for representing the exponential symbol and the square root of minus one in
the output. We shall use e and I for these quantities here in the text.

should be noted that whenever Mathematica produces TraditionalForm output, it automatically inserts hidden "tags", so that the output expression can be interpreted unambiguously if it is used as input in a subsequent command line.

If it is desired to suppress the output, perhaps because it is a lengthy result in an intermediate step of a multi-step calculation, simply end the input command line with a semi-colon (;). On executing the command line, the calculation will have been performed, but the result not displayed.[5]

b. In the next command line, the expression to be differentiated is assigned the name f by typing in f, followed by an equal sign, and then the expression. One can perform subsequent mathematical operations on a named object.

```
f = x^3 y^2 Cosh[2x/y]/(1+x^4 y^2)
```

$$\frac{x^3 y^2 \cosh(\frac{2x}{y})}{1 + x^4 y^2}$$

For user-defined names, it is good practice to generally use lowercase letters so that your named quantities don't become confused with Mathematica's built-in functions which always start with a capital letter. Of course, we need not use only a single letter such as was done here, but can use whole words or convenient acronyms for our named expressions. Notice that in the output only the expression appears, not the name of the expression or the equal sign.

The same command structure is used for ordinary and partial derivatives. Now the second partial derivative of f with respect to y is performed,[6]

```
D[f, {y, 2}]
```

the result being:

$$\frac{2x^3 \cosh(\frac{2x}{y})}{1 + x^4 y^2} + 4x^3 y \left(\frac{-2x^4 y \cosh(\frac{2x}{y})}{(1 + x^4 y^2)^2} - \frac{2x \sinh(\frac{2x}{y})}{y^2 (1 + x^4 y^2)} \right)$$

$$+ x^3 y^2 \left(\left(\frac{8x^8 y^2}{(1 + x^4 y^2)^3} - \frac{2x^4}{(1 + x^4 y^2)^2} \right) \cosh(\frac{2x}{y}) + \frac{8x^5 \sinh(\frac{2x}{y})}{y (1 + x^4 y^2)^2} \right.$$

$$\left. + \frac{\frac{4x^2 \cosh(\frac{2x}{y})}{y^4} + \frac{4x \sinh(\frac{2x}{y})}{y^3}}{1 + x^4 y^2} \right)$$

Clearly, Mathematica is even more useful if higher derivatives have to be carried out. It not only will save you time, but the output will be error free providing

---

[5]This is not the case for graphics commands. We shall have more to say about this later.
[6]One could use D[f,y,y] here, but this form is less convenient for higher derivatives.

that the input is typed correctly. For example, you might try taking the 7th y derivative of the function f, instead of the second. The resulting output would be extremely challenging to obtain quickly and error free with pen and paper.

c. The indefinite integral of $x \tanh(3x)$ is now carried out.

```
Integrate[x Tanh[3 x], x]
```

$$\frac{I}{9} \left(\frac{-9\,I}{2} x^2 - 3\,I\,x\,\log(1 + e^{-6\,x}) + \frac{I}{2} \mathrm{Li}_2(-e^{-6\,x}) \right)$$

If Mathematica cannot do an integral, it simply leaves the integral undone in the output. The answer here involves a function, $\mathrm{Li}_2(-e^{-6\,x})$, which may be unfamiliar to the reader. If the output is expressed in StandardForm, rather than TraditionalForm, this function is given in the form PolyLog[2, $-e^{-6\,x}$]. If you wish to know what PolyLog stands for, simply type in a question mark followed by the word PolyLog,

```
?PolyLog
```

and execute the command line, the result being:

PolyLog[n, z] gives the nth polylogarithm function of z. More...

The word More, which is colored blue on the computer screen and underlined, is a hyperlink to more information about PolyLog. Placing the cursor on the hyperlink and clicking the mouse opens an information window on this function. Further related material is provided by opening the additional hyperlinks inside the PolyLog information window. The same information window can also be opened by highlighting the word PolyLog in the previous input, clicking on the Help button on the tool bar, and then clicking on Find Selected Function.

It would be nice to see the structure of the original integral on the left-hand side and the expanded form of the integral output on the right-hand side. This is accomplished in the following command line by using the HoldForm function to hold the form of the integral, followed by a double equal sign[7], and using the Expand function on the right-hand side. The percent sign, %, is used to refer to the output of the previous command line, two percent signs, %%, to the output two command lines earlier (as is the case here), and so on.

```
HoldForm[Integrate[x Tanh[3 x], x]] == Expand[%%]
```

$$\int x \tanh(3\,x)\,dx == \frac{x^2}{2} + \frac{1}{3} x \log(1 + e^{-6\,x}) - \frac{1}{18} \mathrm{Li}_2(-e^{-6\,x})$$

d. The definite integral between $x = 0$ and $\pi/2$ for the same integrand as in part c is carried out, the numerical evaluation operator N being "attached" to the definite integral input so as to provide us with a decimal answer

---

[7]The = sign is used for assigning a name and the == sign for an equality in an equation.

```
Integrate[x Tanh[3 x], {x, 0, Pi/2}] // N
```

1.18805

Alternatively, one could input `N[%]` after the `Integrate` command line, or use the numerical integration command `NIntegrate`. In all cases, Mathematica prints the output to six digits. If you wish to print all the digits in the answer that Mathematica knows, you can use the `InputForm` command:

```
InputForm[%]
```

1.1880546728479855

Earlier we indicated how to proceed through the notebook file one command line at a time. If it is desired to execute an entire notebook file in one fell swoop, this may be done by successively clicking on Kernel in the tool bar, then on Evaluation, and finally on EvaluateNotebook.

End Example 1-1

Being able to Taylor expand functions is important in nonlinear modeling. For example, at an elementary level the vibrations of the atoms in an atomic lattice are analyzed using Hooke's law, which states that the restoring force on an atom displaced from equilibrium by an amount x is proportional to x. Hooke's law is valid if the displacements are sufficiently small. For larger displacements, higher-order terms in x should be retained. The functional forms of the force laws can be Taylor expanded to obtain these nonlinear terms.

Example 1-2: Taylor Expansion

Taylor expand the following functions:

a. $e^{\cos(2x)}$ about $x = 0$, keeping terms to order x^{10},

b. the general force function $f(x)$ about $x = 0$ to order x^5

Solution:a. Starting with this example, we shall use the following `Clear` command to clear all user-defined non-subscripted symbols prior to beginning the notebook file. With several files open, we may have forgotten that certain symbols which we intend to use in the current file may have already had specific values or functional forms assigned to them.

```
Clear["Global`*"]
```

In the following command line, the nonlinear function $e^{\cos(2x)}$ is Taylor expanded about $x = 0$ to order x^{11}. Use is made of the exponential function, `Exp`.

```
Series[Exp[Cos[2 x]], {x, 0, 11}]
```

$$e - 2ex^2 + \frac{8ex^4}{3} - \frac{124ex^6}{45} + \frac{758ex^8}{315} - \frac{26224ex^{10}}{14175} + O(x^{12})$$

The "order of" term, $O(x^{12})$, is removed with the following `Normal` command.

`Normal[%]`

$$e - 2\,e\,x^2 + \frac{8\,e\,x^4}{3} - \frac{124\,e\,x^6}{45} + \frac{758\,e\,x^8}{315} - \frac{26224\,e\,x^{10}}{14175}$$

In the output we note that there is a common factor e in each term. The `Collect` command, with `E` as the second argument, is used to extract this factor from the previous output and place it outside the series expansion.

`Collect[%, E]`

$$e\left(1 - 2\,x^2 + \frac{8\,x^4}{3} - \frac{124\,x^6}{45} + \frac{758\,x^8}{315} - \frac{26224\,x^{10}}{14175}\right)$$

b. The above Taylor expansion was for a specific function. The formal expansion of a general function may also be generated. Here a general nonlinear force function $f(x)$ is expanded about $x = 0$ to 5th order, given the name `force`,

`force = Series[f[x], {x, 0, 5}]`

$$f(0) + f'(0)\,x + \frac{1}{2}\,f''(0)\,x^2 + \frac{1}{6}\,f^{(3)}(0)\,x^3 + \frac{1}{24}\,f^{(4)}(0)\,x^4 + \frac{1}{120}\,f^{(5)}(0)\,x^5 + O(x^6)$$

and the "order of" term removed.

`Normal[force]`

$$f(0) + f'(0)\,x + \frac{1}{2}\,f''(0)\,x^2 + \frac{1}{6}\,f^{(3)}(0)\,x^3 + \frac{1}{24}\,f^{(4)}(0)\,x^4 + \frac{1}{120}\,f^{(5)}(0)\,x^5$$

The derivation of the nonlinear equation of motion of the vibrating eardrum which is discussed in Chapter 2 makes use of this formal expansion, with the first nonlinear term in x, i.e., the quadratic (x^2) term, being kept.

End Example 1-2

In Equation (1.4), illustrating the breakdown of the linear superposition principle for the simple pendulum equation, a trigonometric identity was employed. If this particular relation had been forgotten or you were unsure of the sign on the right-hand side of the identity, Mathematica's `TrigExpand` command can be of assistance. This command is only one of many trigonometric and algebraic manipulation commands at our disposal. A few of these commands are illustrated in the next example.

Example 1-3: Trigonometric and Algebraic Manipulations

a. Confirm the following trigonometric identities:

- $\sin(x + y) = \sin(x)\,\cos(y) + \cos(x)\,\sin(y)$,
- $\cos(2x) = \cos^2(x) - \sin^2(x)$,
- $\cosh^2(x) - \sinh^2(x) = 1$.

b. Find the factors of the polynomial

$$x^5 - 3x^4 - 23x^3 + 51x^2 + 94x - 120.$$

c. For a certain direct current (dc) circuit, Kirchhoff's loop rules give the following four simultaneous current equations

$$2y_1 + 3y_2 - 6y_3 - 5y_4 = 2$$

$$y_1 - 2y_2 - 4y_3 + 2y_4 = 0$$

$$-3y_1 + 2.5y_2 + y_4 = 5$$

$$y_2 - 23y_3 + 9.3y_4 = 7.2.$$

for the unknown currents y_1, y_2, y_3, and y_4. Solve this set for the currents, giving the numerical answers to 3 digits accuracy.

`Solution:` a. The first identity to be confirmed is the one that was used in

`Clear["Global`*"]`

Equation (1.4). The `TrigExpand` command does the trick in the following input.

`Sin[x+y] == TrigExpand[Sin[x+y]]`

$$\sin(x + y) == \cos(y)\,\sin(x) + \cos(x)\,\sin(y)$$

In the next line, the same command is used to generate the second trig relation.

`Cos[2 x] == TrigExpand[Cos[2 x]]`

$$\cos(2x) == \cos^2(x) - \sin^2(x)$$

The third trig identity is generated by using the `TrigReduce` command.

`Cosh[x]^2 - Sinh[x]^2 == TrigReduce[Cosh[x]^2 - Sinh[x]^2]`

$$\cosh^2(x) - \sinh^2(x) == 1$$

The last output relation could also have been generated by employing the `Simplify` command.

b. The polynomial is entered and given the name `poly`.

`poly = x^5 - 3x^4 - 23x^3 + 51x^2 + 94x - 120`

$$x^5 - 3x^4 - 23x^3 + 51x^2 + 94x - 120$$

The factors of the polynomial are then found.

    `Factor[poly]`

$$(x - 5)\ (x - 3)\ (x - 1)\ (x + 2)\ (x + 4)$$

c. Although the loop equations could be solved by hand, the algebra is tedious and is better tackled with a computer algebra system. The four equations are entered as a list (enclosed in curly brackets) and the `Solve` command applied to determine y1, y2, y3, and y4. The output is suppressed by using a semi-colon at the end of the input line and the name `solution` given to the result.

    `solution = Solve[{2 y1 + 3 y2 - 6 y3 - 5 y4 == 2,`
    `y1 - 2 y2 - 4 y3 + 2 y4 == 0, -3 y1 + 2.5 y2 + y4 == 5,`
    `y2 - 23 y3 + 9.3 y4 == 7.2}, {y1, y2, y3, y4}];`

Three digits are displayed in the `solution` output by using the `SetPrecision` command.

    `SetPrecision[solution, 3]`

$$\{\{y1 \to -2.78,\ y2 \to -0.893,\ y3 \to -0.794,\ y4 \to -1.09\}\}$$

The solution is expressed as a set of "transformation rules" (indicated by the arrows) for the variables. You can think, e.g., of $y3 \to -0.794$ as being a rule in which "y3 goes to -0.794". However, if one now entered the command y3, the symbol $y3$ would appear rather than the numerical value of y3. To apply a transformation rule to a particular Mathematica expression, we can employ the "replacement operator" `/.` to apply the rule to the expression by using the following syntax: `expression /. rule`

In the present case, a list of the numerical values of the four unknowns (displayed as default six digit numbers) is obtained in the following command line.

    `{y1, y2, y3, y4} = {y1, y2, y3, y4} /. solution[[1]]`

$$\{-2.77527, -0.893032, -0.793916, -1.09323\}$$

Notice that we didn't simply use `solution`. In general, the solution of a set of polynomial equations might have several different values for each of the unknowns. The solution then would be given in the form of a "list of lists", viz., $\{\{y1 \to .2, y2 \to .3, ...\}, \{y1 \to -.5, y2 \to .4, ...\}, \{\ \}, ...\}$. In the present example, because the equations are all linear, there is only one list inside the outer list, and this is extracted by using `[[1]]` to pick out this first list in `solution`. In general, for a list v, `v[[i]]` picks out the ith element of the list.

Now, an individual value is easily extracted. For example, if we want y3, then enter

    `y3`

    -0.793916

Implicit numerical schemes for solving nonlinear ODEs and PDEs, to be discussed in Chapters 6 and 11, respectively, make use of this simultaneous equation solving ability.

End Example 1-3

Several nonlinear sports models are discussed in a delightful reprint collection entitled "The Physics of Sports", edited by Angelo Armenti Jr[PLA92]. The next example, taken from this collection, deals with the effect of nonlinear air resistance on a falling badminton shuttlecock, or bird. The sports minded reader can consult Armenti for other articles on such diverse topics as the aerodynamics of a knuckleball (a baseball gripped with the knuckles and thrown so as to produce an erratic trajectory), the physics of drag racing, the stability of a bicycle, and the physics of karate, to name just a few topics that are covered.

Example 1-4: Plotting Data and Functions

Peastrel, Lynch, and Armenti [PLA92] have analyzed the effect of air resistance on a badminton bird falling vertically from rest. Taking y to be the distance fallen and t to be the elapsed time, their experimental data is given in Table 1.1. Using Newton's law of nonlinear air resistance (to be discussed in Chapter 2)

y (meters)	t (sec)	y (meters)	t (sec)	y (meters)	t (sec)
0.61	0.347	2.13	0.717	6.00	1.354
1.00	0.470	2.44	0.766	7.00	1.501
1.22	0.519	2.74	0.823	8.50	1.726
1.52	0.582	3.00	0.870	9.50	1.873
1.83	0.650	4.00	1.031		
2.00	0.674	5.00	1.193		

Table 1.1: Data for the falling badminton bird.

they find that the data can be fitted by the theoretical formula

$$y = \frac{v_T^2}{g} \ln\left(\cosh\left(\frac{gt}{v_T}\right)\right),$$

where $g = 9.81$ m/s^2 is the acceleration due to gravity and $v_T = 6.80$ m/s the terminal velocity (the velocity at which the downward gravitational and upward resistive forces balance). Plot the data and the theoretical formula together in the same graph.

Solution: The time and distance data are entered as separate named lists, the output being suppressed with a semi-colon. Lists are appropriate for plotting data because the order of the elements in a list is always preserved.

```
Clear["Global`*"]
time = {.347, .47, .519, .582, .65, .674, .717, .766, .823,
        .87, 1.031, 1.193, 1.353, 1.501, 1.726, 1.873};
distance = {.61, 1, 1.22, 1.52, 1.83, 2, 2.13, 2.44, 2.74, 3,
            4, 5, 6, 7, 8.5, 9.5};
```

The `Transpose` command is used to join the two lists into a single list of time-distance `data` points for plotting purposes.

```
data = Transpose[{time, distance}];
```

The data points are plotted with the `ListPlot` command. The `Hue[1]` option colors the points red and `PointSize` controls the size of the points. To suppress the picture produced in `gr1`, the option `DisplayFunction -> Identity` is included, because a line-ending semi-colon is not sufficient.

```
gr1 = ListPlot[data, PlotStyle -> {Hue[1], PointSize[.025]},
         DisplayFunction -> Identity];
```

The formula is now entered, with u standing for the terminal velocity.

```
y = (u^2/g) Log[Cosh[g t/u]]
```

$$\frac{u^2 \, \log(\cosh(\frac{g\,t}{u}))}{g}$$

On entering the given values of g and u,

```
g = 9.81; u = 6.80;
```

these numbers are automatically substituted into y.

```
y
```

$$4.71356 \, \log(\cosh(1.44265 \, t))$$

The formula y is now plotted over the time range $t = 0$ to $t = 2$, the curve being given a blue color with the `Hue[.6]` option. The picture is also suppressed.

```
gr2 = Plot[y, {t, 0, 2}, PlotStyle -> Hue[.6],
          DisplayFunction -> Identity];
```

The graphs `gr1` and `gr2` are superimposed in the same plot with the `Show` command. The option `DisplayFunction -> $DisplayFunction` is included to unsuppress the graphs. The plot range is specified and a title added with the `PlotLabel` option. The double quotes enclosing the title indicate a "string". A string is a sequence of characters which has no other value than itself. The ticks, font style, and image size are controlled with appropriate option choices.

```
Show[{gr1, gr2}, DisplayFunction -> $DisplayFunction,
PlotRange -> {{0, 2}, {0, 10}}, PlotLabel -> "distance vs time",
Ticks -> {{{1, "t"}, 2}, {{0.001, "0"}, {5, "y"}, 10}},
TextStyle -> {FontFamily -> "Times", FontSize -> 20},
ImageSize -> {500, 400}];
```

The plot of the data points and formula is shown in Figure 1.2. The theoretical

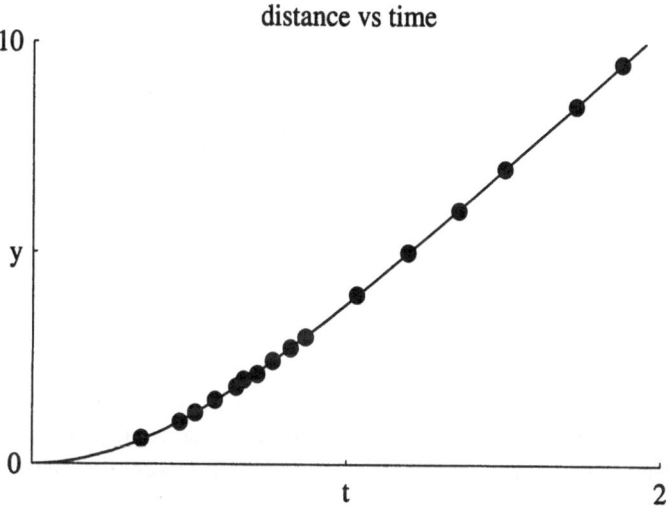

Figure 1.2: Data points and theoretical formula for falling badminton bird.

formula is in excellent agreement with the experimental data. The straight line behavior at larger t is an indication that the badminton bird has essentially reached its terminal velocity.

$\boxed{\textbf{End Example 1-4}}$

In Chapter 8, the behavior of forced nonlinear oscillator systems is examined in great depth. As a preliminary glimpse into this fascinating, complicated, and sometimes bizarre realm, we look at the vibrations of the eardrum driven by a sinusoidal pressure wave. In particular, a numerical solution is obtained for the eardrum displacement and the result plotted.

$\boxed{\textbf{Example 1-5: Numerical Solution of the Eardrum ODE}}$

The eardrum is being driven by a time(t)-dependent sinusoidal pressure wave. Suitably normalized and with arbitrary parameter values, the nonlinear ODE for the eardrum displacement, $y(t)$, from equilibrium is of the form

$$\ddot{y}(t) + y(t) + 0.25y(t)^2 = 0.55\sin(0.1t)$$

with the initial condition $y(0) = 1$ and $\dot{y}(0) = 0$. Solve the ODE numerically and plot the solution $y(t)$ as well as the velocity $v = \dot{y}$ versus the displacement y.

Solution: The ordinary differential equation, labeled ode, is entered using two primes (made with the keyboard apostrophe) for the second derivative.

```
Clear["Global`*"]
ode = y''[t] + y[t] + 0.25 y[t]^2 == 0.55 Sin[0.1 t]
```

$$0.25\, y(t)^2 + y(t) + y''(t) == 0.55\,\sin(0.1\,t)$$

The NDSolve command is used to find a numerical solution of ode, subject to the initial conditions y[0] == 1, y'[0] == 0, over the steady-state time range $t = 350$ to $t = 500$. The option MaxSteps->5000 is included because the default maximum number of 1000 time steps is reached at $t = 240.375$ otherwise. We could employ Mathematica's default numerical ODE solver, but instead opt to use the Rungé–Kutta–Fehlberg 45 (rkf45) method discussed in Chapter 6.

```
sol = NDSolve[{ode, y[0] == 1, y'[0] == 0}, y, {t, 350, 500},
       MaxSteps -> 5000, Method -> RungeKutta];
```

The Plot command is used to graph y[t] over the desired time range. Since the solution sol is expressed as a replacement rule for y, we form y[t] /. sol and apply the Evaluate command. To obtain a reasonably smooth curve the number of plotting points is specified to be 1000. Frame -> True produces a framed picture and FrameLabel allows us to label the frame axes.

```
Plot[Evaluate[y[t] /. sol], {t, 350, 500}, PlotPoints -> 1000,
PlotRange -> {-2.5, 2}, Frame -> True, PlotStyle -> Hue[.9],
FrameLabel -> {"time", "displacement"}, ImageSize->{600,400},
FrameTicks -> {{350, 500}, {-2, 0, 2}, {}, {}},
TextStyle -> {FontFamily -> "Times", FontSize -> 16}];
```

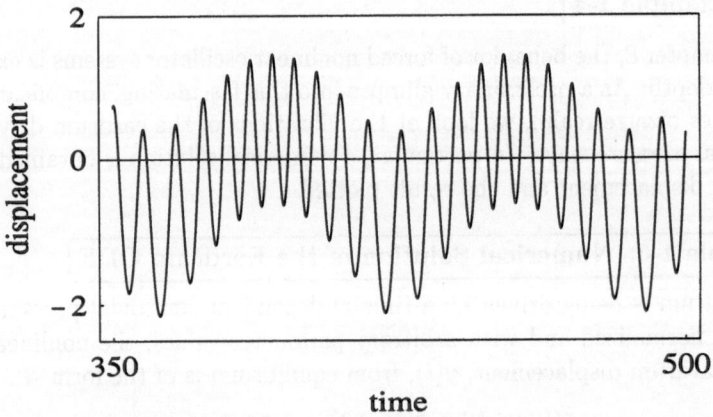

Figure 1.3: Displacement vs. time of the sinusoidally driven eardrum.

The steady-state temporal evolution of the eardrum displacement is shown in Figure 1.3. From the plot, one sees clear evidence of the nonlinearity in the equation as the quadratic term in $y(t)$ is responsible for the asymmetric response, the displacement maximum for negative y being larger than for positive y.

To plot the velocity versus the displacement, the `ParametricPlot` command

```
ParametricPlot[Evaluate[{y[t],y'[t]}/.sol],{t,350,500},
Frame->True,FrameTicks->{{-2,-1,0,{.75,"y"},1.5},
{-1,0,{.5,"v"},1},{},{}},PlotStyle->Hue[.9],ImageSize->
{600,400},TextStyle->{FontFamily->"Times",FontSize->16}];
```

is applied over the same time range as in Figure 1.3, the resulting curve being shown in Figure 1.4. Such a parametric curve is an example of a "phase plane trajectory".

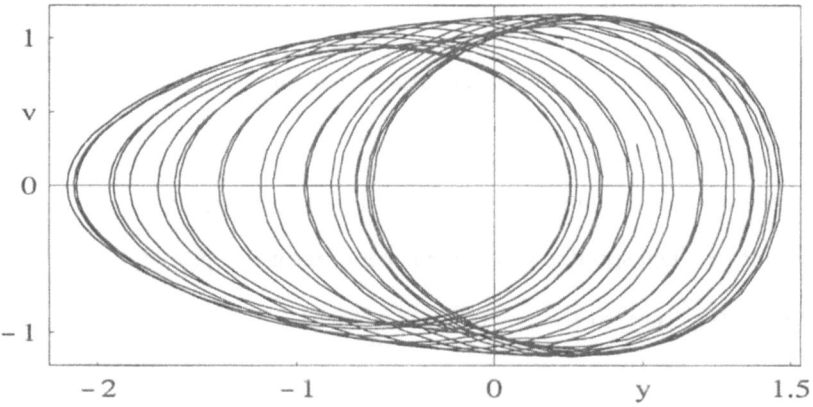

Figure 1.4: Velocity vs. displacement for the driven eardrum.

The coordinates of any point on a curve, e.g., a maximum in Figure 1.3, may be obtained by carrying out the following procedure. Place your cursor on the computer plot and click the mouse button. Then hold the Control key down and move your cursor to the desired location, e.g., one of the maxima in the curve. The coordinate values are then displayed in a small window on the bottom left of the computer screen.

From the ODE, the period of the driving force is $T = 2\pi/\omega = 2\pi/0.1 = 20\pi \simeq 63$. What feature in Figure 1.3 corresponds to this period?

⎡ **End Example 1-5** ⎤

As mentioned at the beginning of this chapter, the linear superposition principle is a cornerstone for the analysis of dynamical systems governed by linear ODEs or PDEs. For example, recognizing that $\psi(x,t) = \sin(x - ct)$ is a solution of the 1-dimensional linear wave equation

$$\frac{\partial^2 \psi(x,t)}{\partial x^2} = \frac{1}{c^2}\frac{\partial^2 \psi(x,t)}{\partial t^2},$$

where c is the wave speed, then, by linear superposition, a function made up of a sum of similar sine terms will also be a solution of the wave equation. In the following example, we consider a Fourier sine series solution of the linear wave equation and animate it. Although linear superposition does not hold for non-linear systems, the subject matter of this text, the animation command will still prove useful in visually clarifying how some special solutions (e.g., "solitons", discussed in Chapters 3 and 10) to nonlinear PDEs of physical interest evolve.

Example 1-6: Animation and Sound

The Fourier sine series

$$\psi(x,t) = \frac{0.4}{\pi} \left(1 + \sum_{j=0}^{n} \frac{\sin((2j+1)(2000)(x-ct))}{(2j+1)} \right)$$

satisfies the linear wave equation. Taking $c = 1$ m/s and $n = 5$ terms,

  a. animate the solution $\psi(x,t)$ over the spatial interval $x = -1/100...1/100$ and the time range $t = 0...10$, and discuss the result,

  b. plot the function $f = \psi(0,t)\,\text{sech}(.5(t-5))\,\sin(2005\,t)$ for $t = 0...10$ and discuss the shape,

  c. use the Play command to produce the sound corresponding to f.

Solution:a. The speed and the number n of terms is entered.

```
Clear["Global`*"]
c=1; n=5;
```

The Sum command is used to add the sine terms in the Fourier series. The index j runs from 0 to n in steps of 1.

```
psi=(0.4/Pi)(1+Sum[Sin[(2j+1)(2000)(x-c t)]/(2j+1),{j,0,n,1}])
```

The output of psi takes the following form.

$$0.127324 \left(1 + \sin(2000\,(x-t)) + \frac{1}{3}\sin(6000\,(x-t)) + \frac{1}{5}\sin(10000\,(x-t)) \right.$$

$$\left. + \frac{1}{7}\sin(14000\,(x-t)) + \frac{1}{9}\sin(18000\,(x-t)) + \frac{1}{11}\sin(22000\,(x-t)) \right)$$

The first sine term contribution has a frequency of $2000/(2\pi) \simeq 318$ Hz, the next sine term a frequency of about 955 Hz, the third term a frequency of 1592 Hz, and so on. The normal range of hearing is from about 20 to 20,000 Hz.

To animate a plot of the series, the Animate command will be used. This command is listed under Animation in Mathematica's standard add-on Graphics package. The following command (<< stands for Get) loads this package.

```
<< Graphics`
```

Then the output of **psi** can be animated with the following command line. The x range is taken to be from $-1/100$ to $1/100$ and the time range from 0 to 10. The animated picture is to consist of 20 frames and 1000 plotting points are used. Suitable tick marks are also chosen.

```
Animate[Plot[psi, {x, -1/100, 1/100}], {t, 0, 10}, Frames -> 20,
PlotPoints -> 1000, Ticks -> {{-.01, {.001, "0"}, {.005, "x"}, .01},
{.1, .2}}, TextStyle -> {FontFamily -> "Times", FontSize -> 12}]
```

On executing the command line, a sequence of 20 frames will result, the wave form in a typical frame being similar to that displayed in Figure 1.5. The Fourier

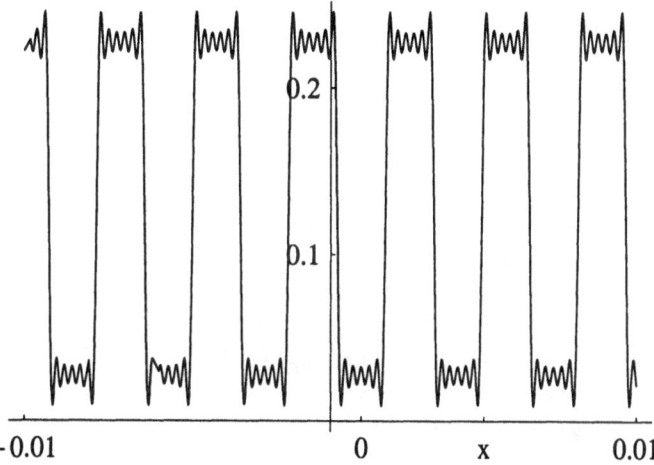

Figure 1.5: A traveling wave of approximately rectangular pulses.

sine series given above approximates a square-pulse wave train. The reader can keep more terms in the series if desired.

Double clicking the mouse button on any plot frame will produce animation. A single click will stop the animation.

b. The function f is entered. Rather than making the assignment x=0, the replacement operator, /., is used to substitute x=0 into **psi**. This approach will prove useful in situations where one would like to substitute a numerical value for, say, x into a particular command line, rather than have x take on the value in all command lines subsequent to the assignment.

```
f = (psi /. x -> 0)(Sech[.5 (t - 5)]) Sin[2005 t]
```

$$0.127324 \operatorname{sech}(0.5\,(t-5))\,\sin(2005\,t)\left(1 - \sin(2000\,t) - \frac{1}{3}\sin(6000\,t)\right.$$

$$\left. -\frac{1}{5}\sin(10000\,t) - \frac{1}{7}\sin(14000\,t) - \frac{1}{9}\sin(18000\,t) - \frac{1}{11}\sin(22000\,t)\right)$$

The function f consists of the square pulse wave train of part (a) evaluated at $x = 0$, which has been given a time-dependent amplitude modulation. The hyperbolic secant term will produce an overall envelope function which peaks at $t = 5$ while the `Sin[2005 t]` term will beat against each term in the sine series to produce periodic drops in intensity.

Although not specifically requested, we shall plot the over all envelope function as well as `f`. To obtain the proper scaling for this function note that when the two sine terms are in phase, the amplitude will be about $2 \times 0.127 \simeq 0.25$. So, we shall create graphs for $\pm.25 \operatorname{sech}(.5(t - 5))$ and superimpose the two graphs in the same plot as `f`.

In Example 1-4, each graph was produced in a separate command line and the output suppressed each time with the `Displayfunction->Identity` option. An alternate approach is to use the following `Block` construct with `$DisplayFunction=Identity` present to suppress the graphical output of the entire block. The first two graphs, `gr1` and `gr2`, will produce red and green colored, dashed, curves that represent the envelope function. The third graph, `gr3`, produces a blue-colored plot of the function `f` with 3000 plotting points being used.

```
Block[{ $DisplayFunction = Identity},
gr1 = Plot[.25 Sech[.5 (t - 5)], {t, 0, 10},
     PlotStyle -> {Hue[.9], Dashing[{.02}]}];
gr2 = Plot[-.25 Sech[.5 (t - 5)], {t, 0, 10},
     PlotStyle -> {Hue[.3], Dashing[{.02}]}];
gr3 = Plot[f, {t, 0, 10}, PlotPoints -> 3000,
     PlotStyle -> Hue[.6]]];
```

The three graphs are superimposed in Figure 1.6 with the `Show` command.

```
Show[{gr1, gr2, gr3}, PlotRange -> {{0, 10}, {-.3, .3}}, Frame -> True,
FrameTicks -> {{{0.01, "0"}, 5, {7.5, "t"}, 10}, {-.2, -.1,
{0, "f"}, .1, .2}, {}, {}}, ImageSize -> {600, 400},
TextStyle -> {FontFamily -> "Times", FontSize -> 18}];
```

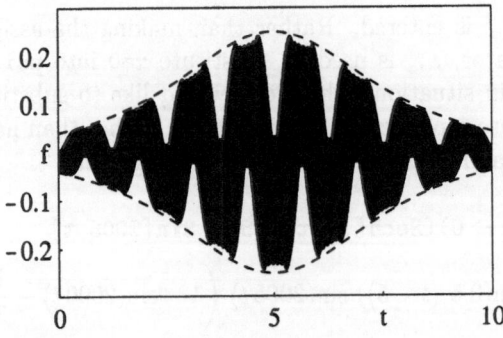

Figure 1.6: The function `f` with the (dashed) envelope curves indicated.

Note that after using the `Block` construct, it was not necessary to include the option `DisplayFunction -> $DisplayFunction` as was done in Example 1-4.

c. To generate the sound associated with the function `f`, the following `Play` command is entered and executed. A pictoral representation of `f` similar to that displayed in Figure 1.6 is also produced.

    Play[f, {t, 0, 10}, PlayRange -> All];

After the initial rendition of the sound, it can be played again by double clicking on the small "speaker box" in the right-hand margin of the cell.

| End Example 1-6 |

As our final introductory example, a complete analytic derivation and numerical solution of a nonlinear mechanics problem will be given. This example is more involved than those presented so far and is intended to show the reader at an early stage what is possible with the assistance of a computer algebra system. We shall have much more to say about solving this class of problems in subsequent chapters.

| Example 1-7: A Classical Mechanics Example |

Consider the experimental set-up shown in Figure 1.7. A mass m_2 is connected

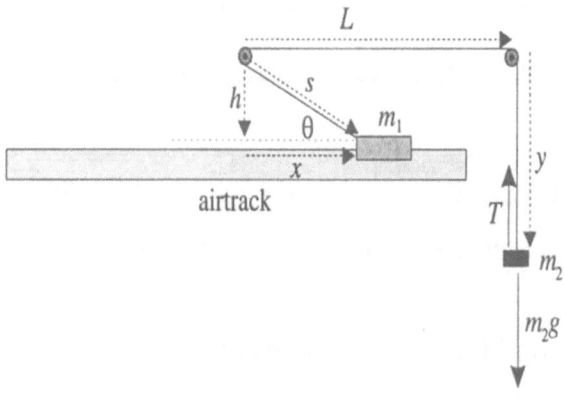

Figure 1.7: Setup for the classical mechanics example.

via a light, nonextensible, string of length c meters passing over two small, frictionless pulleys to a heavier mass m_1 which is allowed to slide freely on a (frictionless) airtrack. The distance between the pulleys is taken to be L meters and the tension in the string to be T Newtons. The origin is chosen to be the center of the left-hand pulley and the locations of m_2 and m_1 are then given by the coordinates $(L, y(t))$ and $(x(t), h)$, respectively. When m_1 is not directly below the left-hand pulley, then the connecting string makes an angle $\theta(t)$ with the horizontal.

a. Analytically determine the equation of motion for the displacement $x(t)$ of the mass m_1.

b. Given $m_1 = 1$ kg, $m_2 = 0.9$ kg, $h = 0.1$ m, $g = 9.81$ m/s^2, $x(0) = 1$ m, $\dot{x}(0) = 0$ m/s, numerically solve the equation of motion and create a 3-dimensional plot showing the displacement $x(t)$ vs. the velocity $\dot{x}(t)$ vs. time t over the time interval $t = 0...10$ s.

Solution:a. Using the theorem of Pythagoras, the variable length $s(t)$ of string between the left-hand pulley and m_1 is given by $s = \sqrt{h^2 + x(t)^2}$.

```
Clear["Global`*"]
```

```
s = Sqrt[h^2 + x[t]^2]
```

$$\sqrt{h^2 + x(t)^2}$$

The string is assumed to have a total fixed length c. A comment to this effect is included in the following command line by using the syntax (* comment *).

```
eq1 = c == s + L + y[t]  (* Fixed string length *)
```

$$c == L + y(t) + \sqrt{h^2 + x(t)^2}$$

By differentiating the output of eq1 twice with respect to time, a constraint equation on the accelerations of m_1 and m_2 is obtained.

```
D[eq1, {t, 2}]  (* Constraint equation for accelerations *)
```

$$0 == \frac{x'(t)^2}{\sqrt{h^2 + x(t)^2}} - \frac{x(t)^2 \, x'(t)^2}{\left(h^2 + x(t)^2\right)^{3/2}} + \frac{x(t) \, x''(t)}{\sqrt{h^2 + x(t)^2}} + y''(t)$$

The acceleration a_2 of mass m_2 must be given by $d^2y(t)/dt^2$. The output of the previous command line is solved for this acceleration.

```
a2 = Solve[%, y''[t]]  (* Acceleration of mass m2 *)
```

$$\left\{\left\{y''(t) \to -\frac{x'(t)^2}{\sqrt{h^2 + x(t)^2}} + \frac{x(t)^2 \, x'(t)^2}{\left(h^2 + x(t)^2\right)^{3/2}} - \frac{x(t) \, x''(t)}{\sqrt{h^2 + x(t)^2}}\right\}\right\}$$

The "replacement operator", /., is used to evaluate a2 at time t.

```
a2 = y''[t] /. a2[[1]]  (* evaluate a2 *)
```

$$-\frac{x'(t)^2}{\sqrt{h^2 + x(t)^2}} + \frac{x(t)^2 \, x'(t)^2}{\left(h^2 + x(t)^2\right)^{3/2}} - \frac{x(t) \, x''(t)}{\sqrt{h^2 + x(t)^2}}$$

The tension T in the string is $T = (m2)\,g - (m2)\,a2$.

```
tension= m2*g - m2*a2 (* Tension in string *)
```

$$g\,m2 - m2\left(-\frac{x'(t)^2}{\sqrt{h^2 + x(t)^2}} + \frac{x(t)^2\,x'(t)^2}{\left(h^2 + x(t)^2\right)^{3/2}} - \frac{x(t)\,x''(t)}{\sqrt{h^2 + x(t)^2}}\right)$$

Since the pulleys are frictionless, and $\cos\theta(t) = x(t)/s(t)$, the mass m_1 in Figure 1.7 experiences a horizontal force component $-T(x(t)/s(t))$, the minus sign appearing because the force is in the negative x direction. Making use of the FullSimplify command, Newton's 2nd law for m_1 then is given by eq2.

```
eq2 = -FullSimplify[tension*x[t]/s] ==m1*D[x[t], {t, 2}]
       (* 2nd law for m1 *)
```

$$-\frac{m2\,x(t)\left(x(t)^3\,x''(t) + h^2\,x(t)\,x''(t) + h^2\,x'(t)^2 + g\left(h^2 + x(t)^2\right)^{\frac{3}{2}}\right)}{\left(h^2 + x(t)^2\right)^2} == m1\,x''(t)$$

The FullSimplify command tries a much wider range of methods than the algebraic and trigonometric transformations used by Simplify. As a consequence, it can sometimes take substantially longer than Simplify.

b. To numerically solve the above second-order nonlinear ODE in $x(t)$, the values of the parameters are entered,

```
m1 = 1; m2 = 0.9; h = 0.1; g = 9.81;
```

and the ODE solved over the time interval $t = 0...10$ for an initial displacement of m_1 of 1.0 meter and zero initial velocity.

```
sol = NDSolve[{eq2, x[0] == 1, x'[0] == 0}, x, {t, 0, 10},
       MaxSteps -> 5000];
```

The ParametricPlot3D command, with suitable plot options, is used to create the 3-dimensional Figure 1.8, showing time vs. displacement vs. velocity.

```
ParametricPlot3D[Evaluate[{t, x[t], x'[t]}/. sol], {t, 0, 10},
   PlotPoints -> 2000, ImageSize->{400, 400}, BoxRatios ->{1.5, 1, 1},
   Ticks -> {{0, {5, "t"}, 10}, {-1, {0, "x"}, 1}, {-4, {0, "v"}, 4}},
   TextStyle -> {FontFamily -> "Times", FontSize -> 18}];
```

If a different view of the 3-dimensional plot is preferred than given by the default orientation, a ViewPoint plot option may be added as follows. Place your cursor at a convenient location inside the ParametricPlot3D command. Then choose Input in the Mathematica tool bar and select 3D ViewPoint Selector. You can then rotate the box shown in the resulting window either by dragging on it with the mouse or using the scroll bars. Once you have the desired orientation of the

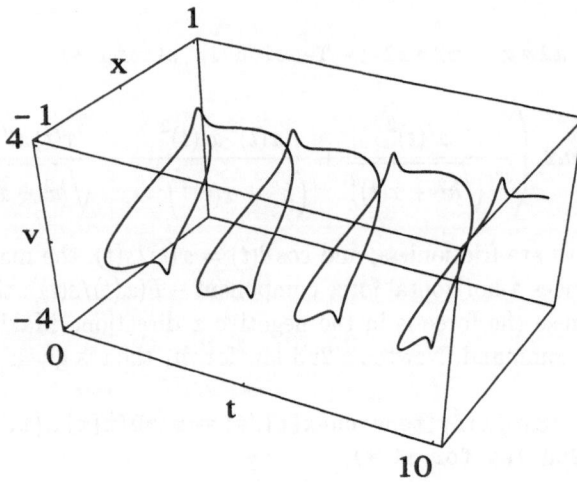

Figure 1.8: 3-dimensional plot of the periodic motion of mass m_1.

box, click Paste and go back to the notebook window. The ViewPoint option, with selected coordinates, will have been pasted into your plot command at the point at which you had placed your cursor. Executing the modified command line will produce a plot with the new orientation.[8]

| End Example 1-7 |

1.2.2   Getting Mathematica Help

In the previous subsection, the reader has seen a very small sampling of the Mathematica commands at his or her disposal. Even if this text were devoted completely to Mathematica, which it is not, we could not cover the thousands of commands and options that are available. Fortunately, Mathematica has an on-line help system which allows Mathematica commands and features, listed by name or subject, to be explored.

Suppose for example, that it is desired to find the command for solving an ODE analytically. Having clicked on Mathematica's Help button, we can proceed by selecting the Help Browser and carrying out the following steps:

- Click on Algebraic Computation. This opens a new list of options.

- Click on Equation Solving. This opens another list of options.

- Click on DSolve. A description of the DSolve command for analytically solving ordinary and partial differential equations then appears along with some examples of how to use the command structure. Click on the arrow adjacent to Further Examples to see these examples.

---

[8]An additional "experimental" feature in Release 4.1 is the real-time rotation of 3-dimensional graphics. Here this may be accomplished by inserting <<RealTime3D` prior to the ParametricPlot3D command and Default3D` afterwards. The default picture may be rotated by dragging on it with the mouse.

The `DSolve` help window is hyperlinked to various related command structures and topics as well as sections of Wolfram's *Mathematica Book*. Clicking with the mouse on such a hyperlink, e.g., NDSolve, takes one to the description of that command (in this case the command structure for numerically solving an ODE). Clicking on Back will take you back to the `DSolve` help window.

If you already know the name of a Mathematica command, e.g., `DSolve`, type DSolve next to the Go To button and then click on that button. This will take you to the same DSolve help window as above.

Even if you are unsure of the complete command name, the Help Browser can still be useful. Remembering that all Mathematica commands must begin with a capital letter, the command for taking the square root could be `Sqr`, `SquareRoot`,.... In this case, type the first couple of letters, Sq, in the Go To box. The Help Browser window then highlights Mathematical Functions in the first options column, Elementary Functions in the second column, and Sqrt in the third column. If you click on Sqrt, a window opens up with a description of the square root command.

1.2.3   Use of Mathematica in Studying Nonlinear Physics

Today, more and more researchers and science educators are using computer algebra tools such as Mathematica, especially in situations where the calculations become excessively complicated or tedious. While the use of such a package is always optional (given sufficient time and patience, one could always reproduce a Mathematica calculation with pen and paper or with a pocket calculator), we strongly encourage students to learn how to use this remarkable tool. As shall be illustrated, by using Mathematica to study nonlinear physics, we

- reduce the amount of time spent on number crunching and plotting, thereby providing more time to spend on the physics;

- permit explorations of how small changes in the controlling parameters or in the initial conditions of the nonlinear systems can result in very different or even chaotic behavior;

- encourage creativity and stimulate the imagination by providing the time and the means to explore new nonlinear models;

- reduce the frustration that results when after completing a lengthy calculation, you have to repeat it because of a computational error;

- reduce mathematical apprehensions and accordingly entice more students into this exciting branch of contemporary physics;

- make nonlinear physics more appealing and accessible to students in other disciplines.

Mathematica's symbolic computational, numerical, plotting, and animating ability provides the user with the means of seeing "new physics", and this should be the main reason and the motivation for investing some time in learning how to use this computer algebra system. By making use of Mathematica the undergraduate student can, to some extent, become a researcher by extending and modifying the many nonlinear systems that will appear in subsequent chapters.

Recognizing that most readers may have little or no expertise with Mathematica, we introduce Mathematica in this book in a very gentle way, basically on a need to use basis as the concepts and methods of nonlinear physics are explored. After all, this is primarily a text on nonlinear physics, not a text on symbolic computation. In addition to the annotated Mathematica examples in the text, a comprehensive set of more than seventy additional Mathematica files or notebooks are also included on the accompanying CD-ROM. These files are designed to supplement, expand and elucidate the concepts presented in the text, as well as to illustrate how to use the relevant Mathematica commands.[9] The placement of all notebook files on the CD is intended to save you the onerous task of typing in the code relevant to our study of nonlinear physics. For the reader's convenience, all Mathematica commands used in the files and text are listed in the index under the heading Mathematica Command.

Brief descriptions of the Mathematica files (MF) provided with this book appear throughout the text, and are placed adjacent to the concept being explored. A Mathematica notebook symbol indicates a provided Mathematica file. The file number (e.g. MF01) appears inside the symbol, and for emphasis is placed in the margin, as shown here. When such an icon appears in the margin, you should run the indicated file. The reader is expected to be an active participant, not just a passive passenger in our exploration of the nonlinear world. For your convenience, all MF notebooks are hyperlinked to the master file **00master.nb**. The Mathematica files are not annotated to the same degree as the examples presented in the text, but comments are included for many of the input command lines, explaining what the command is trying to achieve. If you wish to learn more about the command, simply make use of Mathematica's Help menu as outlined earlier.

A large fraction of the 400 problems in this text have been designed so that it might prove useful to employ Mathematica to carry out part or all of the solution. The reader should use common sense, however, in tackling a given problem. For example, Mathematica should be employed, if necessary, to analytically integrate, differentiate, solve, manipulate, to create plots, to animate, etc. Mathematica is not intended for doing trivial mathematical operations that can be done in your head. Neither is it intended as a substitute for clear thinking. Many of the problems can be tackled by employing techniques similar to those used in the text examples and Mathematica files, so the examples and files should be examined for procedures and clues that might help you do these problems. For some problems the reader is asked to use specific Mathematica commands, while for others hints are given as to what commands might prove helpful in solving the problem. These commands will appear in the following bold type style: **>DSolve(ODE,y(x))**. It should be emphasized that, where hints are given, there is often more than one approach to solving a given problem. In these cases, feel free to use your own approach and to experiment with different combinations of Mathematica commands than those suggested by the authors.

---

[9]Many of the figures in this text are generated with these (and similar) files.

In addition to all of the Mathematica examples and files provided on the CD, we show from time to time in the text how short combinations of analytical steps can be handled using Mathematica symbolic computation. These are indicated by the same bold type style for commands as above. It should be emphasized, however, that almost all of the Mathematica code needed to progress successfully through this text is contained on the provided CD-ROM.

In Chapters 2 and 3, when we present a survey of interesting nonlinear systems, the Mathematica examples and files allow you to "solve" systems of ordinary differential equations for a given set of parameters and visualize the results without any understanding of the underlying mathematics. The numerical and plotting "tools" can be used at an early stage without us worrying about how the tools work. These early examples and files are meant to whet your appetite and to motivate you to learn the methods and concepts of nonlinear physics presented in the rest of the text. In later chapters, the text examples and files will teach you how to perform the theoretical techniques covered in the text using symbolic and numerical computation.

PROBLEMS
Problem 1-1: Arithmetic
Numerically evaluate the following arithmetical operations using Mathematica:

    a. $(245)^3 \div 25 \times 3$,

    b. $e^{2.3} \times (\ln(2))^3 \div \tan(\frac{3\pi}{8})$,

    c. $\pi^5 \times \log_{10}(25) \div (\cosh(1.3))^3$.

For the latter two problems, also use `InputForm` to generate more significant figures than the six given by the default output.

Problem 1-2: Symbolic differentiation
Evaluate the following ordinary and partial derivatives:

    a. $\dfrac{d^5}{dx^5}\left(x^6 \tanh(x) \cos(x) e^{-x^2}\right)$

    b. $\dfrac{d^9}{dx^9}\left(\dfrac{x^{11} \tanh(2x)}{1 + x^4}\right)$

    c. $\dfrac{\partial^2}{\partial x^2}\left(\dfrac{\ln(1 + x^4 + y^4)}{\sqrt{(x^2 + y^2)}}\right)$

    d. $\dfrac{\partial^3}{\partial y^3}\left(\dfrac{\ln(1 + x^4 + y^4)}{\sqrt{(x^2 + y^3)}}\right)$

    e. $\dfrac{\partial^6}{\partial x^4 \partial y^2}\left(\dfrac{x^5 y^3 \tanh(2x/y)}{1 + x^6 y^7}\right)$

Problem 1-3: Indefinite integrals
Evaluate the following indefinite integrals:

    a. $\displaystyle\int \dfrac{x\,dx}{x^5 + 1}$,   **b.** $\displaystyle\int (\sec x)^3\,dx$,   **c.** $\displaystyle\int \dfrac{x\,dx}{1 - x^3}$,

d. $\int \dfrac{x^2 \, dx}{\cosh^4 x}$,   **e.** $\int x e^{ax} \sinh^2(bx) \, dx$,   **f.** $\int (\log(x))^3 / \sqrt{1 - x^2} \, dx$.

Identify any functions in the output that are unfamiliar to you.

Problem 1-4: Definite integrals

Evaluate the following definite integrals, expressing your answers to parts (a) to (f) in both default and in decimal forms:

a. $\displaystyle\int_0^{\pi/2} x^4 \sin x \cos x \, dx$,

b. $\displaystyle\int_0^1 x \arctan x \, dx$,

c. $\displaystyle\int_0^\infty dx/(1 + x^3)$,

d. $\displaystyle\int_0^\infty \sqrt{x} \, dx/(1 + x^2)$,

e. $\displaystyle\int_{-1}^1 dx/(\sqrt{1 - x^2}(1 + x^2))$,

f. $\displaystyle\int_0^1 \ln^{26}(x) \, dx/(\sqrt{1 - x^2})$,

g. $\displaystyle\int_0^\pi d\theta/(a + b \cos \theta)$.

Simplify the output of part **g**.

Problem 1-5: Western European tourists

The number $n(t)$ of foreign visitors to the United States between 1984 and 1991 can be described by the functional form

$$n(t) = \frac{18000}{1 + 36.02 e^{-0.8540t}}$$

where t is the number of years since 1984. The fraction of these visitors who were from Western Europe is given by

$$f(t) = 1.429 \times 10^{-5} t^2 - 2.234 \times 10^{-3} t + 0.08955.$$

a. Plot the number $N(t) = n(t) f(t)$ of Western European visitors for the period 1984 to 1991.

b. Calculate the analytic time derivative of $N(t)$.

c. Determine the inflection point of the curve $N(t)$. This may be done by calculating the second time derivative of $N(t)$ which gives the curvature of the function. The inflection point corresponds to the t value at which the second derivative is zero.

d. How many visitors came from Western Europe in 1989?

e. How rapidly was this number changing in 1989?

Problem 1-6: Hydrogen (H) atom motion

The potential energy of one of the atoms in a hydrogen molecule is given by

$$U(x) = U_0(e^{-2(x-a)/b} - 2e^{-(x-a)/b})$$

with $U_0 = 2.36$ electron volts ($1\text{ev} \equiv 1.6 \times 10^{-19}$ joules), $a = 0.37$ angstroms (1 angstrom $\equiv 1 \times 10^{-10}$ m), and $b = 0.34$ angstroms.

a. Plot the potential energy in ev as a function of x in angstroms.

b. The force F on the atom is given by $F = -\frac{dU}{dx}$. Plot a graph of $F(x)$.

c. Under the influence of this force, the H atom moves back and forth along the x-axis between certain limits, called the turning points, determined by the total energy. If the total energy is $E = -1.15$ ev, find the turning points graphically. This may be done by plotting the total energy and potential energy vs. x on the same graph and using the procedure explained in Example 1-5.

d. For the H atom, is the equation of motion linear or nonlinear? Explain.

Problem 1-7: Projectile motion

The accompanying table shows the horizontal velocity v in meters/second as a function of time t in seconds for a shell fired from a 6-inch naval gun.

Time	Velocity	Time	Velocity	Time	Velocity
0	657	1.20	588	2.40	528
0.30	638	1.50	571	2.70	514
0.60	619	1.80	557	3.00	502
0.90	604	2.10	542		

a. Make a plot of the velocity data as a function of time.

b. The velocity can be approximately represented by the formula

$$v = 655.9 - 61.4t + 3.26t^2.$$

Plot the velocity equation on the same graph as the experimental data.

c. By integrating the area under the analytic velocity curve, calculate the horizontal distance traveled by the shell in the first 3.0 seconds.

d. Plot the analytic form for the acceleration vs. time over the time interval 0 to 3.0 seconds.

Problem 1-8: Taylor expansion

Taylor expand the following functions as indicated:

a. $f(x) = (1/x) - \cot x$ about $x = 0$ to 14th order in x,

b. $f(x) = 1/\sqrt{1 + 3x^2}$ about $x = 0$ to order 10,

   c. $f(x) = x/(e^x - 1)$ about $x = 0$ to order 10,

   d. $f(x) = e^{\arctan(x)}$ about $x = \frac{\pi}{4}$ to order 6,

   e. $f(x) = \ln(1 + \sqrt{1 + x^2})$ about $x = 0$ to order 10,

   f. $f(x) = \ln(\sin(x))$ about $x = 0$ to order 10.

Problem 1-9: How good is the small angle approximation?
For small θ, with θ expressed in radians,

$$\sin \theta = \theta - \frac{\theta^3}{3!} + \frac{\theta^5}{5!} - \cdots.$$

   a. Confirm this expansion out to order x^{12} using the `Series` command.

   b. How big is the cubic correction to the linear term for $\theta = 30\,°$? Express your answer as the numerical ratio of the cubic to the linear term.

   c. How big is the 5th order correction?

   d. For $\theta = 179\,°$, to what term in the series would one have to go to have a 1% correction to the linear term?

Problem 1-10: Small time behavior
For the falling badminton bird in Example 1-4, determine the analytic power law behavior of $y(t)$ for small t by Taylor expanding about $t = 0$ and using the `Normal` command. Keep only the first non-vanishing term in the expansion.

Problem 1-11: Visualizing the small angle approximation
Plot $\sin(x)$ versus x and the series expansions of $\sin(x)$ to 5th, 7th, 9th, and 11th order in x on the same graph over the range $x = 0...2\pi$.

Problem 1-12: Factoring and solving a polynomial
Consider the polynomial

$$f(x) = 35x^9 - 15x^8 + 56x^7 - 17x^6 + 4x^5 + 11x^4 - 20x^3 + 13x^2 - 3x.$$

   a. Factor the polynomial.

   b. Use the `Solve` command to find the roots of the polynomial. How many roots are real? How many roots are complex?

   c. Use the `NSolve` command to find the decimal form of the roots.

Problem 1-13: Roots of an equation
Determine the roots of equations (a) to (c) using the `NSolve` command. Can this command find the roots of the non-polynomial equation (d)? In this case employ the `FindRoot` command using the suggested starting point. Go to the Help Browser to see the syntax for the `FindRoot` command.

   a. $x^3 + 1 = 0$,

   b. $x^5 + 1 = 0$,

c. $x^4 + y^4 = 67$, $x^3 - 3xy^2 = -35$,

d. $x = 0.7 \sin x + 0.2 \cos y$, $y = 0.7 \cos x - 0.2 \sin y$, (root near $(0.5, 0.5)$).

Problem 1-14: Solving a dc circuit problem

Consider the dc electrical circuit shown in the figure. Applying Kirchhoff's rules and using Mathematica to solve the resulting current equations, determine the current flowing through each resistor.

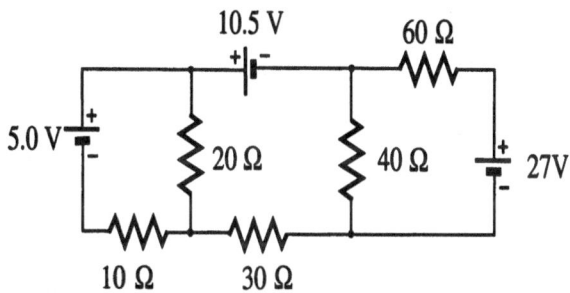

Problem 1-15: Mesh analysis for ac circuits

For the accompanying alternating current (ac) circuit the impedances are $Z_1 = 2 + 5I$, $Z_2 = 8 - I$, $Z_3 = 4 + 3I$, $Z_4 = 5 - 2I$, and $Z_5 = 1 + 5I$, with $I \equiv \sqrt{-1}$, while the voltage amplitudes are $\varepsilon_1 = 10$ volts and $\varepsilon_2 = 3$ volts.

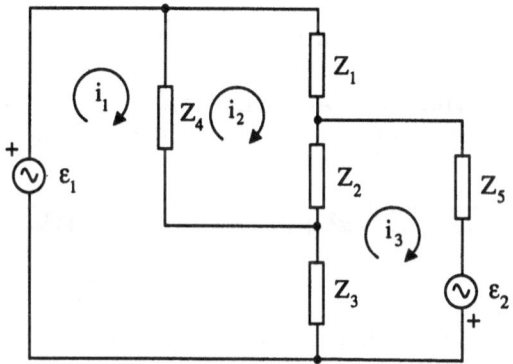

a. Show that the mesh equations for the complex current amplitudes i_1, i_2, i_3 are

$$i_1(Z_3 + Z_4) - i_2 Z_4 - i_3 Z_3 = \varepsilon_1$$

$$-i_1 Z_4 + i_2(Z_1 + Z_2 + Z_4) - i_3 Z_2 = 0$$

$$-i_1 Z_3 - i_2 Z_2 + i_3(Z_2 + Z_3 + Z_5) = \varepsilon_2.$$

b. Use Mathematica to solve for i_1, i_2, and i_3. Note that Mathematica uses I as the input command for $\sqrt{-1}$.

c. What is the net current amplitude through Z_4?

d. What is the meaning of the complex current amplitude?

Problem 1-16: Simultaneous linear equations
Solve the set of linear equations

$$1.1x_1 - x_2 + 2x_3 - 3.55x_4 = -8$$

$$2x_1 - 2.93x_2 + 3x_3 - 3x_4 = -20.9$$

$$x_1 + x_2 + x_3 = -2$$

$$x_1 - x_2 + 4x_3 + 3x_4 = 4.5$$

for x_1, x_2, x_3, and x_4.

Problem 1-17: Trig identities
Using either the `TrigExpand` or `TrigReduce`commands, derive trigonometric
identities for the following functions:

a. $\sin(9x)$,

b. $\cos(7x)$,

c. $2\cos((x+y)/2)\cos((x-y)/2)$,

d. $4\cos^3(x) - 3\cos(x)$,

e. $7\sin(x) - 56\sin^3(x) + 112\sin^5(x) - 64\sin^7(x)$.

Problem 1-18: Polynomial transformation
Making use of the `Factor` command, perform the transform

$$x^2 + 2x + 1 + \frac{1}{x^2 + 2x + 1} \longrightarrow \frac{(x+1)^4 + 1}{(x+1)^2}.$$

Problem 1-19: Sums
Use the `Sum` command to evaluate the following sums:

a. $\displaystyle\sum_{k=1}^{n} k^2$,   **b.** $\displaystyle\sum_{k=1}^{n}(2k-1)$,   **c.** $\displaystyle\sum_{k=1}^{n}(k^2+k-1)/(k+2)!$,

d. $\displaystyle\sum_{k=0}^{\infty} 1/k!$,   **e.** $\displaystyle\sum_{k=1}^{\infty} 1/(2k-1)^2$.

Problem 1-20: Beats
Consider the addition of two waves whose resultant displacement is given by

$$y = A_1 \sin\left(\frac{2\pi}{\lambda_1}(x - v_1 t) + \delta_1\right) + A_2 \sin\left(\frac{2\pi}{\lambda_2}(x - v_2 t) + \delta_2\right)$$

where A is the amplitude, λ the wavelength, x the spatial coordinate in the
direction of wave propagation, v the speed, t the time, and δ a phase factor.
Take $A_1 = A_2 = 1$, $v_1 = v_2 = 1$, $\delta_1 = \delta_2 = 0$, $\lambda_1 = 2\pi$, and $\lambda_2 = 2.05\pi$.

    a. Animate the resultant wave displacement y, taking an x-range from -200 to $+200$, $t = 0...500$, 25 frames, and 600 points.

    b. Describe the envelope of the resultant wave. If these were sound waves, what intensity variation would an observer hear at a fixed spatial point as the resultant wave passed by? This is the phenomenon of beats.

Problem 1-21: Eardrum equation

Consider the eardrum equation of Example 1-5.

    a. Suppose that the quadratic term had a minus sign in front of it instead of a plus sign. What do you intuitively expect to happen? Confirm your intuition by running the code with the above sign change.

    b. If the quadratic term in the eardrum equation were changed to a cubic term, what qualitative change in the solution might you expect? Run the code with this change and see if your intuition was correct.

Problem 1-22: Dispersion

Dispersion refers to a phenomenon in physics where each frequency (Fourier) component of a wave form has a different (phase) velocity. In Example 1-6, replace the speed $c = 1$ in the jth Fourier component with $c = 1/(1 + a\,j)$ where a is a positive constant, e.g., $a = 1$. Execute the modified notebook and discuss the results. Note that the amplitude of the envelope function and vertical plot range will have to be modified.

Problem 1-23: Viscous drag

Suppose that in the classical mechanical Example 1-7 the mass m_1 experienced a viscous drag force $F = -\mu v(t)$, where $v(t)$ is the velocity and μ the positive drag coefficient. Neglecting any drag on mass m_2, determine the equation of motion of m_1 and plot its motion for $\mu = 0.05$ N·s/m and over a time interval of your choosing.

1.3 Nonlinear Experimental Activities

We have also included a variety of "real" experimental activities to round off our multi-prong approach to learning nonlinear physics. These experiments are optional but we strongly encourage instructors in educational institutions to arrange for their students to attempt some of these experiments. The experiments are designed to use the readily available equipment and chemicals found in most science and engineering departments.

    Brief three- or four-line descriptions of the nonlinear experiments appear throughout Part I, the theory portion of this text. These descriptions are adjacent to the relevant text section and are indicated by placing the experiment number inside the symbol of a stopwatch. For emphasis the stopwatch is printed in the margin as shown here. The complete details of each experimental activity are given in Part II of this text. The page number on which an experimental activity begins may be easily obtained by either consulting the table of contents or the Experimental Activity index entry. These experiments have been designed

to complement the material in Part I and are not intended to be long or tedious. The emphasis is on seeing nonlinear physics in the real world, not on gathering elaborate results or generating pretty lab reports. The creative instructor or student will undoubtedly be able to think of modifications or improvements to the experiments. With the mushrooming interest in nonlinear physics, more and more experimental write-ups are appearing in appropriate physics journals such as the *American Journal of Physics (AJP)* and others.

Lastly, many of the suggested experimental activities are open-ended and could serve as student term projects. When nonlinear physics has been taught by the authors, theoretical and experimental term projects have been used and found to be valuable educational alternatives to formal final examinations. On completion of the projects, each student is required to present his or her results in a class talk. With proper supervision and motivation, it has been found that most students do first-rate jobs on their projects and gain valuable experience by standing up in front of their peers and explaining what they have done.

As an introductory sample of the nonlinear experimental activities that will accompany the theoretical development in Part I, the experimentally inclined reader might try the following two related activities. Each of these easy to perform experiments is accompanied with a Mathematica notebook file which aids in the analysis of the data. Remember that the detailed description of each activity can be found in Part II of this text.

Magnetic Force
The mathematical relationship between the force of repulsion and the separation distance of two thin cylindrical magnets is to be determined. Mathematica is used to plot the experimental data and extract the magnetic force law.

Magnetic Tower
A number of thin disk magnets are stacked vertically on a wooden dowel, with the magnetic poles oriented in such a way that a repulsive force exists between each magnet. The equilibrium separation distance between each magnet is measured and Mathematica is used to solve for the predicted distances.

PROBLEMS
Problem 1-24: Looking for a term project?
A question frequently asked by students beginning the nonlinear physics course on which this text is based is: "How do I find a suitable theoretical or experimental topic on which to base a term paper or project?" Of course, the instructor can suggest possible topics and experimentally inclined students can make use of the activities of Part II. But, quite often students want to do their own thing. Assuming that you have had no previous exposure to nonlinear physics, it will probably take you a while to get ideas as you progress through this text.

The next two chapters will survey a wide variety of topics which are referenced in the bibliography with the article title and, in the case of journals, the volume number, first page, and year of publication given. So if a topic looks interesting, go to your college or university library and look up the relevant reference(s). An alternate approach might be to do a search on the Web. Try searching for nonlinear topics that relate to your personal interests. Search for journal names or words such as *nonlinear, chaos, music, art, experiments, electronics,* etc., or various combinations and permutations of these words.

For example, two papers which have served as the basis of term papers or projects are "Population dynamics of fox rabies in Europe", published by Anderson and coworkers[AJMS81] in *Nature*, and "Solitons in the undergraduate laboratory", published in the *American Journal of Physics* by Bettini et al [BMP83]. Look up one of these papers or find one of your own and describe in no more than one page and in your own words what the article is about. Of course, at this stage you will not understand many of the details contained in these papers. Don't worry about it! The understanding will come as you progress through this text.

1.4   Scope of Part I (Theory)

As mentioned, in the following two chapters we shall begin by surveying different classes of nonlinear systems. Such a survey is bound to be uneven and incomplete and to some extent reflect the background and research experiences of the authors. Trying our best to avoid this, we have made the survey fairly long and made an effort to give the student the flavor of not only traditional physics but also examples from mathematical biology, engineering, chemistry, and so on. The equations that crop up in physics have their analogies in these other areas of science, and since the modern physics student often ends up after graduation in non-traditional areas, we feel that it is important to give some examples of this cross-fertilization. If there is a specific topic in Chapters 2 and 3 that you do not fully understand because of your own academic background, don't worry about it. It's not crucial to progressing through the rest of the text. In these chapters, we are trying to convey some of the richness of the subject of nonlinear phenomena. If more information is desired on a given topic, consult the cited references in your college or university library. Keep in mind that many of the topics covered in Chapters 2 and 3 are actually quite complex and we obviously cannot do them full justice in our brief coverage. In particular, the derivation of some of the basic dynamical equations is simply beyond the scope of this text.

To study these nonlinear systems, the student must learn how to use certain mathematical tools. To this end, the following three chapters examine in detail the topological, analytical (exact and approximate), and numerical approaches to deal with systems described by nonlinear ODEs, using the examples of Chapters 2 and 3 and others as illustrations of the techniques. All three approaches will be complemented, but not supplanted by the use of Mathematica. The student may wonder why not. For example, as shall be seen, there are problems, (e.g., the simple pendulum problem) where analytic answers exist but Mathematica can only yield them with considerable "poking and prodding". Similarly, although Mathematica can numerically solve nonlinear ODEs, it is much more limited in dealing with nonlinear PDEs of physical interest. So it's important to see how numerical schemes are conceptually created.

After Chapters 4, 5 and 6 have introduced the necessary mathematical tools, we then set out to explore in greater depth the nonlinear systems and concepts introduced in Chapters 2 and 3, the survey chapters. Subsequent chapters on Limit Cycles (Ch. 7), Forced Oscillators (Ch. 8), and Nonlinear Maps (Ch. 9) are included. The latter chapter involves the use of finite difference equations which are generally easier to analyze than differential equations.

The last three chapters of Part I give the student an introduction to some of the analytic and numerical methods and underlying concepts that are important for the study of nonlinear PDE systems. It is assumed that the student has already encountered the linear wave, diffusion, and Schrödinger equations. In Chapter 10, we explore various nonlinear PDE phenomena and concepts such as nonlinear diffusion, solitary wave solutions, and nonlinear superposition. Because it is so important for modern research into nonlinear PDE systems, we devote Chapter 11 to explaining how numerical simulations may be carried out for nonlinear diffusive and wave equations. Finally, we end the text with an optional chapter illustrating a conceptually powerful analytic technique, the Inverse Scattering Method. The close connection of the method, when applied to the Korteweg–de Vries equation describing shallow water waves, to quantum mechanical scattering is emphasized. This chapter is included to give the reader the flavor of a more advanced topic in nonlinear physics, and to show that in this text we have barely scratched the surface of a nonlinearly growing subject.

Unlike the prevailing situation in many traditional undergraduate physics courses, which deal with subject matter that has been known in some cases for hundreds of years, nonlinear physics gives you the reader a glimpse of the future, a future in which you have the potential to explore and perhaps unravel some of the mysteries of nature. The exciting world of nonlinear physics is beckoning, so let us begin our journey into this new world with due haste. Onward to the future!

Chapter 2

Nonlinear Systems. Part I

His style is chaos illumined by flashes of lightning. As a writer, he has mastered everything except language.
Oscar Wilde (1855-1900), Anglo-Irish Writer

2.1 Nonlinear Mechanics

In physics and engineering, the most readily visualized examples are usually those in classical mechanics. The starting point at the elementary level is Newton's second law, while at the more advanced level one can form the Lagrangian

$L = T - V$ where T is the kinetic energy and V the potential energy. We shall apply both approaches to nonlinear mechanics where the force is a nonlinear function of the displacement and, perhaps, the velocity.

2.1.1 The Simple Pendulum

We start with a familiar example which was mentioned in the introduction. A small mass m on the end of a light (idealized to be weightless) rigid rod of length ℓ is allowed to swing along a circular arc in a vertical plane as shown in Figure 2.1. Friction at the pivot point, air resistance, etc. are neglected. The

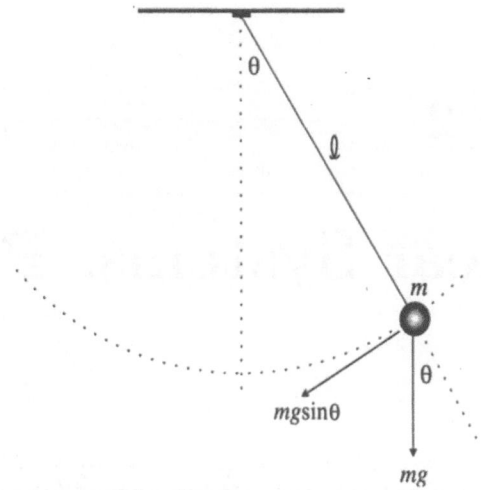

Figure 2.1: Force diagram for the simple plane pendulum.

equation of motion for the simple pendulum is first derived by using Newton's second law, $\vec{F} = m\vec{a}$. Since motion is constrained to be along the circular arc, we resolve the gravitational force $m\vec{g}$ into components parallel and perpendicular to the arc. For the parallel case,

$$m\ell\ddot{\theta} = -mg\sin\theta \qquad (2.1)$$

where $\ddot{\theta}$ is the angular acceleration, or

$$\ddot{\theta} + \omega_0^2 \sin\theta = 0 \qquad (2.2)$$

with $\omega_0 = \sqrt{g/\ell}$. The minus sign appears in the restoring force term of Eq. (2.1) because the force component is in the opposite direction to increasing θ. The pendulum Equation (2.2) is trivial to derive using Newton's second law. However, when the forces and geometry are more complicated, the Lagrangian approach can prove to be simpler to work with as it is often easier to determine the kinetic and potential energies rather than the forces.

As an example, let us rederive the pendulum's equation of motion using the Lagrangian formulation. Taking the zero of potential to be at the bottom of the arc ($\theta = 0$), from Figure 2.1,

$$V = mg\ell(1 - \cos\theta) \qquad (2.3)$$

and

$$T = \frac{1}{2}m(\ell\dot{\theta})^2 \qquad (2.4)$$

yielding the Lagrangian

$$L = T - V = \frac{1}{2}m(\ell\dot{\theta})^2 - mg\ell(1 - \cos\theta) \qquad (2.5)$$

with $\dot{\theta}$ being the angular velocity. Substitution of (2.5) into Lagrange's equation of motion [FC86]

$$\frac{d}{dt}\left(\frac{\partial L}{\partial \dot{\theta}}\right) - \frac{\partial L}{\partial \theta} = 0 \qquad (2.6)$$

yields Equation (2.2) as expected. With either approach, the derivation of the simple pendulum's equation of motion is easy to carry out with pen and paper. For more complicated geometries, one can use Mathematica to mimic the hand derivation with the assurance that careless algebraic mistakes will be eliminated. The following Lagrangian example illustrates the procedure.

Example 2-1: Parametric Excitation

The pivot point O for the simple pendulum is undergoing vertical oscillations given by $A\sin(\omega t)$ as shown. Show that the relevant equation of motion is

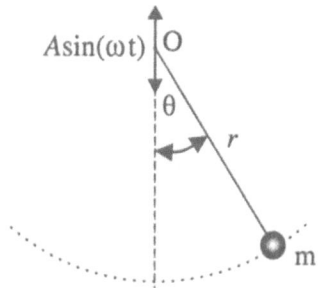

$$\ddot{\theta} + \left[\Omega^2 - \frac{A\omega^2}{r}\sin(\omega t)\right]\sin\theta = 0$$

with $\Omega = \sqrt{g/r}$. This nonlinear ODE with a time-dependent coefficient is an example of a parametric excitation.

Solution: To solve this problem, the `Calculus`VariationalMethods`` package is loaded. We shall be using the `EulerEquations` command, found in this package, to apply the Lagrange equation (2.6).

```
<<Calculus`VariationalMethods`
Clear["Global`*"]
```

First we will derive the Lagrangian. The Cartesian coordinates x, y of the mass m are expressed in terms of the length r and angle $\theta(t)$ at time t.

```
x = r Sin[θ[t]];
```

```
y = r (1 - Cos[θ[t]]) + A Sin[ω t];
```

In the above two input lines we have chosen to enter the symbols θ and ω using the palettes option, so that the final output equation will look like the requested form. To access the relevant palette, successively click on File in the

Mathematica tool bar, then on Palettes, and finally on BasicInput. With your cursor having been placed at the desired location in the command line, clicking on a symbol (e.g., θ) will paste the symbol into your Mathematica notebook at the correct position.

The x and y components of the velocity are calculated

```
v1 = D[x, t]; v2 = D[y, t];
```

and the kinetic energy expression T formed and labeled ke. The Simplify and Expand commands help reduce the output of ke to a more tractable form.

```
ke = Simplify[Expand[(m/2)(v1^2 + v2^2)]]
```

$$\frac{1}{2}m\left(A^2\omega^2\cos(t\omega)^2 + 2Ar\omega\cos(t\omega)\sin(\theta(t))\theta'(t) + r^2\theta'(t)^2\right)$$

The potential energy pe is entered and the lagrangian formed.

```
pe = m g y
```

$$gm\left(r\left(1 - \cos(\theta(t))\right) + A\sin(t\omega)\right)$$

```
lagrangian = ke - pe
```

$$\frac{1}{2}m\left(A^2\omega^2\cos(t\omega)^2 + 2Ar\omega\cos(t\omega)\sin(\theta(t))\theta'(t) + r^2\theta'(t)^2\right)$$
$$-gm\left(r\left(1 - \cos(\theta(t))\right) + A\sin(t\omega)\right)$$

Eq. (2.6) is applied to lagrangian with the EulerEquations command.

```
eq = EulerEquations[lagrangian, θ[t], t]
```

$$-mr\left((g - A\omega^2\sin(t\omega))\sin(\theta(t)) + r\theta''(t)\right) == 0$$

We then replace g with $r\Omega^2$ on the left-hand side of eq, divide by $(-mr^2)$, expand the result, and set it equal to zero.

```
eq2 = Expand[(eq[[1]] /. g -> r Ω^2)/(-m r^2)] == 0
```

$$-\frac{A\omega^2\sin(t\omega)\sin(\theta(t))}{r} + \Omega^2\sin(\theta(t)) + \theta''(t) == 0$$

Collecting $\sin(\theta(t))$ terms in eq2, the parametric excitation equation results.

```
eq3 = Collect[eq2[[1]], Sin[θ[t]]] == 0
```

$$\left(\Omega^2 - \frac{A\omega^2\sin(t\omega)}{r}\right)\sin(\theta(t)) + \theta''(t) == 0$$

End Example 2-1

Once an equation of motion is derived, the next step is to solve it, either analytically or numerically. For our example of parametric excitation, a numerical approach must be used, since an analytic solution does not exist. Being somewhat simpler in form, the reader might ask whether the simple pendulum Equation (2.2) can be solved exactly analytically. The first Mathematica file addresses this question. Run the file and see what happens.

The Simple Pendulum
The symbolic package attempts (and fails) to analytically solve the pendulum Equation (2.2) for $\theta(t)$ using the Mathematica DSolve command. However, a numerical value is found for the period T for large $\theta_{max} = 2$ radians ($\simeq 115°$). Mathematica commands in file: DSolve, Integrate, Solve, Chop, N, /.

According to the first file, the answer to our above question appears to be no. Actually, Equation (2.2) *can* be solved analytically for θ but first it is necessary to introduce a special function, the Jacobian elliptic function. This will be done in Chapter 5. These elliptic functions will also appear in later computer files. Despite this lack of initial success, we shall encounter some other nonlinear ODEs (e.g., nonlinear damping in the next section) for which analytic answers are obtainable with little extra effort.

Regardless of whether an analytic form for the solution can be found or not, a general feature of undriven nonlinear pendulum equations is that the period of oscillation depends on the angular amplitude. The experimentally inclined reader can verify that this is the case for the spin toy pendulum which is the subject of the next experiment.

Spin Toy Pendulum
In this experiment an inexpensive commercial toy is used to investigate how the period of oscillation varies with the angular amplitude. The experiment verifies that the period is substantially different for large angular displacements compared to that measured for small angles.

PROBLEMS
Problem 2-1: Period of the simple pendulum
Determine the period of a simple plane pendulum which is released from rest from an initial angle $\theta(0) = 175°$.

Problem 2-2: Time of descent
Determine the time of descent of a simple pendulum from $\theta = 170°$ to i) $\theta = 25°$, ii) to $\theta = -30°$.

Problem 2-3: Damped simple pendulum
The damped simple pendulum equation is given by

$$\ddot{\theta} + 2\alpha\dot{\theta} + \omega_0^2 \sin\theta = 0$$

where α is the positive damping coefficient. If $\ell = 1$ meter, $g = 9.8$ m/s^2, $\alpha = 0.25$ s^{-1}, and $\theta(0) = 177°$, how long does it take the pendulum swinging from rest to first reach the position $\theta = 0°$. How long does it take the pendulum to reach $\theta = 0°$ a second time?

Problem 2-4: Parametric excitation revisited

In the parametric excitation Example 2-1, suppose that the pivot point O is undergoing horizontal oscillations given by $A\sin(\omega t)$. Using the Lagrangian approach, derive the equation of motion of the mass m.

Problem 2-5: Parametric excitation solution

Numerically solve Example 2-1 by taking $A = 0.20$ m, $g = 9.8$ m/s^2, $r = 1$ m, $\omega = 1$ radian/s, $\theta(0) = 30°$, $\dot{\theta}(0) = 0$. Plot the solution over a suitable time range and determine the period of the repeat pattern. Relate the period to one of the two frequencies in the problem.

Problem 2-6: Example 1-7 revisited

Use the Lagrangian approach to derive the equation of motion of the mass m_1 in Example 1-7.

Problem 2-7: The spherical pendulum

If a single particle of mass m is constrained to move without friction on the surface of a sphere of radius ℓ and is acted upon by a uniform gravitational field, it forms a spherical pendulum. Write down the equations of motion in

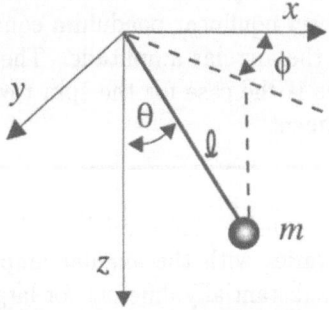

terms of the spherical polar coordinates r, θ, ϕ. Show that $\theta(t)$ satisfies the differential equation

$$\ddot{\theta} - \frac{L^2 \cos\theta}{m^2 \ell^4 \sin^3\theta} + \omega_0^2 \sin\theta = 0$$

where $L = $ constant is a component of the angular momentum that you must identify. You may use either the Newtonian or Lagrangian approach.

Problem 2-8: The rotating pendulum

A vertically oriented circular wire of radius ℓ rotates with angular velocity ω about the z-axis as shown in the accompanying figure. A bead of unit mass ($m = 1$) is allowed to slide along the frictionless wire.

a. If the plane of the circular wire is oriented along the y-axis at $t = 0$, what are the x, y and z coordinates of the bead at time t?

b. Using the Lagrangian approach, show that the equation of motion for the bead is

$$\ddot{\theta} + \omega_0^2 \sin\theta - \frac{1}{2}\omega^2 \sin(2\theta) = 0.$$

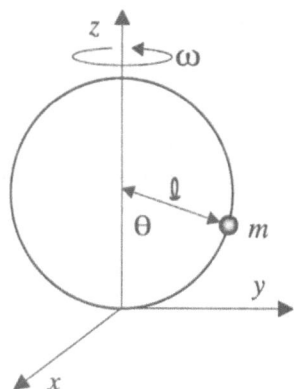

Problem 2-9: The double pendulum

The double pendulum consists of two small masses m_1 and m_2, with light connecting rods of lengths r_1 and r_2, free to execute planar motion about the pivot point O. Derive the equations of motion expressed in terms of the angles θ_1, θ_2 using the Lagrangian approach.

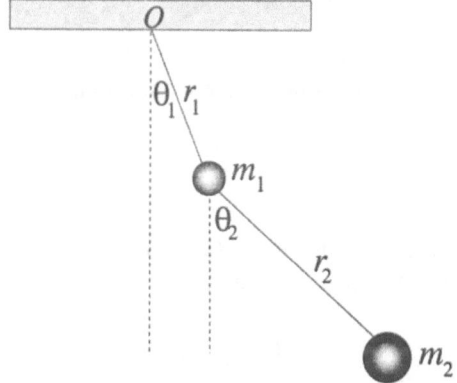

Problem 2-10: The ball bearing pendulum equation

Suppose that a solid ball bearing of radius r rolls back and forth without slipping on a circular track of radius R as illustrated in the figure. Show that the angle θ with the vertical satisfies the simple pendulum equation but with the frequency

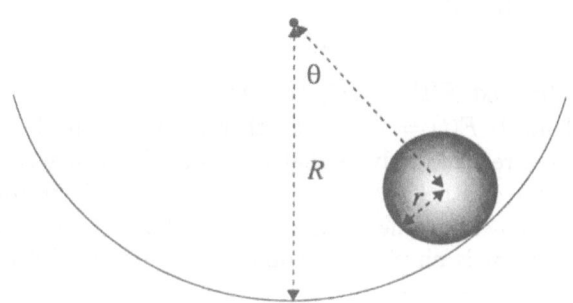

given by $\omega = \sqrt{5g/[7(R-r)]}$ where g is the acceleration due to gravity. Hint: What is the moment of inertia of a solid sphere about an axis through its center?

Problem 2-11: The double pendulum

Derive the equations of motion for the double pendulum of Problem 2-9 using the Newtonian approach.

2.1.2   The Eardrum

Newton's second law can also be applied to biological examples such as the vibrating eardrum. As early as 1895, Helmholtz [Hel95] was aware that the ear perceived frequencies that are not contained in the incident acoustic radiation. To understand this, let us treat the eardrum's tympanic membrane as a mechanical system undergoing 1-dimensional vibrations about its equilibrium position, the displacement being $x(t)$. Quite generally, we can write the restoring force for small x as the Taylor expansion[1]

$$F(x) = F_0 + \left(\frac{dF}{dx}\right)\bigg|_0 x + \frac{1}{2!}\left(\frac{d^2F}{dx^2}\right)\bigg|_0 x^2 + \frac{1}{3!}\left(\frac{d^3F}{dx^3}\right)\bigg|_0 x^3 + \cdots. \quad (2.7)$$

In equilibrium, $x = 0$, and the restoring force $F(x)$ must also vanish so that $F_0 = 0$. Suppose that the term linear in x is dominant so that

$$F(x) \simeq \left(\frac{dF}{dx}\right)\bigg|_0 x. \quad (2.8)$$

To be a restoring force, we must have $(dF/dx)|_0 < 0$ for positive x. Setting $(dF/dx)|_0 = -k$, with the spring constant k positive, gives us the well-known Hooke's law

$$F(x) = -kx \quad (2.9)$$

which is valid for small x.

If $f(t)$ is the driving force on the eardrum produced by the periodically varying pressure of the incoming sound wave and m is the mass of the tympanic membrane, Newton's second law yields

$$m\ddot{x} = -kx + f(t), \quad (2.10)$$

which can be rewritten as

$$\ddot{x} + \tilde{\omega}_o^2 x = F(t) \quad (2.11)$$

with $\tilde{\omega}_o = \sqrt{k/m}$ and $F(t) = f(t)/m$. This is the forced simple harmonic oscillator problem. If $F(t) = A\cos(\omega t)$, then after an initial transient period the eardrum would respond only at that particular frequency ω.

To hear frequencies other than ω requires that the eardrum equation possess some degree of nonlinearity, the nonlinearity coupling the input into other harmonics. When the ear is physically examined, it is observed that the eardrum

---

[1]Recall Example 1-2, part(b).

is asymmetrically loaded and, as a consequence, undergoes asymmetric oscillations. If we keep the quadratic term in the Taylor expansion (2.7) and set $(1/2!) (d^2F/dx^2)|_0 = -\beta m$, Equation (2.11) is replaced with the nonlinear equation

$$\ddot{x} + \tilde{\omega}_o^2 x + \beta x^2 = F(t). \tag{2.12}$$

The quadratic term in (2.12) introduces asymmetry because it does not change sign with x. Equation (2.12) is the eardrum equation deduced by Helmholtz.

The forced oscillation of nonlinear systems such as those described by Equation (2.12), usually with damping included, leads to a wide variety of physical phenomena, including harmonic (multiples of the driving frequency) oscillation, subharmonic oscillations, and chaotic or non-periodic behavior. The eardrum can also hear the sum and difference of two input frequencies. The following diode circuit experiment simulates how the eardrum can produce these frequency combinations.

Driven Eardrum

In this experiment a biased diode driven by two sinusoidal signals is used to model the frequency response of the eardrum. The output signal is converted to audio so that the combination of frequencies can be heard.

EXP 04

PROBLEMS
Problem 2-12: Eardrum potential
Consider the freely vibrating eardrum equation

$$\ddot{x} + x + 0.1x^2 = 0.$$

Plot the potential energy curve for this eardrum over the range $x = -12...6$. Assuming that the eardrum is initially at rest $(\dot{x}(0) = 0)$ at some positive value $x(0)$, what is the maximum value that $x(0)$ can have for the eardrum to oscillate? What happens if this $x(0)$ value is exceeded?

Problem 2-13: Eardrum behavior
Making use of the numerical approach of Example 1-5, explore the solutions of the eardrum equation of the previous problem as a function of increasing driving frequency ω if it is driven by the forcing function $F(t) = 0.6\sin(\omega t)$. Take $x(0) = 1$, $\dot{x}(0) = 0$, and begin with $\omega = 0.2$. Try to identify the repeat pattern of the eardrum oscillations and relate the period of the pattern to that of the driving force. You may have to alter the time range for your plots.

Problem 2-14: Damped eardrum behavior
Making use of the approach of Example 1-5, numerically determine and plot the solution of the damped eardrum equation,

$$\ddot{x} + 0.5\dot{x} + x + 0.1x^2 = 1.0\sin(0.1t),$$

with $x(0) = 1$, $\dot{x}(0) = 0$. What is the repeat pattern of the eardrum oscillations? Relate the period of the pattern to that of the driving force. Take the time range to be $t = 0...400$ and choose an adequate number of plotting points.

2.1.3   Nonlinear Damping

When an object moves through a viscous fluid (e.g., the atmosphere or water), the fluid exerts a retarding or drag force $\vec{F}_{\text{drag}}$ on the object. Drag plays an important role in the flight characteristics of high speed aircraft as well as golf balls. The mathematical form of the drag force is in general quite complicated for aircraft, race cars, America's Cup racing yachts, etc., and is usually experimentally determined through the use of wind tunnels and large water tanks.

If $\vec{v}$ is the instantaneous velocity, $\vec{F}_{\text{drag}} = \vec{F}_{\text{drag}}(\vec{v})$. The simplest model which is usually considered is where

$$\vec{F}_{\text{drag}} \propto |\vec{v}|^{n-1}\vec{v} \tag{2.13}$$

with n an integer. Then, for example, the equation of motion of an object near the earth's surface would be described by

$$m\dot{\vec{v}} = m\vec{g} - mk\vec{v}|\vec{v}|^{n-1} \tag{2.14}$$

with k a positive constant which depends, among other things, on the density and viscosity of the air and on the shape of the projectile. Experimentally, for a pointed military shell moving in air, one finds that $n = 1$ for $v \leq 24$ m/s or 86 km/h. The $n = 1$ case is referred to as Stokes' law of resistance. For higher v, but still below the speed of sound (≈ 330 m/s), $n = 2$. The $n = 2$ case is referred to as Newton's law of resistance. In both cases n is not precisely 1 or 2, but integer powers are convenient as they allow the equations of motion to be analytically integrated. These force laws have been qualitatively confirmed by the U.S. Army Artillery Corps studying the flight of small projectiles. There is a "bump" in the drag force curve just above the speed of sound and then for $v \geq 600$ m/s the drag force becomes approximately linear in v again. See [MT95]. Nonlinear damping is important in other contexts, for example, in the stabilizers that are intended to counteract the rolling of large ships in choppy seas.

On a much smaller scale, the third experiment illustrates how the mathematical form of the drag force due to air resistance may be found.

Nonlinear Damping
This experiment determines the values of the exponent n in the drag force law. A cardboard sail is attached to an airtrack glider which produces nonlinear drag. The drag force is plotted as a function of velocity to find n.

In the second Mathematica file, the equation of motion of a falling sphere, with viscous air drag present, is investigated.

Nonlinear Air Drag on a Sphere
Here, a spherical object of diameter d falling from rest is acted upon by a drag force [FC86]

$$F_{\text{drag}} = -A\,v - B\,v^2$$

with A and B positive. In SI units, $A = 1.55 \times 10^{-4}\,d$ and $B = 0.22\,d^2$. This empirical force law is a combination of Stokes' and Newton's laws of resistance. Mathematica has no difficulty in generating an analytic solution, expressed in the form of a "pure" function, which is then plotted as a function of time. Mathematica commands in file: `D, DSolve, FullSimplify, NSolve, Plot, RGBColor, AxesLabel, PlotLabel, TextStyle, PlotStyle, /`.

We have already mentioned the importance of nonlinear damping for the flight characteristics of a golf ball. Another important feature is the nonlinear lift provided to the ball by the backspin imparted to it by a golf club.

If you have golfed or watched golf on television, you may have wondered why there are "dimples" on a golf ball, instead of it being perfectly smooth like a billiard ball. The dimples help with the lift. To see this lift effect, run Mathematica File MF 03.

Drag and Lift on a Golf Ball
Explore the effects of nonlinear drag and lift on a golf ball. Good golfing! Mathematica commands in file: `Sqrt, Cos, Sin, N, Table, NDSolve, Show, Block, DisplayFunction, ParametricPlot, Evaluate, MaxSteps, Hue`

PROBLEMS

Problem 2-15: Nonlinear air drag on a baseball
A baseball has a diameter of 0.0764 m. Using the formula for the nonlinear

air drag on a sphere of diameter d, plot the magnitude of the drag force on the baseball as a function of the velocity v for $v=0$ to 50 m/s. A major league pitcher can throw a "fast ball" at 95 miles per hour or 42 m/s. What is the drag force in Newtons on the baseball? Which is the dominant term in the drag force formula in this case, linear or quadratic?

Problem 2-16: Falling raindrops and basketballs
A certain raindrop has a diameter of 0.10 mm and a mass of 0.52×10^{-9} kg while a basketball has a diameter of 25 cm and a mass of 0.60 kg.

a. Assuming that each is dropped from rest, modify MF02 to determine $v(t)$ for the raindrop and the basketball.

b. Plot $v(t)$ in each case, showing the approach to the terminal velocity.

   c. Determine the terminal velocity in each case.

   d. Approximately how long does it take the raindrop and the basketball to come within one percent of the terminal velocity?

Problem 2-17: Falling soap bubble

A soap bubble of diameter of 0.01 m and mass 10^{-7} kg falls from rest.

   a. Modify MF02 to determine $v(t)$ for the soap bubble.

   b. Plot $v(t)$ and determine the terminal velocity.

   c. Approximately how long does it take the soap bubble to come within one percent of the terminal velocity?

Problem 2-18: Using a 9-iron

After a good approach shot, an LPGA golfer has placed her golf ball 100 meters from the hole. She selects a 9-iron and asks her caddy the following questions:

   a. At what angle on her next shot should the ball leave the club to hit the ground at the hole? Assuming that all other conditions are the same, run MF03 with lift and drag included to answer her question, finding the angle of ascent to the nearest degree.

   b. To what maximum height will the ball rise for this shot?

   c. If both lift and drag had been neglected, what would have been the horizontal range of the shot?

   d. If the golfer's shot had been from an elevated tee to a green located 50 meters lower, how far would her ball have landed from the point at which it was hit? Include lift and drag.

Problem 2-19: Golf data

Run MF03 and collect data on the maximum height and horizontal range as a function of initial angle and plot the data in separate graphs. Both drag and lift are to be included.

Problem 2-20: American vs British golf ball

In MF03, data for a British golf ball was used. Do a literature search and find the mass and diameter of an American golf ball. Assuming that all other parameters are the same as in the file, and that lift and drag are included, display the trajectories of the American and British balls in the same graph. Is there a significant difference in horizontal range? Could there be other differences between the balls which could be important? Explain.

Problem 2-21: Some "what if" golf questions

Use MF03 to answer all the following questions. Except where otherwise specified, assume that all conditions are as in the file.

   a. As a part of a golf ball testing routine, the golf ball is shot straight upwards. What does the trajectory look like with lift and drag included? Is the result surprising or not? Explain.

b. A golf ball that has been teed up 3 cm above the ground is hit badly and leaves the tee horizontally. Including lift and drag, plot the trajectory of the ball and determine where it strikes the ground, assuming that the fairway is level?

c. At a Rocky Mountain golf resort, the elevation is such that the density of air is $\rho = 1.0$ kg/m^3 and $g = 9.8$ m/s^2. How much further would the ball travel than for the sea level data given in the file?

2.1.4 Nonlinear Lattice Dynamics

Research on nonlinear lattice dynamics began seriously in the 1950s with the numerical work of the Nobel physics laureate Enrico Fermi, J. Pasta, and Stan Ulam (FPU) on the MANIAC I computer at Los Alamos [FPU55]. For a system of N coupled simple harmonic oscillators, it is well known in classical mechanics that the energy of each normal mode of the system will remain constant. FPU intended to verify the widely accepted assumption that the introduction of small nonlinearities would lead to an equipartition of energy, i.e., the small nonlinearities would cause energy to flow from one mode to another until all modes, in a time-averaged sense, would have the same energy. The existence of such an equilibrium is known as *the zeroth law of thermodynamics* and is an important assumption of statistical mechanics.

FPU numerically solved Newton's equations of motion for the $N = 64$ 1-dimensional spring system shown in Figure 2.2. They considered nearest-

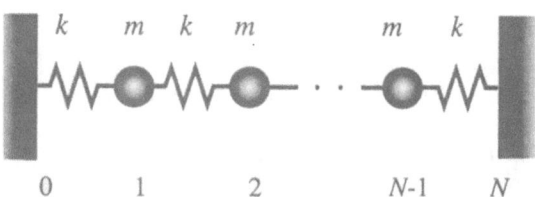

Figure 2.2: Nonlinear spring setup in the FPU problem.

neighbor interactions, taking the interaction potential to be

$$V(r) = \frac{1}{2}kr^2 + \frac{1}{3}k\alpha r^3 \qquad (2.15)$$

with r the relative displacement of nearest neighbors from equilibrium and k and α constants. The force $F(r) = -dV/dr$, so that

$$F(r) = -kr - k\alpha r^2 \qquad (2.16)$$

which is of the same structural form as we had for the eardrum! In lattice dynamics, the first and second terms in (2.15) are referred to as the "harmonic" and "anharmonic" contributions.

FPU found that with $F(r)$ given by (2.16), the nonlinear system displayed non-ergodic behavior, that is, the system did *not* approach equilibrium as expected but instead displayed recurrences of energy in certain modes. This was

referred to as the FPU anomaly. The resolution of this anomaly is beyond
the scope of this text, the interested reader being referred to David Campbell's
Los Alamos report [Cam87] and to Morikazu Toda's *Theory of Nonlinear Lattices* [Tod89].

Toda was able to study the FPU problem analytically instead of numerically by inventing an interaction potential for which analytical solutions could
be derived. The Toda lattice is described by the nearest neighbor interaction
potential

$$V(r) = \frac{a}{b}e^{-br} + ar \tag{2.17}$$

with the product ab positive. $V(r)$ schematically looks like the curves in Figure 2.3

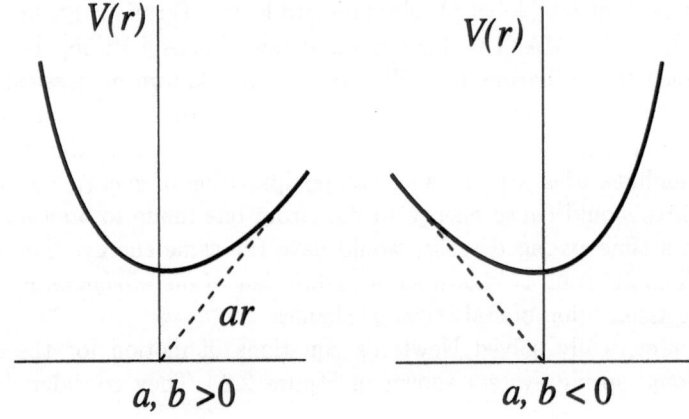

Figure 2.3: The Toda potential.

for $a, b > 0$ and $a, b < 0$. The corresponding force is

$$F(r) = a(e^{-br} - 1). \tag{2.18}$$

For sufficiently small r, the exponential can be expanded yielding

$$F(r) = -abr + \frac{1}{2}ab^2r^2 - \cdots . \tag{2.19}$$

The leading two terms in the expansion are of the same structural form as
in the original FPU problem. The study of nonlinear lattice dynamics is not
only theoretically important in studying the foundations of statistical physics
but is, for example, practically important in trying to understand the onset of
melting in a solid. As the temperature of a solid is increased, the amplitude
of oscillations of the atoms or ions about their equilibrium positions increases
and anharmonic contributions become increasingly more important. At some
critical temperature, the oscillations are so large that the solid transforms into
a liquid.

If the reader wishes to see a simple example of an anharmonic potential,
although not in a lattice dynamics context, consider doing the next experiment.

Anharmonic Potential
This airtrack experiment demonstrates how an anharmonic potential produces nonlinear oscillations. The anharmonic potential is made by tilting the airtrack and attaching magnets to the airtrack glider. A large repelling magnet is placed at the lower end of the airtrack. The nature of the anharmonic potential is explored and the period measured and compared with calculated results.

PROBLEMS
Problem 2-22: The Toda equation of motion
 Consider the infinitely long 1-dimensional Toda lattice shown in the accompanying figure. If x_k is the displacement of the kth atom (all atoms having the

same mass m) from equilibrium, Newton's second law gives

$$m\ddot{x}_k = -F(x_{k+1} - x_k) + F(x_k - x_{k-1}).$$

Explain the structure of the force terms. Setting $y_k = br_k = b(x_{k+1} - x_k)$, show that Toda's equation of motion

$$\ddot{y}_k(\tau) = 2e^{-y_k} - e^{-y_{k+1}} - e^{-y_{k-1}}$$

can be obtained. What is the relation of τ to t?

2.2 Competition Phenomena

In nature, there exist a wide variety of examples of competing groups, species or processes which can be modeled by nonlinear equations. Some important examples are:

- populations of interacting biological species at the macroscopic and microscopic levels;

- interacting laser beams;

- competing political parties, businesses, countries, etc.

One can construct models describing the interactions either from first principles or intuitively with phenomenological equations. A few examples will serve to illustrate these ideas. An ever-increasing number of examples may be found in the current research literature.

2.2.1   Volterra–Lotka Competition Equations

The Volterra–Lotka equations describing the competition between biological species were motivated by the work of Vito Volterra's friend D'Ancona who carried out a statistical analysis of the fish catches in the Adriatic sea in the period 1905–1923. It was observed that the populations of two species of fish (big fish and little fish) varied with the same period, but somewhat out of phase. The big fish survived by eating the little fish. As the big fish ate the little fish, they grew and their numbers increased. Eventually, however, a time was reached when the population of small fish decreased to such a level that some of the large ones could not survive (i.e., they starved) so that their numbers began to decline. With the decline in the number of big fish, the small fish population began to increase again because there were fewer predators. However, with more small fish becoming available, the population of large fish began to grow again, and the cycle was repeated. As a model of the predator–prey relationship, Volterra [Vol26] proposed the following pair of phenomenological equations describing the population numbers N_B and N_L of big and little fish, respectively:

$$\dot{N}_B = (-\alpha_B + g_B N_L)N_B$$
$$\dot{N}_L = (\alpha_L - g_L N_B)N_L, \tag{2.20}$$

with all coefficients positive. The structure of these equations is readily understandable. In the absence of any interaction between the two species of fish, i.e., $g_B = g_L = 0$, the big fish number N_B decays exponentially and the small fish number increases. This assumes that there is an adequate food supply for the small fish. When there is some interaction, a "growth coefficient" $g_B N_L$, depending on the number of little fish present, must appear in the big fish equation. Correspondingly, a "decay coefficient" $-g_L N_B$, appears in the little fish equation. g_B and g_L are not usually the same numerically for the simple reason that one big fish can eat several little fish. As shall be seen, predator–prey equations such as (2.20) can be handled nicely by a combination of topological, analytical and numerical techniques.

This phenomenological approach to predator–prey interactions is quite common and the equations can accurately mimic actual situations by appropriate choices of numerical coefficients. The trading records (Figure 2.4) of fur catches by trappers working for the Hudsons Bay Company in the Canadian north for the period 1845 to 1935 display the predator–prey interaction, the predator being lynx and the prey, snowshoe hares.

R. B. Neff and L. Tillman [NT75] discussed the numerical solution of the rabbits–foxes equations

$$\dot{r} = 2r - \alpha r f$$
$$\dot{f} = -f + \alpha r f. \tag{2.21}$$

Here, r and f refer to the rabbit and fox population numbers respectively. For $\alpha = 0$, there are no encounters and the rabbits, having plenty of vegetation to

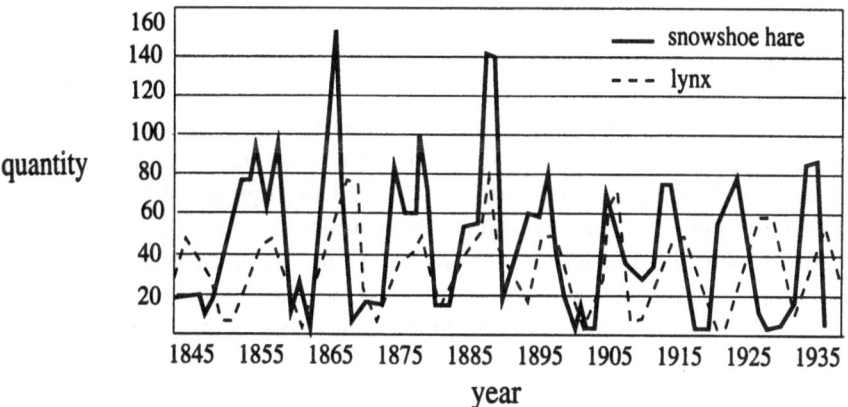

Figure 2.4: Trading records of fur catches for the Hudsons Bay Company.

eat, breed rapidly (note the factor of 2 in the rabbit equation) as rabbits are prone to do while the foxes, whose diet consists mainly of rabbits, starve. The number α, measuring the strength of the interaction, can only be determined by comparison with actual population statistics, but, for numerical illustration, the authors took $\alpha = 0.01$.

An example of the cyclic variation of rabbit and fox population numbers is shown in Figure 2.5 for this α value. This plot was generated for the initial values $r(0) = 100$, $f(0) = 5$ using MF04. Here time runs counterclockwise around the "orbit" in the r versus f plane. Such a plane is an example of a "phase plane" and the orbit is referred to as a "phase plane trajectory".

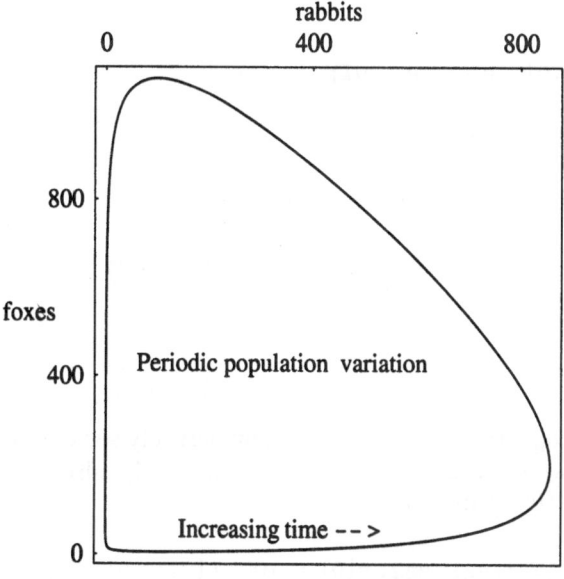

Figure 2.5: Rabbit versus fox numbers.

Rabbits–Foxes Equations

Eqs. (2.21) are solved numerically (as an analytic answer does not exist) using Mathematica's default numerical ODE solving method. Two plots are created with `ParametricPlot3D`, the second plot being reproduced in Figure 2.5. For an initial rabbit number $r(0) = 100$ and fox number $f(0) = 5$, we obtain the periodic behavior shown in the figure where the number of foxes is plotted versus the number of rabbits, and the direction of increasing time indicated. In this notebook, the reader may change the controlling coefficients and experiment with different initial conditions. Mathematica commands in file: `NDSolve`, `Evaluate`, `ViewPoint`, `AxesLabel`, `PlotPoints`, `Epilog`, `ParametricPlot3D`, `BoxRatios`, `Hue`, `PlotRange`, `Ticks`, `Text`

An alternate way to generate the phase plane trajectory of Figure 2.5 is presented in Example 2-2.

Example 2-2: Phase Plane Trajectory

Create a plot showing the evolution of the rabbit and fox numbers with time and then reproduce the phase plane trajectory shown in Figure 2.5. Use the Rungé–Kutta–Fehlberg method in the `NDSolve` command and take the time interval to be $t = 0...20$.

Solution: With the coupled rabbits-foxes equations written in the form

$$\dot{r}(t) = a\,r(t) - b\,r(t)\,f(t),$$
$$\dot{f}(t) = -c\,f(t) + d\,r(t)\,f(t),$$

the coefficient values a, b, c, and d are entered

```
Clear["Global`*"]
a = 2; b = .01; c = 1; d = .01;
```

along with the two equations.

```
eq1 = r'[t] == a r[t] - b r[t] f[t]
```

$$r'(t) == 2\,r(t) - 0.01\,f(t)\,r(t)$$

```
eq2 = f'[t] == -c f[t] + d r[t] f[t]
```

$$f'(t) == -f(t) + 0.01\,f(t)\,r(t)$$

The system of equations, `eq1` and `eq2`, is numerically solved over the time range $t = 0...20$ using the Rungé–Kutta–Fehlberg method, subject to the initial conditions $r(0) = 100$, $f(0) = 5$.

```
sol = NDSolve[{eq1, eq2, r[0] == 100, f[0] == 5}, {r, f}, {t, 0, 20},
    Method -> RungeKutta];
```

Using the `Block` construct, two graphs are produced but not displayed. The `Plot` command is used in `gr1` to plot the rabbit and fox numbers versus time. In the `PlotStyle` option the rabbit curve is given a blue hue with `Hue[.6]`, while the fox curve is dashed with the `Dashing` command and given a red hue with `Hue[1]`. The `ParametricPlot` command produces the phase plane trajectory, which is colored green with `Hue[.3]`, in `gr2`.

```
Block[$DisplayFunction = Identity,
gr1 = Plot[Evaluate[{r[t], f[t]} /. sol], {t, 0, 20},
    PlotStyle -> {Hue[.6], {Dashing[{.02}], Hue[1]}}, Ticks ->
    {{{.01, 0}, {10, "time"}, 20}, {500, {700, "number"}, 1000}},
    TextStyle -> {FontFamily -> "Times", FontSize -> 14},
    PlotLabel -> "solid blue:  rabbits, dashed red:  foxes"];
gr2 = ParametricPlot[Evaluate[{r[t], f[t]} /. sol], {t, 0, 20},
    PlotStyle -> {Hue[.3]}, PlotRange -> {{0, 1000}, {0, 1100}},
    Ticks -> {{{.01, 0}, 500, {700, "rabbits"}, 1000},
    {500, {700, "foxes"}, 1000}},
    TextStyle -> {FontFamily -> "Times", FontSize -> 14}]];
```

The `GraphicsArray` command is used to place the two graphs, `gr1` and `gr2`, side by side and the result displayed in Figure 2.6 with the `Show` command.

```
Show[GraphicsArray[{gr1, gr2}], ImageSize -> {600, 200}];
```

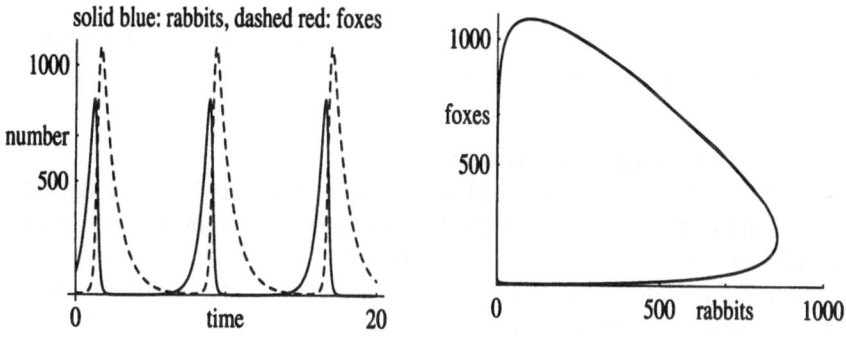

Figure 2.6: Left: Rabbit, fox numbers vs. time. Right: Phase plane trajectory.

The periodic variation of the rabbit and fox numbers with time is shown on the left, the phase plane trajectory on the right. The latter curve is the same as shown earlier in Figure 2.5.

End Example 2-2

As is discussed in a review paper by Goel, Maitra and Montroll [GMM71] many generalizations and modifications of the Volterra–Lotka equations have been suggested to more accurately describe various interacting populations. For example, one aspect which is neglected in the Volterra model, although it was apparently well known to him, is the approach to saturation of a population that must survive on limited resources. If a species is not preyed upon, it is expected that its population will saturate and not continue to grow indefinitely (as it would in the Volterra model). This point was first made by the Belgian mathematician Verhulst in 1845. Thus, for example, the term $\alpha_L N_L$ in the little fish equation (2.20) would be replaced by the so-called Verhulst term, $\alpha_L N_L (\theta - N_L)/\theta$, where θ is called the saturation number.

Still another feature that various investigators have taken into account is the time lag due to reproduction, or due to the reaction of the members of the population to any change in environment, or to changes in populations of other species. For example, in Problem 6-21 (page 247), time delay has been built into the equations modeling the population dynamics of baleen whales.

Of course, the Volterra model and its generalizations are phenomenological models which must be experimentally verified. An interesting discussion of this point may be found in Section 10 of the paper by Goel and coworkers [GMM71].

It should also be pointed out that in the above modeling, we are treating the population numbers as continuous variables, whereas in reality the numbers should in fact be integers. For small population numbers, this could become important. An alternate approach is to make use of number density, e.g., foxes/km^2 as in the example of the next subsection.

PROBLEMS
Problem 2-23: Fur catches
Discuss the behavior of the snowshoe hare and lynx curves in Figure 2.4, using the text discussion of the big fish–little fish interaction as a guide. Compare the trading record curves with the idealized theoretical curves for the rabbits and foxes in the first graph of Figure 2.6 and discuss the differences.

Problem 2-24: Foxes and Rabbits
Taking $r(0) = 50$, $f(0) = 50$, and $\alpha = 0.01$ in the rabbits–foxes equations, calculate and plot the time evolution of the fox and rabbit numbers as well as the phase plane trajectory using (a) MF04, (b) Example 2-2. To what (approximate) maximum number does the rabbit population grow? the fox population? What are the minimum population numbers?

Problem 2-25: Saturable Volterra model
To take into account a saturation effect in the interaction term caused by a large number of rabbits, the rabbits (r)–foxes (f) equations can be written as

$$\dot{r} = 2r - \frac{\alpha r f}{1 + sr}, \quad \dot{f} = -f + \frac{\alpha r f}{1 + sr}$$

with the saturation parameter $s \geq 0$. Modify file MF04 to include the above saturation effect. Taking $\alpha = 0.01$, $r(0) = 100$, $f(0) = 5$, determine f and r vesus time t, and $f(t)$ versus $r(t)$ for $s = 0.001$ and plot the results. Compare your solution with that obtained for $s = 0$ and discuss the effect of nonzero s. Experiment with different values of α, s and $r(0)$ and discuss what happens.

Problem 2-26: Saturable Volterra Problem 2

Repeat the preceding problem, but with the suitably modified code of Example 2-2 and $r(0) = f(0) = 50$. All other parameters are the same. Discuss the behavior in this case.

Problem 2-27: Rabbits and sheep

Some rabbits and sheep are competing with each other in the munching of Farmer Brown's pasture of luscious green grass. Suppose that the equations describing the rabbit number $r(t)$ and sheep number $s(t)$ are

$$\dot{r} = r(3 - r) - 2rs, \qquad \dot{s} = s(2 - s) - rs,$$

these equations being phenomenological in nature.

a. Discuss the interpretation of the various terms in each equation.

b. By modifying the code in Example 2-2, determine the phase plane trajectories for the two initial conditions $r(0) = 5$ rabbits, $s(0) = 5$ sheep and $r(0) = 10$, $s(0) = 5$. Include both trajectories in the same graph. Describe the outcome in each case.

c. Try some other initial conditions. What can you conclude? Can the two animal species competing for the same limited resource both coexist?

Problem 2-28: Big fish–little fish

Choosing appropriate (albeit, artificial) values for the coefficients, confirm that the big fish–little fish population equations can have a cyclic solution.

2.2.2 Population Dynamics of Fox Rabies in Europe

The use of Volterra-like equations to model the spread of diseases, a branch of epidemiology, is discussed in the book *Mathematical Biology* by J.D. Murray [Mur89]. An example of the modeling of the rabies epidemic in Central Europe, which is believed to have originated in Poland in 1939, was given by

Anderson, Jackson, May and Smith [AJMS81]. This epidemic was primarily transmitted by the fox population.

In the model, the fox population is divided into three classes, all of which are measured in population density (foxes/km^2), namely: susceptibles, denoted by X, are foxes that are currently healthy but are susceptible to catching the virus; infected, denoted by Y, are foxes that have caught the virus but are not yet capable of passing on the virus; and finally, infectious, denoted by Z, are foxes that are capable of infecting the susceptibles. Note that the model has no category of recovered immune foxes because very few, if any, survive after acquiring the rabies virus. For other diseases with a lower mortality rate, one would add a recovered category and therefore an additional modeling equation. The relevant equations for the fox rabies epidemic are

$$\dot{X} = aX - (b + \gamma N)X - \beta XZ$$

$$\dot{Y} = \beta XZ - (\sigma + b + \gamma N)Y \tag{2.22}$$

$$\dot{Z} = \sigma Y - (\alpha + b + \gamma N)Z$$

with $N = X + Y + Z$ being the total fox density. The meaning of the various coefficients in (2.22) and their estimated values are given in the following table. Note that the estimated value of γ has quite a wide range.

Symbol	Meaning	Value
a	average per capita birth rate of foxes	1 yr^{-1}
b	average per capita intrinsic death rate	0.5 yr^{-1}
β	rabies transmission coefficient	79.67 km^2 yr^{-1}
σ	$1/\sigma$ is the average latent period ($\sim$ 28 to 30 days)	13 yr^{-1}
α	death rate of rabid foxes (average life expectancy $\sim$ 5 days)	73 yr^{-1}
γ	γN represents increased death rate when N is large enough to deplete the food supply	0.1-5 km^2 yr^{-1}

Figure 2.7 shows a representative result from Mathematica File 05 for $\gamma = 0.1$, all other parameters taken from the above table. Here we have chosen to plot X and Y versus time t in years. The start of the numerical run is indicated. Note how, for example, the population density X of healthy foxes nearly dies away, builds up again, nearly dies away again, and so on. This cyclic behavior is characteristic of many common contagious diseases when left untreated.

The fox rabies model described by Equations (2.22) was formulated to assess the effectiveness of different methods for controlling the spread of rabies such as putting oral vaccines in different baits versus hunting the foxes.

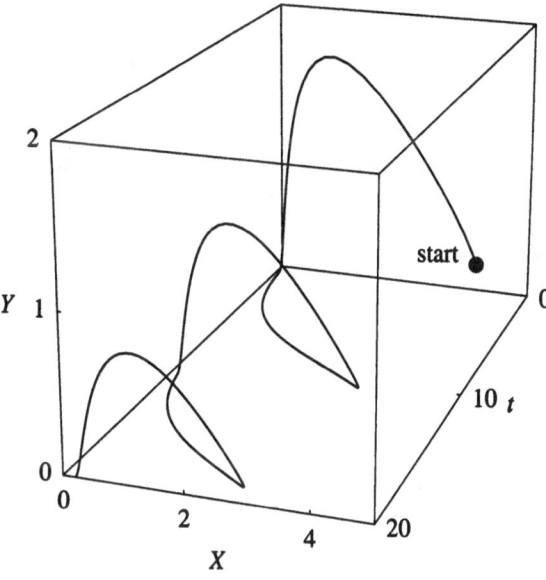

Figure 2.7: Population dynamics for fox rabies (X healthy, Y infected foxes).

Fox Rabies Epidemic Model

Using the parameter values from the table, Equations (2.22) are solved numerically and the results plotted. Figure 2.7 shows one set of results from the file. Here we have taken $\gamma = 0.1$, initial conditions $X(0) = 4$, $Y(0) = 0.2$ and $Z(0) = 0.1$ and allowed 20 years to elapse. The reader may explore other values of γ and various initial population densities. Mathematica commands in file: `NDSolve`, `Method->RungeKutta`, `PrecisionGoal`, `MaxSteps`, `Evaluate`, `ParametricPlot3D`, `AxesLabel`, `PlotPoints`, `PlotRange`, `Text`, `Point`, `ViewPoint`, `BoxRatios`, `Ticks`, `BoxStyle`, `RGBColor`, `Epilog`, `PointSize`, `TextStyle`, `FontSize`, `AxesEdge`, `PlotLabel`, `StyleForm`,

PROBLEMS

Problem 2-29: Epidemic Model

Discuss the mathematical structure of each term in the fox rabies equation.

Problem 2-30: SIR epidemic model

Kermack and McKendrik (KM27) developed the SIR epidemic model to describe the data from the Bombay plague of 1906. The acronym SIR stands for the three groups of people in the model:

- the susceptible (S) group who are healthy but can become infected,

- the infected (I) group who have the disease and are able to infect healthy individuals,

- the removed (R) group who are either dead or have recovered and are permanently immune.

The SIR system of equations is given by

$$\dot{S} = -\alpha SI, \quad \dot{I} = \alpha SI - \beta I, \quad \dot{R} = \beta I,$$

with α and β positive.

   a. Discuss the mathematical structure of these equations.

   b. List and discuss any assumptions that you think have gone into the model.

   c. By modifying MF05, explore and discuss the solutions of this model.

2.2.3   Selection and Evolution of Biological Molecules

According to current theory, in the primordial chemical "soup" where life began, RNA chemistry provided an environment for Darwinian selection and evolution. According to Nobel chemistry laureate Manfred Eigen and his collaborator Peter Schuster [Eig71, EGSWO81], the first carriers of genetic information were probably primitive strands of RNA which could self-replicate, although not perfectly. Mutations caused slight errors in the nucleotide sequence making up a given RNA strand, thus producing other closely related species of RNA. These RNA species competed against each other for energy and food (energy-rich monomers).

As a starting point to understanding what might have occurred, Eigen and Schuster have suggested simple competition models which are built up phenomenologically. We shall outline how one such model is formulated. Let us first define the necessary symbols.

Symbol	Meaning
$X_k(t)$	concentration of species k ($k = 1, 2, \ldots N$)
A_k	total reproduction rate of species k, including the imprecise copies it produces due to mutations
Q_k	fraction of the copies which are precise (the quality factor)
D_k	decomposition ("death") rate of species k
$\phi_{k\ell}$	a mutation coefficient which represents the rate of production of species k due to errors in replication of species ℓ
W_k	$\equiv A_k Q_k - D_k$, net intrinsic rate of producing exact copies

The linear rate equation for producing species k is then

$$\dot{X}_k(t) = W_k X_k(t) + \sum_{\ell \neq k}^{N} \phi_{k\ell} X_\ell(t). \tag{2.23}$$

Now, since $1 - Q_k$ represents the fraction of mutations or imprecise copies of k produced, we must have the conservation relation

$$\sum_k A_k(1 - Q_k)X_k = \sum_k \sum_{\ell \neq k} \phi_{k\ell} X_\ell. \tag{2.24}$$

Selection would occur in nature under certain complex environmental constraints. As an illustrative rather than realistic constraint that leads to selection, Eigen and Schuster imposed the conservation relation

$$\sum_k X_k = n \tag{2.25}$$

with n a constant. That such a constraint leads to selection is clear because an increase in the concentration of one species necessarily implies a decrease in the concentrations of the others. If $\sum_k X_k = $ constant, then

$$\sum_k \dot{X}_k = \frac{d}{dt}(\sum_k X_k) = 0. \tag{2.26}$$

But, on summing over (2.23) and using (2.24), we have

$$\begin{aligned}
\sum_k \dot{X}_k &= \sum_k (A_k Q_k - D_k)X_k + \sum_k A_k(1 - Q_k)X_k \\
&= \sum_k (A_k - D_k)X_k \equiv \sum_k E_k X_k > 0
\end{aligned} \tag{2.27}$$

since we assume that $A_k > D_k$ in general. If we want to keep $\sum_k X_k = $ constant we must "dilute" the soup by removing the excess production of the X_k. This is done by adding a "dilution term" to (2.23)

$$\dot{X}_k(t) = W_k X_k(t) + \sum_{\ell \neq k}^{N} \phi_{k\ell} X_\ell(t) - \Omega X_k \tag{2.28}$$

where we have, for simplicity, taken the same coefficient Ω for each species. If $\Omega_0 \equiv \Omega \sum_k X_k$ is the total dilution, we have, on summing (2.28) and setting the result equal to zero to maintain the constraint, that

$$\Omega_0 = \sum_k (A_k - D_k)X_k. \tag{2.29}$$

The total dilution must be equal to the overall excess production which makes sense. Making use of (2.29), Equation (2.28) becomes

$$\dot{X}_k(t) = [W_k - \overline{E}(t)]X_k(t) + \sum_{\ell \neq k} \phi_{k\ell} X_\ell(t) \tag{2.30}$$

with $\overline{E}(t) \equiv \left(\sum_{i=1}^{N} E_i X_i\right) / \left(\sum_{i=1}^{N} X_i\right)$ where $E_i \equiv A_i - D_i$.

This coupled set (2.30) of nonlinear equations has been exactly solved analytically and discussed by Billy Jones, Richard Enns and Sada Rangnekar [JER76]. The system is referred to as the quasi-species model because for certain choices of parameters, it allows the selection of a distribution of species (the "quasi-species") rather than a single species. For the special case of $N = 2$, the

equations can be combined into the well-known Riccati equation of nonlinear
mathematics which has the generic structure,

$$\dot{X} + aX^2 + f_1(t)X + f_2(t) = 0 \qquad (2.31)$$

where a is a constant and f_1 and f_2 are, in general, functions of t. The Riccati equation is solved in Chapter 5 for the case where f_1 and f_2 are time-
independent.

Eigen has proposed different constraints leading to more complex coupled
nonlinear equations which show important quantitative differences, but the
above equation contains the important qualitative features of selection. For
an understandable discussion of the biological background leading to the quasi-
species model, the reader is referred to the Scientific American article by Eigen
et al [EGSWO81].

PROBLEMS
Problem 2-31: Riccati equation
Proceeding by hand, explicitly show that for the $N = 2$ case, the two equations
can be combined into a single Riccati equation. Identify a, f_1, and f_2.

Problem 2-32: Mathematica derivation of Riccati equation
Making use of the appropriate Mathematica commands, repeat the first part of
the previous problem.

2.2.4   Laser Beam Competition Equations

For certain nonlinear optical problems ("stimulated scattering" processes), one

Stimulated Scattering

can derive, from first principles, competition equations between two intense laser
beams of different frequencies. A typical experimental setup used, for example,
in the study of stimulated thermal scattering, is illustrated schematically in
Figure 2.8. The output of a ruby laser ($\lambda = 6943$ Å) is split into two beams.
One, referred to as the laser beam L with frequency ω_L, is sent directly through
a cell containing liquid carbon tetrachloride (CCl_4) colored with a small amount
of iodine (I_2). The other beam, referred to as the signal beam S with frequency
ω_S, first passes through a frequency converter which changes the frequency from
ω_L to ω_S and then passes through the cell in the opposite direction to the first
beam. The alignment is such that the two beams, which have finite widths, pass
through each other for the entire length of the cell. Briefly what happens inside

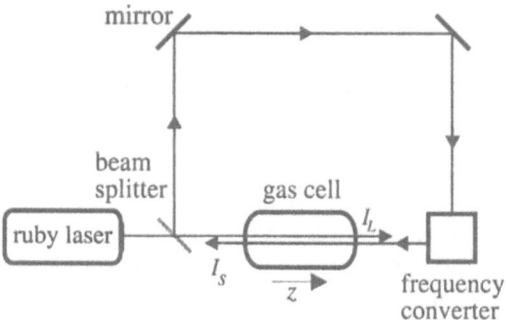

Figure 2.8: Experimental setup for obtaining stimulated thermal scattering.

the cell is as follows. The I_2 molecules absorb electromagnetic energy at the difference frequency, $\omega = \omega_L - \omega_s$, and deliver this energy via collision processes to the host liquid (CCl_4). CCl_4 has a large thermal expansion coefficient. The resulting thermal fluctuations in the CCl_4 produce a modulation of the refractive index in the region of beam overlap, which causes light to be scattered from one beam into the other. Under steady state conditions, which prevail if the duration of the light pulses is long compared to the lifetime of the thermal fluctuations, and making certain approximations, Maxwell's equations for the light beams and the hydrodynamic equations for the liquids may be reduced to yield competition equations similar to those for the biological problem [BEP71, ER79]

$$\frac{dI_L}{dz} = -gI_LI_s - \alpha I_L, \quad \frac{dI_s}{dz} = -gI_LI_s + \alpha I_s. \tag{2.32}$$

Here I_L, I_s are the intensities[2] inside the cell of the laser and signal beams respectively, α is the positive linear absorption coefficient (for I_2) and the gain factor $g = g(\omega)$ depends on the frequency difference between the two beams and the properties of the "host" medium through which the beams are traveling. The analytic form of $g(\omega)$ may be found in the review articles referenced above. g may be positive or negative, depending on the frequency difference ω. The validity of the above nonlinear modeling has been confirmed experimentally for a variety of host and absorbing liquids.

Similarly, one can also have competition between two laser beams traveling through the cell in the *same* direction. In this case, the relevant intensity equations become

$$\frac{dI_L}{dz} = -gI_LI_s - \alpha I_L, \quad \frac{dI_s}{dz} = +gI_LI_s - \alpha I_s. \tag{2.33}$$

The only difference between the two sets of Equations (2.32) and (2.33) is that the right-hand side of the I_s equation differs by a minus sign. This spells out the difference between finding an exact analytic solution and having to resort to numerical techniques! The latter set can be manipulated to yield a Bernoulli equation which has the general structure

$$\frac{dy}{dz} + f_1(z)y = f_2(z)y^n, \tag{2.34}$$

---

[2]intensity = energy crossing a unit area per unit time

with f_1 and f_2, in general, functions of z and n a positive integer, This equation is one of the few exactly solvable, physically important, nonlinear ODEs. The method of solution is discussed in Chapter 5. In contrast, the first set of coupled equations, (2.32), must be solved numerically.

PROBLEMS
Problem 2-33: Bernoulli equation
By adding the two equations in the set (2.33), integrating the result, and eliminating $I_L(z)$, show that I_s satisfies a Bernoulli-type equation. Identify f_1 and f_2, and determine the value of n.

Problem 2-34: Laser Beam Competition
Consider the case above where the two laser beams are traveling in the same direction. Set $I_L = x, I_s = y$ and take $g = 1, \alpha = 0.01$. If $x(0) = y(0) = 1$, numerically solve for $x(z)$ and $y(z)$ over the range $z = 0$ to 30 and plot both solution curves in the same picture. Also plot the trajectory in the phase plane. Repeat with $\alpha = 0.10$ and 1 and discuss the results.

2.2.5   Rapoport's Model for the Arms Race

One last illustration of nonlinear competition equations will be mentioned, namely the competition between nations or groups of nations in the arms race. One possible measure of this competition is the expenditure (X units of cur-

rency) of money by nations in their defense budgets. L. F. Richardson, in his essay "Mathematics of War and Foreign Politics" [Ric60], has reviewed, for example, the defense budgets of France, Germany, Russia and the Austria–Hungarian empire for the pre-World War I years, 1909–1913. Over this time interval he found that the defense budgets of Group 1 (France and Russia, who were allies) and Group 2 (Germany and Austria–Hungary) could be mathematically described by assuming that the rate at which the arms budget X of Group 1 increased was proportional to that Y of Group 2 and vice versa,

$$\dot{X} = a_1 Y, \quad \dot{Y} = a_2 X, \tag{2.35}$$

where a_1, a_2 are positive constants. In fact, Richardson obtained a reasonable fit to the total budget expenditure of both groups by taking $a_1 = a_2 = k$ which leads, on adding the equations, to a solution of (2.35) of the form

$$X(t) + Y(t) = [X(0) + Y(0)]e^{kt} \qquad (2.36)$$

i.e., exponential growth of the total defense budget of the nations involved! Of course, this exponential growth cannot and did not continue forever. The declaration of war changed the situation and thus the structure of the equations.

At times of crisis, nations are not about to tell their opponents how much money is being spent on arms but there is no doubt that the rate of growth accelerates. To take this into account, Rapoport [Rap60] introduced a model with additional nonlinear growth terms. In his model he also introduced decay terms which reflect the pressure within nations to reduce the defense budget and spend the money on non-defense items. Rapoport's model for the arms race between two nations is

$$\dot{X} = -m_1 X + a_1 Y + b_1 Y^2$$
$$\dot{Y} = -m_2 Y + a_2 X + b_2 X^2 \qquad (2.37)$$

all constants being positive. Figure 2.9 shows the behavior of $X(t)$ and $Y(t)$ for

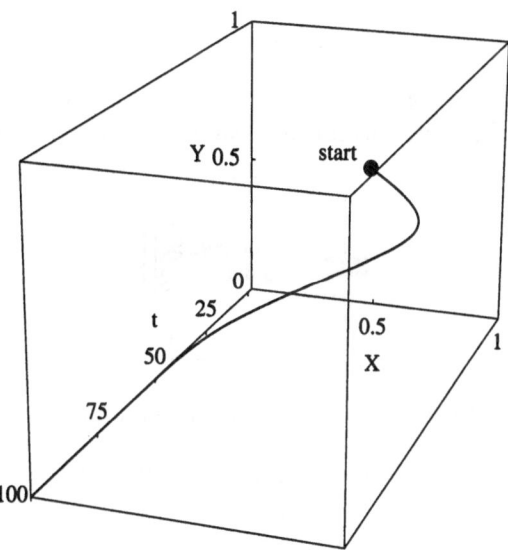

Figure 2.9: Arms expenditures (X, Y) vs time (t) for two competing nations.

a particular (artificial) choice of coefficients used in Mathematica File 06. What can you conclude from this figure?

Still other models have been investigated, for example by Saperstein [Sap84]. Saperstein's thesis was that the appearance of "chaos" (highly irregular behavior) signals the outbreak of war. Along these same lines, Siegfried Grossman and Gottfried Mayer-Kress have written an interesting article entitled "Chaos in the International Arms Race" [GMK89]. As the title suggests, the article discusses the role of chaos in nonlinear models of the arms race.

Arms Race between Two Nations

Equations (2.37) are solved numerically for, e.g, $m_1 = 0.5$, $a_1 = 1$, $b_1 = 0.02$, $m_2 = 0.4$, $a_2 = 0.1$ and $b_2 = 0.05$ and $X(0) = Y(0) = 0.5$. For this choice, we see how X and Y evolve with time in Figure 2.9. The reader may adjust the values of the coefficients to see what affect they have on the plot. Mathematica commands in file: `NDSolve`, `ParametricPlot3D`, `Evaluate`, `ViewPoint`, `PlotPoints`, `/.`, `AxesLabel`, `PlotRange`, `AxesEdge`, `BoxRatios`, `Ticks`, `BoxStyle`, `Epilog`, `Text`, `RGBColor`, `Point`, `PointSize`, `FontSize`, `ImageSize`

PROBLEMS

Problem 2-35: The arms race

Modify MF06 by replacing the quadratic terms with cubic terms. This reflects additional pressure to increase defense spending. Taking the same parameters and initial conditions as in the original Mathematica file, except for reducing m_1 to 0.40 and m_2 to 0.30, determine what happens in the arms race. Discuss your result.

2.3   Nonlinear Electrical Phenomena

2.3.1   Nonlinear Inductance

A simple example of a nonlinear electrical circuit is a charged capacitor C connected to a coil of N turns wrapped around an iron core. The current (i) vs.

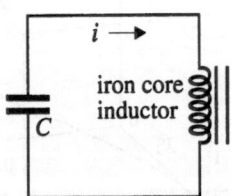

Figure 2.10: Nonlinear inductor circuit.

flux (Φ) relation for the iron core inductor has the form

$$i = \frac{N\Phi}{L_0} + A\Phi^3. \tag{2.38}$$

If the cubic term were not present, we would have the usual linear relation $N\Phi = L_0 i$ where L_0 is the self-inductance of the coil and Φ is the flux threading through one turn of the coil. The cubic term arises from the presence of the iron core. The precise form of the nonlinearity depends on the nature of the core. If q is the charge on the capacitor at time t, Kirchhoff's law for the sum of the potential drops around the circuit yields

$$\frac{q}{C} + N\frac{d\Phi}{dt} = 0. \tag{2.39}$$

Differentiating (2.39), noting that $dq/dt = i$, and using Equation (2.38), we obtain

$$\ddot{\Phi} + \alpha\Phi + \beta\Phi^3 = 0 \qquad (2.40)$$

where $\alpha \equiv 1/L_0 C$ and $\beta \equiv A/NC$.

In the following experiment, the reader can learn more about a nonlinear inductor, or tank, circuit.

Iron Core Inductor
This activity explores the relationship between the amplitude of nonlinear oscillations and the period in a nonlinear electric circuit.

EXP 07

The next activity explores the relationship between the amplitude of nonlinear oscillations and the period when the nonlinear coefficient β in Eq. (2.40) is negative. The equation is then referred to as a "soft spring" oscillator equation.

Nonlinear LRC Circuit
A piecewise linear (nonlinear) LRC circuit that models a soft spring oscillator is produced and the relation between amplitude and period investigated.

EXP 08

PROBLEMS
Problem 2-36: The iron core inductor
Attempt to analytically solve Equation (2.40) using Mathematica for $\alpha = \beta = 1$ and with initial conditions $\Phi(0) = 1$, $\dot{\Phi}(0) = 0$. As for the simple pendulum, a periodic solution exists, expressible in terms of elliptic functions. As was the case in MF01, an integral expression for the time t is easily found from which the period of the motion can be calculated. Evaluate the period. What would the period be if $\beta = 0.1$?

2.3.2   An Electronic Oscillator (the Van der Pol Equation)

In the development of nonlinear theory, the Van der Pol equation [Van26] has played an important role, particularly because it displays the so-called limit cycle, a phenomenon which does not occur in linear problems. A "limit cycle" corresponds to a periodic motion which the system approaches no matter what the initial conditions are. All electronic oscillator circuits display limit cycles as do many acoustical and mechanical systems, the heart, a mechanical pump, being a prime example. The Van der Pol (VdP) equation, which is derived in Example 2-3 for a particular electronic oscillator circuit, can be regarded as a simple harmonic oscillator equation to which a variable damping term has been added, viz.

$$\ddot{X} - \epsilon(1 - X^2)\dot{X} + X = 0. \qquad (2.41)$$

Here ϵ is a positive quantity which is related to the circuit parameters and X is a "displacement" variable which remains to be interpreted for the electronic oscillator circuit. When $X > 1$, the damping term tends to reduce the magnitude of the oscillations, but for $X < 1$, one has a negative damping term which tends to increase the oscillations. It is this latter feature that gives rise to the self-excitation of oscillator circuits because any spontaneous fluctuations (in an electrical circuit, this is due to the thermal noise) of X will tend to grow. It also follows that there must be some amplitude for which the motion neither increases nor decreases with time (it is not $X = 1$!). This situation corresponds to the limit cycle. Systems which display self-sustained oscillations are very common in nature, occurring when a periodic motion is maintained through absorption of energy from a constant energy source. In the electronic oscillator case, the energy is supplied by a battery. Self-excited oscillations also occur frequently in mechanical and acoustical systems. The failure of the Tacoma Narrows Bridge in 1940 is thought to have been due to the onset of a self-excited oscillation in which the constant energy source was the steady wind. When the torsional oscillations became too large, the bridge collapsed.

Example 2-3: Tunnel Diode Oscillator

A typical tunnel diode oscillator circuit [Cho64] is shown in the accompanying figure with the current (I) versus voltage (V) characteristics of the nonlinear diode (D) also sketched. The battery voltage V_b is adjusted to coincide with the

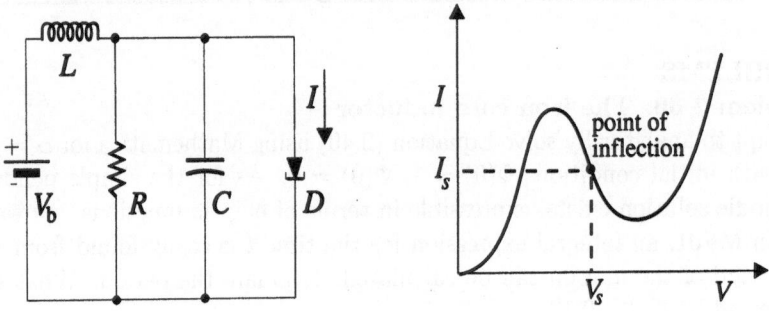

inflection point V_s of the I vs. V curve. In the neighborhood of this operating point, one may write $i = -av + bv^3$, where i and v are the current and voltage relative to values at the operating point ($i = I - I_s$, $v = V - V_s$) and a and b are positive constants. Using Kirchhoff's rules and the i vs. v relation, show that the tunnel diode circuit yields the Van der Pol equation.

Solution: The capital letter I is Mathematica's input symbol for $\sqrt{-1}$, so we will use the small letter i with appropriate subscripts to represent all current variables. Similarly, all voltage variables will be labeled with the lower case symbol v with appropriate subscripts. To enter the subscripted quantities, the Basic Typesetting Palette will be used.

```
Clear["Global`*"]
```

The current-voltage relation, $i(t) = -av(t) + bv(t)^3$, is entered and assigned the name ii. If we label the diode current as i_D, then from the current-voltage curve we have $i_D = i_S + i(t) \equiv i_S + $ ii.

```
ii = -a v[t] + b v[t]^3; iD = iS + ii
```

$$i_S - a\,v(t) + b\,v(t)^3$$

Kirchhoff's current rule states that the current i_L through the inductor L is equal to the sum of the currents through the resistor R, capacitor C, and diode D. This fact is entered as **eq1**.

```
eq1 = -iL[t] + iR + iC + iD == 0
```

$$i_C + i_R + i_S - i_L(t) - a\,v(t) + b\,v(t)^3 == 0$$

From the current-voltage curve, the diode voltage is $v_D = v_S + v(t)$. Further, the voltage drops v_R and v_C across the resistor and capacitor, respectively, are equal to the voltage drop across the diode.

```
vD = vS + v[t]; vR = vD; vC = vD
```

$$v_S + v(t)$$

By Ohm's law, the current through the resistor is $i_R = v_R/R$. The current through the capacitor is $i_C = C\,(dv_C/dt)$.

```
iR = vR/R; iC = C D[vC, t];
```

With all of the above relations entered, **eq1** is given by the output of the following command line.

```
eq1
```

$$i_S - a\,v(t) + b\,v(t)^3 + \frac{v_S + v(t)}{R} - i_L(t) + C\,v'(t) == 0$$

Note how the previous relations were automatically substituted. The time derivative of **eq1** is now taken and the result labeled as **eq2**.

```
eq2 = D[eq1, t]
```

$$-a\,v'(t) + \frac{v'(t)}{R} + 3\,b\,v(t)^2\,v'(t) - i_L'(t) + C\,v''(t) == 0$$

Kirchhoff's potential drop rule is applied to the outer loop of the circuit.

```
vL = vS - vD;
```

From the definition of inductance, one has $di_L/dt = v_L/L$. This relation is substituted into **eq2** and the result labeled as **eq3**.

```
eq3 = eq2 /. iL'[t] -> vL/L
```

$$\frac{v(t)}{L} - a\,v'(t) + \frac{v'(t)}{R} + 3\,b\,v(t)^2\,v'(t) + C\,v''(t) == 0$$

In the following input line the left-hand side of **eq3**, obtained by entering **eq3[[1]]**, is divided by C and the result expanded and set equal to zero.

> **eq4 = Expand[eq3[[1]]/C] == 0**

$$\frac{v(t)}{CL} - \frac{a\,v'(t)}{C} + \frac{v'(t)}{CR} + \frac{3\,b\,v(t)^2\,v'(t)}{C} + v''(t) == 0$$

The coefficients of $dv(t)/dt$ are collected on the lhs of **eq4** and the result set equal to zero. This yields the unnormalized VdP equation.

> **eq5 = Collect[eq4[[1]], v'[t]] == 0**

$$\frac{v(t)}{CL} + \left(-\frac{a}{C} + \frac{1}{CR} + \frac{3\,b\,v(t)^2}{C}\right)v'(t) + v''(t) == 0$$

The following substitution makes use of the identification of $1/\sqrt{LC}$ as a frequency, labeled ω.

> **eq6 = eq5 /. L -> 1/(C ω^2)**

$$\omega^2\,v(t) + \left(-\frac{a}{C} + \frac{1}{CR} + \frac{3\,b\,v(t)^2}{C}\right)v'(t) + v''(t) == 0$$

Noting that $\omega^2 v(t)$ and $v''(t)$ have the same dimensions, it is clear that a dimensionless time variable $\tau \equiv \omega t$ can be introduced. Studying the coefficient of $v'(t)$, we also see that the voltage variable $v(t)$ can be transformed into a dimensionless voltage variable $x(\tau)$, by writing $v(t) \equiv x(\tau)\,\mathrm{tr}$, where

$$\mathrm{tr} \equiv \frac{\sqrt{a - 1/R}}{\sqrt{3\,b}}.$$

The transformation relation **tr** is entered,

> **tr = Sqrt[(a - 1/R)]/Sqrt[3 b];**

and the above variable changes made on the lhs of **eq6**. To be consistent, one must also substitute $v'(t) = \omega\,x'(\tau)\,\mathrm{tr}$ and $v''(t) = \omega^2\,x''(\tau)\,\mathrm{tr}$. Finally, the whole result is divided by $\omega^2\,\mathrm{tr}$ and set equal to zero in **eq7**.

> **eq7 = Expand[(eq6[[1]] /. {v[t] ->x[τ] tr, v'[t] ->ω x'[τ] tr,**
> **   v''[t] ->ω^2 x''[τ] tr})/(ω^2 tr)] == 0**

$$x(\tau) - \frac{a\,x'(\tau)}{C\omega} + \frac{x'(\tau)}{CR\omega} + \frac{a\,x(\tau)^2\,x'(\tau)}{C\omega} - \frac{x(\tau)^2\,x'(\tau)}{CR\omega} + x''(\tau) == 0$$

We then successively collect $x'(\tau)$, ω, $1/C$, and $x(\tau)^2$ on the lhs of **eq7** and set the result equal to zero.

`eq8 = Collect[eq7[[1]], {x'[τ], ω, 1/C, x[τ]^2}] == 0`

$$x(\tau) + \frac{\left(\left(a - \frac{1}{R}\right) x(\tau)^2 - a + \frac{1}{R}\right) x'(\tau)}{C\omega} + x''(\tau) == 0$$

Again, on examining the coefficient of $x'(\tau)$ in eq8, we are led to introduce the dimensionless parameter

$$\epsilon = \frac{(a - 1/R)}{\omega C} = \frac{(aR - 1)}{\omega CR}.$$

Making the replacement $\omega -> (a - 1/R)/(\epsilon C)$ on the lhs of eq8, simplifying,

`eq9 = Simplify[(eq8[[1]] /. ω -> (a - 1/R)/(ε C))] == 0`

$$x(\tau) - \epsilon x'(\tau) + \epsilon x(\tau)^2 x'(\tau) + x''(\tau) == 0$$

and collecting terms, the dimensionless Van der Pol equation results.

`vanderpolequation = Collect[eq9[[1]], {x'[τ], ε}] == 0`

$$x(\tau) + \epsilon \left(x(\tau)^2 - 1\right) x'(\tau) + x''(\tau) == 0$$

Since a, ω, and C are positive, the parameter $\epsilon > 0$ provided that the resistance $R > 1/a$. It should be noted that subscripted names, e.g., the diode voltage v_D used earlier, will not be cleared by a subsequent `Clear["Global`*"]` command.

`Clear["Global`*"]`
v_D

$$v_S + v(t)$$

The following command line will clear v_D.
$v_D =.$

v_D

$$v_D$$

Alternatively, one can clear all subscripted names by quitting the Mathematica kernel. Click on Kernel in the tool bar, then on Quit Kernel, and on Local.

End Example 2-3

If you wish to learn how the current-voltage curve for a tunnel diode can be experimentally determined, try the following activity.

Tunnel Diode Negative Resistance Curve
The I-V curve of a tunnel diode is determined and plotted. The region of negative resistance (the negative slope region of the I-V curve) is determined.

The next experiment involves building a tunnel diode oscillator circuit and studying its behavior.

Tunnel Diode Self-Excited Oscillator
In this activity a tunnel diode circuit similar to that in Example 2-3 is used to produce self-excited oscillations. By varying the resistance R, the value of the parameter ϵ may be changed, producing different types of oscillations.

PROBLEMS
Problem 2-37: The tunnel diode oscillator
For the tunnel diode 1N3719, $i = -0.050v + 1.0v^3$, with v measured in volts and i measured in amperes. For what range of resistance R is the parameter ϵ positive? Evaluate ϵ for $R = 50$ ohms, inductance $L = 25$milli-henries, and capacitance $C = 1$ micro-farad. Would you expect the variable damping term to be important or negligible in this case? Explain. Confirm your opinion by obtaining a numerical solution and plotting the result for $x(0) = 0$, $\dot{x}(0) = 0.1$. Use the same approach as in Example 1-5 and compare the shape of the oscillations with what they would be for very small ϵ.

Problem 2-38: Limit cycle
By making a suitable plot, determine the shape of the limit cycle for the VdP equation for $\epsilon = 5$. Hint: Rewrite the second-order equation as two first-order equations by setting $\dot{x} = y$, $\dot{y} = \epsilon(1 - x^2)y - x$. Take $x(0) = y(0) = 0.1$ and let $t = 0...15$. What does the limit cycle look like for $\epsilon = 0.10$? Take $x(0) = y(0) = 1.5$ and let $t = 0...30$. Discuss the effects of increasing ϵ.

Problem 2-39: Another tunnel diode oscillator circuit: Part 1
Consider the tunnel diode circuit shown in the accompanying figure with the

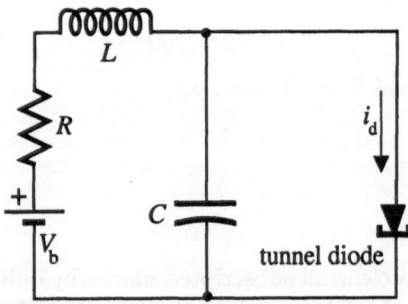

diode current i_d (in amperes) related to the potential drop V (in volts) across the diode by the nonlinear relation $i_d = a(V^3 - \frac{3}{4}V^2 + \frac{1}{7}V)$, where a is positive.

a. Plot i_d/a versus V over the range 0 to 0.5 volts and confirm that the shape of the curve is similar to that in Example 2-3.

b. By differentiating i_d/a once with respect to V and setting the result equal to zero, find the voltage range over which the slope of the i_d vs. V curve is negative. Explain why this is called the negative resistance region.

c. If the battery voltage, V_b, is set equal to the voltage at the inflection point in the negative resistance region, what is the value of V_b? Hint: Differentiate i_d/a twice with respect to V and set the result equal to zero.

Problem 2-40: Another tunnel diode oscillator circuit: Part 2
Consider the tunnel diode circuit of the previous problem with the same nonlinear $i_d(V)$ relation.

a. Using Kirchhoff's rules, show that the circuit is governed by the dimensionless Van der Pol-like ODE

$$\ddot{x} - \epsilon(1 - x^2)\dot{x} + f(x) + x = 0$$

with $b = \sqrt{\frac{5}{336} - \frac{RC}{3La}}$, $x = \frac{V - \frac{1}{4}}{b}$, $\tau = \frac{t}{\sqrt{LC}}$, $\epsilon = 3ab^2\sqrt{\frac{L}{C}}$, and the form of the function $f(x)$ to be determined. Assume that $V_b = \frac{1}{4}$ volts here.

b. Taking $a = 1.2$ amperes/volt, $L = 20$ henry, $C = 0.01$ farad, and $R = 1$ ohm, numerically solve the ODE for x vs. τ and plot the result over a suitable time range. Take $x(0) = \dot{x}(0) = 0$. The oscillation shape is characteristic of a relaxation oscillation, to be discussed in Chapter 7.

c. Determine analytically and confirm numerically that there is a critical value of R above which the oscillations die away.

Problem 2-41: A neon tube circuit
A neon tube is inserted into the circuit as shown in the accompanying figure. The voltage versus current characteristics of the neon tube are as shown and

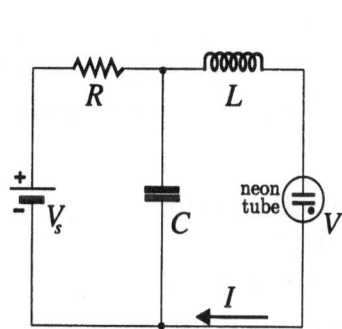
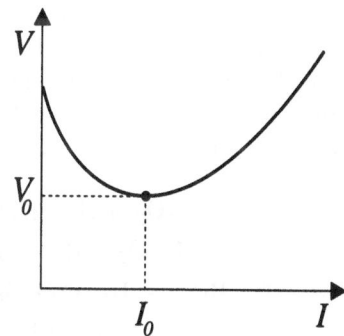

may be idealized by the parabolic relation $V = V_0 + a(I - I_0)^2$. Using Kirchhoff's rules, show that the equation of motion for the circuit is

$$\ddot{x} + (b + dx)\dot{x} + ex + fx^2 = g,$$

where $x = I - I_0$, and b, d, e, f and g are constants which you must identify in terms of the circuit parameters.

Problem 2-42: Rayleigh's equation
Show that Rayleigh's equation

$$\ddot{y} - \epsilon(\dot{y} - \frac{1}{3}\dot{y}^3) + y = 0,$$

first derived by Lord Rayleigh [Ray83] for certain nonlinear acoustical phenomena, is easily transformed into the Van der Pol equation.

2.4   Chemical and Other Oscillators

2.4.1   Chemical Oscillators

The Belousov–Zhabotinski (BZ) chemical reaction is the best known of the chemical oscillators. To achieve the BZ reaction, John Tyson [Tys76] suggests using the ingredients[3] listed in the following table:

	Ingredients	Initial Molar Concentration
150 ml	sulphuric acid(H_2SO_4)	1 M
0.175 g	cerium ammonium nitrate ($Ce(NO_3)_6$ $(NH_4)_2$)	0.002 M
4.292 g	malonic acid ($CH_2(COOH)_2$)	0.28 M
1.415 g	sodium bromate ($NaBrO_3$)	0.063 M

When the malonic acid and cerium ammonium nitrate are dissolved and stirred in the sulfuric acid, the solution will initially be yellow, then turn clear after a few minutes. Then, on adding the sodium bromate, the solution will oscillate between yellow and clear with a period of about one minute. A more dramatic color change can be achieved by adding a few ml of 0.025 M Ferroin (1, 10 phenanthroline iron). The periodic color change then will be between red and blue.

Field, Körös and Noyes [FKN72] were able to measure the periodic oscillations in the Br^- concentration and the Ce^{4+}/Ce^{3+} ratio in the BZ reaction. They then isolated the following five important reactions in the complicated chemistry that was taking place,

$$BrO_3^- + Br^- + 2H^+ \rightarrow HBrO_2 + HOBr$$
$$HBrO_2 + Br^- + H^+ \rightarrow 2HOBr$$
$$2Ce^{3+} + BrO_3^- + HBrO_2 + 3H^+ \rightarrow 2Ce^{4+} + 2HBrO_2 + H_2O$$
$$2HBrO_2 \rightarrow BrO_3^- + HOBr + H^+$$
$$4Ce^{4+} + BrCH(COOH)_2 + 2H_2O \rightarrow 4Ce^{3+} + Br^- + HCOOH +$$
$$2CO_2 + 5H^+$$

Field and Noyes [FN74] invented a kinetic model called[4] the "Oregonator". They labeled the concentrations of BrO_3^-, etc., as follows:

---

[3]Belousov [Bel58] originally used citric acid, but malonic acid is now commonly substituted. Because it was contrary to the then current belief that all solutions of reacting chemicals must go monotonically to equilibrium, Belousov could not initially get his discovery published in any Soviet journal. Only years later, when his results were confirmed by Zhabotinski, was he given due recognition for his discovery. For his pioneering research work he was awarded, along with Zhabotinski, the Soviet Union's highest medal but, unfortunately, he had passed away 10 years earlier.

[4]The name of their model reflects the location in which their research was carried out.

Chemical Species	Symbol
BrO_3^-	A
$HBrO_2$	X
Br^-	Y
Ce^{4+}	Z
all others	$*$

Then, for the five reactions shown above,

$$
\begin{aligned}
A + Y + * &\overset{k_1}{\to} X + * \\
X + Y + * &\overset{k_2}{\to} * \\
A + X + * &\overset{k_3}{\to} 2X + 2Z + * \\
2X &\overset{k_4}{\to} A + * \\
Z + * &\overset{k_5}{\to} hY + *
\end{aligned}
\tag{2.42}
$$

where $h \sim 1/2$ is a fudge factor introduced because of the severe truncation of the full set of equations describing the complicated chemistry and the k_i denote the various rates of reaction. Further, the depletion of A is neglected so that A is treated as a constant. We can write down the rate equations for the production of X, Y, Z as

$$
\dot{X} = k_1 AY - k_2 XY + k_3 AX - k_4 X^2
$$

$$
\dot{Y} = -k_1 AY - k_2 XY + hk_5 Z
\tag{2.43}
$$

$$
\dot{Z} = 2k_3 AX - k_5 Z.
$$

In writing down (2.43), we have made use of the empirical rule: When two substances react to produce a third, the reaction rate is proportional to the product of the concentrations of the two substances. Note that $2X$ in the fourth reaction is treated as $X + X$, thus the X^2 term in the $\dot{X}$ equation. The factor of 2 in the $\dot{Z}$ equation appears because in the third reaction, two of Z appear for each net $(2X - X)$ one of X. Equations (2.43) are nonlinear with terms present which are structurally similar to those that we saw earlier for some of the competition equations.

Equations (2.43) can be converted into a dimensionless form, a process known as normalization, by setting

$$
x = \frac{k_2}{k_1 A} X, \quad y = \frac{k_2}{k_3 A} Y, \quad z = \frac{k_2 k_5}{2k_1 k_3 A^2} Z, \quad \tau = (k_1 A)t,
$$

and

$$
\epsilon = \frac{k_1}{k_3}, \quad p = \frac{k_1 A}{k_5}, \quad q = \frac{k_1 k_4}{k_2 k_3}.
$$

The Oregonator equations then reduce to the form

$$\epsilon \dot{x}(\tau) = x + y - qx^2 - xy$$

$$\dot{y}(\tau) = -y + 2hz - xy \qquad\qquad (2.44)$$

$$p\dot{z}(\tau) = x - z$$

with the parameters ϵ, q, p and h all positive. By normalizing the rate equations, the number of independent parameters has been reduced, thus making for easier analysis.

Figure 2.11 shows how the Ce^{4+} (z) concentration varies with time for some specific values of the coefficients and initial concentrations used in file MF07.

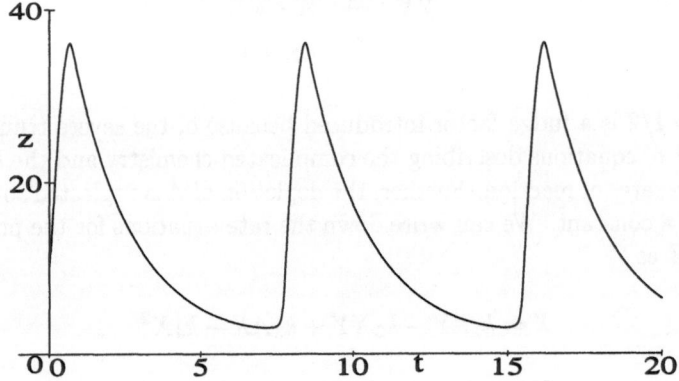

Figure 2.11: Oscillatory behavior of Ce^{4+} (z) concentration vs time.

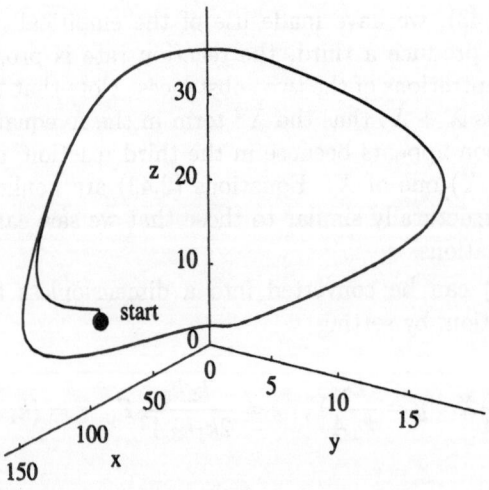

Figure 2.12: Evolution of chemical system onto limit cycle in 3D phase space.

The chemical oscillator behavior is quite evident. The $HBrO_2$ (x) and Br^- (y) concentrations show similar oscillatory behavior. Figure 2.12 shows the x vs. y vs. z plot (a 3-dimensional "phase space") for the same conditions. Note how the chemical system winds onto a closed trajectory. A closed loop is approached no matter what the initial conditions, so one has a 3-dimensional "limit" "cycle".

The Oregonator Model
Equations (2.44) are solved numerically for $\epsilon = 0.03$, $p = 2$, $q = 0.006$, $h = 0.75$, $x(0) = 100$, $y(0) = 1$ and $z(0) = 10$. By varying the initial conditions, 3-dimensional limit cycle behavior can be confirmed. Mathematica commands in file: NDSolve, MaxSteps, ParametricPlot3D, Evaluate, AxesLabel, Hue, PlotPoints, PlotRange, ViewPoint, Boxed->False, BoxRatios, Ticks, AxesStyle, AxesEdge, Epilog, RGBColor, PointSize, Point, Plot, TextStyle, PlotLabel, FontWeight, ImageSize

MF07

PROBLEMS
Problem 2-43: The Oregonator model
Confirm that a limit cycle results in MF07 regardless of the initial conditions.

Problem 2-44: Chemical oscillator
The rate equations for a certain chemical oscillator are

$$A \xrightarrow{k_1} X$$
$$B + X \xrightarrow{k_2} Y + *$$
$$2X + Y \xrightarrow{k_3} 3X$$
$$X \xrightarrow{k_4} *$$

where the concentrations of species A and B are held constant.

a. Using the empirical rule for chemical reactions, write down the rate equations for X and Y.

b. Convert the rate equations into a normalized form by setting
$x = \sqrt{\frac{k_3}{k_4}}X$, $y = \sqrt{\frac{k_3}{k_4}}Y$, $a = \sqrt{\frac{k_3}{k_4}}\frac{k_1}{k_4}A$, $b = \frac{k_2}{k_4}B$, and $\tau = k_4 t$.

c. Taking $a = 1$, $b = 2.5$, $x(0) = y(0) = 0.1$, produce a 3-dimensional plot showing $x(t)$ vs. $y(t)$ vs. t.

d. Choose an orientation which clearly shows that a periodic orbit is achieved.

e. Confirm the limit cycle nature by trying a few different initial normalized concentrations.

Problem 2-45: Adjusting the fudge factor
In MF07 the fudge factor h was taken to be $h = 0.75$. Exploring the range $h = 0.1$ to $h = 1$, with all other conditions the same, determine whether or not a limit cycle occurs. Comment on the sensitivity of the model on h.

2.4.2   The Beating Heart

Figure 2.13 shows an electrocardiogram recording the electrical activity associated with contractions of a normal heart. As early as 1928, Van der Pol and Van der Mark [VV28] considered the heartbeat as a "relaxation oscillator" (one for which the periodic spikes are separated by relatively long periods of inactivity) and proposed an electric model of the heart. More recently, Glass and

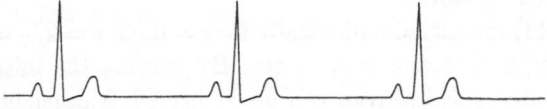

Figure 2.13: Electrocardiogram recording for a normal heart.

coworkers [GSB86] have experimentally studied electrical stimulation of spontaneously beating cells from embryonic chick hearts. Chaotic (irregular) heart action is a precursor of sudden death and understanding the onset of chaos is the motivation for their research.

PROBLEMS
Problem 2-46: Relaxation oscillators
Suggest some other possible examples of relaxation oscillator behavior in the natural world.

Problem 2-47: The teeter-totter
Consider a teeter-totter, or seesaw, with a weight B on the end initially touching the ground as shown in the figure. Let water drip into the container on the other end A until the weight of the water is sufficient that B suddenly rises and point

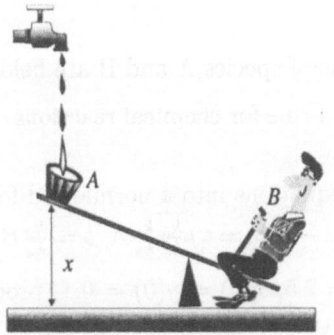

A touches the ground. The container is allowed to empty quickly and the teeter-totter returns to its original position, and so on. Sketch the distance x of A from the ground as a function of t. Explain why this is an example of a relaxation oscillator.

Chapter 3

Nonlinear Systems. Part II

But you will ask, how could a uniform chaos coagulate at first irregularly in heterogeneous veins or masses to cause hills--- Tell me the cause of this, and the answer will perhaps serve for the chaos.
Isaac Newton (1642-1727)

3.1 Pattern Formation

Patterns pervade the world of nature as well as the world of the intellect. In the biological realm we are quite familiar with the stripes on a zebra, the spots on a leopard, and the colorful markings of certain birds, fish, and butterflies. In the physical world we may have noticed the pretty fringe patterns which occur when thin films of oil spread on a road surface or the wonderful shapes that ice crystals can assume when trees are coated after an ice storm. If we go into a wallpaper shop, we can be overwhelmed by the wide variety of patterns available, the patterns created by someone's artistic imagination. If we talk to a scientist we will soon find that his or her goal in life is usually to discover (impose?) some underlying pattern to the phenomena under investigation and then attempt to mathematically model that pattern. In this section, we shall look at some attempts to understand or create patterns through the use of nonlinear modeling and concepts. Our first example is from the world of chemistry.

3.1.1 Chemical Waves

Arthur Winfree [Win72] has given a convenient list of chemical ingredients for observing so-called target or bulls-eye patterns in a petri dish, viz.

- 3 ml of concentrated sulphuric acid and 10 g of sodium bromate dissolved in 134 ml of water

- 1 g of sodium bromide dissolved in 10 ml of water

- 2 g of malonic acid dissolved in 20 ml of water

On adding 0.5 ml of the second solution to 6 ml of the first, and then adding 1 ml of the third solution, the resulting mixture becomes clear after a few minutes. On mixing in 1 ml of 0.025 M (standard) Ferroin, stirring well, and pouring into a shallow petri dish, colorful and intriguing patterns will be seen to form.

As first reported by Zaikin and Zhabotinski [ZZ70], the solution will initially be uniformly red, but in a few minutes blue dots will appear and spread out in rings (target patterns) as schematically illustrated in Figure 3.1. The blue dots correspond to spontaneous chemical oscillator sites which jiggle the liquid surface up and down at each point and thus produce circular wave patterns. Unlike small amplitude water waves which would run through each other and linearly superimpose, the nonlinear waves from each nucleation site in this chemical mixture come to a "crashing halt" when they encounter each other, eventually leaving a static pattern in the petri dish. When the dish is shaken, the process

Figure 3.1: "Target" patterns in a petri dish; an example of chemical waves.

will start over again with a new final pattern being observed. The mathematically inclined reader is referred to John Tyson's book [Tys76] for a theoretical discussion of this fascinating nonlinear pattern formation. The above chemical mixture is an example of an "excitable medium", i.e., a nonlinear medium in which colliding wave fronts annihilate each other and stop, and for which there is a "refractory time" during which no further wave action is possible.

Spiral waves can be generated by tilting the petri dish in order to break some of the blue wave fronts. The free ends of the wave fronts wrap around into spirals. Spiral waves can also occur in biological examples of excitable media, for example, in cardiac tissue [KG95].

PROBLEMS
Problem 3-1: Chemical Waves
Go to one of your friendly chemistry department's instructors and acquire the chemicals listed in Winfree's recipe. Experimentally confirm the behavior of the target patterns and spiral waves discussed above. See if you are able to influence the behavior, e.g., by clapping two dusty chalk board brushes together above the chemical mixture.

Problem 3-2: Forest fires
Consider the occurrence of several lightning strikes each of which initiates a forest fire at different points in a forest. If there is no wind and the forest is more or less homogeneous in its content, terrain, dryness, etc., describe the

spreading of the fires. If each fire is pretty thorough in consuming the trees, what happens when the fire fronts meet? What is the refractory time if the forest is regarded as an example of an excitable medium?

3.1.2 Snowflakes and Other Fractal Structures

Snowflakes, such as those shown in Figure 3.2, are an example of so-called frac-

Figure 3.2: Snowflakes as examples of fractal structure.

tal structures occurring in nature [BH62]. The term "fractal" was introduced by the mathematician Benoit Mandelbrot [Man75, Man77] in the 1970s to describe geometrical shapes such as coastlines, clouds, etc., having highly complex boundaries. To give a more precise meaning, we must briefly explore the concept of dimension.

There are several different types of dimension, the most familiar being the dimension of Euclidean space. In this case, the dimension is the minimum number of coordinates needed to specify the location of a point uniquely. In the last section of this chapter, we shall refer to the dimension of a dynamical system, i.e., the number of state variables (temperature, pressure, or whatever) needed to describe the dynamics of the system. In both cases, the dimension is an integer; non-integer values cannot occur.

The concept of dimension can be generalized to allow non-integer values. A fractal shape is characterized by a non-integer dimension. There are at least five different types of fractal dimension, the most wellknown of which is called the capacity dimension, D_C. In this introduction, we shall only examine D_C, the interested reader being referred to the excellent discussion of other types in the text by Parker and Chua [PC89]. Keep in mind that D_C should reduce to integer values in those situations where we would expect it to do so.

Consider a 1-dimensional figure such as the straight line, or more generally a smooth curve, of length L such as is shown in Figure 3.3. This line is covered by $N(\epsilon)$ 1-dimensional segments, each of length ϵ. Dots are used to represent the boundaries of each of the segments. On the top line, $\epsilon = L$ and $N(\epsilon) = 1$. At the next level, we have divided the line into three parts. Here, $\epsilon = L/3$ and $N(\epsilon) = 3 = L/\epsilon$. Quite generally, for any line subdivided in the same manner, $N(\epsilon) = L/\epsilon$.

Next, consider a 2-dimensional square of side L, as shown in Figure 3.4. We

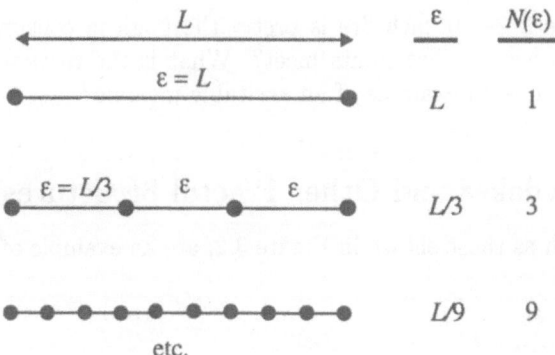

Figure 3.3: Covering a line of length L with line segments of length ϵ.

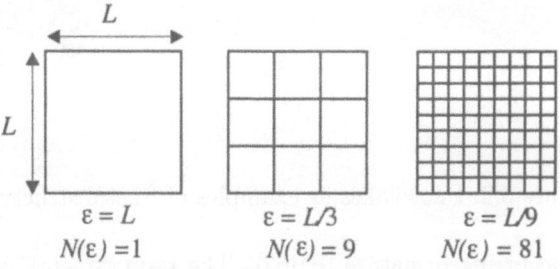

Figure 3.4: Covering a square of side L with boxes of side ϵ.

will cover the square with identical boxes of side ϵ and again determine $N(\epsilon)$, the number of boxes needed to fill the square. In this case, we find that in general $N(\epsilon) = L^2/\epsilon^2$.

In three dimensions, we would obviously obtain $N(\epsilon) = L^3/\epsilon^3$ and, generalizing, in D-dimensions

$$N(\epsilon) = \frac{L^D}{\epsilon^D}. \tag{3.1}$$

Taking the logarithm and solving for D, we obtain

$$D = \frac{\ln N(\epsilon)}{\ln L + \ln(1/\epsilon)}. \tag{3.2}$$

Taking the limit as $\epsilon \to 0$, $\ln(1/\epsilon) \gg \ln L$ and we define the capacity dimension

$$D_c = \lim_{\epsilon \to 0} \frac{\ln N(\epsilon)}{\ln(1/\epsilon)}. \tag{3.3}$$

So, D_c agrees with our "normal" concept of dimension for the examples above. Let us now return to Figure 3.3 and throw away the middle third at each step as in Figure 3.5. We take $L = 1$ for simplicity and count the number of line segments $N(\epsilon)$ needed to cover the unit interval, i.e., we do not count the empty segments. On the kth step we have $\epsilon = (1/3)^k$ and $N(\epsilon) = 2^k$, yielding a

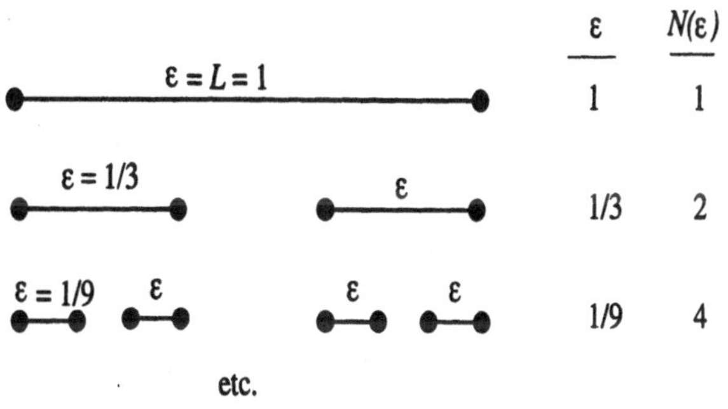

Figure 3.5: The Cantor set.

capacity dimension

$$D_c = \lim_{k \to \infty} \frac{\ln 2^k}{\ln 3^k} = \frac{\ln 2}{\ln 3} = 0.6309\ldots . \tag{3.4}$$

The segmented line with gaps in Figure 3.5 is referred to as a Cantor set. It has a fractal dimension intermediate between a point (a 0-dimensional object) and a continuous line (a 1-dimensional object). The Cantor set has a fractal dimension, which makes intuitive sense as it is "more" than a point but not quite a solid line.

What has all this to do with snowflakes? It happens that the jagged boundary of a snowflake can be characterized by a fractal dimension. The problem on the Koch triadic curve at the end of this section illustrates this point. The Koch curve and the Cantor set are examples of self-similar fractals. On each step, the new line segment is a scaled-down version of the old segment.

The basic physics underlying the formation of so-called dendritic structures such as snowflakes has been explored, e.g., by Ben-Jacob, Goldenfeld, Langer and Schön [BJGLS84]. Other examples of dendritic structures in nature having fractal dimensions are lightning bolts and ferns. In the following example, a fractal fern is "grown" mathematically from a single input point.

Example 3-1: Barnsley's Fractal Fern

By generating a random number r between 0 and 1, taking the starting point to be $x_0 = y_0 = 0$, and iterating the following 2-d piecewise relation (due to Michael Barnsley [Bar88]) from $n = 0$ to $n = 40000$,

$$(x_{n+1}, y_{n+1}) = \begin{cases} (0, 0.16y_n), & 0.0 < r < .01 \\ (0.2x_n - 0.26y_n, 0.23x_n + 0.22y_n + 0.2), & .01 < r < .08 \\ (-0.15x_n + 0.28y_n, 0.26x_n + 0.24y_n + 0.2), & .08 < r < .15 \\ (0.85x_n + 0.04y_n, -0.04x_n + 0.85y_n + 0.2), & .15 < r < 1.0 \end{cases}$$

create a plot of a green fern. Is the fern a self-similar fractal? Explain.

Solution: For programming convenience, the relation will be expressed as

$$x_{n+1} = a_i\, x_n + b_i\, y_n + e_i, \quad y_{n+1} = c_i\, x_n + d_i\, y_n + f_i,$$

where the first branch corresponds to $i = 1$ and is selected if the random number $r < p_1 = 0.01$, the second branch corresponds to $i = 2$, and so on.

```
Clear["Global`*"]
```

The coefficients for each branch of the piecewise relation are entered as indexed quantities as are the boundaries for the random number r.

```
a[1] = 0;  a[2] = 0.2;  a[3] = -0.15;  a[4] = 0.85;
b[1] = 0;  b[2] = -0.26;  b[3] = 0.28;  b[4] = 0.04;
c[1] = 0;  c[2] = 0.23;  c[3] = 0.26;  c[4] = -0.04;
d[1] = 0.16;  d[2] = 0.22;  d[3] = 0.24;  d[4] = 0.85;
e[1] = 0;  e[2] = 0;  e[3] = 0;  e[4] = 0;
f[1] = 0;  f[2] = 0.2;  f[3] = 0.2;  f[4] = 0.2;
p[1] = 0.01;  p[2] = 0.08;  p[3] = 0.15;  p[4] = 1
```

The total number of iterations and the starting coordinates are specified.

```
total = 40000;  x[0] = 0;  y[0] = 0;
```

The following command line produces a random number r from the continuous uniform distribution of real decimal numbers in the range 0 to 1. A different random number will be generated on each successive iteration of the governing piecewise relations.

```
r: = Random[]
```

The piecewise relations are entered as Mathematica functions. A Mathematica function $f(x)$ is defined by the following syntax: `f[x_]:=expression`. Whenever an argument is given to `f`, then the rule contained in `expression` is applied to it. For example, if we define the function `f[x_]:=x^2`, then the output of `f[2]` is 4, and so on. Functions can be defined with more than one argument, e.g., `f[x_,y_]:=x^2+y^2`. Then `f[3,2]` generates the number 13. In the following two input lines, the piecewise functions `xx[i_, n_]` and `yy[i_, n_]` are defined with two arguments, one being the branch index i and the other the iteration number n. The right-hand sides of the functional definitions are just the piecewise relations expressing the rule for advancing from n to n+1.

```
xx[i_, n_] : = x[n+1] = a[i]  x[n] + b[i]  y[n] + e[i]
yy[i_, n_] : = y[n+1] = c[i]  x[n] + d[i]  y[n] + f[i]
```

One of the most useful commands in Mathematica is the `Table` command, which has the syntax `Table[expression, {i, min, max}]`. This command applies `Do` to `expression` while i goes from `min` to `max` in steps of 1, and forms a list of the results. In the following input line, `Table` is applied to the piecewise functions. To pick out the appropriate branch on each iteration, the `Which` command is used inside `Table`. The syntax is `Which[test 1, then 1, test 2, then 2, ...]`, i.e., If *test 1* is true, then do *then 1*, otherwise if *test 2* is true, then do *then 2*, otherwise.... Here, if $r<p[1] = 0.01$, then the coefficients corresponding to $i = 1$ are selected, etc. The resulting list of points is assigned the name pts.

```
pts = Table[Which[r < p[1], {xx[1, n], yy[1, n]},
      r < p[2], {xx[2, n], yy[2, n]}, r < p[3], {xx[3, n], yy[3, n]},
      r < p[4], {xx[4, n], yy[4, n]}], {n, 0, total}];
```

The list of points is now graphed using the ListPlot command, with various plotting options being chosen so as to produce a nice picture.

```
ListPlot[pts, AspectRatio -> 1, Axes -> False, Frame -> True,
PlotRange -> {{-.75, .75}, {0, 1.5}}, FrameTicks ->
{{-.5, {0, "x"}, .5}, {{.001, "0"}, .5, {.75, "y"}, 1, 1.5}, {}, {}},
PlotStyle -> {RGBColor[.1, 1, .1], PointSize[.007]}, TextStyle ->
{FontFamily -> "Times", FontSize -> 16}, ImageSize->{400,400}];
```

The resulting fern is shown in Figure 3.6. Although presented in black and white here, the fern is an appropriate shade of green when viewed on the computer screen. You can experiment in the RGBColor command with the fraction of red, green, and blue that you prefer.

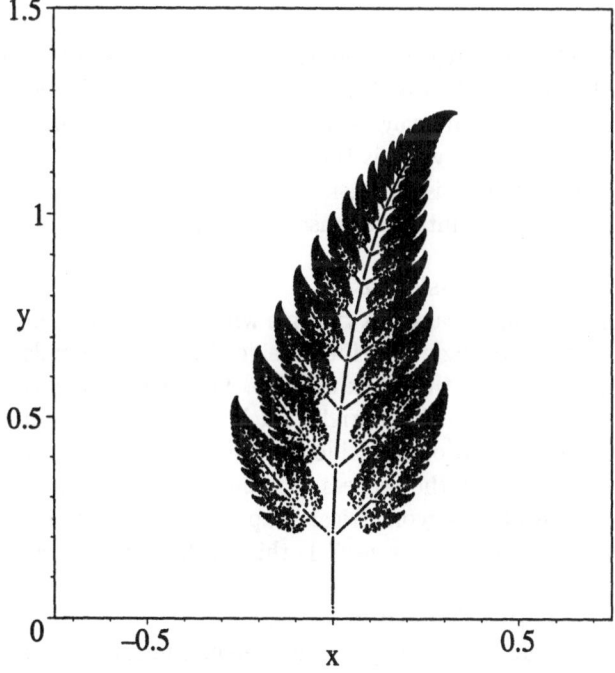

Figure 3.6: Barnsley's fractal fern.

It should be noted that Barnsley's fern resembles a real fern, the Black Spleenwort (Asplenium adiantum-nigrum). Barnsley's fern is self-similar in the sense that each frond is a miniature version of the whole fern, and each frond branch is similar to the whole frond, and so on.

End Example 3-1

PROBLEMS
Problem 3-3: The Koch triadic curve
Consider a line of length 1. Instead of throwing away the middle third as in the Cantor set, form an equilateral triangle in the middle third. Each line segment is $\epsilon = 1/3$. Repeat the process with each new line segment in step 1 to

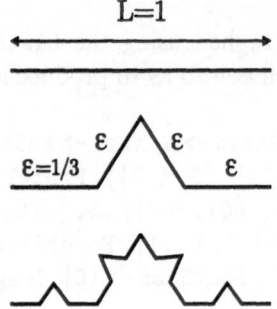

produce step 2. Each line segment now has length 1/9. Repeating this process indefinitely, determine D_C. Does your answer make intuitive sense? Explain.

Problem 3-4: The middle-half Cantor set
The Cantor set is also known as the middle-third Cantor set as on each step the middle third of each remaining line segment is thrown away. In the middle-half Cantor set, the line is initially divided into quarters and the inner two quarters (the middle half) are thrown away. If this action is repeated indefinitely with the remaining line segments, what is the capacity dimension of the middle-half Cantor set? If you compare this dimension with that for the middle-third Cantor set, does your answer make intuitive sense? Explain.

Problem 3-5: The Sierpinski gasket
Consider an upright equilateral black triangle with sides of unit length. Remove an inverted equilateral triangle inscribed inside the black triangle with vertex points bisecting the sides of the black triangle. One will now have an inverted white triangle with three smaller upright black triangles adjacent to its three sides. Repeat this removal process inside each of the three new black triangles. Sketch the result. Repeating the process as many times as necessary, determine the fractal dimension of this geometrical shape, known as Sierpinski's gasket. Does your answer make intuitive sense? Is this a self-similar fractal? Explain.

Problem 3-6: Sierpinski's carpet
A black square with sides of unit length is divided into nine smaller equal squares and the central square is colored white. Then this process is repeated for each of the eight remaining black squares, and so on. Sketch the Sierpinski "carpet" which results after five such iterations. Determine the fractal dimension of Sierpinski's carpet and comment on whether the answer makes intuitive sense. Is this a self-similar fractal? Explain.

Problem 3-7: A fractal tree
By generating a random number between 0 and 1, show that the following

piecewise expression produces a fractal appearing tree when plotted.

$$(x_{n+1}, y_{n+1}) = \begin{cases} (0.05x_n, 0.60y_n), & 0.0 < r < 0.1 \\ (0.05x_n, -0.50y_n + 1.0), & 0.1 < r < 0.2 \\ (0.46x_n - 0.15y_n, 0.39x_n + 0.38y_n + 0.60), & 0.2 < r < 0.4 \\ (0.47x_n - 0.15y_n, 0.17x_n + 0.42y_n + 1.1), & 0.4 < r < 0.6 \\ (0.43x_n + 0.28y_n, -0.25x_n + 0.45y_n + 1.0), & 0.6 < r < 0.8 \\ (0.42x_n + 0.26y_n, -0.35x_n + 0.31y_n + 0.70), & 0.8 < r < 1.0 \end{cases}$$

Take $x_0 = 0.5$, $y_0 = 0.0$, $N = 25000$, and a color of your own choice.

3.1.3 Rayleigh–Bénard Convection

This phenomenon, originally studied by Lord Rayleigh [Ray83], is concerned with the flow of heat energy upward through a horizontal fluid layer of infinite extension and thickness d when heated from below, the lower surface being held at a temperature ΔT above the upper surface. For small ΔT, the fluid is at rest and the transfer of energy is via heat conduction. However, as ΔT is increased above a critical value, fluid convection occurs. Rayleigh observed the formation of cylindrical "rolls" as schematically indicated in Figure 3.7. Hot fluid rises

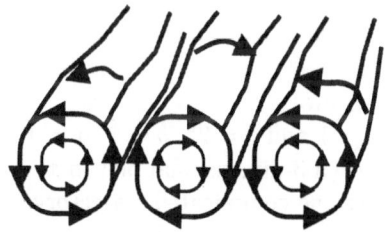

Figure 3.7: Cylindrical rolls in Rayleigh–Bénard convection.

(depicted by the upright arrows) along a boundary between a pair of rolls, cools at the top surface and then drops along the boundaries of adjacent rolls. As ΔT is further increased, more complex behavior occurs and ultimately chaotic convection is observed. The governing nonlinear hydrodynamic equations for this system can be found in an article by Saltzman [Sal62].

Figure 3.8: Rolls and hexagonal cells for cylindrical geometry.

Similar behavior has been observed in a cylindrical geometry. Rolls and hexagonal cells have been seen [Kos74] as shown in Figure 3.8. The lighter region corresponds to hotter rising fluid while the darker region corresponds to cooler descending fluid. The formation of hexagonal cells in a cylindrical geometry is an example of "symmetry breaking", the system being forced to decide on a particular orientation as a stability threshold (in this case the temperature) is crossed.

In an attempt to understand weather patterns, Edward Lorenz tackled a similar heat driven thermal convection problem for the atmosphere. Out of this research evolved the famous Lorenz model [Lor63] which, although not an accurate description of the atmosphere, shows how complex behavior can arise from three simple coupled nonlinear differential equations. The Lorenz equations [Lor63] are

$$\dot{x} = \sigma(y - x), \quad \dot{y} = rx - y - xz, \quad \dot{z} = xy - bz, \tag{3.5}$$

with x, y, z real and σ, r and b real, positive constants. Note that there are two nonlinear terms, the structure of each similar to that seen previously in competition equations. Physically, x, y and z are proportional to the convective velocity, the temperature difference between ascending and descending flows, and the mean convective heat flow. The coefficients σ and r are the Prandtl and reduced Rayleigh numbers, respectively, and b is related to the wave number.

PROBLEMS
Problem 3-8: Lorenz model
Given $x(0) = y(0) = z(0) = 1$ and $r = 25$, $b = \frac{8}{3}$, $\sigma = 10$, numerically solve the Lorenz equations and create a 3-dimensional plot in $x - y - z$ space. Choose an orientation that best shows the complicated trajectory.

3.1.4   Cellular Automata and the Game of Life

In using the routine NDSolve to numerically solve nonlinear ODEs using Mathematica, we have already mentioned that the Rungé–Kutta–Fehlberg algorithm may be used. We shall discuss numerical schemes in some detail in Chapter 6. All such schemes are based on replacing continuous variables such as time by small but finite intervals. For nonlinear PDEs, one does the same thing for the spatial part, replacing for example, the continuous variable x by small increments Δx. The state variable, temperature $T(x, t)$ for example, will be evaluated at "grid points" (x_i, t_i) with $i = 0, 1, \ldots$. As the grid is made finer and finer, it is hoped that the "correct" temperature distribution will be calculated.

There are, however, nonlinear dynamical problems where everything is discrete from the beginning. Instead of writing down differential equations, rules are postulated which are intended to capture the underlying physics. Such nonlinear dynamical systems are referred to as cellular automata. Cellular automata systems were first investigated by John Von Neumann and Stan Ulam. Later, interest was revived through the work of Stephen Wolfram [Wol86]

Interesting pattern formations can occur for 2-dimensional or higher cellular automata depending on the initial configurations and the underlying rules which

I wonder why I wonder why
I wonder why I wonder why ...

HINTS FOR
SURVIVAL
Game
of
Life

The game is to correctly punctuate the above sentence.

are specified. As an example, we consider the "Game of Life" invented by
John Conway in the 1970s [Gar70, Gar71a, Gar71b]. We begin by dividing 2-
dimensional space into discrete cells as in Figure 3.9. Each cell is either dead,

Figure 3.9: A dark (live) cell has eight nearest neighbor (n) white (dead) cells.

in which case it takes on the value 0, or alive, in which case it takes on the value
1. To visualize the system, let us color a live cell black and a dead cell white.
One starts with some initial configuration of live and dead cells (one live cell
in Figure 3.9). For this square lattice, each live or dead cell has eight nearest
neighbors, while for other lattices shapes, the number of nearest neighbors would
be different and would lead to different behavior. In the Game of Life, the rules
are as follows:

- Each cell is only allowed to interact with its nearest neighbors.

- A cell that is alive at one time-step dies at the next time-step if

  1. it has only 0 or 1 live neighbors; i.e., it perishes from "loneliness"

  2. it has 4 or more live neighbors; i.e., it succumbs to "overcrowding"

  A live cell will stay alive if it has 2 or 3 live neighbors.

- A dead cell becomes alive on the next step if it has exactly 3 live neighbors
  (clearly the live trio needed a fourth for bridge!)

Obviously other rules could be postulated and other lattices used.

The challenge is to find initial cell cluster configurations which lead to interesting behavior. Some cell configurations simply die out, such as that illustrated

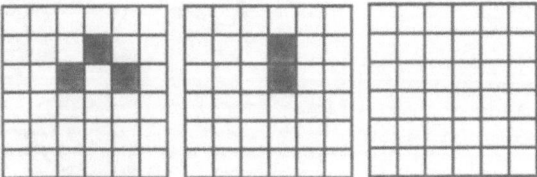

Figure 3.10: Death of an initial configuration.

in Figure 3.10. In the first time-step, from t_0 to t_1, two of the live cells die, but a dead one comes to life. On the next step, both live cells die of loneliness. By time-step 2, all cells are dead.

However, the initial live cell configuration shown in Figure 3.11 tends to a stationary pattern while that shown in Figure 3.12 oscillates. Clusters which oscillate are known as "blinkers".

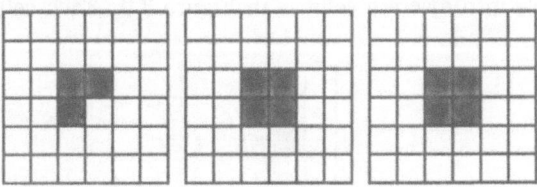

Figure 3.11: Evolution toward a stationary pattern.

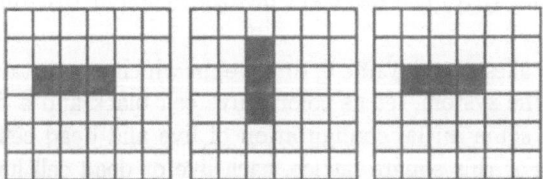

Figure 3.12: Example of a "blinker" pattern.

The above example appears to be more a game than real science. However, cellular automata concepts *have* been applied to real problems. For example, Murray and Paola [MP94] have applied a cellular model to the geophysical phenomena of stream braiding. The authors studied how a broad sheet of water flowing over non-cohesive sediment breaks up into a network of interconnected channels forming the "braided" stream. They postulate rules of nonlinear interaction of the cells based on the underlying physical mechanisms. Based on the patterns formed, they were able to determine which mechanisms are the most important for stream braiding.

In a delightful book, aimed at students in the biological sciences, Daniel Kaplan and Leon Glass [KG95] discuss how cellular automata concepts can be applied to the problem of generating patterns in biological systems, for example the exotic triangular pattern seen on the conus sea shell.

Example 3-2: The One-Out-of-Eight Rule

Using the rule that a cell becomes alive on a two-dimensional square lattice if exactly one of its eight neighbors is alive, otherwise it remains unchanged, and an initial configuration of one lone live cell, plot the pattern which results after twenty-nine steps.

Solution: Although the pattern that evolves after only a few steps can be easily generated by hand, it is quite tedious and challenging to produce the correct pattern when the number of steps is large. The following simple Mathematica code can be easily modified to handle any number of steps.

Live cells will be assigned the value one and dead cells the value zero. Using the `ListDensityPlot` command, the pattern of live cells will be plotted as white squares on a black background. We will start with one live cell (white square) placed at the center of a square black lattice which has $L = 62$ rows and 62 columns. This size of lattice is required if we are to see the entire pattern of live cells produced after a `total` number of 29 steps. If more steps are desired, the value of L must be increased.

```
Clear["Global`*"]

L = 62; total = 29;
```

Using the `Table` command, two null[1] matrices, each of dimension L by L, are created. One matrix is labeled "new", the other "old".

```
new = Table[0, {L}, {L}];

old = Table[0, {L}, {L}];
```

In the old matrix, the zero at (L/2, L/2) is replaced with the value 1. The following "old" matrix now represents the initial configuration with a single live cell placed at the center of a matrix of dead cells.

```
old = ReplacePart[old, 1, {{L/2, L/2}}];
```

Each time a step is executed a new matrix will be calculated, making use of the one-out-of-eight rule. To begin with, the new matrix has already been "initialized" to zero.

To understand how the one-out-of-eight rule is to be implemented, consider Figure 3.13 which shows a representative cell (i, j) and its eight immediate neighbors. Here i labels the row and j the column. In row $i - 1$, there are three immediate neighbors, namely $(i - 1, j - 1)$, $(i - 1, j)$, and $(i - 1, j + 1)$. In the same row, the neighbors are $(i, j - 1)$ and $(i, j + 1)$, while in row $i + 1$ they are

---

[1]Matrices with all elements equal to zero.

$(i + 1, j - 1)$, $(i + 1, j)$, and $(i + 1, j + 1)$.

The following function will check the nearest neighbors to (i, j) listed above and determine the number that are alive in the old matrix. For example, if three

$i\text{-}1, j\text{-}1$	$i\text{-}1, j$	$i\text{-}1, j\text{+}1$
$i, j\text{-}1$	i, j	$i, j\text{+}1$
$i\text{+}1, j\text{-}1$	$i\text{+}1, j$	$i\text{+}1, j\text{+}1$

Figure 3.13: Immediate neighbors of cell (i, j).

neighbors are alive, then s[i,j] will have the value 3. Note how the double square brackets are used to pick out the matrix elements in "old".

```
s[i_, j_] := old[[i + 1, j - 1]] + old[[i + 1, j]] + old[[i + 1, j + 1]]
          + old[[i, j - 1]] + old[[i, j + 1]]
          + old[[i - 1, j - 1]] + old[[i - 1, j]] + old[[i - 1, j + 1]]
```

To create the pattern generated by the one-out-of-eight rule, a double Do loop command structure is used. The syntax for Do is Do[body,{i, min, max}], i.e., Do body while i goes from min to max in steps of 1. In the inner Do loop, body consists of two commands. First s[i, j] is calculated to determine the number of nearest neighbors to cell (i, j) that are alive. This is followed by an "If...then...else" command, the syntax being If[test, then, else]. Here, if s[i, j] is equal to 1, then the matrix element new[[i, j]] is set equal to one (i.e., comes alive), otherwise it is set equal to the "old" value of the matrix element. To avoid difficulties in evaluating s[i,j] at the outer edges of the square lattice, the indices i and j are only allowed to run from 2 to L-1=61. After the "new" matrix has been evaluated, the process is repeated with the outer Do loop. The "old" matrix is set equal to the "new" one and the index n iterated from 1 to total.

```
Do[Do[s[i, j]; If[s[i, j] == 1, new[[i, j]] = 1, new[[i, j]] =
   old[[i, j]]], {i, 2, L - 1}, {j, 2, L - 1}]; old = new, {n, 1, total}]
```

Using the ListDensityPlot command, the new matrix is plotted after 29 steps,

```
ListDensityPlot[new, ImageSize -> {500, 500},
TextStyle -> {FontFamily -> "Times", FontSize -> 16}];
```

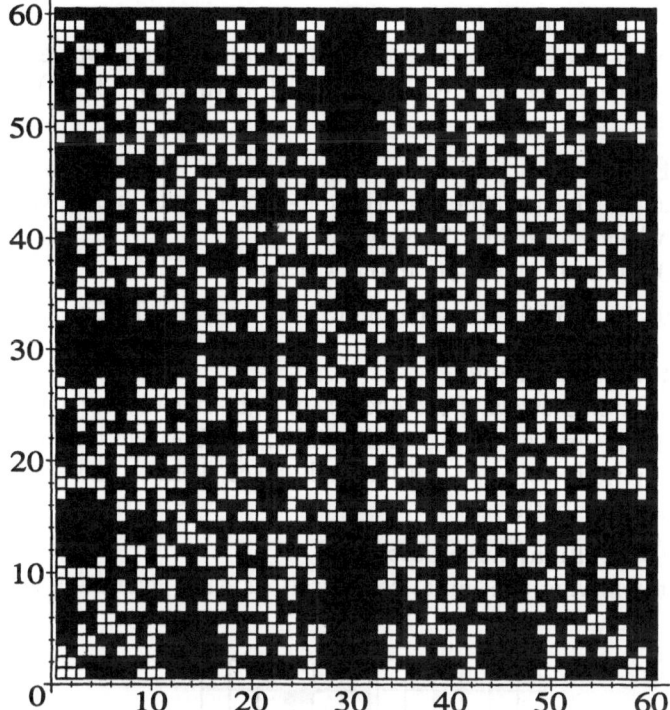

Figure 3.14: Complex pattern generated by the one-out-of-eight rule.

the resulting complex pattern being shown in Figure 3.14. Starting with one white live cell (the center cell of the central white square), the one-out-of-eight rule has generated a "family" of live cells, their spatial distribution creating a rich geometric pattern.

As one changes the rule structure, many interesting cellular patterns may be generated. Clearly, by the proper choice of rules, fabric designs and wallpaper patterns could be created, not to mention some patterns that are observed in nature.

| End Example 3-2 |

It should be noted that Mathematica has illustrations in the Demos section of its online Help of how to code the Game of Life and produce various cellular automata patterns. The Game of Life is contained in Programming Sampler and the cellular automata patterns in Cellular Automata.

PROBLEMS
Problem 3-9: The one-out-of-eight pattern in color

A colored one-out-of-eight pattern can be produced by using, e.g., the `ListPlot3D` command. Experiment with this command and any other plot command or plot options that you can find to produce colored versions of the one-out-of-eight rule pattern.

Problem 3-10: Four live cells
Using the one-out-of-eight rule, determine the pattern that evolves after 29 steps if there are initially four live cells located at $(30, 25)$, $(30, 35)$, $(25, 30)$, and $(35, 30)$.

Problem 3-11: Lace
Modify the one-out-of-eight rule to a two-out-of-eight rule and determine the lacy patterns which evolve after 29 steps if

   a. Cells $(30, 30)$ and $(31, 31)$ are initially alive.

   b. Cells $(29, 30)$, $(30, 30)$, $(31, 30)$, and $(32, 30)$ are initially alive.

Problem 3-12: Exploration
Explore what patterns evolve if the one-out-of-eight rule is changed to a two-out-of-eight rule, a three-out-of-eight rule, and so on. You will have to increase the number of initially live cells and judicially place them on the grid in order to see any interesting patterns.

Problem 3-13: Cellular automata
Show that the following two initial configurations each generate a stationary structure.

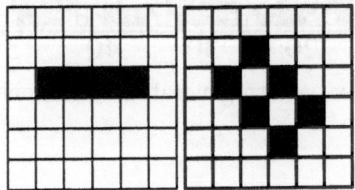

3.2   Solitons

To introduce the concept of a soliton, it is convenient to first consider the linear dispersionless wave equation which describes small amplitude vibrations,

$$\frac{\partial^2 \psi}{\partial x^2} - \frac{1}{c^2}\frac{\partial^2 \psi}{\partial t^2} = 0. \tag{3.6}$$

Physics students know that a localized traveling wave obeying Equation (3.6) travels along unchanged in shape in the x-direction with speed c. A typical example is illustrated schematically in Figure 3.15 with the pulse traveling in the positive x direction. This pulse has a shape $\psi(x, t) = \psi(x - ct)$, i.e., it looks exactly the same in a coordinate system moving to the right with velocity c. At $t = t_0$, the shape[2] has been translated without distortion to the right, the maximum moving from $x = 0$ to $x_0 = ct_0$.

---

[2]The general solution of the linear wave equation (3.6) is $\psi(x, t) = f_1(x + ct) + f_2(x - ct)$, where f_1 and f_2 are arbitrary functions. The first term represents a wave form traveling to the left, the second a wave traveling to the right. The general solution may be derived with the Mathematica commands:

```
waveeq=D[psi[x,t],{x,2}] - (1/c^2) D[psi[x,t],{t,2}] ==0
DSolve[waveeq, psi[x,t],{x,t}]
Simplify[%,c>0]
```

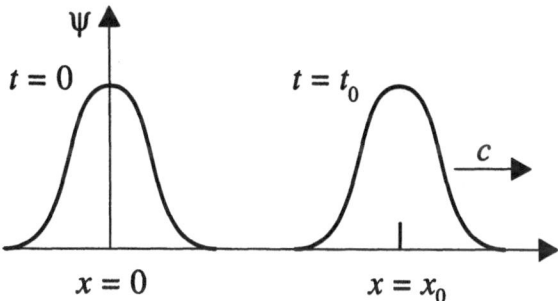

Figure 3.15: Propagation of a solitary pulse solution to the linear wave equation.

This localized pulse is an example of a "solitary" wave. A "soliton" is a solitary wave solution of a wave equation that asymptotically preserves its shape and velocity upon collision with other solitary waves. Thus, if a solution $\psi(x,t)$ is composed only of solitary waves for large negative time, i.e.,

$$\psi(x,t) = \sum_{j=1}^{N} \psi_{\text{sol}}(x - c_j t) \tag{3.7}$$

as $t \rightarrow -\infty$, such solitary waves will be called solitons if they emerge from the interaction with no more than a phase shift, i.e.,

$$\psi(x,t) \sim \sum_{j=1}^{N} \psi_{\text{sol}}(x - c_j t + \delta_j) \tag{3.8}$$

as $t \rightarrow +\infty$, δ_j being the phase shift of the jth pulse. The definition of a soliton clearly implies something about the stability of the solitary waves. Since pulses propagating according to the linear dispersionless wave equation can pass through one another completely unaffected, they are, according to our definition, solitons. However, they are a trivial example of a soliton.

Now introduce some dispersion, i.e., we let $c = c(\omega)$ where ω is frequency. The localized pulse shape in Figure 3.15 may be thought of as the Fourier sum of different frequency components. Since c is different for different ω, the Fourier components will travel at different speeds and the pulse will change shape. Dispersion, without any other compensation, tends to spread the pulse and destroy the possibility of a solitary wave.

Suppose, on the other hand, that some nonlinearities are introduced into the wave equation. Introducing nonlinear terms again removes the possibility of solitary waves because the effect of nonlinearities is to redistribute the energy via harmonic generation into higher frequency modes. Because there is effectively an increase in frequency spread, the well-known Heisenberg uncertainty principle of physics tells us that the pulse should be squeezed in the time domain (and in the spatial domain as well). An optical example [RE76] of spatial squeezing is shown in Figure 3.16 where the transient behavior of two short, initially rectangular envelope, light pulses is shown as they pass through each other. The relevant equations are the laser competition equations (2.32) with first time derivatives

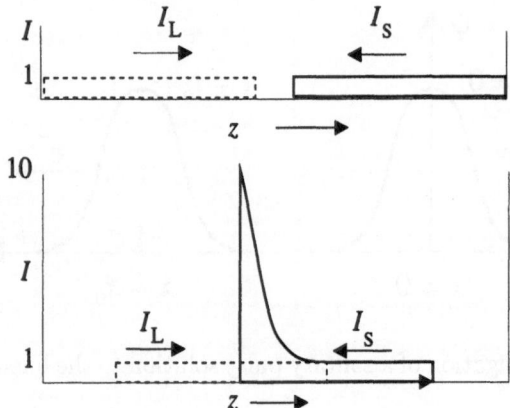

Figure 3.16: Transient behavior of two interacting laser pulses. The intensities are normalized to their input values.

included and $\alpha = 0$, namely,

$$\frac{\partial I_{\mathrm{L}}}{\partial z} + \frac{1}{v}\frac{\partial I_{\mathrm{L}}}{\partial t} = -gI_{\mathrm{L}}I_{\mathrm{s}}, \quad \frac{\partial I_{\mathrm{s}}}{\partial z} - \frac{1}{v}\frac{\partial I_{\mathrm{s}}}{\partial t} = -gI_{\mathrm{L}}I_{\mathrm{s}}, \tag{3.9}$$

where v is the speed of light in the medium. If we set $g = 0$, i.e., no interaction, and use DSolve, the general solutions of (3.9) are of the structure

$$I_{\mathrm{L}}(z,t) = I_{\mathrm{L}}(z - vt), \ I_{\mathrm{s}}(z,t) = I_{\mathrm{s}}(z + vt). \tag{3.10}$$

The rectangular shapes would simply translate to the right and the left, respectively, without distortion. Here we have a nonlinear term $I_{\mathrm{L}}I_{\mathrm{s}}$ but no dispersion. In Chapter 11, we shall reduce the nonlinear PDEs (3.9) to ODEs, whereupon we can numerically solve for I_{L} and I_{s} using Mathematica.

By now the student might ask: Can one have a situation where both dispersion and nonlinearities are present and the spreading and squeezing can balance each other? The answer is yes provided that the structure of the dispersive and

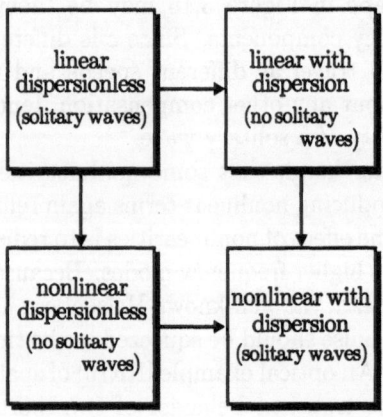

Figure 3.17: Necessary ingredients for solitary waves to exist.

nonlinear terms is just right. Figure 3.17 summarizes our discussion.

Some specific examples will now be discussed. Because the derivations are in most cases too lengthy, involve approximations and boundary conditions which have to be carefully examined, and involve normalization through suitable scale transformations, no attempt will be made to derive the soliton equations.

Although the balancing scenario between nonlinearity and dispersion that we have just outlined is one of the most common ways for solitary waves to occur, they can also arise as a consequence of other balancing acts. For example, a candle flame maintains its solitary pulse nature by balancing the diffusion of heat from the flame into the wax and the nonlinear energy release of the vaporizing wax [Sco81]. For the 3-wave interaction discussed in Chapter 10, three co-existing solitary waves (two electromagnetic and a sound wave) are possible due to the presence of several "competing" nonlinear terms.

3.2.1 Shallow Water Waves (KdV and Other Equations)

The one-dimensional Boussinesq equation [Bou72],

$$\frac{\partial^2 \psi}{\partial x^2} - \frac{\partial^2 \psi}{\partial t^2} + 6\frac{\partial^2 (\psi^2)}{\partial x^2} + \frac{\partial^4 \psi}{\partial x^4} = 0 \tag{3.11}$$

first derived in an attempt to describe shallow-water waves (ψ is the surface displacement) propagating in both directions, possesses soliton solutions. If the nonlinear (third) term is absent, Equation (3.11) is structurally the same as the equation of motion for a vibrating wire [Mor48]. A wire is a string with stiffness added. If you hold the end of a string, the rest of the string flops down due to gravity. For a wire of reasonable thickness, the wire bends downward but doesn't immediately flop down at the juncture with your fingers. Detailed analysis of the wire [Mor48] leads to the inclusion of a fourth derivative term as in Equation (3.11) to account for stiffness.

If one assumes a plane wave solution $\psi \sim \exp(i(kx - \omega t))$ in the wire equation, the dispersion relation $k^4 - k^2 + \omega^2 = 0$ results. The phase velocity $v = \omega/k(\omega)$ is frequency dependent, so the wire equation displays dispersion. Putting the nonlinear term back in allows the possibility of soliton solutions. In Chapters 10 and 11, we shall outline some methods for finding these solutions.

An even more famous example involving water waves is the Korteweg–deVries equation

$$\frac{\partial \psi}{\partial t} + \alpha \psi \frac{\partial \psi}{\partial x} + \frac{\partial^3 \psi}{\partial x^3} = 0 \tag{3.12}$$

with α a numerical factor and ψ again the surface displacement. Since α can be scaled out of the equation by letting $\psi \to \frac{\psi}{\alpha}$, the particular choice of α used doesn't matter. Common choices appearing in the literature are $\alpha = 1$, $\alpha = 6$, and, by mathematicians, $\alpha = -6$. Note that the latter negative value inverts ψ relative to the physical displacement. Equation (3.12) was derived by Korteweg and de Vries in 1895 [KdV95] to explain earlier observations by the Scottish engineer[3] John Scott Russell [Rus44].

---

[3]John Scott Russell evidently was one of the great naval architects of the 19th century, for example introducing the concept of "solid of least resistance" to radically alter thinking on the design of a ship's bow [Emm77].

In the less formal style of scientific reporting of the day, Scott Russell wrote:

> I was observing the motion of a boat which was rapidly drawn along a
> narrow channel by a pair of horses, when the boat suddenly stopped
> – not so the mass of water in the channel which it had put in motion;
> it accumulated round the prow of the vessel in a state of violent
> agitation, then suddenly leaving it behind, rolled forward with great
> velocity, assuming the form of a large solitary elevation, a rounded
> smooth and well-defined heap of water, which continued its course
> along the channel apparently without change of form or diminution
> of speed. I followed it on horseback, and overtook it still rolling on
> at a rate of some eight or nine miles an hour, preserving its original
> figure some thirty feet long and a foot to a foot and a half in height.
> Its height gradually diminished, and after a chase of one or two miles
> I lost it in the windings of the channel. Such, in the month of August
> 1834, was my first chance interview with that singular and beautiful
> phenomenon . . .

The KdV equation allows soliton solutions for the uni-directional propagation of lossless shallow water waves in a rectangular canal. The KdV equation doesn't look like a wave equation, having a first time derivative and a third spatial derivative term. It does describe shallow water waves quite well, however. Solitons are again possible because the dispersion is counterbalanced by the nonlinearity. The KdV equation turns out to be very important in the study of solitons because, under suitable approximations, it arises in many different physical problems [SCM73], including

- ion-acoustic waves in a plasma

- magnetohydrodynamic waves in a plasma

- anharmonic lattices

- longitudinal dispersive waves in elastic rods

- pressure waves in liquid-gas bubble mixtures

- rotating flow down a tube

- thermally excited phonon packets in low temperature nonlinear crystals.

The derivation of the KdV equation for shallow water waves in a rectangular canal is beyond the scope of this text. The interested reader is referred to the book by Leibovich and Seebass [LS74]. Although Korteweg and DeVries showed that the KdV equation has solitary wave solutions, it was not until 1965 that Norm Zabusky and Martin Kruskal [ZK65] published numerical results indicating the formation of solitons. In 1971, Fred Tappert, of Bell Laboratories, obtained the following explicit analytic expression describing two interacting solitons of the KdV equation:

$$\psi = \left(\frac{72}{\alpha}\right) \frac{3 + 4\cosh(2x - 8t) + \cosh(4x - 64t)}{[3\cosh(x - 28t) + \cosh(3x - 36t)]^2}. \tag{3.13}$$

Figure 3.18, obtained by running MF08, shows a plot of Equation (3.13) just before, during, and after the collision. The two solitons are traveling to the right, the taller one having the greater velocity. Look at the plot representing the collision process. Does the linear superposition principle hold here?

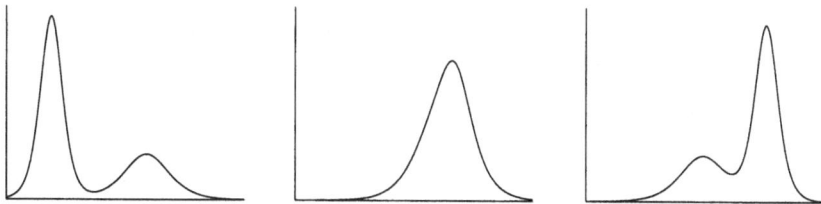

Figure 3.18: Two-soliton solution (3.13) before, during, and after collision.

Water tank experiments demonstrating the validity of the KdV equation and the soliton solutions have been carried out by Bettini, Minelli, and Pascoli[BMP83] and Olsen, Smith, and Scott[OSS84].

The KdV Two-Soliton Solution

Equation (3.13) is animated. Figure 3.18 shows "snapshots" of the motion before, during, and after the collision. Both solitons are traveling to the right, the taller soliton moving faster than the shorter one. As with the target patterns, we see here another example of nonlinear superposition during the collisions. Mathematica commands in file: `Cosh, Animate, Plot, PlotStyle, Hue, Frames, PlotPoints, Ticks, PlotRange,TextStyle, <<Graphics`

The two-dimensional generalization of the KdV equation is the Kadomtsev–Petviashvili (KP) equation [KP70]

$$(\psi_t + \alpha\psi\psi_x + \psi_{xxx})_x + 3\psi_{yy} = 0 \tag{3.14}$$

where ψ is the surface displacement, and subscripts denote partial derivatives.

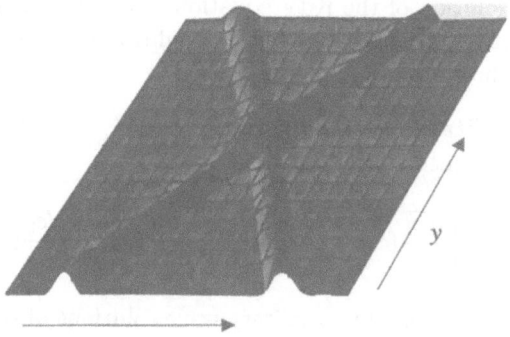

Figure 3.19: Schematic representation of the KP two-soliton solution.

Except where confusion may arise, we shall generally use this more compact subscript notation for partial differentiation from now on.

Figure 3.19 schematically shows a 2-soliton solution of the KP equation. Such an "X-shaped" soliton configuration has actually been observed in shallow ocean water off of the coast of Oregon, the event captured in a famous photograph which appeared in the March 1991 issue of *Physics Today* [Kru91]. In this case, x and y in Figure 3.19 are directions normal and parallel to the shoreline, respectively.

PROBLEMS
Problem 3-14: A derivation of the KdV equation
Consider a small amplitude wave $U = A\cos(kx - \omega t)$ traveling through a medium characterized by the dispersion relation

$$\omega = sk(1 - \frac{1}{2}\epsilon k^2 + O(\epsilon^2))$$

where s is speed, ϵk^2 is small, and O( ) means "of the order of." Show that to order ϵ, U satisfies the linear differential equation

$$U_t + sU_x + \frac{s\epsilon}{2}U_{xxx} = 0.$$

Now suppose that s depends weakly on the small displacement U, viz.

$$s = s_0 + s_1 U + O(U^2).$$

Neglecting terms of order U^2 and ϵU, derive the resulting nonlinear differential equation. Finally, by setting $X = x - s_0 t$, $T = t$ and introducing suitable scaling, show that the KdV equation results.

Problem 3-15: Solitary waves of the KdV equation
By direct substitution, use Mathematica to prove that

$$\psi = \frac{3c}{\alpha}\,\text{sech}^2\left[\frac{\sqrt{c}}{2}(x - ct)\right]$$

is a solitary wave solution of the KdV equation. The `TrigReduce` command is useful here. Use the `Animate` command to confirm that taller solitary waves travel faster than shorter ones.

Problem 3-16: Solitary wave collision
The following displacement formula

$$\psi(x,t) = \frac{3c_1}{\alpha}\,\text{sech}^2\left[\frac{\sqrt{c_1}}{2}(x - x_1 - c_1 t)\right] + \frac{3c_2}{\alpha}\,\text{sech}^2\left[\frac{\sqrt{c_2}}{2}(x - x_2 - c_2 t)\right]$$

represents the superposition of two solitary wave solutions of the KdV equation if the separation $|x_1 - x_2|$ is sufficiently large initially. Animate $\psi(x, t)$ for $c_1 = 5$, $c_2 = 1$, $\alpha = -6$, $x_1 = -10$, $x_2 = 10$. Discuss your results.

3.2.2 Sine-Gordon Equation

The sine-Gordon equation

$$\psi_{xx} - \psi_{tt} = \sin\psi \tag{3.15}$$

supports so-called "topological solitons". (The KdV solitons are referred to as "non-topological solitons".) Such solitons, as shown in Figure 3.20, which

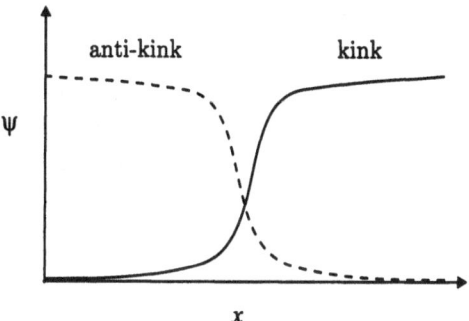

Figure 3.20: Kink-antikink soliton solutions to the sine-Gordon equation.

look like "kinks" and "anti-kinks", can propagate without distortion. A good physical example of a kink solution is a Bloch[4] wall between two magnetic domains in a ferromagnet. The magnetic spins rotate from, say, spin down in one domain to spin up in the adjacent domain (Figure 3.21). The transition

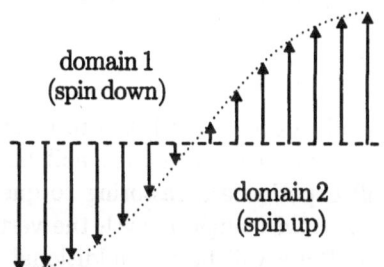

Figure 3.21: Bloch wall between two ferromagnetic domains.

region between down and up is called the Bloch wall. Under the influence of an applied magnetic field, the Bloch wall can propagate according to the sine-Gordon equation (3.15).

If one introduces the linear approximation $\sin\psi = \psi$, the resulting equation is known as the Klein–Gordon equation, an equation which appears in particle physics

$$\psi_{xx} - \psi_{tt} = \psi. \tag{3.16}$$

The Klein–Gordon equation is dispersive and thus the $\sin\psi$ term in (3.15) contains both dispersion and nonlinearity, allowing (3.15) to have soliton solutions.

A mechanical model of the sine-Gordon equation can be constructed as shown in Figure 3.22, the details being found in the review paper by A. Barone

---

[4]Named after the theoretical physicist and Nobel Laureate Felix Bloch.

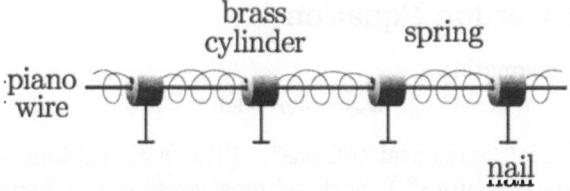

Figure 3.22: A mechanical model of the sine-Gordon equation.

et al. [BEMS71]. Here, ψ is the twist angle of the nails from the vertical. The derivation of the sine-Gordon equation for the mechanical model is straightforward and the mathematical form of the kink solution easily confirmed.

Example 3-3: Derivation and Solution of Sine-Gordon Equation

For the mechanical model of Figure 3.22,

a. derive the sine-Gordon equation, stating what assumptions are made,

b. confirm by direct substitution that $\psi(x,t) = 4\arctan\left(e^{(x-ct)/\sqrt{1-c^2}}\right)$, with $c < 1$, is a solution of this equation,

c. by animating $\psi(x,t)$ with $c = 0.5$, show that it is a kink solitary wave traveling in the positive x direction.

Solution:a. The derivation of the sine-Gordon equation is a straightforward generalization of the simple pendulum equation (2.2).

```
Clear["Global`*"]
```

The brass cylinders and nail thickness are taken to be small and it is assumed that all of the mass m is concentrated in each nail head. As with the pendulum, the nail head will experience a restoring torque due to gravity when the nail of length L makes a twist angle ψ with the vertical. On a given nail, say the one at position x, there will be two additional forces due to the two springs connecting that nail to its two nearest neighbors. If the spacing between nails is h and Hooke's law is assumed, the force exerted on the nail at x at time t due to the springs connecting it to the nails at $x+h$ and $x-h$ will be $k(\psi(x+h,t) - \psi(x,t)) - k(\psi(x,t) - \psi(x-h,t))$, where k is the torque constant for the springs. Putting these ideas all together, Newton's 2nd law, in the form mass times acceleration minus total force equals zero, is entered.

```
eq1 = m L D[ψ[x, t], {t, 2}] + m g Sin[ψ[x, t]]
      - k (ψ[x+h, t] - ψ[x, t]) + k (ψ[x, t] - ψ[x-h, t]) == 0
```

$$g\, m \, \sin(\psi(x, t)) + k \, (\psi(x, t) - \psi(x - h, t)) - k \, (\psi(x + h, t) - \psi(x, t))$$
$$+ L\, m\, \psi^{(0,\, 2)}(x, t) == 0$$

In the output, the notation $\psi^{(0,2)}(x, t)$ denotes the second partial time derivative of ψ with respect to time t. The lhs of **eq1** output is now divided by the weight $m\,g$, and the result expanded and set equal to zero.

```
eq2 = Expand[eq1[[1]]/(m g)] == 0
```

$$\sin(\psi(x,\,t)) + \frac{2\,k\,\psi(x,\,t)}{g\,m} - \frac{k\,\psi(x-h,\,t)}{g\,m} - \frac{k\,\psi(x+h,\,t)}{g\,m} + \frac{L\,\psi^{(0,\,2)}(x,\,t)}{g} == 0$$

Since h is small, the lhs of eq2 is Taylor expanded about $x = 0$ to order h^2.

```
eq3 = Series[eq2[[1]], {h, 0, 2}]
```

$$\sin(\psi(x,\,t)) + \frac{L\,\psi^{(0,\,2)}(x,\,t)}{g} - \frac{k\,\psi^{(2,\,0)}(x,\,t)\,h^2}{g\,m} + O(h)^3$$

The $O(h^3)$ term is removed in eq3 with the **Normal** command and the result set equal to zero.

```
eq4 = Normal[eq3] == 0
```

$$\sin(\psi(x,\,t)) + \frac{L\,\psi^{(0,\,2)}(x,\,t)}{g} - \frac{h^2\,k\,\psi^{(2,\,0)}(x,\,t)}{g\,m} == 0$$

On making the following self-evident substitutions, the required form of the sine-Gordon equation results.

```
sgEq = eq4 /. {(L D[ψ[x, t], {t, 2}]/g) -> D[Ψ[X, T], {T, 2}],
       (k D[ψ[x, t], {x, 2}] h^2/(m g)) -> D[Ψ[X, T], {X, 2}],
       Sin[ψ[x, t]] -> Sin[Ψ[X, T]]}
```

$$\sin(\Psi(X,\,T)) + \Psi^{(0,\,2)}(X,\,T) - \Psi^{(2,\,0)}(X,\,T) == 0$$

b. The proposed solution is entered, the independent variables being written as capital letters to match the above output notation.

```
Ψ = 4 ArcTan[Exp[(X - c T)/Sqrt[1 - c^2]]]
```

$$4 \tan^{-1}\left(e^{\frac{X-cT}{\sqrt{1-c^2}}}\right)$$

On entering the lhs of the sine-Gordon equation the above solution is automatically substituted,

```
sgeqlhs = D[Ψ, {T, 2}] - D[Ψ, {X, 2}] + Sin[Ψ]
```

producing the following output.

$$-4\left(\frac{-2\,e^{\frac{3(X-cT)}{\sqrt{1-c^2}}}}{(1-c^2)\left(1+e^{\frac{2(X-cT)}{\sqrt{1-c^2}}}\right)^2} + \frac{e^{\frac{(X-cT)}{\sqrt{1-c^2}}}}{(1-c^2)\left(1+e^{\frac{2(X-cT)}{\sqrt{1-c^2}}}\right)}\right)$$

$$+4\left(\frac{-2\,c^2\,e^{\frac{3\,(X-c\,T)}{\sqrt{1-c^2}}}}{(1-c^2)\left(1+e^{\frac{2\,(X-c\,T)}{\sqrt{1-c^2}}}\right)^2}+\frac{c^2\,e^{\frac{(X-c\,T)}{\sqrt{1-c^2}}}}{(1-c^2)\left(1+e^{\frac{2\,(X-c\,T)}{\sqrt{1-c^2}}}\right)}\right)$$

$$+\sin(4\tan^{-1}(e^{\frac{(X-c\,T)}{\sqrt{1-c^2}}}))$$

Applying the `FullSimplify` command reduces the above output to zero, thus confirming the suggested solution.

    FullSimplify[sgeqlhs]

$$0$$

c. To animate the solution, the `Graphics` package is loaded,

    << Graphics`

and the velocity $c = 0.5$ is entered and automatically substituted into Ψ.

    c = 0.5; Ψ

$$4\tan^{-1}(e^{1.1547\,(X-0.5\,T)})$$

The function Ψ is now animated, producing a kink profile traveling to the right on execution of the command line.

    Animate[Plot[Ψ, {X, -20, 20}, PlotStyle -> {Hue[1]}], {T, 0, 20},
    Frames -> 50, PlotPoints -> 500, ImageSize->{600,400}, AxesLabel ->
    {"X", "Ψ"}, TextStyle -> {FontFamily -> "Times", FontSize -> 16}]

    End Example 3-3

The review article of Barone et al. also describes how the sine-Gordon equation occurs for

- magnetic flux propagation in Josephson junctions

- propagation of crystal dislocations in a solid

- propagation of ultra-short optical pulses

- a unitary theory of elementary particles

PROBLEMS
Problem 3-17: Kink–kink collision
The following expression describes the collision of two sine-Gordon kink solitons,

$$\psi = 4\arctan(c\,\frac{\sinh(x/\sqrt{1-c^2})}{\cosh(ct/\sqrt{1-c^2})}),$$

where c is the velocity. Taking $c = 0.5$, animate the kink–kink solution over a suitable range of x and t. What does negative c correspond to? Animate the kink–antikink collision by replacing the first c by $1/c$, x by ct, and ct by x.

Problem 3-18: Kink–kink solution
By direct substitution, confirm that the kink–kink formula given in the previous problem is a solution of the sine-Gordon equation.

Problem 3-19: Sine-Gordon breather mode
The sine-Gordon equation permits a moving (velocity v) "breather mode" solution, which is localized in space but oscillatory in time, of the form

$$\psi = 4 \arctan\left(\sqrt{\frac{m}{(1-m)}} \frac{\sin((t - vx)\gamma\sqrt{(1-m)})}{\cosh((x - vt)\gamma\sqrt{m})}\right)$$

with $\gamma = 1/\sqrt{1-v^2}$, $-1 < v < 1$, and $0 < m < 1$. Animate this solution for $m = 1/2$ and (a) $v = 0$, (b) $v = 0.5$, (c) $v = -0.9$. The factor γ is the special Lorentz transformation of relativity with the speed of light equal to one.

3.2.3 Self-Induced Transparency

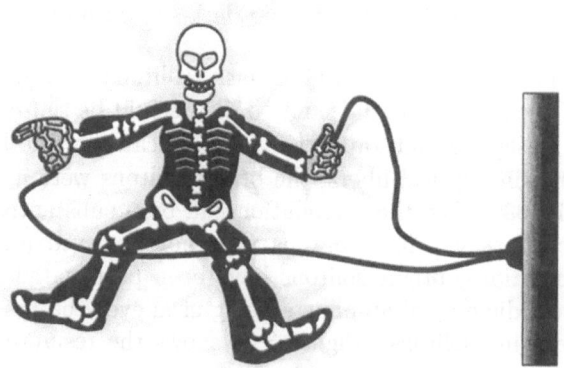

Using numerical techniques, McCall and Hahn [MH67, MH69] discovered that ultra-short pulses of light can travel through a resonant two-level optical medium as if it were transparent. Under certain conditions it is found that the leading edge of the pulse inverts the atomic population while the trailing edge returns it to the ground state via stimulated emission. Thus the energy absorbed by the quantum system from the leading edge of the pulse is recaptured by the trailing edge. Under proper conditions of coherence and intensity, a steady pulse (i.e., a solitary wave) can propagate without damping at a velocity two or three orders of magnitude less than the phase velocity of light in the medium.

3.2.4 Optical Solitons

The nonlinear Schrödinger equation (NLSE) can be used to describe ultra-short (picosecond) optical envelope solitons in transparent optical fibers. It has the dimensionless (normalized) form

$$iE_z \pm \frac{1}{2}E_{\tau\tau} + |E|^2E = 0 \qquad (3.17)$$

where the plus–minus sign corresponds to anomalous or normal dispersion respectively. Here, E is proportional to the electric field amplitude ϕ, z to the distance coordinate Z in the direction of propagation, and $\tau \propto t - Z/v_g$, where t is the time and v_g is the group velocity. Equation (3.17) may be derived[5] from Maxwell's wave equation for the optical Kerr nonlinearity which models a medium possessing an intensity dependent refractive index $n = n_0 + n_2|\phi|^2$ with $n_2 > 0$ [Has90, Agr89]. With profiles as shown in Figure 3.23, "bright" solitons

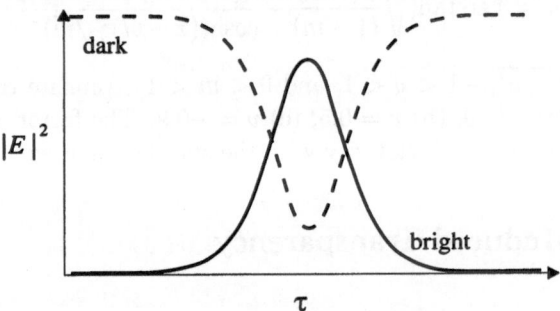

Figure 3.23: Bright and dark soliton profiles.

occur for the plus sign in (3.17) while "dark" solitons occur for the minus sign. The origin of the names "bright" and "dark" should be self-evident. Although predicted by Hasegawa and Tappert in 1973, because of the large attenuation in the then existing optical fibers, the bright solitons were not experimentally observed until 1980 when the attenuation had been substantially reduced. For other technical reasons, dark solitons were not observed until 1987. Under laboratory conditions, bright solitons have been propagated for thousands of kilometers, periodic optical amplifiers being used every 60 km or so to reboost the slowly decaying solitons. Figure 3.24 shows the results of an experiment

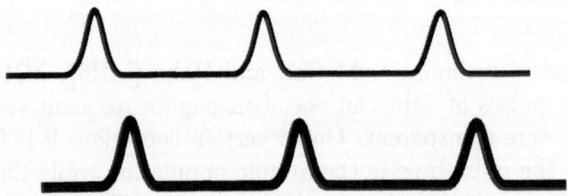

Figure 3.24: Optical soliton profiles transmitted at 5 gigabits (top) and received (bottom) after 10,000 kilometers. The transmissions were error-free.

carried out at AT&T Bell Laboratories that illustrates this feature.

Solitons are envisioned by telecommunications engineers as high bit-rate[6]

---

[5] A simpler alternate approach is left as a problem.

[6] A pulse, representing a 1, represents one bit of information. The narrower the pulse, the higher is the bit rate. Present optical cable technology uses light pulses which are very much wider than optical solitons.

carriers of digitized information, a bright soliton, e.g., being used to represent a digital "1" and a blank space being used to represent a "0" in the binary number system. There is also considerable interest in using optical solitons in all optical logic gates.

Edmundson and Enns [EE95] have numerically investigated 3-dimensional spherical optical solitons ("light bullets") for the generalized NLSE

$$iE_z + \frac{1}{2}(E_{xx} + E_{yy} + E_{\tau\tau}) + \left(\frac{|E|^2}{1 + a|E|^2}\right)E = 0. \qquad (3.18)$$

Here, x and y refer to directions transverse to the propagation direction. The nonlinearity arises from saturation of the Kerr refractive index. For small $|E|$, the nonlinear coefficient in (3.18) is as in (3.17) but for large $|E|$, the coefficient approaches the constant value $1/a$. Three-dimensional spherical soliton-like solutions to (3.18) can exist because the nonlinear term can balance not only the dispersion in the propagation (z) direction, but also the diffraction or spreading in the transverse directions due to the E_{xx} and E_{yy} terms. Figure 3.25 shows a

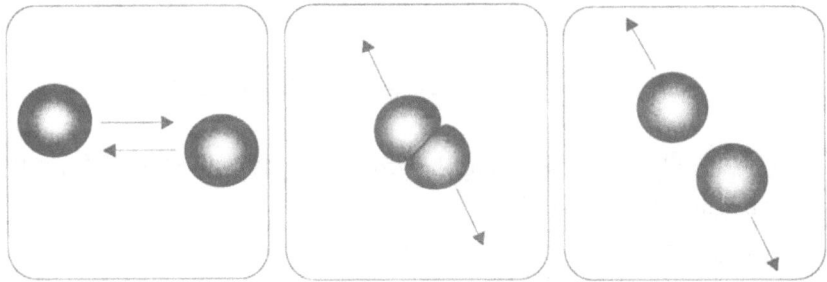

Figure 3.25: Billiard ball behavior of colliding (repulsive) light bullets.

representative numerical simulation in which "out of phase" light bullets bounce off of one another in a repulsive fashion, behaving much like billiard balls. Notice the flattening of the light bullets along their contact faces in the middle frame. For "in phase" light bullets, there is a tendency for them to attract each other. If they have equal but opposite velocities and a non-zero impact parameter as in the first frame of Figure 3.26, they can begin to rotate about each other (center

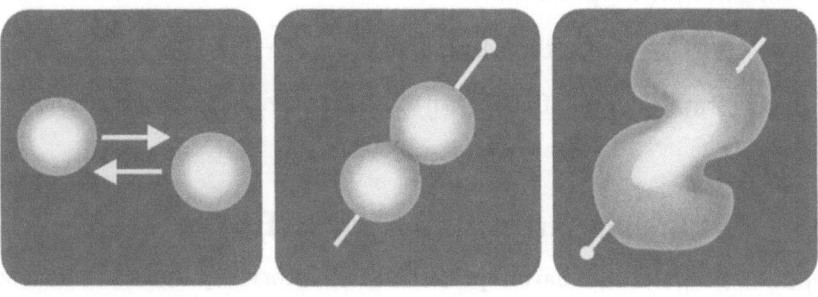

Figure 3.26: Coalescence and spiral galaxy formation of attractive light bullets. A reference line has been added to indicate the sense of rotation.

frame) before coalescing. As coalescence takes place, the light bullet duo shed radiative arms, at this stage resembling a spiral galaxy (third frame). When the galactic disc arms have been shed, what is left appears to be a single rotating soliton (not shown)[7]. The confirmation of these numerical light bullet scenarios awaits experimental confirmation or repudiation.

PROBLEMS
Problem 3-20: Derivation of the NLSE
As a simple model of 1-dimensional bright optical soliton propagation in a dielectric fiber, consider an electromagnetic field $\mathcal{E}(x,t) = \phi(x,t)e^{i(k_0x-\omega_0t)}$ propagating in a medium characterized by a refractive index of the form $n = \frac{ck}{\omega} = n_1(\omega) + n_2|\mathcal{E}|^2$. Here ϕ is a (slowly varying) complex amplitude, k_0 the central wave number, $\omega_0 = \frac{2\pi c}{\lambda_0}$ the central frequency with corresponding wavelength λ_0, c the speed of light, n_1 the linear refractive index, and n_2 a frequency-independent constant.

  a. By expanding the wave number $k = k(\omega, |\mathcal{E}|^2)$ around ω_0 and zero electric field and making use of the Fourier integral representation $\mathcal{E}(x,t) = \frac{1}{2\pi}\int_{-\infty}^{\infty}\int_{-\infty}^{\infty} e(k,\omega)e^{i(kx-\omega t)}dkd\omega$, show that (keeping appropriate terms) $\phi(x,t)$ satisfies the equation

$$i\left(\phi_x + k'\phi_t\right) - \frac{1}{2}k''\phi_{tt} + \frac{2\pi n_2}{\lambda_0}|\phi|^2\phi = 0.$$

  Here $k' \equiv \frac{\partial k}{\partial \omega}|_0 \equiv \frac{1}{v_g}$, v_g being the group velocity, and $k'' \equiv \frac{\partial^2 k}{\partial \omega^2}|_0 < 0$ for the bright soliton case.

  b. Introducing the dimensionless quantities $E = 10^{4.5}\sqrt{2\pi n_2}\phi$, $z = 10^{-9}\left(\frac{x}{\lambda_0}\right)$, and $\tau = \frac{10^{-4.5}}{\sqrt{-\lambda_0 k''}}\left(t - \frac{x}{v_g}\right)$, show that the above equation reduces to the NLSE

$$iE_z + \frac{1}{2}E_{\tau\tau} + |E|^2E = 0.$$

  c. For a glass fiber at $\lambda = 1.5\mu m$, $n_2 = 1.2 \times 10^{-22}$ m^2/V^2 and $-\lambda k'' = 3.23 \times 10^{-32}$ s^2. How many km, V/m, and picoseconds (ps) do $z = 1$, $E = 1$, and $\tau = 1$ correspond to?

Problem 3-21: Solitary waves of the NLSE
By direct substitution, prove that

$$E = A_0\,\text{sech}(A_0\tau)e^{\frac{1}{2}iA_0^2z}$$

is a solitary wave solution of (3.17) with the plus sign, A_0 being a real positive constant.

3.2.5   The Jovian Great Red Spot (GRS)

Antipov et al. [ANST86] have reported on laboratory experiments which support the theory that the Jovian GRS (Figure 3.27) is a solitary wave vortex

---

[7]To learn more about light bullets and see color rendered movie versions, the reader can go to the light bullet home page (http://www.sfu.ca/renns/lbullets.html) on the Internet.

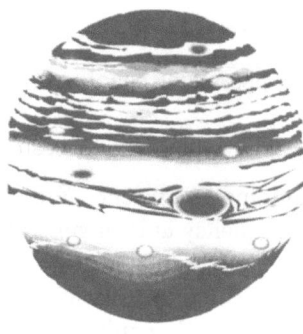

Figure 3.27: The Jovian Great Red Spot.

known as a Rossby soliton, which is kept stationary by counter-streaming zonal winds. The GRS was first observed by astronomers in the 17th century and has remained remarkably stable over the intervening centuries. The Red Spot is extremely large, measuring approximately 400,000 km across its diameter (about the same as the earth–moon distance!). The interested reader is also referred to an article by Ingersoll [Ing73].

3.2.6   The Davydov Soliton

In molecular biology, the understanding of the mechanism of energy transfer in proteins is a long-standing problem. Davydov and Kislukha [DK76] have proposed a model in which the hydrolysis of ATP (Adenosine triphosphate) leads to the production of amide I vibrations in the hydrogen-bonded spines of protein α helixes. A nonlinear interaction and soliton states arise in the model because of the interaction of the amide I vibration with the hydrogen bonds. In the literature, there has been an ongoing debate over the thermal stability of the Davydov soliton, a relatively recent paper by Cruzeiro–Hansson [CH94], for example, advancing reasons why this soliton may be thermally stable.

3.3   Chaos and Maps

Finally, in our introduction to some of the phenomena and their associated nonlinear equations, equations that will be explored in more depth in subsequent chapters, we shall look at a few systems that permit "chaotic" behavior. Chaotic dynamical systems obey deterministic nonlinear differential or finite-difference equations. Chaos[8] refers to the irregular and unpredictable time evolution that can occur. In the chaotic regime, there is also an extreme sensitivity to initial conditions. In addition to requiring the presence of some nonlinearity, for chaos to occur, the dynamical system must have at least three dependent state variables.

---

[8]Chaos, in Greek mythology, is the vacant, unfathomable space from which everything arose. In the Olympian myth Gaea sprang from Chaos and became the mother of all things.

3.3.1 Forced Oscillators

A simple mechanical example possessing chaotic solutions is the driven, damped, simple pendulum,

$$\ddot{\theta} + 2\gamma\dot{\theta} + \omega_0^2 \sin\theta = F\cos(\omega t) \qquad (3.19)$$

where γ is the positive damping coefficient, F is the amplitude of the periodic force, and the driving frequency ω is in general different than the natural frequency ω_0. Letting $\tau = \omega_0 t$, $b = 2\gamma/\omega_0$, $\mathcal{F} = F/\omega_0^2$ and $a = \omega/\omega_0$, Equation (3.19) can be rewritten as three coupled first-order ODEs with three state variables, the angle θ, the normalized angular velocity $\mu = \dot{\theta}(\tau)$, and the phase $\phi = a\tau$,

$$\dot{\theta}(\tau) = \mu$$

$$\dot{\mu}(\tau) = -b\mu - \sin\theta + \mathcal{F}\cos\phi \qquad (3.20)$$

$$\dot{\phi}(\tau) = a$$

Of course, this rescaling could have been easily avoided by simply choosing $\omega_0 = 1$. Whether the driven pendulum system (3.20) has periodic or chaotic solutions depends on the values of the parameters b, $\mathcal{F}$ and a and the choice of initial conditions for the state variables.

Another mechanical example of great historical interest which displays a rich variety of periodic as well as chaotic solutions is the forced spring equation, more commonly known as the Duffing [Duf18] equation, named after the pioneer who studied its behavior,

$$\ddot{X} + 2\gamma\dot{X} + \alpha X + \beta X^3 = F\cos(\omega t). \qquad (3.21)$$

The Duffing equation is further categorized according to the signs and values of the parameters α and β. For $\alpha > 0$ and $\beta > 0$, it is known as the "hard spring" Duffing equation, a name which should be self-evident. A hard spring becomes harder to stretch for larger displacements from equilibrium. On the other hand, if one expands the $\sin\theta$ term in (3.19), and keeps terms out to θ^3, Duffing's equation with $\beta < 0$ would result. This is referred to as the "soft spring" case. Two other important categories that have been extensively studied are the "nonharmonic [Ued79]" ($\alpha = 0$) and "inverted" ($\alpha = -1$) cases. As with the driven pendulum equation, Duffing's equation can be re-expressed as three first-order nonlinear ODEs. The reader can explore all categories of the Duffing equation in Mathematica File MF09.

Duffing's Equation
Duffing's equation is numerically integrated. With all other parameters fixed, the force amplitude F can be varied producing periodic and chaotic solutions. Figure 3.29 shows an example of a period two and a chaotic orbit produced with this notebook. The initial conditions may be altered to check the sensitivity of the chaotic solution to the starting point. Mathematica commands in file: `NDSolve, ParametricPlot, Evaluate, AspectRatio, PlotLabel`

Figure 3.28 shows a plot of $X(t)$, over the time interval $t = 0...200$, derived from the Mathematica file for the inverted Duffing equation. Here $\alpha = -1$, $\beta = 1$, $\gamma = 0.25$, $\omega = 1$ and a) $F = 0.34875$, b) $F = 0.420$. The initial conditions were taken to be $X(0) = 0.090$ and $\dot{X}(0) = 0$. The left frame of Figure 3.28,

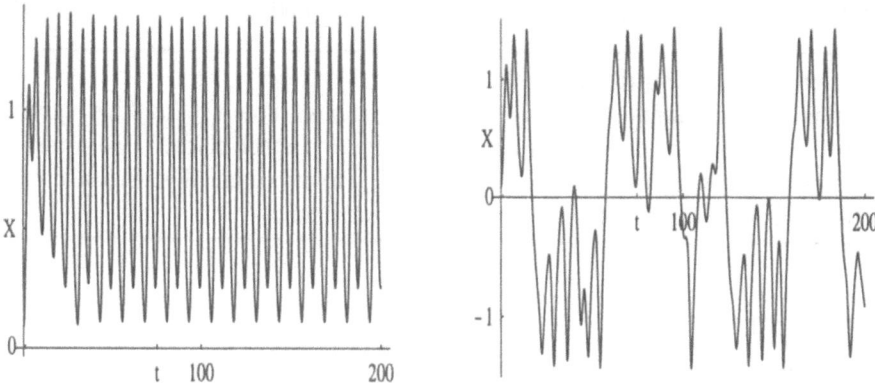

Figure 3.28: Period-two and chaotic behavior for the Duffing equation.

corresponding to $F = 0.34875$, is an example of a "period-two" solution as the pattern repeats every two oscillations when steady-state is achieved. The right frame, for $F = 0.420$, is quite irregular with no obvious pattern emerging, even if a longer time range is chosen. It is an example of a chaotic solution.

An alternate way to plot the solution is to make a phase plane trajectory picture by plotting the velocity $(V = \dot{X})$ versus the displacement X. Figure 3.29 shows such pictures for the period two and chaotic solutions respectively over the time interval $t = 100...200$. The transient portion of the numerical solution has been removed. For the period two solution on the left, the trajectory has wound

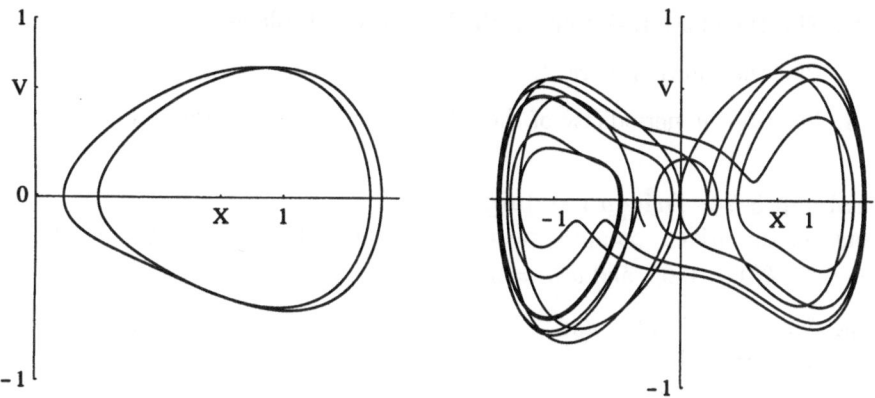

Figure 3.29: Phase trajectories of the solutions shown in Figure 3.28.

onto a closed (periodic) orbit which appears to cross itself. These crossings are an artifact resulting from the fact that we have projected a three dimensional phase trajectory onto a two dimensional plane. We shall discuss this point further in Chapter 8. For the solution on the right of the figure, the trajectory does indeed look chaotic. An important aspect of chaos is the extreme sensitivity to initial conditions, a feature that may be explored in MF09.

The Duffing equation is more than just a theoretical equation on a piece of paper or simulated on the computer. The following experiment illustrates how an inverted steel blade, when driven, can be described by the Duffing equation.

Forced Duffing Equation

This is an experimental activity which investigates how the Duffing equation can be used to model a driven oscillating inverted steel blade. Two alternate methods of driving the blade are presented: (a) mechanically with a driving motor, (b) electromagnetically with a Helmholtz coil.

PROBLEMS

Problem 3-22: Three state variables

Rewrite the Duffing oscillator equation as three coupled, normalized, first-order ODEs involving three dependent state variables.

Problem 3-23: Duffing equation

Consider the Duffing equation with the following input parameters:

- $\alpha = 1$, $\beta = -1$, $\gamma = 0.25$, $\omega = 1$, $F = 0.34875$, $x(0) = 0.09$, $\dot{x}(0) = 0$;

- $\alpha = 1$, $\beta = 0.2$, $\gamma = 0.2$, $\omega = 1$, $F = 4.0$, $x(0) = 4.0$, $\dot{x}(0) = 0$;

- $\alpha = 0$, $\beta = 1$, $\gamma = 0.04$, $\omega = 1$, $F = 0.2$, $x(0) = 0.28$, $\dot{x}(0) = 0$;

For each case,

a. Categorize the Duffing equation according to the α, β values.

b. Plot the phase trajectory in the $V = \dot{X}$ vs. X plane.

c. Plot the solution X vs. t.

d. Identify the periodicity of the solution in the steady-state regime.

Problem 3-24: Driven pendulum

Consider the driven, damped, pendulum equation (3.19) with $\omega_0 = 1$, $\omega = \frac{2}{3}$, $\gamma = 0.25$, $\theta(0) = 0.09$, and $\dot{\theta}(0) = 0$.

a. For $F = 1.0$, 1.06, and 1.25, plot $V = \dot{\theta}$ vs. θ. Choose a suitable time range in each case.

b. For the same F values, plot θ vs. t.

c. Identify the periodicity for each F value.

d. Explore the range between $F = 1.06$ and 1.25 and identify the periodicity in each case.

3.3.2 Lorenz and Rössler Systems

For the forced pendulum problem, we have seen how a single second-order equation could be rewritten as three first-order equations. The Duffing equation could be treated similarly. Still other systems are already in this latter form, e.g., the Lorenz equations (3.5). The Lorenz system has two nonlinear cross terms and as a consequence can have complicated behavior, this behavior depending on the values of the coefficients. A traditional choice, one used by Lorenz, is to take $\sigma = 10$, $b = 8/3$ and allow the so-called bifurcation parameter r to increase from zero. As r is increased, the Lorenz system undergoes sudden changes in behavior as certain critical or bifurcation values of r are passed. For $r = 28$ and $x(0) = 2$, $y(0) = 5$, $z(0) = 5$, one obtains the beautiful chaotic trajectory shown in Figure 3.30. Qualitatively resembling the gossamer wings

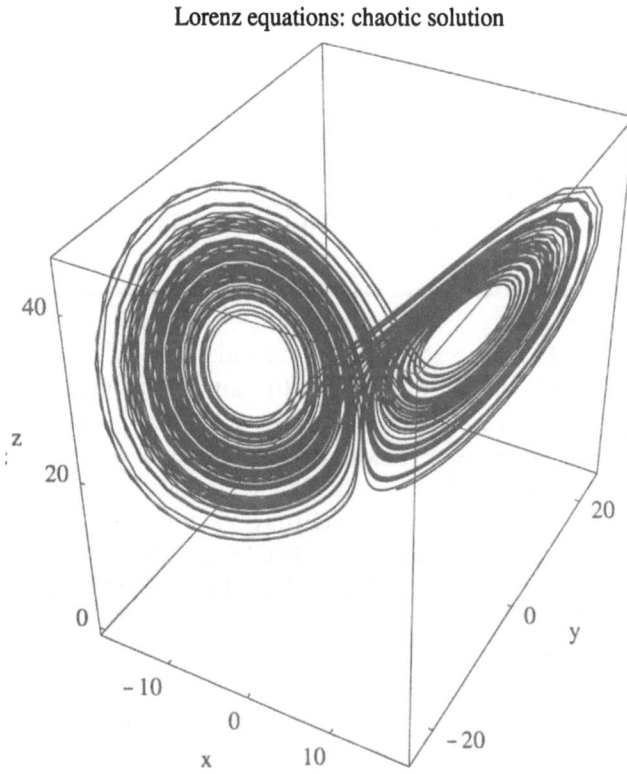

Lorenz equations: chaotic solution

Figure 3.30: "Butterfly wings" chaotic trajectory for the Lorenz system.

of a butterfly, the trajectory may be viewed in color in the notebook file MF10. Although the trajectory remains confined to a given region as the representative point evolves with time, it continually traces out new paths, a feature of chaotic behavior. This may also be seen by plotting one of the dependent variables, e.g., $x(t)$ as in Figure 3.31. Up until $t \simeq 7$, one probably would feel that the future behavior could be predicted. Suddenly, however, $x(t)$ shoots upward and again settles down to a pattern of increasing oscillations reminiscent of the behavior up until $t \sim 7$. Then, unpredictably, $x(t)$ shoots downward where you might

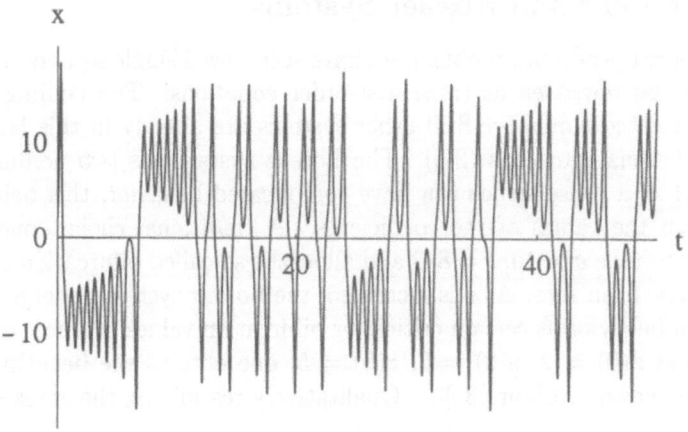

Figure 3.31: $x(t)$ for the butterfly trajectory.

intuitively guess that it would repeat the first pattern. Surprisingly, it doesn't, instead undergoing large irregular oscillations. Again at even larger t, there are hints of the initial patterns of oscillation but nothing is ever the same as time evolves.

The Lorenz System
The NDSolve and ParametricPlot3D commands are applied to (3.5) with the above specified initial conditions $x(0)$, $y(0)$ and $z(0)$ to generate Figure 3.30. Although the student can produce plots for different values of r, experimentally searching for bifurcation values, a "neater" way of determining these special points analytically is discussed in Chapter 4. Mathematica commands in file: NDSolve, MaxSteps, ParametricPlot, ParametricPlot3D, Evaluate, PlotPoints, Background, BoxStyle, RGBColor, AxesLabel, AxesEdge, Ticks, PlotLabel, Hue, ImageSize, TextStyle, PlotRange

Unlike the Lorenz model which had its origin in an attempt to describe a real physical system (the atmosphere), a simple artificial 3-dimensional system was introduced by Rössler [Rö76] with only one nonlinear cross term. The Rössler system is described by the following three ODEs,

$$\dot{x} = -(y+z), \quad \dot{y} = x + ay, \quad \dot{z} = b + z(x-c), \qquad (3.22)$$

with x, y, z real and a, b and c positive constants. Despite its simple appearance, (3.22) can still yield complicated trajectories. This is left as a problem.

PROBLEMS
Problem 3-25: The Rössler system
Modify the Lorenz equations (3.5) in the file MF10 to obtain the Rössler system (3.22). Numerically solve and plot the Rössler system with $a = 0.2$, $b = 0.2$, $c = 5.7$ and initial conditions $x(0) = -1$, $y(0) = 0$ and $z(0) = 0$. Investigate the effect of changing the parameter c, holding a and b fixed at the previous values.

With $a = 0.2$, $b = 0.2$ and $c = 5.7$, investigate the effect of changing the initial conditions. Comment on your results.

Problem 3-26: Chua's butterfly
In Experimental Activity 30 (Chua's Butterfly), Chua's electrical circuit (see Figure 30.3) is used to produce a chaos-exhibiting double scroll or butterfly wings trajectory. From the equivalent circuit of Figure 30.2, the dynamical equations are found to be

$$\dot{I} = \frac{V_1}{L}, \quad \dot{V}_1 = \frac{(V_2 - V_1)}{C_1 R} - \frac{I}{C_1}, \quad \dot{V}_2 = \frac{(V_1 - V_2)}{C_2 R} - \frac{V_2}{C_2 r}$$

where I is the current through the inductor $L = 0.0040$ H, V_1 is the potential drop across the capacitor $C_1 = 4.7 \times 10^{-8}$ F, and V_2 is the potential drop across the capacitor $C_2 = 1.1 \times 10^{-9}$ F. R is a variable resistor while r is a piecewise linear negative resistance function given by $r = -2000$ ohms for $|V_2| < 1.7$ volts and $r = -2015$ ohms, otherwise.

By modifying MF10, produce a 3-dimensional plot of Chua's butterfly, taking $R = 2011$ and initial conditions $I(0) = 0$, $V_1(0) = 0$, $V_2(0) = 0.2$. Take the time interval $t = 0$ to 0.01 s and choose an orientation which best shows the trajectory. Explore the behavior of the trajectory in the neighborhood of $R = 2011$ ohms and comment on the results.

3.3.3 Poincaré Sections and Maps

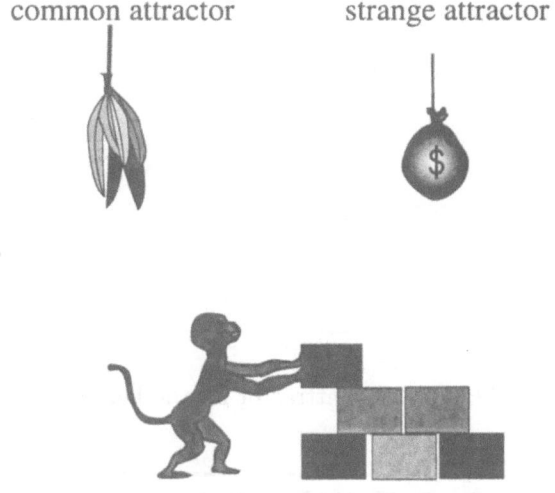

common attractor   strange attractor

Let us return to the Duffing equation (3.21) with $\alpha = -1$, $\beta = 1$, $\gamma = 0.25$ and $\omega = 1$ so that

$$\ddot{x} + 0.5\dot{x} - x + x^3 = F\cos(t) \tag{3.23}$$

or, equivalently (with $\phi(0) = 0$),

$$\dot{x} = y, \quad \dot{y} = -0.5y - x^3 + x + F\cos\phi, \quad \dot{\phi} = 1. \tag{3.24}$$

Still another way of viewing the dynamical behavior of this (or any) system is to take a "snapshot" of the y vs. x phase plane at each period $T = 2\pi/\omega$ of the driving force. This is called a Poincaré section. One looks at the Poincaré section at time intervals $t = nT$ with n a positive integer. After an initial transient period, the dynamical system will settle down to either a periodic or chaotic motion. If the system evolves to a periodic solution of frequency ω ($\omega = 1$ in (3.23) or (3.24)), the same as the driving frequency, the Poincaré section will consist of a single dot which is reproduced at each multiple of the period T. If the system settles down to a periodic solution of frequency $\omega/2$ ($\omega/2 = 1/2$ in our example), the Poincaré section will have two points between which the system oscillates as multiples of T elapse. This is a period two solution.

For chaotic motion, on the other hand, each period produces a point at a different location and the "sum" of the individual snapshots can produce strange patterns of dots with complex boundaries in the x vs. y plane. These geometrical shapes are called strange attractors and are characterized by fractal dimensions. Figure 3.32 shows the strange attractor (70 points are plotted here) corresponding to the chaotic solution of Figure 3.29.

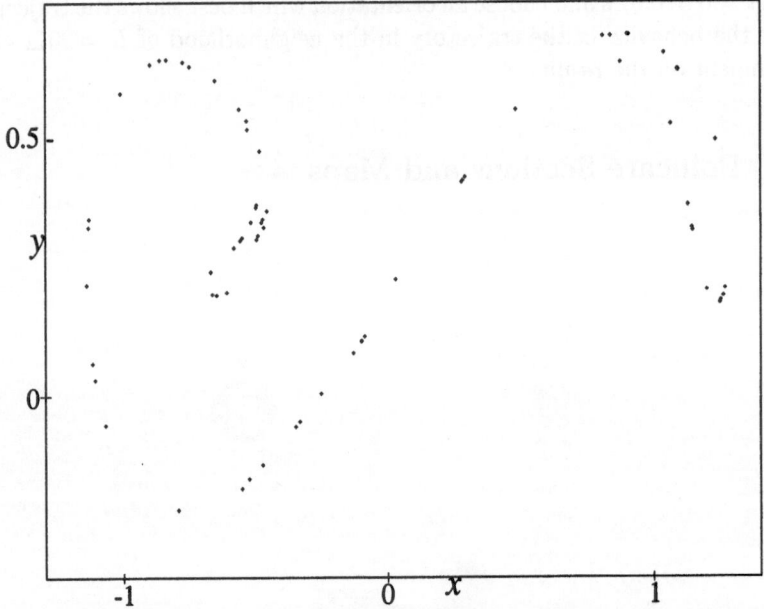

Figure 3.32: A strange attractor for the Duffing equation.

Poincaré Sections for Duffing's Equation
Poincaré sections corresponding to (3.24) are obtained for some different F values. By showing the sequential time development of points, one can clearly see the periodicity. For $F = 0.34875$, for example, the system "hops" back and forth between two points. Mathematica commands in file: Do, NDSolve, Flatten, Evaluate, Table, MaxSteps, Epilog, Point, Text, PlotRange, Frame, ListPlot, PlotStyle, RGBColor, PointSize, FrameTicks, PlotLabel

Because the differential equation is deterministic, a definite relationship must exist between the phase plane coordinates (X_n, Y_n) at the end of the nth period of the driving force, and the coordinates (X_{n+1}, Y_{n+1}) at the end of the $(n+1)$st period. Symbolically

$$X_{n+1} = f_1(X_n, Y_n), \quad Y_{n+1} = f_2(X_n, Y_n), \tag{3.25}$$

where the f_i in general are nonlinear. For most dynamical systems it has proved impossible to find analytic forms of f_1 and f_2. Therefore, mathematicians have reversed the process. They postulate (usually simple) analytic forms for the f_i and study the behavior of the resulting difference equations. This can shed light, e.g., on the onset of chaos observed in the full differential system of equations. Equations (3.25) represent a 2-dimensional "map". A 1-dimensional map is described by

$$X_{n+1} = f(\lambda, X_n) \tag{3.26}$$

where we have inserted a "control parameter" λ. Control parameters may also be inserted into 2-dimensional maps.

PROBLEMS
Problem 3-27: Duffing oscillator
Consider the Duffing oscillator equation with the following input parameters:

- $\alpha = 1$, $\beta = -1$, $\gamma = 0.25$, $\omega = 1$, $F = 0.34875$, $x(0) = 0.09$, $\dot{x}(0) = 0$;

- $\alpha = 0$, $\beta = 1$, $\gamma = 0.04$, $\omega = 1$, $F = 0.2$, $x(0) = 0.28$, $\dot{x}(0) = 0$;

Obtain the Poincaré section in each case and identify the periodicity. You may have to alter the range of the plot.

Problem 3-28: Driven pendulum
Consider the driven pendulum system (3.20) with $a = \frac{2}{3}$, $b = 0.5$, $\theta(0) = 0.09$, $\mu(0) = 0$, $\phi(0) = 0$, and (i) $\mathcal{F} = 1.0$, (ii) $\mathcal{F} = 1.07$, (iii) $\mathcal{F} = 1.47$. Modify MF11 and obtain the Poincaré section for this system for each $\mathcal{F}$ value. Identify the periodicity in each case. Hint: Note that the period of the cosine driving term is $T = 2\pi/a$. To keep the angular coordinate in a limited range, divide θ by 2π and make use of the `FractionalPart` command.

3.3.4   Examples of One- and Two-Dimensional Maps

In mathematical biology, one naturally encounters difference equations and maps. The biologist measures, e.g., the population of insects from one generation to the next, each generation undergoing seasonal breeding and then dying before the next generation hatches. Salmon populations behave in the same way, coming back to their original streams every four years before laying their eggs and subsequently dying.

Let N_n be the population number corresponding to the nth generation. Since it takes a certain time interval for each generation to occur, n plays the role of a time interval. One famous model, pioneered by Robert May [May76], that has been discussed in great detail in the literature, is the logistic model

$$N_{n+1} = \left(1 + r - \frac{r}{k}N_n\right) N_n \tag{3.27}$$

where r is the real positive "growth coefficient" and k is also a positive constant. Note that when $r = 0$, $N_{n+1} = N_n$ and no growth occurs as successive generations pass. The $(-r/k)N_n^2$ term reflects a tendency for the population to decrease as available food and space becomes inadequate. This is reminiscent of our discussion of saturation effects in Chapter 2 for predator–prey differential equation models.

Setting $X_n = rN_n/k(r+1)$ and $a = 1 + r$ as the control parameter, Equation (3.27) becomes

$$X_{n+1} = aX_n(1 - X_n). \tag{3.28}$$

Confining our interest to $0 \leq X \leq 1$, this one-dimensional mapping is referred to as the logistic map. We shall explore the properties of this map in detail in Chapter 9. A preliminary idea of the nature of the solutions can be obtained by running the logistic map code given in the following example for different values of the control parameter.

| Example 3-4: The Logistic Map |

Iterate the logistic map for $a = 3.05$ and $X_0 = 0.1$, taking $N = 120$ iterations. Plot the output.

Solution: The input parameters are entered along with the logistic function.

```
Clear["Global`*"]

a = 3.05;  x[0] = 0.1;  total = 120;

f[x_] := a x(1 - x)
```

Using the Nest command, the logistic function f is applied n times to the input value x[0]. Plotting points $\{n, x[n]\}$ are created with the Table command, with n running from 0 to total.

```
pts = Table[{n, Nest[f, x[0], n]}, {n, 0, total}];
```

The points are plotted with the ListPlot command, the result being shown in Figure 3.33.

```
ListPlot[pts, PlotRange -> {{0, 120}, {0, 1}},
  PlotStyle -> {Hue[0], PointSize[.015]}, ImageSize -> {600, 400},
  Ticks -> {{{0.01, "0"}, 20, 40, 60, {70, "n"}, 80, 100, 120},
  {.2, .4, {.5, "Xn"}, .6, .8, 1}},
  TextStyle -> {FontFamily -> "Times", FontSize -> 16}];
```

After a few iterations the transient dies away and the solution hops back and forth every two steps between the two branches of Figure 3.33, located at $X \simeq 0.59$ and $X \simeq 0.74$. Thus, a period-2 solution occurs for $a = 3.05$. As the control parameter a is further increased, the logistic map solution will display a series of period doublings at certain bifurcation values of a, ultimately leading

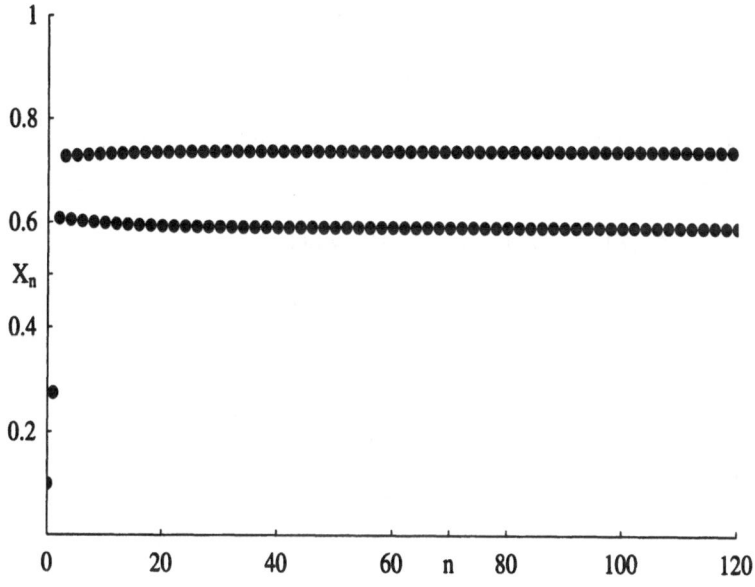

Figure 3.33: A period two solution of the logistic map.

to a chaotic regime. A detailed discussion of bifurcations of the logistic map will be given in Chapter 9.

End Example 3-4

A 2-dimensional predator–prey type of map of the form

$$X_{n+1} = aX_n \left(1 - X_n - Y_n\right), \quad Y_{n+1} = bX_nY_n, \quad (3.29)$$

with $2 < a \leq 4$ and $2 < b \leq 4$, has been discussed by the mathematician H. A. Lauwerier [Lau86]. If the Y variable is not present, this is just the logistic map again. One obtains a predator–prey interaction by inserting the XY terms. Which variable refers to the predator? A phase plane picture showing Y_n vs. X_n is easily generated as demonstrated in the next example.

Example 3-5: Predator–prey Map

Iterate the predator–prey map for $a = 3.0$, $b = 3.5$, $X_0 = 0.1$, $Y_0 = 0.2$, taking $N = 2000$ iterations. Create a phase-plane plot by showing Y_n vs. X_n.

Solution: The parameters are specified,

```
Clear["Global`*"]

a=3; b=3.5; x[0]=0.1; y[0]=0.2; total=2000;
```

and Equations (3.29) entered (letting $n \to n - 1$) as Mathematica functions.

```
x[n_]:=x[n]=a x[n-1](1-x[n-1]-y[n-1])
```

```
y[n_]:=y[n]=b x[n-1] y[n-1]
```

The `Table` command is used to iterate these functions and create the plotting points,

```
pts=Table[{x[n],y[n]},{n,0,total}];
```

which are graphed with the `ListPlot` command.

```
ListPlot[pts,PlotRange->{{0,.7},{0,.7}},AspectRatio->1,
Ticks->{{{0.001,"0"},.3,{.4,"Xn"},.6},{{0.001,"0"},.3,
{.4,"Yn"},.6}},PlotStyle->{Hue[0],PointSize[.015]},TextStyle->
{FontFamily->"Times",FontSize->16},ImageSize->{400,400}];
```

The resulting plot of Y_n versus X_n is shown in Figure 3.34. Starting at the

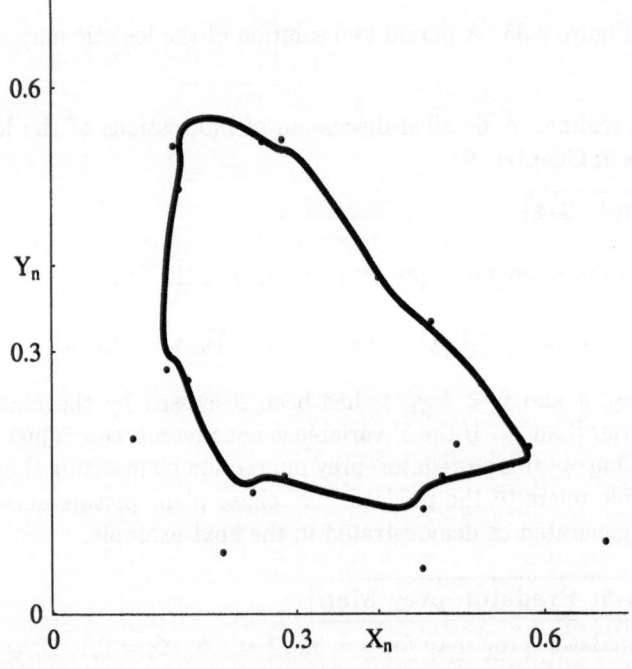

Figure 3.34: A cyclic solution of the predator–prey map.

point ($X_0 = 0.1$, $Y_0 = 0.2$), the predator–prey system evolves after a transient interval onto a loop, indicating some sort of cyclic variation in the population (density) numbers. The code is easily modified to show either X_n or Y_n as a function of n. This is left as a problem.

End Example 3-5

Perhaps one of the most famous maps is the complex Mandelbrot map

$$Z_{n+1} = Z_n^2 + C, \tag{3.30}$$

where $Z = X + iY$ and $C = p + iq$ are complex numbers. Separating real and imaginary parts, (3.30) can be rewritten as the 2-dimensional map

$$X_{n+1} = X_n^2 - Y_n^2 + p, \qquad Y_{n+1} = 2X_nY_n + q, \tag{3.31}$$

with two real control parameters p and q. Such maps as the Mandelbrot map (3.30) can be used to generate intriguing 2-dimensional fractal pictures, as illustrated by Heinz-Otto Peitgen and Peter H. Richter [PR86]. The colored centerpiece fractal pictures in their book look like works of art and, indeed, have been presented as such at public exhibitions. To give the reader some flavor of what these pictures look like, the Mathematica File MF12 is provided, this notebook generating a colored plot of the so-called "Mandelbrot set".

The Mandelbrot Set

A Mandelbrot function is defined with the following `Module` construct:

```
mandelbrot[p_, q_] := Module[{n = 0, z = 0 + 0 I},
        While[Abs[z] < 2 && n < 20, z = z^2 + (p + I q); n++]; n];
```

Inside `Module` a list of the initial values of the "local" variables, the iteration number **n** and the complex **z**, is given. Here we have taken $X = Y = 0$ initially in (3.30). This complex map is iterated for a range of p and q, the iterated values either diverging to infinity or being attracted to a finite value of X_n, Y_n. To decide which occurs, the iteration is continued while the absolute value of **z** < 2 and (`&&` is the logical "and" symbol) **n** < 20. Following the `While` command, the **n** values are recorded. If $n = 20$, convergence occurs, while $n < 20$ indicates divergence. The region in the p–q plane corresponding to attraction (convergence) is colored red using the `DensityPlot` command, and is referred to as the Mandelbrot set of points. The boundary of the Mandelbrot set has a fractal structure which is colored blue. All other colors in the plot indicate different rates of divergence. Mathematica commands in file: `Module`, `While`, `Abs`, `DensityPlot`, `Mesh`, `PlotPoints`, `ColorFunction`, `Hue`, `FrameTicks`, `TextStyle`

Chaotic behavior can also be observed in dynamical systems described by PDEs. These tend to be more complicated to analyze and we shall not deal with any problems of this type in this text. As an example, the interested reader is referred to the numerical simulation of the driven optical ring cavity by McAvity, Enns, and Rangnekar [MER88]. As the reader will probably have noticed in these introductory chapters, optics is a particularly good field for demonstrating nonlinear behavior. (We leave the authors' objectivity as a subject for coffee room debate!)

PROBLEMS

Problem 3-29: The standard map

A perfectly elastic ball bounces vertically on a horizontal vibrating plate whose velocity is given by $v_{plate} = A\sin(\omega t)$. Let V_n be the speed of the ball prior to the nth bounce at time t_n. Neglecting the vertical displacement of the plate relative to the flight of the ball, show that

$$V_{n+1} = V_n + 2A\sin\theta_n, \qquad \theta_{n+1} = \theta_n + 2\frac{\omega}{g}V_{n+1},$$

where $\theta_n = \omega t_n$ is the phase at the nth bounce. This system is an example of the so-called "standard map".

Problem 3-30: The logistic map

Consider the logistic map (3.28) with $X_0 = 0.1$ and the following values of a: i) $a = 2.75$, ii) $a = 3.10$, iii) $a = 3.50$, iv) $a = 3.70$, and v) $a = 3.83$. Calculate X_n up to $n = 120$. Use the code in Example 3-4, inserting the appropriate a value. If, after the transient has died away, X takes on n different values before repeating, the solution is referred to as a period-n solution. Identify the periodicity for each of the a values. Which one of the a values probably corresponds to chaos? Confirm your conclusion by choosing a larger N.

Problem 3-31: Predator–prey map 1

In the predator–prey example, experiment with other a and b values in the range $2 < a \leq 4$, $2 < b \leq 4$, keeping the starting values the same as in the example. Discuss the observed behavior in the Y_n vs. X_n plane.

Problem 3-32: Predator–prey map 2

Modify the code in Example 3-5 to display X_n vs. n and Y_n vs. n. Interpret the observed behavior of the population numbers. Experiment with other a and b values in the range $2 < a \leq 4$, $2 < b \leq 4$ and interpret the behavior.

Problem 3-33: Another 1-d map

Taking $X_0 = 0.1$, iterate the following map up to $n = 150$ and plot the output,

$$X_{n+1} = -(1+a)X_n + X_n^3,$$

for i) $a = 0.5$, ii) $a = 1.2$, iii) $a = 1.25$, iv) $a = 1.30$, and v) $a = 1.5$. Identify the periodicity of the solution in each case. If necessary, take larger N values.

Problem 3-34: Map exploration

Taking $X_0 = 0.1$, explore the behavior of the map

$$X_{n+1} = a + X_n - X_n^2$$

over the range $a = 0.5...1.8$. You will want to take smaller steps in a nearer the upper end of the range. Identify the periodicity for each a value chosen.

Problem 3-35: Tapestries

Consider the 2-dimensional map

$$X_{n+1} = X_n + a\sin(Y_n), \qquad Y_{n+1} = Y_n + bX_{n+1}$$

with $b = 1$, $X_0 = Y_0 = 0.1$. Iterate the map up to $N = 3000$ for $a = 0.1, 0.2, 0.3, ..., 1.0$. You should see some examples of artistic tapestries. Try some other a values (e.g., $a = 0.43$) and see what interesting patterns you can discover. You might also wish to try other a and b values.

Chapter 4

Topological Analysis

For all men strive to grasp what they do not
know, while none strive to grasp what they
already know; and all strive to discredit what
they do not excel in, while none strive to
discredit what they do excel in. This is why
there is chaos.
Chuang-tzu (368-286 BC)

4.1   Introductory Remarks

Most nonlinear systems cannot be solved exactly so we must resort to a variety
of approaches in order to obtain an approximate solution. Where applicable,
the phase plane portrait can serve as a valuable tool for qualitatively determining the types of possible solutions before resorting to numerical or (usually

The unexamined life is not worth
living.   Socrates (470–399 BC)

approximate) analytical methods for specific initial conditions. In this chapter,
the concept of phase plane analysis will be examined in some depth, not for
a specific problem, but for a wide class of physical problems described by the

following system of first order equations:

$$\frac{dx}{dt} = P(x,y), \quad \frac{dy}{dt} = Q(x,y), \tag{4.1}$$

where P, Q are, in general, nonlinear functions of x and y and the independent variable has been taken here to be time t. In the laser competition equations (2.32), t would, of course, be replaced with z, the spatial coordinate. The mathematician would refer to this set of equations as being autonomous, meaning that P and Q do not explicitly depend on t. Why it is desirable to restrict the discussion for the moment to autonomous equations will become readily apparent.[1]

All mechanical problems arising from Newton's second law possessing the structure

$$\ddot{x} = F(x, \dot{x}) \tag{4.2}$$

where the force F is in general a nonlinear function of the position x and the speed $\dot{x}$ can be cast into what will be referred to hereafter as the "standard form" of Equations (4.1), simply by setting

$$\dot{x} \equiv \frac{dx}{dt} = y \tag{4.3}$$

so that Equation (4.2) becomes

$$\dot{y} \equiv \frac{dy}{dt} = F(x,y). \tag{4.4}$$

This last pair of equations then is clearly in standard form with $P(x,y) = y$ and $Q(x,y) = F(x,y)$. Thus, for example, the equations for the simple pendulum and the nonlinear hard spring as well as Van der Pol's equation can be written in standard form with $Q(x,y) = -\omega_0^2 \sin x$, $Q(x,y) = (-k/m)(1 + a^2 x^2)x$, and $Q(x,y) = -x + \epsilon(1 - x^2)y$, respectively. Of course, there are those problems which are already described by equations in standard form, e.g., the big fish–little fish problem, Equation (2.20); the quasi-species evolutionary Equations[2] (2.30) for $N = 2$; the laser competition Equations (2.32) or (2.33), to mention just a few.

Returning now to the basic autonomous equations (4.1), the independent variable t is readily eliminated by dividing one equation by the other:

$$\frac{dy}{dx} = \frac{Q(x,y)}{P(x,y)}. \tag{4.5}$$

The solution of this equation obviously yields a phase plane diagram of y versus x with certain characteristic curves or "phase trajectories" along which

---

[1] Non-autonomous equations (P, Q contain t explicitly) arise when time-dependent driving forces are present, e.g., the problem of forced oscillations of a simple pendulum described by

$$\ddot{x} = -\omega_0^2 \sin x + F \sin \omega t.$$

Of course, as seen in Section 3.3, this equation can be made autonomous by increasing the dimensionality of the state variable space from 2 to 3 dimensions.

[2] For $N > 2$, e.g., the Lorenz Equations (3.5), the topological analysis would involve an N-dimensional phase space. Even for $N = 3$, the general topological analysis is difficult [Hay64, Jac90].

the system will evolve as t increases. For Newtonian mechanics problems, the phase plane corresponds to plotting the velocity of the system versus the displacement, while for, say, the rabbits–foxes system, we plot the rabbit number against the fox number (or vice versa). In general the integration to determine $y(x)$ may not be possible or the result so complex that it is difficult to analyze. The topological approach is to show that there exist certain special ("stationary" or "singular") points x_0, y_0 in the phase plane in the region of which the nature of the solutions can be readily determined. The entire phase plane can then, in principle, be pieced together by connecting all these regions. In practice, particularly when the phase plane portraits are complicated, portraits are more easily and speedily obtained by using Mathematica.

Equation (4.5) gives the slope of the tangent to the trajectory passing through the point (x, y). Let's consider a specific easily visualized example, namely the undamped simple pendulum swinging back and forth between θ_{max} and $-\theta_{max}$ with $\theta_{max} < \pi$. At $\pm\theta_{max}$, the angular velocity $\dot\theta$ is zero, while taking on its maximum positive or negative value at $\theta = 0$. For a given θ_{max}, or total energy E, the pendulum system can be described by a closed trajectory (actually an ellipse) in the $\dot\theta$ versus θ phase plane. For different θ_{max} or E values, different ellipses are generated as indicated in Figure 4.1. For a given trajec-

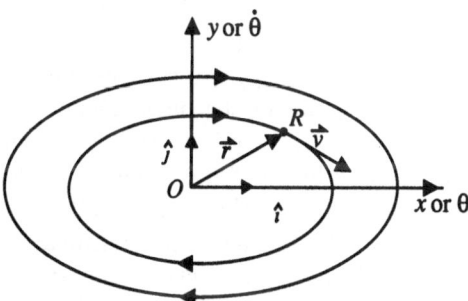

Figure 4.1: Phase plane portrait of simple pendulum for $\theta_{max} < \pi$.

tory, a representative phase point R moves along that trajectory as a function of time. For the pendulum, time clearly evolves clockwise as shown by the sense of the arrow heads. If $\vec{r}$ is the radius vector from O to R, then the velocity of the point R, called the phase velocity, is given by

$$\vec{v} = \frac{d\vec{r}}{dt} = \hat{\imath}\dot{x} + \hat{\jmath}\dot{y} \tag{4.6}$$

where $\hat{\imath}$, $\hat{\jmath}$ are unit vectors along the x, y axes respectively. For $\theta_{max} \ll \pi$, the motion of the simple pendulum is simple harmonic and the phase velocity, as the reader may verify, is

$$\vec{v} = \theta_{max}\omega_0[\hat{\imath}\cos\omega_0 t - \hat{\jmath}\sin\omega_0 t]. \tag{4.7}$$

Assuming that ω_0 is not zero, this phase velocity is never zero unless $\theta_{max} = 0$. The latter situation corresponds to the mass m being placed at $\theta = 0$ with zero angular velocity, i.e., the representative point R being placed at O in Figure 4.1. The point O corresponds to an equilibrium point because the net force on the mass m vanishes there. At this point the representative point R is at rest. O is one type of stationary or singular point, a singular point being defined as a point (x_0, y_0) for which $P(x_0, y_0) = Q(x_0, y_0) = 0$ simultaneously. The point O is an example of a stable equilibrium point because if damping were added to the pendulum equation, the pendulum would evolve toward O as $t \to \infty$ for $\theta_{max} < \pi$. Since P, Q are nonlinear[3] there may be a number of singular points present.

Phase Plane Portrait of Undamped Pendulum
The phase plane portrait, showing all possible types of trajectories, is generated for the undamped simple pendulum. Mathematica commands in file: `Solve`, `Plot`, `Evaluate`, `Table`, `Re`, `PlotPoints`, `Hue`, `Ticks`, `AxesLabel`, `Text`, `ImageSize`, `Epilog`, `PointSize`, `Point`, `RGBColor`

For the simple pendulum, not only is O a singular point in the phase plane but also the points $\dot{\theta} = 0$, $\theta = \pm\pi$, $\pm2\pi$, etc. This may be seen physically. If, for example, $\theta = \pi, \dot{\theta} = 0$, the pendulum would be standing on end with zero velocity. This is an example of an unstable equilibrium point in mechanics, as the slightest "nudge" would cause the mass m to move away from $\theta = \pi$. The other singular points correspond to the mathematical periodicity of the $\sin\theta$ function. At these various singular points, $P = y = 0$ corresponding to zero angular velocity, and $Q = F(x, y) = 0$ corresponding to no net force (i.e., no acceleration.) Any other point of the phase plane which is not a singular point is, naturally enough, called an "ordinary point". From Equation (4.2), an ordinary point is characterized by a definite value of the slope of the tangent to the phase trajectory passing through that point, while for a singular point (for example, O) the direction of the tangent is indeterminate. It also follows from the fundamental theorem[4] of Cauchy, for the existence of solutions to differential equations, that through every ordinary point of the phase plane there passes one and only one phase trajectory. A little thought should convince the reader that if (x_0, y_0) is a singular point, a trajectory passing through an ordinary point (x, y) at some instance will never reach (x_0, y_0) in a finite time.

With these preliminary comments and definitions dispensed with, we shall now classify the types of singular points that can occur and sketch the phase trajectories in the neighborhood of these points. Given this information, it is possible to construct the entire phase plane diagram and, consequently, interpret the dynamics of the physical problem.

---

[3]For example, they could be polynomials in x and y as in the arms race system.

[4]For a differential equation of nth order

$$x^{(n)} \equiv \frac{d^n x}{dt^n} = F\left(t, x, \dot{x}, \ldots, x^{(n-1)}\right)$$

there exists a unique analytic (analytic at a point means that one can Taylor expand in a series about that point) solution in the neighborhood of $t = t_0$ such that the function x and its $n - 1$ derivatives acquire for $t = t_0$ the set of prescribed values x_0, $\dot{x}_0 \ldots x_0^{(n-1)}$ provided that the function $F\left(t, x, \dot{x} \ldots x^{n-1}\right)$ is analytic in the neighborhood of this set of values. For the problem at hand, $n = 2$ and F is not a function of t.

4.2   Types of Simple Singular Points

At this stage it is useful to list the four types of "simple" singularities that can occur, postponing for the moment the precise meaning of the word "simple" as well as the criterion for the existence of a given singularity.

Vortex Point (V)

The vortex point (often called a center) has already made its debut in our brief discussion of the simple pendulum problem, e.g., at $\theta = 0$, $\dot{\theta} = 0$. The trajectories consist of a continuum of closed curves, elliptical in the pendulum problem, enclosing the singularity as shown in Figure 4.2.

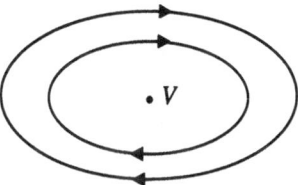

Figure 4.2: Trajectories in the neighborhood of a vortex point V.

Saddle Point (S)

The points $\theta = \pm n\pi$ with n = odd integer, for the simple pendulum, are examples of saddle points, the behavior of the trajectories in the neighborhood of which is schematically illustrated in Figure 4.3, the arrows indicating the

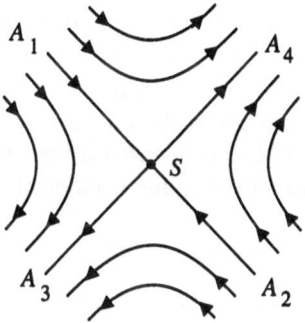

Figure 4.3: Trajectories in the neighborhood of a saddle point S.

direction of increasing t. The four trajectories A_1S, A_2S, A_3S and A_4S divide the area about the saddle point S into four regions. A representative point only approaches S along A_1S and A_2S as $t \rightarrow \infty$, while it takes an infinite time to move away from S along A_3S and A_4S. The nature of these trajectories is quite apparent for the simple pendulum. If the pendulum were initially at $\theta = \pi$, for example, an infinitesimal "nudge" in a clockwise (counterclockwise) sense would cause the undamped pendulum to go by energy conservation to $\theta = -\pi(+3\pi)$. The trajectories A_1S, A_2S, etc., connect adjacent saddle points. In the limit that the nudge goes to zero, it would take an infinite time for the pendulum to swing from one saddle point to another. On the other hand, if the nudge is of finite magnitude the pendulum will clearly swing past the next saddle point as well as all subsequent saddle points. This represents a different motion than in the vicinity of a vortex. The saddle point "trajectories" separate two different motions and are referred to as "separatrixes". Mathematica File MF13 illustrates these various features.

Focal Point (F)

For a focal or spiral point F a representative point R approaches the singularity along a spiral path, e.g., along P_1 in Figure 4.4. Since the spiral winds

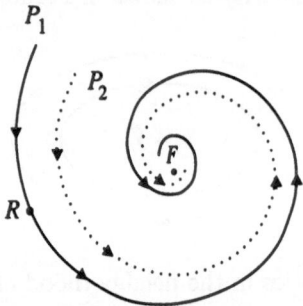

Figure 4.4: Trajectories in the neighborhood of a focal point F.

around F an infinite number of times, the direction of approach to F is indeterminate. There are an infinite number of such spirals approaching F, the curve P_2 being another. Through each ordinary point of the phase plane passes only one such spiral. For the situation as depicted in Figure 4.4 where R approaches F as $t \to \infty$, F is called a stable focal point. It is also an example of a "point attractor". As a simple physical example of how a stable focal point can occur, imagine that a simple pendulum has been immersed in a viscous medium which offers a resistive force proportional to the velocity. If the pendulum is pulled out through a small angle $\theta(0) = A$ and released from rest, the pendulum would be described by the linear damped (damping coefficient $\gamma > 0$) harmonic oscillator equation

$$\ddot{\theta} + 2\gamma\dot{\theta} + \omega_0^2\theta = 0 \tag{4.8}$$

which, for the underdamped case ($\gamma < \omega_0$), has the solution

$$\theta(t) = Ae^{-\gamma t}\left(\cos\left(\sqrt{\omega_0^2 - \gamma^2}\ t \right) + \frac{\gamma \sin\left(\sqrt{\omega_0^2 - \gamma^2}\ t \right)}{\sqrt{\omega_0^2 - \gamma^2}} \right). \tag{4.9}$$

This solution is confirmed in Example 4-1.

Example 4-1: Damped Harmonic Oscillator Solution

Derive the underdamped harmonic oscillator solution (4.9).

Solution: Omitting the subscript on ω for convenience, the damped harmonic oscillator equation is entered using the BasicInput palette to input θ, γ, and ω.

```
Clear["Global`*"]
```

```
osc = θ''[t] + 2 γ θ'[t] + ω^2 θ[t] == 0
```

$$\omega^2\, \theta(t) + 2\gamma\theta'(t) + \theta''(t) == 0$$

The ODE osc is analytically solved for the function θ for $\theta(0) = A$ and $\dot{\theta}(0) = 0$.

```
sol = DSolve[{osc, θ[0] == A, θ'[0] == 0}, θ, t]
```

The lengthy output of this command line and the following one have been omitted here in the text, but can be viewed on the computer screen by executing the example on the CD. The solution given by the output of sol is expressed in terms of exponential functions. To convert the solution into a trigonometric structure, the ExpToTrig command is employed.

```
sol2 = ExpToTrig[θ[t] /. sol[[1]]]
```

The output of sol2 is given in terms of the hyperbolic functions sinh and cosh. To obtain a solution in terms of sines and cosines, the FullSimplify command is applied to sol2 subject to the condition $\omega > \gamma$.

```
sol3 = FullSimplify[sol2, ω > γ]
```

$$A\,e^{-t\gamma}\left(\cos(t\,\sqrt{\omega^2 - \gamma^2}) + \frac{\gamma\,\sin(t\,\sqrt{\omega^2 - \gamma^2})}{\sqrt{\omega^2 - \gamma^2}}\right)$$

The output of sol3 is the underdamped solution (4.9) that we were after.

$\boxed{\textbf{End Example 4-1}}$

The underdamped pendulum would oscillate about $\theta = 0$ with the amplitude decreasing exponentially with time. In terms of a phase plane diagram (Fig. 4.5), the system starting at, say, $\theta = \theta_0$, $\dot{\theta} = 0$ would converge on $\theta = 0$, $\dot{\theta} = 0$ as $t \to \infty$ so that the origin behaves as a stable focal point. If the pendulum

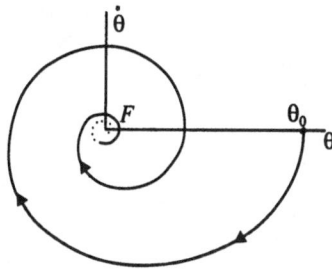

Figure 4.5: Phase trajectory for an underdamped simple harmonic oscillator.

is pulled out through a large angle, the phase trajectory far from F would be described by the full nonlinear equation. However, when the amplitude decreases sufficiently so that the linear approximation holds, then the above picture is valid, i.e., near the origin, the trajectories are spirals. Thus, adding a small amount of damping changes a vortex point into a focal point.

For an unstable focal point F, the spiral trajectories leave F (the arrows in Figure 4.4 would be reversed) starting out with an indeterminate direction. As we shall see later in the chapter, the origin $x = 0$, $\dot{x} = 0$ is an unstable focal point for the Van der Pol equation. The next experimental activity also features an unstable focal point.

Focal Point Instability

This experimental activity investigates an electrical circuit which produces an unstable focal point by introducing "negative resistance" through the use of an operational amplifier.

Nodal Point (N)

A nodal point is a singularity that is approached by trajectories having the following property. Referring to Figure 4.6, as $r \to 0$ the directions of the tangents to the trajectories approach definite limits.

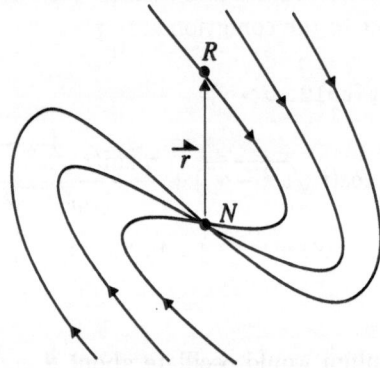

Figure 4.6: Trajectories in the neighborhood of a nodal point N.

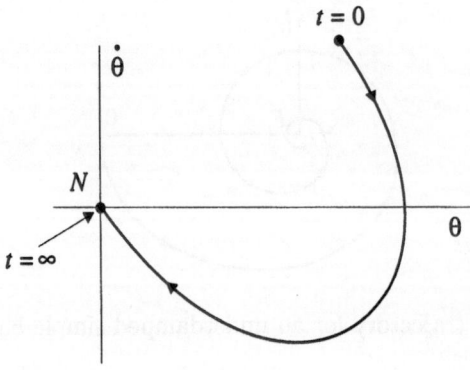

Figure 4.7: Phase trajectory for an overdamped harmonic oscillator.

The nodal point in Figure 4.6 is called a stable nodal point because a representative point R approaches N as $t \to \infty$. For an unstable nodal point, the arrows would be reversed.

Again referring to the example of a damped pendulum undergoing small oscillations, for overdamping the general solution is $\theta = Ae^{-r_1 t} + Be^{-r_2 t}$ with $r_1, r_2 > 0$ and real. If initially, for example, $\theta > 0$, $\dot{\theta} > 0$, the evolution of the pendulum would be as schematically illustrated in Figure 4.7, the origin now being a stable nodal point.

4.3 Classifying Simple Singular Points

To study the nature of trajectories passing through ordinary points (x, y) in the neighborhood of a singular point (x_0, y_0), we must return to the general expression for the slope of a trajectory,

$$\frac{dy}{dx} = \frac{Q(x, y)}{P(x, y)}. \tag{4.10}$$

At the singular point, $Q(x_0, y_0) = P(x_0, y_0) = 0$, while at the ordinary points, although either Q or P may be zero (corresponding to zero slope or infinite slope), they are not zero simultaneously. For ordinary points near the singular point, it is convenient to write

$$x = x_0 + u$$

$$y = y_0 + v \tag{4.11}$$

where u and v are small. Then, at an ordinary point

$$\frac{dy}{dx} = \frac{Q(x_0 + u, y_0 + v)}{P(x_0 + u, y_0 + v)} \tag{4.12}$$

and on Taylor-expanding[5] both the numerator and denominator of the right-hand side for small u and v

$$\frac{dy}{dx} = \frac{dv}{du} = \frac{cu + dv + c'u^2 + d'v^2 + f'uv + \cdots}{au + bv + a'u^2 + b'v^2 + e'uv + \cdots} \tag{4.13}$$

where we have made use of the fact that $Q(x_0, y_0) = P(x_0, y_0) = 0$ and set

$$a \equiv \left(\frac{\partial P}{\partial x}\right)_{x_0, y_0}, \quad b \equiv \left(\frac{\partial P}{\partial y}\right)_{x_0, y_0}, \quad c \equiv \left(\frac{\partial Q}{\partial x}\right)_{x_0, y_0}, \quad d \equiv \left(\frac{\partial Q}{\partial y}\right)_{x_0, y_0},$$

etc. The coefficients $a, b, c, \ldots$ are real if Q, P are real. A "simple singularity" is one in the neighborhood of which the behavior of the trajectories is correctly described by retaining only the linear terms in u and v in the denominator and numerator of Equation (4.13). Then, for a simple singularity,

$$\frac{dv}{du} = \frac{cu + dv}{au + bv}. \tag{4.14}$$

---

[5]If Q and P are polynomials, the power series follows by direct expansion.

Clearly, if a, b, c, and d are nonzero and u and v are sufficiently small, then Equation (4.14) should be a good approximation to Equation (4.13), the higher order terms involving u^2, uv, etc., making only small corrections which, except for the vortex, do not qualitatively change the nature of the trajectories. If, on the other hand, c and d (or a and b) both vanish, then higher order terms should be kept in the numerator (or denominator). Even for a, b, c, and d all nonzero, one can have $au + bv = 0$ and $cu + dv = 0$ for u and $v \neq 0$ (u, $v = 0$ corresponds to the singular point (x_0, y_0) of interest) in which case higher order terms should be kept in both the numerator and denominator. A nontrivial solution of $cu + dv = 0$, $au + bv = 0$, can only occur if the determinant condition

$$\begin{vmatrix} c & d \\ a & b \end{vmatrix} = 0 \tag{4.15}$$

or $bc - ad = 0$. If this occurs, the singularity is no longer simple (i.e., it is not determined by linear terms in u and v alone). Since setting either c and d or a and b equal to zero also makes $bc - ad$ vanish, it follows that a simple singularity can occur if

$$bc - ad \neq 0. \tag{4.16}$$

In the neighborhood of such a singularity, the trajectories are described by

$$\frac{dv}{du} = \frac{cu + dv}{au + bv}, \tag{4.17}$$

i.e., their nature is completely determined by the coefficients a, b, c, and d. We shall now outline how it is established that there are only four types of simple singularities for the 2-dimensional phase plane.

The expression for dv/du can be thought of as resulting from a pair of simultaneous first-order coupled linear equations

$$\begin{aligned} \dot{u} = \frac{du}{dt} &= au + bv \\ \dot{v} = \frac{dv}{dt} &= cu + dv. \end{aligned} \tag{4.18}$$

Solving for v in the first equation of (4.18) and substituting into the second yields

$$\ddot{u} + p\dot{u} + qu = 0 \tag{4.19}$$

with

$$\begin{aligned} p &\equiv -(a + d) \\ q &\equiv ad - bc. \end{aligned} \tag{4.20}$$

Since the above linear ODE has constant coefficients, a solution is sought of the form $u \sim e^{\lambda t}$. There are two roots, λ, obtained by solving

$$\lambda^2 + p\lambda + q = 0 \tag{4.21}$$

namely

$$\lambda_{1,2} = -\frac{p}{2} \pm \frac{1}{2}\sqrt{p^2 - 4q}. \tag{4.22}$$

Although hardly necessary, the mathematical steps leading from Equations (4.18) to (4.22) are checked in the following example. The Mathematica commands shown here will prove very useful in the linear stability analysis of 3-dimensional nonlinear systems, which will be discussed at the end of this chapter.

Example 4-2: Roots

Derive the two roots of λ starting with the first-order coupled Equations (4.18) and using a matrix approach.

Solution: The matrix containing the coefficients of u and v is formed. This matrix is known as the "Jacobian matrix" and therefore we label it with this name.

```
Clear["Global`*"]

jacobianMatrix = {{a, b}, {c, d}}
```

$$\begin{pmatrix} a & b \\ c & d \end{pmatrix}$$

If proceeding by hand, one would assume that both $u, v \sim e^{\lambda t}$ and rewrite Equations (4.18) in the matrix form

$$\begin{bmatrix} a & b \\ c & d \end{bmatrix} \begin{bmatrix} u \\ v \end{bmatrix} = \lambda \begin{bmatrix} u \\ v \end{bmatrix} = \begin{bmatrix} \lambda & 0 \\ 0 & \lambda \end{bmatrix} \begin{bmatrix} u \\ v \end{bmatrix}.$$

Thus,

$$\begin{bmatrix} a - \lambda & b \\ c & d - \lambda \end{bmatrix} \begin{bmatrix} u \\ v \end{bmatrix} = 0,$$

so that for a nontrivial solution the determinant condition

$$\begin{vmatrix} a - \lambda & b \\ c & d - \lambda \end{vmatrix} = 0$$

must be imposed. Expansion of the determinant will lead to the quadratic equation in λ which can be solved for the roots. All of this procedure may be accomplished with the following **Eigenvalues** command.

```
eivs = Eigenvalues[jacobianMatrix]
```

$$\left\{ \frac{1}{2}\left(a + d - \sqrt{a^2 - 2da + d^2 + 4bc}\right), \frac{1}{2}\left(a + d + \sqrt{a^2 - 2da + d^2 + 4bc}\right) \right\}$$

To obtain the roots λ_1 and λ_2, the substitutions $a = -p - d$ and $bc = -q + ad$ are made in each result of **eivs** and the **ExpandAll** command applied.

`lambda1 = ` λ_1 ` == ExpandAll[eivs[[2]]] //. {a->-p-d, b c->-q+a d}]`

$$\lambda_1 == \frac{1}{2}\sqrt{p^2 - 4q} - \frac{p}{2}$$

`lambda2 = ` λ_2 ` == ExpandAll[eivs[[1]]] //. {a->-p-d, b c->-q+a d}]`

$$\lambda_2 == -\frac{p}{2} - \frac{1}{2}\sqrt{p^2 - 4q}$$

To obtain the desired forms, note that the "repeated replacement" rule `//.` was used here, not `/.`, the replacement rule. The latter applies the rules (substitutions) once on the expression, whereas the former applies the rules repeatedly until the result no longer changes.

The Mathematica approach employed in this example will be used to later carry out the stability analysis of the 3-dimensional Lorenz system.

$\boxed{\textbf{End Example 4-2}}$

The analysis of the λ roots is straightforward, so we only sketch the main points here. For simple singularities, $q \neq 0$, so a zero root of (4.22) is not possible. The case $q = 0$ corresponds to "higher order singular points" which will be examined later. The possible roots λ_1, λ_2 depend on the relative size and signs of p and q.

First, we consider $q > 0$ and $p \neq 0$. For $p^2 > 4q$, the roots λ_1 and λ_2 are real and of the same sign, while for $p^2 < 4q$ the roots are complex conjugate. For $p^2 = 4q$ (or $p^2 - 4q = 0$), $\lambda_1 = \lambda_2 = -\frac{1}{2}p$ so the roots are degenerate and obviously of the same sign. In this case, a second linearly independent solution $te^{\lambda t}$ must be introduced. For nonzero p, the singular points are clearly stable for $p > 0$ and unstable for $p < 0$. This latter condition is a special case of *Lyapunov's Theorem*, which applies to an N-dimensional system. For $N = 3$, for example, one would solve a cubic equation for λ.

Lyapunov's theorem states that if the real parts of the characteristic roots obtained from the linearized equations in the neighborhood of the singular point are not zero, then the stationary point is stable if all real parts are negative and unstable if at least one real part is positive.

Now, the classification of the stationary points for $q > 0$ and $p \neq 0$ is made and summarized in the p-q plane picture of Figure 4.8.

- Nodal points occur when the roots λ_1 and λ_2 are real and of the same sign, i.e., for $p^2 - 4q > 0$. That this is so is apparent from our earlier discussion of the overdamped harmonic oscillator. Nodal points also occur for the degenerate root case, i.e., along the parabolic curve $p^2 - 4q = 0$ in Figure 4.8. For the harmonic oscillator, this would correspond to critical damping. From Lyapunov's theorem, stable nodal points occur for $p > 0$, unstable for $p < 0$.

- Focal or spiral points occur when the roots are complex conjugate, i.e., for $p^2 - 4q < 0$. This case corresponds to that for the underdamped oscillator. Stable focal points occur for $p > 0$, unstable for $p < 0$.

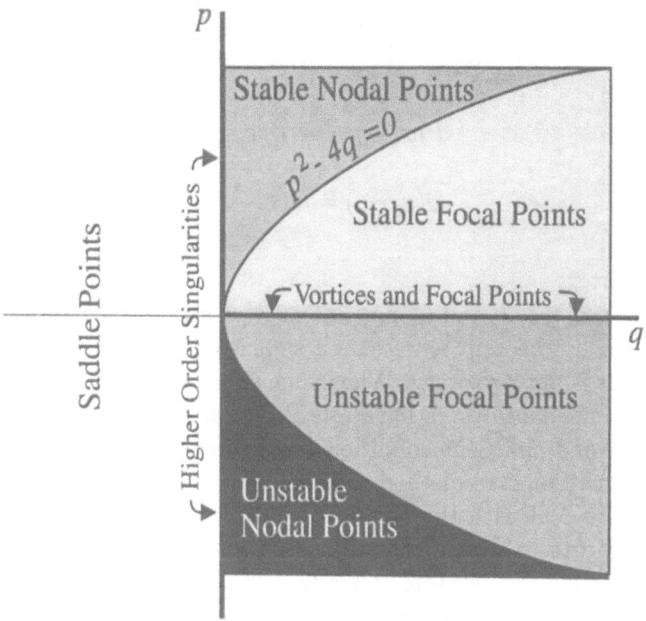

Figure 4.8: The p–q plane for establishing types of simple singular points.

For $q > 0$, now consider the case when $p = 0$. The two roots, λ_1 and λ_2, are now purely imaginary. This is the case for the undamped pendulum. The stationary point is a vortex point.

Finally, we examine the situation when $q < 0$. Independent of the value or sign of p, the roots λ_1 and λ_2 are real but of the opposite sign. Because the roots are of opposite signs, it follows from Lyapunov's theorem that the associated stationary points are always unstable. The region $q < 0$ in Figure 4.8 corresponds to saddle points. Since all possible roots of λ have been examined, it follows that there are only four types of simple singular points.

However, it is clear that the analysis for the vortices is not definitive because we have kept only first-order terms in u and v in the Taylor expansion (4.13). Higher order terms in the expansion may turn vortices into focal points. This is why the positive q axis in Figure 4.8 is labeled vortices and focal points. With the Taylor expansion option available in Mathematica, we could, of course, keep higher order terms in an attempt to distinguish between the two types of singular points. This can be done for individual cases, but it is difficult to make "global" statements that apply to all nonlinear systems. A simple global theorem is due to Poincaré.

4.3.1 Poincaré's Theorem for the Vortex (Center)

Suppose that for the system of equations

$$\dot{x} = P(x, y), \qquad \dot{y} = Q(x, y), \tag{4.23}$$

the functions $P(x,y)$, $Q(x,y)$ satisfy in the neighborhood of O the conditions for O to be a vortex or a focus. If $P(x,y)$, $Q(x,y)$ satisfy the conditions

$$P(x,-y) = -P(x,y),$$

$$Q(x,-y) = Q(x,y),$$

(4.24)

then O is a vortex.

In some situations, P and Q may not satisfy the above conditions yet O is a vortex. Poincaré's theorem represents a sufficiency condition for the existence of a vortex, but is not necessary. When such situations arise, life becomes more difficult. Poincaré and others have established a complex method that leads to a necessary and sufficient condition for a singularity to be a vortex. Two references on the Poincaré method can be found in [SC64] and [Hay64].

We shall not deal with this method here but illustrate in the next section the use of Poincaré's theorem in conjunction with the analysis of singularities for a few of the physical systems already mentioned. If Poincaré's theorem is inconclusive, use can be made of Mathematica's numerical options.

PROBLEMS
Problem 4-1: Poincaré's theorem
Give a geometrical argument that supports Poincaré's theorem.

4.4   Examples of Phase Plane Analysis

To firmly fix the ideas just presented, the following examples will be analyzed in some detail. To complement our analysis, the "tangent field" will be introduced and plotted with Mathematica. We have already made use of Mathematica's "phase portrait" plotting ability in Chapter 2, which allowed us to generate phase plane trajectories corresponding to specified initial conditions. By specifying many initial conditions, one could clearly check our topological analysis. As shall be seen, the tangent field approach can also be very illuminating.

4.4.1   The Simple Pendulum

Recall the simple pendulum equation

$$\ddot{x} = -\omega^2 \sin x.$$

(4.25)

Putting $\dot{x} = y$, then $\dot{y} = -\omega^2 \sin x$ and

$$\frac{dy}{dx} = \frac{-\omega^2 \sin x}{y} \equiv \frac{Q(x,y)}{P(x,y)}.$$

(4.26)

The singular points occur at $y_0 = 0$ and $x_0 = n\pi$ ($n = 0, \pm 1, \pm 2, \ldots$). Setting $x = x_0 + u$, $y = y_0 + v$, then

$$\frac{dy}{dx} = \frac{dv}{du} = \frac{-\omega^2(\sin(x_0 + u))}{y_0 + v}.$$

(4.27)

First, consider the singular point $x_0 = 0$, $y_0 = 0$ for which

$$\frac{dv}{du} = \frac{-\omega^2 \sin u}{v} = \frac{-\omega^2 \left(u - \frac{u^3}{3!} + \dots \right)}{v} \simeq \frac{-\omega^2 u}{v} \qquad (4.28)$$

keeping only the first-order term in the numerator. Comparing with Equation (4.14), we identify $a = 0$, $b = 1$, $c = -\omega^2$, $d = 0$, so that $p = -(a + d) = 0$ and $q = ad - bc = \omega^2 > 0$. Referring to the p–q diagram, Figure 4.8, the singularity is either a vortex or a focal point. Let's apply Poincaré's theorem to see if the singularity could be a vortex.(We know that it is!) Since $P(x, y) = y$, $Q(x, y) = -\omega^2 \sin x$, then in the vicinity of $x_0 = 0$, $y_0 = 0$ we have

$$P(x, -y) = -y = -P(x, y) \qquad (4.29)$$

$$Q(x, -y) = -\omega^2 \sin x = Q(x, y).$$

As Poincaré's theorem is satisfied, the singularity $x_0 = 0, y_0 = 0$ is correctly identified as a vortex.

Now consider the singularity $x_0 = \pi$, $y_0 = 0$. The numerator of (4.27) becomes $\sin(\pi + u) = -\sin u \simeq -u$ so

$$\frac{dv}{du} \simeq \frac{\omega^2 u}{v} \qquad (4.30)$$

from which we identify $a = 0$, $b = 1$, $c = +\omega^2$, $d = 0$ and therefore $p = 0$, $q = -\omega^2 < 0$. Referring to the p–q diagram (Figure 4.8) and noting that $q < 0$, this singularity corresponds to a saddle point.

The other singularities corresponding to different values of n can be analyzed in the same manner. Vortices occur for n even and saddle points for n odd. The phase plane can now be qualitatively constructed as shown in Figure 4.9. It

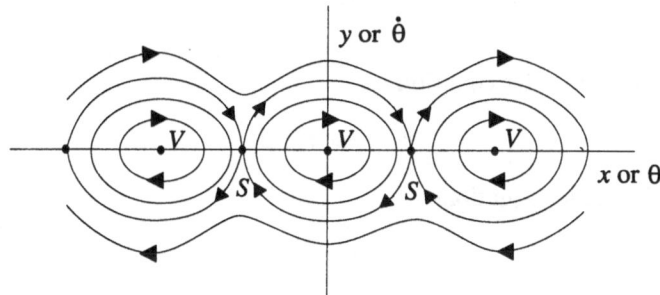

Figure 4.9: Phase plane portrait for undamped simple pendulum.

agrees with the results from Mathematica File 13. The saddle point "trajectories" connect adjacent saddle points and the sense of the time arrows must be the same as the enclosed vortices. Remember that motion doesn't actually occur along such a trajectory since it would take an infinite time for the pendulum system to leave a saddle point and move to another saddle point. As mentioned earlier the saddle point trajectories are referred to as separatrixes since they separate two qualitatively different types of pendulum motion. Inside the separatrixes are the back and forth motion characterized by vortices. Outside corresponds to situations where the pendulum has sufficient energy to go over the top and, in the absence of damping, continue to advance in the direction of ever increasing θ. Above the top separatrixes, $\dot{\theta}$ is positive and the motion is in the counterclockwise sense, while below the bottom separatrixes the motion is clockwise. It is clear from our physical understanding of the pendulum system that the phase plane portrait displays all possible motions of the undamped pendulum.

Mathematica can also be employed to confirm our qualitative picture in a different manner. Equation (4.5) gives the exact slope relation at any phase plane point x, y, that is to say the tangent to the trajectory at that point. With the proper Mathematica command, the "tangent field" can be plotted by drawing short line segments of appropriate slope at equally spaced points and attaching arrow heads to indicate the sense of increasing time. This is illustrated for the simple pendulum in the following example.

Example 4-3: Tangent Field

Plot the tangent field for the simple pendulum, taking the characteristic frequency to be $\omega = 1$.

Solution: The Graphics package is loaded,

```
Clear["Global`*"]
```

```
<< Graphics`
```

and the value of the frequency entered. The Greek letter ω is formed with the BasicInput palette.

```
ω = 1;
```

Writing the second-order equation

$$\ddot{x} = -\omega^2 \sin x$$

as the two first-order equations

$$\dot{x} = y, \quad \dot{y} = -\omega^2 \sin x,$$

the rhs are entered as components of a list and given the name simpend.

```
simpend = {y, -ω^2 Sin[x]};
```

The tangent field graph gr1 is produced with the PlotVectorField command, the length and shape of the arrows being controlled with the ScaleFunction, ScaleFactor, and HeadLength options. The density of arrows is controlled with PlotPoints. The output of gr1 is suppressed.

```
gr1 = PlotVectorField[simpend, {x, -7, 7}, {y, -3, 3},
    PlotPoints -> 15, Frame -> True, FrameTicks -> {{-2 Pi, -Pi,
    0, {1.5, "x"}, Pi, 2 Pi}, {-3, 0, {1.5, "y"}, 3}, {}, {}},
    ScaleFunction -> (1 &), ScaleFactor -> .5, HeadLength -> 0.01,
    DisplayFunction -> Identity];
```

Using the Show command, the graph gr1 is now displayed along with colored filled circles locating the stationary points of the simple pendulum system.

```
Show[gr1, PlotRange -> {{-7.5, 7.5}, {-3.3, 3.3}},
Epilog -> PointSize[.03], {RGBColor[0, 1, 0], Point[{-2 Pi, 0}],
Point[{0, 0}], Point[{2 Pi, 0}], {RGBColor[1, 0, 0], Point[{-Pi, 0}],
Point[{Pi, 0}]}}, DisplayFunction -> $DisplayFunction,
TextStyle -> {FontFamily -> "Times", FontSize -> 16},
ImageSize -> {600, 300}];
```

Figure 4.10 shows the results of applying the above Mathematica code. The tangent field plot confirms our phase plane analysis of the simple pendulum and gives us a quantitatively accurate picture.

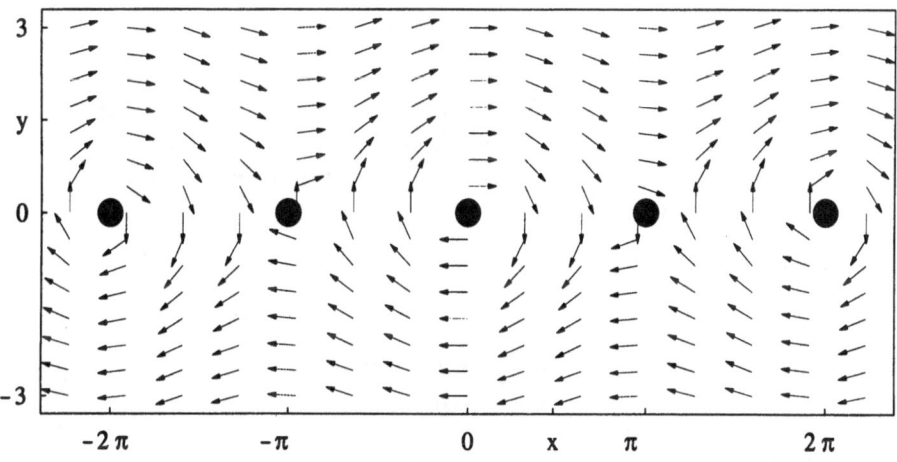

Figure 4.10: Tangent field for simple pendulum with stationary points indicated.

End Example 4-3

4.4.2 The Laser Competition Equations

Recall, from Chapter 2, that for beams traveling in the opposite direction, the laser beam competition equations were

$$\frac{dI_L}{dz} = -gI_L I_s - \alpha I_L$$

$$\frac{dI_s}{dz} = -gI_L I_s + \alpha I_s$$

$$(4.31)$$

where the absorption coefficient α is intrinsically positive and the gain coefficient g can be of either sign. In contrast to the simple pendulum problem which will be solved in the next chapter, this set of equations cannot be solved exactly in analytic form. To conform with the general analysis just presented, it is convenient to relabel the intensities, $x \equiv I_L$, $y \equiv I_s$, so that the equation for the slope of the trajectories is

$$\frac{dy}{dx} = \frac{(-gxy + \alpha y)}{(-gxy - \alpha x)} = \frac{y(-gx + \alpha)}{x(-gy - \alpha)}. \tag{4.32}$$

The singular points of Equation (4.32) are $x_0 = 0$, $y_0 = 0$ and $x_0 = \alpha/g$, $y_0 = -\alpha/g$. For $\alpha \neq 0$, the latter singular point is in the 4th quadrant of the x–y plane for $g > 0$ and in the 2nd quadrant for $g < 0$.

Let's examine each of these singularities:

(a) $x_0 = 0$, $y_0 = 0$:

Writing $x = 0 + u$, $y = 0 + v$ and linearizing, Equation (4.32) becomes

$$\frac{dv}{du} = \frac{\alpha v}{-\alpha u} \tag{4.33}$$

from which we identify $a = -\alpha$, $b = 0$, $c = 0$, $d = \alpha$ so that $p = -(a+d) = 0$ and $q = ad - bc = -\alpha^2$. Since $q < 0$, the origin is a saddle point in the neighborhood of which the trajectories are hyperbolic. (Integration of Equation (4.33) gives $uv = $ const., a rectangular hyperbola.)

(b) $x_0 = \alpha/g$, $y_0 = -\alpha/g$:

Setting $x = \alpha/g + u$, $y = -\alpha/g + v$, Equation (4.32) yields

$$\frac{dv}{du} = \frac{\alpha u - guv}{-\alpha v - guv} \tag{4.34}$$

or, on retaining the first-order (linear) terms,

$$\frac{dv}{du} \simeq \frac{\alpha u}{-\alpha v}. \tag{4.35}$$

Therefore, $a = 0$, $b = -\alpha$, $c = +\alpha$, $d = 0$, $p = 0$ and $q = \alpha^2 > 0$. Since this singularity must be either a vortex or a focal point, let's appeal to Poincaré's theorem to decide which it is. Identifying

$$P(u, v) \equiv -\alpha v - guv, \quad Q(u, v) \equiv \alpha u - guv, \tag{4.36}$$

then

$$P(u, -v) = \alpha v + guv = -P(u, v)$$
$$\tag{4.37}$$
$$Q(u, -v) = \alpha u + guv \neq Q(u, v).$$

Thus Poincaré's theorem isn't satisfied and doesn't help in deciding whether the singularity is a vortex or a focal point. However, if it is remembered that we should ultimately be interested in the physics of the problem, this need not bother us. Physically, the intensities of both beams must always remain positive. Therefore the solution of physical interest must lie in the 1st quadrant of the x–y plane. Since the singularity at the origin is a saddle point and the other singularity lies in the 4th or 2nd quadrant (depending on the sign of g) and further remembering that "real" trajectories cannot cross, the trajectories relevant to the physical problem must be as schematically illustrated in Figure 4.11. The direction of increasing z for $g > 0$ is indicated by the arrows, the direction following from the original Equations (4.31).

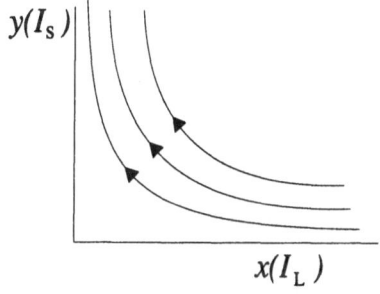

Figure 4.11: Physical region of laser competition phase plane portrait.

Although the identification of the singular or stationary points in this example was easily done by hand, Mathematica can be extremely useful when the algebra becomes tedious or a large number of singular points are involved. The Mathematica code for the previous example, which is now provided in the following file, is easily modified to handle other more complex coupled systems, polynomial or otherwise.

Phase Plane Analysis
This file shows how Mathematica can be used to perform the phase plane analysis for the laser competition equations. It analytically locates all the stationary points and calculates the mathematical forms of p, q, and $p^2 - 4q$ for each such point. The numerical values of these quantities are calculated for $\alpha = 1$ and plotting points formed for each stationary point. These singular points are then placed on a colored plot similar to Figure 4.8, from which the nature of the stationary points can be determined. To distinguish between vortices and focal points in situations where $p = 0$ and $q > 0$, Poincaré's theorem is included in the file. Poincaré's theorem is not satisfied for the laser competition equations. The file can be modified to handle other systems of nonlinear equations. Mathematica commands in file: Factor, Solve, Simplify, D, Plot, Sqrt, AspectRatio, PlotStyle, Background, PointSize, RGBColor, Text, Point, Hue, AxesLabel, PlotRange, Epilog, ImageSize, TextStyle, FontFamily.

To see how MF14 can be applied to other nonlinear systems, consider the following oversimplified model of the competition between two warring nations.

Example 4-4: Competing Armies

The armies of two warring countries are modeled by the competition equations

$$\frac{dC_1}{dt} = \alpha C_1 - \beta C_1 C_2$$

$$\frac{dC_2}{dt} = (\alpha + 1)C_2 - \gamma \beta C_1 C_2$$

with α and β both positive and $\gamma > 1$. Here C_1 and C_2 are the numbers of individuals in the armies of countries 1 and 2.

    a. Discuss the structure of the equations and suggest how improvements could be made in the model.

    b. Locate and identify all stationary points of this system.

    c. Make a tangent field plot which includes both stationary points and some representative trajectories. Take $\alpha = 5$, $\gamma = 1.15$, and $\beta = 1/2500$.

    d. Discuss possible outcomes on the basis of the above picture.

Solution:a. This oversimplified model has linear growth terms in each equation to reflect recruitment and nonlinear loss terms to account for battle casualties. Since the parameters are all positive, country number 2 recruits more rapidly than country number 1, but also has a higher casualty rate since $\gamma > 1$. For a sufficiently long war, the recruitment terms should be modified to include saturation effects because the rate would decrease as the total number of individuals in each country available to fight is limited. What suggestions do you have?

b. We make use of MF14. The Graphics package is loaded first.

```
Clear["Global`*"]; << Graphics`
```

We use the symbol x for C_1 and y for C_2. The rhs of the ODEs are entered and assigned the names eqP and eqQ, respectively.

```
eqP = α x - β x y;
```

```
eqQ = (α + 1) y - γ β x y;
```

The slope is calculated by dividing eqQ by eqP. Applying the **Factor** command and inspecting the output reveals that there are two stationary points.

```
slope = Factor[eqQ/eqP]
```

$$-\frac{y\,(-1 - \alpha + x\beta\gamma)}{x\,(\alpha - y\beta)}$$

The location of the two stationary points is extracted by applying the **Solve** command to eqP == 0 and eqQ == 0.

```
statpts = Solve[{eqP == 0, eqQ == 0}, {x, y}] // Simplify
```

$$\{\{x \to 0, y \to 0\}, \{x \to \frac{1+\alpha}{\beta\gamma}, y \to \frac{\alpha}{\beta}\}\}$$

The stationary points are $(x = 0, y = 0)$ and $(x = \frac{\alpha+1}{\beta\gamma}, y = \frac{\alpha}{\beta})$.

In the following command line, p and q, as well as $p^2 - 4q$, are evaluated for both stationary points.

```
pandq = {p = -(D[eqP, x] + D[eqQ, y]),
         q = D[eqP, x] D[eqQ, y] - D[eqP, y] D[eqQ, x],
         Simplify[p^2 - 4q]} /. statpts
```

$$\begin{pmatrix} -2\alpha - 1 & \alpha(\alpha + 1) & 1 \\ 0 & -\alpha(\alpha + 1) & 2\alpha^2 + 2(\alpha + 2)\alpha \end{pmatrix}$$

The top row of the output yields $p = -2\alpha - 1$, $q = \alpha(\alpha+1)$, and $p^2 - 4q = 1$ for the first stationary point at the origin. From the bottom row, we have $p = 0$, $q = -\alpha(\alpha+1)$, and $p^2 - 4q = 2\alpha^2 + 2(\alpha+2)\alpha$ for the nonzero stationary point. Since α is positive, the stationary point at the origin must be an unstable nodal point, since $p < 0$, $q > 0$, and $p^2 - 4q > 0$. On the other hand, the nonzero stationary point is a saddle point since $q < 0$.

c. The parameter values are entered.

```
α = 5; γ = 1.15; β = 1/2500;
```

A graph of the tangent field is created with the `PlotVectorField` command, the output being suppressed.

```
gr1 = PlotVectorField[{eqP, eqQ}, {x, -2500, 30000}, {y, -2500, 30000},
    PlotPoints -> 20, Frame -> True, FrameTicks -> {{0, 10000, {15000,
    "x"}, 20000, 30000}, {0, 10000, {15000, "y"}, 20000, 30000},
    {}, {}}, ScaleFunction -> (.4 &), DisplayFunction -> Identity];
```

To superimpose some representative trajectories in the same plot as the tangent field, the time-dependent ODEs are now entered, subject to the initial conditions x[0] == x0, y[0] == y0.

```
eq = {x'[t] == (eqP /. x -> x[t] /. y -> y[t]),
      y'[t] == (eqQ /. x -> x[t] /. y -> y[t]),
      x[0] == x0, y[0] == y0};
```

For the initial x value, we take x0 = 4000.

```
x0 = 4000;
```

Using the `Table` command, the ODE system contained in `eq` is numerically solved over the time interval $t = 0$ to 1 for initial y values varying from 1000 to 5000 in steps of 500. Nine different trajectories are produced.

```
sol = Table [NDSolve [eq /. y0 -> yy, {x[t], y[t]}, {t, 0, 1}],
      {yy, 1000, 5000, 500}];
```

`ParametricPlot` is used to create a graph of the trajectories. The `Compiled -> False` option is included so as to obtain high-precision numbers.

```
gr2 = ParametricPlot [Evaluate [{x[t], y[t]} /. sol], {t, 0, 1},
      DisplayFunction -> Identity, PlotStyle -> Hue [.6],
      Compiled -> False, Frame -> True];
```

The graphs `gr1` and `gr2` are superimposed in Fig. 4.12 with the `Show` command. The (colored) stationary points are included by using the `Epilog` option.

```
Show [gr1, gr2, Epilog -> {PointSize [.03], {RGBColor [0, 1, 0],
Point [{0, 0}]}, {RGBColor [1, 0, 0], Point [{(α + 1)/(β γ), α/β}]}},
PlotRange -> {{-4000, 32000}, {-4000, 32000}}, DisplayFunction ->
$DisplayFunction, Frame -> True, ImageSize -> {500, 500},
TextStyle -> {FontFamily -> "Times", FontSize -> 16}];
```

d. The trajectories displayed in Figure 4.12 show country 1 (x) winning out in one case and country 2 (y) in the other. From the plot, the reader should be able to deduce the general rules for a given country to prevail.

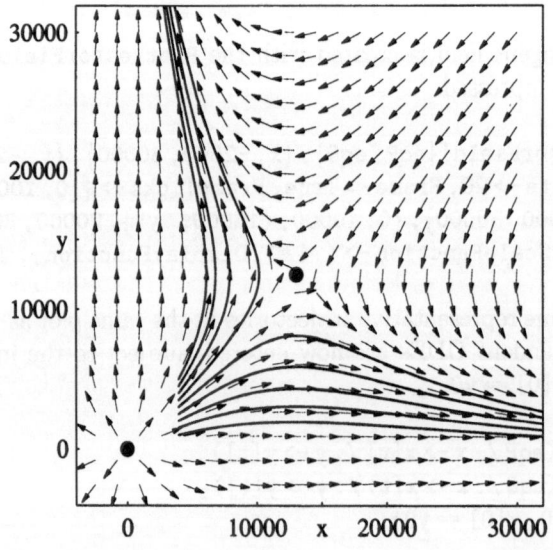

Figure 4.12: Phaseportrait plot for armies example.

End Example 4-4

PROBLEMS

Problem 4-2: Laser Competition Equations
Plot the tangent field picture for the laser competition equations for $g = 1$, $\alpha = 0.1$, and locate the stationary points with suitably-sized colored points. Superimpose some representative trajectories in the same figure.

Problem 4-3: Laser Competition 2
By hand, find and determine the nature of the singular points for the laser competition equations when the laser and signal beams are propagating in the same direction. Sketch the phase plane portrait for the region of physical interest for $g = 1$, $\alpha = 0.1$ and discuss the result. Check your result by using Mathematica to perform the singular point analysis and to plot the tangent field picture.

4.4.3   Example of a Higher Order Singularity

Consider a light (weightless) spring suspended from one end, with a mass m attached to the other. Suppose that the force F required to deflect the mass m a distance x downward from the equilibrium position displays the behavior shown in Figure 4.13. At $x = \tilde{x}$ (assumed to be nonzero), the force $F = 0$ and

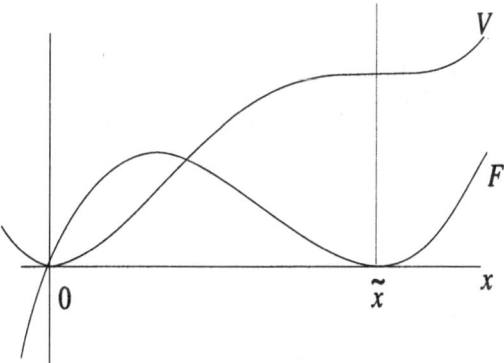

Figure 4.13: Force (F) law and potential energy (V) for a hypothetical spring.

$dF/dx = 0$. The corresponding potential energy (V) curve is also shown in the figure. ($F = +dV/dx$ since F is the deflecting force). The analytical form of F may be written as

$$F = kx - gx^2 + hx^3 \qquad (4.38)$$

with k, g, h, all positive. For sufficiently small x, the force displays a linear Hooke's law behavior, the corresponding potential energy being parabolic in this region. As x is increased, F decreases, so we include $-gx^2$ as the next term to accomplish this. At still larger x, F turns up again. We achieve this by including the $+hx^3$ term. This phenomenological third-order polynomial force law would have three zeros in general but here two of the roots are degenerate at $\tilde{x}$. At $x = \tilde{x}$

$$F = 0 = (k - g\tilde{x} + h\tilde{x}^2)\tilde{x} \qquad (4.39)$$

$$\frac{dF}{dx} = 0 = k - 2g\tilde{x} + 3h\tilde{x}^2 \qquad (4.40)$$

so g and h must be related to k as follows,

$$g = \frac{2k}{\tilde{x}}, \quad h = \frac{k}{\tilde{x}^2}. \tag{4.41}$$

The equation of motion is

$$m\ddot{x} + (kx - gx^2 + hx^3) = 0. \tag{4.42}$$

On setting $y = \dot{x}$,

$$\dot{y} = \frac{1}{m}(-kx + gx^2 - hx^3) \tag{4.43}$$

and

$$\frac{dy}{dx} = \frac{\frac{1}{m}(-kx + gx^2 - hx^3)}{y}. \tag{4.44}$$

We shall not completely analyze this problem here but only examine the trajectories in the neighborhood of the singular point $x_0 = \tilde{x}$, $y_0 = 0$. Setting $x = \tilde{x} + u$, $y = 0 + v$, and using the relations for g and h, Equation (4.44) reduces to

$$\frac{dv}{du} = -\frac{(k/m\tilde{x})\, u^2\, [(1 + u/\tilde{x})]}{v}. \tag{4.45}$$

Since the lowest order term in the numerator is not linear, but quadratic, in u, the singular point $x_0 = \tilde{x}$, $y_0 = 0$ is not a simple singularity but, instead, an example of a higher order singularity.

What do the trajectories look like close to this singularity? Sufficiently close to the singular point the cubic contribution in (4.45) is negligible, so

$$\frac{dv}{du} \simeq -\frac{(k/m\tilde{x})\, u^2}{v} \tag{4.46}$$

which on cross-multiplication can be integrated to give

$$v^2 + au^3 = C \tag{4.47}$$

with $a \equiv 2k/3m\tilde{x}$ and the constant C determined by initial conditions.

By choosing various values of C, the trajectories in the neighborhood of the singularity are determined and may be plotted as in Figure 4.14. The tangent field is also shown. This plot was produced with the Mathematica File MF15. The higher order singularity in this case looks like the coalescence of a saddle point and a vortex, having saddle point trajectories to the left of the singularity and vortex trajectories to the right.

Higher Order Singular Point

This file generates the trajectories in the neighborhood of the higher order singular point as well as displaying the tangent field and the stationary point. Mathematica commands in file: `PlotVectorField`, `ImplicitPlot`, `RGBColor`, `PlotPoints`, `Show`, `Frame -> True`, `FrameTicks`, `Hue`, `Epilog`, `PointSize`, `DisplayFunction -> Identity`, `Point`, `ImageSize`, `AspectRatio`.

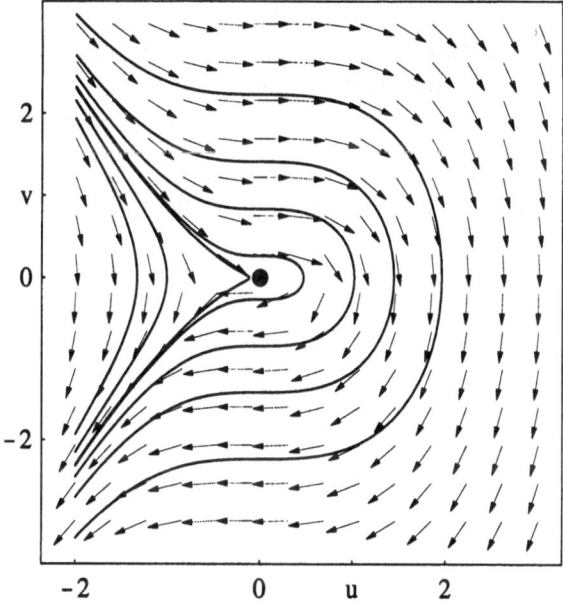

Figure 4.14: Phase plane portrait near the higher order singular point.

PROBLEMS

In the following problem set it is suggested that you use Mathematica to carry out part or all of each problem, depending on the complexity of the singular point analysis and whether tangent field and trajectory plots are requested.

Problem 4-4: Van der Pol equation

Locate all the singular points for the Van der Pol equation as ϵ is increased from zero and determine the nature of the trajectories near each singularity point. Confirm your analysis with a tangent field plot, including some representative trajectories.

Problem 4-5: An eardrum equation

For the following eardrum equation locate and identify the stationary points.

$$\ddot{x} + x - \frac{1}{2}x^2 = 0.$$

Plot the tangent field and some representative trajectories.

Problem 4-6: Hard and soft springs

Locate and identify all the singular points of the following spring equations. Create a phase plane portrait in each case, showing the tangent field and some reprentative trajectories.

1. Hard spring: $\ddot{x} + \omega_0^2(1 + a^2x^2)x = 0$

2. Soft spring: $\ddot{x} + \omega_0^2(1 - b^2x^2)x = 0$

Problem 4-7: Rapoport's model for the arms race

Discuss in detail the nature of the singular points for Rapoport's model, Equations (2.37), of the arms race between two nations. Confirm your analysis by choosing suitable parameter values and making appropriate plots.

Problem 4-8: Damped simple pendulum

The motion of a simple pendulum with viscous damping is described by

$$\ddot{\theta} + b\dot{\theta} + k\sin\theta = 0$$

with b, $k > 0$. Ascertain the nature of the singular points and comment on them. Determine the bifurcation values of b at which the physical solutions change. Confirm your results with suitable plots.

Problem 4-9: Artificial examples

Consider the following two systems of coupled first-order nonlinear ODEs:

1. $\dot{x} = y + 2y^3, \quad \dot{y} = -x - 2x^3.$

2. $\dot{x} = y + x(x^2 + y^2), \quad \dot{y} = -x + y(x^2 + y^2).$

By the method of analysis of singularities, demonstrate that the origin in each case is either a vortex or a focal point. Show that one case will definitely have a vortex. Actually, both systems of equations can be integrated by eliminating the time in the first system and changing to polar coordinates in the second. Determine the analytic solutions and plot them in the phase plane for constants of your own choosing. Identify the singularities at the origin. Comment on the results.

Problem 4-10: Rabbits and foxes

Locate and identify the singularities for the rabbits–foxes competition equations:

$$\dot{r} = 2r - 0.01rf, \quad \dot{f} = -f + 0.01rf.$$

Does the Poincaré criterion help in this problem? Explain. By making a suitable plot, show that Mathematica is very useful in identifying the singularities.

Problem 4-11: Verhulst equations 1

Consider the following predator–prey equations:

$$\dot{x} = x - ax^2 - bxy, \quad \dot{y} = y - cy^2 + dxy,$$

where x and y are the population densities of the prey and predators respectively. Here a, b, c, d are positive. What do the terms involving a and c represent physically? Find the singular points of this system and determine their nature. Confirm the analysis by plotting the tangent field and some trajectories.

Problem 4-12: Verhulst equations 2

Suppose that in the preceding problem $d = -b$ so that the interaction is disadvantageous to both species. Taking $a = c = 1$, find the singular points of the new system and determine their nature. Confirm by plotting the tangent field and some representative trajectories.

Problem 4-13: Nonlinearly damped pendulum

Consider a simple pendulum immersed in a medium that exerts a force proportional to the square of its velocity and in a direction opposite to the velocity. Let x be the angle of swing measured from the stable equilibrium position. Starting with Newton's second law, analyze the singularities of the resulting equation. Obtain an exact analytic expression for the trajectories in the x–y plane where $y \equiv v/\sqrt{k}$, k being the coefficient of the $\sin x$ terms and $v \equiv \dot{x}$. Plot and discuss the trajectories. Confirm your results by creating a plot with the tangent field and several trajectories present .

Suppose that the pendulum is given an impulse when in the position $x = 0$ so that it acquires the initial angular velocity $\tilde{v}$. If $\tilde{v}$ is sufficiently large, the pendulum will go over the top one or more times before finally oscillating about the stable equilibrium position. Show that the pendulum will execute $(n+1)/2$ full revolutions (n is odd) if $\tilde{v}$ lies in the range $\tilde{v}_n < \tilde{v} < \tilde{v}_{n+2}$ where

$$\tilde{v}_n^2 = \left(\frac{2}{1 + 4c^2} \right) \left(1 + e^{2cn\pi} \right)$$

c being the coefficient of the damping term when the equation is in the form $\ddot{x} + \cdots = 0$.

Problem 4-14: Completing text example

For the text example of the higher order singularity, the analysis was only carried out in the neighborhood of the higher order singular point. Locate and identify any further singular points and produce the complete phase plane portrait using Mathematica.

Problem 4-15: Current-carrying wires

The force F per unit length between two parallel current-carrying wires is

$$F = \frac{\mu_0}{4\pi} \frac{I_1 I_2}{d},$$

where I_1 and I_2 are the currents, d the separation of the wires, and μ_0 the permeability of free space. The force is attractive (repulsive) if the currents are in the same (opposite) direction. In the accompanying figure, the short wire

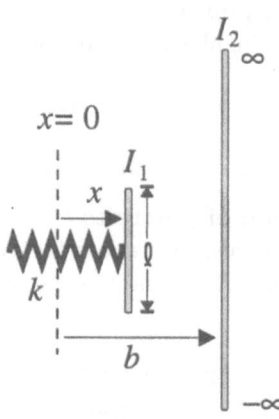

(length ℓ), which is connected to a spring (spring constant k) and is carrying current I_1, is displaced a distance x from equilibrium and allowed to vibrate. The other infinitely long wire carrying current I_2 is fixed in position.

    a. Assuming that the two wires remain parallel, derive the equation of motion for the short wire for parallel and for anti-parallel currents.

    b. For each case in (a), find and identify the singular points.

    c. Confirm your analysis with suitable phase plane portraits

Problem 4-16: Double-well potential

Consider a particle of mass $m = 1$ moving in a double-well potential $V(x) = -\frac{1}{2}x^2 + \frac{1}{4}x^4$.

    a. Derive the equation of motion.

    b. Plot the potential energy and discuss the possible motions.

    c. Find and identify the singular points.

    d. Plot the tangent field and some representative trajectories.

Problem 4-17: Rotating pendulum

The equation of motion for the rotating pendulum (See Chapter 2) is

$$\ddot{\theta} + \omega_0^2 \sin\theta - \frac{1}{2}\omega^2 \sin(2\theta) = 0.$$

Find and identify the stationary points. Confirm your analysis by making a phase plane picture.

Problem 4-18: Spherical pendulum

The equation of motion for the spherical pendulum (See Chapter 2) is

$$\ddot{\theta} - C\frac{\cos\theta}{\sin^3\theta} + \omega_0^2 \sin\theta = 0$$

where C is a positive constant related to a component of the angular momentum. Find and identify all stationary points. Create a phase plane picture.

Problem 4-19: Four fixed points

Locate and identify the four singular points of the system

$$\dot{x} = 2\cos x - \cos y, \quad \dot{y} = 2\cos y - \cos x.$$

Plot the tangent field and some representative trajectories.

Problem 4-20: A collection of portraits

For each of the following systems, locate and identify all singular points, and plot the tangent field and some representative trajectories.

    a. $\dot{x} = 2xy, \quad \dot{y} = y^2 - x^2,$

    b. $\dot{x} = y + y^2, \quad \dot{y} = -\frac{1}{2}x + \frac{1}{5}y - xy + \frac{6}{5}y^2,$

    c. $\dot{x} = y + y^2, \quad \dot{y} = -x + \frac{1}{5}y - xy + \frac{6}{5}y^2.$

Problem 4-21: Another collection
For each of the following three systems, locate and identify all singular points, and plot the tangent field and some representative trajectories.

    a. $\dot{x} = y(1 - x^2)$, $\dot{y} = 1 - y^2$,

    b. $\dot{x} = \sin y$, $\dot{y} = \sin x$,

    c. $\ddot{x} + (\dot{x})^2 + x = 3$.

Problem 4-22: Squid and herring
The major food source for squid is herring. If S and H are the numbers of squid and herring, respectively, per acre of seabed, the interaction between the two species can be modeled [Sco87] by the system (with time in years)

$$\dot{H} = k_1 H - k_2 H^2 - k_3 HS, \quad \dot{S} = -k_4 S - k_5 S^2 + bk_3 HS,$$

with $k_1 = 1.1$, $k_2 = 10^{-5}$, $k_3 = 10^{-3}$, $k_4 = 0.9$, $k_5 = 10^{-4}$, and $b = 0.02$.

    a. Locate and identify all stationary points of the squid–herring system.

    b. Plot the tangent field showing all the stationary points and include some representative trajectories.

    c. Discuss possible outcomes for different ranges of initial populations.

    d. Suppose that every last squid were removed from the area occupied by the herring and from all surrounding areas. Would the herring population increase indefinitely without bound or would there be an upper limit on the number of herring per unit area? If you believe the latter would occur, what is that number?

    e. If the squid free situation just described had persisted for many years, how many squid would there be two years later if a pair of fertile squid is introduced into the area?

Problem 4-23: Tenured engineering faculty [Sco87]
A particular engineering faculty is made up of x untenured and y tenured professors. Suppose that the engineering school has the following policy. Each year a number of new untenured professors are hired equal to 10% of the entire engineering faculty. Also 10% of the untenured professors are given tenure and 10% are thrown out each year. Historically, for this engineering school, 5% of the tenured professors retire or leave each year.

    a. Write down the rate equations for the untenured and tenured professors.

    b. Locate and identify all stationary points.

    c. Create a phase plane portrait with two trajectories starting at $x(0) = 30$, $y(0) = 10$ and $x(0) = 10$, $y(0) = 20$. Take $t = 0..30$. Show the tangent field on the same plot. Comment on the long time behavior shown in the plot.

    d. In the long run, tenured professor growth will be what percentage greater than untenured professor growth?

    e. Suggest a simple variation which would make this model nonlinear and then analyze your model.

4.5 Bifurcations

In doing the problems of the previous section, the reader will have noticed that in some cases the character of the stationary points changes as one or more parameters are varied. As the stationary points change, then so do the solutions as reflected by the changing nature of the trajectories in the phase plane. The value of the parameter at which the topological change occurs is called the bifurcation value or point. Mathematicians have classified the bifurcation values according to the change in topological behavior which is observed. We shall now briefly discuss some of the standard classifications. The mathematically inclined reader should consult either Verhulst [Ver90] or Strogatz [Str94] for a more detailed account.

a. **Saddle–Node Bifurcation:** Consider the simple system

$$\dot{x} = \epsilon - x^2, \qquad \dot{y} = -y \qquad (4.48)$$

with the single parameter ϵ. For $\epsilon > 0$, there are two stationary points, $(\sqrt{\epsilon}, 0)$ and $(-\sqrt{\epsilon}, 0)$. Carrying out the standard singular point analysis, for the first singular point we have $a = -2\sqrt{\epsilon}$, $b = 0$, $c = 0$, $d = -1$, $p = -(a + d) = 1 + 2\sqrt{\epsilon} > 0$, $q = ad - bc = 2\sqrt{\epsilon} > 0$, and $p^2 - 4q = (1 - 2\sqrt{\epsilon})^2 > 0$. Thus, referring to Figure 4.8, the stationary point $(\sqrt{\epsilon}, 0)$ is identified as a stable nodal point. For the second stationary point, $(-\sqrt{\epsilon}, 0)$, $q = -2\sqrt{\epsilon} < 0$ so it is an unstable saddle point. As ϵ is decreased towards zero, the stationary points move towards each other, coalescing into a higher-order singular point (since then $q = 0$) when $\epsilon = 0$. For $\epsilon < 0$, there are no real stationary points. The critical value $\epsilon = 0$ is an example of a saddle–node bifurcation point. As ϵ increases through this value, a saddle point and node are created for $\epsilon > 0$.

b. **Transcritical Bifurcation:** Consider the system

$$\dot{x} = x(\epsilon - x), \qquad \dot{y} = -y \qquad (4.49)$$

which has two stationary points, $(0, 0)$ and $(\epsilon, 0)$. For the 1st stationary point, we can identify $a = \epsilon$, $b = 0$, $c = 0$, $d = -1$, $p = 1 - \epsilon$, $q = -\epsilon$, and $p^2 - 4q = (1 + \epsilon)^2$. For $\epsilon < 0$, then $q > 0$ and $p > 0$, so the stationary point is stable. Indeed, since $p^2 - 4q > 0$, it is a stable nodal point. On the other hand, for $\epsilon > 0$, we have $q < 0$, so the stationary point loses its stability, becoming a saddle point.

For the 2nd stationary point, $a = -\epsilon$, $b = 0$, $c = 0$, $d = -1$, $p = 1 + \epsilon$, $q = \epsilon$, and $p^2 - 4q = (\epsilon - 1)^2$. For $\epsilon < 0$, $q < 0$ so this singular point is unstable, being a saddle point. For $\epsilon > 0$, we have $q > 0$ and $p > 0$, so the singular point is stable.

As the parameter ϵ passes through zero, the two stationary points exchange stability. The critical value $\epsilon = 0$ is an example of a transcritical bifurcation point.

c. **Pitchfork Bifurcation:** Consider the system

$$\dot{x} = x(\epsilon - x^2), \qquad \dot{y} = -y \qquad (4.50)$$

which has one stationary point, $(0,0)$, for $\epsilon < 0$ and three stationary points, $(0,0)$, $(\sqrt{\epsilon},0)$, and $(-\sqrt{\epsilon},0)$ for $\epsilon > 0$. For the singular point, $(0,0)$, one finds that $a = \epsilon$, $b = 0$, $c = 0$, $d = -1$, $q = -\epsilon$, $p = 1 - \epsilon$, and $p^2 - 4q = (\epsilon + 1)^2$. For $\epsilon < 0$, then $q > 0$ and $p > 0$, so the stationary point is stable. For $\epsilon > 0$, we have $q < 0$, so the stationary point is an unstable saddle point.

For both singular points $(\sqrt{\epsilon},0)$ and $(-\sqrt{\epsilon},0)$, which only exist for $\epsilon > 0$, one obtains $a = -2\epsilon$, $b = 0$, $c = 0$, $d = -1$, $q = 2\epsilon$, $p = 2\epsilon + 1$, and $p^2 - 4q = (2\epsilon - 1)^2$. For $\epsilon > 0$, one has $q > 0$ and $p > 0$, so they are both stable. Thus, as ϵ increases from a negative value and passes through zero, the stationary point $(0,0)$ loses its stability and two additional symmetrically located stable stationary points are born. Since the x-coordinate of the singular point $(0,0)$ is zero for all $\epsilon < 0$, and the x-coordinates of the other two stationary points are $\sqrt{\epsilon}$ and $-\sqrt{\epsilon}$ for all $\epsilon > 0$, a sketch of these x coordinates vs. ϵ for the stable branches qualitatively resembles the handle and two symmetric prongs of a pitchfork. For this reason, the critical value $\epsilon = 0$ is an example of a pitchfork bifurcation. More precisely, it is referred to as a "supercritical" pitchfork bifurcation. For a "subcritical" bifurcation, which occurs if the term $x(\epsilon - x^2)$ in the x equation is replaced with $x(\epsilon + x^2)$, the two prongs exist for $\epsilon < 0$.

d. Hopf Bifurcation: Consider the Van der Pol equation

$$\ddot{x} - \epsilon(1 - x^2)\dot{x} + x = 0 \tag{4.51}$$

with the parameter ϵ allowed to be negative, zero, or positive. There is only one stationary point, namely the origin $(x = 0, y = 0)$. If you did the relevant problem, you would have found that for this stationary point $a = 0$, $b = 1$, $c = -1$, $d = \epsilon$, $q = 1$, $p = -\epsilon$, and $p^2 - 4q = \epsilon^2 - 4$. Since $q > 0$, the origin is a stable focal or nodal point for $\epsilon < 0$ and an unstable focal or nodal point for $\epsilon > 0$. In the neighborhood of the origin, as ϵ is increased from a negative value through zero, the phase plane trajectory changes from that characteristic of an oscillatory solution decaying to the origin to one associated with a growing oscillatory solution which eventually winds onto a limit cycle. The Van der Pol equation is said to have undergone a Hopf bifurcation at $\epsilon = 0$, changing from a stable spiral for small negative ϵ to an unstable spiral for small positive ϵ.

PROBLEMS
Problem 4-24: Mathematica confirmation
Make tangent field plots for each of the systems (4.48), (4.49), and (4.50) for appropriate values of ϵ and confirm the bifurcation analysis given in the text.

Problem 4-25: Hopf bifurcation
Consider the system

$$\dot{x} = \epsilon x - y + xy^2, \qquad \dot{y} = x + \epsilon y + y^3$$

   a. Locate and identify the stationary point of this system.

   b. Show that a Hopf bifurcation occurs as the real parameter ϵ is varied. What is the critical value of ϵ?

   c. Make tangent field plots for ϵ close to the bifurcation value and confirm your analysis.

Problem 4-26: Pitchfork bifurcation

Consider the following coupled system,

$$\dot{x} = \epsilon x + y + \sin x, \qquad \dot{y} = x - y.$$

   a. Locate and identify the singular points of this system.

   b. Show that a pitchfork bifurcation occurs at the origin.

   c. Determine the critical or bifurcation value ϵ_c.

   d. Is the bifurcation supercritical? Explain.

   e. Make a tangent field plot for an ϵ value slightly larger than ϵ_c.

Problem 4-27: A genetic control system

Griffith [Gri71] has discussed a genetic control system modeled by the dimensionless equations

$$\dot{x} = -\alpha x + y, \qquad \dot{y} = \frac{x^2}{1 + x^2} - \beta y$$

with α and β both positive. Here x is proportional to the concentration of a certain protein and y is proportional to the concentration of messenger RNA from which it is translated. The activity of a particular gene is directly proportional to the concentration of protein present.

   a. Show that below a critical value α_c, there are three stationary points. Locate and identify these points.

   b. Determine α_c.

   c. Show that two of the stationary points coalesce in a saddle–node bifurcation at $\alpha = \alpha_c$.

   d. Make a tangent field plot for $\alpha < \alpha_c$ and discuss the result.

Problem 4-28: Biased Van der Pol oscillator

For the biased Van der Pol equation,

$$\ddot{x} - \epsilon(1 - x^2)\dot{x} + x = c,$$

with c a constant, determine the curves of ϵ versus c at which a Hopf bifurcation occurs. Confirm your analysis by plotting the tangent field and some representative trajectories for $\epsilon = 1$ and varying c.

4.6 Isoclines

As has been demonstrated, the appropriate Mathematica command allows us to plot the tangent field corresponding to the exact slope relation

$$\frac{dy}{dx} = \frac{Q(x,y)}{P(x,y)} = f(x,y). \tag{4.52}$$

If one is forced to construct the tangent fields by hand, as was the case in the precomputer days, the method of isoclines can be used. In this method, set $f(x,y) = $ const. $= C$. Then construct curves (isoclines) corresponding to a few different choices of C. Along each of these curves, the slope is C. Draw equally spaced line segments ("hash marks") with appropriate slope along the various isoclines. Although the modern computer approach will be used throughout this text, it is instructive to illustrate an example of the isocline method using the Van der Pol equation

$$\ddot{x} - \epsilon(1 - x^2)\dot{x} + x = 0 \tag{4.53}$$

for $\epsilon = 1$. Introducing $y = \dot{x}$, then

$$\frac{dy}{dx} = \frac{(1 - x^2)y - x}{y}. \tag{4.54}$$

The isoclines are given by

$$\frac{dy}{dx} = \frac{(1 - x^2)y - x}{y} = C \tag{4.55}$$

or

$$(1 - x^2)y - x = Cy. \tag{4.56}$$

Choosing $C = 1$, corresponding to isoclines of slope $+1$ or $45\,°$, Equation (4.56) reduces to

$$x(1 + xy) = 0 \tag{4.57}$$

so that $x = 0$ and $xy = -1$, the latter being a rectangular hyperbola. These $C = 1$ isoclines are shown in Figure 4.15. The rectangular hyperbola has two branches. On each branch draw equally spaced hash marks of slope $+1$ as well as on the y axis. Next, choose $C = 0$ corresponding to slope 0. Then

$$y = \frac{x}{1 - x^2} \tag{4.58}$$

which is also plotted in Figure 4.15 with slope zero hash marks added. From (4.55), $C = \infty$ corresponds for $x \neq 0$ to $y = 0$. One can already get some preliminary idea of how the trajectories look. With a little imagination, you may already be able to visualize a spiral unwinding from the origin. As more isoclines are drawn, the complete tangent field would emerge. This precomputer approach to drawing the complete tangent field and deducing the shape of the limit cycle is extremely tedious.

In Figure 4.16 the tedious hand-drawing procedure has been avoided by using the computer to generate the isoclines for $C = -5, -3, -1, 1$, and 3. The

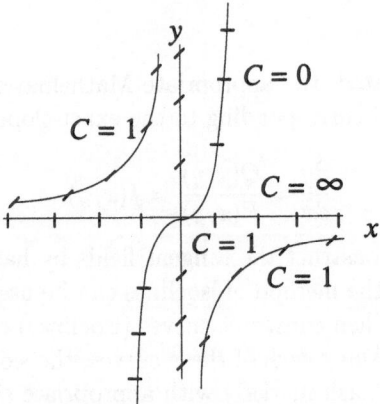

Figure 4.15: Some isoclines for the Van der Pol equation.

tangent field is also plotted. You should be able to determine by looking at the isoclines and the tangent field which isoclines correspond to which C values. The tangent field shows the sense of the trajectories in the phase plane. A trajectory starting at the origin spirals outward from the origin, a result which is consistent with the singular point analysis of Problem 4-4. As $t \to +\infty$, the spiral evolves into a closed trajectory (heavy curve). This behavior may be confirmed by going back to the limit cycle Problem 2-38 (page 74) and redoing it with $\epsilon = 1$. This closed trajectory however differs from the vortex trajectory. The latter occurs physically for conservative systems, and there is a continuum of such trajectories

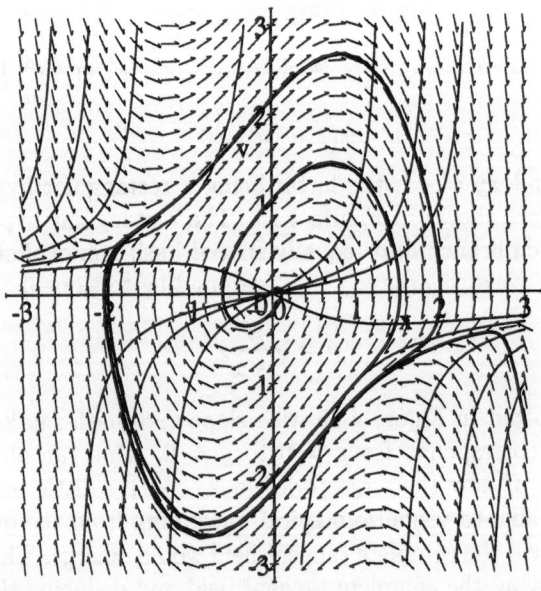

Figure 4.16: Isoclines (thin solid lines), tangent field (arrows), and approach of two trajectories (heavy solid lines) to the Van der Pol limit cycle for $\epsilon = 1$.

surrounding a vortex point. In certain non-conservative systems (of which the Van der Pol equation is an example), there exists a particular closed trajectory, the limit cycle, which is determined only by the properties of the equation itself and is independent of how the oscillation began. A stable limit cycle is another example of an attractor. As ϵ is increased, the shape of the Van der Pol limit cycle is altered substantially.

PROBLEMS
Problem 4-29: Do by hand at your own peril!
Take $\epsilon = 5$ and use the isocline method to determine the shape of the Van der Pol limit cycle. Compare your result with that obtained in the limit cycle Problem 2-38 on page 74. How does the amount of effort compare?

4.7   3-Dimensional Nonlinear Systems

The analysis of the general autonomous 3-dimensional nonlinear $(P, Q, R$ are arbitrary nonlinear functions) system,

$$\dot{x} = P(x, y, z), \quad \dot{y} = Q(x, y, z), \quad \dot{z} = R(x, y, z), \qquad (4.59)$$

can be tackled in a manner similar to the 2-dimensional case but is much too complicated to present here. Although the location and stability of the singular points is easily achieved, the identification of their nature is nontrivial [Jac90, Hay64].

Instead of discussing the general case, we shall look briefly at a specific example, the Lorenz equations,

$$\dot{x} = \sigma(y - x)$$

$$\dot{y} = rx - y - xz \qquad (4.60)$$

$$\dot{z} = xy - bz$$

with σ, r, b positive and x, y, z real. To find the stationary or singular points, x_0, y_0, z_0, set $\dot{x}$, $\dot{y}$, $\dot{z} = 0$ in the Lorenz equations. From the first equation, $x_0 = y_0$. Then, the second equation yields $x_0(r - 1 - z_0) = 0$ so that either $x_0 = 0$ or $z_0 = r - 1$. From the third equation, $x_0 = y_0 = 0$ means that $z_0 = 0$, while for $z_0 = r - 1$ we have $x_0 = y_0 = \pm\sqrt{b(r - 1)}$. For $r < 1$, the latter is imaginary and must be rejected. So for $r < 1$, the only stationary point is $(0, 0, 0)$ while for $r > 1$ there are three stationary points, namely $(0, 0, 0)$, $(\sqrt{b(r - 1)}, \sqrt{b(r - 1)}, r - 1)$, and $(-\sqrt{b(r - 1)}, -\sqrt{b(r - 1)}, r - 1)$. The parameter r is another example of a bifurcation parameter and $r = 1$ is the bifurcation value or point. As one increases r through one, the nature of the solutions changes. In physical terms, $r < 1$ corresponds to heat conduction in the Lorenz model and $r > 1$ to heat convection.

For $r < 1$, set $x = 0 + u$, $y = 0 + v$, $z = 0 + w$ in the Lorenz equations and linearize in u, v, w so that

$$\dot{u} = \sigma(v - u)$$

$$\dot{v} = ru - v \tag{4.61}$$

$$\dot{w} = -bw.$$

Assuming solutions of the form $e^{\lambda t}$, the coupled u, v equations yield the two roots $\lambda_{1,2} = -(1/2)(\sigma + 1) \pm (1/2)\sqrt{(\sigma + 1)^2 - 4\sigma(1 - r)}$, while the uncoupled w equation gives the 3rd root $\lambda_3 = -b$. For $r < 1$, $\lambda_{1,2}$ are both real and negative. Since λ_3 is also real and negative, the origin is a stable singular point. All nearby solutions go to zero as $t \to \infty$. This singular point is identifiable as a 3-dimensional nodal point, which may be confirmed by using Mathematica.

For $r > 1$, the linear analysis in the neighborhood of $(0, 0, 0)$ is identical to that for $r < 1$, so the same expressions for the λ roots result. For $r > 1$, all three roots are real with λ_2 and λ_3 negative, but λ_1 positive. So the origin is now an unstable point. For the other pair of stationary points, the following cubic equation in λ results:

$$\lambda^3 + (\sigma + b + 1)\lambda^2 + b(\sigma + r)\lambda + 2(r - 1)b\sigma = 0. \tag{4.62}$$

We shall not attempt to analyze the cubic equation for general values of the parameters. Choosing Lorenz's "standard" values $\sigma = 10$ and $b = 8/3$, the cubic equation can be solved using Mathematica. There is a bifurcation point at $r = \bar{r} = 24.73684....$ For $r < \bar{r}$, all three roots of the cubic equation have negative real parts so that the corresponding two stationary points are stable. For $r > \bar{r}$, one root is real and negative but the other two roots have positive real parts so the corresponding two singular points are unstable. In file MF10, the value of r was chosen to be 28. In this case the three singular points are all unstable and are at $(0, 0, 0)$, $(6\sqrt{2} \simeq 8.49, 6\sqrt{2}, 27)$ and $(-6\sqrt{2}, -6\sqrt{2}, 27)$. The reader should go back to MF10 and try to locate the latter two singular points in relation to the "wings" of the "butterfly" trajectory.

In closing, it should be noted that Mathematica can plot a vector field in 3 dimensions indicating the directions of the trajectories but, without the actual trajectories present, the configuration of arrows can often turn out to be quite confusing. In MF16, an illustration of the superposition of the butterfly trajectory on the 3-dimensional vector field is given for the Lorenz equations.

Butterfly Trajectory and 3-Dimensional Tangent Field

Using the `PlotVectorField3D` command, a three-dimensional tangent field is plotted for the Lorenz system. To guide the eye, the butterfly trajectory is superimposed on the vector field. Mathematica commands in file: `NDSolve`, `Graphics`, `Block`, `ParametricPlot3D`, `PlotVectorField3D`, `Evaluate`, `PlotPoints`, `AxesLabel`, `AxesEdge`, `StyleForm`, `PlotLabel`, `Background`, `Show`, `ViewPoint`, `VectorHeads`, `ScaleFunction`, `ScaleFactor`

Although one can debate the visual effectiveness of creating a 3-dimensional tangent field, there is no doubt that Mathematica is extremely useful in carrying out the stability analysis for 3-dimensional systems, as demonstrated in the following example for the Lorenz system. The techniques illustrated here can be applied even to 4-dimensional systems such as found in the hyperchaos Problem 4-37 at the end of this chapter.

Example 4-5: Stability Analysis for the Lorenz System

Derive the cubic Equation (4.62) for λ. For $\sigma = 10$, $b = \frac{8}{3}$, and $r = 28$, numerically solve the cubic equation for the three roots. Confirm the location of the 2nd stationary point and its stability.

Solution: The right-hand sides of the three Lorenz equations are entered with a more computationally convenient index notation, x[1], x[2], x[3] being used to denote the variables x, y, and z.

```
Clear["Global`*"]

eq[1] = σ (x[2] - x[1]);

eq[2] = r x[1] - x[2] - x[1] x[3];

eq[3] = x[1] x[2] - b x[3];
```

Setting the right-hand sides of the three equations equal to zero, the singular points are determined using the **Table** and **Solve** commands.

```
singpoints = Solve[Table[eq[i] == 0, {i, 3}], Table[x[i], {i, 3}]]
```

$$\{\{x(2) \to 0,\ x(3) \to 0,\ x(1) \to 0\},$$
$$\{x(2) \to -\sqrt{br - b},\ x(3) \to r - 1,\ x(1) \to -\sqrt{br - b}\},$$
$$\{x(2) \to \sqrt{br - b},\ x(3) \to r - 1,\ x(1) \to \sqrt{br - b}\}\}$$

The singular point coordinates agree with the values previously obtained by hand. We now concentrate on the third solution (which is the 2nd stationary point in the earlier discussion) and select it in the following input line.

```
sp3 = singpoints[[3]]
```

$$\{x(2) \to \sqrt{br - b},\ x(3) \to r - 1,\ x(1) \to \sqrt{br - b}\}$$

A function is created to differentiate eq[i] with respect to x[j],

```
a[i_, j_] := D[eq[i], x[j]]
```

and the Jacobian matrix A of coefficients in the linearized equations formed with the **Table** command. The coordinate values obtained in sp3 are substituted into the result.

```
A = Table[a[i, j], {i, 3}, {j, 3}] /. sp3
```

$$\begin{pmatrix} -\sigma & \sigma & 0 \\ 1 & -1 & -\sqrt{br-b} \\ \sqrt{br-b} & \sqrt{br-b} & -b \end{pmatrix}$$

The determinant command, Det, is used to obtain the characteristic polynomial cp in λ from the matrix A. The command IdentityMatrix[3] creates a 3 by 3 identity matrix with 1s along the diagonal and all off-diagonal elements equal to zero.

```
cp = Det[A - λ IdentityMatrix[3]]
```

$$-\lambda^3 - b\lambda^2 - \sigma\lambda^2 - \lambda^2 - br\lambda - b\sigma\lambda + 2b\sigma - 2br\sigma$$

Equation (4.62) follows on collecting with respect to λ, b, and σ in -cp.

```
cp2 = Collect[-cp, {λ, b, σ}] == 0
```

$$\lambda^3 + (b + \sigma + 1)\,\lambda^2 + b\,(r+\sigma)\,\lambda + b\,(2r-2)\,\sigma == 0$$

In order to numerically solve the cubic λ equation, the parameter values are entered.

```
σ = 10;  b = 8/3;  r = 28;
```

The characteristic polynomial then takes the form given in the output of cp2,

```
cp2
```

$$\lambda^3 + \frac{41\lambda^2}{3} + \frac{304\lambda}{3} + 1440 == 0$$

which is numerically solved for the roots.

```
lambdavalues = NSolve[cp2]
```

$$\{\{\lambda \to -13.8546\}, \{\lambda \to 0.0939556 - 10.1945\,I\}, \{\lambda \to 0.0939556 + 10.1945\,I\}\}$$

The roots can alternately be obtained from the matrix A by applying the Eigenvalues command.

```
lambdavalues2 = Eigenvalues[A] // N
```

$$\{-13.8546,\ 0.0939556 + 10.1945\,I,\ 0.0939556 - 10.1945\,I\}$$

The first root is real and negative, while the other two are complex with positive real parts. By Lyapunov's theorem, the stationary point is unstable. Finally, the location of the 2nd stationary point is confirmed for the given parameters.

sp3

$$\{x(2) \rightarrow 6\sqrt{2},\ x(3) \rightarrow 27,\ x(1) \rightarrow 6\sqrt{2}\}$$

The coordinates are just those of the 2nd stationary point in the text discussion.

$\boxed{\textbf{End Example 4-5}}$

PROBLEMS
Problem 4-30: Bifurcation type
For the Lorenz system, what type of bifurcation occurs at $r = 1$?

Problem 4-31: Bifurcation point
For $\sigma = 10$ and $b = 8/3$, show that $\bar{r} = 24.73684$ is indeed a bifurcation point for the Lorenz system, with the nature of the roots changing at this r value.

Problem 4-32: Butterfly trajectory
In the analysis of the Lorenz equations, the reader was asked to locate two of the singular points in relation to the wings of the butterfly trajectory. By choosing initial conditions near the singular points and taking the time run to be short, identify the nature of the singular points.

Problem 4-33: The Oregonator
Consider the Oregonator system

$$\epsilon\dot{x} = x + y - qx^2 - xy, \quad \dot{y} = -y + 2hz - xy, \quad p\dot{z} = x - z,$$

with $\epsilon = 0.03$, $p = 2$, $q = 0.006$, $h = 0.75$. Show analytically that the origin is an unstable singular point. Locate any other stationary points and determine their stability.

Problem 4-34: Magnetic field reversal
To account for the geologically observed changes and reversals of the earth's magnetic field as deduced from magnetized rock strips on the earth's ocean floor, a modified disc dynamo model of the earth's interior has been proposed

[CH80]. When the relevant magnetohydrodynamic equations are truncated, the following system of equations is obtained:

$$\dot{x} = a(y - x), \quad \dot{y} = zx - y, \quad \dot{z} = b - xy - cz,$$

with a, b, c positive real parameters and $b > ac(a + c + 3)/(a - 1 - c)$. Here, x is related to a poloidal potential, y to a toroidal magnetic field component, and z to the moment of angular momentum.

Taking $a = 3$, $b = 25$, and $c = 1$, locate the singular points and determine their stability. Using Mathematica, explore the solutions of this system for different initial values. Relate the singular points to your plots. Qualitatively, relate the behavior of $y(t)$ to reversals of the earth's magnetic field.

Problem 4-35: The Rössler system
Consider the Rössler system

$$\dot{x} = -(y + z), \quad \dot{y} = x + ay, \quad \dot{z} = b + z(x - c),$$

with $a, b, c > 0$. Analytically show that there are no singular points for $c < \sqrt{4ab}$ and two singular points for $c > \sqrt{4ab}$. Find analytic expressions for the latter points. Linearizing the system in the vicinity of these singular points, find the cubic equation for the roots λ. Taking $a = b = 0.2$ and allowing c to take on the values $c = 2.4, 3.5, 4.0, 5.0, 8.0$, solve the cubic equation numerically for the roots and determine the stability in each case.

Taking $x(0) = 0.1$, $y(0) = 0.1$, $z(0) = 0$, show by calculating the trajectories numerically and plotting them in the x–y plane that each c value leads to a qualitatively different behavior. Identify the behavior by plotting $y(t)$.

Problem 4-36: Two predators and a single prey
Consider the 3-species system

$$\dot{x} = ax - xy - xz, \quad \dot{y} = -by + xy, \quad \dot{z} = -cz + xz,$$

with $a, b, c > 0$. Which dependent variables refer to the predators and which to the prey? Find the stationary points of this system and determine their stability. Explore this system numerically by choosing different initial values $x(0)$, $y(0)$, $z(0)$ and identify the nature of the singular points. Is it possible to have all nonzero initial values and have one predator vanish? Illustrate your answer with an appropriate trajectory in 3-dimensional phase space, remembering that x, y, z cannot be negative.

Problem 4-37: Hyperchaos
Rössler [Rö79] has studied the 4-dimensional system

$$\dot{x} = -y - z, \quad \dot{y} = x + 0.25y + r, \quad \dot{z} = 3 + xz, \quad \dot{r} = -0.5z + 0.05r,$$

which can display solutions more irregular than chaos (hyperchaos). Locate the stationary points and determine their stability. Explore the solutions of this system, plotting the trajectories in x–y–z space.

Problem 4-38: Multiple stationary points
The following 3-dimensional system

$$\dot{x} = (1 - z)[(4 - z^2)(x^2 + y^2 - 2x + y) + 4(2x - y) - 4],$$

$$\dot{y} = (1 - z)[(4 - z^2)(xy - x - zy) + 4(x + zy) - 2z],$$

$$\dot{z} = z^2(4 - z^2)(x^2 + y^2),$$

has several stationary points.

a. Locate all of the stationary points.

b. Determine the stability of each stationary point.

c. Explore the solutions of this system by making a 3-d x–y–z plot.

Problem 4-39: Owls

Let x, y, and z be the population numbers for tawny owls, long-eared owls, and little owls, respectively which co-exist in some forest. The population equations are given by

$$\dot{x} = x(1 - a_1 x) - x(a_2 y + a_3 z),$$

$$\dot{y} = y(1 - b_1 y) + y(b_2 x - b_3 z),$$

$$\dot{z} = z(1 - c_1 z) + z(c_2 x + c_3 y)$$

with all coefficients positive.

a. Interpret the structure of each equation.

b. Locate the singular points of this system.

c. Discuss the stability of these points.

Problem 4-40: More multiple singular points

Consider the coupled system of equations

$$\dot{x} = x - xy - y^3 + z(x^2 + y^2 - 1 - x + xy + y^3),$$

$$\dot{y} = x - z(x - y + 2xy),$$

$$\dot{z} = (z - 1)(z + 2zy^2 + z^3)$$

which has several singular points.

a. Locate all of the stationary points.

b. Determine the stability of each stationary point.

c. Explore the nature of the solutions to this system by making suitable 3-dimensional plots.

Problem 4-41: The Bombay plague

In a pioneering work in epidemiology, Kermack and McKendrick [KM27] developed the SIR model to describe the data from the Bombay plague of 1906. The acronym SIR stands for the three groups of people in the model:

- the susceptible (S) group who are healthy but can become infected,

- the infected (I) group who have the disease and are able to infect healthy individuals,

- the removed (R) group who are either dead or have recovered and are permanently immune.

The SIR system of equations is given by

$$\dot{S} = -\alpha SI, \quad \dot{I} = \alpha SI - \beta I, \quad \dot{R} = \beta I$$

with α and β positive.

a. Setting $u \equiv S/S_0$, $v \equiv I/S_0$, $w \equiv R/S_0$, $\tau \equiv \beta t$, $b \equiv \alpha S_0/\beta$, with $S_0 \equiv S(0)$, show that the SIR system may be rewritten in the normalized form

$$\dot{u}(\tau) = -buv, \quad \dot{v}(\tau) = buv - v, \quad \dot{w}(\tau) = v.$$

b. Show that there is a conservation of total number $N = S + I + R$ which can be re-expressed as $u + v + w = a$, with the constant a to be determined. What is the range of a? The conservation of total number reflects the assumption that the epidemic evolves so rapidly that slower changes in the total population number due to births, deaths due to other causes, emigration, etc., can be ignored.

c. Show that solving the normalized SIR system can be reduced to solving the single first-order nonlinear ODE

$$\dot{w}(\tau) = a - w(\tau) - e^{-bw(\tau)}.$$

This equation must be solved numerically. Relate $u(\tau)$ and $v(\tau)$ to $w(\tau)$.

d. Confirm that the maximum in $\dot{w}(\tau)$ occurs at the same time as the maximum in $v(\tau)$. This time is referred to as the "peak" time τ_{peak} of the epidemic. At $\tau = \tau_{peak}$, there are more sick (infected) people and a higher daily death rate than at any other time.

e. Show that the value of w at τ_{peak} is given by $w = \ln(b)/b$. Confirm that for $b > 1$, $\tau_{peak} > 0$ while for $b < 1$, $\tau_{peak} = 0$. The condition $b = 1$ is referred to as the "threshold" for an epidemic, an epidemic situation occuring if $b > 1$. In an epidemic, things get worse before they get better.

f. Take $u(0) = 1$, $v(0) = 0.001$, $w(0) = 0$, and $b = 2$. What is the stationary value of $w(\tau)$? Numerically solve the first-order nonlinear ODE for $w(\tau)$ and show that w approaches this stationary value. What fraction of the susceptible group survive the epidemic in this case?

g. For the same parameter values as above, numerically solve the normalized SIR system of equations and make 3-dimensional plots of u vs. v vs. w and τ vs. v vs. w. For the latter choose an orientation which shows the maximum in $v(\tau)$. What is the numerical value of τ_{peak} in this case?

Chapter 5

Analytic Methods

Chaos often breeds life, when order breeds habit.
Henry B. Adams, American Historian (1838-1918)

5.1   Introductory Remarks

The topological approach allows us to easily determine the types of solutions
that a given physically interesting nonlinear ODE system will permit. With
its qualitative nature established, the next step is to ascertain whether a given
solution can be described by some sort of analytic expression.

By Jove, Watson,
by careful analysis
we should be able
to solve the mystery
of who made these
footprints.

   In this chapter, a few methods of generating exact (when they exist) and
approximate analytical solutions will be developed. The bulk of this chapter
will be devoted to the latter because there are relatively few nonlinear differential equations of physical interest that are exactly solvable. It follows therefore
that Mathematica's capability of exactly solving nonlinear ODEs is limited. In
MF01 we saw Mathematica fail to tell us that the solution to the simple pendulum equation with a specified initial condition is a Jacobian elliptic function.

That the solution has this form will be shown in the next section. More commonly, however, an explicit analytic solution simply doesn't exist and we are forced to be satisfied with some approximate expression. Most methods, e.g., perturbation theory, of generating approximate solutions are, by their very nature, rather tedious to implement. It is here that the analytic manipulation capability of Mathematica can prove to be particularly valuable. Even if only some of the steps are done with a computer algebra system, this often can save considerable time and more importantly avoid algebraic and sign errors. To this end, we shall show in this chapter how approximate, as well as some exact, analytic solutions may be generated using Mathematica. Some of the code will be slightly more sophisticated and involved than that seen so far, but it is hoped that the Mathematica text examples and files will be clear and that the reader will derive intellectual pleasure from applying them to the task of solving the text problems. Without the aid of the Mathematica system, some of the problems would be quite difficult or very tedious to carry out by hand.

Before proceeding with this chapter on analytic methods, some excellent useful references should be mentioned. The reader who wants to learn more about different analytic and numerical methods for solving ODEs and PDEs should consult the *Handbook of Differential Equations* by Daniel Zwillinger [Zwi89]. As the name implies, this is a handbook which gives a quick synopsis of a given method with usually one or two examples. It is not a textbook, but gives references to more detailed explanations of the topics. If one has a burning desire or serious need to learn more about the properties of special functions such as Bessel functions, elliptic functions, etc., then we recommend consulting the *Handbook of Mathematical Functions* by Milton Abramowitz and Irene A. Stegun [AS72].

5.2   Some Exact Methods

In Mathematica File MF02 the equation of motion for a sphere of mass m falling from rest near the earth's surface and acted on by a nonlinear drag force $F_{\text{drag}} = -Av - Bv^2$ was analytically solved for the velocity v. The DSolve command was applied to the relevant nonlinear equation

$$\dot{v} = g - av - bv^2, \qquad (5.1)$$

where $a = A/m$, $b = B/m$, yielding a general solution involving one arbitrary constant. On evaluating the constant and substituting representative values for the coefficients, a plot of $v(t)$ was created, the plot showing the sphere's approach to the terminal velocity.

Not surprisingly, Mathematica will also analytically solve the above nonlinear ODE with the initial condition included in the DSolve command. This is demonstrated in the following example.

Example 5-1: Nonlinear Drag on a Falling Sphere

Determine the analytic solution $v(t)$ of (5.1) for $v(0) = 0$. Taking the same parameter values, confirm that the solution is identical in form to that in MF02. Plot the $v(t)$ curve with the terminal velocity clearly indicated and labeled.

Solution: Equation (5.1) is entered and labeled as **de**.

```
Clear["Global`*"]
```

```
de = v'[t] == g - a v[t] - b v[t]^2
```

$$v'(t) == g - a v(t) - b v(t)^2$$

The ODE **de** is analytically solved for the function v, subject to $v(0) = 0$.

```
vel = DSolve[{de, v[0] == 0}, v, t]
```

The lengthy output, which has been suppressed here in the text, yields four functional forms and a warning message that inverse functions are being used, so some solutions may not be found. After some experimentation, the fourth solution is chosen and simplified by assuming that a, b, and g are all positive. This assumption is entered in the following **Simplify** command as an option with the logical "and" function (&&).

```
vel2 = Simplify[v[t] /. vel[[4]], a > 0 && b > 0 && g > 0]
```

$$\frac{\sqrt{a^2 + 4bg} \tanh\left(\frac{1}{2}\sqrt{a^2 + 4bg}\,t - I \cos^{-1}\left(\frac{1}{2\sqrt{\frac{bg}{a^2+4bg}}}\right)\right) - a}{2b}$$

The analytical form of $v(t)$ is given by the output of **vel2**. Since the answer must be real, the imaginary factor must ultimately vanish. To see that this does happen, the same values are entered for the parameters as in MF02,

```
a = .1; b = .05; g = 9.8;
```

and **vel2** then simplified.

```
vel3 = Simplify[vel2]
```

$$14.0357 \tanh(0.701783\,t + (0.071368 + 0.\,I)) - 1.$$

The coefficient of I should be identically zero, but is not precisely zero because a finite number of digits are being used. The **Chop** command is used to replace the coefficient of I by the exact integer 0. Chop uses a default tolerance of 10^{-10}.

```
vel4 = Chop[vel3]
```

$$14.0357 \tanh(0.701783\,t + 0.071368) - 1.$$

The output expression generated in **vel4** is identical with that obtained in MF02. Although the terminal velocity is easily extracted by hand, let's use Mathematica's **Limit** command to do the job.

```
terminalvelocity = Limit[vel4, t -> ∞]
```

13.0357

The terminal velocity is slightly more than 13 m/s. A horizontal line corresponding to the terminal velocity and the analytic result vel4 are now plotted together in Figure 5.1 using the Plot command. The Epilog option is used to add the words "terminal velocity" to the graph just above the horizontal line.

```
Plot[{terminalvelocity, vel4}, {t,0,5}, PlotStyle -> Hue[.6],
PlotRange -> {{0,5}, {0,15}}, ImageSize -> {600,400},
Ticks -> {{{.01,"0"},1,2,{2.5,"t"},3,4,5},{5,{7.5,"v"},10,15}},
Epilog -> Text["terminal velocity", {1,13.5}],
TextStyle -> {FontFamily -> "Times",FontSize -> 16}];
```

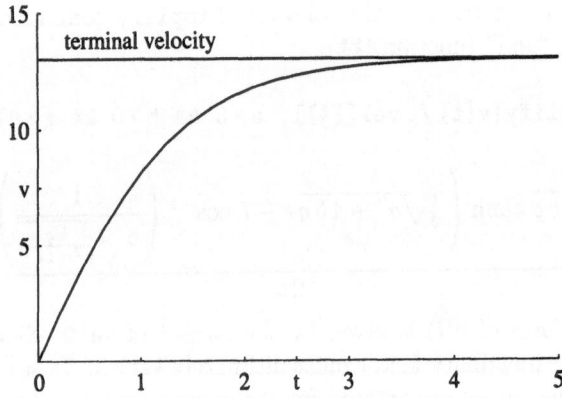

Figure 5.1: Velocity as a function of time for the falling sphere.

The falling sphere effectively reaches the terminal velocity in about four seconds.

End Example 5-1

Since Mathematica was successful in solving Equation (5.1), its method of attack must be based on known nonlinear ODE solving techniques. What methods are available and who cares as long as Mathematica gives us an answer? As to the former, we shall provide one simple method in the next subsection. As to why bother, remember that the primary goal of this text is to introduce the reader to the concepts and methods of nonlinear physics, using Mathematica as an *auxiliary tool*. Also, the DSolve command is not always successful even if an analytic answer exists. This was the case in MF01. So it is necessary to know some of the analytic methods that are available, so that we can guide Mathematica in the ODE solving process when necessary. We shall not present all methods for analytically solving nonlinear ODEs in this section, just highlight a few of the more important ones. We begin by examining the separation of variables method for solving first-order nonlinear ODEs.

5.2.1 Separation of Variables

A first-order ODE of the structure

$$dy/dx = f(x)/g(y) \qquad (5.2)$$

is trivially solved, even if f and g are not linear in x and y respectively, by cross-multiplying and integrating:

$$\int g(y)dy = \int f(x)dx + \text{constant}. \qquad (5.3)$$

An obvious candidate for the variable separation method is the nonlinear drag equation (5.1). Assuming that $v = 0$ at $t = 0$ and $v = V$ at $t > 0$, we separate variables and obtain

$$\int_0^V \frac{dv}{(1 - \frac{a}{g}v - \frac{b}{g}v^2)} = gt. \qquad (5.4)$$

The lhs of (5.4) may now be analytically integrated and the resulting equation solved for $V(t)$. Although this can be done by hand by consulting integral tables, we prefer to use Mathematica as in the following example.

Example 5-2: Solving for V

Perform the integration and solve (5.4) for the velocity V at time t. Taking $g = 9.8$ m/s^2, $a = 0.1$ s^{-1}, and $b = 0.05$ m^{-1}, confirm that this solution is the same as in MF02 and Example 5-1.

Solution: Equation (5.4) is entered and the integration on the lhs performed.

```
Clear["Global`*"]
```

```
eq = Integrate[1/(1 - (a/g) v - (b/g) v^2), {v, 0, V}] == g t
```

$$\frac{2g \tan^{-1}(\frac{a}{\sqrt{-a^2-4bg}})}{\sqrt{-a^2 - 4bg}} - \frac{2g \tan^{-1}(\frac{a+2bV}{\sqrt{-a^2-4bg}})}{\sqrt{-a^2 - 4bg}} == g t$$

Then, **eq** is analytically solved for V,

```
vel = Solve[eq, V]
```

$$\{\{V \to \frac{\sqrt{-a^2 - 4bg} \tan(\frac{1}{2}(2 \tan^{-1}(\frac{a}{\sqrt{-a^2-4bg}}) - \sqrt{-a^2 - 4bg}\, t)) - a}{2b}\}\}$$

and the result simplified with the **FullSimplify** command, subject to a, b, and g all being positive.

```
vel2 = FullSimplify[V /. vel[[1]], a > 0 && b > 0 && g > 0]
```

$$\frac{\sqrt{a^2 + 4bg} \tanh(\frac{1}{2}\sqrt{a^2 + 4bg}\, t + \tanh^{-1}(\frac{a}{\sqrt{a^2+4bg}})) - a}{2b}$$

The parameter values are entered,

    g = 9.8; a = 0.1; b = 0.05;

and the Expand command applied to vel2.

    vel3 = V == Expand[vel2]

$$V == 14.0357 \tanh(0.701783\,t + 0.071368) - 1$$

The analytic form of V is the same as in MF02 and Example 5-1.

End Example 5-2

The variable separation procedure may also be applied to equations of the form

$$\frac{dy}{dx} = \frac{f(x,y)}{g(x,y)} \tag{5.5}$$

provided that $f(x,y)$ and $g(x,y)$ are homogeneous polynomials of x and y of the same degree n, i.e., every term in $f(x,y)$ or $g(x,y)$ is of the form $x^m y^{n-m}$ with $m = 0, 1, \ldots n$. A separable equation follows by setting $y(x) = z(x)x$.

Nonlinear equations that are separable and for which the resulting integrals have analytic answers could be directly solved using Mathematica's analytic ODE solver. Nevertheless, we would like the reader to do the following problems without using DSolve, unless otherwise indicated, so that the basic variable separation technique is well understood.

PROBLEMS
Problem 5-1: A mathematical example
Consider the equation

$$\frac{dy}{dx} = \frac{2x^3 y - y^4}{x^4 - 2xy^3}.$$

Assuming $y(x) = z(x)x$, find the differential equation for $z(x)$. Separate variables and integrate to show that the general solution is $x(1 + z^3) = Cz$, where C is an arbitrary constant. Substitute $z(x) = y/x$ and simplify to find the final implicit solution.

Problem 5-2: An alternate approach
Solve Problem 5.1 explicitly for $y(x)$ using the DSolve command. Because the explicit form is "ugly", also obtain the implicit general solution of the previous problem by performing appropriate Mathematica manipulations.

Problem 5-3: Newton's law of resistance
A particle is projected vertically upward with an initial speed v_0 near the earth's surface. Show that if Newton's law of resistance applies, then the speed of the particle when it returns to its initial position is

$$v_0 v_{\text{term}} / \sqrt{v_0^2 + v_{\text{term}}^2}$$

where v_{term} is the terminal speed.

Problem 5-4: Nonlinear friction
A small block of mass m slides down an inclined plane (angle θ with horizontal) under the pull of gravity. If $F_{\text{drag}} = -kmv^2$, show that the time T required to move a distance D from rest is

$$T = \frac{\cosh^{-1}\left[e^{kD}\right]}{\sqrt{kg\sin\theta}}.$$

Hint: $\operatorname{arctanh}(x) = \operatorname{arccosh}\left(\frac{1}{\sqrt{1-x^2}}\right)$.

Problem 5-5: Linear pursuit
An unarmed merchant ship is being pursued by a dastardly pirate vessel. The merchant ship is moving vertically along the straight line $x = a$ of the following figure. The pirate ship moves in such a manner that its direction of motion is always towards the merchant craft. Both vessels are traveling at constant speed, the speed of the pirate being n times that of the merchant ship.

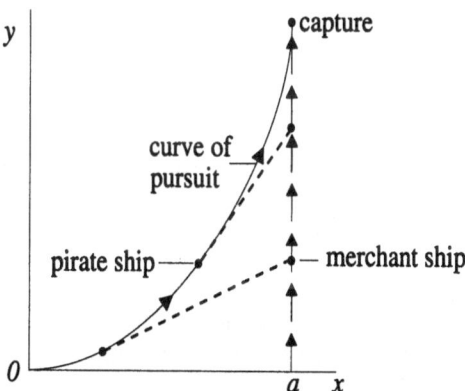

a. If the merchant ship is initially at $x = a$, $y = 0$ and the pirate vessel at the origin, show that the "curve of pursuit" for the pirate to capture the merchant ship is determined by the second-order nonlinear equation

$$n^2(a - x)^2\left(\frac{d^2y}{dx^2}\right)^2 - \left(\frac{dy}{dx}\right)^2 - 1 = 0. \tag{5.6}$$

b. For $n = 2$, prove that the merchant ship will be captured at $y = 2a/3$. [Hint: To solve the nonlinear equation, set $p = dy/dx$, separate variables, and use the DSolve command on the first-order equation.]

This problem, referred to in the literature as the problem of linear pursuit, was first proposed and solved by the French hydrographer Pierre Bougner in 1732.

Problem 5-6: A beagle gives chase
A loyal beagle, named Patches, is out in a field when she spots her beloved mistress Heather walking along a straight road. Patches runs towards Heather in such a way as to always aim at her. With distances in km, Patches is initially at $(x = 1, y = 0)$, Heather at $(0,0)$, the road is described by the straight line equation $x = 0$, and the ratio of Heather's speed to the dog's speed is r.

a. Show that Patches' path is described by the nonlinear ODE

$$x\frac{d^2y}{dx^2} = r\sqrt{1 + \left(\frac{dy}{dx}\right)^2}.$$

b. Show that the solution of the ODE, subject to the initial conditions, is

$$y = \frac{1}{2}\left(\frac{x^{1+r}}{1+r} - \frac{x^{1-r}}{1-r}\right) + \frac{r}{1-r^2}.$$

c. If Heather is walking at 3 km/h and Patches runs at 8 km/h, how many minutes does it take the dog to reach her mistress? Plot Patches' path.

Problem 5-7: Minimum surface of revolution
In the calculus of variations, it is proved that the form of the function $y(x)$ that minimizes or maximizes the integral

$$I = \int_a^b F(x, y(x), y'(x))\, dx,$$

with $y' \equiv dy/dx$, may be found by solving the Euler–Lagrange equation

$$\frac{\partial F}{\partial y} - \frac{d}{dx}\frac{\partial F}{\partial y'} = 0,$$

subject to the boundary conditions at the end points a and b. In most physical problems it is usually obvious whether a minimum or a maximum has been achieved.

Consider a surface of revolution generated by revolving a curve $y(x)$, stretching between the points $(0, y_0)$ and (x_1, y_1), about the x-axis.

a. Show that the surface area A is given by

$$A = 2\pi \int_0^{x_1} y\sqrt{(1 + (y')^2)}\, dx.$$

b. Identifying F and substituting it into the Euler–Lagrange equation, show that $y(x)$ satisfies the nonlinear ODE

$$yy'' - (y')^2 = 1.$$

c. Setting $p = dy/dx$, separating variables, and integrating, show that the analytic form of $y(x)$ that minimizes the area A is $y(x) = \frac{1}{a}\cosh(ax + b)$, where a, b are determined from the boundary conditions.

d. Taking $y_0 = 1$, $x_1 = 1$, and $y_1 = 2$, plot the curve $y(x)$.

e. Plot the 3-dimensional surface obtained by rotating $y(x)$ about the x-axis.

5.2.2 The Bernoulli Equation

Recall the laser beam competition equations, Equation (2.33), for light beams traveling in the same direction through an absorbing fluid,

$$\frac{dI_\mathrm{L}}{dz} = -gI_\mathrm{L}I_\mathrm{s} - \alpha I_\mathrm{L}$$
$$\frac{dI_\mathrm{s}}{dz} = +gI_\mathrm{L}I_\mathrm{s} - \alpha I_\mathrm{s}. \tag{5.7}$$

Adding these two equations yields

$$\frac{d}{dz}\left[I_\mathrm{L}(z) + I_\mathrm{s}(z)\right] = -\alpha(I_\mathrm{L} + I_\mathrm{s}) \tag{5.8}$$

which is readily integrated to give

$$I_\mathrm{L}(z) + I_\mathrm{s}(z) = \left[I_\mathrm{L}(z=0) + I_\mathrm{s}(z=0)\right]e^{-\alpha z} \equiv I_0 e^{-\alpha z}. \tag{5.9}$$

If the absorption coefficient $\alpha = 0$, then $I_\mathrm{L} + I_\mathrm{s}$ is constant corresponding to energy conservation for the two light beams. For $\alpha \neq 0$, the medium absorbs energy from the light beams according to the well-known exponential decay law (5.9). Using Equation (5.9), I_L can be eliminated from the second equation in (5.7) resulting in the equation

$$\frac{dI_\mathrm{s}}{dz} + \left[\alpha - gI_0 e^{-\alpha z}\right]I_\mathrm{s} = -gI_\mathrm{s}^2. \tag{5.10}$$

This is a special case of the Bernoulli equation

$$\frac{dy}{dz} + f_1(z)y = f_2(z)y^n \tag{5.11}$$

with $y \equiv I_\mathrm{s}$, $f_1(z) \equiv \alpha - gI_0 e^{-\alpha z}$, $f_2(z) \equiv -g$ and $n = 2$. Although Bernoulli's equation is nonlinear, it can be converted into a linear differential equation by a simple change of dependent variable, viz., by introducing $p = 1/y^{n-1}$. For example, for our laser equation (5.10), since $n = 2$, we have $p = 1/I_\mathrm{s}$. Then, $dI_\mathrm{s}/dz = -(1/p^2)dp/dz$ and Equation (5.10) may be rewritten as

$$-\frac{1}{p^2}\frac{dp}{dz} + \left[\alpha - gI_0 e^{-\alpha z}\right]\frac{1}{p} = -\frac{g}{p^2} \tag{5.12}$$

or, on multiplying through by $-p^2$,

$$\frac{dp}{dz} + \left[gI_0 e^{-\alpha z} - \alpha\right]p = g \tag{5.13}$$

a first-order linear ODE which is solved according to a standard mathematics procedure. Labeling the coefficient of p as $f(z)$, one multiplies both sides of (5.13) by the integrating factor $\exp(\int f(z)\,dz)$ and rewrites the equation as

$$\frac{d}{dz}\left(e^{\int f(z)\,dz}p\right) = ge^{\int f(z)\,dz}. \tag{5.14}$$

Equation (5.14) may be integrated to find $p(z)$ and then $I_s(z) = 1/p(z)$ is the desired solution. The detailed derivation of the analytic form is left as a problem which we are asking the student to carry out and compare with the answer found in file MF17. Not surprisingly, the original Bernoulli equation (5.10) is directly analytically solvable with Mathematica. This is also left as a problem

---

Laser Beam Competition

Equation (5.13) is analytically solved with the `DSolve` command and applied to the `Laser beam competition` problem given below. Mathematica commands in file: `DSolve, Plot, Evaluate, PlotStyle, Thickness, PlotLabel, Block, $DisplayFunction=Identity, AxesLabel, Hue, ImageSize, Background`

PROBLEMS

Problem 5-8: Laser beam competition
Solve Equation (5.13) for p by carrying out the step (5.14) and inverting the result to obtain $I_s(z)$. Plot $\ln(I_s(z)/I_s(0))$ for $0 \le z \le 10$ cm given $I_s(0)/I_L(0) = 0.01$, $gI_L(0) = 1$ cm^{-1} for the two cases $\alpha = 0$ and 0.5 cm^{-1}. Discuss and compare the two cases. Check your answer against file MF17.

Problem 5-9: Bernoulli equation
Solve the Bernoulli equation (5.10) using the Mathematica `DSolve` command and plot $I_s(z)$ from $z = 0$ to 15 for $I_0 = 1.01$, $I_s(0) = 0.01$, $g = 1$, and $\alpha = 0.01$.

Problem 5-10: Nonlinear diode circuit
Consider a linear capacitor C connected in series with a diode as shown. Suppose

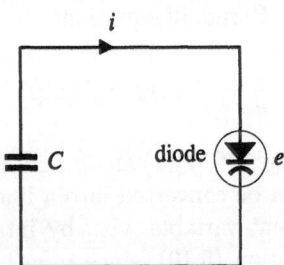

that the relation between the instantaneous values of the current i and the voltage e of this diode can be written as $i = ae + be^2$ where a, b are positive constants. At $t = 0$, the voltage across the capacitor is E. Show that $e(t)$ satisfies Bernoulli's equation. Then solve for $e(t)$ and plot (e/E) versus (at/C) for $(bE/a) = 1/2$.

Problem 5-11: Logistic equation
In 1838, Verhulst suggested the "logistic equation" (with $a, b > 0$)

$$\dot{N} = aN(1 - bN)$$

as a model for the growth of the human population. Eliminate a and b by normalizing the variables. Analytically solve the resulting equation by (a) recognizing that it is a Bernoulli equation, (b) separating variables, (c) using Mathematica's `DSolve` command.

5.2.3 The Riccati Equation

For $N = 2$, the quasi-species evolutionary Equations (2.30) can be reduced to the Riccati equation, i.e., for $N = 2$ we have

$$\dot{X}_1 = W_1 X_1 - \left(\frac{E_1 X_1 + E_2 X_2}{n} \right) X_1 + \phi_{12} X_2$$
$$\dot{X}_2 = W_2 X_2 - \left(\frac{E_1 X_1 + E_2 X_2}{n} \right) X_2 + \phi_{21} X_1$$

$$(5.15)$$

subject to the constraint $X_1 + X_2 = n$. Using this condition, X_2 can be eliminated from Equation (5.15) resulting in an equation for X_1 alone:

$$\dot{X}_1 + \frac{1}{n} [E_1 - E_2] X_1^2 + [E_2 - W_1 + \phi_{12}] X_1 - n\phi_{12} = 0. \qquad (5.16)$$

This is a special case of the Riccati equation which has the general form

$$\frac{dy}{dt} + ay^2 + f_1(t)y + f_2(t) = 0 \qquad (5.17)$$

where a is a constant. The Riccati equation may be reduced to a linear equation by a change of dependent variable, viz., introducing

$$z = e^{a \int_0^t y \, dt}. \qquad (5.18)$$

It follows from (5.18) that

$$\frac{dz}{dt} = ayz, \quad \text{or} \quad y = \frac{1}{az} \frac{dz}{dt}. \qquad (5.19)$$

Then

$$\frac{dy}{dt} = \frac{1}{az} \frac{d^2 z}{dt^2} - \frac{1}{az^2} \left(\frac{dz}{dt} \right)^2 = \frac{1}{az} \frac{d^2 z}{dt^2} - ay^2. \qquad (5.20)$$

Substituting (5.20) into Equation (5.17) yields the second-order linear ODE

$$\frac{d^2 z}{dt^2} + f_1(t) \frac{dz}{dt} + af_2(t)z = 0. \qquad (5.21)$$

In our biological example, Equation (5.16), the coefficients are assumed to be constants so the resulting second-order linear equation is easily solved. One assumes that $z = \exp(\lambda t)$. Substitution into (5.21) yields a quadratic equation for λ, i.e., two roots λ_1 and λ_2. The general solution is then a linear combination of $\exp(\lambda_1 t)$ and $\exp(\lambda_2 t)$. More generally, f_1, f_2 in Equation (5.21) are functions of t. If the student is lucky, the equation might still have an exact analytic solution, as in the following example. Consider the differential equation

$$\frac{dy}{dx} + ay^2 + \frac{1}{x} y + \frac{1}{a} = 0. \qquad (5.22)$$

Making the standard substitution (5.18), Equation (5.22) reduces to

$$\frac{d^2 z}{dx^2} + \frac{1}{x} \frac{dz}{dx} + z = 0, \qquad (5.23)$$

which can be identified as a zeroth order Bessel equation with solution

$$z = C_1 J_0(x) + C_2 Y_0(x). \tag{5.24}$$

The solution of (5.22) is then

$$y = \frac{1}{az}\frac{dz}{dx} = \frac{C_1 J_0'(x) + C_2 Y_0'(x)}{a[C_1 J_0(x) + C_2 Y_0(x)]} = -\frac{1}{a}\left[\frac{C_1 J_1(x) + C_2 Y_1(x)}{C_1 J_0(x) + C_2 Y_0(x)}\right] \tag{5.25}$$

or

$$y = \frac{-[C_3 J_1(x) + Y_1(x)]}{a[C_3 J_0(x) + Y_0(x)]}, \tag{5.26}$$

with $C_3 \equiv C_1/C_2$. The solution of the first-order ODE (5.22) can have only one arbitrary constant. This solution can also be derived directly from Equation (5.22) with Mathematica's DSolve command.

Example 5-3: Riccati Example

Analytically solve the Riccati equation (5.22) using the DSolve command.

Solution: The relevant equation is entered and given the name ode.

```
Clear["Global`*"]

ode = y'[x] + a y[x]^2 + y[x]/x + 1/a == 0
```

$$\frac{1}{a} + \frac{y(x)}{x} + a y(x)^2 + y'(x) == 0$$

Applying DSolve to ode,

```
sol = DSolve[ode, y, x]
```

$$\left\{\left\{y \to \text{Function}\left[\{x\}, \frac{-Y_1(x) - J_1(x)\,c_1}{a\,(Y_0(x) + J_0(x)\,c_1)}\right]\right\}\right\}$$

```
y = y[x] /. sol[[1]]
```

$$\frac{-Y_1(x) - J_1(x)\,c_1}{a\,(Y_0(x) + J_0(x)\,c_1)}$$

yields the solution $y(x)$, with a single arbitrary constant c_1. The answer is the same as was obtained in the hand derivation.

End Example 5-3

Although Mathematica was able to directly solve still another nonlinear ODE, we are running out of physical systems for which this is true.

PROBLEMS
Problem 5-12: Quasi-species Riccati equation
Show that Equation (5.16) can be solved with the DSolve command.

Problem 5-13: Another Riccati problem

An equidimensional linear differential equation

$$x^n \frac{d^n z}{dx^n} + b_1 x^{n-1} \frac{d^{n-1} z}{dx^{n-1}} + \cdots + b_{n-1} x \frac{dz}{dx} + b_n z = h(x),$$

where the b's are constants, can be reduced to a linear differential equation with constant coefficients by making the substitution $x = \exp(t)$. Use this approach to solve the equation

$$x^2 \frac{dy}{dx} + xy + x^2 y^2 = 1.$$

Problem 5-14: Mathematica approach to the Riccati problem

Solve the previous problem using Mathematica's `DSolve` command, expressing the solution as a ratio of two simple polynomials. The `TrigToExp` and `Simplify` commands could be useful.

5.2.4 Equations of the Structure $d^2 y/dx^2 = f(y)$

Equations of this structure can arise in mechanical problems formulated in terms of Newton's second law. Of course, in this case, the independent variable will be the time t. The simple plane pendulum and hard spring equations, viz.,

$$\ddot{\theta} = -\omega_0^2 \sin \theta$$

$$\ddot{x} = -\omega^2 (1 + a^2 x^2) x \tag{5.27}$$

are two obvious examples. To solve

$$\frac{d^2 y}{dx^2} = f(y) \tag{5.28}$$

we multiply the equation by $2(dy/dx)\, dx = 2\, dy$ and integrate, viz.,

$$\int 2 \frac{dy}{dx} \frac{d^2 y}{dx^2}\, dx = \int \frac{d}{dx} \left(\frac{dy}{dx} \right)^2 dx = \int 2 f(y)\, dy + C \tag{5.29}$$

where C is a constant, or

$$\left(\frac{dy}{dx} \right)^2 = 2 \int f(y)\, dy + C \tag{5.30}$$

or

$$\frac{dy}{dx} = \left[2 \int f(y)\, dy + C \right]^{1/2} \equiv g(y). \tag{5.31}$$

In mechanical problems, where x is replaced with t, Equation (5.30) is just a statement about the conservation of energy with C setting the total energy. Continuing with the calculation, y is determined as a function of x from

$$\int \frac{dy}{g(y)} = x + \text{constant} \tag{5.32}$$

provided that the integration can be carried out. The last constant merely shifts $y(x)$ along the x-axis.

The Simple Pendulum Solution

As our first example, let us solve the simple plane pendulum equation

$$\ddot{\theta} = -\omega_0^2 \sin\theta \tag{5.33}$$

using the above procedure. Multiplying (5.33) by $2\dot{\theta}\,dt$ and integrating yields

$$\dot{\theta}^2 = 2\omega_0^2 \cos\theta + C \tag{5.34}$$

where $\dot{\theta}$ is the angular velocity. Assuming that the pendulum is swinging back

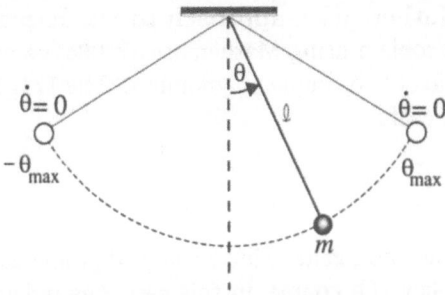

Figure 5.2: Oscillations of a simple pendulum.

and forth (as in Figure 5.2) between some maximum angle of deflection $\theta_{\max}$ (in magnitude), which implies $\dot{\theta} = 0$ at $\theta = \pm\theta_{\max}$ since the mass m is momentarily at rest there, one obtains $C = -2\omega_0^2 \cos(\theta_{\max})$. Then Equation (5.34) becomes

$$\dot{\theta}^2 = 2\omega_0^2(\cos\theta - \cos\theta_{\max}) = 4\omega_0^2 \left[\sin^2\left(\frac{1}{2}\theta_{\max}\right) - \sin^2\left(\frac{1}{2}\theta\right)\right] \tag{5.35}$$

on using the trigonometric identity $\cos\theta = 1 - 2\sin^2(\theta/2)$.

It is convenient to set $k \equiv \sin\left(\frac{1}{2}\theta_{\max}\right)$ and introduce a new angular variable ϕ (which has no direct geometric meaning in Figure 5.2) through the relation

$$\sin\left(\frac{1}{2}\theta\right) = k\sin\phi. \tag{5.36}$$

Differentiating the last line, squaring, and substituting for $\dot{\theta}^2$, we obtain

$$\dot{\phi}^2 = \omega_0^2 \cos^2\left(\frac{1}{2}\theta\right) = \omega_0^2\left(1 - \sin^2\left(\frac{1}{2}\theta\right)\right) = \omega_0^2\left[1 - k^2\sin^2\phi\right]. \tag{5.37}$$

Then, taking the positive square root to give the pendulum a positive (i.e., counterclockwise, according to the usual convention) angular velocity at $t = 0$, we have

$$\dot{\phi} = \omega_0\sqrt{\left(1 - k^2\sin^2\phi\right)}. \tag{5.38}$$

On separating variables and integrating, Equation (5.38) yields

$$\omega_0(t - t_0) = \int_{\phi_0}^{\phi} \frac{d\phi'}{\sqrt{\left(1 - k^2\sin^2\phi'\right)}} \tag{5.39}$$

where ϕ_0 is the value of ϕ at some instant t_0. We now choose $\phi_0 = 0$ at $t_0 = 0$ which (from Equation (5.36)) corresponds to having $\theta = 0$ at time $t = 0$. With this choice Equation (5.39) becomes

$$w_0 t = u \equiv F(\phi, k) \tag{5.40}$$

where

$$F(\phi, k) = \int_0^\phi \frac{d\phi'}{\sqrt{(1 - k^2 \sin^2 \phi')}} \tag{5.41}$$

is called the elliptic integral of the first kind.[1] The quantity k is called the modulus ($k = \text{mod } u$) of the integral, while ϕ is referred to as the amplitude ($\phi = \text{am } u$) of the integral. The elliptic integral of the first kind[2] may be evaluated numerically using the Mathematica command `EllipticF[ϕ,m]` with ϕ in radians and $m \equiv k^2$ and the appropriate values of the arguments being inserted. Thus, for example, $F(\phi = \pi/6, k = 0.5) = 0.529429$ to 6-figure accuracy on noting that $m = (0.5)^2 = 0.25$ and using `EllipticF[Pi/6, 0.25]`.

The solution described by Equations (5.40) and (5.41) shall be discussed in a moment. Let us first derive the period T of the simple pendulum. Recalling the relation

$$\sin \left(\frac{1}{2}\theta \right) = \sin \left(\frac{1}{2}\theta_{\text{max}} \right) \sin \phi \tag{5.42}$$

we can see that as θ increases from 0 to θ_{max}, ϕ must increase from 0 to $\pi/2$. Since this takes place in a quarter of a period, the period T must be given by

$$T = \frac{4}{w_0} F \left(\frac{\pi}{2}, k \right) \equiv \frac{4}{w_0} K(k) \tag{5.43}$$

where

$$K(k) \equiv \int_0^{\pi/2} \frac{d\phi'}{\sqrt{(1 - k^2 \sin^2 \phi')}} \tag{5.44}$$

is called the complete elliptic integral of the first kind. $K(k)$ can be evaluated numerically with the Mathematica command `EllipticK[m]`, again with $m \equiv k^2$. Then, for example, $K(0.5) = 1.68575$. For $k < 1$ (i.e., $\theta_{\text{max}} < \pi$), the integrand of (5.44) can be Taylor expanded and integrated term by term using Mathematica, the result being

$$K(k) = \int_0^{\pi/2} \left[1 + \frac{1}{2}k^2 \sin^2 \phi' + \cdots \right] d\phi' = \frac{\pi}{2} \left[1 + \frac{k^2}{4} + \frac{9}{64}k^4 + \cdots \right]. \tag{5.45}$$

Setting $T_0 = 2\pi/w_0$ and $k = \sin(\frac{1}{2}\theta_{\text{max}})$, (5.43) then becomes

$$\frac{T}{T_0} = 1 + \frac{1}{4}\sin^2 \left(\frac{1}{2}\theta_{\text{max}} \right) + \frac{9}{64}\sin^4 \left(\frac{1}{2}\theta_{\text{max}} \right) + \cdots. \tag{5.46}$$

---

[1]Although it doesn't arise in this physical problem, it should be mentioned that there is an elliptic integral of the second kind $E(\phi, k)$ defined by $E(\phi, k) = \int_0^\phi \sqrt{(1 - k^2 \sin^2 \phi')} \, d\phi'$.

[2]Setting $y = \sin(\phi')$, $F(\phi, k) = \int_0^{\sin(\phi)} \frac{dy}{\sqrt{1-y^2}\sqrt{1-k^2 y^2}}$ is a common alternate form of the elliptic integral.

T_0 is the familiar period that would be obtained in the linear approximation, i.e., for small θ_{max}. When θ is not restricted to small angles, the period becomes increasingly dependent on the angular amplitude θ_{max} as θ_{max} is increased. However, except at angles θ_{max} approaching π, the correction to T_0 is small. For example, for $\theta_{max} = 30°$, the correction is 1.7% and for $\theta_{max} = 60°$, it is about 7%. For $\theta_{max} = \pi$, we physically expect that the period should be infinite. That this is mathematically so is readily verified. For $\theta_{max} = \pi$, $k = 1$ and

$$K(1) = \int_0^{\pi/2} \frac{d\phi'}{\left[1 - \sin^2 \phi'\right]^{\frac{1}{2}}} = \int_0^{\pi/2} \frac{d\phi'}{\cos \phi'}.$$

The latter integral may be evaluated[3] with Mathematica as follows:

```
Limit[Integrate[1/Cos[x], {x, 0, a}], a -> Pi/2]
```

$$\infty$$

The series expression that was written down earlier for $K(k)$ is only valid for $k < 1$. The variation of T/T_0 with θ_{max} up to $\theta_{max} = \pi$ must be obtained from Equation (5.43) and (5.44) and (using MF18) is shown in Figure 5.3.

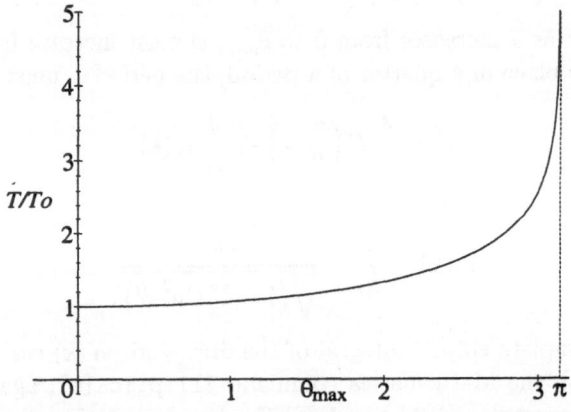

Figure 5.3: Normalized period T/T_0 versus maximum angular amplitude θ_{max}.

Period of the Simple Pendulum
The period of the simple plane pendulum is calculated for various θ_{max} and the plot of Figure 5.3 is created. Mathematica commands in file: Sqrt, EllipticK, Series, Block, $DisplayFunction=Identity, Plot, ListPlot, Show, PlotStyle, PlotJoined->True, RGBColor, PlotRange, TextStyle, Ticks, FontFamily, FontSize, ImageSize

In the next experiment the period of a meter stick, which is an example of a compound pendulum, is measured and compared with the values predicted

---

[3]The command EllipticK[1] yields the answer ComplexInfinity, indicating an infinite magnitude but undetermined complex phase.

theoretically. If the meter stick of length L is pivoted at a distance r from the center of mass, the equation of motion is the same as for the simple pendulum, but with the period given by

$$T = (2\sqrt{3}/3)\sqrt{(L^2 + 12r^2)/rg} \, K(k) \qquad (5.47)$$

where $K(k)$ is the complete elliptic integral. The derivation of the theoretical formula (5.47) is left as a problem.

Period of a Compound Pendulum
In this experiment the period of a meter ($L = 1$ m) stick pivoted at a point a distance r from the center of mass is measured. For varying $\theta_{\text{max}} < \pi$, the experimental results are compared with those predicted by the formula (5.47).

EXP 13

Although our interest in this section is in the nonlinear behavior of the simple pendulum, it should be noted that careful experimental observation of the decreasing oscillations of a simple pendulum for $\theta_{\text{max}} \ll \pi$ can be used to determine the analytic form of the nonlinear drag force due to the surrounding air. This is the subject of the following easy-to-carry-out experiment.

Damped Simple Pendulum
Assuming that the drag force due to the surrounding air on a large simple pendulum bob of mass m moving with speed v is given by $F = -bmv^n$, the decrease in amplitude for small oscillations can be used to determine the values of the exponent n and the constant b.

EXP 14

Let's now continue with our mathematical exploration of the behavior of the pendulum when the angular amplitude is not necessarily small. To express the behavior of θ as a function of time t for the simple pendulum, a new special function, the elliptic function, is introduced by writing

$$\sin\left(\frac{\theta}{2}\right) = k\sin\phi = k\sin(\text{am}(u)) \equiv k\,\text{sn}(u) \qquad (5.48)$$

where sn (u) is called the Jacobian elliptic sine function of the integral u. Since $u = \omega_0 t$ (from Equation (5.40)), then

$$\theta = 2\sin^{-1}(k\,\text{sn}(\omega_0 t)) \qquad (5.49)$$

gives the desired $\theta(t)$ relation for the simple pendulum. This, finally, is the elusive analytic solution that we were unable to find in our first Mathematica File MF01. Of course, to fully appreciate the temporal behavior of θ, we must learn a little bit about elliptic integrals and elliptic functions.

PROBLEMS
Problem 5-15: Period of a meter stick
Derive the period (in the form (5.47)) of a meter stick of length L pivoted at a point a distance r from the center of mass.

Problem 5-16: Evaluating elliptic integrals

Evaluate $F(\phi, k)$ for $\theta_{max} = 170°$ and $\phi = 3\pi/8$. Also calculate the corresponding complete elliptic integral of the first kind as well as the complete elliptic integral of the second kind.

Problem 5-17: Period of the simple pendulum

Plot the Taylor series representation for the normalized period T/T_0 for a large $\theta_{max} < \pi$ and several different orders of approximation. Hint: You can Taylor expand the integrand of (5.44) and derive the series representation of the normalized period to any order.

Problem 5-18: Period of oscillation in an anharmonic potential

Show that the period of oscillation of a particle of mass m in a potential $U = A \mid x \mid^n$ is given by

$$T = (2/n)\sqrt{2\pi m/E}(E/A)^{1/n}\Gamma(1/n)/\Gamma(1/2 + 1/n)$$

where E is the total energy and Γ is the Gamma function. Take $n = 2$ and evaluate the Γ functions using Mathematica. Then show that T reduces to the normal expression for the parabolic potential. Evaluate T for $n = 4$. To learn more about the period of oscillation in an anharmonic potential, run the Mathematica File MF19.

Aspects of the Anharmonic Period

This file explores the period of oscillation of the anharmonic potential for different values of n and different total energies E. Mathematica commands in file: Abs, Sqrt, Pi, Gamma, Block, $DisplayFunction=Identity, Do, Plot, PlotStyle, Hue, Show, Table, PlotRange, Ticks, PlotLabel, Epilog, StyleForm, Text, TextStyle, ImageSize

---

Elliptic Integral of the First Kind

The elliptic integral of the first kind

$$u = F(\phi, k) = \int_0^\phi \frac{d\phi'}{\sqrt{(1 - k^2 \sin^2 \phi')}} \tag{5.50}$$

is usually only tabulated for the range $\phi = 0$ to $\pi/2$. The behavior of $F(\phi, k)$ in this range is plotted with MF20 and shown in Figure 5.4.

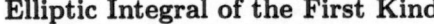

Elliptic Integrals

This file creates three plots, the first showing $u = F(\phi, k)$ versus ϕ for different k values. The second plot is of u versus $\theta_{max}/2$ for different values of ϕ. The final plot produces Figure 5.5. Mathematica commands in file: EllipticF, EllipticK, Table, Block, $DisplayFunction=Identity, Plot, Epilog, PlotStyle, Show, ListPlot, PlotJoined->True, PlotRange, Ticks, TextStyle, RGBColor, ImageSize, Evaluate, Thickness

---

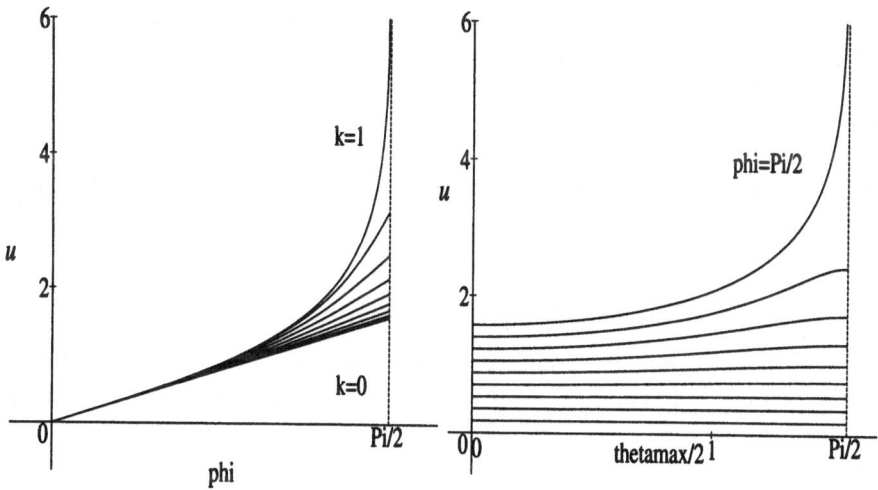

Figure 5.4: First plot: $u = F(\phi, k)$ vs ϕ for different k values; Second plot: u vs $\theta_{\max}/2$ for different ϕ.

The first graph in Figure 5.4 shows u as a function of ϕ for different k values. For $\theta_{\max} = 0$, i.e., $k=0$, F is linear in ϕ, but as $\theta_{\max}$ increases, F deviates more and more from a straight line and asymptotically approaches the vertical at $\phi = \pi/2$ as $\theta_{\max}/2 \to \pi/2$ ($k \to 1$). The second graph shows u as a function of $\theta_{\max}/2$ for $\phi = 0...\pi/2$. Outside the "fundamental" range $\phi = 0$ to $\pi/2$, values of $F(\phi, k)$ may be determined from the relations

$$F(-\phi, k) = -F(\phi, k), \quad F(\phi + n\pi, k) = 2nK(k) + F(\phi, k) \tag{5.51}$$

with $n = 0, 1, 2\ldots$, which the student may easily verify from the definition of $F(\phi, k)$. In Figure 5.5, the second relation in (5.51) has been used in MF20 to

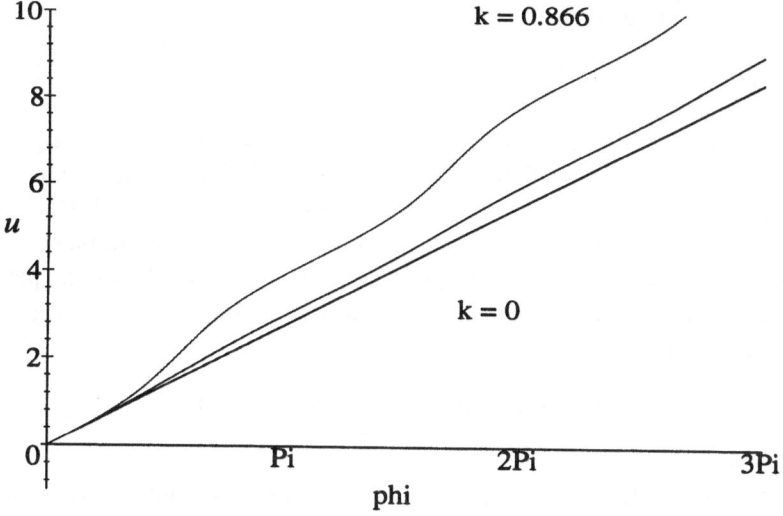

Figure 5.5: $F(\phi, k)$ over the range $\phi = 0...3\pi$ for $k = 0$, 0.5, and 0.866.

plot the elliptic integral over the range $\phi = 0$ to 3π. The curves are for $\theta_{\max} = 0$, $60°$, and $120°$ $(k = \sin(120°/2) = \sin(60°) = 0.866)$.

PROBLEMS
Problem 5-19: Extending the range
Verify the relations (5.51) in the text. Evaluate $F(\phi, k)$ for $\theta_{\max} = 170°$ and $\phi = 2.375\pi$.

Jacobian Elliptic Functions

In analogy with the familiar circular sine and cosine functions, there exist elliptic sine and cosine functions. We have just encountered $\operatorname{sn} u$ in our pendulum problem. The elliptic cosine function $\operatorname{cn} u$ is defined, not surprisingly, by $\operatorname{cn} u = \cos \phi = \cos(\operatorname{am} u)$.

How are the elliptic functions determined? For $k = 0$, $u = F(0, \phi) = \phi$ and therefore $\operatorname{sn} u = \sin u$, $\operatorname{cn} u = \cos u$. As k is increased from zero, the shapes of $\operatorname{sn} u$ and $\operatorname{cn} u$ change. What do they look like? Referring to Figure 5.5, consider the $\theta_{\max} = 120°$ curve $(k = 0.866)$. If we choose, say, $\phi = \phi_1$ the projection off this curve yields the corresponding u value u_1. Taking the sine

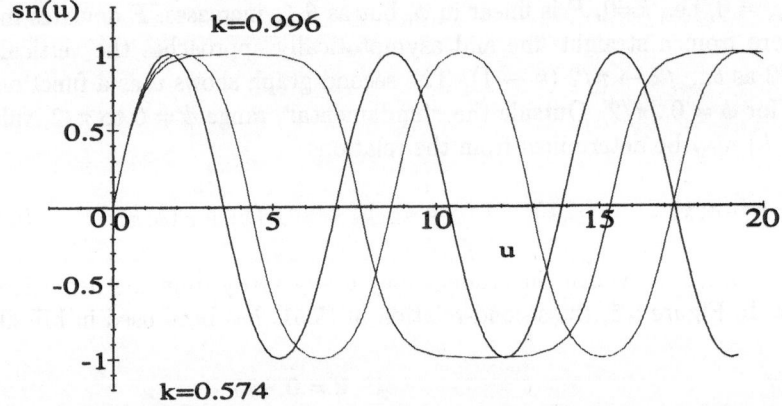

Figure 5.6: $\operatorname{sn} u$ vs u for $k = 0.574$ $(\theta_{\max} = 70°)$, 0.866 $(120°)$, 0.996 $(170°)$.

and cosine of ϕ_1, we have from the basic definitions, $\operatorname{sn}(u_1) = \sin(\phi_1)$ and $\operatorname{cn}(u_1) = \cos(\phi_1)$. Figure 5.6 (produced with MF21) shows $\operatorname{sn} u$ obtained by this procedure for $k = 0.866$ (middle curve) as well as $k = 0.574$ and 0.996. For $k = 1$, $\operatorname{sn} u = \tanh u$ and $\operatorname{cn} u = \operatorname{sech} u$, the proof being left as a problem.

From our earlier discussion of the period of the pendulum, clearly the period of both $\operatorname{sn} u$ and $\operatorname{cn} u$ is $4K(k)$. This conclusion also follows from, e.g., the relation $\operatorname{sn} u = \sin \phi$, since

$$\operatorname{sn}(u + 4K) = \sin(\phi + 2\pi) = \sin \phi = \operatorname{sn} u$$

and similarly for $\operatorname{cn} u$.

The elliptic functions have properties similar to the circular functions, e.g.,

$$\operatorname{cn}^2 u + \operatorname{sn}^2 u = \cos^2 \phi + \sin^2 \phi = 1 \tag{5.52}$$

$$\frac{d}{du}\, \text{sn}\, u = \text{cn}\, u\, \text{dn}\, u \tag{5.53}$$

where dn u is another elliptic function defined to be

$$\text{dn}\, u = \sqrt{(1 - k^2 \sin^2 \phi)} = \sqrt{(1 - k^2 \,\text{sn}^2\, u)}. \tag{5.54}$$

The reader is referred to [AS72] for a complete list of the properties of elliptic functions.

Mathematica may be used to evaluate and manipulate elliptic functions. For example, in MF21 the derivative relation (5.53) is confirmed and the integral of sn u performed. Unlike what we have done here in the text for notational convenience, the second argument k is not suppressed in the Mathematica input command or in the output. For example, the input command for the Jacobian elliptic sine function takes the form JacobiSN[u,m], with $m \equiv k^2$. For, say, $u = 4$ and $k = 0.996$ ($\theta_{\text{max}} = 170°$), JacobiSN[4,0.996^2) yields the value 0.99985.

Exploring the Jacobian Elliptic Functions
This file creates the elliptic sine function sn u for arbitrary k values and then produces Figure 5.6. The analytic derivative and integral of sn u are also carried out. Mathematica commands in file: JacobiSN, Sin, Plot, Evaluate, PlotRange, RGBColor, Epilog, ImageSize, D, Integrate, Simplify

The Hard Spring Solution

As a contrast to our hand calculation for the simple pendulum equation, where one of our priorities was to establish the properties of elliptic integrals and functions, we now show how Mathematica can readily solve the hard spring equation

$$\ddot{x}(\tau) + \left(1 + a^2 x^2(\tau)\right) x(\tau) = 0 \tag{5.55}$$

for $x(0) = A$, $\dot{x}(0) = 0$. Here, $\tau \equiv \omega t$.

Example 5-4: Analytical Hard Spring Solution

Given the initial conditions specified above, derive

   a. the analytic formula for the hard spring period,

   b. the analytic solution, $x(t)$, to the hard spring ODE (5.55).

Solution: The left-hand side of the hard spring ODE is entered

```
Clear["Global`*"]
```

and labeled eq1. The BasicInput palette is used to input τ.

```
eq1 = x''[τ] + (1 + a^2 x[τ]^2) x[τ]
```

$$x(\tau) \left(a^2 x(\tau)^2 + 1\right) + x''(\tau)$$

Then `eq1` is multiplied by the velocity and the result integrated with respect to the (normalized) time τ, yielding the energy (potential plus kinetic). The multiplication operator has been included here to avoid possible confusion.

>   energy = Integrate[eq1*x'[τ], τ]

$$\frac{1}{4}a^2\,x(\tau)^4 + \frac{x(\tau)^2}{2} + \frac{1}{2}x'(\tau)^2$$

The initial amplitude will be taken to be A and the initial velocity equal to zero.

>   amplitude = A; velocity = 0;

Substituting these values into the energy expression will yield the total energy, labeled `e`.

>   e = energy /. {x[τ] -> amplitude, x'[τ] -> velocity}

$$\frac{a^2\,A^4}{4} + \frac{A^2}{2}$$

The complete energy expression is then obtained by equating `energy` to `e`.

>   eq2 = energy == e

$$\frac{1}{4}a^2\,x(\tau)^4 + \frac{x(\tau)^2}{2} + \frac{1}{2}x'(\tau)^2 == \frac{a^2\,A^4}{4} + \frac{A^2}{2}$$

The energy expression `eq2` is solved for the velocity, yielding two solutions in `eq3`. The positive square root is selected in `eq4`.

>   eq3 = Solve[eq2, x'[τ]]

$$\{\{x'(\tau) \to -\sqrt{2}\,\sqrt{\frac{a^2\,A^4}{4} + \frac{A^2}{2} - \frac{a^2\,x(\tau)^4}{4} - \frac{x(\tau)^2}{2}}\},$$

$$\{x'(\tau) \to \sqrt{2}\,\sqrt{\frac{a^2\,A^4}{4} + \frac{A^2}{2} - \frac{a^2\,x(\tau)^4}{4} - \frac{x(\tau)^2}{2}}\}\}$$

>   eq4 = x'[τ] /. eq3[[2]]

$$\sqrt{2}\,\sqrt{\frac{a^2\,A^4}{4} + \frac{A^2}{2} - \frac{a^2\,x(\tau)^4}{4} - \frac{x(\tau)^2}{2}}$$

To obtain the integrand for calculating the period, we replace $x(\tau)$ with a new variable y and form the reciprocal 1/eq4.

>   integrand = (1/eq4) /. x[τ] -> y

$$\frac{1}{\sqrt{2}\,\sqrt{\dfrac{a^2\,A^4}{4} + \dfrac{A^2}{2} - \dfrac{a^2\,y^4}{4} - \dfrac{y^2}{2}}}$$

The period is obtained by integrating the integrand from $y = 0$ to $y = A$ and multiplying by $4/\omega$.

```
period = (4/ω) Integrate[integrand, {y, 0, amplitude}]
```

$$\frac{4\sqrt{\dfrac{2\,a^2\,A^2 + 2}{a^2\,A^2 + 2}}\,K\left(-\dfrac{a^2\,A^2}{a^2\,A^2 + 2}\right)}{\sqrt{\dfrac{1}{A}}\,\sqrt{a^2\,A^3 + A}\,\omega}$$

The analytic result for the period is expressed in terms of the complete elliptic integral K of the first kind. By using the FullSimplify command with the assumptions that $a > 0$ and $A > 0$, the period formula can be simplified.

```
period2 = FullSimplify[period, a > 0 && A > 0]
```

$$\frac{4\,K\left(\dfrac{2}{a^2\,A^2 + 2} - 1\right)}{\sqrt{\dfrac{a^2\,A^2}{2} + 1}\,\omega}$$

As a partial check on the period2 expression, it can be Taylor expanded about $a = 0$ to second order.

```
linearperiod = Series[period2, {a, 0, 2}]
```

$$\frac{2\pi}{\omega} - \frac{3\,A^2\,\pi\,a^2}{4\omega} + O(a)^3$$

The first term in the output is the usual linear expression for the period, the second term being the correction for small a. To obtain $x(t)$, the integrand is now integrated from some point x to $y = A$ and the result equated to $\omega\,t$. Because of its length, the resulting output is suppressed here in the text.

```
eq5 = ω t == Integrate[integrand, {y, x, amplitude}]
```

The output of eq5 is simplified subject to $a > 0$ and $A > 0$ and $x < A$.

```
eq6 = FullSimplify[eq5, a > 0 && A > 0 && x < A]
```

$$t\,\omega == \frac{\sqrt{2}\left(K\left(\dfrac{2}{a^2\,A^2 + 2} - 1\right) - F\left(\sin^{-1}\left(\dfrac{x}{A}\right)\Big|\,\dfrac{2}{a^2\,A^2 + 2} - 1\right)\right)}{\sqrt{a^2\,A^2 + 2}}$$

The implicit form of the solution is given in terms of the elliptic integral F of the first kind. The first argument of F contains the x that we are seeking. To extract an explicit expression for x, eq6 is now solved for x and simplified.

```
eq7 = Solve[eq6, x]
```

```
x = Simplify[x /. eq7[[1]]]
```

$$A \,\text{JacobiSN} \left(K \left(\frac{2}{a^2 \, A^2 + 2} - 1 \right) - \sqrt{\frac{a^2 \, A^2}{2} + 1} \, t\omega, \; \frac{2}{a^2 \, A^2 + 2} - 1 \right)$$

The solution is expressed in terms of an elliptic sine function, with a complicated argument. As a partial check, let's look at the leading term in a series expansion of x for small a.

```
xlinear = Series[x, {a, 0, 1}]
```

$$A \, \cos(t\omega)$$

The above result is exactly what would be expected in the linear limit for the given initial conditions. To see the effect of a nonzero value of a, let's plot the exact result x for $a = 3$, $A = 3$, and $\omega = 1$, and compare with xlinear.

```
a = 3;  A = 3;  ω = 1;
```

The first graph plots $x(t)$ over the range, $t = 0...6$, coloring the curve red, while the second graph plots the linear solution, coloring the curve blue. The two graphs are then superimposed with the Show command in Figure 5.7.

```
gr1 = Plot[Evaluate[x], {t, 0, 6}, PlotPoints -> 1000,
      PlotStyle -> Hue[1], DisplayFunction -> Identity];

gr2 = Plot[xlinear {t, 0, 6}, PlotStyle -> Hue[.6],
      DisplayFunction -> Identity];

Show[gr1, gr2, Ticks -> {{2, {2.85, "t"}, 4, 6}, {-3, {1.5, "x"}, 3}},
DisplayFunction -> $DisplayFunction, TextStyle -> {FontFamily ->
"Times", FontSize -> 16}, ImageSize -> {500, 250}];
```

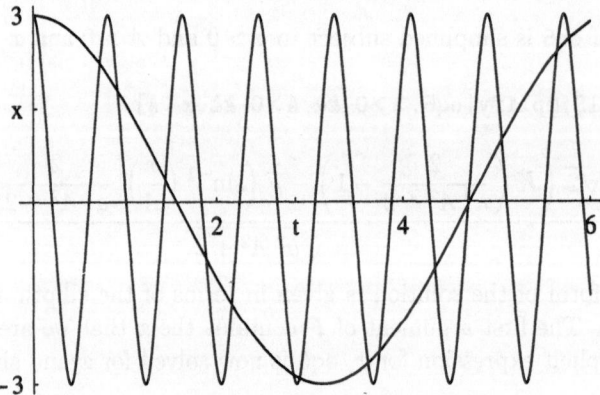

Figure 5.7: Fast oscillation: hard spring. Slow oscillation: linear spring.

The slowly varying oscillation, with period 2π, is the linear solution `xlinear`. The effect of a nonzero value of a is to increase the restoring force and therefore speed up the oscillations. This is clearly seen in the figure. The hard spring period may be either extracted from the figure or by numerically evaluating `period2` as follows.

```
N[period2]
```

0.816726

For our choice of parameters, the hard spring period is about one-eighth that of the linear period.

> **End Example 5-4**

By suitably modifying the code, other equations of motion involving polynomial force laws are also easily solved. Even the simple pendulum problem can be explicitly solved with Mathematica, if some mathematical guidance is provided. This is left as a problem for the reader.

PROBLEMS

Problem 5-20: Interpretation of k
Analytically show that for $\theta_{\max} < \pi$, $k^2 = E/E_0$, where E is the total energy and $E_0 = 2mgl$ is the maximum potential energy.

Problem 5-21: Over the top
Analytically show that, when the pendulum has sufficient energy to go "over the top", the solution is $\theta = 2\arcsin(\mathrm{sn}(\omega_0 t/k'))$ where $(k')^2 = E_0/E$. Here E_0 is the maximum potential energy and E is the total energy.

Problem 5-22: The separatrix
Analytically derive the separatrix equation describing the curves that separate the two types of solutions in the simple pendulum problem. To do this, put $\theta_{\max} = \pi$ in the energy equation. Plot the separatrix equation and attach arrowheads to indicate the sense of motion.

Problem 5-23: Elliptic cosine function
Modify file MF21 to calculate and plot $\mathrm{cn}\, u$ for $\theta_{\max} = 120\,^\circ$ and $170\,^\circ$.

Problem 5-24: Properties of Jacobian elliptic functions
Verify the following properties of the Jacobian elliptic functions: a) by hand, b) by making use of Mathematica.

- $\frac{d}{du}(\mathrm{cn}\, u) = -\,\mathrm{sn}\, u \,\mathrm{dn}\, u$

- $\frac{d^2}{du^2}(\mathrm{cn}\, u) = (2k^2 - 1)\,\mathrm{cn}\, u - 2k^2\,\mathrm{cn}^3 u$

- $\int \mathrm{cn}\, u \, du = \frac{1}{k}\arccos(\mathrm{dn}\, u)$

Problem 5-25: Simple pendulum solution
Taking $\theta_{max} = 170°$, $\ell = 1$ m, $g = 9.8$ m/s^2, and using Mathematica, plot

$$\theta(t) = 2\sin^{-1}[k\,\text{sn}(\omega_0 t)]$$

for one cycle of the simple pendulum. How long does it take the pendulum to descend from 170° to 30°? Confirm your answer by making use of file MF01.

Problem 5-26: Free vibrations of the eardrum
In the eardrum equation (2.12), set $F(t) = 0$. Taking $\tilde{\omega}_0 = 1$, $\beta = 1$, $x(0) = \frac{1}{3}$, and $\dot{x}(0) = 0$, analytically determine the period and the solution $x(t)$. Determine the numerical value of the period.

Problem 5-27: $k = 1$ limits
Prove analytically that for $k = 1$, $\text{sn}\,u = \tanh u$ and $\text{cn}\,u = \text{sech}\,u$.

Problem 5-28: Lennard–Jones potential
The Lennard–Jones potential

$$V(r) = V_0(1/r^{12} - 2/r^6)$$

is a well-known molecular potential in physics and chemistry.

   a. Plot $V(r)/V_0$ and physically interpret the structure.

   b. Calculate the distance r_0 at which the minimum in $V(r)$ occurs.

   c. Write $r = r_0 + x$ with x small and Taylor expand $V(r)$ out to x^3.

   d. Calculate the force $F(x) = -dV/dx$ and equate it to $m\ddot{x}$ to obtain an equation of motion for a particle of mass m moving in the potential well near r_0. Take $V_0 = 1$ and $m = 4$.

   e. If $x(0) = -1/30$ and $\dot{x}(0) = 0$, determine the solution in terms of an elliptic function. Calculate the period analytically and numerically.

Problem 5-29: Planetary motion in general relativity
The normalized equation describing the trajectory of a planet about a star (e.g., our sun) under the assumptions of general relativity is found to be of the form (with a, b positive)

$$d^2y/d\theta^2 + y = a + by^2.$$

Here y is the normalized (normalized to the minimum distance) inverse distance of the planet from the star and θ is the angular coordinate. For $b = 0$, the usual Keplerian equation of motion results. The nonlinear term arises from general relativity.

   Consider the mythical planet of Zabuzzywuk for which $a = 4/5$. If $b = 0$, solve the equation analytically for the initial conditions $y(0) = 1$, $y'(0) = 0$ and show by plotting $1/y$ that the orbit is elliptical. Now, take $b = 1/100$ and the same initial conditions. Discuss the effect of the parameter b on the elliptical orbit by, say, calculating the angle through which the planet moves as y decreases to the value $y = 0.7$.

Problem 5-30: Mathematica and the simple pendulum
Introducing suitable transformations that mimic the hand calculation, show that the simple undamped pendulum equation can be explicitly solved for $\theta(t)$ using Mathematica.

5.3 Some Approximate Methods

Approximate analytic solutions can be readily generated when the nonlinear terms are small compared to the linear part of the differential equation. Perturbation techniques are the most common and two methods, Poisson's and Linstedt's, will be discussed at some length. Linstedt's perturbation approach can be used, for example, to derive the shape of the limit cycle for the Van der Pol equation when the nonlinear parameter ϵ is small. On the other hand, it cannot tell us about the transient evolution to the limit cycle. To handle this problem, we can use a much more complicated perturbation approach called the method of multiple scales [NB95] or, provided we stop at lowest order, a relatively simple alternative method due to Krylov and Bogoliubov [KB43] which is based on the variation of parameters method for solving linear ODEs. In this text, we shall only deal with the Krylov–Bogoliubov method.

When the nonlinearity is not small, one can attempt to find approximate analytic solutions by introducing "trial wave functions" and minimizing the "error" arising from the fact that the trial functions are not exact solutions. The Ritz method and its close "cousin" the Galerkin method are based on this approach and will be discussed in the last section of this chapter.

Before tackling these approximate methods, we shall first demonstrate that Mathematica can directly generate a series solution which will be valid over some range of the independent variable.

5.3.1 Mathematica Generated Taylor Series Solution

To see how Mathematica may be used to generate a series solution, let's consider a concrete example, the simple pendulum equation with $\omega = 1$,

$$\ddot{x} + \sin x = 0 \qquad (5.56)$$

and initial conditions $x(0) = \pi/3$, $\dot{x} = 0$.

Example 5-5: Series Solution

Derive the series solution of the simple pendulum equation, subject to the given initial conditions, out to $O(t^{10})$.

Solution: We begin by expanding $x(t)$ in a series about $t = 0$, keeping terms out to order t^9. The initial conditions are substituted into the series, the label s being assigned to the result.

```
Clear["Global`*"]
```

```
s = Series[x[t], {t, 0, 9}] /. {x[0] -> Pi/3, x'[0] -> 0}
```

$$\frac{\pi}{3} + \frac{x''(0)\, t^2}{2} + \frac{x^{(3)}(0)\, t^3}{6} + \frac{x^{(4)}(0)\, t^4}{24} + \frac{x^{(5)}(0)\, t^5}{120} + \frac{x^{(6)}(0)\, t^6}{720}$$

$$+ \frac{x^{(7)}(0)\, t^7}{5040} + \frac{x^{(8)}(0)\, t^8}{40320} + \frac{x^{(9)}(0)\, t^9}{362880} + O(t)^{10}$$

The simple pendulum equation is entered, the series being automatically substituted and terms grouped according to powers of t.

$$ode = D[s, \{t, 2\}] + Sin[s] == 0$$

$$\left(x''(0) + \frac{\sqrt{3}}{2}\right) + x^{(3)}(0)\, t + \left(\frac{x''(0)}{4} + \frac{x^{(4)}(0)}{2}\right) t^2 + \left(\frac{x^{(3)}(0)}{12} + \frac{x^{(5)}(0)}{6}\right) t^3$$

$$+ \left(\frac{-\sqrt{3}\, x''(0)^2}{16} + \frac{x^{(4)}(0)}{48} + \frac{x^{(6)}(0)}{24}\right) t^4 + \left(\frac{-x''(0)\, x^{(3)}(0)}{8\sqrt{3}} + \frac{x^{(5)}(0)}{240} + \frac{x^{(7)}(0)}{120}\right) t^5$$

$$+ \left(\frac{-x''(0)^3}{96} + \frac{\sqrt{3}}{4}\left(\frac{-x^{(3)}(0)^2}{36} - \frac{x''(0)\, x^{(4)}(0)}{24}\right) + \frac{x^{(6)}(0)}{1440} + \frac{x^{(8)}(0)}{720}\right) t^6$$

$$+ \left(\frac{-x''(0)^2\, x^{(3)}(0)}{96} + \frac{\sqrt{3}}{4}\left(\frac{-\left(x^{(3)}(0)\, x^{(4)}(0)\right)}{72} - \frac{x''(0)\, x^{(5)}(0)}{120}\right)\right.$$

$$\left. + \frac{x^{(7)}(0)}{10080} + \frac{x^{(9)}(0)}{5040}\right) t^7 + O(t)^8 == 0$$

For arbitrary time t, the only way that the above output relation can be true is if the coefficient of each power of t is separately set equal to zero. This may be implemented by applying the `LogicalExpand` command to `ode`.

    eqs = LogicalExpand[ode]

$$x''(0) + \frac{\sqrt{3}}{2} == 0 \ \bigwedge \ x^{(3)}(0) == 0 \ \bigwedge \ \frac{x''(0)}{4} + \frac{x^{(4)}(0)}{2} == 0$$

$$\bigwedge \ \frac{x^{(3)}(0)}{12} + \frac{x^{(5)}(0)}{6} == 0 \ \bigwedge \ \frac{-\sqrt{3}\, x''(0)^2}{16} + \frac{x^{(4)}(0)}{48} + \frac{x^{(6)}(0)}{24} == 0$$

$$\bigwedge \ \frac{-x''(0)\, x^{(3)}(0)}{8\sqrt{3}} + \frac{x^{(5)}(0)}{240} + \frac{x^{(7)}(0)}{120} == 0$$

$$\bigwedge \ \frac{-x''(0)^3}{96} + \frac{\sqrt{3}}{4}\left(\frac{-x^{(3)}(0)^2}{36} - \frac{x''(0)\, x^{(4)}(0)}{24}\right) + \frac{x^{(6)}(0)}{1440} + \frac{x^{(8)}(0)}{720} == 0$$

$$\bigwedge \ \frac{-x''(0)^2\, x^{(3)}(0)}{96} + \frac{\sqrt{3}}{4}\left(\frac{-\left(x^{(3)}(0)\, x^{(4)}(0)\right)}{72} - \frac{x''(0)\, x^{(5)}(0)}{120}\right)$$

$$+ \frac{x^{(7)}(0)}{10080} + \frac{x^{(9)}(0)}{5040} == 0$$

The symbol $\bigwedge$ appearing between each equation in the output indicates the logical "and". That is to say, the first equation must hold, "and" the second equation, "and" the third, and so on. The equations `eqs` are then solved for $x''(0)$, $x^{(3)}(0)$, $x^{(4)}(0)$, etc.

`Solve[eqs]`

$$\{\{x^{(8)}(0) \to \frac{-49\sqrt{3}}{8},\ x^{(9)}(0) \to 0,\ x^{(6)}(0) \to \sqrt{3},\ x^{(7)}(0) \to 0,$$

$$x^{(5)}(0) \to 0,\ x^{(4)}(0) \to \frac{\sqrt{3}}{4},\ x^{(3)}(0) \to 0,\ x''(0) \to \frac{-\sqrt{3}}{2}\}\}$$

The series solution then follows by replacing $x^{(8)}(0)$, $x^{(9)}(0)$, etc., in **s** with the above values.

`sol = s /. %[[1]]`

$$\frac{\pi}{3} - \frac{\sqrt{3}\,t^2}{4} + \frac{t^4}{32\sqrt{3}} + \frac{t^6}{240\sqrt{3}} - \frac{7\,t^8}{15360\sqrt{3}} + O(t)^{10}$$

Note that there are no odd powers of t in this series. To complete the derivation, the order of term is removed from **sol** with the **Normal** command.

`Normal[sol]`

$$\frac{\pi}{3} - \frac{\sqrt{3}\,t^2}{4} + \frac{t^4}{32\sqrt{3}} + \frac{t^6}{240\sqrt{3}} - \frac{7\,t^8}{15360\sqrt{3}}$$

The above series solution will be accurate only over a limited time range. This range can be determined by comparing it with either the exact analytic solution of the pendulum equation or the "exact" numerical solution.

$\boxed{\textbf{End Example 5-5}}$

Of course the series solution could also be obtained by hand, but by using Mathematica we avoid a tedious task, and are assured that no algebraic or arithmetic mistakes are made. If we suddenly decide for some strange reason to go out to 20th order, we have to change only one number in our code. Or, if the nonlinear term in the equation has some other structure, again the change is easily made.

PROBLEMS
Problem 5-31: 20th order
For the text example, calculate the series solution out to $O(t^{20})$. Discuss the practicality of doing this calculation by hand.

Problem 5-32: Comparison with exact solution
For the above example, plot the series solution for different orders against the exact solution and determine over what times the different orders are valid.

Problem 5-33: Mathematica eases pain
Suppose that in our above example the $\sin x$ term were replaced with $x^2 \sin x$. Find the series solution to 10th order in t using the Mathematica code. Do you think that you would want to do this by hand?

5.3.2   The Perturbation Approach: Poisson's Method

Perturbation techniques are applicable when the nonlinear terms are small. The parameter ϵ, which may or may not be dimensionless, will be used to characterize the size of the nonlinear terms as, for example, in the Van der Pol equation

$$\ddot{x} - \epsilon(1 - x^2)\dot{x} + x = 0. \tag{5.57}$$

The basic Poisson procedure is to assume that the solution may be written as a power series in ϵ. If ϵ is small, only a few terms in the expansion are usually necessary; if not, the series may converge very slowly or not at all. There are certain technical pitfalls that can arise but may be circumvented by appropriate modifications.

Let's see how Poisson's method works. The solution $x(t)$ for a time-dependent problem is written in the form

$$x(t) = x_0(t) + \epsilon x_1(t) + \epsilon^2 x_2(t) + \cdots \tag{5.58}$$

where the subscripts refer to the zeroth order, first order, etc., in ϵ. If more than one equation is present as, for example, in the laser competition problem, each solution may be expanded in the same manner.

It is standard practice to adjust the solution so that when $\epsilon \to 0$ (the nonlinear terms vanish), the lowest order or "generating solution" x_0 is the exact solution of the remaining linear equation subject to the initial conditions. A few examples will suffice to illustrate Poisson's method. We first select a problem, viz. the nonlinear diode circuit introduced in Problem 5-10 (page 176), which has an exact solution so that some feeling for the accuracy of the perturbation expansion can be gained as different orders in ϵ are retained.

Nonlinear Diode Circuit

A capacitor of capacitance C is connected in series to a nonlinear diode which has the current (i)–voltage (e) relation of the form

$$i = ae + be^2 \tag{5.59}$$

with the voltage across the capacitor at $t = 0$ having the value $e = E$. The relevant circuit equation is

$$C\frac{de}{dt} + ae + be^2 = 0. \tag{5.60}$$

Introducing dimensionless variables $x \equiv e/E$ and $\tau \equiv at/C$, (5.60) becomes

$$\dot{x}(\tau) = -x - \frac{bE}{a}x^2. \tag{5.61}$$

This Bernoulli-type equation is easily solved by hand or by using Mathematica. Pretending to be ignorant of this fact, we shall solve it using the Poisson procedure. The form of the equation clearly suggests that we choose $\epsilon \equiv bE/a$ to be the dimensionless parameter characterizing the size of the nonlinear term. ϵ shall be assumed to be small but otherwise arbitrary. To solve

$$\dot{x}(\tau) = -x - \epsilon x^2 \tag{5.62}$$

assume the standard expansion

$$x(\tau) = x_0 + \epsilon x_1 + \epsilon^2 x_2 + \cdots. \qquad (5.63)$$

The differential equation (5.62) then yields

$$\dot{x}_0 + \epsilon \dot{x}_1 + \epsilon^2 \dot{x}_2 + \cdots = -(x_0 + \epsilon x_1 + \epsilon^2 x_2 + \cdots) - \epsilon(x_0 + \epsilon x_1 + \epsilon^2 x_2 + \cdots)^2. \qquad (5.64)$$

Since ϵ is arbitrary, equal powers of ϵ can be equated, and writing down the various orders of ϵ,

$$\epsilon^0 : \qquad \dot{x}_0 + x_0 = 0 \qquad (5.65)$$

$$\epsilon^1 : \qquad \dot{x}_1 + x_1 = -x_0^2 \qquad (5.66)$$

$$\epsilon^2 : \qquad \dot{x}_2 + x_2 = -2x_0 x_1. \qquad (5.67)$$

Because the nonlinear term in this example was a low order polynomial, the rhs of each order equation was easy to obtain. For more complicated nonlinearities or if a higher order is desired, it is better to automate the process using Mathematica's symbolic capabilities. This is illustrated in the following example.

Example 5-6: Generating Perturbation Equations

Generate the Poisson perturbation equations out to 5th order in ϵ for

$$\dot{x}(\tau) + x(\tau) + \epsilon\, x(\tau)^2 = 0.$$

Solution: The maximum order $n = 5$ is specified.

```
Clear["Global`*"]

n = 5;
```

The Poisson perturbation expansion of $x(\tau)$ is entered, terms being kept up to order n.

```
x[τ] = Sum[xᵢ[τ] ε^i, {i, 0, n}]
```

$$x_5(\tau)\,\epsilon^5 + x_4(\tau)\,\epsilon^4 + x_3(\tau)\,\epsilon^3 + x_2(\tau)\,\epsilon^2 + x_1(\tau)\,\epsilon + x_0(\tau)$$

The nonlinear diode equation is given and the output result expanded. Remove the semicolon if you wish to see the lengthy result, which contains terms out to order ϵ^{11}.

```
eq = Expand[D[x[τ], τ] + x[τ] + ε x[τ]^2];
```

Then the various powers of ϵ are collected in eq.

```
eq2 = Collect[eq, ε];
```

For arbitrary ϵ, the coefficient of each power of ϵ must be set equal to zero.

The command `CoefficientList` gives a list of coefficients of powers of ϵ in eq2. The `Table` command is used to create the list of perturbation equations. The index must run from $i = 1$ to $i = n + 1 = 6$ to pick up the six equations corresponding to ϵ^0, ϵ^1, ϵ^2,...,ϵ^5. `TableForm` is used to place each equation on a separate line, instead of running the equations sequentially one after the other in the same line.

```
Table[CoefficientList[eq2, ϵ][[i]] == 0, {i, 1, n + 1}] // TableForm
```

$$x_0(\tau) + x_0{'}(\tau) == 0$$

$$x_0(\tau)^2 + x_1(\tau) + x_1{'}(\tau) == 0$$

$$2\,x_0(\tau)\,x_1(\tau) + x_2(\tau) + x_2{'}(\tau) == 0$$

$$x_1(\tau)^2 + 2\,x_0(\tau)\,x_2(\tau) + x_3(\tau) + x_3{'}(\tau) == 0$$

$$2\,x_1(\tau)\,x_2(\tau) + 2\,x_0(\tau)\,x_3(\tau) + x_4(\tau) + x_4{'}(\tau) == 0$$

$$x_2(\tau)^2 + 2\,x_1(\tau)\,x_3(\tau) + 2\,x_0(\tau)\,x_4(\tau) + x_5(\tau) + x_5{'}(\tau) == 0$$

The code in this example is easily adapted to handle any finite order of ϵ or any polynomial nonlinearity.

End Example 5-6

Now let's continue with the perturbation development to 2nd order in ϵ for the nonlinear diode circuit, which will suffice for this particular problem. Noting that the rhs of Equations (5.66) and (5.67) involve known solutions of previous orders, let's find the detailed solution. At $t = 0$, $e = E$, so that

$$x(0) = 1 = x_0(0) + \epsilon x_1(0) + \epsilon^2 x_2(0) + \cdots. \tag{5.68}$$

Choosing $x_0(0) = 1$, then, since ϵ is arbitrary, $x_1(0) = x_2(0) = \cdots = 0$.

Zeroth Order (ϵ^0)

Making use of the condition $x_0(0) = 1$, the generating solution of (5.65) is

$$x_0(\tau) = e^{-\tau}. \tag{5.69}$$

First Order (ϵ^1)

With x_0 known, the first-order equation (5.66) becomes

$$\dot{x}_1 + x_1 = -x_0^2 = -e^{-2\tau} \tag{5.70}$$

which is easily integrated, subject to $x_1(0) = 0$, to yield

$$x_1(\tau) = e^{-2\tau} - e^{-\tau} = e^{-\tau}(e^{-\tau} - 1). \tag{5.71}$$

Again, because the nonlinear ODE was quite simple, the integration was easy. In the more complex examples which follow in this and the following section, more reliance will be placed on Mathematica's analytic ODE solver.

Second Order (ϵ^2)

Substituting for $x_0(\tau)$ and $x_1(\tau)$, Equation (5.67) becomes

$$\dot{x}_2 + x_2 = -2x_0 x_1 = -2e^{-\tau}\left(e^{-2\tau} - e^{-\tau}\right) \tag{5.72}$$

which, on using Mathematica's `DSolve` command with $x_2(0) = 0$, has the solution

$$x_2(\tau) = e^{-3\tau} - 2e^{-2\tau} + e^{-\tau} = e^{-\tau}\left(e^{-\tau} - 1\right)^2. \tag{5.73}$$

Putting all of the pieces together, the complete solution to second order is

$$x(\tau) = e^{-\tau}\left[1 + \epsilon\left(e^{-\tau} - 1\right) + \epsilon^2\left(e^{-\tau} - 1\right)^2\right]. \tag{5.74}$$

What would the solution be to third order in ϵ?

The reader who has done the nonlinear diode circuit Problem 5-10 will know that the exact solution is

$$x(\tau) = \frac{e^{-\tau}}{1 - \epsilon\left(e^{-\tau} - 1\right)}. \tag{5.75}$$

If this exact result is expanded to second order in ϵ for ϵ small, then the perturbation result (5.74) is obtained. In Figure 5.8, the contributions of the terms x_0, ϵx_1, $\epsilon^2 x_2$ are plotted for $\epsilon = \frac{1}{2}$. To the accuracy of the figure, these terms add up to the exact solution (5.75). It is left as an exercise for you to identify the curves that correspond to the individual perturbation terms and the line that represents their sum.

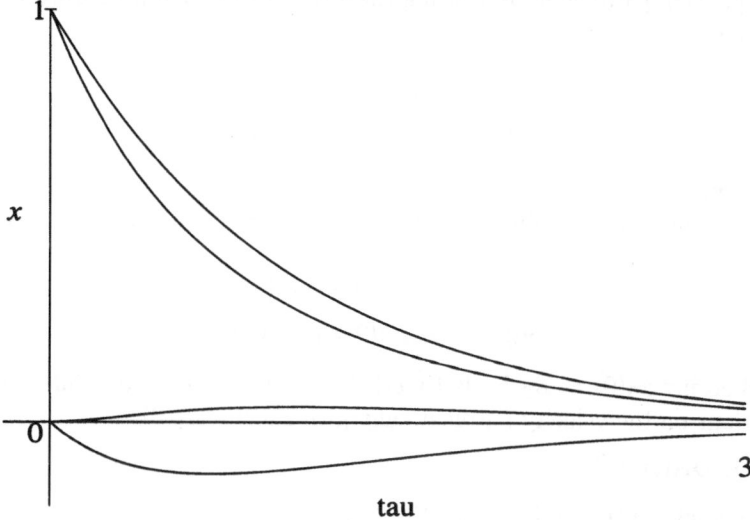

Figure 5.8: Perturbation solution for $\epsilon = 1/2$ for the nonlinear diode circuit.

This was an easy example, and although some Mathematica code was introduced, this was not really necessary. For most realistic problems involving more complex nonlinear terms and where an exact solution is not possible, the perturbation approach can prove quite tedious to apply and mistakes are apt to occur. Mathematica can then help to alleviate the pain. The next few Mathematica files on the CD-ROM will illustrate how this is done, complementing the textual material which follows. For completeness' sake, the above example is done in file MF22. The reader can use it, for example, to determine how good the perturbation result fits the exact solution for, say, $\epsilon = 1$.

Poisson Perturbation Method: The Nonlinear Diode
This file shows how Mathematica may be applied to generate a perturbation solution to the nonlinear diode circuit problem. The student can explore how good the perturbation expansion is for a given order in ϵ as the numerical value of ϵ is increased. Mathematica commands in file: Sum, Expand, Collect, Table, CoefficientList, TableForm, DSolve, Plot, Hue, Thickness, Block, $DisplayFunction=Identity, PlotStyle, Show, AxesLabel, PlotLabel, Text, Epilog, RGBColor, StyleForm, Ticks, FontFamily, ImageSize

Nonlinear Hard Spring

As another example, consider a mass on a nonlinear hard spring described by the equation of motion
$$\ddot{x} + \omega_0^2 x + \epsilon x^3 = 0 \tag{5.76}$$
with ϵ small. Physically the motion will be nearly simple harmonic and mathematically we know how to obtain an exact solution in terms of elliptic functions whether ϵ is small or not. The purpose of this example is to illustrate the occurrence of so-called secular terms which arise in nonlinear oscillatory problems.

Although ϵ is not dimensionless here, nothing prevents us from using it as the expansion parameter. Repeating the standard Poisson procedure yields the equations
$$\epsilon^0 : \qquad \ddot{x}_0 + \omega_0^2 x_0 = 0 \tag{5.77}$$
$$\epsilon^1 : \qquad \ddot{x}_1 + \omega_0^2 x_1 = -x_0^3, \tag{5.78}$$

and so on.

Assuming initial conditions $x(0) = A_0$, $\dot{x}(0) = 0$ (i.e., zero initial velocity) then
$$x(0) = A_0 = x_0(0) + \epsilon x_1(0) + \cdots \tag{5.79}$$
$$\dot{x}(0) = 0 = \dot{x}_0(0) + \epsilon \dot{x}_1(0) + \cdots \tag{5.80}$$

and choosing $x_0(0) = A_0$ leads to $x_1(0) = x_2(0) = \cdots = 0$, while the initial conditions on the velocity automatically give $\dot{x}_0(0) = \dot{x}_1(0) = \cdots = 0$.

Zeroth Order (ϵ^0)

The general solution of Equation (5.77) is
$$x_0(t) = a_0 \cos \omega_0 t + b_0 \sin \omega_0 t \tag{5.81}$$

where a_0, b_0 are arbitrary constants. Since $x_0(0) = A_0$ and $\dot{x}_0(0) = 0$, we have $a_0 = A_0$ and $b_0 = 0$ so that

$$x_0(t) = A_0 \cos \omega_0 t. \tag{5.82}$$

First order (ϵ^1)

Substituting for x_0 into Equation (5.78), we have

$$\ddot{x}_1 + \omega_0^2 x_1 = -A_0^3 \cos^3 \omega_0 t = -\frac{3}{4} A_0^3 \cos \omega_0 t - \frac{1}{4} A_0^3 \cos 3\omega_0 t \tag{5.83}$$

where the trigonometric identity $\cos 3\theta = 4 \cos^3 \theta - 3 \cos \theta$ has been used on the rhs. Alternatively, the `TrigReduce` command could be employed on $\cos^3 \omega_0 t$.

On integrating, the first-order solution is

$$x_1(t) = a_1 \cos \omega_0 t + b_1 \sin \omega_0 t - \frac{3}{8} \frac{A_0^3 t}{\omega_0} \sin \omega_0 t + \frac{A_0^3}{32 \omega_0^2} \cos 3\omega_0 t. \tag{5.84}$$

Since $x_1(0) = \dot{x}_1(0) = 0$, then $a_1 = -A_0^3/(32\omega_0^2)$, $b_1 = 0$ and thus

$$x_1(t) = -\frac{3}{8} \frac{A_0^3 t}{\omega_0} \sin \omega_0 t + \frac{A_0^3}{32 \omega_0^2} (\cos 3\omega_0 t - \cos \omega_0 t). \tag{5.85}$$

To first order in ϵ, the complete solution is

$$x(t) = A_0 \cos \omega_0 t + \frac{\epsilon A_0^3}{32 \omega_0^2} \left[-12\omega_0 t \sin \omega_0 t + (\cos 3\omega_0 t - \cos \omega_0 t) \right]. \tag{5.86}$$

For a sufficiently small ϵ, we would normally expect the solution calculated to first order to be a reasonably accurate description of the actual motion (i.e., oscillatory in time). However, as $t \to \infty$, the term (referred to as the "secular term") $\omega_0 t \sin \omega_0 t$ will cause $x(t)$ to apparently "blow up" instead of displaying well-behaved oscillatory motion.[4] Furthermore, the secular term destroys periodicity at any time t.

Such secular terms will appear in every order when the Poisson method is applied to an oscillatory problem. Clearly, this is undesirable. What is needed is a solution that displays the correct oscillatory behavior in every order. Therefore, a procedure must be devised to remove the secular terms as they arise, such a procedure having been developed by Lindstedt. Before outlining this procedure, we might ask, "why does the difficulty occur in the first place?" A simple linear differential equation illustrates what is happening. Let us consider the equation

$$\ddot{x} + x = -\epsilon x \tag{5.87}$$

with initial conditions $x(0) = A_0$, $\dot{x}(0) = 0$. Since the equation is linear, the exact solution, whether ϵ is small or not, is easily found to be

$$x_{\text{exact}}(t) = A_0 \cos(\sqrt{1 + \epsilon}\, t). \tag{5.88}$$

---

[4]Of course, there can exist problems where the solution actually does "blow up" as $t \to \infty$. However, here the apparent divergence is an artifact of the Poisson perturbation procedure.

Let's now assume ϵ to be small and solve (5.87) by the Poisson procedure. To first order in ϵ, the equations to be solved are

$$\ddot{x}_0 + x_0 = 0$$

$$\ddot{x}_1 + x_1 = -x_0.$$

(5.89)

With $x_0(0) = A_0$, $\dot{x}_0(0) = 0$, then

$$x_0(t) = A_0 \cos t$$

(5.90)

and the first-order equation becomes

$$\ddot{x}_1 + x_1 = -A_0 \cos t$$

(5.91)

which has a solution

$$x_1(t) = a_1 \cos t + b_1 \sin t - \frac{1}{2} A_0 t \sin t.$$

(5.92)

Since $x_1(0) = \dot{x}_1(0) = 0$, then $a_1 = 0$, $b_1 = 0$, and the complete solution to first order is

$$x(t) = A_0 \cos t - \frac{1}{2} \epsilon A_0 t \sin t$$

(5.93)

the last term being the secular one. If the exact solution (5.88) is expanded for ϵ small, we obtain to first order in ϵ

$$x_{\text{exact}}(t) = A_0 \cos \left[\left(1 + \frac{1}{2}\epsilon + \cdots \right) t \right]$$

$$\simeq A_0 \cos t \cos \left(\frac{1}{2}\epsilon t \right) - A_0 \sin t \sin \left(\frac{1}{2}\epsilon t \right)$$

$$= A_0 \cos t - \frac{1}{2}\epsilon A_0 t \sin t + \cdots$$

(5.94)

which is exactly the same as the perturbation result to first order in ϵ. Although the exact solution is periodic, this fact would not be immediately apparent from the series representation (5.94), the secular term masking the periodicity. The problem is that in the series representation we are forcing our solution into a form with a generating solution at the original frequency (1 here), while the frequency is actually $\sqrt{1+\epsilon}$. Lindstedt has used this idea to modify Poisson's method in such a way that secular terms are eliminated as they arise, i.e., order by order. His approach, guided by examples such as this one, is to let the frequency change and let the "new" frequency be a perturbation expansion around the "old" one.

PROBLEMS
Problem 5-34: Perturbation equations
For the nonlinear ODE

$$\dot{X}(t) + X + \epsilon X^4 = 0$$

write out the equations that result in a Poisson perturbation expansion to 10th order in ϵ.

5.3.3   Lindstedt's Method

To solve the nonlinear hard spring equation

$$\ddot{x} + \omega_0^2 x + \epsilon x^3 = 0 \tag{5.95}$$

for ϵ small, we shall assume that the frequency is changed when the nonlinear terms are present and can be represented by a perturbation expansion. Letting Ω be the unknown frequency of the periodic solution of the spring equation, it is convenient to introduce a new independent time variable $\tau \equiv \Omega t$. Then, independent of what value Ω turns out to have, the solution $x(\tau)$ will be periodic with period 2π, i.e.,

$$x(\tau + 2\pi) = x(\tau). \tag{5.96}$$

With the understanding that all derivatives are now with respect to τ, the nonlinear spring equation (5.95) becomes

$$\Omega^2 \ddot{x} + \omega_0^2 x + \epsilon x^3 = 0. \tag{5.97}$$

Lindstedt's procedure is to assume that

$$x(\tau) = x_0(\tau) + \epsilon x_1(\tau) + \cdots, \quad \text{and} \quad \Omega = \Omega_0 + \epsilon \Omega_1 + \cdots. \tag{5.98}$$

Since as $\epsilon \to 0$, we must have $\Omega \to \omega_0$, then $\Omega_0 = \omega_0$. If the reader has run the file MF22, which showed how Mathematica could be used to handle the Poisson method, the advantage of using computer algebra for more complex perturbation problems should be self-evident. For the Lindstedt method, the algebra tends to be even more involved because of the double perturbation expansion. In the following example, the hard spring perturbation solution is obtained using Mathematica.

Example 5-7: Hard Spring Perturbation Solution

Derive the Lindstedt solution of the hard spring ODE to first order in ϵ.

Solution: The maximum order, $n = 1$, is specified,

```
Clear["Global`*"]
```

```
n = 1;
```

and the perturbation expansions for x and Ω are entered using the BasicInput palette.

```
x[τ] = Sum[xᵢ[τ] εⁱ, {i, 0, n}]
```

$$x_0(\tau) + \epsilon x_1(\tau)$$

```
Ω = Sum[ωᵢ εⁱ, {i, 0, n}]
```

$$\omega_0 + \epsilon \omega_1$$

The hard spring equation (5.97) is inputted and the result expanded.

```
eq = Expand[Ω^2 D[x[τ], τ, τ] + ω₀^2 x[τ] + ε x[τ]^3];
```

We collect powers of ϵ in eq,

```
eq2 = Collect[eq, ε];
```

and use the `CoefficientList` and `Table` commands to generate the zeroth and first order equations.

```
odesystem = Table[Expand[CoefficientList[eq2, ε][[i]]/ω₀^2] == 0,
              {i, 1, n+1}] // TableForm
```

$$x_0(\tau) + x_0''(\tau) == 0$$

$$\frac{x_0(\tau)^3}{\omega_0{}^2} + x_1(\tau) + \frac{2\,\omega_1\,x_0''(\tau)}{\omega_0} + x_1''(\tau) == 0$$

The above two ODEs are extracted from `odesystem` and displayed separately.

```
ode0 = odesystem[[1, 1]]
```

$$x_0(\tau) + x_0''(\tau) == 0$$

```
ode1 = odesystem[[1, 2]]
```

$$\frac{x_0(\tau)^3}{\omega_0{}^2} + x_1(\tau) + \frac{2\,\omega_1\,x_0''(\tau)}{\omega_0} + x_1''(\tau) == 0$$

In the next two input lines, ode0 is analytically solved for $x_0(\tau)$, subject to the initial conditions $x(0) = A$, $\dot{x}(0) = 0$.

```
sol0 = DSolve[{ode0, x₀[0] == A, x₀'[0] == 0}, x₀, τ];
```

```
x0 = x₀[τ] /. sol0[[1]]
```

$$A\cos(\tau)$$

The zeroth order solution is substituted into ode1,

```
ode1 = ode1 /. sol0[[1]]
```

$$\frac{A^3\cos^3(\tau)}{\omega_0{}^2} - \frac{2\,A\,\omega_1\,\cos(\tau)}{\omega_0} + x_1(\tau) + x_1''(\tau) == 0$$

and the lhs of ode1 simplified with the `TrigReduce` and `Expand` commands.

```
lhsode1 = Expand[TrigReduce[ode1[[1]]]]
```

$$\frac{3\cos(\tau)\,A^3}{4\omega_0{}^2} + \frac{\cos(3\tau)\,A^3}{4\omega_0{}^2} - \frac{2\cos(\tau)\,\omega_1\,A}{\omega_0} + x_1(\tau) + x_1{}''(\tau)$$

The $\cos(\tau)$ terms are collected in lhsode1 and ode1 reformed.

```
ode1 = Collect[lhsode1, Cos[τ]] == 0
```

$$\frac{\cos(3\tau)\,A^3}{4\omega_0{}^2} + \cos(\tau)\left(\frac{3A^3}{4\omega_0{}^2} - \frac{2A\omega_1}{\omega_0}\right) + x_1(\tau) + x_1{}''(\tau) == 0$$

Then, ode1 is analytically solved for x_1, subject to the initial condition $x_1(0) = 0$, $\dot{x}(0) = 0$.

```
sol1 = DSolve[{ode1, x₁[0] == 0, x₁'[0] == 0}, x₁, τ];
```

The solution $x_1(\tau)$ is simplified,

```
x1 = Simplify[x₁[τ] /. sol1[[1]]]
```

$$\frac{-A\sin(\tau)\left(A^2\left(6\tau + \sin(2\tau)\right) - 16\tau\omega_0\omega_1\right)}{16\omega_0{}^2}$$

and terms involving τ are collected.

```
x1 = Collect[x1, τ]
```

$$-\frac{\sin(\tau)\sin(2\tau)\,A^3}{16\omega_0{}^2} - \frac{\tau\sin(\tau)\left(6A^2 - 16\omega_0\omega_1\right)A}{16\omega_0{}^2}$$

The coefficient of $\tau\sin(\tau)$ is set equal to zero in x1 and the first order frequency correction obtained by solving the resulting output.

```
freq = Solve[Coefficient[x1, τ Sin[τ]] == 0, ω₁]
```

$$\left\{\left\{\omega_1 \to \frac{3A^2}{8\omega_0}\right\}\right\}$$

The frequency, to first order in ϵ, is given by the output of the following command line.

```
Ω = Ω /. freq[[1]]
```

$$\frac{3\epsilon A^2}{8\omega_0} + \omega_0$$

In the first order solution x_1, we set $\tau\sin(\tau)$ equal to zero, and simplify the result with the TrigReduce command.

```
x1 = TrigReduce[(x1 /. τ Sin[τ] -> 0)]
```

$$\frac{A^3\cos(3\tau) - A^3\cos(\tau)}{32\omega_0{}^2}$$

On collecting with respect to A^3 in $\mathbf{x_1}$,

$\mathbf{x1 = Collect[x1, A\char`\^3]}$

$$\frac{A^3 \, (\cos(3\tau) - \cos(\tau))}{32\,{\omega_0}^2}$$

the complete solution to first order in ϵ is obtained.

$\mathbf{x = x0 + \epsilon\, x1}$

$$A\cos(\tau) + \frac{\epsilon\, A^3\, (\cos(3\tau) - \cos(\tau))}{32\,{\omega_0}^2}$$

$\boxed{\textbf{End Example 5-7}}$

Recall that, in Example 5-4, the nonlinear spring equation, now stated in the form

$$\ddot{x} + \omega_0^2(1 + a^2 x^2)x = 0, \tag{5.99}$$

was solved exactly. The exact period expression was expanded in powers of a to second order, yielding

$$T = \frac{2\pi}{\omega_0} - \frac{3\pi a^2 A^2}{4\omega_0} = T_0(1 - \frac{3}{8}a^2 A^2), \tag{5.100}$$

with $T_0 = 2\pi/\omega_0$. Comparing Equations (5.99) and (5.95), we can make the identification $\epsilon \equiv \omega_0^2 a^2$. How does the Lindstedt perturbation result for the period compare with that in (5.100)? From Example 5-7, we have

$$T = \frac{2\pi}{\Omega} = \frac{2\pi}{\omega_0 + 3\epsilon A^2/8\omega_0} = \frac{2\pi/\omega_0}{1 + 3a^2 A^2/8} \simeq T_0(1 - \frac{3}{8}a^2 A^2). \tag{5.101}$$

The Lindstedt perturbation result agrees with the exact result to order ϵ.

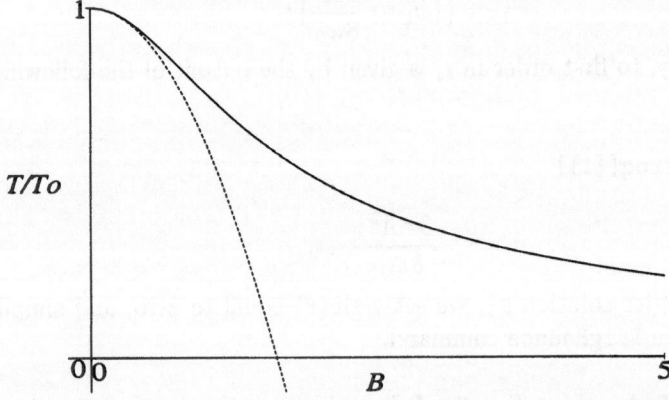

Figure 5.9: Normalized period: first-order perturbation (dashed curve) and exact (solid curve) formulae.

The first-order formula is only valid for sufficiently small amplitude. Its range of validity can be determined by setting $B \equiv aA$ and plotting T/T_0 versus B for the exact and perturbation formulae. Figure 5.9 shows that the first-order perturbation result diverges from the exact result for $B \geq 0.5$, and thus higher order terms in the perturbation expansion must be retained for larger B.

Van der Pol's Equation

As the next example, let's look for a periodic solution (the limit cycle) of Van der Pol's equation

$$\ddot{x} - \epsilon\left(1 - x^2\right)\dot{x} + x = 0 \tag{5.102}$$

with $\epsilon > 0$ and small. Only the main results are outlined here, a Mathematica treatment being given in Mathematica File MF23. Once again introducing $\tau = \Omega t$, noting that $\Omega_0 = 1$ here, and using the Lindstedt perturbation expansions (5.98) for $x(\tau)$ and Ω, we obtain the following equations to second order in ϵ:

$$\ddot{x}_0 + x_0 = 0$$

$$\ddot{x}_1 + x_1 = -2\Omega_1 \ddot{x}_0 + (1 - x_0^2)\dot{x}_0 \tag{5.103}$$

$$\ddot{x}_2 + x_2 = -(2\Omega_2 + \Omega_1^2)\ddot{x}_0 - 2\Omega_1 \ddot{x}_1 + \Omega_1(1 - x_0^2)\dot{x}_0 - 2x_0 x_1 \dot{x}_0 + (1 - x_0^2)\dot{x}_1.$$

Defining the zero of time to be when the "velocity" $\dot{x}$ is zero[5], i.e., $\dot{x}(0) = 0$, then $\dot{x}_i(0) = 0$ while the periodicity condition on $x(\tau)$ gives $x_i(\tau + 2\pi) = x_i(\tau)$ for $i = 0, 1, 2, \ldots$.

Zeroth order(ϵ^0):

Subject to the single initial condition $\dot{x}_0(0) = 0$, the zeroth-order solution is

$$x_0(\tau) = a_0 \cos\tau \tag{5.104}$$

the constant a_0 being still undetermined.

First order(ϵ^1):

Using the zeroth-order solution, the first-order equation in (5.103) simplifies to

$$\ddot{x}_1 + x_1 = 2\Omega_1 a_0 \cos\tau - (1 - a_0^2 \cos^2\tau)a_0 \sin\tau$$

$$= 2\Omega_1 a_0 \cos\tau + (a_0^3 - a_0)\sin\tau - a_0^3 \sin^3\tau \tag{5.105}$$

or, on making use of the identity $\sin 3\tau = 3\sin\tau - 4\sin^3\tau$,

$$\ddot{x}_1 + x_1 = 2\Omega_1 a_0 \cos\tau + a_0\left(\frac{a_0^2}{4} - 1\right)\sin\tau + \frac{a_0^3}{4}\sin 3\tau \tag{5.106}$$

---

[5]Since the solution is periodic, $\dot{x}$ has to be zero at some time. We can always shift the origin of time to be when $\dot{x} = 0$, since the result cannot depend on how we choose $t = 0$.

which can be readily integrated (making use of $\dot{x}_1(0) = 0$), yielding

$$x_1(\tau) = a_1 \cos\tau + \frac{3}{4}\sin\tau + \frac{1}{2}\tau(2\Omega_1 a_0)\sin\tau$$
$$- \frac{1}{2}\tau\left[a_0\left(\frac{a_0^2}{4} - 1\right)\right]\cos\tau - \frac{a_0^3}{32}\sin 3\tau. \tag{5.107}$$

There are two secular terms in this equation which must be eliminated in order that $x_1(\tau + 2\pi) = x_1(\tau)$. Since $a_0 \neq 0$ for a nontrivial solution, then $\Omega_1 = 0$ and $a_0^2/4 - 1 = 0$, for the $\tau\sin\tau$ and $\tau\cos\tau$ terms, respectively, to vanish. Thus[6], $a_0 = 2$ which means that the zeroth-order solution is finally completely determined,

$$x_0(\tau) = 2\cos\tau. \tag{5.108}$$

With the secular terms removed, the first-order solution is

$$x_1(\tau) = a_1 \cos\tau + \frac{3}{4}\sin\tau - \frac{1}{4}\sin 3\tau \tag{5.109}$$

where a_1 will be determined by imposing the periodicity condition on the second-order solution $x_2(\tau)$.

Second order(ϵ^2):

With $\Omega_1 = 0$, the second-order equation in (5.103) reduces to

$$\ddot{x}_2 + x_2 = -2\Omega_2\ddot{x}_0 - 2x_0 x_1\dot{x}_0 + (1 - x_0^2)\dot{x}_1 \tag{5.110}$$

which, on using the identity

$$\cos 5\tau = 16\cos^5\tau - 20\cos^3\tau + 5\cos\tau \tag{5.111}$$

as well as the identity for $\cos 3\tau$, becomes

$$\ddot{x}_2 + x_2 = \left(4\Omega_2 + \frac{1}{4}\right)\cos\tau + 2a_1\sin\tau - \frac{3}{2}\cos 3\tau + 3a_1\sin 3\tau + \frac{5}{4}\cos 5\tau. \tag{5.112}$$

Experience with earlier examples has taught us that the $\cos\tau$ and $\sin\tau$ terms will result in secular terms on integration. The periodicity condition $x_2(\tau + 2\pi) = x_2(\tau)$ forces us to conclude that $4\Omega_2 + 1/4 = 0$ and $a_1 = 0$. Therefore, the second-order correction to the frequency is $\Omega_2 = -1/16$ and with $a_1 = 0$ the first-order solution (5.109) reduces to

$$x_1(\tau) = \frac{3}{4}\sin\tau - \frac{1}{4}\sin 3\tau. \tag{5.113}$$

Using the condition $\dot{x}_2(0) = 0$, the second-order solution is

$$x_2(\tau) = a_2 \cos\tau + \frac{3}{16}\cos 3\tau - \frac{5}{96}\cos 5\tau. \tag{5.114}$$

---

[6]The other choice of $a_0 = -2$ does not yield a new solution, but simply changes the phase by π radians.

The reader can check that the constant a_2, which must be determined by imposing the periodicity condition on the third-order solution, turns out to have the value $a_2 = -1/8$, so

$$x_2 = -\frac{1}{8}\cos\tau + \frac{3}{16}\cos 3\tau - \frac{5}{96}\cos 5\tau. \qquad (5.115)$$

Thus, to second order in ϵ, the periodic solution of Van der Pol's equation is

$$x(\tau) = 2\cos\tau + \epsilon\left(\frac{3}{4}\sin\tau - \frac{1}{4}\sin 3\tau\right) + \epsilon^2\left(-\frac{1}{8}\cos\tau + \frac{3}{16}\cos 3\tau - \frac{5}{96}\cos 5\tau\right) \qquad (5.116)$$

with $\tau = \Omega t$ where

$$\Omega = 1 - \frac{1}{16}\epsilon^2. \qquad (5.117)$$

For $\epsilon = 0.7$, e.g., the frequency is lowered by 3%. As the lowest-order solution is $x(\tau) = 2\cos\tau = 2\cos t$, then, to this order, $x^2 + \dot{x}^2 = 4$ which is the equation of a circle in the phase plane of radius $r = 2$. Thus, the limit cycle for small ϵ is nearly circular with the corrections to order ϵ^2 being given by (5.116). It is interesting to evaluate the time average over one period of the nonlinear terms in Van der Pol's equation, e.g., to zeroth order

$$\int_0^{2\pi} \left[-\epsilon\Omega(1 - x^2)\dot{x}\right] d\tau = \int_0^{2\pi} 2\epsilon(1 - 4\cos^2\tau)\sin\tau\, d\tau = 0. \qquad (5.118)$$

For part of the cycle one has a negative resistance where the circuit picks up energy from the battery, while for the other part the damping is positive and the circuit loses energy, the time average over the cycle being zero consistent with stable operation of an electronic oscillator circuit. The reader may check that the higher-order terms also average out to zero over a period.

In the following file, a Lindstedt perturbation solution is produced for the Van der Pol equation out to order ϵ^3 for x and to order ϵ^4 for the frequency. This result would be very challenging to obtain quickly and accurately by hand.

Lindstedt Method: The Van der Pol Equation
In this file, a Mathematica treatment of the Lindstedt perturbation method, which may be extended to arbitrary order in ϵ, is carried out for the Van der Pol equation. The perturbation result is plotted for several different orders of ϵ. Mathematica commands in file: Sum, Expand, Collect, Table, CoefficientList, TableForm, DSolve, TrigReduce, Collect, Sin, Cos, Solve, Coefficient, Block, $DisplayFunction=Identity, PlotStyle, Plot, RGBColor, Show, AxesLabel, PlotLabel, StyleForm, ImageSize

PROBLEMS
Problem 5-35: Trigonometric identities
Analytically prove the following trigonometric identities:

- $\cos 5\tau = 16\cos^5\tau - 20\cos^3\tau + 5\cos\tau$,

- $\sin 3\tau = 3\sin\tau - 4\sin^3\tau$.

Problem 5-36: Hard spring 3rd-order solution
Generate the 3rd-order perturbation solution to the hard spring equation. By
plotting the 3rd-order and exact results for the period, determine the value of
aA at which the 3rd-order result for the period begins to seriously break down.
Is there much improvement over the first-order result discussed in the text?

Problem 5-37: A modified Van der Pol equation
Consider the modified Van der Pol equation

$$\ddot{x} - \epsilon^2(1 - x^2)\dot{x} + x - \epsilon x^3 = 0$$

with $\epsilon > 0$ and small. Find the periodic solution to this equation to order ϵ^2
using perturbation theory.

Problem 5-38: The eardrum
Consider the equation of motion of the eardrum

$$\ddot{x} + \omega_0^2 x + \epsilon x^2 = 0$$

subject to the initial conditions $x(0) = A_0$, $\dot{x}(0) = 0$. Assuming ϵ to be small,
solve this equation to 2nd order in ϵ using the appropriate perturbation method.

Problem 5-39: SHO with nonlinear "damping"
Consider the following simple harmonic oscillator equation with nonlinear damping

$$\ddot{x} + \epsilon \dot{x}^2 + \omega_0^2 x = 0.$$

Assuming $x(0) = A_0$, $\dot{x}(0) = 0$, find $x(t)$ to order ϵ^2. Plot for $A_0 = 1$, $\omega_0 = 1$,
and $\epsilon = 1/2$ and 1 and comment on the results.

5.4   The Krylov–Bogoliubov (KB) Method

An approximation technique, which has as its starting point the variation of
parameters approach used in solving linear ODEs, has been developed by Krylov
and Bogoliubov [KB43] for nonlinear equations of the structure

$$\ddot{x} + \omega_0^2 x + \epsilon f(x, \dot{x}) = 0, \tag{5.119}$$

where f is a nonlinear function and ϵ is a parameter. The Van der Pol equation,
the eardrum equation, and the nonlinear spring are examples of equations of
this structure. The KB procedure is to assume a solution that retains the same
structure for the "displacement" $x(t)$ and the "velocity" $\dot{x}(t)$ as in the absence
of the nonlinear term, viz.,

$$x = a \sin(\omega_0 t + \phi)$$

$$\dot{x} = a\omega_0 \cos(\omega_0 t + \phi) \tag{5.120}$$

but with the amplitude a and phase angle ϕ now functions of t. They are no longer constant. The differential equations for a and ϕ can be readily found. Differentiating the assumed form of $x(t)$ and equating the result to the assumed form of $\dot{x}(t)$ yields

$$\dot{a}\sin(\omega_0 t + \phi) + a\dot{\phi}\cos(\omega_0 t + \phi) = 0. \tag{5.121}$$

Then, differentiating $\dot{x}(t)$ and substituting the resulting expression for $\ddot{x}(t)$ as well as the assumed forms for $x(t)$ and $\dot{x}(t)$ into the general structure above yields the second equation relating a and ϕ,

$$\dot{a}\omega_0\cos(\omega_0 t + \phi) - a\omega_0\dot{\phi}\sin(\omega_0 t + \phi) + \epsilon f(a\sin(\omega_0 t + \phi), a\omega_0\cos(\omega_0 t + \phi)) = 0. \tag{5.122}$$

Finally, separating into equations for the rate of change of a and ϕ, we obtain

$$\dot{a} = -\frac{\epsilon}{\omega_0} f\left(a\sin(\omega_0 t + \phi), a\omega_0\cos(\omega_0 t + \phi)\right)\cos(\omega_0 t + \phi)$$

$$\dot{\phi} = \frac{\epsilon}{a\omega_0} f\left(a\sin(\omega_0 t + \phi), a\omega_0\cos(\omega_0 t + \phi)\right)\sin(\omega_0 t + \phi). \tag{5.123}$$

To this point, no approximations have been made. Depending on the functional form of f, these equations may be difficult or impossible to solve exactly for $a(t)$ and $\phi(t)$. However, when ϵ is small, a and ϕ are clearly slowly varying functions of t and approximate solutions may be generated. The KB method is based on this fact. If a and ϕ vary slowly with time, then in a time $T_0 = 2\pi/\omega_0$, a and ϕ will not have changed much, remaining essentially constant.[7] Therefore, it is a reasonable first approximation to consider a and ϕ as constant on the rhs during the time T_0. For notational convenience, let's put $\psi \equiv \omega_0 t + \phi$ and expand the right-hand sides of these equations in a general Fourier series,

$$\dot{a} = -\frac{\epsilon}{\omega_0}\left(K_0(a) + \sum_{n=1}^{\infty}\{K_n(a)\cos(n\psi) + L_n(a)\sin(n\psi)\}\right)$$

$$\dot{\phi} = \frac{\epsilon}{a\omega_0}\left(M_0(a) + \sum_{n=1}^{\infty}\{M_n(a)\cos(n\psi) + N_n(a)\sin(n\psi)\}\right) \tag{5.124}$$

where

$$K_0(a) = \frac{1}{2\pi}\int_0^{2\pi} f(a\sin\psi, a\omega_0\cos\psi)\cos\psi\, d\psi$$

$$K_n(a) = \frac{1}{\pi}\int_0^{2\pi} f(a\sin\psi, a\omega_0\cos\psi)\cos\psi\cos(n\psi)\, d\psi \tag{5.125}$$

$$L_n(a) = \frac{1}{\pi}\int_0^{2\pi} f(a\sin\psi, a\omega_0\cos\psi)\cos\psi\sin(n\psi)\, d\psi$$

and similarly for M_0, M_n and N_n. Now, integrate these equations between some

---

[7] T_0 is, of course, the period when the nonlinear terms are absent. Since these terms are assumed to be small, the period will not change much either.

instant t and $t + T_0$ assuming that a and ϕ remain constant under the integral sign during the interval.[8] Noting that for nonzero integral values of n,

$$\int_t^{t+T} \cos[n(\omega_0 t + \phi)]\, dt = \frac{1}{\omega_0} \int_0^{2\pi} \cos(n\psi)\, d\psi = 0,$$

and similarly

$$\int_0^{2\pi} \sin(n\psi)\, d\psi = 0,$$

we obtain

$$
\begin{aligned}
\frac{a(t + T_0) - a(t)}{T_0} &= -\frac{\epsilon}{\omega_0} K_0[a(t)] \\
\frac{\phi(t + T_0) - \phi(t)}{T_0} &= \frac{\epsilon}{a\omega_0} M_0[a(t)].
\end{aligned}
\tag{5.126}
$$

If T_0 is small compared to the total time duration involved, T_0 may be replaced with dt. Similarly, since the changes

$$
\begin{aligned}
\Delta a &= a(t + T_0) - a(t) \\
\Delta \phi &= \phi(t + T_0) - \phi(t)
\end{aligned}
\tag{5.127}
$$

are small, they may also be replaced with differentials. The difference equations (5.126) are then transformed into the differential equations

$$
\begin{aligned}
\dot{a} &= -\frac{\epsilon}{\omega_0} K_0[a(t)] \\
\dot{\phi} &= \frac{\epsilon}{a\omega_0} M_0[a(t)]
\end{aligned}
\tag{5.128}
$$

or

$$
\begin{aligned}
\dot{a} &= -\frac{\epsilon}{2\pi\omega_0} \int_0^{2\pi} f(a \sin\psi, a\omega_0 \cos\psi) \cos\psi\, d\psi \\
\dot{\phi} &= \frac{\epsilon}{2\pi a\omega_0} \int_0^{2\pi} f(a \sin\psi, a\omega_0 \cos\psi) \sin\psi\, d\psi.
\end{aligned}
\tag{5.129}
$$

Since $\psi \equiv \omega_0 t + \phi$, the last equation could also be written as

$$\dot{\psi} = \omega_0 + \frac{\epsilon}{2\pi a\omega_0} \int_0^{2\pi} f(a \sin\psi, a\omega_0 \cos\psi) \sin\psi\, d\psi. \tag{5.130}$$

The solution $x = a \sin(\omega_0 t + \phi)$ with a and ϕ determined by the above equations is referred to as the first approximation of Krylov and Bogoliubov. If the approximate equations (5.129) for a and ϕ are compared with the exact ones (5.123), it is clear that the right-hand sides have been replaced with the average

---

[8]That is, they retain the values that they had at the instant t.

values over one period (2π) of ψ. The assumption about the quantity a remaining constant during the interval T_0 can only be a good approximation if (the rate of change of a) $\times$ (time for one period) $\ll$ (value of a), i.e.,

$$|\dot{a}|\frac{2\pi}{\omega_0} \ll |a|, \quad \text{or} \quad \left|\frac{\dot{a}}{a}\right|\frac{2\pi}{\omega_0} \ll 1, \tag{5.131}$$

and, for slowly varying ϕ (i.e., for $\dot{\psi} \simeq \omega_0$), we must have $|\dot{\phi}/\omega_0| \ll 1$.

The Nonlinear Spring

Let's test the KB method on an old friend, the nonlinear spring equation

$$\ddot{x} + \omega^2 x + \epsilon x^3 = 0 \tag{5.132}$$

with initial conditions $x(0) = A_0$, $\dot{x}(0) = 0$. Identifying $f = x^3 = a^3 \sin^3 \psi$ and using the expressions (5.129),

$$\dot{a} = -\frac{\epsilon a^3}{2\pi\omega_0}\int_0^{2\pi}(\sin^3\psi)\cos\psi\,d\psi = 0$$
$$\dot{\phi} = \frac{\epsilon a^3}{2\pi a\omega_0}\int_0^{2\pi}(\sin^3\psi)\sin\psi\,d\psi = \frac{3\epsilon a^2}{8\omega_0}. \tag{5.133}$$

On integrating,

$$a(t) = C_1$$
$$\phi(t) = \frac{3\epsilon a^2}{8\omega_0}t + C_2 \tag{5.134}$$

where the constants C_1, C_2 remain to be evaluated. The KB solution is

$$x(t) = a\sin\psi = a\sin(\omega_0 t + \phi) \tag{5.135}$$

or

$$x(t) = C_1\sin\left[\left(\omega_0 + \frac{3}{8}\frac{\epsilon C_1^2}{\omega_0}\right)t + C_2\right]. \tag{5.136}$$

Then,

$$\dot{x}(0) = 0 = C_1(\omega_0 + \frac{3}{8}\frac{\epsilon C_1^2}{\omega_0})\cos C_2 \tag{5.137}$$

which implies that $C_2 = \pi/2$ while $x(0) = A_0 = C_1\sin\pi/2$, so $C_1 = A_0$. Noting that $\sin(\theta + \pi/2) = \cos\theta$, the first-order KB solution is

$$x(t) = A_0\cos\left[(\omega_0 + \frac{3}{8}\frac{\epsilon A_0^2}{\omega_0})t\right]. \tag{5.138}$$

For the approximation to be valid, $|\dot{\phi}/\omega_0| \ll 1$ or $3\epsilon A_0^2/8\omega_0 \ll \omega_0$. Since a is constant, the inequality on $\dot{a}$ is automatically satisfied. In the KB first approximation, the frequency is

$$\omega = \omega_0 + \frac{3}{8}\frac{\epsilon A_0^2}{\omega_0} \tag{5.139}$$

which agrees exactly with the first-order perturbation result using Lindstedt's method. The KB solution for x differs slightly, however, from the perturbation result, the latter (see Example 5-7) having the additional small term

$$\frac{\epsilon A_0^3}{32\omega_0^2}(\cos 3\tau - \cos \tau)$$

where $\tau \equiv \omega t$.

Actually, Krylov and Bogoliubov developed an improved first approximation, which for this particular example will yield exact agreement of the solution with the perturbation result to order[9] ϵ. For our purposes, the "unimproved" first KB approximation shall suffice. It is easy to apply and has the important virtue that for some problems it can readily yield information which it is not possible to obtain via the Lindstedt perturbation technique. The Van der Pol equation for ϵ small provides a good example. The Lindstedt perturbation method gave us the limit cycle solution but did not provide us with the complete time evolution of a representative point toward the limit cycle, i.e., it was not conducive to studying the transient growth of the solution where the amplitude changes with time. The KB first approximation, on the other hand, can handle this situation very nicely without too much algebraic effort as is demonstrated in file MF24.

Nonlinear Damping

Let us consider a linear spring system with nonlinear damping

$$\ddot{x} + \omega_0^2 x + \epsilon f(\dot{x}) = 0 \tag{5.140}$$

where $\epsilon > 0$ and small. The KB approximate solution is

$$x = a\sin(\omega_0 t + \phi) \tag{5.141}$$

with a, ϕ determined by

$$\dot{a} = -\frac{\epsilon}{2\pi\omega_0}\int_0^{2\pi} f(a\omega_0\cos\psi)\cos\psi\, d\psi$$
$$\dot{\phi} = \frac{\epsilon}{2\pi a\omega_0}\int_0^{2\pi} f(a\omega_0\cos\psi)\sin\psi\, d\psi. \tag{5.142}$$

Let's examine the integral in $\dot{\phi}$. It may be rewritten as

$$-\frac{1}{a\omega_0}\int_0^{2\pi} f(a\omega_0\cos\psi)\,d(a\omega_0\cos\psi) = -\frac{1}{a\omega_0}\,\Phi(a\omega_0\cos\psi)\big|_{\psi=0}^{2\pi} = 0. \tag{5.143}$$

Since $\dot{\phi} = 0$, then $\phi = \phi_0$, a constant. So there is no change in frequency in the first KB approximation. The frequency correction turns out to be of second order so does not appear here. The solution is then

$$x(t) = a(t)\sin(\omega_0 t + \phi_0) \tag{5.144}$$

---

[9]See Section 68 of Minorsky [Min64]. The higher order approximations, which shall not be examined here, are discussed in Section 70 of the same reference.

where $a(t)$ remains to be determined.

Clearly, the nonlinear damping analysis can be applied to the Van der Pol equation, which is of the structure (5.140) with $\omega_0 = 1$ and $f(x, \dot{x}) = (x^2 - 1)\dot{x}$. Although the form of $a(t)$ can be readily determined by hand, the determination of how accurate the KB solution is demands that a comparison be made with the "exact" numerical solution. In the following file, Mathematica is used to calculate the KB solution relevant to the Van der Pol transient growth problem and the approximate analytic formula is compared with the exact result.

Krylov–Bogoliubov Method

This file demonstrates the application of the KB method to the Van der Pol transient growth problem. The KB solution is compared to the numerical solution by plotting both results in the same figure for $\epsilon = 1/2$. Mathematica commands in file: `HoldForm`, `Integrate`, `Sin`, `Cos`, `Pi`, `ReleaseHold`, `DSolve`, `Block`, `NDSolve`, `$DisplayFunction=Identity`, `Plot`, `Evaluate`, `Hue`, `PlotRange`, `Show`, `Ticks`, `PlotPoints`, `PlotLabel`, `StyleForm`, `ParametricPlot`, `ImageSize`, `Frame->True`, `AspectRatio`, `FrameTicks`

Reproduced from MF24, Figure 5.10 shows the KB analytic solution (dashed curves) for the Van der Pol equation for $\epsilon = 1/2$, as well as the exact numerical solution (solid curves). On the left, x has been plotted as a function of t, while on the right, the phase plane portrait ($v = \dot{x}$ versus x) is presented. In this case, the KB solution for $x(t)$ is reasonably close to the exact result, but does less well in producing the correct shape of the limit cycle at large t.

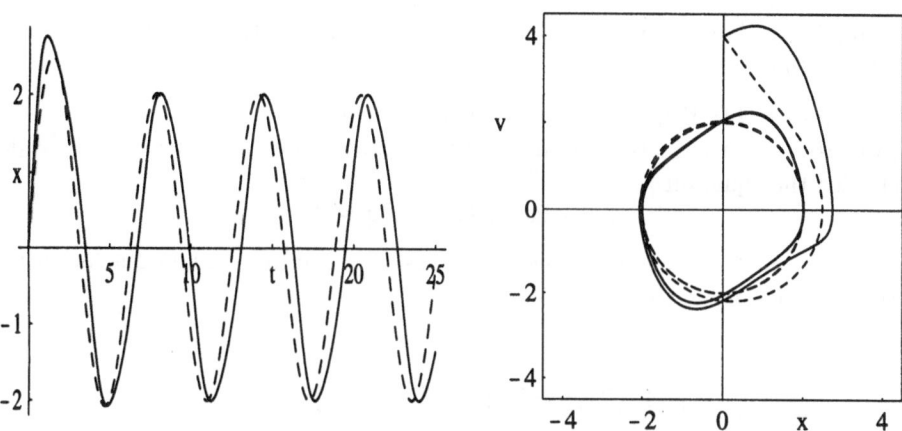

Figure 5.10: KB solution: dashed curves; numerical solution: solid curves.

PROBLEMS
Problem 5-40: Quadratic damping
Consider quadratic damping $f(\dot{x}) = |\dot{x}|\dot{x}$ for the linear spring system (5.140). Determine the first KB approximate solution. Take $\epsilon = 0.5$, $\omega_0 = 1$, $a(0) = 1$, $\phi(0) = \pi/2$ and plot $x(t)$ and $\dot{x}$ vs. x up to $t = 20$. Plot the exact numerical result in the same figures and discuss the accuracy of the KB solution.

Problem 5-41: Coulomb, or dry, friction
Coulomb, or dry, friction refers to the type of friction which occurs when two dry solid surfaces slide over each other. To a crude approximation, the frictional force is found to be a constant (f_0, say) independent of the speed and proportional to the normal force between the surfaces. Of course, the frictional force must change sign with a change of direction of motion. With Coulomb friction present, Newton's second law for 1-d motion of a mass m attached to a spring with spring constant k yields

$$m\ddot{x} + f_0\, sgn(\dot{x}) + kx = 0,$$

where sgn is the signum function

$$sgn(\dot{x}) = +1, \dot{x} > 0; \ = -1, \dot{x} < 0; \ = 0, \dot{x} = 0.$$

Analyze the behavior of the above oscillatory system using phase plane techniques. Solve the equation exactly for each half cycle, making sure to then join the solutions at the end of each half cycle to get the complete solution. Solve the equation using the first KB approximation and compare the result with the exact solution.

Problem 5-42: Van der Pol transient growth
Making use of MF24, compare the KB solution for the Van der Pol transient growth with the numerical solution for different initial radii and phase angles. Take $\epsilon = 0.1, 1, 2$.

Problem 5-43: Hard spring with damping
Add a linear damping term $2\epsilon\gamma\dot{x}$ to the hard spring equation and determine the KB solution.

Problem 5-44: Electrically driven tuning fork
In considering the motion of an electrically driven tuning fork, Lord Rayleigh studied the equation
$$\ddot{x} - 2a\dot{x} + bx^3 + \omega_0^2 x = 0,$$

where a and b are small positive constants. Find an approximate solution to this equation using the KB method.

5.5　Ritz and Galerkin Methods

The perturbation and Krylov–Bogoliubov approaches depended on the nonlinear terms being small compared to the linear part. The Ritz and Galerkin methods are more general and, at least in principle, may be applied to the general ODE

$$f(x, \dot{x}, \ddot{x}, \ldots, t) = 0 \tag{5.145}$$

with f a nonlinear function of its arguments. One assumes a solution of the form

$$\Phi(t) = \sum_{i=0}^{N} C_i \phi_i(t), \tag{5.146}$$

The Ritz Method

where the ϕ_i are appropriately chosen linearly independent functions and the C_i's are constants which are to be adjusted so as to make $\Phi(t)$ as close as possible to the exact solution. There are only loose guidelines to choosing the ϕ_i:

1. Choose simple functions.

2. Keep the number of terms as small as possible to ease the labor involved. (If Mathematica is used, this stipulation can be somewhat relaxed.)

3. Use your intuition, e.g., for an oscillatory problem you might choose a combination of sinusoidal functions.

4. Remember that the assumed solution $\Phi(t)$ must satisfy the same initial conditions as the exact solution $x(t)$. This can be arranged, for example, by insisting that ϕ_0 satisfy the initial conditions, thus fixing C_0 and having all other ϕ_i vanish at $t = 0$.

What criterion can be established to estimate the accuracy of the assumed solution? If Φ were the exact solution, then $f(\Phi, \dot{\Phi}, \ddot{\Phi}, \ldots, t)$ would be identically zero. Since, in general, Φ will only be an approximate solution, $f(\Phi, \dot{\Phi}, \ddot{\Phi}, \ldots, t) \neq 0$, i.e.,

$$f(\Phi, \dot{\Phi}, \ddot{\Phi}, \ldots, t) = e(t) \qquad (5.147)$$

where $e(t)$ is referred to as the "residual". The size of e (whether it is positive or negative) is a measure of the error in the approximate solution, i.e., how much it differs from the exact solution. Since e will generally be a function of t, it may be small over certain ranges of t indicating that the approximate solution is reasonably accurate over those ranges, but large over other ranges of t suggesting a poor approximation. It may therefore be necessary to approximate the exact solution by different assumed solutions $\Phi(t)$ over different ranges of t.

In analogy with the method of least squares for fitting the "best" straight line to a set of experimental points,[10] we shall actually take $e(t)^2$ to be a convenient and simple measure of accuracy. If the range of interest is from a to b, say, then a measure of the total error over that interval is

$$E = \int_a^b e^2(t)\, dt. \qquad (5.148)$$

---

[10]Recall that in the method of least squares, it is desired to fit the best straight line $y = C_1 x + C_2$ to a set of experimental points. This is done by adjusting C_1 and C_2 so as to minimize $E \equiv \sum_{i=1}^{N} e_i^2$ where the residuals $e_1, e_2, \ldots, e_N$ are the differences (positive or negative) between the y coordinates of the experimental points and the y coordinates (at the same x values) of the straight line.

Since a continuous function is involved instead of discrete points, the sum which appears in the least squares method must be replaced with an integral in (5.148). The integral E is to be minimized by adjusting the C_i's. Setting

$$dE = \frac{\partial E}{\partial C_0}dC_0 + \cdots + \frac{\partial E}{\partial C_N}dC_N = 0 \qquad (5.149)$$

and noting that the C_i are independent, we have

$$\frac{\partial E}{\partial C_i} = \frac{\partial}{\partial C_i}\int_a^b e^2(t)\,dt = 0 \qquad (5.150)$$

or

$$\int_a^b e(t)\frac{\partial e(t)}{\partial C_i}\,dt = 0 \qquad (5.151)$$

with $i = 0, 1, 2, \ldots, N$. The resulting N equations can be solved for the coefficients C_i.[11] The procedure that we have outlined is called the Ritz method and (5.151) is called the Ritz condition.

The Galerkin method is identical to the Ritz method, except that the Ritz condition is replaced by the Galerkin condition

$$\int_a^b e(t)\phi_i(t)\,dt = 0 \qquad (5.152)$$

where the $\phi_i(t)$ are the same linearly independent functions used in the trial solution in (5.146). In (5.152), the residual $e(t)$ is said to be "orthogonal" to all of the ϕ_i ($i = 0, 1, 2, 3, \ldots, N$).

The Galerkin condition can be understood from a physical point of view. Consider, for example, a nonlinear mechanical problem described by the $F = ma$ equation

$$m\ddot{x} + f(x, \dot{x}) = 0 \qquad (5.153)$$

where x is the displacement. Each term in this equation represents a force. When the approximate solution $\Phi(t)$ is substituted into the left-hand side of the equation, the residual $e(t)$ results. Clearly $e(t)$ has the dimensions of a force and can be thought of as the "excess force" existing at any instant t because of the error inherent in the approximate solution. The product of this force $e(t)$ and the assumed displacement $\Phi(t)$ has the dimensions of work or energy, and we can interpret $e\Phi$ as a kind of error in the work done by the system.

Galerkin made the reasonable assumption that the average of this work over the time interval of interest (e.g., over a period for an oscillatory problem) should be zero. In mathematical terms this requirement is

$$\int_a^b e(t)\Phi(t)\,dt = 0 \qquad (5.154)$$

or

$$\sum_i C_i \int_a^b e(t)\phi_i(t)\,dt = 0 \qquad (5.155)$$

---

[11]One must of course check that the result corresponds to a minimum and not a maximum.

which, since the ϕ_i are assumed to be linearly independent, yields the Galerkin condition

$$\int_a^b e(t)\phi_i(t)\,dt = 0. \tag{5.156}$$

Lest the reader think that, because of this mechanical example, Galerkin's method is restricted to second-order equations, consider the first-order nonlinear equation describing the capacitor (C) discharge through a nonlinear diode with current–voltage relation $I = aV + bV^2$,

$$C\dot{V} + aV + bV^2 = 0. \tag{5.157}$$

In this case, each term of the equation is a current contribution, so $e(t)$ can be interpreted as an "excess current". The approximate solution $\Phi(t)$ will be a voltage, so $e\Phi$ has the dimensions of power. Galerkin's condition for this situation is then just a statement that the total excess work over the time interval must add up to zero, which again is not unreasonable. The reader might argue that the physical interpretation in this example is slightly different than in the previous one. If we were to keep exactly the same physical interpretation in each case, the mathematical condition would have to be changed accordingly. It is simpler to retain the same mathematical condition irrespective of whether the equation is first or second order. In the Galerkin method, this is precisely what is done.

Finally, before applying Ritz's and Galerkin's methods to some examples, it should be mentioned that for the (linear) Sturm–Louiville equation discussed in standard mathematics texts, it can be proven that the Ritz and Galerkin methods must yield exactly the same result. An excellent discussion of this point may be found on page 264 of the book by Kantorovich and Krylov [KK58].

Example of the Ritz Method

The familiar example of the capacitor discharge through a nonlinear diode is considered. Although this example illustrates how the Ritz method is applied, it also shows how ugly the procedure can become for a problem which was previously solved easily and exactly. It has the virtue, however, of showing the power of Mathematica as seen in the accompanying file MF25.

Example of the Ritz Method
This file shows how Mathematica's symbolic capabilities may be applied in the context of the Ritz method. Mathematica commands in file: `Exp`, `D`, `Integrate`, `NSolve`, `Block`, `$DisplayFunction=Identity`, `Plot`, `Ticks`, `PlotRange`, `PlotStyle`, `RGBColor`, `PlotLabel`, `StyleForm`, `TextStyle`, `Expand`, `Collect`, `Show`, `ImageSize`

The relevant normalized equation is

$$\dot{x}(\tau) = -x - \epsilon x^2 \tag{5.158}$$

with $x = 1$ at $\tau = 0$. The student has seen the perturbation method applied to (5.158) and also has solved it exactly.

If the nonlinear term were not present, the exact solution satisfying the initial condition would be

$$x(\tau) = e^{-\tau}. \tag{5.159}$$

To apply the Ritz method, let's choose

$$\phi_0(\tau) = e^{-\tau}. \tag{5.160}$$

To find an approximate solution to (5.158), assume it to be of the form

$$\Phi(\tau) = e^{-\tau} + C_1 \phi_1(\tau). \tag{5.161}$$

What form should be chosen for ϕ_1? Since ϕ_0 already satisfies the initial condition, ϕ_1 must vanish at $\tau = 0$. From the structure of Equation (5.158), clearly ϕ_1 should go to zero as $\tau \to \infty$. So, let's choose the simplest thing that we can think of, namely

$$\phi_1(\tau) = \tau e^{-\tau}, \tag{5.162}$$

where ϕ_1 is linearly independent of ϕ_0. Then

$$e(\tau) = \dot{\Phi} + \Phi + \epsilon\Phi^2 = C_1 e^{-\tau} + \epsilon(1 + C_1\tau)^2 e^{-2\tau}. \tag{5.163}$$

Since the range of physical interest is $\tau = 0$ to ∞, the Ritz condition (5.151) becomes

$$\int_0^\infty e\frac{\partial e}{\partial C_1}d\tau = 0$$

or

$$\int_0^\infty \left[C_1 e^{-\tau} + \epsilon(1 + C_1\tau)^2 e^{-2\tau}\right]\left[e^{-\tau} + 2\epsilon(1 + C_1\tau)\tau e^{-2\tau}\right]d\tau = 0$$

which, on integrating with Mathematica, yields

$$C_1^3 + \left(3 + \frac{128}{27\epsilon}\right)C_1^2 + \left(4 + \frac{256}{27\epsilon} + \frac{32}{3\epsilon^2}\right)C_1 + \frac{64}{9\epsilon} + \frac{8}{3} = 0. \tag{5.164}$$

Although it is possible to solve this messy cubic equation analytically for arbi-

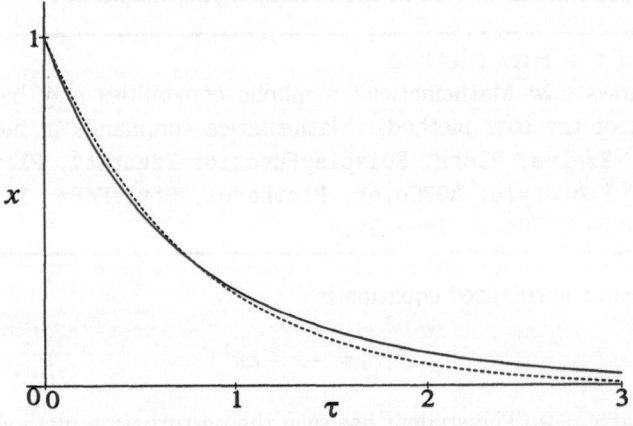

Figure 5.11: Comparison of Ritz (dashed) and exact (solid curve) solutions.

trary ϵ, it is easier to proceed by specifying numerical values for ϵ and using the numerical root solving capability of Mathematica. For example, for $\epsilon = 1/2$, one obtains the pair of complex conjugate roots $(-6.10524 \pm 5.00466\,I)$ and the real root (-0.27100). In this case there is only one real root. If more than one root is real, one must check to see which root gives a minimum value to the total error E, given by (5.148).

The Ritz solution for $\epsilon = 1/2$ is plotted in Figure 5.11 along with the exact solution. The Ritz curve is slightly too high at small t and too low at large t. Other trial functions Φ could be chosen to improve the agreement.

PROBLEMS

Problem 5-45: Ritz approximate solution
Taking $\epsilon = 1, 5$ in the above example, determine the corresponding roots of the cubic equation. Then plot the Ritz approximate solution and compare it with the perturbation result derived earlier. By comparing with the exact solution, comment on the accuracy of the Ritz solution for larger values of ϵ.

Problem 5-46: Other trial functions
Repeat the previous problem, but try several different trial wave functions than the one used in the text example. Compare the accuracy attained with the different wave functions.

Problem 5-47: Galerkin solution
Solve the above example in the text using Galerkin's condition. Take the same $\phi_0(t)$ and $\phi_1(t)$. Compare the result for $\epsilon = 1/2$ with that obtained by the Ritz method and the exact result.

Problem 5-48: Ritz method applied to hard spring
Apply Ritz's method to the nonlinear hard spring equation

$$\ddot{x} + \omega_0^2 x + \epsilon x^3 = 0$$

with $x(0) = A_0$, $\dot{x}(0) = 0$. Assume a solution of the form $\Phi(t) = C_0 \cos(\omega t)$ with ω an unknown frequency. Show that the roots of ω are given by

$$\omega^2 = \omega_0^2 + \left(\begin{array}{c} 0.89 \\ 2.11 \end{array} \right) \epsilon C_0^2.$$

Show that the first root of ω^2 yields the minimum value for E. For ϵ small, calculate the period $T = 2\pi/\omega$ to first order in ϵ and compare your answer with the perturbation result.

Problem 5-49: A different nonlinear spring
Use the Ritz method to find an approximate solution to the following equation describing a certain nonlinear spring,

$$\ddot{x} + a^2 x |x| = 0$$

where $a^2 > 0$ with $x = A_0$, $\dot{x} = 0$ at $t = 0$. Choose $\Phi(t) = C_0 \cos(\omega t)$. Can this equation be solved using the Krylov–Bogoliubov procedure? Explain.

Problem 5-50: Ritz method applied to rotating pendulum

Using a trial wave function of your own choice, use the Ritz method to solve the rotating pendulum equation

$$\ddot{\theta} + \omega_0^2 \sin\theta - \frac{1}{2}\omega^2 \sin(2\theta) = 0,$$

with $\omega_0 = 1$, $\omega = 1$, $\theta(0) = 30°$, and $\dot{\theta} = 0$. Compare your answer with the numerical solution for $\theta(t)$. Plot the two results.

Problem 5-51: Galerkin method applied to rotating pendulum

Using a trial wave function of your own choice, use the Galerkin method to solve the rotating pendulum equation

$$\ddot{\theta} + \omega_0^2 \sin\theta - \frac{1}{2}\omega^2 \sin(2\theta) = 0,$$

with $\omega_0 = 1$, $\omega = 2$, $\theta(0) = 45°$, $\dot{\theta} = 0$. Compare your answer with the numerical solution for $\theta(t)$. Plot the two results.

Chapter 6

The Numerical Approach

What a chimera then is man! What a novelty! What a
monster, what a chaos, what a contradiction, what a
prodigy! Judge of all things, feeble earthworm,
depostitory of truth, a sink of uncertainty and error, the
glory and the shame of the universe.
Blaise Pascal (1623-1662)

The combination of finite-difference approximations to the derivatives and the
use of a high speed digital computer leads to a very powerful approach to solving
the nonlinear ordinary and partial differential equations of physics. For many
nonlinear systems, particularly those where the nonlinear terms are not small
corrections to an otherwise linear behavior, the numerical route may be the best
or only feasible way to travel. For the nonlinear ODEs encountered earlier in

The First Numerical Approach

the text, the student has been allowed to use the Mathematica numerical ODE
solver without any explanation provided of the principles on which it is based.
In this chapter, we would like to partially fill that void by briefly describing
how some of the common numerical schemes for solving nonlinear ODEs are
derived. Our aim is to provide a simple conceptual framework that will make

the reader more comfortable with the numerical approach while progressing through the rest of the topics that lie ahead. It should be emphasized that we are not attempting to explain the code which underlies Mathematica's `NDSolve` command which is about 500 pages long.

Since many of the underlying numerical concepts do not depend on whether the system is linear or nonlinear, occasionally a linear ODE shall be chosen for illustrative purposes because an exact analytic solution can then be generated. This will allow us to readily gauge the success, or lack thereof, of a particular numerical approach. For the same reason, the numerical schemes will be tested on nonlinear systems already solved in earlier Mathematica files. It should be strongly emphasized that the treatment presented in this chapter is no substitute for taking a relevant formal numerical analysis course. At best, we can only provide a very small glimpse of a vast and rapidly growing subject. It should be noted, however, that all of the concepts developed in this chapter, although expressed in terms of Mathematica, are relevant to programming in other computer languages. In a later chapter, the concepts developed here will be extended and some algorithms for numerically solving nonlinear PDEs derived.

Before plunging into our exploration of the numerical realm, we would be remiss if the excellent reference book on numerical techniques entitled *Numerical Recipes* by Press, Flannery, Teukolsky, and Vetterling [PFTV89] were not mentioned. This book has rapidly become the main source used by researchers looking for the best numerical methods to use in a given situation. For students who are studying computer programming, the book is available in different versions, namely in Fortran, in C, and in Pascal.

6.1   Finite-Difference Approximations

To numerically solve, for example, the laser competition equations or the Van der Pol equation, we must first learn how to represent derivatives by their finite difference approximations. To see how this works, consider the arbitrary function $y(x)$ sketched in Figure 6.1 whose first and second derivatives with respect to x, viz., y' and y'', are desired at the point P. The notationally convenient prime notation is used here to denote derivatives. The independent variable has been taken to be x, but it could just as well be time t. The standard mathematical convention is to use primes for spatial derivatives and, as has been done many times already in the text, dots for time derivatives.

Let R and Q be two neighboring points on $y(x)$ very close to P (i.e., h in the figure is supposed to be small). Taylor expanding $y(x)$ in powers of h at R and Q yields

$$y(x+h) = y(x) + hy'(x) + (1/2!)h^2y''(x) + (1/3!)h^3y'''(x) + \cdots \qquad (6.1)$$

$$y(x-h) = y(x) - hy'(x) + (1/2!)h^2y''(x) - (1/3!)h^3y'''(x) + \cdots \qquad (6.2)$$

Subtracting the second expansion from the first and keeping only the lowest-order term on the right-hand side (rhs) gives

$$y'(x) = (1/2h)[y(x+h) - y(x-h)] + O(h^2) \equiv \Delta y/\Delta x + O(h^2), \qquad (6.3)$$

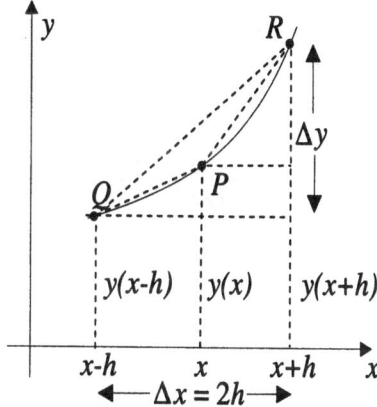

Figure 6.1: Obtaining finite difference approximations to derivatives.

where $O(h^2)$ indicates that the error in this approximation to the first derivative is of order h^2. Equation (6.3) is just the slope of the chord QR which, in the limit $\Delta x = 2h \to 0$, approaches the slope of the tangent to the curve $y(x)$ at P (i.e., the derivative at P). For small, but finite, h Equation (6.3) is called the "central-difference approximation" to the first derivative.

Two alternate approximations to the first derivative may also be derived. The "forward-difference approximation" results on dropping terms in (6.1) of order h^2, h^3, etc.,

$$y'(x) = (1/h)[y(x + h) - y(x)] + O(h). \tag{6.4}$$

This result corresponds to approximating the derivative at P by the slope of the chord PR.

The "backward-difference approximation" is obtained in a similar manner from Equation (6.2):

$$y'(x) = (1/h)[y(x) - y(x - h)] + O(h). \tag{6.5}$$

This is just the slope of the chord QP. For a given size of h, the backward-and forward-difference approximations are clearly not as accurate as the central-difference approximation. However, as shall be seen, the forward-difference approximation is commonly used to represent the time derivative in explicit schemes because it allows us to advance forward in time from $t = 0$.

How is the second derivative y'' represented? Adding Equations (6.1) and (6.2) produces the finite-difference formula

$$y''(x) = (1/h^2)[y(x + h) - 2y(x) + y(x - h)] + O(h^2). \tag{6.6}$$

Just as with the first derivative, other possible finite difference approximations may be obtained for the second derivative. The following example uses Mathematica's **Series** command to confirm an alternate finite difference approximation for the second derivative. This new approximation, containing five y terms instead of the three y terms occuring in Equation (6.6), is of $O(h^4)$ accuracy.

Example 6-1: Alternate Difference Approximation for y''

By also Taylor expanding $y(x + 2h)$ and $y(x - 2h)$ show that

$$y''(x) = \frac{-y(x + 2h) + 16y(x + h) - 30y(x) + 16y(x - h) - y(x - 2h)}{12\,h^2} + O(h^4).$$

Solution: To verify the above difference approximation of $y''(x)$,

```
Clear["Global`*"]
```

the function $f(i) = y(x + i\,h)$ is created, with i the variable.

```
f[i_]:=y[x+ih]
```

The five y terms in the numerator of the $y''(x)$ approximation correspond to f[2], f[1], f[0], f[-1], and f[-2]. The numerator is now entered.

```
num=-f[2]+16f[1]-30f[0]+16f[-1]-f[-2]
```

$$-30\,y(x) - y(x - 2\,h) + 16\,y(x - h) + 16\,y(h + x) - y(2\,h + x)$$

The left-hand side of eq below is just the formal sum given above, while on the right-hand side we expand num in powers of h, keeping terms to order h^5.

```
eq=num==Series[num, {h, 0, 5}]
```

$$-30\,y(x) - y(x - 2\,h) + 16\,y(x - h) + 16\,y(h + x) - y(2\,h + x) == 12\,y''(x)\,h^2 + O(h^6)$$

In the output, we see that only a term of order h^2 survives on the right-hand side, the next term being of order h^6. Since $y''(x)$ appears in this h^2 term, we can extract the second derivative (labeled secder) by applying Normal to eq to remove $O(h^6)$ and solving for $y''(x)$.

```
sol=Solve[Normal[eq], y''[x]]
```

$$\left\{\left\{y''(x) \rightarrow -\frac{30\,y(x) + y(x - 2\,h) - 16\,y(x - h) - 16\,y(h + x) + y(2\,h + x)}{12\,h^2}\right\}\right\}$$

```
secder=y''[x] /. sol[[1]]
```

$$-\frac{30\,y(x) + y(x - 2\,h) - 16\,y(x - h) - 16\,y(h + x) + y(2\,h + x)}{12\,h^2}$$

The output in secder agrees with the suggested finite difference formula for $y''(x)$. The error in this approximation is clearly $O(h^6)/h^2 = O(h^4)$.

End Example 6-1

As mentioned earlier, this alternate finite approximation for $y''(x)$ contains two more terms than in the approximation given by Equation (6.6). Additional terms mean that for the same step size more computing time would be needed to evaluate the second derivative. But, on the other hand, a higher order in h allows one to use larger h steps to achieve the same numerical accuracy and this reduces the computing time. In assessing any numerical scheme built up out of finite difference approximations, it becomes a trade-off between these two competing aspects because there is a limit to how big h can be and still obtain meaningful results. This latter point follows from the fact that the numerical finite difference methods are built on the Taylor series expansion with a finite number of terms being retained.

The finite-difference approximation concept is easily extended to situations where y is a function of more than one independent variable, e.g., x and t, and partial differential equations are to be solved. This conceptually simple extension is left until Chapter 11.

PROBLEMS
Problem 6-1: Comparison of approximations for $y''(x)$
Consider the function $y(x) = x^{5.1}$. Taking $h = 0.1$ and Mathematica's default accuracy, evaluate the second derivative of $y(x)$ at $x = 4.95$ using the approximation of Example 6-1 as well as the approximation given by Equation (6.6). Compare your two answers with the exact result and comment on their accuracy. Repeat the problem with increasingly smaller step sizes and discuss the results.

Problem 6-2: Fourth derivative
Show that

$$y''''(x) = (1/h^4)[y(x + 2h) - 4y(x + h) + 6y(x) - 4y(x - h) + y(x - 2h)] + O(h^2).$$

Suggest a physical nonlinear equation from Chapter 3 for which this representation of the fourth spatial derivative might prove useful. Consider the function $y = x^{5.2}$. Taking $h = 0.1$ and Mathematica's default accuracy, plot the difference between the exact fourth derivative and this approximation over the range $x = 0...20$. Repeat with smaller and smaller step sizes and discuss the results.

Problem 6-3: Comparison of first derivative approximations
Consider the function $y(x) = x^5$. By plotting on the same graph the difference between the exact first derivative and each of the approximations, compare the central, forward, and backward approximations over the range $x = 0...20$ with each other and with the exact derivative for $h = 0.1, 0.01, 0.001, 0.0001$. Discuss the results.

6.2   Euler and Modified Euler Methods

Although it is not employed for serious research calculations because very small steps in h must be used to gain acceptable accuracy thus causing it to be slow, the Euler method is historically and conceptually important. It is the starting point for understanding all explicit single step schemes.

6.2.1 Euler Method

To illustrate the Euler method, an example which is already quite familiar to the reader is chosen, namely the rabbits–foxes equations which were encountered in the first survey chapter and which the student has explored in MF04. Since it is known what the solution should look like, this will help us to get a feeling for how well the Euler method is working as, for example, the step size h is changed. The rabbits–foxes equations are

$$\dot{r} = 2r - arf, \quad \dot{f} = -f + arf. \tag{6.7}$$

This set of equations, with $a = 0.01$, was solved numerically in MF04 using the default implementation of Mathematica's numerical ODE solving command NDSolve. As mentioned earlier, the explanation of the underlying code is beyond the scope of this chapter. One option which is available in NDSolve is Method ->RungeKutta which is a variation on the Rungé–Kutta (RK) scheme due to Fehlberg. We shall be content here to explain how Rungé–Kutta schemes work, including the Rungé–Kutta–Fehlberg 4-5 (RKF45) method. We can use Mathematica's programming ability to implement the various RK schemes.

We begin by first examining the Euler method. This latter approach involves using the forward-difference approximation for the time derivative, thus connecting, say, the kth time step to the step $k + 1$, and approximating the rhs of the equation(s) by its value at the kth step. Starting at the zero time step in an initial value problem, one progresses from one temporal step to another, a single step at a time. The new value of the dependent variable is expressed explicitly in terms of the old one. For the rabbits–foxes equations, on setting $t_{k+1} = t_k + h$, this procedure gives us

$$r_{k+1} = r_k + h[2r_k - ar_k f_k]$$

$$f_{k+1} = f_k + h[-f_k + ar_k f_k]. \tag{6.8}$$

Thus, assuming that r and f have been evaluated at the kth time step, the rhs of the equations is determined and the values of r and f may be explicitly calculated at the $k + 1$ step. For example, taking $h = 0.02$, $a = 0.01$, and the initial populations to be $r(t = 0) = 300$, $f(t = 0) = 150$, the evolution of the populations for, say, $n = 1000$ time steps can be calculated by writing the Euler algorithm in terms of Mathematica as follows.

Example 6-2: Euler Algorithm

Numerically solve the rabbits–foxes equations (6.8) for the specified conditions.

Solution: The ScatterPlot3D command will be used. This command is found

```
Clear["Global`*"]
```

in Mathematica's Graphics package which is now loaded.

```
<< Graphics`
```

The initial time, initial rabbit and fox numbers, interaction coefficient, step size, and number of time steps are entered.

```
t[0] = 0; r[0] = 300; f[0] = 150; a = .01; h = .02; n = 1000;
```

The `TimeUsed[]` command below gives the total number of seconds of cpu time used so far in the current Mathematica session. The cpu starting time for implementation of the Euler algorithm is recorded, but not displayed.

```
cpustart = TimeUsed[];
```

Functions are created for the relevant Euler difference equations (replacing k with $k-1$ in (6.8) and the time increment relation).

```
r[k_] := = r[k] = r[k-1] + h (2 r[k-1] - a r[k-1] f[k-1])
```

```
f[k_] := f[k] = f[k-1] + h (-f[k-1] + a r[k-1] f[k-1])
```

```
t[k_] := = t[k] = t[k-1] + h
```

The following `Table` command iterates the equations, producing a list of three numbers (time, rabbit number, and fox number) at each k value up to $k = n$. These are the plotting point coordinates in the 3-dimensional t–r–f space.

```
pts = Table[{t[k], r[k], f[k]}, {k, 0, n}];
```

The "current" cpu time is again recorded and the starting cpu time subtracted,

```
cpuend = (TimeUsed[] - cpustart) seconds
```

0.16 seconds

yielding the cpu time to execute the Euler algorithm. In this case it took about 0.16 seconds on a 1 GHz PC. This number will vary slightly from one run to the next. Can you suggest why? How much computer time is used to go out to the same total time (e.g., $t = 1000 \times 0.02 = 20$, above) in obtaining the numerical solution is an important issue when comparing numerical schemes for dealing with complex nonlinear systems. A 3-dimensional picture of all of the plotting points is created by applying the `ScatterPlot3D` command, the resulting 3-dimensional trajectory being shown in Figure 6.2.

```
ScatterPlot3D[pts, BoxRatios -> {1.6, 1, 1}, ViewPoint -> {3, 2, 1},
   PlotStyle -> {RGBColor[0, 0, 1]}, Boxed -> False, AxesEdge -> {{-1,-1},
   {-1, -1}, {-1, -1}}, Ticks -> {{0, {8, "t"}, 20}, {200, {275, "r"}, 400},
   {200, {350, "f"}, 400, 600}}, TextStyle -> {FontFamily -> "Times",
   FontSize -> 16}, ImageSize -> {400, 400}];
```

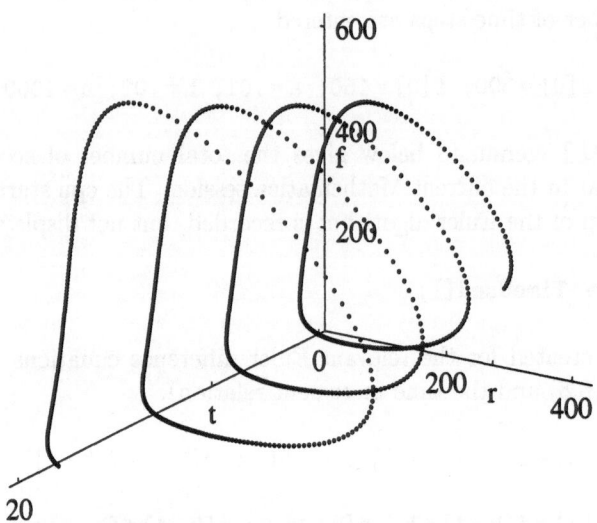

Figure 6.2: Euler solution of rabbits (r)–foxes (f) equations.

Each successive dot in Figure 6.2 represents a time step so one can see where the nonlinear system moves quickly with time along the trajectory and where it doesn't. To compare with Figure 2.5 generated from MF04, the `Table` command can be used again to create plotting points in the 2-dimensional r–f space,

```
pts2 = Table[{r[k], f[k]}, {k, 0, n}];
```

and the points plotted with the `ListPlot` command.

```
ListPlot[pts2, PlotStyle -> {RGBColor[0, 0, 1]}, AspectRatio -> 1,
Ticks -> {{{.01, "0"}, 200, {300, "r"}, 400}, {200, {300, "f"}, 400,
600}}, TextStyle -> {FontFamily -> "Times", FontSize -> 16},
ImageSize -> {400, 400}];
```

The resulting phase plane picture is shown in Figure 6.3. Time runs counterclockwise in the figure from the starting point $r[0] = 300$, $f[0] = 150$.

End Example 6-2

The phase plane trajectory in Figure 6.3 is consistent with what is predicted by phase plane analysis, viz., a saddle point at the origin and a vortex or focal point at $r = 100$, $f = 200$. If Mathematica File MF04 had not been included, it would be tempting to conclude from this figure that the second singular point is an unstable focal point. This conclusion would be wrong. Cumulative

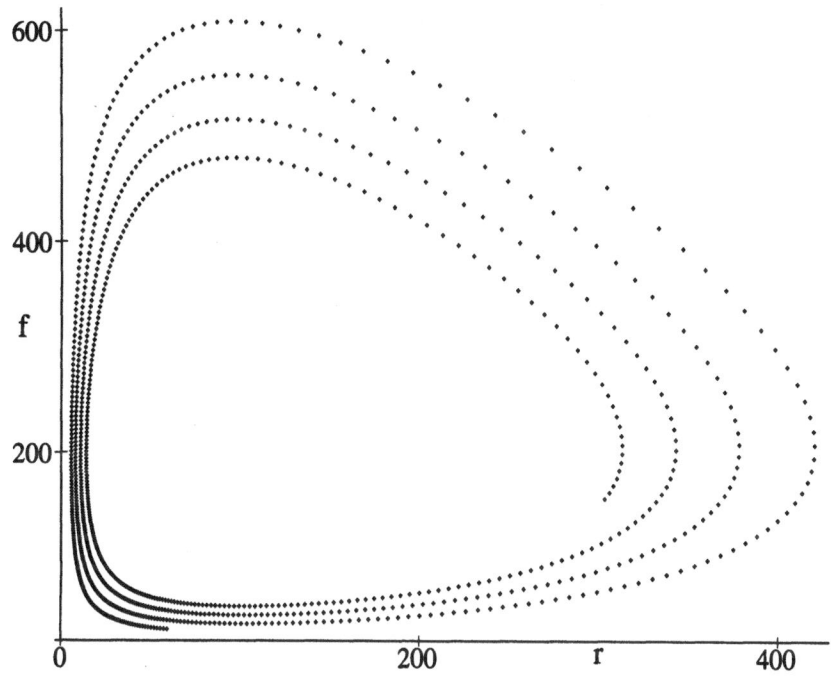

Figure 6.3: Phase plane picture for r–f equations generated by Euler method.

numerical error over the 1000 time steps involved has prevented the trajectory from being a closed loop or vortex. How does such cumulative error arise? On each time step, there is some inherent error due to our choice of algorithm, the step size specified, and the fact that a finite number of digits is used. For a given step size and number of digits, any algorithm will be characterized by a so-called truncation error. The Euler method is of order h accuracy and is said to have a truncation error of order h. As shall be seen, this corresponds to a Taylor series expansion of the rhs of our equations to this order. The truncation error arises because all higher order terms in the Taylor series have been neglected. The fourth-order RK algorithm which shall be encountered in a later section corresponds to a Taylor series expansion to order h^4, and therefore has a truncation error of this order. For a given step size, the fourth-order RK method will be more accurate than the Euler method because it keeps more terms in the Taylor series. It will have a smaller truncation error.

Given that the Euler method has been used for our rabbits and foxes example, how do we know that our step size is small enough to give as accurate an answer as is practically possible for the chosen method? Although $h = 0.02$ sounds pretty small, it really isn't when applying the Euler method. A standard check on the adequacy of the step size for any numerical scheme is to cut the step size in half and double the number of steps to achieve the same total time. If this produces a noticeable difference in the answer, repeat this step doubling procedure a number of times until no appreciable difference is observed.

In the next Figure 6.4, the step size has been cut by four to $h = 0.005$, but $n = 4000$ so that the total time is still 20. Notice that the trajectory comes closer to closing. So, you might say, the step size should be cut even further.

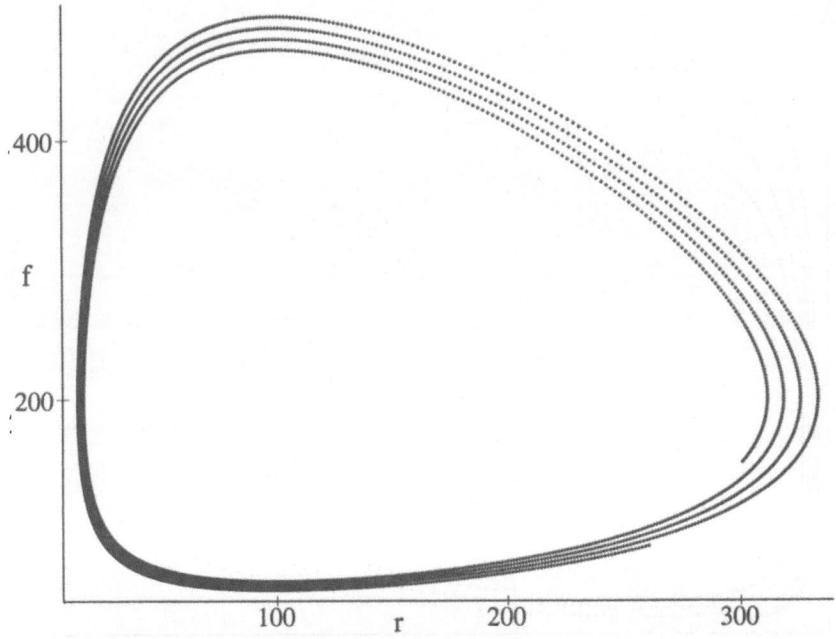

Figure 6.4: Applying the step doubling procedure to the Euler method.

How many times this step size halving procedure[1] can be carried out depends
on the speed and memory of the computer that is being used. A much better
approach is to devise a numerical scheme for a given step size h which is of higher
order accuracy in h. This will enable us to attain reasonable accuracy without
taking the time step to be extremely small. Sticking with the forward-difference
approximation for the time derivative which allows us to advance forward in
time from $t = 0$, this means that we have to improve our approximation of the
rhs of our equations. A systematic approach involving the Taylor series concept
will be presented in the next section, but first let's see how a slight alteration of
the Euler method leads to an algorithm of $O(h^2)$ accuracy, viz. the "modified"
(or "improved") Euler method.

6.2.2   The Modified Euler Method

So that a comparison can be made with the Euler method, the rabbits–foxes
equations[2] are tackled once more. Let's write our nonlinear system with the rhs
of the two equations represented for notational brevity by $R(r, f)$ and $F(r, f)$,
respectively, viz.,

$$\dot{r} = R(r, f), \quad \dot{f} = F(r, f). \tag{6.9}$$

Setting $t_{k+1} = t_k + h$, $R_1[k] \equiv R(r_k, f_k)$, $F_1[k] \equiv F(r_k, f_k)$, $R_2[k] \equiv R(r_k + hR_1[k], f_k + hF_1[k])$, and $F_2[k] \equiv F(r_k + hR_1[k], f_k + hF_1[k])$, the modified Euler

---

[1]Even if it were practical to do so, one cannot make h too small or error will occur because
of the finite number of digits being used.

[2]We hope that the reader hasn't tired of this ongoing saga of the wild kingdom.

equations are

$$r_{k+1} = r_k + (1/2)h(R_1[k] + R_2[k])$$

$$f_{k+1} = f_k + (1/2)h(F_1[k] + F_2[k]).$$

(6.10)

That this scheme is of $O(h^2)$ accuracy may be confirmed by writing the rhs of the equations out in detail and comparing with the Taylor series expansion (see Problem 6.5). Assuming that $h = 0.02$ and that the initial conditions and parameter values are the same as before, the algorithm for the modified Euler method applied to the rabbits–foxes equations is as follows.

Example 6-3: Modified Euler Algorithm

Apply the modified Euler algorithm to the rabbits–foxes equations for the same initial conditions, input parameters, and step size as previously, but take twice as many steps, ie., take $n = 2000$.

Solution: Except for the change from $n = 1000$ to $n = 2000$, the input

```
Clear["Global`*"]
```

parameters are the same as in Example 6-1.

```
t[0] = 0;  r[0] = 300;  f[0] = 150;  a = .01;  h = .02;  n = 2000;
```

The starting cpu time is recorded,

```
cpustart = TimeUsed[];
```

and functions created for the modified Euler algorithm.

```
r1[k_] := 2 r[k] - a r[k] f[k]

f1[k_] := -f[k] + a r[k] f[k]

r2[k_] := 2 (r[k] + h r1[k]) - a (r[k] + h r1[k]) (f[k] + h f1[k])

f2[k_] := -(f[k] + h f1[k]) + a (r[k] + h r1[k]) (f[k] + h f1[k])

r[k_] := r[k] = r[k - 1] + 0.5 h (r1[k - 1] + r2[k - 1])

f[k_] := f[k] = f[k - 1] + 0.5 h (f1[k - 1] + f2[k - 1])

t[k_] := t[k] = t[k - 1] + h
```

The modified Euler equations are iterated with the Table command, creating the plotting points in the two-dimensional r–f space.

    pts = Table[{r[k], f[k]}, {k, 0, n}];

The current cpu time is again recorded and the starting cpu time subtracted,

    cpuend = (TimeUsed[] - cpustart) seconds

    0.82 seconds

yielding a cpu time for implementation of the modified Euler algorithm of about 0.82 seconds, again performed on a 1 GHz PC. The ListPlot command is used to plot the points and produce the phase plane picture shown in Figure 6.5.

    ListPlot[pts, PlotStyle -> {RGBColor[0, 0, 1]}, AspectRatio -> 1,
    Ticks -> {{{-5, "0"}, 100, {150, "r"}, 200, 300}, {100, 200, {250,"f"},
    300, 400}}, TextStyle -> {FontFamily -> "Times", FontSize -> 16},
    ImageSize -> {400, 400}];

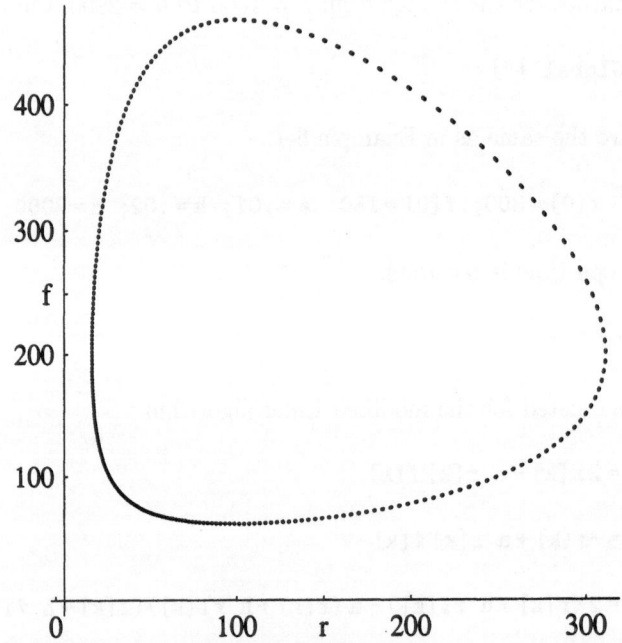

Figure 6.5: Modified Euler algorithm applied to r–f equations.

End Example 6-3

A significant improvement in the accuracy of the solution compared to the Euler method for the same step size is gained. Even though the run is twice as long as in Figure 6.3, the trajectory in Figure 6.5 looks more like that to be expected around a vortex point. It is even better than the numerical result obtained in Figure 6.4 which used a time step four times smaller. However, the modified Euler algorithm tends to be slower than the Euler algorithm because

of the extra evaluations which must be performed on the rhs. As one goes to
higher and higher order accuracy, more and more evaluations must be carried
out on the rhs. It is natural to ask if there is a certain optimum order beyond
which the number of evaluations of the rhs to be carried out cancels the time
advantage that can be gained by using larger h steps due to increased accuracy.
Remember that h cannot be increased indefinitely. In fact, for the rabbits–foxes
system, trouble arises using the modified Euler method even for h below 1.

Figure 6.6 shows the results obtained for the rabbits–foxes system with the

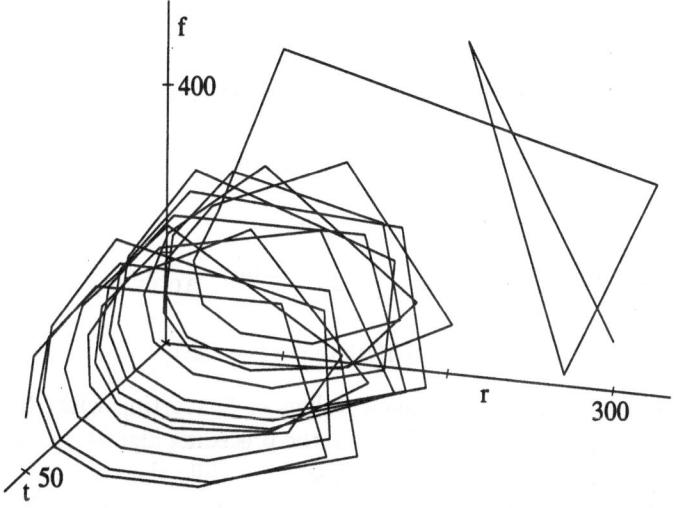

Figure 6.6: Behavior of r–f equations for $h = 0.565$ in modified Euler code.

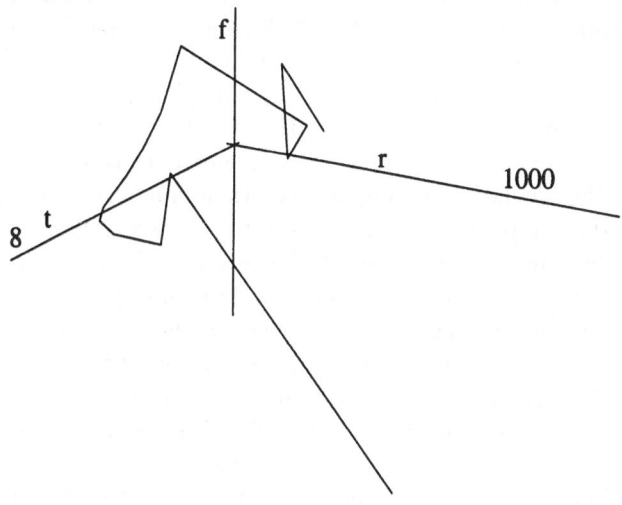

Figure 6.7: Onset of numerical instability in modified Euler code for $h = 0.570$.

modified Euler algorithm using $h = 0.565$, $n = 100$, and the `ScatterPlot3D` command. For viewing convenience, the `PlotJoined -> True` option has been used to join the plotting points by straight lines. If the points are not connected, the figure turns out to be rather confusing. Although the behavior in Figure 6.6 is qualitatively correct, the curve has become quite jagged and quantitatively inaccurate when compared with the smaller step size result in Figure 6.5.

Now, increase the step size slightly to $h = 0.570$. For $n = 14$, the trajectory shown in Figure 6.7 results. The run has been stopped at a small n value because, as the reader may confirm, for larger n, the curve does not close on itself but instead diverges to infinity extremely rapidly. Completely incorrect behavior has occurred, and we say that "numerical instability" has set in. Although the detailed form of the numerical instability and at what step size it occurs depends on the ODE being solved and the algorithm being used, it is always a problem when the step size is too large.

A combination of a trivial linear ODE and the Euler method reinforces this point. Consider the ODE

$$y' = -10y \tag{6.11}$$

with $y(0) = 1$. The exact solution is $y(x) = e^{(-10x)}$. Applying the Euler method yields

$$y_{k+1} = y_k - h(10y_k) = (1 - 10h)y_k. \tag{6.12}$$

Let's see what happens for increasing values of h, starting with the relatively small value $h = 1/20 = 0.05$. In this case, the preceding algorithm yields $y_0 = 1$, $y_1 = 0.5$, $y_2 = 0.25$, $y_3 = 0.125$, $y_4 = 0.0625,...$ while the corresponding exact values (quoted to three figures) are $1, 0.607, 0.368, 0.223, 0.135,....$ Although the numerical values are not too close to the exact values because the step size is not small enough, the numerical scheme is said to be stable because the numerical results qualitatively mimic the behavior of the exact solution.

For $h = 1/10$, the numerical scheme produces $y_1 = 0$, $y_2 = 0$, $y_3 = 0,....$ The numerical solution immediately drops to zero and remains there. Not surprisingly, numerical instability sets in for larger h values, the numerical solution displaying incorrect behavior. For $h = 3/20$, the Euler method gives $y_0 = 1$, $y_1 = -0.5$, $y_2 = 0.25$, $y_3 = -0.125$, $y_4 = 0.0625,....$ The numerical solution has begun to show spurious oscillatory behavior, although the oscillations are at least correctly decaying. For $h = 3/10$, the numerical values are $y_0 = 1$, $y_1 = -2$, $y_2 = 4$, $y_3 = -8$, $y_4 = 16$, ... Any resemblance to the correct solution has been lost, the oscillations diverging, even though h is still well below $h = 1$. So, based on this simple example and the rabbits–foxes example treated with the modified Euler method, the student should always be careful not to end up in too large a step size regime where numerical instability can occur.

To this point, our approach to developing a higher-order accuracy solution has not been systematic. To attempt to answer the question of what is the optimum algorithm to use, a systematic procedure of generating higher-order accuracy algorithms must be developed. The Rungé–Kutta (RK) approach discussed in the next section provides such a treatment.

Before tackling this topic, the reader is encouraged to try some of the following problems which involve modifying the code in the examples to handle the given ODE systems

PROBLEMS

Problem 6-4: Van der Pol equation

Choosing $h = 0.01$, solve the Van der Pol equation for $\epsilon = 5$ and $x(0) = \dot{x}(0) = 0.1$ using both the Euler and modified Euler algorithms. Let t run up to 15, and by suitably orienting your plot compare your answers with the phase portrait solution of the limit cycle Problem 2-38 (page 74).

Problem 6-5: Accuracy of modified Euler method

Confirm explicitly that the modified Euler method is of order h^2 accuracy in a Taylor series expansion for the rabbits–foxes equations. To do this, show, for example, that the Taylor series expansion for the rabbits equation is

$$r_{k+1} = r_k + h(\dot{r})_k + \frac{h^2}{2!}(\ddot{r})_k + \cdots$$

or

$$r_{k+1} = r_k + hR_k + \frac{h^2}{2}(\frac{\partial R}{\partial r}R + \frac{\partial R}{\partial f}F)_k + \cdots$$

with the subscript $k+1$ referring to $t_{k+1} = t_k + h$. Evaluate the terms in this expansion and show that they exactly agree with the expansion of Equation (6.10) in the text to order h^2. (The two results do not agree to order h^3.)

Problem 6-6: White dwarf equation

In his theory of white dwarf stars, Chandresekhar [Cha39] introduced the non-linear equation

$$x(d^2y/dx^2) + 2(dy/dx) + x(y^2 - C)^{3/2} = 0,$$

with the boundary conditions $y(0) = 1$, $y'(0) = 0$. Making use of the Euler method with $h = .01$, numerically compute $y(x)$ over the range $0 \leq x \leq 4$ with $C = 0.1$ and plot the result. (Hint: Start at $x = 0.01$ to avoid any problem at the origin.)

Problem 6-7: Thomas–Fermi equation

A nonlinear ODE due to Thomas [Tho27] and Fermi [Fer28] which appears when determining the effective nuclear charge in heavy atoms is

$$y'' = (1/\sqrt{x})y^{3/2}$$

with boundary conditions $y(0) = 1$, $y(x) \to 0$ as $x \to \infty$. Numerically determine $y(x)$ using either the Euler or modified Euler methods and plot the result. You are free to choose the h value. (Hint: To achieve the asymptotic boundary condition, your initial slope at the origin should be about -1.59. You also have to start slightly outside $x = 0$ to avoid the singularity. This problem will probably involve considerable fiddling of the input slope condition and step size.) As a check, note that [BC31] the numerical values at $x = 1.0$, 2.0, 4.0, 10.0, 20.0 should be 0.425, 0.247, 0.106, 0.0244, 0.0058, respectively.

Problem 6-8: Spruce budworm infestation

The sudden outbreak of the spruce budworm which can rapidly defoliate a forest and kill the trees can be described [LJH78] by the dimensionless equation

$$\dot{x}(\tau) = rx\left(1 - \frac{x}{k}\right) - \frac{x^2}{1 + x^2}.$$

Here $x(\tau)$ is proportional to the budworm population number at time τ and the growth coefficient r and carrying capacity parameter k are positive constants. The first term describes the growth of the budworm population with a saturation effect included due to the finite forest available, while the last term models the decrease in population due to bird predation. Taking $k = 300$, use the Euler method with $h = 0.01$ and $x(0) = 0.5$ to determine the time evolution of the budworm population for $r = 0.1$, $r = 0.5$, and $r = 1.0$. Discuss the results.

Problem 6-9: Biochemical switch

Explaining biological pattern formation, such as the stripes on a zebra and the spots on a leopard, is an area of much interest [Mur89]. A simple dimensionless model of a biochemical switch for turning on a gene, which is normally inactive, to produce a pigment is as follows:

$$\dot{x}(\tau) = s - rx + \frac{x^2}{1 + x^2}.$$

Here $x(\tau)$ is proportional to the concentration of pigment produced by the gene, $s > 0$ proportional to a fixed concentration of biochemical signal substance which activates the gene, and $r > 0$ a degradation coefficient. The last term in the equation represents a positive feedback process which stimulates the production of pigment. Explore this model for different values of s and r using the modified Euler method with $h = 0.01$ and $x(0) = 0$. Discuss your results.

Problem 6-10: Artificial example

Consider the nonlinear equation

$$\frac{dy}{dx} = xy(y - 2)$$

with $y(0) = 1$. Taking $h = 0.02$, solve for $y(x)$ out to $y = 3$ using Euler's method. Solve the equation exactly with Mathematica and compare the numerical and analytic results by plotting them in the same figure.

6.3　Rungé–Kutta (RK) Methods

6.3.1　The Basic Approach

The Euler and modified Euler approaches are the simplest examples of single-step explicit methods for tackling nonlinear ODEs. A broad class of such methods can be developed, referred to as the Rungé–Kutta methods [see, e.g., L. Lapidus and J.H. Seinfeld, *Numerical Solution of Ordinary Differential Equations* [LS71]], in which the Euler and modified Euler methods correspond to the lowest two approximations.

Keeping in mind that a second-order differential equation can always be rewritten as two equivalent first-order equations, let us consider the first-order nonlinear ODE

$$y' = f(x, y) \tag{6.13}$$

with $f(x, y)$ a nonlinear function. Our discussion of Equation (6.13) can be easily extended to a coupled set of such equations.

Rungé was the first to point out that it was possible to avoid the successive differentiation in a Taylor series expansion while still preserving the accuracy. In the Rungé–Kutta approach, one bypasses the derivatives in a Taylor series expansion by introducing undetermined parameters and making the result as high-order accuracy as desired by including additional evaluations of $f(x,y)$ within the interval (x_k, y_k) to (x_{k+1}, y_{k+1}).

The general single-step equations,

$$y_{k+1} = y_k + \sum_{i=1}^{v} w_i K_i \tag{6.14}$$

are set up, where the w_i are referred to as the "weighting coefficients", v is the order of accuracy of the RK method, and

$$K_i = hf(x_k + c_i h, y_k + \sum_{j=1}^{i-1} a_{ij} K_j) \tag{6.15}$$

with $c_1 = 0, 0 < c_i \leq 1$ for $i \geq 2$, and h the step size. Explicitly, Equation (6.15) yields

$$K_1 = hf(x_k, y_k)$$

$$K_2 = hf(x_k + c_2 h, y_k + a_{21} K_1) \tag{6.16}$$

$$K_3 = hf(x_k + c_3 h, y_k + a_{31} K_1 + a_{32} K_2),$$

and so on. For $v = 1$ and $w_1 = 1$, the Euler formula is regained. For $v = 2$, $w_1 = w_2 = 1/2$, $c_2 = 1$, and $a_{21} = 1$, the modified Euler formula results. Since it is desired to go beyond these lowest-order algorithms, we shall consider higher values of v and proceed in a more general manner. The parameters $w_1, w_2, ..., w_v$; $c_2, ..., c_v$; and the a_{ij} must be determined. Each set of parameters (when obtained) will specify the points (x, y) between (x_k, y_k) and (x_{k+1}, y_{k+1}) at which $f(x, y)$ is to be evaluated.

To determine these unknown parameters, expand y_{k+1} in powers of h such that it agrees with the solution of the differential equation to a specified number of terms in a Taylor series. As a concrete example, take $v = 3$. Looking back at our general formulas (6.14), (6.15), there are eight undetermined parameters, namely $w_1, w_2, w_3, c_2, c_3, a_{21}, a_{31}, a_{32}$.

Now, noting that f is a function of x and $y(x)$ and using subscripts to denote partial derivatives, we have

$$y' = f(x, y) \equiv f$$

$$y'' = f_x + f_y f \tag{6.17}$$

$$y''' = f_{xx} + 2f_{xy}f + f_{yy}f^2 + f_y f_x + f_y^2 f.$$

Labeling $f(x_k, y_k) \equiv f_k$ and $y(x_k + h) \equiv y_{k+1}$, a Taylor expansion to third order in the derivatives of the following form can then be written

$$y_{k+1} = y_k + h(y')_k + (h^2/2!)(y'')_k + (h^3/3!)(y''')_k$$

$$= y_k + hf_k + (1/2)h^2(f_x + f_y f)_k \tag{6.18}$$

$$+ (1/6)h^3(f_{xx} + 2f_{xy}f + f^2 f_{yy} + f_y f_x + f_y^2 f)_k.$$

Now expand the expressions for K_1, K_2, K_3 in powers of h out to the same order h^3:

$$K_1 = hf_k$$

$$K_2 = hf_k + h^2(c_2 f_x + a_{21} f_y f)_k$$

$$+ (h^3/2)(c_2^2 f_{xx} + 2c_2 a_{21} f f_{xy} + a_{21}^2 f_{yy} f^2)_k$$

$$K_3 = hf_k + h^2(c_3 f_x + a_{31} f_y f + a_{32} f_y f)_k \tag{6.19}$$

$$+ h^3((1/2)c_3^2 f_{xx} + c_3(a_{31} + a_{32})f f_{xy}$$

$$+ (1/2)(a_{31} + a_{32})^2 f_{yy} f^2 + a_{32}(c_2 f_x + a_{21} f f_y)).$$

Making use of these latter expressions, we obtain

$$y_{k+1} = y_k + \sum_{i=1}^{3} w_i K_i$$

$$= y_k + w_1 hf_k + w_2 hf_k + w_2 h^2(c_2 f_x + a_{21} f f_y)_k$$

$$+ w_2 h^3((1/2)c_2^2 f_{xx} + c_2 a_{21} f f_{xy} + (1/2)a_{21}^2 f_{yy} f^2)_k \tag{6.20}$$

$$+ w_3 hf_k + w_3 h^2(c_3 f_x + a_{31} f_y f + a_{32} f_y f)_k$$

$$+ w_3 h^3((1/2)c_3^2 f_{xx} + c_3(a_{31} + a_{32})f f_{xy}$$

$$+ (1/2)(a_{31} + a_{32})^2 f^2 f_{yy} + a_{32}(c_2 f_x + a_{21} f f_y)).$$

Comparing coefficients term by term in (6.18) and (6.20) yields the following eight equations for the unknown parameters:

$$w_1 + w_2 + w_3 = 1, \quad c_2 w_2 + c_3 w_3 = 1/2, \quad a_{21} w_2 + (a_{31} + a_{32})w_3 = 1/2,$$

$$(1/2)c_2^2 w_2 + (1/2)c_3^2 w_3 = 1/6, \quad c_2 a_{21} w_2 + c_3(a_{31} + a_{32})w_3 = 1/3, \tag{6.21}$$

$$(1/2)a_{21}^2 w_2 + (1/2)(a_{31} + a_{32})^2 w_3 = 1/6, \quad c_2 a_{32} w_3 = 1/6, \quad a_{21} a_{32} w_3 = 1/6.$$

On comparing the last two equations in (6.21), note that $c_2 = a_{21}$. Then, from the second and third equations, $c_3 = a_{31} + a_{32}$, and we are left with only four other independent equations,

$$w_1 + w_2 + w_3 = 1$$

$$a_{21}w_2 + (a_{31} + a_{32})w_3 = 1/2$$

$$a_{21}^2 w_2 + (a_{31} + a_{32})^2 w_3 = 1/3 \tag{6.22}$$

$$a_{21}a_{32}w_3 = 1/6.$$

Since there are only six independent relationships for the eight unknown parameters, we have two "free" parameters which can be taken to be c_2 and c_3. These parameters aren't completely free in that they are restricted to lie between 0 and 1 so as to subdivide the interval. For specified values of c_2, c_3, all other parameters are uniquely determined. In the next subsection, some choices which lead to RK algorithms commonly appearing in the numerical analysis literature shall be indicated.

The above derivation was carried out for $v = 3$. The derivation for larger v becomes quite tedious to do by hand, and one should attempt to make some use of symbolic computation. We shall be content here to quote some common higher-order formulas, leaving the fourth-order analysis as a problem. It should be mentioned that to any order v, one finds the identities $\sum_{i=1}^{v} w_i = 1$ and $c_i = \sum_{j=1}^{i-1} a_{ij}$ for $i \neq 1$.

6.3.2 Examples of Common RK Algorithms

For $v = 2$, results can be obtained from the analysis of the previous subsection by truncating the equations at order h^2. In this case, the three relations $w_1 + w_2 = 1$, $c_2 = a_{21}$, and $c_2 w_2 = 1/2$ result. Taking c_2 as the free parameter, we can then solve for w_1 and w_2. For example, for $c_2 = 1$, $w_1 = w_2 = 1/2$ and the modified Euler algorithm drops out. Another well-known result, called Heun's algorithm, follows on choosing $c_2 = 2/3$. These two results represent the "standard" choices quoted in the numerical analysis literature. Obviously, other choices could be made.

For $v = 3$, a standard choice is to take $c_2 = 1/2$ and $c_3 = 1$. Then, using Equations (6.22), we find that $a_{21} = 1/2$, $a_{31} = -1$, $a_{32} = 2$, $w_1 = w_3 = 1/6$ and $w_2 = 2/3$. The corresponding third-order RK formula is then given by

$$y_{k+1} = y_k + (1/6)[K_1 + 4K_2 + K_3] \tag{6.23}$$

with $K_1 = hf(x_k, y_k)$, $K_2 = hf(x_k + h/2, y_k + K_1/2)$, and $K_3 = hf(x_k + h, y_k - K_1 + 2K_2)$. This formula which is accurate to order h^3 involves three "f substitutions".

For $v = 4$, one of the most often quoted RK formulas in the numerical analysis literature is

$$y_{k+1} = y_k + (1/6)[K_1 + 2K_2 + 2K_3 + K_4] \tag{6.24}$$

with $K_1 = hf(x_k, y_k)$, $K_2 = hf(x_k + h/2, y_k + K_1/2)$, $K_3 = hf(x_k + h/2, y_k + K_2/2)$, and $K_4 = hf(x_k + h, y_k + K_3)$. Since this is a fourth-order RK formula, the error is of $O(h^5)$. Four "f substitutions" are involved.

As an example of applying the fourth-order RK formula, consider the following first-order nonlinear ODE [Dav62]

$$\frac{dy}{dx} = xy(y - 2) \tag{6.25}$$

with $y(0) = 1$. This equation is exactly solvable by using the separation of variables technique or by making use of the Mathematica command

```
sol = DSolve[{y'[x] == x y[x] (y[x] - 2), y[0] == 1}, y, x];
```

the analytic solution being

```
y[x] /. sol[[1]]
```

$$\frac{2}{1 + e^{x^2}} \tag{6.26}$$

The Mathematica code for the fourth-order RK algorithm applied to this nonlinear example is as follows:

Example 6-4: Fourth-Order RK Algorithm

Solve Equation (6.25) using the fourth-order RK formula and compare the numerical solution with the analytic result (6.26). Take the step size to be $h = 0.1$ and $n = 25$ steps. Display the solutions in the same figure.

Solution: The initial values of x and y, the step size h, and the number

```
Clear["Global`*"]
```

of steps n are specified.

```
x[0] = 0;  y[0] = 1;  h = 0.1;  n = 25;
```

The fourth order RK functions are entered,

```
f1[k_] := h x[k] y[k] (y[k] - 2)

f2[k_] := h (x[k] + 0.5 h) (y[k] + 0.5 f1[k]) (y[k] + 0.5 f1[k] - 2)

f3[k_] := h (x[k] + 0.5 h) (y[k] + 0.5 f2[k]) (y[k] + 0.5 f2[k] - 2)

f4[k_] := h (x[k] + h) (y[k] + f3[k]) (y[k] + f3[k] - 2)

y[k_] := y[k] = N[y[k-1] + (f1[k-1] + 2 f2[k-1] + 2 f3[k-1]
                + f4[k-1])/6]
```

```
x[k_] := x[k] = N[x[k-1]+h]
```

and iterated with the `Table` command to produce the plotting points.

```
pts = Table[{x[k], y[k]}, {k, 0, n}];
```

The exact analytic expression is given,

```
yy[x_] := 2/(1+Exp[x^2])
```

and graphs of the numerical and exact solutions created and then superimposed with the `Show` command.

```
Block[{$DisplayFunction = Identity},
gr1 = ListPlot[pts,PlotStyle->{PointSize[.015],RGBColor[0,0,1]}];
gr2 = Plot[yy[x], {x, 0, 2.5}, PlotStyle -> {RGBColor[1,0,0]}];]

Show[gr1, gr2, Ticks -> {{{.01, "0"}, 1, {1.5, "x"}, 2}, {.5,{.75, "y"},
1}}, TextStyle -> {FontFamily -> "Times", FontSize -> 16},
ImageSize->{600,400}];
```

The resulting picture is shown in Figure 6.8. The numerical points in the figure

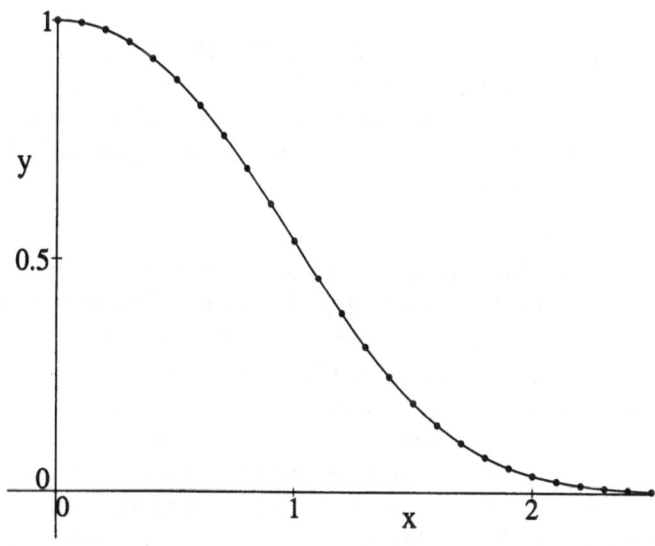

Figure 6.8: Fourth-order RK solution (circles) of (6.25) and exact result (curve).

are in good agreement with the exact solution. How good? If we apply the `Precision` command to the analytic expression at, say, $x = 1$,

```
Precision[yy[1]]
```

∞

Mathematica yields the value ∞. The exact solution has infinite precision. On the other hand, applying Precision to the numerical result at $x = 1$ (10 steps of size $h = 0.1$),

    Precision[y[10]]

    16

informs us that the numerical answer has been calculated to 16 digits. (This is the "machine precision" of the computer.) The analytic and numerical answer can be compared to the same number of digits by using the SetPrecision command. For example, taking 10 digit accuracy,

    SetPrecision[yy[1], 10]

    0.5378828427

the exact solution at $x = 1$ is 0.5378828427,

    SetPrecision[y[10], 10]

    0.5378836571

while the fourth-order RK result is 0.5378836571. The two answers differ in the 6th decimal place. If the step size in the RK algorithm is reduced to $h = 0.01$ and 100 steps are taken, the numerical value of y at $x = 1$ is 0.5378828428 which differs from the exact result in the 10th decimal place. (Remember that only 10-digit accuracy was specified.)

| End Example 6-4 |

If the reader was delinquent and didn't do the "artificial example" Problem 6-10 (page 238) using the Euler method, perhaps it would be instructive to solve it now for $h = 0.1$. This will give the reader a clear idea of how much better the fourth order RK method is than the Euler method.

For orders $v = 5$ and higher, detailed examination of the RK method reveals that it takes more "f substitutions" than the order of the accuracy. For example, a sixth-order formula requires a minimum of seven substitutions. As a compromise between computing time (more f substitutions require more time) and high accuracy, fourth- and fifth-order RK schemes are generally the most popular. Mathematica's RungeKutta option in the DSolve command uses a variation due to Fehlberg [Feh70] involving these two orders. The Mathematica numerical ODE solver also makes use of the idea of an adaptive step size, a concept which is discussed in the next section.

PROBLEMS
Problem 6-11: Separation of variables
Use the separation of variables method to confirm the analytic solution given in Example 6-4.

Problem 6-12: Heun's method
Explicitly write out Heun's algorithm. Then use it to solve the rabbits–foxes equations for the same step size, parameters, etc., as in the text for the modified Euler method. How does your solution compare with the modified Euler solution? Is there much difference between the two results?

Problem 6-13: Third-order RK formula for a second-order ODE
Expressing a general second-order ODE as two coupled first-order ODEs, explicitly write out a generalization of the third-order RK formula given in the text for this situation. Write out the Mathematica code for this algorithm for a second-order nonlinear system of your choice. Numerically explore the solution to your system as a function of step size for a given set of parameter values.

Problem 6-14: Fourth-order Rungé–Kutta parameters
For the general fourth-order RK formula, what is the total number of parameters $w_1...$, $c_2...$, $a_{21}...$ which have to be determined by comparing with the Taylor expansion to the fourth order? For the specific form of the fourth-order RK formula quoted in the text, what are the values of these parameters? Comment on the feasibility of deriving the general relations for the parameters by hand.

Problem 6-15: Chemical reaction [BF89]
Consider the irreversible chemical reaction

$$2K_2Cr_2O_7 + 2H_2O + 3S \rightarrow 4KOH + 2Cr_2O_3 + 3SO_2$$

with initially N_1 molecules of potassium dichromate ($K_2Cr_2O_7$), N_2 molecules of water, and N_3 atoms of sulphur. The number X of potassium hydroxide (KOH) molecules at time t seconds is given by the rate equation

$$dX/dt = k(2N_1 - X)^2(2N_2 - X)^2(4N_3/3 - X)^3$$

with $k = 1.64 \times 10^{-20} \ s^{-1}$ and $X(0) = 0$. Making use of the discussion on chemical reactions in Chapter 2, explain the three exponents on the rhs. Also explain, for example, the factor $(4N_3/3 - X)$. Suppose $N_1 = N_2 = 2000$ and $N_3 = 3000$. Determine $X(t)$ using the fourth order RK method with a step size $h = 0.001$. How many KOH molecules are present at $t = 0.1$ seconds? at $t = 0.2$ seconds?

Problem 6-16: Competition for the same food supply
Two biological species competing for the same food supply are described by the population equations

$$\dot{N}_1 = (4 - 0.0002N_1 - 0.0004N_2)N_1$$

$$\dot{N}_2 = (2 - 0.00015N_1 - 0.00005N_2)N_2.$$

a. Find and identify the stationary points of this system.

b. Suppose that $N_1(0) = 8000$ and $N_2(0) = 4000$. Using the fourth-order RK method with $h = 0.05$ and taking $t = 0..25$, plot $N_1(t)$, $N_2(t)$, and N_1 versus N_2. Discuss the results.

Problem 6-17: Orbital motion

With a suitable choice of values for the constants, the Newtonian orbit of a particle in an inverse square law gravitational field is governed by the system of equations

$$\ddot{r} = \frac{9}{r^3} - \frac{2}{r^2}, \quad \dot{\theta} = \frac{3}{r^2},$$

where r and θ are the radial and angular coordinates, respectively. At time $t = 0$, the particle is located at the minimum radial distance $r(0) = 3$ with $\theta(0) = 0$ and $\dot{r}(0) = 0$.

   a. Using the fourth-order RK method with $h = 0.1$, numerically integrate the equations to determine the particle's orbit.

   b. Show that the exact analytic solution is the ellipse $r = 9/(2 + \cos\theta)$.

   c. Plot the analytic and numerical solutions in the same graph.

   d. The exact period is $T = 12\sqrt{3}\pi$. How does the numerical value of the period compare to the exact value?

Problem 6-18: Cancerous tumor growth

Using the fourth-order RK method, solve

$$\dot{x} = -x \ln x$$

for the initial condition $x(0) = 0.00001$ and plot the result. Discuss the nature of the curve. This equation, referred to as Gompertz law, has been used to model the growth of cancerous tumors [ALSB73] [New80].

Problem 6-19: Hortense, the duck

Hortense, a duck with a gimpy wing, attempts to swim across a river by steadily aiming at a target point directly across the river. The river is 1 km wide and has a speed of 1 km/hour while Hortense's speed is 2 km/hour. In Cartesian coordinates, Hortense is initially at $(x = 1, y = 0)$ while the target point is at $(0, 0)$. Hortense has not yet learned to handle the drift of the river and is initially swept downstream slightly. Her equations of motion are

$$\dot{x} = -\frac{2x}{\sqrt{x^2 + y^2}}, \quad \dot{y} = 1 - \frac{2y}{\sqrt{x^2 + y^2}}.$$

   a. Justify the structure of the equations.

   b. Using the fourth-order RK method with $h = 0.01$, determine how long it takes Hortense to reach the target point.

   c. Determine the analytic solution $y(x)$ for Hortense's path across the river.

   d. Plot the analytic and numerical solutions together for Hortense's path.

Problem 6-20: Harvesting of fish

To take into account the effect of fishing on a single species of fish, a "harvesting term" can be added to the normalized logistic equation describing population growth, viz.,

$$\dot{x} = x(1 - x) - Hx/(a + x).$$

Using the fourth-order RK method with $h = 0.1$, numerically investigate this equation for $x(0) = 0.1$, $a = 0.2$, and various "harvesting coefficients" $H = 0.1, 0.2, 0.3, \ldots$ and plot your results. Discuss the change in behavior of your answer as H is increased.

Problem 6-21: Population dynamics of baleen whales
May [May80] has discussed the solution of the following normalized equation describing the population of sexually mature adult baleen whales,

$$\dot{x}(t) = -ax(t) + bx(t - T)(1 - (x(t - T))^N).$$

Here $x(t)$ is the normalized population number at time t, a and b the mortality and reproduction coefficients, T the time lag necessary to achieve sexual maturity, and N a positive parameter. If the term $1 - (x(t - T))^N < 0$, then this term is to be set equal to zero. Taking $a = 1$, $b = 2$, $T = 2$, step size $h = 0.01$, and 4000 time steps, use the Euler method to numerically solve for $x(t - T)$ versus $x(t)$ and for $x(t)$ for (a) $N = 3.0$, (b) $N = 3.5$. Plot your results. For (a) you should observe a period one solution, and for (b) a period two solution. Discuss how this interpretation can be made from your plots.

6.4 Adaptive Step Size

6.4.1 A Simple Example

In solving certain nonlinear ODEs numerically, regions of "solution space" can be encountered where the solution is changing relatively slowly and we could get away with a larger step size while maintaining reasonable accuracy, while in other regions the solution is rapidly changing and a smaller step size should be taken. Such is the situation for the Van der Pol equation when $\epsilon \gg 1$. One would like to have a simple automatic way of changing the step size as the nature of the solution changes. We would like to "race" through flat, uninteresting, regions but should "tiptoe" through precipitous areas. What is needed in these situations is an adaptive step size that will adjust to the terrain being encountered. In many numerical problems, a properly designed adaptive step size can lead to greatly reduced computing time without sacrificing accuracy, or to improved accuracy without too much additional time. There are many approaches to this problem in the numerical analysis literature and two of the more important ones will be discussed in the ensuing subsections. First, however, the concept of an adaptive step size will be illustrated with a simple example.

We look at a minor variation on the example that was studied in the previous section with the fourth-order RK method and a fixed step size h. The nonlinear equation to be solved is

$$\frac{dy}{dx} = x(y - 3)(y - 2) \tag{6.27}$$

subject to the initial condition $y(0) = 1$. By separating variables or using Mathematica's DSolve command, the exact solution is found to be

$$y(x) = \frac{-3 + 4e^{\frac{1}{2}x^2}}{-1 + 2e^{\frac{1}{2}x^2}} \tag{6.28}$$

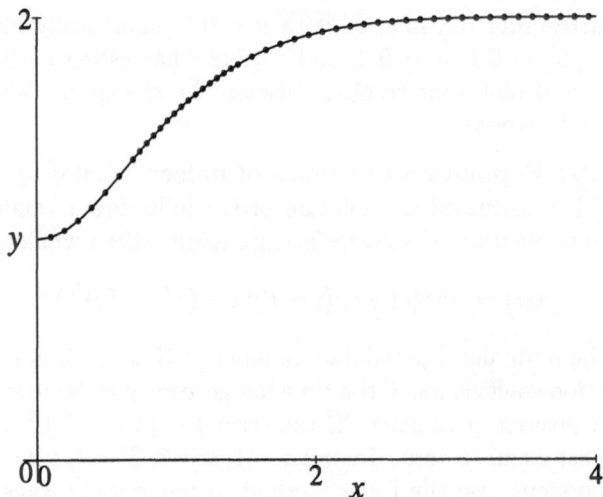

Figure 6.9: Applying the adaptive step fourth-order RK algorithm to (6.27).

which is the solid curve in Figure 6.9. The solution starts out relatively flat near $x = 0$, begins to rise rapidly, and then eventually saturates to $y = 2$. How could a simple scheme be devised that alters the step size according to the terrain? In analogy to hiking up a mountain trail, one possible approach might be to make use of the slope to set the step size. If the slope becomes too steep, smaller steps are taken, whereas if the slope is small, larger steps can be used. For the latter, a maximum step size can be imposed so that accuracy is not lost. Similarly, a lower bound can be set on the step size, so that the calculation doesn't bog down on a huge number of small steps. By adding appropriate "conditional" statements to the fourth-order RK code, this adaptive step size approach has been carried out in the next file MF26.

Adaptive Step Sizing

A simple adaptive step size scheme for the fourth-order RK method is illustrated. Use is made of the Mathematica Which and For commands. The syntax for the latter is For[start,test,increment,body]. The command executes "start", then repeatedly evaluates "body" and "increment" until "test" fails to give True. In the increment, k++ is used to increment the value of k by 1. As a measure of the error in the numerical scheme, the y values calculated from the exact analytic solution are subtracted from the numerical y values. Mathematica commands in file: Abs, Which, For, Table, Exp, Block, $DisplayFunction=Identity, ListPlot, PlotStyle, PointSize, RGBColor, PlotLabel, Show, Ticks, PlotRange −> All, Plot, TextStyle, ImageSize

---

Using the file MF26, the numerical points shown in Figure 6.9 are generated and superimposed on the exact solution. The variation in the step size is clearly seen. Now this file was only intended to illustrate how one would go about programming an adaptive step scheme, rather than being a serious attempt to improve either the accuracy or the speed of the numerical calculation. What adaptive step schemes do the numerical analysis experts recommend?

Two methods that are used in serious nonlinear ODE solving are the step doubling approach and the Rungé–Kutta–Fehlberg fourth–fifth (RKF 45) algorithm. These schemes are briefly discussed in the following two subsections.

PROBLEMS
Problem 6-22: Variation on Text Example
Consider the nonlinear ODE

$$\frac{dy}{dx} = x(y - x)(y - 2)$$

with $y(0) = 1$. Can this equation be solved analytically using the separation of variables method? Why not? How about with Mathematica's analytic ODE solving capability? Solve the equation using Mathematica's numerical ODE solver and plot the result out to $x = 4.0$. Modify the adaptive step scheme in MF26 to numerically solve the equation out to $x = 4.0$. Plot the adaptive scheme result in the same figure as the Mathematica result. Discuss the accuracy of your scheme.

Problem 6-23: Rabbits and Foxes
Combine an adaptive step scheme similar to that in MF26 with the Euler method to solve the rabbits-foxes equations for $\alpha = 0.01$, $r(0) = 300$, and $f(0) = 150$. Plot the phase plane trajectory and compare the result with that in Figure 6.3.

6.4.2 The Step Doubling Approach

A recommended approach for the fourth order Rungé–Kutta formula may be found in the reference book *Numerical Recipes*. It is based on the "step doubling" idea for checking accuracy that was mentioned earlier in the chapter.

For a given algorithm, one can check if the step size h is sufficiently small by running the calculation again with, say, half the step size and monitoring in what decimal place the two answers differ. In analogy with what was done in our illustrative example, the difference between the two answers, $\Delta = y_2 - y_1$, can be used as a measure of how accurate the answer is and a tolerance level specified.

In the fourth-order RK algorithm, for each step four evaluations (four f substitutions), i.e., K_1, K_2, K_3, K_4, are required. We could imagine doing our calculation with steps of size $2h$ or, alternatively, with twice as many steps of size h. In a single step of size $2h$ (advance from x to $x + 2h$, say), it takes four RK evaluations. For two steps of size h it takes eight evaluations. So let's write our program to calculate in steps of h, but also to calculate in steps of $2h$ (i.e., do both calculations).

The first evaluation in the $2h$ step approach involves the same $f(x_k, y_k)$ as in the h step approach, so it only costs three extra evaluations to do the $2h$ case after the h case (involving eight evaluations) has been done. The extra work involved is $3/8 = 0.375$. That is to say, the extra computing time is increased by roughly this amount.

What does this extra work buy us? Label the exact solution for an advance from x to $x + 2h$ by $Y(x + 2h)$. Let the approximate numerical solution for the single step $2h$ be denoted by y_1 and for two steps of size h by y_2. Then, remembering how the fourth-order RK formula was obtained by comparing with the Taylor series to the fourth order, we can write

$$Y(x + 2h) = y_1 + (2h)^5\Phi + O(h^6) + \cdots$$
$$Y(x + 2h) = y_2 + 2(h)^5\Phi + O(h^6) + \cdots \tag{6.29}$$

where Φ is a number whose order of magnitude is $y''''' (x)/5!$. To order h^5, Φ remains constant over the step.

As a measure of the error, calculate $\Delta = y_2 - y_1$. By adjusting h, we shall endeavor to keep a desired degree of accuracy Δ. Now that the error is known, at least approximately, we need to know how to keep it within desired bounds. Since Δ scales as h^5, then $h \sim \Delta^{0.2}$. If Δ_0 and Δ_1 are the errors corresponding to step sizes h_0 and h_1, respectively, then

$$\frac{h_0}{h_1} = \left|\frac{\Delta_0}{\Delta_1}\right|^{0.2} \tag{6.30}$$

or

$$h_0 = h_1 \left|\frac{\Delta_0}{\Delta_1}\right|^{0.2}. \tag{6.31}$$

Now let Δ_0 represent the desired accuracy which must be specified. Then, if Δ_1 is larger than Δ_0 in magnitude, the above equation tells us how much to decrease the step size when the present (failed) step is retried. (Note that in our simple scheme, we simply moved onto the next step, whereas here the move only occurs if the desired tolerance is met.) If Δ_1 is smaller than Δ_0, the equation tells us how much the step size can be safely increased for the next step.

Our argument did not take cumulative error into consideration. When this is done, *Numerical Recipes* recommends a slight alteration to (6.31), replacing it with

$$h_0 = Sh_1 \left|\frac{\Delta_0}{\Delta_1}\right|^{0.20} \tag{6.32}$$

for $\Delta_0 \geq \Delta_1$, and

$$h_0 = Sh_1 \left|\frac{\Delta_0}{\Delta_1}\right|^{0.25} \tag{6.33}$$

for $\Delta_0 < \Delta_1$. The parameter S is a safety factor which is taken to be a few percent less than 1. It is inserted because the estimates of error are not exact. The interested reader can find the complete adaptive step fourth-order RK code in *Numerical Recipes*.

6.4.3   The RKF 45 Algorithm

As a compromise between computing time and high accuracy, fourth- and fifth-order RK schemes have proven to be quite popular among researchers. The RungeKutta option of Mathematica's numerical ODE solver uses a variation due to Fehlberg [Feh70], viz., the Rungé–Kutta–Fehlberg fourth-fifth (RKF 45) order algorithm. To understand how it works, the concept of the local truncation error which was briefly mentioned earlier in the chapter must be fleshed out.

If $y(x_k + h)$ is the exact solution and if we subtract from it its Taylor expansion to order h^p, then

$$
\begin{aligned}
y(x_k + h) - [y_k + hf(x_k, y_k) + \frac{h^2}{2!}f'(x_k, y_k) + \cdots + \frac{h^p}{p!}f^{(p-1)}(x_k, y_k)] \\
= \frac{h^{p+1}}{(p+1)!}f^p(x_k, y_k) + \cdots
\end{aligned}
\tag{6.34}
$$

In an RK scheme of order p, the series expansion is truncated at order h^p, the remaining terms being thrown away. The dominant term for $h \ll 1$ on the rhs is of order h^{p+1}. The local truncation error is defined as

$$
\frac{y(x_k + h) - \tilde{y}}{h} = \frac{h^p}{(p+1)!}f^p(x_k, y_k) + \cdots
\tag{6.35}
$$

where $\tilde{y}$ is the series expansion out to order h^p. Thus, the local truncation error for a pth order RK scheme is $O(h^p)$.

The RKF 45 method uses an RK scheme with local truncation error of order h^5, viz.,

$$
\bar{y}_{k+1} = y_k + \frac{16}{135}K_1 + \frac{6656}{12825}K_3 + \frac{28561}{56430}K_4 - \frac{9}{50}K_5 + \frac{2}{55}K_6
\tag{6.36}
$$

to estimate the local error in an RK method of order h^4, viz.,

$$
y_{k+1} = y_k + \frac{25}{216}K_1 + \frac{1408}{2565}K_3 + \frac{2197}{4104}K_4 - \frac{1}{5}K_5
\tag{6.37}
$$

where

$$
\begin{aligned}
K_1 &= hf(x_k, y_k) \\
K_2 &= hf(x_k + \frac{1}{4}h, y_k + \frac{1}{4}K_1) \\
K_3 &= hf(x_k + \frac{3}{8}h, y_k + \frac{3}{32}K_1 + \frac{9}{32}K_2) \\
K_4 &= hf(x_k + \frac{12}{13}h, y_k + \frac{1932}{2197}K_1 - \frac{7200}{2197}K_2 + \frac{7296}{2197}K_3) \\
K_5 &= hf(x_k + h, y_k + \frac{439}{216}K_1 - 8K_2 + \frac{3680}{513}K_3 - \frac{845}{4104}K_4) \\
K_6 &= hf(x_k + \frac{1}{2}h, y_k - \frac{8}{27}K_1 + 2K_2 - \frac{3544}{2565}K_3 + \frac{1859}{4104}K_4 - \frac{11}{40}K_5).
\end{aligned}
\tag{6.38}
$$

The fourth-order RK method involves four f substitutions while, in general, the fifth-order RK method involves six such substitutions. So, generally the two methods together would cost 10 evaluations. The RKF scheme, as can be seen from above, is quite clever in that it involves a total of only six f substitutions.

As an estimate of the local truncation error on the $k+1$st step in the fourth-order scheme, one calculates the difference

$$R = \left|\overline{y}_{k+1} - y_{k+1}\right| /h$$
$$= \left|\frac{1}{360}K_1[k] - \frac{128}{4275}K_3[k] - \frac{2197}{75240}K_4[k] + \frac{1}{50}K_5[k] + \frac{2}{55}K_6[k]\right| /h. \tag{6.39}$$

If R exceeds the tolerance, a smaller step size is taken in the program. Conversely, if the difference is well below the tolerance, a larger step size can be taken. The step size adjustments are the same as in the previous section with the safety factor S often taken to be 0.84. Once the tolerance is satisfied on a given step, $y[k+1]$ is calculated using Equation (6.37).

The RungeKutta option of Mathematica's numerical ODE solver has automated this RKF 45 procedure so we do not have to generate our own code. The student may use this Mathematica option when numerical solutions are required in the following problems, even though most do not involve rapidly varying solutions.

PROBLEMS
Problem 6-24: Exploring the quasispecies model for N=4
Consider the quasispecies model of biological selection discussed in Chapter 2 with $N = 4$ species and an initial distribution $x_1(0) = 4$, $x_2 = x_3 = x_4 = 0$. Take $W_1 = 1$, $W_2 = 2$, $W_3 = 3$, $W_4 = 4$, $E_1 = 1$, $E_2 = 2$, $E_3 = 3$, and $E_4 = 4$. Find mutation coefficients ϕ_{kl} for which only species $k = 4$ survives as $t \to \infty$. Also find mutation coefficients for which all species survive with roughly comparable x_k values as $t \to \infty$. Plot your results.

Problem 6-25: Parametric excitation
Consider the parametric excitation equation of Example 2-1 with $\omega_0 = 1$, $A = r$, $\theta(0) = \pi/3$, and $\dot{\theta}(0) = 0$. Investigate the solution for various values of the driving frequency ω and plot the results in a format of your own choice. Discuss your results.

Problem 6-26: Emden's equation
Consider a spherical cloud of gas of radius R. The gravitational attraction of the molecules is balanced by the pressure p. At a radius $r \leq R$, Newton's second law applied to gravitational attraction gives for the acceleration g of gravity, $g = GM(r)/r^2 = -d\phi/dr$. Here G is the gravitational constant, $M(r)$ the mass of cloud inside r, and ϕ the gravitational potential. As one moves outward from r to $r + dr$, the decrease in pressure is $dp = -\rho g dr$ where ρ is the density. Assume:

- an adiabatic equation of state $p = k\rho^\gamma$ prevails where k is a positive constant and γ is the ratio of specific heat at constant pressure to that at constant volume;

- ϕ satisfies Poisson's equation in spherical coordinates

$$\nabla^2\phi = d^2\phi/dr^2 + (2/r)d\phi/dr = -4\pi G\rho;$$

- the boundary conditions are $\phi = 0$, $\rho = 0$, $p = 0$ at $r = R$ and $\phi = \phi_0$, $g = 0$ at $r = 0$.

a. Show that the various equations above can be combined to yield Emden's nonlinear equation

$$\frac{d^2y}{dx^2} + \frac{2}{x}\frac{dy}{dx} + y^n = 0.$$

Identify y, x, and n in terms of the original variables. What are the values of y and y' at $x = 0$?

b. Taking $y(0) = 1$ and $y'(0) = 0$, determine the analytic solution for $n = 0$ and $n = 1$. Show that $n = 2$ does not have an explicit analytic solution.

c. Solve the $n = 2$ case numerically and plot $y(x)$. On the same diagram, also plot the $n = 3, 4, 5$ cases.

Problem 6-27: Fitzhugh–Nagumo model of nerve cell firing
A simple model system that captures the important aspects [KG95] of electrical impulse transmission in nerve cells are the Fitzhugh–Nagumo equations

$$\frac{dv(t)}{dt} = i(t) - v(t)(v(t) - a)(v(t) - 1) - w(t)$$
$$\frac{dw(t)}{dt} = \epsilon(v(t) - \gamma w(t)),$$

where a, ϵ, γ are parameters, $v(t)$ is the voltage across the cell membrane, $w(t)$ a recovery variable, and $i(t)$ is the stimulus current injected into the cell. Taking $a = 0.139$, $\epsilon = 0.008$, $\gamma = 2.54$, $v(0) = 0$, $w(0) = 0$, $i(t) = 0.10$ for $10 \leq t < 20$ and $i(t) = 0$ otherwise, determine the behavior of $v(t)$ vs t for t up to 120 seconds and plot your result. Also try the values 0.02 and 0.03, instead of 0.10, in $i(t)$ and comment on the change in behavior as the values are varied. Biologists refer to the sequence of firing and returning to rest in the 0.03 and 0.10 cases as examples of an "action potential". The biologically inclined reader can go to the cited reference to learn more about action potentials in nerve cells.

6.5 Stiff Equations

As soon as the student has to deal with nonlinear systems involving more than one first order ODE or a single equation of second order or higher, the possibility of "stiffness" occurs. Stiffness refers to a situation where there are two or more very different time or spatial scales of the independent variable. The phrase "stiff equations" reputedly [BF89] had its origin in the study of differential equations governing spring and mass systems with the springs having large spring constants. Such springs are referred to as stiff springs because it requires a large force to stretch them. A stiff ODE system can also arise for a given mass and spring constant if the damping is sufficiently large to put the system in the overdamped regime.

To flesh out the concept of stiffness, consider the following strongly over-damped simple harmonic oscillator equation [HP95],

$$\ddot{x}(t) + 20\dot{x}(t) + x(t)/100 = 0 \qquad (6.40)$$

with initial conditions $x(0) = 0$ and $\dot{x}(0) = 10$. The oscillator is initially at the origin with nonzero velocity. Although this equation can be readily solved analytically by hand, the solution is such that a loglog plot should be used. Thus, we use Mathematica for the entire treatment which follows.

Example 6-5: Loglog Plot of Stiff Equation Solution

Derive the analytic solution of the stiff equation (6.40), with $x(0) = 0$ and $\dot{x}(0) = 10$. Make a loglogplot of the solution.

Solution: The Graphics package is loaded,

```
Clear["Global`*"]
```

```
<< Graphics`
```

and the stiff equation (6.40) is entered,

```
ode = x"[t] + 20 x'[t] + 0.01 x[t] == 0
```

$$0.01\, x(t) + 20\, x'(t) + x''(t) == 0$$

and analytically solved for $x(t)$.

```
sol1 = DSolve[{ode, x[0] == 0, x'[0] == 10}, x[t], t]
```

$$\{\{x(t) \to 0.500025\, e^{-20.\,t}\left(-1.\,e^{0.000500013\,t} + 1.\,e^{19.9995\,t}\right)\}\}$$

```
sol2 = Expand[x[t] /. sol1[[1]]]
```

$$-0.500025\, e^{-19.9995\,t} + 0.500025\, e^{-0.000500013\,t}$$

The solution given by sol2 may be written in the equivalent form

$$x(t) = .500025\left(e^{-t/\tau_1} - e^{-t/\tau_2}\right) \qquad (6.41)$$

with the characteristic times $\tau_1 \simeq 1/0.0005 \simeq 2000$ and $\tau_2 \simeq 1/20 \simeq 0.05$. The widely differing values of τ_1 and τ_2 set the time scales for the two exponentials. To see the two disparate time scales, a loglog plot of the solution, sol2, is produced over the time range $t = 0.01$ to $t = 1000$.

```
LogLogPlot[sol2, {t, 0.01, 1000}, PlotPoints->5000,PlotRange->All,
   PlotStyle->{RGBColor[1, 0, 0]}, Ticks -> {{.01,.1,1,10,100,1000},
   {.1, .5}}, AxesLabel -> {"t", "x"},TextStyle -> {FontFamily ->
   "Times", FontSize -> 16}, ImageSize -> {600,400}];
```

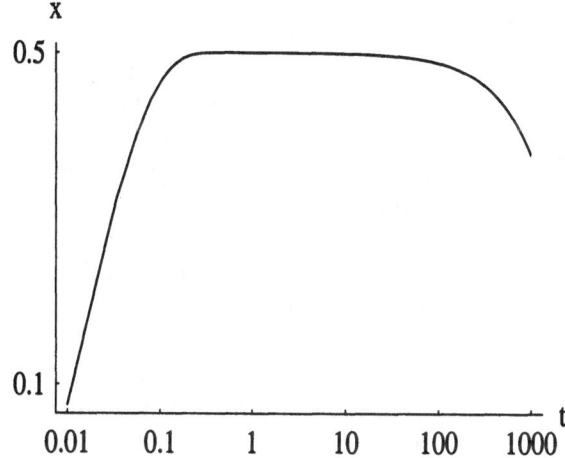

Figure 6.10: Loglog plot of the analytic solution to the stiff equation.

Figure 6.10, which spans five decades of time, clearly shows the rise and fall of the stiff equation solution with time. Because τ_1 is so large, the first term in the solution (6.41) remains essentially constant at about 0.50, while the second term decays from -0.50 to zero on a time scale of the order of τ_2. Thus the latter characteristic time governs the rapid rise of the solution to about 0.5. Thereafter, the solution decays very slowly until t becomes comparable to τ_1, and then the first term also rapidly decays to zero.

$\boxed{\textbf{End Example 6-5}}$

Now consider what happens if one starts with the exact solution at $t = 10$ and numerically integrates forward in time using, for example, the Euler method with constant step size $h = 0.2$. To six digits, the input values are

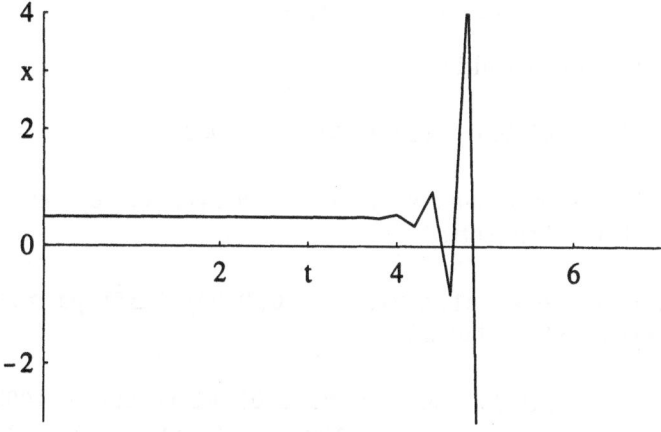

Figure 6.11: Euler method applied to stiff equation for $h = 0.2$.

$x(0) = 0.497531$, $\dot{x}(0) = -0.000248772$. As the reader may easily verify, the numerical solution follows the analytic solution for a short while, but then goes unstable, displaying increasing oscillations as shown in Figure 6.11. By systematically decreasing the step size in our example, one finds that an accurate and stable solution requires that the step size be smaller than the shortest time scale (here $\tau_2 \simeq 0.05$) in the problem. This is a somewhat surprising result as our numerical integration began well past the rapid initial rise of the solution and on the portion of the solution curve whose behavior is governed by the slow decay time τ_1. That the shortest spatial or time scale in the problem dictates the step size which can be safely used is a feature of all stiff equations. Of course, the Euler method is not used in serious calculations, but even with more accurate methods the same sort of numerical instability problem can occur for stiff systems. Because quite small step sizes may be required when applying explicit numerical schemes with fixed h to stiff equations, thus increasing the computing time and in some situations introducing roundoff error, either adaptive step size schemes or implicit and semi-implicit schemes are used to cure the numerical instability. The latter schemes are discussed in the next section.

That the adaptive step size approach works quite well can be verified by using, for example, the RKF 45 option of NDSolve to solve the above example for which the Euler method with $h = 0.2$ failed. The relevant code is given in the following example.

Example 6-6: RKF Method Applied to Stiff Equation

Use the RungeKutta option of NDSolve to solve the stiff equation problem, with $x(0) = 0.497531$, $\dot{x}(0) = -0.000248772$, and plot the numerical solution.

Solution: The relevant ODE is entered,

```
Clear["Global`*"]

ode = x''[t] + 20 x'[t] + 0.01 x[t] == 0
```

$$0.01\, x(t) + 20\, x'(t) + x''(t) == 0$$

along with the initial conditions.

```
ic1 = x[0] == 0.497531;  ic2 = x'[0] == -0.000248772;
```

The stiff ODE is solved numerically with the RungeKutta option of NDSolve, and the solution is then plotted.

```
sol = NDSolve[{ode, ic1, ic2}, x, {t, 0, 2500}, MaxSteps -> 20000,
    Method -> RungeKutta];

Plot[Evaluate[x[t] /. sol], {t, 0, 2500}, PlotPoints -> 1000,
PlotRange -> {0, .6}, Ticks -> {{1000, {1500, "t"}, 2000}, {{0.01, "0"},
{.25, "x"}, .5}}, PlotStyle -> Hue[.3], TextStyle -> {FontFamily ->
"Times", FontSize -> 18}, ImageSize -> {600, 400}];
```

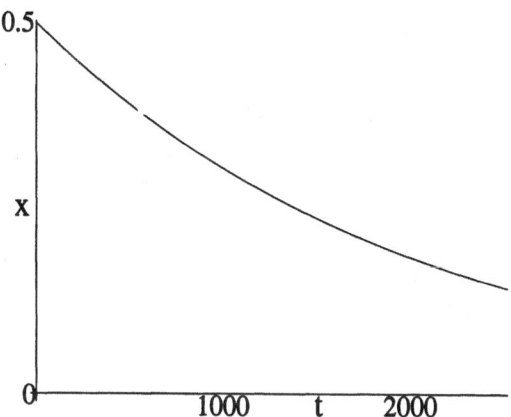

Figure 6.12: RKF 45 method successfully applied to stiff equation.

Figure 6.12 shows the correct smooth decrease of the solution towards zero.

End Example 6-6

PROBLEMS
Problem 6-28: Stiff harmonic oscillator 1
Verify the numerical result obtained in Figure 6.11. Show that the shortest time scale in the problem approximately sets the step size which can be safely used.

Problem 6-29: Stiff harmonic oscillator 2
For the stiff ODE discussed in the text, start with an initial value of y which has been slightly perturbed at $t = 10$ away from the exact value. Using the Euler method and an appropriate step size, show that the shorter time scale predominates for a while, and then the solution displays the same long time scale as the exact (unperturbed) solution.

Problem 6-30: A stiff system
Consider the system [PFTV89]

$$u' = 998u + 1998v$$

$$v' = -999u - 1999v$$

with $u(0) = 1$ and $v(0) = 0$. Solve this set of equations exactly and show that it is a stiff set. What condition should be imposed on the step size to avoid numerical instability? Apply the modified Euler method to the problem and by choosing different step sizes test the condition that you have stated.

Problem 6-31: A forced stiff system

Consider the following system [BF89] of linear forced ODEs

$$\dot{u} = 9u + 24v + 5\cos t - \frac{1}{3}\sin t$$

$$\dot{v} = -24u - 51v - 9\cos t + \frac{1}{3}\sin t$$

with $u(0) = 4/3$ and $v(0) = 2/3$. Determine the exact analytic solution for $u(t)$ and $v(t)$ and show that the system is stiff. Apply the fourth-order RK method to this problem taking step sizes of 0.1 and 0.05. Discuss the stability of your results and relate your discussion to the shortest time scale in the problem.

6.6   Implicit and Semi-Implicit Schemes

To illustrate the underlying concept of an implicit scheme, once again consider the trivial linear ODE

$$y' = -10y \tag{6.42}$$

with $y(0) = 1$. We modify the Euler method by evaluating the rhs at the "new" step instead of the "old" one, i.e.,

$$y_{k+1} = y_k - h(10y_{k+1}). \tag{6.43}$$

This procedure is referred to as the backward Euler method because (replacing k with $k - 1$) it makes use of the backward-difference approximation (Section 6.1) to the first derivative in the Euler scheme. Because this ODE is so simple, y_{k+1} can be solved for analytically,

$$y_{k+1} = y_k/(1 + 10h). \tag{6.44}$$

In general, this analytic inversion is not possible, particularly for nonlinear ODEs. For example, consider the nonlinear equation

$$y'' = 2y^5 - 6y. \tag{6.45}$$

Setting $y' = z$ and applying the same procedure as above, we obtain

$$y_{k+1} = y_k + hz_{k+1}$$

$$z_{k+1} = z_k + h(2y_{k+1}^5 - 6y_{k+1}). \tag{6.46}$$

This set cannot be analytically inverted to obtain y_{k+1}. Because this cannot be explicitly done, the procedure is also referred to as the implicit Euler method. From now on this phrase shall be used, regardless of the structure of our ODE. Since our ultimate goal is to solve nonlinear equations, we shall show later how to handle an equation such as the one above.

Returning to our pedagogical example, take $h = 3/10$. When this step size was chosen in the Euler scheme, the numerical solution bore no resemblance to the exact solution, displaying rapidly diverging oscillations. Here, numerically $y_0 = 1$, $y_1 = 0.25$, $y_2 = 0.0625$, $y_3 = 0.0156$,..., while the analytic solution yields (to three figure accuracy) 1, 0.607, 0.368, 0.223,.... Although the new scheme is not too accurate for the step size chosen, it is stable and displays the correct qualitative behavior. Indeed, the new algorithm is stable for any step size h. The implicit (backward) Euler method is only $O(h)$ accurate, which may be verified by expanding for h small.

Let's now revisit the stiff harmonic oscillator equation (6.40) of the previous section and apply the implicit Euler method to it. Because the equation is linear, an exact inversion is again possible so that this procedure (setting $\dot{x} \equiv y$) yields the simple algorithm

$$y_{k+1} = (y_k - 0.01hx_k)/(1 + 20h + 0.01h^2)$$

$$x_{k+1} = ((1 + 20h)x_k + hy_k)/(1 + 20h + 0.01h^2). \tag{6.47}$$

If, as before, we start at $t = 10$ and take $h = 0.2$, the well-behaved and qualitatively correct numerical solution shown in Figure 6.13 is obtained. Of course, h

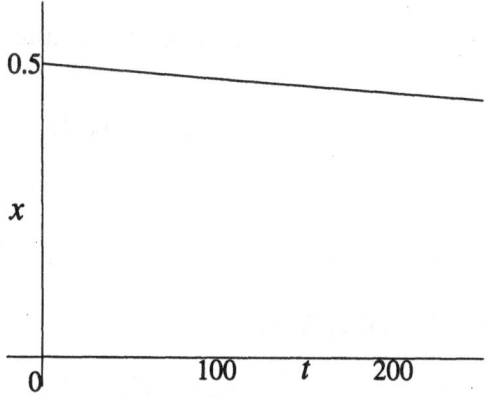

Figure 6.13: Implicit Euler method applied to stiff equation.

can be decreased to improve the accuracy, but a better approach is to come up with an implicit scheme of higher order.

To create a stable second-order accurate implicit scheme, we can average the "old" and the "new" on the rhs. (This is also the conceptual basis for the semi-implicit Crank–Nicolson scheme for solving nonlinear PDEs which is discussed in Chapter 11 .) For example, for our simple pedagogical ODE,

$$y_{k+1} = y_k - 10h \times \frac{1}{2}[y_k + y_{k+1}]. \tag{6.48}$$

Again, one can solve analytically for y_{k+1}, yielding

$$y_{k+1} = \frac{(1 - 5h)}{(1 + 5h)}y_k. \tag{6.49}$$

To see that this scheme is second-order accurate, simply expand the denominator for $5h \ll 1$, the result being

$$y_{k+1} = (1 - 10h + O(h^3))y_k, \tag{6.50}$$

i.e., the error is of order h^3.

How are these ideas on implicit schemes applied to nonlinear systems? As an example, suppose that we want to apply a first-order implicit method to the fox rabies system of Section 2.2.2. This system has already been solved in MF05 using the Mathematica NDSolve command with the RungeKutta option, so the behavior of the solution is known. The student will thus be able to gain some idea of how well the implicit scheme works. Rewrite the original coupled fox rabies equations in the following expanded form so that the nonlinear terms are clearly displayed:

$$\dot{X} = (a - b)X - \gamma X^2 - \gamma XY - (\beta + \gamma)XZ$$

$$\dot{Y} = -(\sigma + b)Y + \beta XZ - \gamma XY - \gamma Y^2 - \gamma YZ \tag{6.51}$$

$$\dot{Z} = \sigma Y - (\alpha + b)Z - \gamma XZ - \gamma YZ - \gamma Z^2.$$

To first-order accuracy, the corresponding implicit algorithm is

$$X_{k+1} = X_k + h((a - b)X_{k+1} - \gamma X_{k+1}^2 - \gamma X_{k+1}Y_{k+1}$$

$$-(\beta + \gamma)X_{k+1}Z_{k+1})$$

$$Y_{k+1} = Y_k + h(-(\sigma + b)Y_{k+1} + \beta X_{k+1}Z_{k+1} - \gamma X_{k+1}Y_{k+1} \tag{6.52}$$

$$-\gamma Y_{k+1}^2 - \gamma Y_{k+1}Z_{k+1})$$

$$Z_{k+1} = Z_k + h(\sigma Y_{k+1} - (\alpha + b)Z_{k+1} - \gamma X_{k+1}Z_{k+1}$$

$$-\gamma Y_{k+1}Z_{k+1} - \gamma Z_{k+1}^2).$$

For three linear coupled equations with constant coefficients, we could readily solve for the unknowns X_{k+1}, Y_{k+1}, Z_{k+1} at each time step by using Mathematica, as illustrated in Example 1-3. However, for the fox rabies system above, our equations are nonlinear in the unknowns, involving quadratic terms such as $X_{k+1}Z_{k+1}$. How does one deal with such terms?

For a general nonlinear function $f(x_{k+1}, z_{k+1})$, the standard procedure in the literature is to expand thus,

$$f(x_{k+1}, z_{k+1}) = f(x_k, z_k) + (x_{k+1} - x_k)\left(\frac{\partial f}{\partial x}\right)_{x_k, z_k} + (z_{k+1} - z_k)\left(\frac{\partial f}{\partial z}\right)_{x_k, z_k} + \cdots. \tag{6.53}$$

For our fox rabies example, this gives us, for example,

$$f \equiv (XZ)_{k+1} = X_k Z_k + Z_k(X_{k+1} - X_k) + X_k(Z_{k+1} - Z_k). \qquad (6.54)$$

Applying this procedure to all of the quadratic terms on the rhs of our system and gathering all the "new" values X_{k+1}, etc., on the lhs, we obtain

$$(1 + hA_{1k})X_{k+1} + hB_{1k}Y_{k+1} + hC_{1k}Z_{k+1} = X_k + hD_{1k}$$

$$hA_{2k}X_{k+1} + (1 + hB_{2k})Y_{k+1} + hC_{2k}Z_{k+1} = Y_k + hD_{2k} \qquad (6.55)$$

$$hA_{3k}X_{k+1} + hB_{3k}Y_{k+1} + (1 + hC_{3k})Z_{k+1} = Z_k + hD_{3k}$$

with $N_k = X_k + Y_k + Z_k$ and

$$A_{1k} = (b - a) + \gamma(N_k + X_k) + \beta Z_k, \quad A_{2k} = \gamma Y_k - \beta Z_k, \quad A_{3k} = \gamma Z_k,$$

$$B_{1k} = \gamma X_k, \quad B_{2k} = (\sigma + b) + \gamma(N_k + Y_k), \quad B_{3k} = \gamma Z_k - \sigma,$$

$$C_{1k} = (\beta + \gamma)X_k, \quad C_{2k} = \gamma Y_k - \beta X_k, \quad C_{3k} = (\alpha + b) + \gamma(N_k + Z_k),$$

$$D_{1k} = \gamma N_k X_k + \beta X_k Z_k, \quad D_{2k} = \gamma N_k Y_k - \beta X_k Z_k, \quad D_{3k} = \gamma N_k Z_k.$$

We now have three linear equations (6.55), with known numerical coefficients determined from the previous step, for the three unknowns X_{k+1}, Y_{k+1}, Z_{k+1}. The problem can be finished by solving the three simultaneous equations on each new time step. For the student's convenience, the lengthy code has been written out in Mathematica File MF27.

Semi-Implicit Scheme Applied to Fox Rabies System

In this file, the semi-implicit Euler scheme developed in the text is applied to the fox rabies system for the same initial values and parameters as in MF05. The student may compare the results obtained and the computing time involved for different h sizes with those obtained in MF05 where the Mathematica numerical ODE solver (with the RungeKutta option) was used. Mathematica commands in file: NSolve, Graphics, ScatterPlot3D, Boxed->False, PlotRange, BoxRatios, AxesLabel, Do, Table, AxesEdge, ImageSize, PlotStyle, RGBColor, TextStyle

Taking the same initial values and parameters as in MF05, a step size $h = 0.004$, and 5000 steps, the trajectory shown in Figure 6.14 is produced which, aside from orientation, closely resembles that in Figure 2.7. Because the semi-implicit method used here is not of high-order accuracy and doesn't have adaptive step size capability, the computing time involved is much longer than in file MF05. The computing time in MF27 may be reduced by taking a much larger h value, but becomes increasingly more inaccurate as h is increased. The semi-implicit scheme remains stable, however. To confirm these remarks, the reader should, e.g., take $h = 0.5$ and 40 steps.

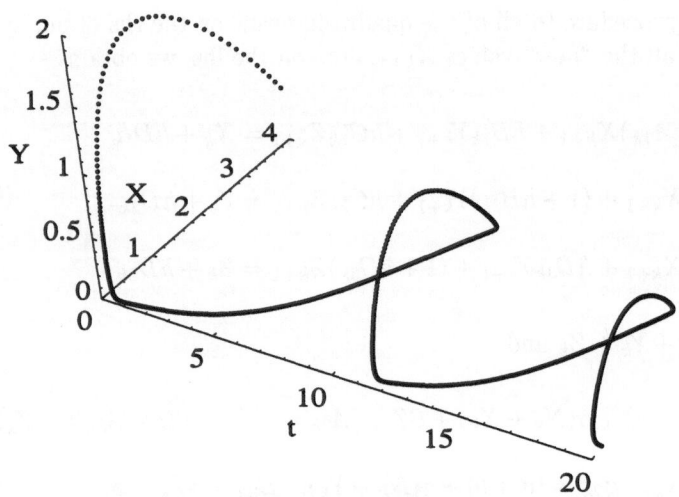

Figure 6.14: Semi-implicit solution of fox rabies equations.

The above modification of an implicit scheme to handle nonlinear ODEs is an example of a semi-implicit method. Semi-implicit schemes are not guaranteed to be stable but usually are. It should also be mentioned that it is quite complicated to design higher-order implicit schemes with automatic step size adjustment. We shall not pursue the idea of implicit and semi-implicit algorithms any further, being content to give the student a conceptual idea of what is involved. After all, this is not a text on numerical analysis, and it's about time to apply the various mathematical tools that have been developed to the exploration of nonlinear systems and concepts. As a general rule, the Mathematica numerical ODE solver shall be used in ensuing chapters unless otherwise stated. Given that this is our intention, the reader might argue that all this discussion on implicit and semi-implicit methods was perhaps a waste of time. In addition to the reasons already advanced earlier, the numerical concepts presented here can be readily generalized to nonlinear PDEs, the subject matter of Chapters 10 and 11. Understanding the underlying concepts of how to create numerical PDE solving schemes will deepen the reader's understanding of that subject matter.

PROBLEMS
Problem 6-32: Semi-implicit scheme
In the text, applying the backward Euler method to the nonlinear ODE

$$y'' = 2y^5 - 6y,$$

yielded the following implicit scheme for solving the differential equation.

$$y_{k+1} = y_k + hz_{k+1}, \qquad z_{k+1} = z_k + h(2y_{k+1}^5 - 6y_{k+1}).$$

Derive a first-order semi-implicit scheme for this set of equations, with all the unknowns on the lhs of the equations and the known quantities on the rhs.

Problem 6-33: Stiff equation
Derive an analytic solution of the following stiff ODE, subject to the given initial condition.

$$\frac{dy}{dt} = \frac{50}{y} - 50\,y, \quad y(0) = \sqrt{2}.$$

Numerically solve the ODE over the range $t = 0..1$ using

  a. Euler's method,

  b. the fourth-order RK method,

  c. a first-order semi-implicit scheme based on the backward Euler method.

Use a step size $h = 0.05$ for $0 \le t \le 0.2$ and $h = 0.1$ for $0.2 \le t \le 1$. Compare the numerical solutions with the exact solution and discuss.

Problem 6-34: The Rössler system
Derive a second-order accurate semi-implicit algorithm for the Rössler system (Equations (3.22). Taking $a = b = 0.2$, $c = 5.0$, $x(0) = 4.0$, $y(0) = z(0) = 0$, $h = 0.05$, determine the behavior of the system up to $t = 100$ and plot the trajectory in x–y–z space. By proper orientation of the viewing box, your result should look like that in Figure 8.31 of Chapter 8.

Problem 6-35: The arms race revisited
Apply the first-order accurate semi-implicit Euler method to Rapoport's model (Section 2.2.5) for the arms race, using the same initial values and parameters as in file MF06. Try different step sizes.

6.7   Some Remarks on NDSolve

Since we intend to generally use NDSolve to solve nonlinear ODEs of interest from now on, a few concluding remarks should be made about some of the optional numerical methods available besides Method -> RungeKutta. With Method -> Automatic, NDSolve switches between a non-stiff Adams method and a stiff Gear method. With Method -> Adams, an implicit Adams method with order between 1 and 12 is used, while for Method -> Gear a backward difference formula with order between 1 and 5 is employed. The interested reader is referred to standard numerical analysis books for a discussion of the Adams and Gear numerical methods.

Aside from the choice of method, various other options such as MaxStepSize, AccuracyGoal, Compiled, etc. are also available. These options are described in the Help entry for NDSolve.

Lastly, if you have ever omitted the semi-colon in executing the NDSolve command, you will have seen that an "InterpolatingFunction" is produced over the interval requested for the independent variable (e.g., the time interval). The underlying Mathematica code generates approximate functions to represent the numerical solution. Quoting from Page 897 of the *Mathematica Book* [Wol99a], "The InterpolatingFunction object contains a representation of the approximate function based on interpolation. Typically, it contains values and possibly derivatives at a sequence of points. It effectively assumes that the function varies smoothly between these points. As a result, when you ask for the value of the function with a particular argument, the `InterpolatingFunction` object can interpolate to find an approximation to the value you want."

PROBLEMS
Problem 6-36: Adams Methods
By consulting an appropriate numerical analysis text, briefly discuss the Adams methods for numerically solving ODEs. Illustrate your discussion by choosing a particular nonlinear ODE, writing a code (not involving the `NDSolve` command) which implements an Adams method to solve the ODE, and plotting the resulting solution.

Problem 6-37: Gear Method
By consulting an appropriate numerical analysis text, briefly discuss the Gear methods for numerically solving ODEs. Illustrate your discussion by choosing a particular nonlinear ODE, writing a code (not involving the `NDSolve` command) which implements a Gear method to solve the ODE, and plotting the resulting solution.

Chapter 7

Limit Cycles

Excellent wretch! Perdition catch my soul
But I do love thee! and when I love thee not,
Chaos has come again.
William Shakespeare (1564-1616), Othello, Act III, Scene iii

7.1 Stability Aspects

For autonomous nonlinear and non-conservative systems described by the differential equations

$$\frac{dx}{dt} = P(x,y), \quad \frac{dy}{dt} = Q(x,y) \tag{7.1}$$

a new kind of trajectory, the limit cycle, has been briefly encountered at various

points in the preceding chapters. The Van der Pol (VdP) electronic oscillator with $P(x,y) = y$ and $Q(x,y) = \epsilon(1 - x^2)y - x$, for example, made its debut in Chapter 2. In this chapter we would like to explore some of the more important properties of limit cycles in greater depth.

A limit cycle C is an isolated closed trajectory having the property that all other trajectories C' in its neighborhood are spirals winding themselves onto C for $t \to +\infty$ (a stable limit cycle) or $t \to -\infty$ (an unstable limit cycle). All trajectories approach the limit cycle independent of the choice of initial conditions. Semistable limit cycles displaying a combination of both stable and unstable behaviors can also occur. Some simple examples and problems, for which the limit cycle can be either analytically determined or found numerically using Mathematica, illustrate these concepts.

As an example of an analytically derivable limit cycle, consider the set of coupled nonlinear ODEs

$$\dot{x} = -y + \frac{x}{\sqrt{x^2 + y^2}}(1 - x^2 - y^2)$$
$$\dot{y} = x + \frac{y}{\sqrt{x^2 + y^2}}(1 - x^2 - y^2). \tag{7.2}$$

To solve this system, introduce the plane polar coordinates $x = r\cos\theta$ and $y = r\sin\theta$. Then, on multiplying the first and second equations by x and y respectively and adding, we obtain

$$x\dot{x} + y\dot{y} = \frac{1}{2}\frac{d}{dt}(x^2 + y^2) = \frac{1}{2}\frac{d}{dt}(r^2) = r\dot{r} = r(1 - r^2) \tag{7.3}$$

or

$$\dot{r} = 1 - r^2. \tag{7.4}$$

On the other hand, multiplying the first and second equations in (7.2) by y and x, respectively, and subtracting produces

$$x\dot{y} - y\dot{x} = r^2\dot{\theta} = x^2 + y^2 = r^2 \tag{7.5}$$

or

$$\dot{\theta} = 1. \tag{7.6}$$

If $r = r_0$, $\theta = \theta_0$ at $t = 0$, we obtain on integration[1] for[2] $r_0 \neq 1$

$$r = \frac{Ce^{2t} - 1}{Ce^{2t} + 1}$$
$$\theta = \theta_0 + t, \tag{7.7}$$

where $C \equiv \frac{1+r_0}{1-r_0}$. Regardless of whether the initial radius $r_0 < 1$ or $r_0 > 1$, as $t \to +\infty$, $r \to 1$, i.e., all trajectories that start initially from inside or outside the circle of radius $r = 1$ spiral in a counterclockwise sense (from (7.7)) onto this circle as $t \to +\infty$. This is demonstrated in Figure 7.1 where two different starting points, $(r_0, \theta_0) = (2, \pi/4)$ and $(0.1, \pi/4)$ have been taken. The PolarPlot plotting command was used, the command structure being given as part of the following example which shows how the analytical steps in the above derivation

---

[1] Noting that $\int \frac{dr}{1-r^2} = \frac{1}{2}\ln\left(\frac{1+r}{1-r}\right)$.
[2] For $r_0 = 1$, $\dot{r} = 0$ and r doesn't change.

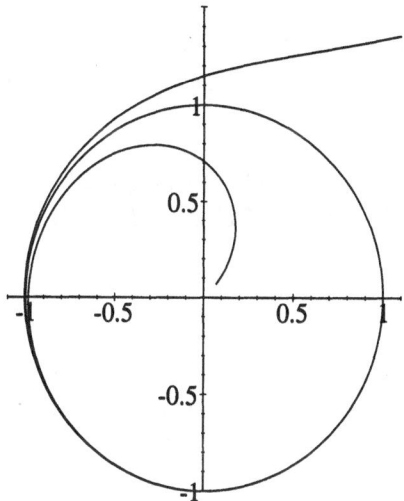

Figure 7.1: Two trajectories winding onto a stable circular limit cycle.

are carried out with the Mathematica computer algebra system.

Example 7-1: Approach to Stable Limit Cycle

Using Mathematica,

   a. Derive Equations (7.4) and (7.6).

   b. Solve these ODEs to obtain Equations (7.7) and produce the polar plot shown in Figure 7.1.

Solution: A call is made to the `Graphics` package,

```
Clear["Global`*"]; << Graphics`
```

and Equations (7.2) are entered.

```
eq1 = D[x[t],t] == -y[t]+x[t](1-x[t]^2-y[t]^2)/Sqrt[x[t]^2+y[t]^2]
```

$$x'(t) == -y(t) + \frac{x(t)\left(1 - x(t)^2 - y(t)^2\right)}{\sqrt{x(t)^2 + y(t)^2}}$$

```
eq2 = D[y[t],t] == x[t]+y[t](1-x[t]^2-y[t]^2)/Sqrt[x[t]^2+y[t]^2]
```

$$y'(t) == x(t) + \frac{y(t)\left(1 - x(t)^2 - y(t)^2\right)}{\sqrt{x(t)^2 + y(t)^2}}$$

Functions are created relating Cartesian coordinates to polar coordinates.

```
x[t_]:=r[t] Cos[θ[t]];  y[t_]:=r[t] Sin[θ[t]];
```

Carrying out the same manipulations as in the earlier hand derivation and applying the FullSimplify command, subject to the condition $r(t) > 0$, yields Equations (7.4) and (7.6).

```
eq3 = FullSimplify[x[t] eq1[[1]] + y[t] eq2[[1]]
      == x[t] eq1[[2]] + y[t] eq2[[2]], r[t] > 0]
```

$$r(t)^2 + r'(t) == 1$$

```
eq4 = FullSimplify[y[t] eq1[[1]] - x[t] eq2[[1]]
      == y[t] eq1[[2]] - x[t] eq2[[2]], r[t] > 0]
```

$$\theta'(t) == 1$$

The radial equation, eq3, is analytically solved subject to the initial condition $r(0) = r0$,

```
DSolve[{eq3, r[0] == r0}, r, t];
```

and the answer simplified.

```
r = Simplify[r[t] /. %[[1]]]
```

$$\frac{r0 + e^{2t}\,(r0+1) - 1}{-r0 + e^{2t}\,(r0+1) + 1}$$

The output is of the same structure as the radial solution in Equation (7.7). The reader can manipulate the Mathematica result into exactly the same form if so desired. The angular ODE, eq4, is also readily solved,

```
DSolve[{eq4, θ[0] == θ0}, θ, t];
```

yielding the desired result. To make a polar plot, the "dummy" angle variable ϕ is introduced,

```
θ = φ == θ[t] /. %[[1]]
```

$$\phi == t + \theta0$$

and the θ equation solved for the time t.

```
Solve[θ, t]
```

$$\{\{t \to \phi - \theta0\}\}$$

Then the time is eliminated from the radial solution by substituting the above result into r. The polar form $r(\phi)$ results.

```
r=r/.%[[1]]
```

$$\frac{r0 + e^{2(\phi-\theta0)}(r0+1) - 1}{-r0 + e^{2(\phi-\theta0)}(r0+1) + 1}$$

Two trajectories will be plotted with different initial radii ($r0 = 0.1$ and $r0 = 2$) but with the same initial angle $\theta0 = \pi/4$.

```
r1=r/.{r0->.1,θ0->Pi/4}; r2=r/.{r0->2,θ0->Pi/4};
```

Figure 7.1 then results on plotting r1 and r2 with the `PolarPlot` command.

```
PolarPlot[{r1,r2},{φ,Pi/4,10},PlotStyle ->Hue[.6],TextStyle->
{FontFamily->"Times",FontSize ->12},ImageSize->{400,400}];
```

End Example 7-1

In Figure 7.1 one trajectory winds onto the circular limit cycle from the inside, the other from the outside. This is an example of a stable limit cycle, such cycles being of importance for the operation of all electronic oscillator circuits. The Van der Pol equation, for example, displays a stable limit cycle which must be found either using an approximate analytical method (e.g., Lindstedt's perturbation method) or numerically. For $\epsilon \ll 1$, Lindstedt's method allowed us to establish that the limit cycle for the Van der Pol case is also (almost) circular in the phase plane with a radius $r = 2$. Study of the transient growth, using

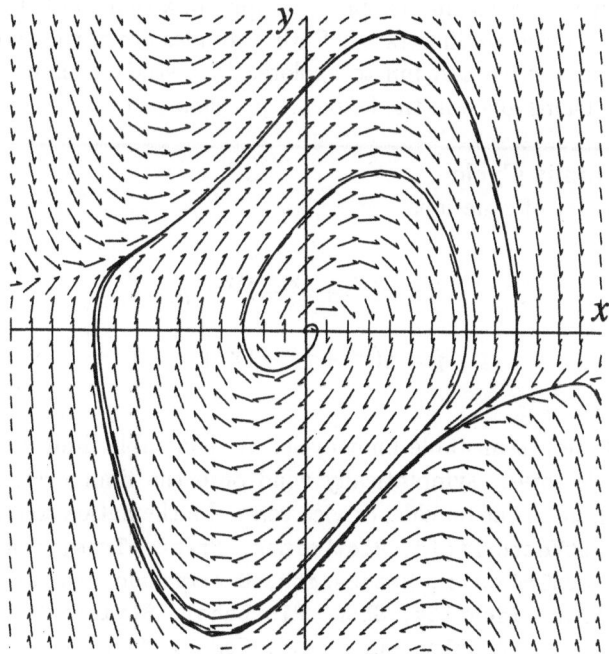

Figure 7.2: Two trajectories winding onto the $\epsilon = 1$ Van der Pol limit cycle.

the KB method, confirmed that the limit cycle is stable. For large ϵ, the above approximate analytic methods break down and a numerical approach is needed. As ϵ becomes large, the Van der Pol limit cycle becomes distorted away from the circular shape. We have already observed this distortion in Figure 4.16 where the limit cycle was numerically generated for $\epsilon = 1$. For viewing convenience, the plot has been reproduced in Figure 7.2 but with the isoclines which appeared previously left out. Referring to either one of our stable limit cycle examples, as illustrated in Figures 7.1 and 7.2, it is not too surprising that a hydrodynamical language has evolved in discussing limit cycles. The stable limit cycle is referred to as a "sink" for the trajectories and the unstable focal point at the origin a "source". Limit cycles may be studied in the laboratory by making use of the Wien bridge oscillator as illustrated in the next two experimental activities. In the first experiment the signal has its origin in an unstable focal point and evolves towards a stable limit cycle, while in the second experiment stable VdP limit cycle oscillations are produced.

Stable Limit Cycle
In this activity the Wien bridge oscillator makes use of a counterbalancing negative and positive feedback to produce a very stable nonlinear limit cycle. The signal is studied in its transient and steady state regimes.

Van der Pol Limit Cycle
This activity uses the Wien bridge oscillator to investigate and model the limit cycle predicted by the Van der Pol equation.

In the next two Mathematica files, examples of a semistable limit cycle and a multiple limit cycle are illustrated. Both examples are of the artificial variety but are still instructive and fun to explore. The student should also try the associated problems given in the following problem section.

Semistable Limit Cycle
In this file the set of equations

$$\dot{x} = x(x^2 + y^2 - 1)^2 - y,$$

$$\dot{y} = y(x^2 + y^2 - 1)^2 + x$$

is shown to have a semistable limit cycle. A semistable limit cycle is one for which the trajectories asymptotically wind onto the limit cycle from the inside and wind off on the outside, or vice versa. Mathematica commands in file: `Graphics, PlotVectorField, PlotPoints, Frame->True, ScaleFunction, FrameTicks, ScaleFactor, HeadLength, Background, ParametricPlot, DisplayFunction->Identity, Show, PlotLabel, StyleForm, TextStyle, ImageSize, Table, NDSolve, Evaluate, Compiled->False, PlotStyle, DisplayFunction->$DisplayFunction, RGBColor, Hue, PolarPlot`

Multiple Limit Cycles
This file numerically explores the stability of the multiple limit cycles which occur for the nonlinear system given in Problem 7-3 by integrating trajectories with different initial conditions. Mathematica commands in file: Graphics, PlotVectorField, PlotPoints, Frame->True, ScaleFunction, Evaluate, PlotLabel, ScaleFactor, HeadLength, Background, ParametricPlot, DisplayFunction->Identity, Show, StyleForm, TextStyle, ImageSize, NDSolve, Compiled->False, PlotStyle, RGBColor, FrameTicks, Table, Hue, DisplayFunction->$DisplayFunction,

Before concluding this section, it should be stressed that a limit cycle represents a new kind of periodic motion and should not be confused with the vortex trajectories discussed in Chapter 4. In the first of the following problems, the reader is asked to list as many differences as possible.

PROBLEMS

Problem 7-1: Limit cycles versus vortices
Enumerate and discuss as many differences between vortices and limit cycles as you can. Can vortices exist in the real world?

Problem 7-2: An unstable limit cycle
Consider the set of equations

$$\dot{x} = y + x(x^2 + y^2 - 1),$$

$$\dot{y} = -x + y(x^2 + y^2 - 1).$$

Paralleling the procedure for the stable limit cycle example in the text, solve these equations exactly and show that an unstable limit cycle of radius 1 exists. Confirm your answer numerically by considering several different initial conditions and plotting the resulting trajectories. What type of singular point exists at the origin? Confirm your answer by carrying out a topological analysis.

Problem 7-3: Multiple nested limit cycles
Given the pair of equations

$$\dot{x} = -y + xf(\sqrt{x^2 + y^2}),$$

$$\dot{y} = x + yf(\sqrt{x^2 + y^2}),$$

with $f(\sqrt{x^2 + y^2}) = f(r) = r^2 \sin(1/r)$, analytically show that all of the circles of radius $r = 1/(n\pi)$ ($n = 1, 2, 3 \ldots$) are limit cycles, unstable for odd n and stable for even n. (Hint: To establish stability, let $1/r = n\pi + \epsilon(t)$ and study the temporal behavior of ϵ for $\epsilon \ll n\pi$.)

Problem 7-4: The pursuit problem of A.S. Hathaway [Hat21]
A playful dog d initially at the center O of a large pond swims straight for a
duck D, which is swimming at constant speed in a circle of radius a centered on
O and taunting the dog. The dog is swimming k times as fast as the duck.

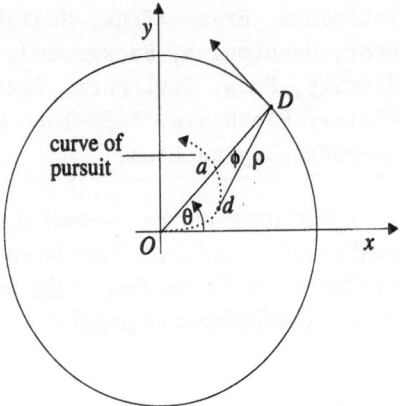

a. Show that the path of the pursuing dog is determined by the following
   coupled nonlinear equations:

   $$\frac{d\phi}{d\theta} = \frac{a}{\rho}\cos\phi - 1, \qquad \frac{d\rho}{d\theta} = a\sin\phi - ka,$$

   where the angles θ and ϕ are defined in the figure and ρ is the distance
   between the instantaneous positions of the dog and the duck.

b. The duck is safe! For $k = 2/3$, numerically show that the dog never reaches
   the duck, but instead traces out a path which asymptotically approaches a
   circle of radius $2a/3$ about the origin. By choosing other starting positions
   for the dog, show by plotting out the pursuit trajectories that the circle
   is a stable limit cycle. Take $a = 1$ and start the duck out at $x = 1, y = 0$
   and swimming counterclockwise. By choosing some other k values, show
   that a similar limit cycle behavior prevails for all $k < 1$, the value of k
   dictating the size of the limit cycle.

c. Fly, duck, fly! For $k > 1$, numerically demonstrate that capture will take
   place. Choose several different starting positions for the dog. Taking
   $k = 3/2, a = 1$, the dog at O, and the same initial conditions for the duck
   as in (b), determine through what angular displacement θ the duck has
   swum before being captured?

Problem 7-5: What's the stability?
By carrying out an exact analytic solution, show that the equations

$$\dot{x} = -x - y + (x - y)(x^2 + y^2), \qquad \dot{y} = x - y + (x + y)(x^2 + y^2)$$

have a limit cycle and determine whether it is stable, unstable, or semistable.
Confirm your analysis by plotting representative trajectories in the phase plane.

Problem 7-6: A semistable limit cycle
Demonstrate analytically that the nonlinear system in Mathematica File MF28 has a semistable limit cycle.

Problem 7-7: Cepheid type variable stars
Krogdahl [Kro55] has used a variation on Van der Pol's equation to explain the pulsation of variable stars of the Cepheid type. The time-dependent radius is given by $r(\tau) = r_0[1 + x(\tau)]^{1/3}$ where τ is the normalized time, r_0 is the mean radius, and $x(\tau)$ satisfies the nonlinear equation

$$\ddot{x} - b(1 - x^2)\dot{x} - a(1 - \frac{3}{2}a)\dot{x}^2 + x - ax^2 + \frac{7}{6}a^2x^3 = 0$$

where a and b are positive constants. Taking $a = 0.2/3$, $b = 0.5$, $x(0) = 1.1$, and $y(0) \equiv \dot{x}(0) = 0$, plot the phase plane trajectory and show that a stable limit cycle exists. Try some other initial conditions to confirm that it is a limit cycle. Also examine how the shape of the limit cycle changes with differing values of a and b.

7.2 Relaxation Oscillations

As mentioned in the previous section, for $\epsilon \ll 1$, we have seen in our study of the Lindstedt perturbation method that the limit cycle for the Van der Pol oscillator is nearly circular with a radius $r = 2$. As ϵ is increased, the circle becomes more and more distorted as higher harmonics are added in. Figure 7.3, which can be reproduced from file MF30, shows the limit cycles obtained numerically for $\epsilon = 1, 2, ..5$. The shapes of the limit cycles for even larger values of ϵ may be generated by using the file.

Relaxation Oscillations for Van der Pol (VdP) Oscillator
This file generates the limit cycles in the phase plane for the Van der Pol oscillator as a function of large and increasing ϵ. The file's limit cycles are color coded for viewing convenience. For large ϵ, for example $\epsilon = 10$, $x(t)$ displays relaxation oscillations. Mathematica commands in file: `Table`, `NDSolve`, `MaxSteps`, `ParametricPlot`, `Evaluate`, `PlotPoints`, `PlotRange`, `Hue`, `Frame->True`, `Compiled->False`, `FrameTicks`, `Epilog`, `Plot`, `Thickness`, `PlotLabel`

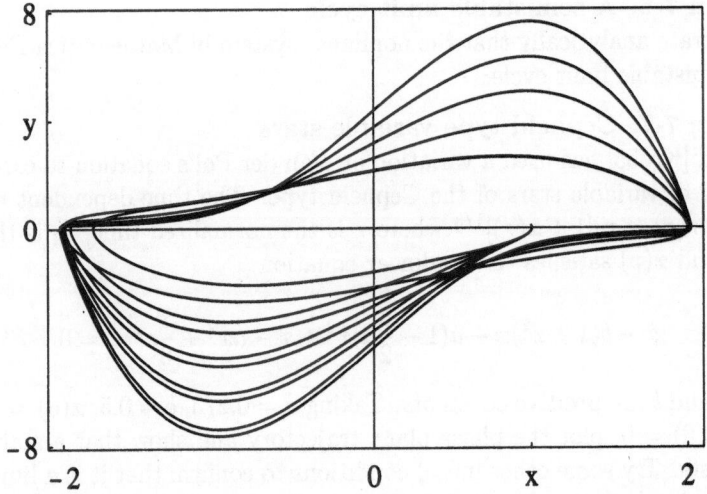

Figure 7.3: VdP limit cycles for $\epsilon = 1$ (inner curve), 2,..5 (outer curve).

As ϵ becomes large compared to 1, the Van der Pol solution $x(t)$ displays so-called relaxation oscillations. Figure 7.4 shows the behavior for $\epsilon = 10, 20$. There are fast changes of $x(t)$ near certain values of t with relatively slowly

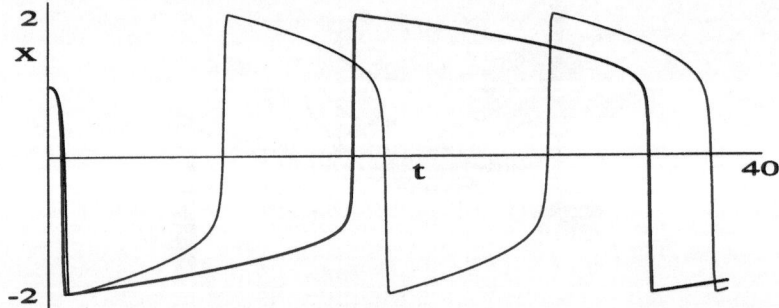

Figure 7.4: Relaxation oscillations of the VdP eq. for $\epsilon =10$ (thin), 20 (thick).

varying regions in between. As ϵ is further increased, the slowly varying regions span longer and longer time intervals.

To understand what is going on for large ϵ, let's rewrite Van der Pol's equation in the form

$$\dot{x} = \epsilon(y - f(x)), \quad \dot{y} = -x/\epsilon, \tag{7.8}$$

with $f(x) \equiv -x + (1/3)x^3$. (Van der Pol's equation results on differentiating the first equation and using the second.) Then, dividing the second equation by the first and rearranging, we obtain

$$(y - f(x))\frac{dy}{dx} = -\frac{x}{\epsilon^2}. \tag{7.9}$$

For $\epsilon \gg 1$, the rhs is nearly zero. In this limit, either $y = f(x)$ or $dy/dx = 0$, i.e., $y =$constant. Sketching $y = f(x)$, the behavior is as schematically indicated

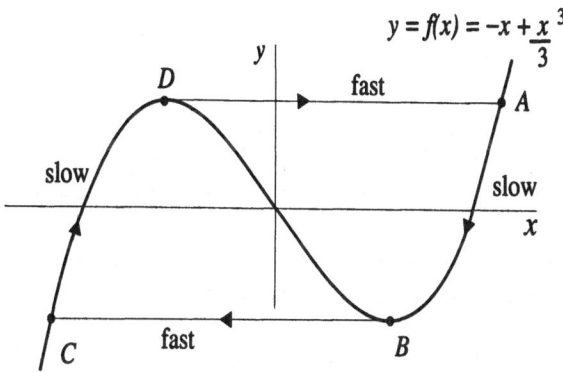

Figure 7.5: Origin of fast and slow time scales for relaxation oscillations.

in Figure 7.5. The system slowly traverses the $f(x)$ curve in the sense of the arrows from A to B, jumps horizontally and quickly from B to C, again slowly moves along the $f(x)$ curve to D, and then jumps quickly back to A. This behavior is readily confirmed by creating the phase plane portrait for the Van der Pol system (7.8) for $\epsilon = 10$.

Example 7-2: Phase Plane Trajectory for Relaxation Oscillation

---

Plot the phase plane trajectory for the VdP system (7.8) for $\epsilon = 10$ and $t = 0..20$. Take $x(0) = y(0) = 0.8$ as the initial values.

Solution: After loading the **Graphics** package,

```
Clear["Global`*"]; << Graphics`
```

the value of ϵ and the form of $f(x)$ are specified.

```
ε = 10; f = -x[t] + x[t]^3/3
```

$$\frac{x(t)^3}{3} - x(t)$$

The two equations making up the VdP system (7.8) are entered,

```
eq1 = x'[t] == ε (y[t] - f)
```

$$x'(t) == 10 \left(-\frac{1}{3}x(t)^3 + x(t) + y(t) \right)$$

```
eq2 = y'[t] == -x[t]/ε
```

$$y'(t) == \frac{-x(t)}{10}$$

and numerically solved, subject to the initial conditions and specified time range.

```
sol = Table[NDSolve[{eq1, eq2, x[0] == .8, y[0] == .8}, {x[t], y[t]},
    {t, 0, 20}]];
```

The `ParametricPlot` command is used to produce a graph of the phase plane trajectory. The curve is colored blue and the labels A, B, C, and D are placed at similar points on the trajectory to those shown in Figure 7.5.

```
gr1 = ParametricPlot[Evaluate[{x[t], y[t]} /. sol], {t, 0, 20},
    PlotStyle -> Hue[.6], PlotRange -> {{-2.5, 2.5}, {-1.1, 1.1}},
    Ticks -> {{-2, {1, "x"}, 2}, {-1, {.5, "y"}, 1}}, Epilog ->
    {Text["A", {2.2, .6}, {0,-1}], Text["B", {1.3, -.9}, {0,-1}],
    Text["C", {-2.2, -.9}, {0,-1}], Text["D", {-1.4, .6}, {0,-1}]},
    DisplayFunction -> Identity];
```

The curve $f = -x + x^3/3$ is plotted as a reddish dashed line,

```
gr2 = Plot[-x + x^3/3, {x, -2.1, 2.1}, DisplayFunction -> Identity,
    PlotStyle -> {Hue[.9], Dashing[{.02}]}];
```

and the two graphs are superimposed with the `Show` command.

```
Show[gr1, gr2, DisplayFunction -> $DisplayFunction, TextStyle ->
{FontFamily -> "Times", FontSize -> 20}, ImageSize -> {600, 400}];
```

The resulting Figure 7.6 agrees with our earlier analysis. By reducing the

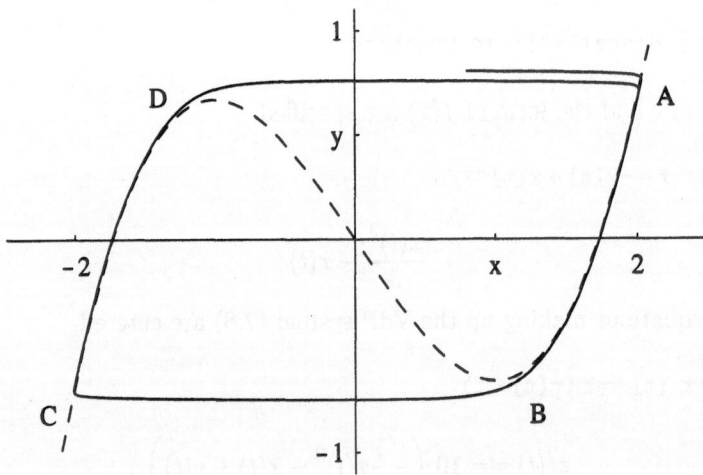

Figure 7.6: Mathematica confirmation of the trajectory in Figure 7.5.

elapsed time from 20 to smaller values, the reader may readily confirm the qualitative comments previously made on which legs of the trajectory are fast and which are slow.

End Example 7-2

Although they differ in their detailed behavior, other examples of relaxation oscillations are also characterized by long "dormant" periods interspersed with very rapid changes. Familiar examples in nature are the periodic bursts of boiling water in geysers such as Old Faithful and the sudden movements of the Earth's tectonic plates resulting in earthquakes. Relaxation oscillations are easily demonstrated in the laboratory as the student will realize if he or she tries one or more of Experimental Activities 17 to 19. Can you think of any other possible examples of relaxation oscillations, either in nature or which could be the subject of an experimental activity?

Relaxation Oscillations: Neon Bulb
Relaxation oscillations are produced by a fixed voltage source in a circuit that contains a neon glow lamp in parallel with a capacitor.

EXP 17

Relaxation Oscillations: Drinking Bird
An example of a relaxation oscillation is that of the commercially available toy "drinking bird" which, when placed perched on the edge of a beaker of water, bobs rapidly with its beak to the water and back again, followed by a long delay time before the next bobbing action. This experimental activity is a variation on the drinking bird.

EXP 18

Relaxation Oscillations: Tunnel Diode
In this activity, self-excited relaxation oscillations produced by an electric circuit that consists of a constant energy source connected to a tunnel diode, inductor, and resistor, are investigated. The tunnel diode's current-voltage curve is deduced from the relaxation oscillations.

EXP 19

PROBLEMS
Problem 7-8: Fast and slow legs
Determine the approximate times for each leg of the trajectory in Figure 7.6 and thus confirm the identification of the fast and slow legs for the trajectory.

Problem 7-9: A modified VdP equation
What effect does altering the variable damping term in the VdP equation to the form $(1 - x^2 - \dot{x}^2)\dot{x}$ have on the limit cycle curves of MF30 for $\epsilon = 1, 2, 3, ..., 10$? Do relaxation oscillations occur for $\epsilon = 10$? Explain.

Problem 7-10: The Rayleigh equation
For $\epsilon \gg 1$, analytically show that the period T of the relaxation oscillations for the Rayleigh equation

$$\ddot{y} - \epsilon(\dot{y} - \frac{1}{3}\dot{y}^3) + y = 0$$

is given approximately by the formula $T = (3 - 2\ln 2)\epsilon = 1.61\epsilon$. Compare T calculated from the formula with the value obtained numerically for $\epsilon = 5, 10, 15$ and 20. Comment on the accuracy.

Problem 7-11: Another relaxation oscillator
For $\epsilon \gg 1$, show that the variable damping oscillator equation

$$\ddot{x} - \epsilon(1 - |x|)\dot{x} + x = 0$$

has the approximate period $T = 2(\sqrt{2} - \ln(1 + \sqrt{2}))\epsilon = 1.07\epsilon$.

7.3   Bendixson's First Theorem

The literature dealing with the existence and properties of limit cycles is vast, so only a few of the more important aspects will be highlighted in this chapter. A great deal of mathematical effort and brain power has gone into trying to establish simple, yet general, theorems which will allow one to decide whether a given set of nonlinear equations has a limit cycle or not. One such theorem, referred to as Bendixson's first theorem or more commonly as Bendixson's negative criterion, can sometimes be used to establish the nonexistence of limit cycles for the basic system of equations

$$\dot{x} = P(x,y), \quad \dot{y} = Q(x,y). \tag{7.10}$$

7.3.1   Bendixson's Negative Criterion

Bendixson's negative criterion (first theorem) may be stated as follows:
If the expression

$$\frac{\partial P}{\partial x} + \frac{\partial Q}{\partial y} \neq 0$$

does not change its sign within a simply connected domain D of the phase plane, no periodic motions can exist in that domain.
   A simply connected domain (or "simple domain") is a region having the property that any closed curve or surface lying in the region can be shrunk continuously to a point without passing outside of the region. Thus a plane annular region between two concentric circles in the phase plane, for example, is not a simply connected domain, i.e., a simple domain has no holes.

7.3.2   Proof of Theorem

Assume that a closed curve Γ enclosing the area D' exists in the domain D of

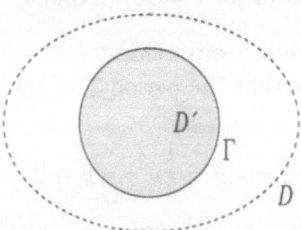

Figure 7.7: Closed curve Γ enclosing area D' in domain D of phase plane.

the phase plane as in Figure 7.7. We make use of Stokes' theorem from vector calculus. For any continuously differentiable vector field $\vec{A}$

$$\int\int_S (\nabla \times \vec{A}) \cdot \hat{n} ds = \oint_\Gamma \vec{A} \cdot d\vec{r} \tag{7.11}$$

where S is a surface in three dimensions (Figure 7.8) having a closed curve Γ as its boundary. The direction of the unit vector $\hat{n}$ is related to the direction of the line integral around the contour Γ as shown. If one curls the fingers of the right hand in the sense of going around Γ, the thumb points in the direction of $\hat{n}$. Stokes' theorem is only true if S is a simply connected domain. It turns

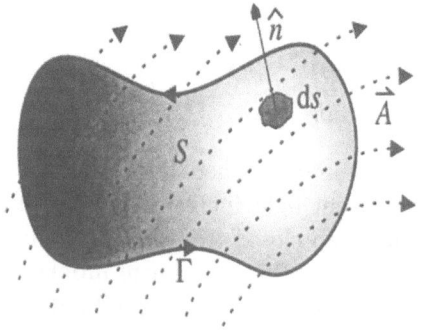

Figure 7.8: Geometrical meaning of symbols in Stokes' theorem.

out that this is necessary to guarantee the single-valuedness of the line integral. The student is referred to any standard vector calculus text for a discussion of this point.

If the surface S is in the x–y plane (our phase plane) with Γ bounding D' (Figure 7.7) and $\vec{A} = -Q\hat{i} + P\hat{j}$, then Stoke's theorem reduces to

$$\int\int_{D'} \left(\frac{\partial P}{\partial x} + \frac{\partial Q}{\partial y} \right) dx dy = \oint_\Gamma (P dy - Q dx). \tag{7.12}$$

If $\partial P/\partial x + \partial Q/\partial y$ is nonzero and does not change sign within D', then the lhs of (7.12) cannot be zero and thus

$$\oint_\Gamma (P dy - Q dx) \neq 0. \tag{7.13}$$

However, if the closed path Γ is to be a phase trajectory, then from Equations (7.10) we have

$$dt = \frac{dx}{P} = \frac{dy}{Q} \tag{7.14}$$

so that $P dy - Q dx = 0$ and

$$\oint_\Gamma (P dy - Q dx) = 0. \tag{7.15}$$

But the results (7.15) and (7.13) contradict each other. The closed curve Γ cannot be a phase trajectory, which implies that no periodic motion can exist in D. The converse of this theorem does not hold, however. I.e., if $\partial P/\partial x + \partial Q/\partial y$ does change sign nothing can be concluded about the existence of a limit cycle.

7.3.3   Applications

As the first application of the theorem, recall the laser beam competition equations (with $I_{\rm L} \equiv x$, $I_{\rm s} \equiv y$) from Chapter 2 for beams traveling in the opposite directions from each other:

$$\frac{dx}{dz} = -gxy - \alpha x \equiv P(x,y)$$
$$\frac{dy}{dz} = -gxy + \alpha y \equiv Q(x,y). \tag{7.16}$$

To apply Bendixson's negative criterion, we calculate

$$\frac{\partial P}{\partial x} + \frac{\partial Q}{\partial y} = -g(x+y). \tag{7.17}$$

For the physical problem of interest where both x and y are restricted to positive values because they represent light intensities, there can be no change of sign of (7.17). So there can be no closed trajectories in the first quadrant, i.e., no cyclic solutions are possible for this laser beam problem.

As a second example, consider the unstable limit cycle of Problem 7-2. In this case, we have

$$P(x,y) \equiv y + x(x^2 + y^2 - 1), \quad Q(x,y) \equiv -x + y(x^2 + y^2 - 1), \tag{7.18}$$

so

$$\frac{\partial P}{\partial x} + \frac{\partial Q}{\partial y} = 4\left(x^2 + y^2 - \frac{1}{2}\right). \tag{7.19}$$

Since this expression does not change sign for

$$\sqrt{(x^2 + y^2)} \equiv r < \frac{1}{\sqrt{2}}$$

no closed trajectory can exist lying inside a circle of this radius. The quantity $\partial P/\partial x + \partial Q/\partial y$ also does not change sign outside the radial distance $r = 1/\sqrt{2}$. However, we know that there is a limit cycle at $r = 1$. What has gone wrong? The problem is that in excluding the region $r < 1/\sqrt{2}$ in the second case, one is no longer dealing with a simply connected domain (i.e., the domain has a circular hole cut out of it) and Bendixson's theorem should not be applied! On the other hand, if the whole phase plane is considered there is a change of sign and no conclusions can be drawn about the existence of a limit cycle. Thus, Bendixson's first theorem, although simple to apply, is not always decisive in answering the question of whether or not a limit cycle exists.

PROBLEMS
Problem 7-12: Van der Pol oscillator
Apply Bendixson's first theorem to the Van der Pol equation. What conclusion would you draw as to the existence of a limit cycle on the basis of this theorem?

Problem 7-13: A successful application
Show that the nonlinear system [Ver90]

$$\dot{x} = -x + y^2, \quad \dot{y} = -y^3 + x^2$$

has no periodic solutions by using the negative criterion.

Problem 7-14: No periodic solutions
By using the negative criterion, show that the following nonlinear systems [Str94] have no periodic solutions.

   a. $\dot{x} = -x + 4y, \quad \dot{y} = -x - y^3,$

   b. $\dot{x} = -2xe^{(x^2+y^2)}, \quad \dot{y} = -2ye^{(x^2+y^2)},$

   c. $\dot{x} = y - x^3, \quad \dot{y} = -x - y^3.$

Problem 7-15: Variable damping Duffing equation
Consider the Duffing equation

$$\ddot{x} - (1.8 - x^2)\dot{x} - 2x + x^3 = 0$$

with variable damping.

   a. Apply Bendixson's negative criterion to this problem and state what conclusion you would come to as to the existence of a limit cycle.

   b. Find and identify the stationary points of this system. Confirm the nature of two of these stationary points by producing a phase plane portrait for the initial values $x = 0.75, y = \dot{x} = 0$ and $x = -0.75, y = 0$. Take $t = 0..30$.

   c. With the same time interval as in (b), take the initial phase plane coordinates to be $x = 0.1, y = 0$. What conclusion can you draw from this numerical result? How do you reconcile your answer with (a)?

7.4 The Poincaré–Bendixson Theorem

Another theorem due to Poincaré and Bendixson gives the necessary and sufficient conditions for the existence of a limit cycle. Unfortunately, the theorem is often difficult to apply because it requires a preliminary knowledge of the nature of trajectories. The Poincaré–Bendixson theorem (sometimes called the second theorem of Bendixson) involves the concept of a half-trajectory which is now explained. For the sake of argument, consider the stable limit cycle schematically depicted in Figure 7.9 with trajectories inside and outside winding onto the limit cycle as $t \to \infty$. Suppose that the representative point at some instant of time is at A. The time origin can be arbitrarily chosen at A, so that A divides the trajectory from $t = -\infty$ to $t = +\infty$ into two half-trajectories, that is, a half-trajectory for $t > 0$ describing the future history of the system and a half-trajectory for $t < 0$ describing the past. In physical problems, one is interested in phenomena starting at a particular instant of time which is usually taken to be zero. Therefore, it is only the half-trajectory $t \geq 0$ which is of importance. With this elementary concept in mind, the second theorem of Bendixson can now be stated.

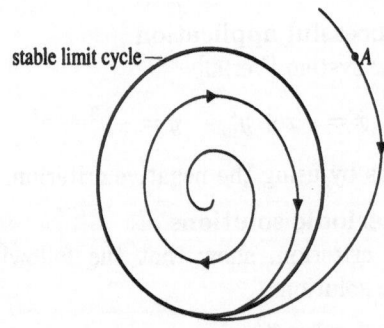

Figure 7.9: Concept of a half-trajectory.

7.4.1 Poincaré–Bendixson Theorem

Let $x(t)$, $y(t)$ be the parametric equations of a half-trajectory C which remains inside the finite domain D for $t \to +\infty$ without approaching any singularity. Then only two cases are possible, either C is itself a closed trajectory or C approaches such a trajectory.

For a proof of this intuitively plausible theorem, the reader is referred to the text by Sansone and Conti [SC64].

7.4.2 Application of the Theorem

Recall that in the limit cycle example with which we began this chapter the polar form of the equations was

$$\dot{r} = 1 - r^2, \quad \dot{\theta} = 1. \tag{7.20}$$

Then $dr/d\theta = 1 - r^2$ with $dr/d\theta > 0$ for $r < 1$ and $dr/d\theta < 0$ for $r > 1$. If an annular domain D of inner radius $r = 1/2$ and outer radius $r = 3/2$ is drawn, the trajectories must cross the inner circle from the region $r < 1/2$ to the region $r > 1/2$, as schematically shown in Figure 7.10 since $dr/d\theta > 0$.

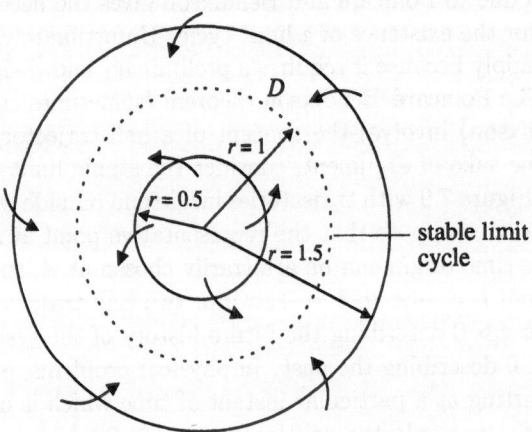

Figure 7.10: Application of the Poincaré–Bendixson theorem.

For the outer circle, the trajectories must cross from the outside $r > 3/2$ to the inside $r < 3/2$ since $dr/d\theta < 0$. It follows from $dr/d\theta = 1 - r^2$ that there are no singularities within D or on its circular boundaries. Since none of the arrows leave the annular domain D, all the trajectories cutting the circles will be trapped inside D. Clearly there exists a half-trajectory that remains inside D as $t \to \infty$ without approaching any singularity. Thus, according to the Poincaré–Bendixson theorem, there is at least one limit cycle inside D. Recall from our previous exact analysis that there is a stable limit cycle at $r = 1$.

PROBLEMS

Problem 7-16: The theorem fails

For $\epsilon \ll 1$, the Van der Pol equation has a limit cycle differing very little from a circle of radius $r = 2$. Show explicitly that if one chooses an annular domain defined by circles of radius $r = 1$ and $r = 3$, the Poincaré–Bendixson theorem is unable to predict the existence of a limit cycle in the annular domain.

Problem 7-17: Is there or isn't there?

Consider the nonlinear system [Ver90]

$$\dot{x} = x(x^2 + y^2 - 2x - 3) - y, \quad \dot{y} = y(x^2 + y^2 - 2x - 3) + x,$$

of coupled ODEs.

   a. Locate and identify all the stationary points.

   b. Use polar coordinates to transform the equations in terms of r and θ.

   c. Choose an annular domain defined by circles of radius $r = 1$ and $r = 3$. Applying the Poincaré–Bendixson theorem, determine if there is a limit cycle inside the annular domain and if so what is its stability.

   d. Confirm your answer by plotting the tangent field for $x = -2..2$, $y = -3..3$.

Problem 7-18: Periodic solution

Consider the nonlinear system [Str94]

$$\dot{x} = -x - y + x(x^2 + 2y^2), \quad \dot{y} = x - y + y(x^2 + 2y^2),$$

of coupled equations.

   a. Locate and identify all the stationary points.

   b. Use polar coordinates to transform the equations in terms of r and θ.

   c. By choosing appropriate concentric circles of different radii centered on the origin and applying the Poincaré–Bendixson theorem to the annular region, show that the system has at least one periodic solution.

   d. Make a tangent field plot over the range $x = -2..2$, $y = -2..2$ and identify the nature of the periodic solution.

Problem 7-19: Limit cycle

Consider the nonlinear system [Str94]

$$\dot{x} = x - y - x^3, \quad \dot{y} = x + y - y^3,$$

of coupled ODEs.

a. Locate and identify all the stationary points.

b. Use polar coordinates to transform the equations in terms of r and θ.

c. By choosing appropriate concentric circles of different radii centered on the origin and applying the Poincaré–Bendixson theorem to the annular region, show that the system has a limit cycle inside the annular region.

d. What is the stability of the limit cycle?

e. By plotting phase plane trajectories for suitable initial conditions, as well as the tangent field, confirm your analysis.

Problem 7-20: Another limit cycle
Consider the nonlinear system [Str94]

$$\dot{x} = x - y - x(x^2 + 5y^2), \quad \dot{y} = x + y - y(x^2 + y^2),$$

of coupled equations.

a. Locate and identify all the stationary points.

b. Use polar coordinates to transform the equations in terms of r and θ.

c. By choosing circles centered on the origin of radii $r = \frac{1}{\sqrt{2}}$ and $r = 1$ and applying the Poincaré–Bendixson theorem to the annular region, show that the system has a limit cycle inside the annular region.

d. Is the limit cycle stable or unstable?

e. By plotting the tangent field as well as phase plane trajectories for suitable initial conditions, confirm your analysis.

Problem 7-21: Still another limit cycle
Consider the nonlinear system

$$\dot{x} = y + (x/2)(1/2 - x^2 - y^2), \quad \dot{y} = -x + (y/2)(1 - x^2 - y^2),$$

of coupled ODEs.

a. Locate and identify all the stationary points.

b. Use polar coordinates to transform the equations in terms of r and θ.

c. By applying the Poincaré–Bendixson theorem to an appropriate annular region establish that there is as at least one limit cycle.

d. By plotting the tangent field as well as phase plane trajectories for suitable initial conditions, determine the number and stability of the limit cycle(s) and confirm your analysis.

7.5 The Brusselator Model

7.5.1 Prigogine–Lefever (Brusselator) Model

A hypothetical set of chemical reactions (the Brusselator model) due to Prigogine and Lefever [PL68] which displays stable limit cycle behavior is the following:

$$A \overset{k_1}{\to} X$$
$$B + X \overset{k_2}{\to} Y + D$$
$$2X + Y \overset{k_3}{\to} 3X$$
$$X \overset{k_4}{\to} E.$$

We have already encountered the chemical oscillator model referred to as the Oregonator. It doesn't take much imagination to guess in which city the present chemical oscillator model originated. In the Brusselator model, it is assumed that the reactions are all irreversible and the concentrations of species A and B are held constant by having very large reservoirs of these species. Then, recalling the empirical rule for chemical reactions discussed in Chapter 2, the rate equations for X and Y are

$$\dot{X} = k_1 A - k_2 B X + k_3 X^2 Y - k_4 X$$
$$\dot{Y} = k_2 B X - k_3 X^2 Y. \tag{7.21}$$

Introducing the new variables

$$x \equiv \sqrt{\frac{k_3}{k_4}} X, \quad y \equiv \sqrt{\frac{k_3}{k_4}} Y, \quad \tau \equiv k_4 t$$

and positive real parameters

$$\bar{a} \equiv \sqrt{\frac{k_3}{k_4}} \frac{k_1 A}{k_4}, \quad \bar{b} \equiv \frac{k_2 B}{k_4},$$

the rate equations become

$$\dot{x}(\tau) = \bar{a} - \bar{b} x + x^2 y - x$$
$$\dot{y}(\tau) = \bar{b} x - x^2 y. \tag{7.22}$$

To find the singular points of (7.22) form the ratio

$$\frac{dy}{dx} = \frac{x(\bar{b} - xy)}{\bar{a} - \bar{b} x + x^2 y - x}. \tag{7.23}$$

There is just one singular point, located at $x_0 = \bar{a}$, $y_0 = \bar{b}/\bar{a}$. Setting $x = \bar{a} + u$, $y = \bar{b}/\bar{a} + v$ and linearizing, (7.23) becomes

$$\frac{dv}{du} = \frac{-\bar{b}u - \bar{a}^2 v}{(\bar{b} - 1)u + \bar{a}^2 v}. \qquad (7.24)$$

Using the standard notation of the topological chapter, we identify $a = \bar{b} - 1$, $b = \bar{a}^2$, $c = -\bar{b}$, $d = -\bar{a}^2$, $p = -(a+d) = -(\bar{b} - 1 - \bar{a}^2)$ and $q = ad - bc = +\bar{a}^2 > 0$. Therefore, making use of the p–q figure of Chapter 4, the singular point is a stable nodal or focal point if $\bar{b} < 1 + \bar{a}^2$ and an unstable nodal or focal point if $\bar{b} > 1 + \bar{a}^2$. The Poincaré–Bendixson theorem can be used to show that, for the unstable case, the trajectory winds onto a stable limit cycle. The detailed supporting argument is presented next.

7.5.2   Application of the Poincaré–Bendixson Theorem

Since we must have x and $y > 0$, i.e., positive chemical concentrations, let's consider the quarter plane shown in Figure 7.11 and choose the domain D to have the appearance of a "lean-to shack" enclosing the singular point at $\bar{a}$, $\bar{b}/\bar{a}$ (with $\bar{b} > 1 + \bar{a}^2$). While the floor and left wall of the shack seem a natural choice,

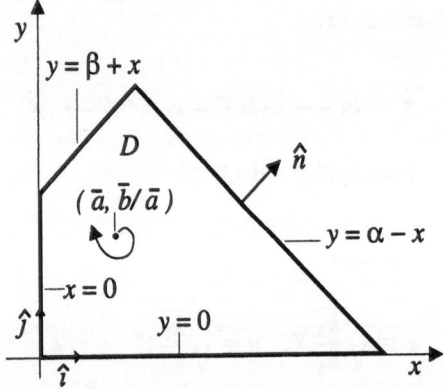

Figure 7.11: Domain needed to apply the Poincaré–Bendixson theorem to the Prigogine–Lefever model.

the slanted roof and other slanted wall are not obvious choices to complete the domain. They arise from a trial and error approach, and are dictated by the fact that we want all trajectories to be crossing a domain wall (or roof) in the same direction. Properly choosing the domain D is the greatest difficulty in applying the Poincaré–Bendixson theorem.

Next, define the velocity vector field

$$\vec{v} = \hat{\imath}\dot{x} + \hat{\jmath}\dot{y} = \hat{\imath}(\bar{a} - \bar{b}x + x^2 y - x) + \hat{\jmath}(\bar{b}x - x^2 y)$$

and let $\hat{n}$ be a unit vector pointing out of the region D. We want to prove that $\vec{v} \cdot \hat{n} < 0$ everywhere along the boundary of D (α, β are not fixed for the moment), i.e., $\vec{v}$ is directed into D at each point on the boundary.

Consider each boundary of D in turn and calculate $\vec{v} \cdot \hat{n}$ for each. Along the boundary line $x = 0$ (the left wall of our shack), since $\hat{n} = -\hat{\imath}$, then

$$\vec{v} \cdot \hat{n}|_{x=0} = -\bar{a} < 0$$

as required.

Along the boundary line $y = 0$ $(x > 0)$ (the "floor"), $\hat{n} = -\hat{\jmath}$, so

$$\vec{v} \cdot \hat{n}|_{y=0} = -\bar{b}x < 0.$$

At $x = 0$, $y = 0$, $\vec{v} = \bar{a}\hat{\imath}$, i.e., the vector field $\vec{v}$ is tangent to the boundary. However, at $x = 0$ and $y = 0$, since $\dot{x} = \bar{a}$, $\ddot{y} = \bar{a}\bar{b} > 0$, no trajectory can leave D through the origin. That is, if $x(0) = y(0) = 0$, then $x(\tau) > 0$ and $y(\tau) > 0$ over some time interval $\tau > 0$.

Along the slanted boundary line $y = \alpha - x$, $\hat{n} = (\hat{\imath} + \hat{\jmath})/\sqrt{2}$ and

$$\vec{v} \cdot \hat{n}|_{y=\alpha-x} = (\bar{a} - x)/\sqrt{2} < 0$$

for $x > \bar{a}$. Thus, $y = \alpha - x$ is a suitable boundary for any $\alpha > 0$ as long as $x > \bar{a}$.

For $x < \bar{a}$, take the boundary to be $y = \beta + x$ (the slanted roof in Figure 7.11). Along $y = \beta + x$, $\hat{n} = (-\hat{\imath} + \hat{\jmath})/\sqrt{2}$ and we want

$$\vec{v} \cdot \hat{n}|_{y=\beta+x} = \frac{-\bar{a} + 2\bar{b}x - 2x^2y + x}{\sqrt{2}} < 0.$$

For this last result to be negative we must have

$$y > \frac{(2\bar{b} + 1)x - \bar{a}}{2x^2}. \tag{7.25}$$

The maximum value of the rhs of (7.25), which is obtained by differentiating, occurs when $x = 2\bar{a}/(2\bar{b} + 1)$, and at this x value, the rhs equals $(2\bar{b} + 1)^2/8\bar{a}$. For $\bar{b} > 1 + \bar{a}^2$, $x = 2\bar{a}/(2\bar{b} + 1) < \bar{a}$, so we are in the correct region. β must be chosen so that the line $y = \beta + x$ goes through the point, $x = 2\bar{a}/(2\bar{b} + 1)$, $y = (2\bar{b} + 1)^2/8\bar{a}$. Therefore

$$\beta = \frac{(2\bar{b} + 1)^3 - 16\bar{a}^2}{8\bar{a}(2\bar{b} + 1)}$$

which is positive for $\bar{b} > 1 + \bar{a}^2$.

Finally, choose α such that at $x = \bar{a}$, $y = \alpha - x = \beta + x$, so $\alpha = 2\bar{a} + \beta$. Since no trajectories can leave region D and there are no other singularities inside D besides the unstable focal (or nodal) point, there must be a stable limit cycle inside D to "absorb" the trajectories. This conclusion is confirmed in Mathematica File 31. Figure 7.12 shows three different trajectories for $\bar{a} = 1$, $\bar{b} = 3$ winding onto the limit cycle which lies inside the boundary of domain D. So the Brusselator model is a success story as far as applying the Poincaré–Bendixson theorem.

Poincaré has also established a series of necessary, but not sufficient, criteria for the existence of limit cycles, based on a theory of characteristic indices associated with closed curves [Min64]. For particular nonlinear equations various

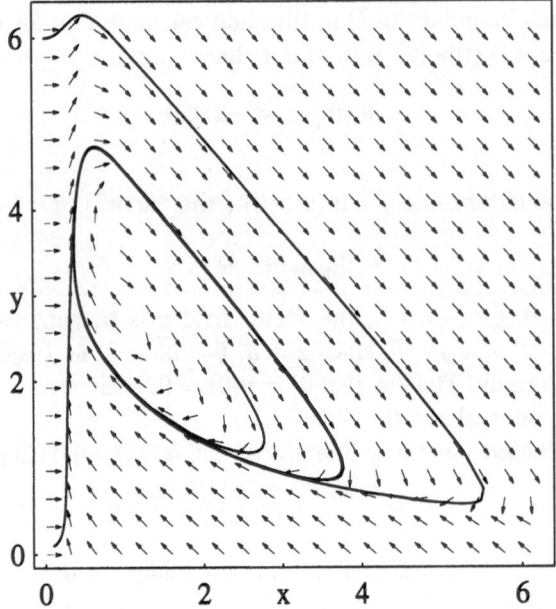

Figure 7.12: Trajectories winding onto stable limit cycle for Brusselator model.

theorems can be proved about the limit cycles of those equations. Sansone and Conti [SC64] discuss such theorems in great detail.

Brusselator Limit Cycle
A stable limit cycle is obtained inside the domain chosen for the Brusselator model for $\bar{b} > 1 + \bar{a}^2$. Mathematica commands in file: Graphics, ScaleFactor, PlotVectorField, PlotPoints, Frame->True, FrameTicks, Background, ScaleFunction, HeadLength, Hue, NDSolve, ParametricPlot, Evaluate, PlotStyle, RGBColor, Compiled->False, Frame->True, Show, ImageSize

---

PROBLEMS
Problem 7-22: Glycolytic oscillator
Living cells obtain energy by breaking down sugar, a process known as glycolysis. In yeast cells, this process proceeds in an oscillatory way with a period of a few minutes. A model proposed by Sel'kov [Sel68] to describe the oscillations is

$$\dot{x} = -x + \alpha y + x^2 y,$$

$$\dot{y} = \beta - \alpha y - x^2 y.$$

Here x and y are the normalized concentrations of adenosine diphosphate (ADP) and fructose-6-phosphate (F6P), and α and β are positive constants.

   a. Show that the nonlinear system has a stationary point at $x = \beta$, $y = \beta/(\alpha + \beta^2)$. Show that the stationary point is an unstable focal or nodal point if $(\beta^2 + \alpha)^2 < (\beta^2 - \alpha)$.

b. Take $\alpha = 0.05$ and $\beta = 0.5$ and check that the inequality in (a) is satisfied. To apply the Poincaré–Bendixson theorem, choose a domain D such

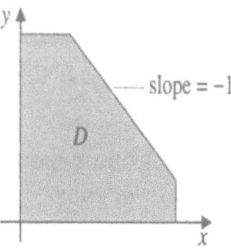

as schematically illustrated in the figure, the slanted wall having a slope of
-1. By calculating $\vec{v} \cdot \hat{n}$ on each boundary of the domain, determine a domain of the indicated shape, such that all trajectories cross the boundaries
from the outside to the inside. Is there a limit cycle inside D? Explain.

c. Confirm that the system with $\alpha = 0.05$, $\beta = 0.5$ has a limit cycle by plotting the tangent field and trajectories corresponding to the initial phase
plane coordinates $(1, 1)$ and $(3, 3)$. Take $t = 0..60$.

Problem 7-23: Why not vertical walls and a flat roof?
Show that if one attempted to make the right wall of the lean-to shack vertical
and the roof flat, the Poincaré–Bendixson theorem would be unable to predict
the limit cycle inside the domain.

Problem 7-24: Indices of Poincaré
Consider a closed curve C, which is to be traversed in a clockwise sense by a
point P, in a vector field $\vec{A}$. Attach to P coordinate axes x', y', whose directions
are to remain parallel to a fixed frame x, y as P transverses C. Let the vector
$\vec{PR}$ represent the direction of the vector field $\vec{A}$ at P. As P moves around C, the

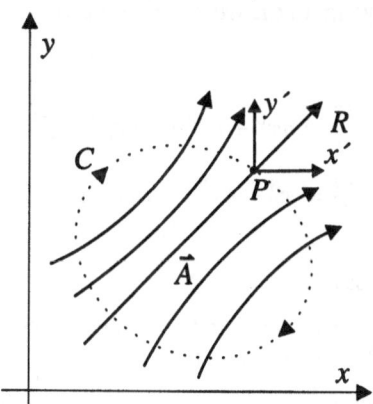

vector $\vec{PR}$ will rotate relative to the coordinate axes. Of course, $\vec{PR}$ will point
in the same direction as it originally did when P returns to the starting point.
Poincaré's *index* of a closed curve C with respect to a vector field is defined to
be the number of complete revolutions of $\vec{PR}$ when P completes one cycle of
C. The index of a singular point is the index of a closed curve in the phase

plane surrounding the singular point and lying in a vector field determined by the phase plane trajectories. Prove the following theorems:

1. *The index of an ordinary point is zero.*

2. *The index of a vortex, nodal or focal point is +1, whereas the index of a saddle point is −1.*

3. *The index of a closed curve containing more than one singular point is the algebraic sum of the indices of the enclosed singular points.*

Problem 7-25: Frommer's theorem

Consider the nonlinear system

$$\dot{x} = P(x, y), \quad \dot{y} = Q(x, y)$$

where $P(x, y)$ and $Q(x, y)$ are homogeneous polynomials of odd degree. Writing $p(u) = P(1, u)$ and $q(u) = Q(1, u)$, prove Frommer's theorem [Fro34]:
A necessary and sufficient condition that the equation

$$\frac{dy}{dx} = \frac{Q(x, y)}{P(x, y)}$$

has a closed cycle about the origin is that the integral

$$I = \int_{-\infty}^{\infty} \frac{p}{q - up} du = 0.$$

(Hint: Set $y = ux$ and show that $\ln(Cx) = \int p \, du/(q - up)$ where C is an integration constant. To complete the proof, consider a closed cycle in the x–y plane and note that u plays the role of slope in the suggested transformation.)

Problem 7-26: Applying Frommer's theorem

Consider the system

$$\frac{dx}{dt} = ax^3 + bx^2y + cxy^2 - y^3$$

$$\frac{dy}{dt} = x^3 + ax^2y + bxy^2 + cy^3.$$

of coupled nonlinear ODEs.

a. Use Frommer's theorem to prove that a necessary and sufficient condition for a closed cycle is that $a + c = 0$.

b. Confirm your result for $a = 1$, $c = -1$, and $b = 2$ by plotting the phase plane trajectory, taking $t = 0..50$ and the initial values $x = 1$, $y = 1$.

c. Cowhands put brands as identifying marks on their cattle. If you were to use the closed cycle generated above as your brand, what would you call it?

7.6 3-Dimensional Limit Cycles

In Mathematica File MF07, the student saw an example of a 3-dimensional stable limit cycle, the nonlinear system being the Oregonator model of the Belousov–Zhabotinski chemical reaction. Trying to use global existence theorems to establish the existence of limit cycles in three dimensions is a difficult task which is left to the student's future education. It should be noted, for example, that Bendixson's first theorem doesn't generalize into three dimensions [Ver90]. On the other hand, individual nonlinear systems of potential physical interest have been exhaustively studied. One such system is the Lorenz model,

$$\dot{x} = \sigma(y - x), \quad \dot{y} = rx - y - xz, \quad \dot{z} = xy - bz. \tag{7.26}$$

For the standard values $\sigma = 10$ and $b = 8/3$, the Lorenz system displays stable limit cycle behavior for extremely large values of r. To see that this is possible, we transform the equations by setting

$$r = 1/\sigma\epsilon^2, \quad x = X/\epsilon, \quad y = Y/\sigma\epsilon^2, \quad z = Z/\sigma\epsilon^2, \quad t = \epsilon T$$

so that

$$\dot{X}(T) = Y - \sigma\epsilon X, \quad \dot{Y}(T) = X - \epsilon Y - XZ, \quad \dot{Z}(T) = XY - \epsilon bZ. \tag{7.27}$$

The zeroth order solution ($\epsilon = 0$), corresponding to the limit $r \to \infty$, can be analytically solved.

Differentiating the first and third equations in Equation (7.27) for $\epsilon = 0$ and substituting for Y and $\dot{Y}$ yields

$$\ddot{X} = X - XZ, \quad \ddot{Z} = \dot{X}^2 + X^2 - X^2 Z. \tag{7.28}$$

Then, multiplying the first equation in (7.28) by X, subtracting the result from the second equation, and integrating twice, we obtain the conserved quantity,

$$Z - \frac{1}{2}X^2 = C_1. \tag{7.29}$$

where C_1 is a constant. Eliminating Z in the first equation of (7.28) yields an equation for X alone,

$$\ddot{X} = (1 - C_1)X - \frac{1}{2}X^3. \tag{7.30}$$

Multiplying this latter result by $2\dot{X}dT$ and integrating produces

$$\dot{X}^2 = (1 - C_1)X^2 - \frac{1}{4}X^4 + C_2 \tag{7.31}$$

where C_2 is the second integration constant. The solution $X(T)$ is expressible in terms of an elliptic function. From the conservation relation, so is $Z(T)$, as is $Y(T)$ since $Y^2 = \dot{X}^2$. Thus, a periodic solution is possible.

To see whether a periodic solution still exists for very large, but finite, r one has to go to higher order in the perturbation expansion. This nontrivial task has been carried out by Robbins [Rob79] and Sparrow [Spa82]. Sparrow has

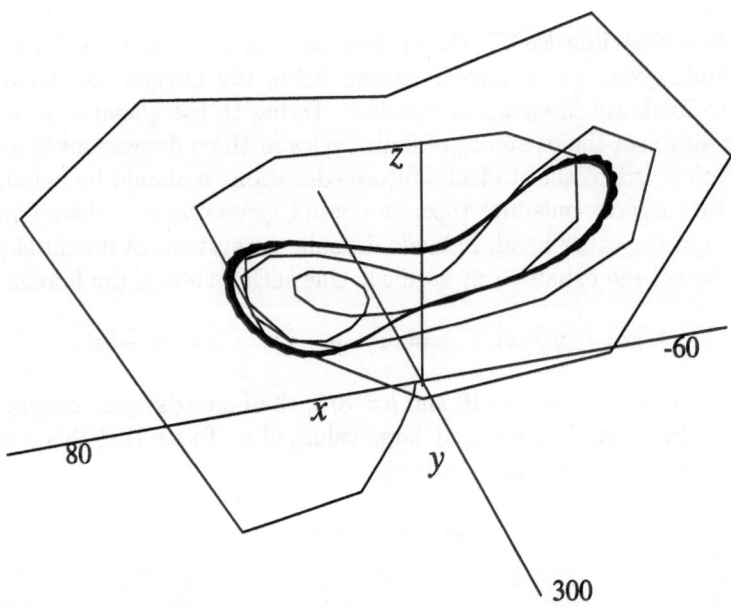

Figure 7.13: Limit cycle solution of the Lorenz system for $r = 320$.

confirmed that stable limit cycles exist for r above about $r = 214$, between $r = 145$ and $r = 166$, and in the narrow range $r = 99.5..100.8$. Setting MaxSteps->∞ in MF10 and adjusting the plotting style options, one can obtain, e.g., the limit cycle shown in Figure 7.13 for $r = 320$.

PROBLEMS
Problem 7-27: Limit cycles for the Lorenz system
Confirm that the Lorenz system has limit cycles for $r = 250$, $r = 163$, and $r = 100.0$. Show that in between the limit cycle regions cited in the text are apparently chaotic bands.

Problem 7-28: The Rössler system
Consider the Rössler system (Equations (3.22)) with $a = 0.2$ and $b = 0.2$. Confirm that 3-dimensional periodic solutions exist for $c = 2.5$, $c = 3.5$, and $c = 4.0$. Identify the periodicity in each case. Are these periodic solutions limit cycles? Explain.

Problem 7-29: Another Rössler limit cycle
The following model due to Rössler produces a 3-dimensional limit cycle for $\epsilon = 0.1$ and $a = 1.5$ which is a distortion in the z direction of the circular loop $x^2 + y^2 = a^2$.

$$\dot{x} = -y + (a^2 - x^2 - y^2)x, \quad \dot{y} = x + (a^2 - x^2 - y^2), \quad \epsilon\dot{z} = (1 - z^2)(x - 1 + z) - \epsilon z$$

Confirm that this model yields a 3-dimensional limit cycle of the indicated shape. Explore the effect of choosing other values of ϵ and a.

Chapter 8

Forced Oscillators

For he being dead, with him is beauty slain,
And, beauty dead, black chaos comes again.
William Shakespeare (1564-1616), Venus and Adonis

The nonlinear physicist (NLP)
producing forced oscillations.

8.1 Duffing's Equation

In Mathematica File MF09, the student has already seen some of the exciting possible solutions that can occur for a forced oscillator depending on the amplitude F chosen for the forcing term. The nonlinear system in that file is the Duffing oscillator

$$\ddot{x} + 2\gamma\dot{x} + \alpha x + \beta x^3 = F\cos(\omega t) \qquad (8.1)$$

with γ the damping coefficient and ω the driving frequency. In mechanical terms, the lhs of the Duffing equation can be thought of as a damped nonlinear spring. With the forcing term on the rhs included, the following special cases have been extensively studied in the literature:

1. Hard spring Duffing oscillator: $\alpha > 0$, $\beta > 0$,

2. Soft spring Duffing oscillator: $\alpha > 0$, $\beta < 0$,

3. Inverted Duffing oscillator: $\alpha < 0$, $\beta > 0$,

Ladies and Gentlemen, I give you an inverted Duffing.

4. Nonharmonic Duffing oscillator: $\alpha = 0$, $\beta > 0$.

The α and β parameters in file MF09 can be readily changed to study all of these cases. As we develop the concepts in this chapter, using the Duffing equation extensively as our illustrative example, the student should reinforce his or her understanding of the ideas by going to file MF09 and carrying out the numerical runs to which we will be referring. They are easy to do and generally don't take much time. In this chapter a conceptual framework will be developed which will help us to understand many of the features that are observed numerically. In the examples and problems, other illustrations of forced oscillator systems shall be introduced. Forced oscillators exhibit a wide variety of solutions depending on the ranges of the parameters and the initial conditions.

In a typical introductory classical mechanics course, the physics student will have studied forced oscillations of the linear spring system, i.e., Duffing's equation with $\alpha > 0$ and $\beta = 0$. The same damped simple harmonic oscillator equation is also encountered in electromagnetics for forced oscillations of the LCR electrical circuit. The standard analytical approach used to solve the forced linear spring ODE is to assume that the transient solution has died away and seek the steady-state periodic solution vibrating at the same frequency ω as that of the oscillatory driving term. As we shall discuss in detail in a later section, a nonlinear spring system can respond in many additional ways that are not possible for the linear system. Nevertheless, our study of forced nonlinear oscillators begins by paralleling the treatment of the linear oscillator and seeking the "harmonic" or period-1 solution which has the same frequency as the driving frequency. As a concrete example, our analysis will concentrate on the hard spring Duffing oscillator, but the other cases will be referred to as well. The following experimental activity shows one approach to determining the coefficients α and β for a hard spring.

Hard Spring

Using an airtrack, the relationship between the extension of a hard spring and the applied force is determined by using a force meter. Assuming a force law of the form $F(x) = \alpha x + \beta x^3$, the values of the constants α and β are found. Then the theoretical values of the period for different amplitudes are calculated and compared with the measured values.

EXP 20

Before any hard spring analysis is begun, go to file MF09 and set $\alpha = 1$, $\beta = 0.2$, $\gamma = 0.2$, $\omega = 1$, and $F = 4.0$ with the initial conditions $x(0) = 0.09$, $\dot{x}(0) = 0$. Taking $t = 100$ to 200 to eliminate the transient, the reader may readily verify that this choice of parameters generates the harmonic solution shown in Figure 8.1. On the left, the steady-state phase plane trajectory is seen

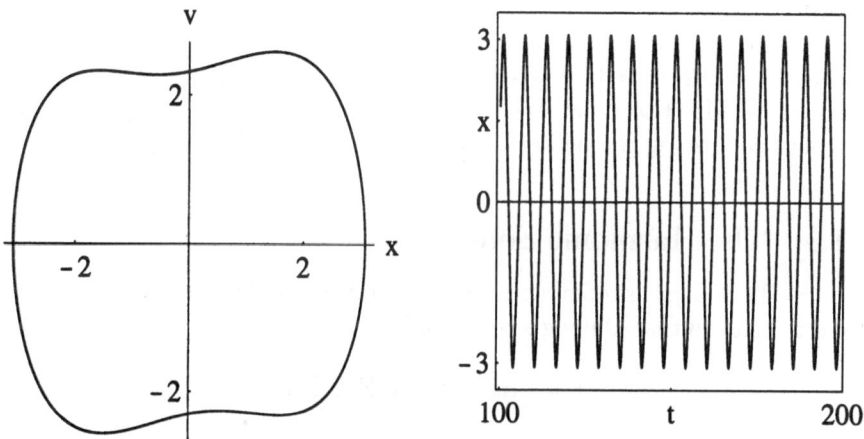

Figure 8.1: Harmonic solution for the hard spring Duffing equation.

to consist of a closed loop indicating a periodic response to the driving force. On the right, we can see from the x versus t figure that a period one solution occurs. The period is $T = 2\pi$, exactly the same as that of the driving force ($T = 2\pi/\omega$ with $\omega = 1$).

Another way of confirming the harmonic (period-1) solution is to make use of the Poincaré section capability of Mathematica File MF11. Recall that the Poincaré section corresponds to taking a "snapshot" of the phase plane at integral multiples of the period of the driving force. If one takes numpts=60 as in the file, the maximum number of steps in the NDSolve command must be increased to, say, MaxSteps->10000. Also the plot range should be changed to PlotRange->{{2.5,3.5},{1,2}}. As the reader can easily verify, the Poincaré section in this case consists of a single dot which may seem rather boring[1]. This steady state situation is rapidly achieved, and watching a stationary dot almost makes you believe that nothing is happening.

---

[1]Don't fall asleep!

To obtain an approximate analytic form[2] for the harmonic solution, another approximation technique will be used, the "iteration method". The basic idea underlying the iteration method as applied to the Duffing equation written in the form

$$\ddot{x} = -2\gamma\dot{x} - \alpha x - \beta x^3 + F\cos(\omega t) \tag{8.2}$$

is to assume a lowest (zeroth, say) order approximation $x_0 = A\cos(\omega t)$ to the solution and substitute it into the rhs of the equation. Then, solving the resulting differential equation will generate a first-order solution $x_1(t)$ which should represent a better approximation to $x(t)$, provided that a certain condition on β is met. The procedure can then be repeated until a sufficiently accurate solution for x has been achieved. After n such iterations, $x(t) \simeq x_n(t)$. This is the procedure that was used originally by Duffing to obtain the harmonic solution.

PROBLEMS
Problem 8-1: Figure 8.1
Confirm the harmonic solution plotted in Figure 8.1. Use MF09 with $\alpha = 1$, $\beta = 0.2$, $\gamma = 0.2$, $\omega = 1$, $F = 4.0$, $x(0) = 0.09$, and $\dot{x}(0) = 0$. Then, use the Poincaré file MF11 to show that the steady-state consists of a single stationary dot, thus confirming the period-1 solution. (Hint: Recall the text comments.)

8.1.1 The Harmonic Solution

To keep the analysis simple, we shall temporarily set the damping coefficient γ equal to zero, so that Duffing's equation becomes

$$\ddot{x} = -\alpha x - \beta x^3 + F\cos(\omega t). \tag{8.3}$$

For our zeroth-order solution, we assume a harmonic solution exists of the form

$$x_0(t) = A\cos(\omega t) \tag{8.4}$$

with the constant A to be determined. Substituting $x_0(t)$ into the rhs of Equation (8.3) and using the trigonometric identity $\cos(3\omega t) = 4\cos^3(\omega t) - 3\cos(\omega t)$, we obtain

$$\ddot{x}_1 = -\left[\alpha A + \frac{3}{4}\beta A^3 - F\right]\cos(\omega t) - \frac{1}{4}\beta A^3\cos(3\omega t). \tag{8.5}$$

This equation is easily integrated twice to yield

$$x_1(t) = A_1\cos(\omega t) + \frac{1}{36}\frac{\beta A^3}{\omega^2}\cos(3\omega t), \tag{8.6}$$

where

$$A_1 \equiv \frac{1}{\omega^2}\left[\alpha A + \frac{3}{4}\beta A^3 - F\right].$$

---

[2]The reader might say, "Who cares if we can find an analytic solution?" You will be surprised and delighted by the interesting phenomena that can be predicted once the analytic solution is developed, phenomena which could prove hard to discover numerically.

As was done in Lindstedt's version of perturbation theory, the integration constants have been set equal to zero to avoid the appearance of secular terms which would spoil the periodicity of our desired solution.

Now using x_1, x_2 can be generated, and so on. These messy higher-order approximations will not be derived here. Instead we shall stop at first order and adopt a procedure historically attributed to Duffing. For a forced linear ($\beta = 0$) oscillator, it is well known in classical mechanics that a characteristic "response" or "resonance" curve (Figure 8.2) can be constructed showing the relation between the magnitude of the amplitude,[3] $|A|$, and the frequency ω for

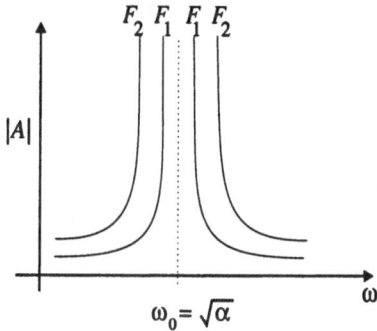

Figure 8.2: Linear response curves for different F values and zero damping.

each specified value of F. A typical set of resonance curves results, each curve "blowing up" at $\omega_0 = \sqrt{\alpha}$ because damping has been completely neglected. For small nonzero β, we would expect a corresponding set of nonlinear response curves lying in the vicinity of the linear response curves. To achieve this, we shall argue that if $x_0(t) = A\cos(\omega t)$ is truly a reasonable zeroth-order approximation to $x(t)$, then the constant A_1 in $x_1(t)$ should differ very little from A. So, we set $A_1 = A$, i.e.,

$$A_1 \equiv \frac{1}{\omega^2}\left[\alpha A + \frac{3}{4}\beta A^3 - F\right] = A \tag{8.7}$$

or

$$\omega^2 = \omega_0^2 + \frac{3}{4}\beta A^2 - \frac{F}{A}. \tag{8.8}$$

Then

$$x_1(t) = A\cos(\omega t) + \frac{1}{36}\frac{\beta A^3}{\left[\omega_0^2 + \frac{3}{4}\beta A^2 - \frac{F}{A}\right]}\cos(3\omega t). \tag{8.9}$$

That this procedure is reasonable is readily confirmed on setting $\beta = 0$. Then (8.8) reduces to

$$\omega^2 = \omega_0^2 - \frac{F}{A} \tag{8.10}$$

or

$$A = \frac{F}{(\omega_0^2 - \omega^2)}. \tag{8.11}$$

Thus

$$|A| = \frac{F}{|\omega_0^2 - \omega^2|} \tag{8.12}$$

---

[3]Defined in the usual way as the maximum numerical value of x.

which yields the linear response curves sketched in Figure 8.2. The solution

$$x(t) = \left[\frac{F}{\omega_0^2 - \omega^2} \right] \cos(\omega t) \tag{8.13}$$

is the well-established steady-state response for the forced undamped linear oscillator problem.

When $\beta > 0$, but sufficiently small, the first order iteration result might be expected to yield a reasonably accurate result. As β is increased, higher order iterations must be included. Equation (8.8) will yield the first-order approximation to the nonlinear response curves. To first order, the harmonic term in $x_1(t)$ will dominate over the third harmonic contribution provided that $\beta A^2 / 36\omega^2 \ll 1$. This puts a restrictive condition on the size of β.

PROBLEMS
Problem 8-2: Second-order nonlinear response curves
Derive the second-order nonlinear response curves for zero damping which correspond to going to second order in the iteration procedure.

Problem 8-3: Perturbation Approach
To first order in the small parameter β, use the Lindstedt method to solve

$$\ddot{x} + \alpha x + \beta x^3 = \beta F_0 \cos(\omega t).$$

Compare your answer with the iteration result derived in the text and discuss. In particular, how does the nonlinear resonance response curve compare?

8.1.2   The Nonlinear Response Curves

Let's now analyze the nonlinear response curves given by

$$\omega^2 = \omega_0^2 + \frac{3}{4}\beta A^2 - \frac{F}{A} \tag{8.14}$$

or, equivalently

$$F = (\omega_0^2 - \omega^2)A + 0.75\beta A^3. \tag{8.15}$$

In analogy with the linear ($\beta = 0$) situation,[4] it is desirable to plot $|A|$ against ω for $\beta > 0$. For example, choosing $\omega_0 = 1$, $\beta = 0.2$, and $F = 0, 1, 2, 4$, a plot of A versus ω can be generated using Mathematica's ImplicitPlot command.

Example 8-1: Zero Damping Nonlinear Response Curves

Plot A vs. ω using Equation (8.15) for $\omega_0 = 1$, $\beta = 0.2$, and $F = 0, 1, 2, 4$.

Solution: After loading the Graphics package, the values of the parameters

```
Clear["Global`*"]; << Graphics`
```

---

[4]It should be noted that A in the nonlinear case ($\beta \neq 0$) is not, strictly speaking, the amplitude of x_1, but the coefficient of the first Fourier term ($\cos(\omega t)$). However, we are assuming that this term is the dominant contribution to the solution for β sufficiently small.

$\omega[0]$ and β are specified using the BasicInput palette to input the symbols.

```
ω[0] = 1;  β = 0.2;
```

The function for generating Equation (8.15) for different F values is entered,

```
f[F_] := (ω[0]² - ω²) A + .75 β A³ == F
```

and checked for a typical F value, viz., $F = 2$.

```
f[2]
```

$$0.15 A^3 + \left(1 - \omega^2\right) A == 2$$

The ImplicitPlot command is used to create a plot of A vs. ω for $F = 0, 1, 2, 4$, the result being shown in Fig. 8.3.

```
ImplicitPlot[{f[0],f[1],f[2],f[4]}, {ω,0,3}, PlotRange -> {{0,3},
{-4,4}}, AspectRatio -> 1, Ticks -> {{1,2,3}, {-4,-2,{0.01,"0"},
{2,"A"},4}}, PlotStyle -> Hue[.9], ImageSize -> {600,400}, Epilog ->
{Text["F = 1",{.5,.4}, {0,-1}], Text["F = 4",{.6,2.8}, {0,-1}],
Text["F = 0",{.95,-1.8},{0,-1}],Text["F = 4", {2.25,-1.8}, {0,-1}],
Text["ω", {1.5,-.5}, {0,-1}]}, TextStyle -> {FontFamily -> "Times",
FontSize -> 16}];
```

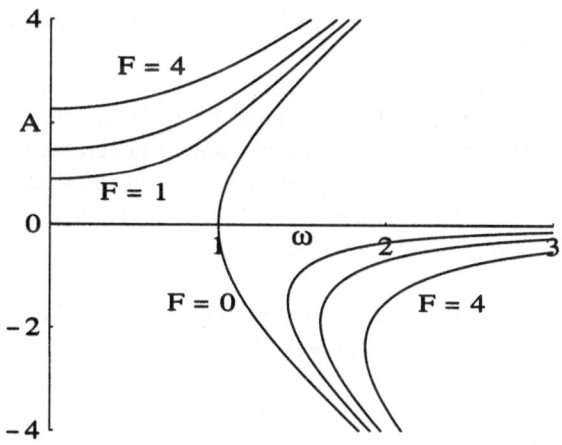

Figure 8.3: A versus ω for $\omega_0 = 1$, $\beta = 0.2$, $\gamma = 0$ and $F = 0, 1, 2, 4$.

$F = 0$ produces a horizontal parabola, while for each nonzero F value there are two branches, one with $A > 0$ and the other with $A < 0$. The former corresponds to $x(t)$ being in phase with the driving force, while the minus sign means that the motion is $180°$ out of phase with the force.

End Example 8-1

To plot $|A|$ versus ω, the code in Example 8-1 can be modified as follows:

Example 8-2: Plot of $|A|$ versus ω

Plot $|A|$ versus ω for the same parameter values as in the previous example.

Solution: The values of $\omega[0]$ and β are given,

```
Clear["Global`*"]; << Graphics`
```

```
ω[0] = 1;  β = 0.2;
```

and the function f[F_] needed for plotting $|A|$ versus ω for different F values is formed.

```
f[F_] := Abs[(ω[0]² - ω²) A + .75 β A³] - F
```

For $F = 0$, the equation $f(0) = 0$ can be solved explicitly for A.

```
sol = Solve[f[0] == 0, A]
```

$$\{\{A \to 0.\}, \{A \to -2.58199\,\sqrt{\omega^2 - 1.}\}, \{A \to 2.58199\,\sqrt{\omega^2 - 1.}\},$$
$$\{A \to -2.58199\,\sqrt{1.\,\omega^2 - 1.}\}, \{A \to 2.58199\,\sqrt{1.\,\omega^2 - 1.}\}\}$$

The third solution is selected and plotted as a blue curve over the range $\omega = 1...4$.

```
gr[0] = Plot[A /. sol[[3]], {ω,1,4}, DisplayFunction -> Identity,
        PlotRange -> {{0,4},{0, 4}}, PlotStyle -> Hue[.6]];
```

To plot the curves for nonzero F, we can employ a different approach than in Example 8-1, using the ContourPlot command with the options Contours -> {0} and ContourShading -> False. The former option generates the contour lines corresponding to $f[F] = 0$. The latter option turns off the shading which would otherwise appear. See what happens if this option is omitted. The Do command is then used to generate graphs for $F = 1...4$.

```
Do[gr[F] = ContourPlot[Evaluate[f[F]], {ω,0,4}, {A,0,4},
Contours -> {0}, ContourShading -> False, PlotPoints -> 100,
DisplayFunction -> Identity], {F,1,4}];
```

The $F = 0, 1, 2, 4$ graphs are plotted in the same figure with the Show command,

```
Show[gr[0],gr[1],gr[2],gr[4], DisplayFunction->$DisplayFunction,
Ticks->{{{.005,"0"},1,2,{2.5,"ω"},3,4},{1,{2,"|A|"},3,4}},
Epilog -> {Text["F = 0",{.85,.5},{0,-1}],Text["F = 4",{.6,2.7},
{0,-1}],Text["F = 4",{2.1,2.7},{0,-1}]},TextStyle->{FontFamily->
"Times",FontSize->16,FontColor->Hue[1]},ImageSize->{600,400}];
```

the result being displayed in Figure 8.4.

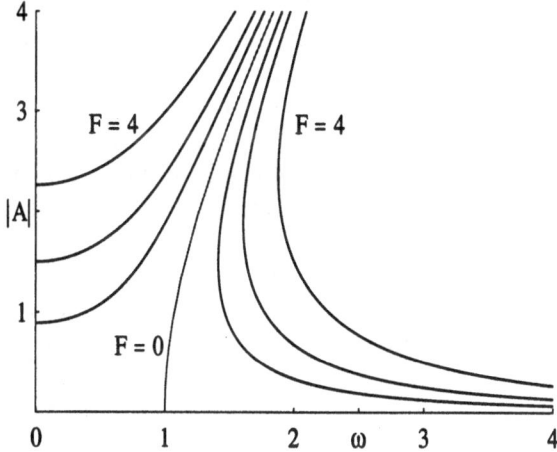

Figure 8.4: $|A|$ versus ω for $\omega_0 = 1$, $\beta = 0.2$, and $F = 0, 1, 2, 4$.

The inner pair of nonlinear resonance curves in the figure correspond to $F = 1$, the outer pair to $F = 4$. For the hard nonlinear spring, the nonlinear response curves look like the linear ones except they are tilted to the right. For the soft nonlinear spring, they would be tilted to the left. For the linear case, it is clear from Figure 8.2 that for a fixed F value, a unique value of $|A|$ results for a given ω. Examination of the nonlinear response or resonance curves in Figure 8.4 reveals that the nonlinear case is more complex. For, say, $F = 1$, there is only one $|A|$ value for $\omega = 0.7$. This $|A|$ value is extracted as follows:

```
sol2 = Solve[(f[1] /. ω->0.7) == 0,A]; A /. sol2[[6]]
```

1.30585

On the other hand, from Figure 8.4 it is clear that there are three possible $|A|$ values for, say, $\omega = 1.7$, which again are easily determined.

```
sol3 = Solve[(f[1] /. ω->1.7) == 0, A];
```

```
{A /. sol3[[4]], A /. sol3[[5]], A /. sol3[[6]]}
```

$\{0.541717, 3.24765, 3.78937\}$

Two of these values correspond to $A < 0$ and the other to $A > 0$. Can you locate them in Figure 8.3?

With the possibility of three $|A|$ values for $F = 1$, $\omega = 1.7$, how does the system know which value to choose? Stability analysis shows that not all of these three values of A correspond to stable oscillations.

$\boxed{\text{End Example 8-2}}$

With zero damping, it has been noted that A, and therefore $x_1(t)$, is either in phase or 180° (π radians) out of phase with the driving force. When damping is included in the forced linear oscillator problem, it is well established in classical mechanics that the displacement and driving force will be out of phase, with the phase angle lying between 0 and π. The exact value of the phase angle depends on the parameters of the problem. A similar behavior can be expected for the nonlinear problem. Further, for the linear oscillator, the resonance curves are "rounded off" to finite values when damping is included, no longer blowing up at ω_0. For small β, a similar rounding off would be expected to occur. These qualitative ideas shall now be confirmed.

Although it might seem more natural to prescribe the driving force and leave the phase of the displacement to be determined, it turns out to be more convenient to fix the latter phase, leaving the phase of the driving force to be ascertained. With this in mind, let's write our equation of motion, with damping included, in the form

$$\ddot{x} = -2\gamma\dot{x} - \alpha x - \beta x^3 + F\cos(\omega t + \phi). \tag{8.16}$$

The amplitude F of the driving force is fixed, but the phase angle ϕ is undetermined. Using the trigonometric identity

$$\cos(\omega t + \phi) = \cos(\omega t)\cos\phi - \sin(\omega t)\sin\phi,$$

and setting $F\cos\phi \equiv H$, $F\sin\phi \equiv G$ (so that $\tan\phi = G/H$, $F = \sqrt{H^2 + G^2}$), our equation may be rewritten as

$$\ddot{x} = -2\gamma\dot{x} - \alpha x - \beta x^3 + H\cos(\omega t) - G\sin(\omega t). \tag{8.17}$$

Assuming that β is small, the iteration procedure can once again be applied with the zeroth-order solution taken to be $x_0 = A\cos(\omega t)$. Substituting x_0 into the rhs of the ODE and integrating twice, with the integration constants set equal to zero, yields

$$x_1(t) = A_1\cos(\omega t) + B_1\sin(\omega t) + \frac{1}{36}\frac{\beta A^3}{\omega^2}\cos(3\omega t) \tag{8.18}$$

with

$$A_1 \equiv (\alpha A + \frac{3}{4}\beta A^3 - H)/\omega^2$$
$$B_1 \equiv -(2\gamma\omega A - G)/\omega^2. \tag{8.19}$$

If the zeroth-order solution is a reasonable starting point, then generalizing our earlier argument leads us to set $A_1 = A$ and $B_1 = 0$. Then, with $\alpha \equiv \omega_0^2$, we have

$$(\omega_0^2 - \omega^2)A + \frac{3}{4}\beta A^3 = H$$
$$2\gamma\omega A = G. \tag{8.20}$$

To this order, the phase angle ϕ is thus given by

$$\tan\phi = \frac{G}{H} = \frac{2\gamma\omega A}{(\omega_0^2 - \omega^2)A + \frac{3}{4}\beta A^3}. \tag{8.21}$$

As a partial check, we note that if $\gamma = 0$, then $\tan \phi = 0$, so that $\phi = 0$ or π as expected.

Recalling that $F = \sqrt{(H^2 + G^2)}$, the desired relation between A and ω for a given F is obtained,

$$\{[(\omega_0^2 - \omega^2)A + \frac{3}{4}\beta A^3]^2 + [2\gamma \omega A]^2\}^{1/2} = F. \tag{8.22}$$

For $\gamma = 0$, this reduces to our earlier expression for A versus ω. If $\beta = 0$, the correct form for the linear case results. Figure 8.5 shows some representative

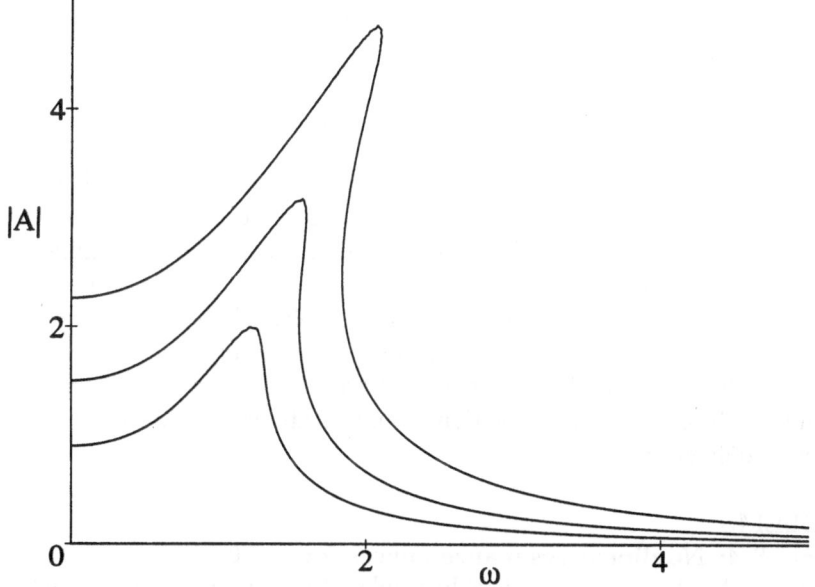

Figure 8.5: First order nonlinear response curves for nonzero damping.

nonlinear resonance curves generated from Equation (8.22) for $\omega_0 = 1$, $\beta = 0.2$, $\gamma = 0.2$, and $F = 1, 2, 4$ using a simple extension of the zero damping Mathematica code from Example 8-1. Again, the nonlinear resonance curves for $\beta > 0$ and $\gamma > 0$ look like the corresponding linear resonance curves "tilted" to the right. All three curves satisfy the convergence condition $\beta A^2 / 36\omega^2 \ll 1$ mentioned earlier. If the approximate A value when $F = 4$ and $\omega = 1$ is read off the plot, the student may verify that it agrees with the A value obtained numerically in Figure 8.1. This gives us further confidence in our iteration procedure. We shall see further supporting evidence shortly.

Note that above a critical F value each curve in Figure 8.5 has two points at which the slope $d|A|/d\omega$ is infinite. It is straightforward to show, although very lengthy [Cun64], that the region of the nonlinear resonance curve lying between these two points (see the dashed region in Fig. 8.6) corresponds to an unstable solution while the two regions (solid curves in Fig. 8.6) outside these two points correspond to stable solutions. The basic approach in the proof is to take the solution $x(t)$ (which may be only approximately known as in the problem being considered here) and let it change by a small amount $u(t)$ to $x(t) + u(t)$. On substituting $x + u$ into the original differential equation and

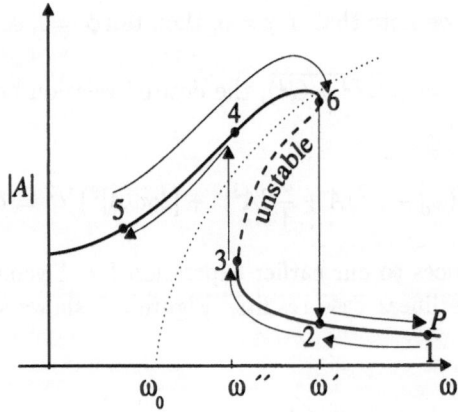

Figure 8.6: Jump phenomenon and hysteresis.

linearizing in u, a new linear equation results for $u(t)$. By studying the behavior of $u(t)$ and seeing whether it grows with time (unstable situation) or decays with time (stable situation), the stability of the solution $x(t)$ may be determined.

For the above procedure, it should be noted that if x were known exactly, then precise stability criteria could be readily established. If x is only approximately known, the stability criteria will also only be approximately known. Thus, depending on the problem, one might have to carry out several iterations to obtain a sufficiently accurate solution to make statements about the stability with some confidence.

PROBLEMS

Problem 8-4: Nonlinear resonance curves for $\gamma > 0$

Generalizing the Mathematica code of Example 8-1 to nonzero damping, confirm the nonlinear resonance curves in Figure 8.5. Take $\omega = 0..5$, $A = 0..5$, and a sufficient number of PlotPoints to obtain smooth curves. Repeat with a generalization of the code in Example 8-2.

Problem 8-5: Soft spring

For the same parameters as in Figure 8.5 but with $\omega_0 = 2$ and $\beta = -0.2$, plot the nonlinear resonance curves for a soft nonlinear spring. Discuss the results.

8.2   The Jump Phenomenon and Hysteresis

Let's now examine how $|A|$ changes as ω is continuously varied, the amplitude F of the driving force being held fixed at some particular value. Suppose a representative point P is followed along the nonlinear resonance curve of Figure 8.6 as ω is decreased from its value at point 1. As P moves from 1 to 2, $|A|$ is smoothly increasing and single-valued. As ω decreases below ω', P must continue to move along the lower stable branch of the resonance curve until it reaches point 3. Even though $|A|$ is no longer single-valued in the frequency range from ω' to ω'', stability considerations dictate that P must continue to move along the same stable segment of the resonance curve that it started out on. Since the resonance curve that is drawn corresponds to a given (fixed) value of F, as ω is decreased further P must "jump" vertically upward to point 4 and then follow the stable upper branch to, say, 5. So $|A|$ jumps discontinuously[5] from its value at 3 to a higher value at 4, and then decreases smoothly as P goes to point 5. If, on the other hand, ω is then increased from 5, P moves along the stable upper branch ($|A|$ increasing smoothly) through the point 4 to 6, whereupon it jumps vertically downward ($|A|$ decreases discontinuously) to 2 and then moves ($|A|$ decreasing smoothly) back to 1 again. If ω is alternately decreased and increased over the frequency range ω' to ω'', a hysteresis cycle[6] $2 \rightarrow 3 \rightarrow 4 \rightarrow 6 \rightarrow 2$ is traced out for $|A|$ versus ω. Numerical confirmation of the jump phenomena and the hysteresis aspects for the hard spring Duffing oscillator using file MF09 is left to the student as a problem. Experimental confirmation can be achieved by carrying out the following mechanical forced oscillator experiment.

Nonlinear Resonance Curve: Mechanical
The nonlinear resonance curve is generated using a driving motor to produce forced vibrations of an airtrack glider. The glider is connected to a hard spring arrangement consisting of two elliptical steel tapes. The jump phenomena and hysteresis are observed as the frequency is increased and decreased.

A nonlinear resonance curve can also be generated both electrically and magnetically as demonstrated in the following two experiments.

Nonlinear Resonance Curve: Electrical
In this activity, the reader may investigate how the resonance frequency changes as a function of the forcing amplitude (voltage) for a piecewise linear (nonlinear) circuit. The nonlinearity is similar to that of a "soft spring" oscillator.

Nonlinear Resonance Curve: Magnetic
In this experiment, the nonlinear resonance curve is produced by using a small magnet attached to a steel ribbon. The system models a hard spring. The magnet is driven by a Helmholtz coil or solenoid connected to a frequency generator.

Instead of performing an experiment in which F is held fixed and ω varied, imagine varying F while holding ω constant. Electrical circuits can actually be constructed where this is possible. Of course, in this case the mechanical variables are replaced with the appropriate electrical ones.

---

[5]The astute reader is going to object at this point and argue that $|A|$ cannot really jump discontinuously because then $x(t)$ (which represents a displacement in the mechanical problem) would also change discontinuously, which is unphysical. Actually, the discontinuous jumps appear because we are looking at the steady-state solution and not including transient terms. If these transient terms were included, $|A|$ would change continuously but very rapidly over a narrow range of ω in the neighborhood of ω''.

[6]The student has probably already encountered the concept of a hysteresis cycle in connection with the B–H curves of ferromagnets. In this case the hysteresis cycle arises from the nonlinear phenomenon of magnetic domain formation.

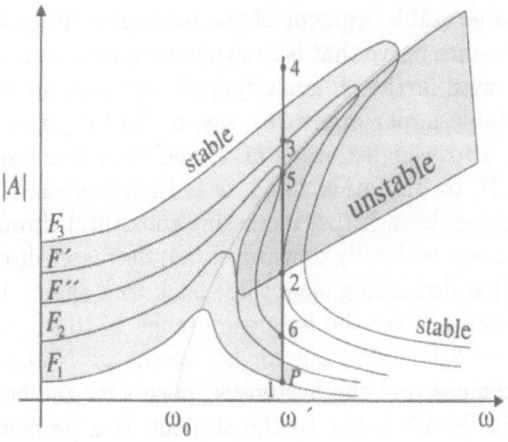

Figure 8.7: Origin of hysteresis cycle as F is varied at a fixed frequency ω'.

Referring to Figure 8.7, choose a frequency ω' as shown and consider a representative point P which moves vertically upward from point 1 as F is increased. It should be kept in mind that there is a continuous set of resonance curves corresponding to a continuous change of F. For reasons of clarity, only a few curves have been drawn. As F is increased, P moves vertically upward to the point 2 at the bottom of the unstable region where F has the value F', say. $|A|$ has increased smoothly up to this point. When F increases infinitesimally above F', P must jump vertically to point 3, and with further increase of F then move continuously to, say, 4. If F is now decreased, P will decrease continuously to 5 (where F has the value F'') at the top of the unstable region and then jump discontinuously downward to point 6. Further decrease of F causes P to move continuously downward to 1. Again a hysteresis cycle results, but this time for $|A|$ versus F.

Examination of Figure 8.7 allows us to sketch this hysteresis cycle as shown in Figure 8.8. The existence of two stable values of $|A|$ for a given F (ω held

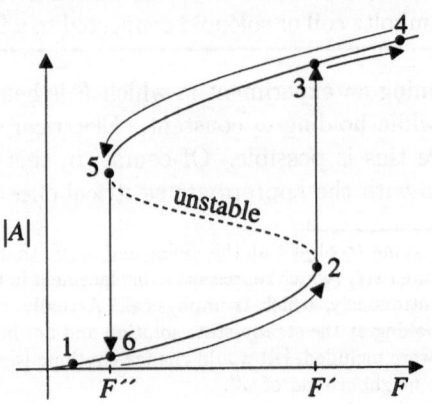

Figure 8.8: Hysteresis cycle when $|A|$ is plotted versus F.

fixed) is an example of bistability.[7] The jump phenomena and accompanying hysteresis cycles of Figures 8.8 and 8.6 were first observed in electrical and mechanical systems some 80 or 90 years ago by Martienssen [Mar10] and Duffing [Duf18]. A modern version of the jump phenomena is contained in the prediction of optical switching between bistable soliton states by Richard Enns and Sada Rangnekar [ER87b], [ER87a] in saturable Kerr materials.

PROBLEMS

Problem 8-6: Jump phenomena, hysteresis, and bistability

Consider the Duffing oscillator with $\alpha = 1$, $\beta = 0.2$, $\gamma = 0.2$, and $F = 4.0$. Confirm the jump phenomena and the hysteresis aspects using file MF09. As input values, take $\dot{x}(0) = 0$, and for $x(0)$ use values close (e.g., slightly above) to the stable branches of the $F = 4$ nonlinear resonance curve in Figure 8.5. In each numerical run choose a value for ω and allow the steady state to evolve so that $|A|$ may be determined. Then sketch or plot $|A|$ versus ω. How do your numerical results compare with what would be predicted from the $F = 4$ resonance curve? Confirm that there is bistability for $\omega = 2.002$ and determine the two $|A|$ values.

Problem 8-7: $|A|$ versus F jump phenomena

Consider the Duffing oscillator with $\alpha = 1$, $\beta = 0.2$, $\gamma = 0.2$, $\omega = 2$, $x(0) = 0.09$, and $\dot{x}(0) = 0$. Starting with $F = 0.1$, determine the steady state amplitude $|A|$ as F is increased and locate the critical value of F at which there is a large jump in amplitude. Determine this critical value to two significant figures.

Problem 8-8: Soft spring jump and hysteresis

Consider the forced soft spring oscillator with $\alpha = 1$, $\beta = -1$, $\gamma = 0.25$, and $F = 0.3$. Taking $\dot{x}(0) = 0$ and $x(0) = 0.09$, solve the oscillator equation and determine $|A|$ for ω varying from 0.3 up to $\omega = 2.0$. Show that there is an upward jump in $|A|$ for ω between 0.58 and 0.59. Demonstrate that there is hysteresis in the system by numerically determining $|A|$ for $\omega = 0.58, 0.577, 0.573$ for the two sets of initial conditions: (a) $\dot{x}(0) = 0$, $x(0) = 0.09$, (b) $\dot{x}(0) = 0$, $x(0) = 0.8$. Identify the nature of the solution in each case, using the Poincaré section Mathematica File MF11 if necessary.

Problem 8-9: Soft spring oscillator

Plot the nonlinear resonance curve $|A|$ versus ω for the forced soft spring oscillator with $\alpha = 1$, $\beta = -0.2$, $\gamma = 0.2$, and $F = 0.8$. Discuss the expected behavior of the oscillator as ω is varied from small values to large values and vice versa. Taking $\dot{x}(0) = 0$ and $x(0) = 1$, calculate the numerical value of the amplitude $|A|$ obtained by solving the Duffing equation for $\omega = 0.16, 0.35, 0.40, 0.90, 1.0, 1.35, 2.0, 3.0$. Compare with the values of $|A|$ predicted from the nonlinear resonance curve. Is the agreement reasonable? Try running Duffing's equation for $\omega = 0.5, 0.6, 0.7, 0.8$. What happens? Can you offer any explanation?

---

[7]Or, alternately, when we have two stable values of $|A|$ for a given ω, F held fixed.

8.3   Subharmonic & Other Periodic Oscillations

Up until now, only the harmonic solution of the hard spring Duffing equation has been studied, i.e., the steady-state periodic solution whose dominant part has the same frequency ω as the external driving force. Permanent oscillations whose frequency $\omega_n = \omega/n$, $n = 2, 3, 4 \ldots$ can also occur in nonlinear systems such as the Duffing equation, hard spring or otherwise. These are referred to as subharmonic oscillations.[8] Since the period of the nth subharmonic oscillation is $T = (\frac{2\pi}{\omega})n$, it is commonly referred to as a period n solution. This subharmonic phenomenon was first studied by Helmholtz in connection with nonlinear vibrations of the eardrum (see Chapter 2). It is beyond the scope of this text to present a complete treatment of subharmonic and other periodic responses for Duffing's equation. Instead, we shall give only a brief overview of the topic.

It is once again useful to go back to the forced linear oscillator problem and point out some features that are usually ignored in the linear case but become important for the forced nonlinear oscillator. Considering only the periodic solutions of the undamped forced linear oscillator equation

$$\ddot{x} + \omega_0^2 x = F \cos(\omega t),   \tag{8.23}$$

we can, without any loss of generality, choose the zero of time such that $\dot{x}(0) = 0$. If $x(0) = A$, the steady-state solution of Equation (8.23) is

$$x(t) = \tilde{A} \cos(\omega_0 t) - \left(\frac{F}{\omega^2 - \omega_0^2} \right) \cos(\omega t)   \tag{8.24}$$

where

$$\tilde{A} = A + \frac{F}{(\omega^2 - \omega_0^2)}.   \tag{8.25}$$

The first and second terms in (8.24) are referred to as the "free oscillation" and "forced oscillation" terms respectively. The solution $x(t)$ is periodic only if one of the following situations prevails:

a. Harmonic Oscillations ($\tilde{A} = 0$):
This choice yields the harmonic solution

$$x(t) = \frac{F}{(\omega_0^2 - \omega^2)} \cos(\omega t)   \tag{8.26}$$

which is usually the only case mentioned in introductory classical mechanics. In principle there are other possibilities which are now enumerated.

b. Subharmonic Oscillations ($\tilde{A} \neq 0$, $\omega_0 = \omega/n$, $n = 2, 3, 4, \ldots$):
As an example, for $n = 3$ we have $\omega_0 = \omega/3$ and the periodic solution is

$$x(t) = (A + \frac{9F}{8\omega^2}) \cos(\frac{\omega t}{3}) - (\frac{9F}{8\omega^2}) \cos(\omega t),   \tag{8.27}$$

the first term being a subharmonic term. The minimum period occurs when ωt changes by 6π, i.e., 3 times the period of the forcing term. This is a period-3 solution, which is plotted in Fig. 8.9 for $A = F = \omega = 1$.

---

[8]In the literature, this topic is often referred to as frequency demultiplication.

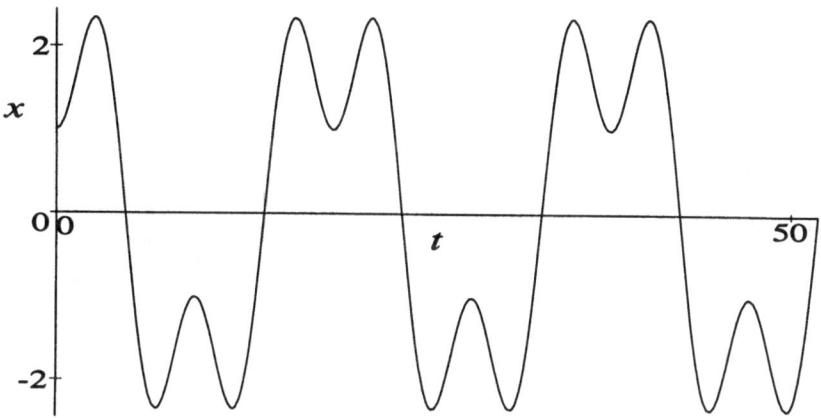

Figure 8.9: $n = 3$ subharmonic (or period 3) solution.

c. **Ultraharmonic Oscillations** ($\tilde{A} \neq 0$, $\omega_0 = n\omega$, $n = 2, 3, 4, ...$):
In this case, for $n = 3$ we have $\omega_0 = 3\omega$ and the periodic solution is

$$x(t) = (A - \frac{F}{8\omega^2}) \cos(3\omega t) + (\frac{F}{8\omega^2}) \cos(\omega t), \qquad (8.28)$$

the first term involving a third harmonic of the driving frequency. The solution is shown in Figure 8.10 for $F = A = \omega = 1$.

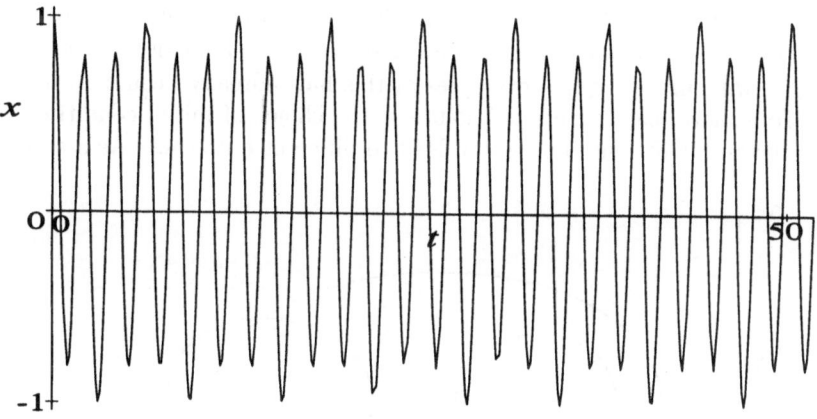

Figure 8.10: $n = 3$ ultraharmonic solution.

d. **Ultrasubharmonic Oscillations** ($\tilde{A} \neq 0$, $\omega_0 = m(\omega/n)$):
In this case one has a term involving an integral multiple of a subharmonic. For example, for $m = 4$, $n = 3$, we have $\omega_0 = 4\left(\frac{\omega}{3}\right)$ and

$$x(t) = (A - \frac{9F}{7\omega^2}) \cos(\frac{4\omega t}{3}) + \frac{9F}{7\omega^2} \cos(\omega t). \qquad (8.29)$$

A plot of this solution for $A = F = \omega = 1$ is shown in Figure 8.11. What is the period of the repeat pattern here?

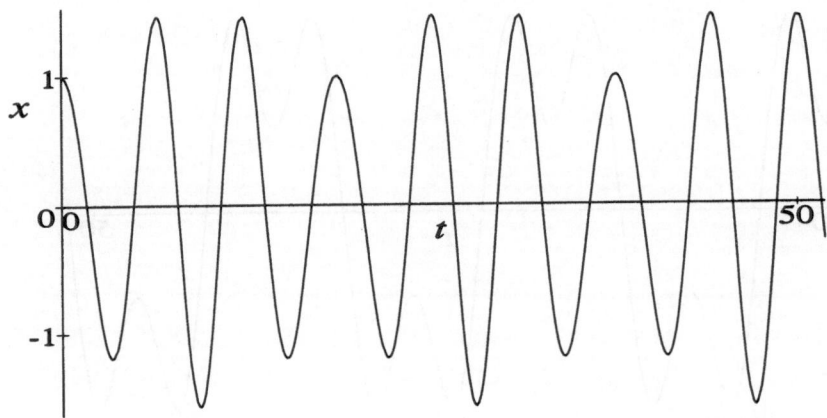

Figure 8.11: $m = 4$, $n = 3$ ultrasubharmonic solution.

In all likelihood, the student has never encountered these other possible periodic solutions in previous mechanics courses. Why not? Well, if damping is included in the forced linear oscillator problem, it is found that only the harmonic case can actually occur because the harmonic oscillation receives energy from the driving force which is at the same frequency, while the subharmonic oscillations, etc., do not receive any energy and thus die away because of the damping. However, for nonlinear systems, even with damping present, these other cases can occur and survive since energy can be fed into other frequency modes than the harmonic because of the coupling of modes due to the presence of nonlinear terms. If one of these nonharmonic modes receives sufficient energy to overcome damping it will survive. Some examples of this have already been seen in file MF09. To achieve, e.g., a period-3 solution, the choice of initial conditions is most important. Different choices can lead to different outcomes. As a simple

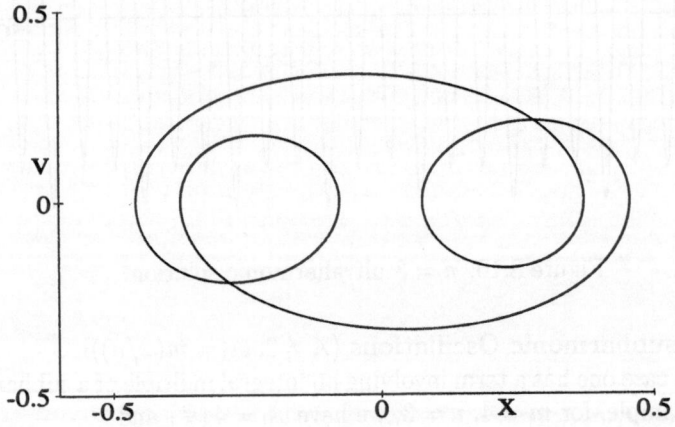

Figure 8.12: Period-3 solution for the nonharmonic Duffing equation.

illustration of this point, go to file MF09 and consider the nonharmonic Duffing oscillator with $\alpha = 0$, $\beta = 1$, $\gamma = 0.04$, $F = 0.2$, $\omega = 1$, and $\dot{x}(0) = 0$. For $x(0) = 0.20$ and $x(0) = 0.30$, a period-1 (harmonic) solution emerges after the transients have died away. If the file is run with the intermediate value $x(0) = 0.25$, we obtain the phase plane trajectory shown in Figure 8.12. Here, the transient has been eliminated by plotting the time range $t = 150...250$. The reader should not be perturbed by the crossing of trajectories in the figure which apparently violates our phase plane discussion in Chapter 4 for "real" trajectories. The crossings are an artifact of presenting the information on a plane. Our phase plane analysis was for 2-dimensional autonomous systems. Duffing's equation is non-autonomous in two dimensions but is autonomous in three dimensions since it can be rewritten as

$$\dot{x} = y$$

$$\dot{y} = -2\gamma y - \alpha x - \beta x^3 + F \cos z \tag{8.30}$$

$$\dot{z} = \omega$$

with $z(0) = 0$.

In three dimensions real trajectories do not cross. The apparent crossings in Figure 8.12 are due to the projection of a 3-dimensional system onto a 2-dimensional phase plane. So what type of solution arises? Again making use of MF09, $x(t)$ behaves as shown in Figure 8.13. Examining the repeat pattern

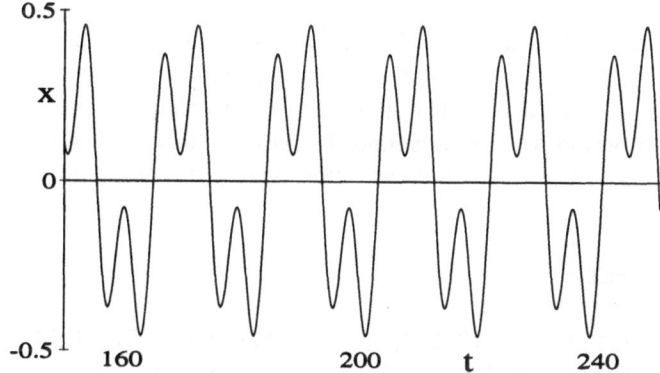

Figure 8.13: $x(t)$ for a period-3 solution.

and the period, a period-3 solution or, equivalently, an $n = 3$ subharmonic can be identified. The picture is qualitatively similar to the $n = 3$ subharmonic solution seen earlier in Figure 8.9. Our identification may also be confirmed by running file MF11 (with the PlotRange adjusted) for the same parameters. The Poincaré section corresponding to steady-state consists of three dots which the system visits consecutively in a regular sequence.

The period-3 solution occurred for $x(0) = 0.25$, while period-1 solutions are observed for $x(0) = 0.20$ and 0.30. One can ask what is the range of $x(0)$ for which the nonharmonic Duffing oscillator evolves into a period-3 solution? This is easily answered by running the file with other $x(0)$ values. We find that the range stretches (to 2-figure accuracy) from $x(0) = 0.25$ to 0.29. Still other ranges of $x(0)$ for which period-3 solutions emerge can be found.

We have taken $\dot{x}(0) = 0$ in all of our runs to this point. Let's hold $x(0) = 0.25$ and start increasing $\dot{x}(0)$. Using file MF11 a period-3 solution is found to persist up to $\dot{x}(0) = 0.06$, but for $\dot{x}(0) = 0.07$ a period-2 solution results. Period 2 lasts until $\dot{x}(0) = 0.10$, period 3 then occurring again for $\dot{x}(0) = 0.11, 0.12$ and $0.14, 0.15$. At $\dot{x}(0) = 0.13$, a period-1 solution is observed. It is clear from all of this that the outcome is somewhat sensitive to initial conditions, a general feature of forced nonlinear systems. Extreme sensitivity to initial conditions occurs in the chaotic regime as will be demonstrated later.

Moving over from $x(0) = 0.25$ to 0.26 and again changing $\dot{x}(0)$, we could begin to map out a region in $\dot{x}(0)$ versus $x(0)$ space for which only period-3 solutions occur. Such a region is called a "basin of attraction" for period-3 solutions. Similarly, other basins of attraction for period 1, period 2, etc., could be mapped out. In the following example, we show the graphical result of carrying out this process for the nonharmonic Duffing oscillator.

Example 8-3: Basins of Attraction

Map out the basins of attraction for the nonharmonic Duffing oscillator with $\alpha = 0$, $\beta = 1$, $\gamma = 0.04$, $F = 0.2$, and $\omega = 1$. Consider the phase space region of initial values $x(0) = 0.20...0.30$, $\dot{x}(0) = 0.00...0.10$, taking 0.01 increments in both directions. Create a figure in which the periodicity is indicated by suitably colored squares.

Solution: Using MF09 and MF11 to solve for the steady-state oscillations of the nonharmonic Duffing oscillator, the periodicity is determined and put into the following **data** array. The numbers refer to the observed periodicity, e.g., 3 refers to period 3. The first inner list is for $\dot{x}(0) = 0$, the second inner list is for $\dot{x}(0) = 0.01$, etc. Within each inner list, $x(0)$ runs from 0.20 to 0.30 in steps of 0.01.

```
Clear["Global`*"]

data={{1,1,1,1,1,3,3,3,3,3,1},{1,1,1,1,3,3,3,3,3,1,3},
      {1,1,1,3,3,3,3,3,1,3,3},{1,1,3,3,3,3,3,3,1,3,2,2},
      {1,3,3,3,3,3,3,3,2,2,2},{3,3,3,3,1,3,3,2,2,2,2},
      {3,3,3,1,3,3,2,2,2,2,2},{3,3,1,3,2,2,2,2,2,2,3},
      {1,3,3,2,2,2,2,2,2,3,3},{3,2,2,2,2,2,2,2,2,1,3,1},
      {2,2,2,2,2,2,3,3,3,1,1}};
```

The data is plotted using the **ListDensityPlot** command, which generates a density plot from the above array of "height" values. Colors are assigned to the different values by using the **ColorFunction** option. The "slot" symbol # represents the first (and, in this case, only) argument supplied to the "pure" **Hue[ ]&** function. The combination .6#^(.5) takes the square root of any supplied number and multiplies the result by 0.6. In the present case, this will color the period one squares red, period two squares green, and period three squares blue. The reader can adjust the colors to taste, if so desired. A plot label is added to the figure indicating this coloring scheme.

```
ListDensityPlot[data, MeshRange -> {{.195, .305}, {-.005, .105}},
ColorFunction -> (Hue[.6#^(.5)]&), PlotLabel -> "Red: Period 1,
Green: Period 2, Blue: Period 3", ImageSize -> {500, 500},
FrameTicks->{{.20, .21, .22, .23, .24, .25, .26, .27, .28, .29, .30},
{0.00, .01, .02, .03, .04, .05, .06, .07, .08, .09, 0.10}, { }, { }},
TextStyle -> {FontFamily -> "Times", FontSize -> 16}];
```

The resulting graph, with $\dot{x}(0)$ plotted vertically and $x(0)$ horizontally, is shown in Figure 8.14. In this text reproduction, black corresponds to period 1, gray to

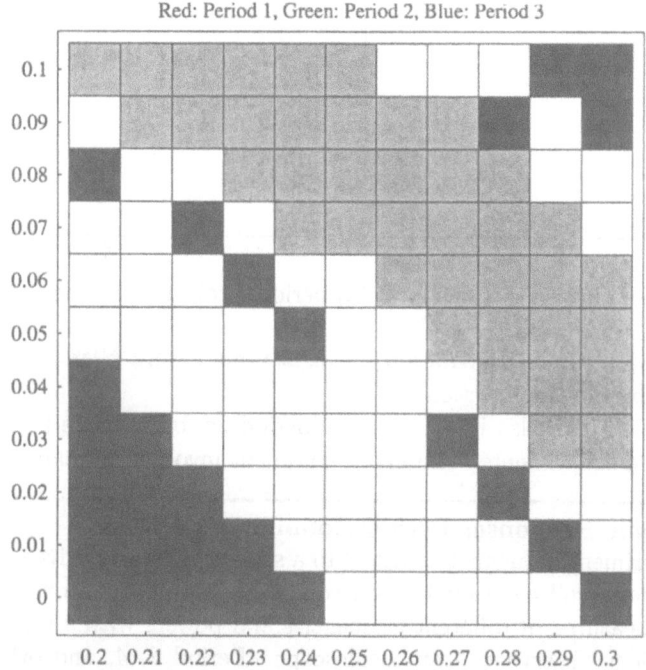

Red: Period 1, Green: Period 2, Blue: Period 3

Figure 8.14: Basins of attraction for the nonharmonic Duffing oscillator.

period 2, and white to period 3. The true colors may be seen on the computer screen.

| End Example 8-3 |

The boundaries of the basins can be smoothed by taking smaller increments, but one should automate the process and use a more sophisticated basin-generating algorithm. Also, by taking smaller increments, one can check whether or not there are even smaller basins of attraction that may have been missed because of the size of our search grid.

In addition to changing the initial conditions, for the Duffing equation one has many other "knobs" to twiddle, viz., the parameters α, β, γ, ω and F. Many different outcomes are possible, including the occurrence of chaotic oscillations, which we discuss in a later section. A common route to chaos is through period doubling when the force amplitude F is increased. An example of period doubling occurs for the inverted Duffing oscillator with $\alpha = -1$, $\beta = 1$, $\omega = 1$, $\gamma = 0.25$, $x(0) = 0.09$, and $\dot{x}(0) = 0$ and a driving term $F\cos(\omega t + 1)$. Using the Poincaré section capability of file MF11 with $\phi(0) = 1$, a period-1 solution occurs for $F = 0.325$, period 2 ($n = 2$ subharmonic) for $F = 0.34875$, period 4 ($n = 4$ subharmonic) for $F = 0.3570$, and period 8 for $F = 0.35797$. The Poincaré sections of the latter two consist of four dots and eight dots respectively as shown in Figure 8.15. It is left as a challenging exercise for the student

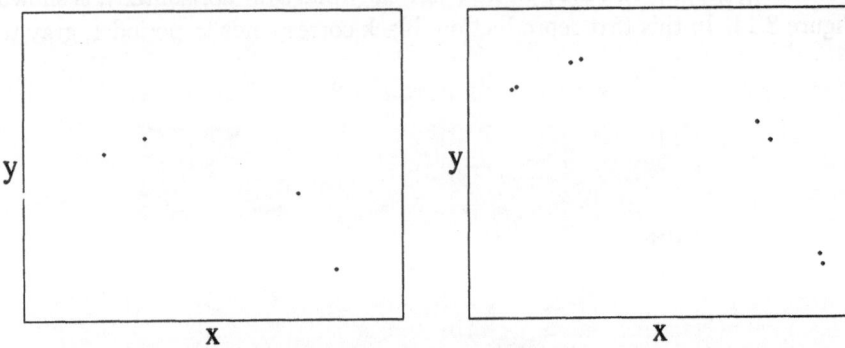

Figure 8.15: Poincaré sections for a (a) period 4 solution, (b) period 8 solution.

to try to find a period 16 solution by slightly increasing F and looking at the corresponding Poincaré section.

The student can also investigate the period doubling route to chaos in the following two experiments. The first experiment involves mechanical motion.

Subharmonic Response: Period Doubling

In this experiment, a magnet attached to a stiff steel tape is driven periodically by a Helmholtz coil connected to a frequency generator. The response of the magnet is measured at different frequencies and powers with a pickup coil and the oscillations observed on an oscilloscope. Period 2, 4, and other periodic solutions are observed.

The second activity involves the nonlinear properties of the diode.

Diode: Period Doubling

In this activity, an ordinary diode is used to produce period doubling.

The magnetic basin of attraction problem which follows in the problem section is the inspiration for this next experimental activity.

Five-well magnetic potential

A commercial toy spherical pendulum has a magnetic bob which swings over a symmetric planar arrangement of five small magnets sitting in the base of the toy. The pendulum is given a kick each time it passes through the vertical position, thus causing the pendulum to execute erratic (chaotic) motion. The motion of the pendulum is investigated.

PROBLEMS

Problem 8-10: Figure 8.9
Plot Equation (8.27) for $A = F = \omega = 1$ and confirm the period-3 solution shown in Figure 8.9.

Problem 8-11: Figure 8.10
Plot Equation (8.28) for $A = F = \omega = 1$ and confirm the $n = 3$ ultraharmonic solution shown in Figure 8.10.

Problem 8-12: Figure 8.11
Plot Equation (8.29) for $A = F = \omega = 1$ and confirm the $m = 4$, $n = 3$, ultrasubharmonic solution shown in Figure 8.11.

Problem 8-13: Period-3 solution
Confirm the period-3 solution shown in Figure 8.12 by

    a. Setting $\alpha = 0$, $\beta = 1$, $\gamma = 0.04$, $F = 0.2$, $\omega = 1$, $x(0) = 0.25$, and $\dot{x}(0) = 0$ in MF09. Take $t = 150..250$ to eliminate the transient. Approximately how long does it take for steady-state to be achieved?

    b. Reproducing $x(t)$ shown in Figure 8.13.

    c. Using MF11 with $\phi(0) = 0$ to confirm that the steady-state Poincaré section consists of three dots, thus confirming the period-3 solution.

Problem 8-14: Figure 8.15
Confirm the two Poincaré sections shown in Figure 8.15 by running MF11 with $\alpha = -1$, $\beta = 1$, $\omega = 1$, $\gamma = 0.25$, $x(0) = 0.09$, $\dot{x}(0) = 0$, $\phi(0) = 1$, and (a) $F = 0.3570$, (b) $F = 0.35797$.

Problem 8-15: Basins of attraction
Making use of Mathematica File MF09 and, if necessary, file MF11, determine the basins of attraction for the nonharmonic Duffing oscillator ($\alpha = 0$, $\beta = 1$, $\gamma = 0.04$, $\omega = 1$, $F = 0.2$) in the range $x(0) = 0.31$ to 0.35 and $\dot{x}(0) = 0.00$ to 0.05. Use increments of 0.01. Draw a picture similar to the one in the text and discuss how your results connect onto the text figure.

Problem 8-16: Unforced underdamped simple pendulum
For $\gamma < \omega_0$, the simple pendulum equation

$$\ddot{x} + 2\gamma\dot{x} + \omega_0^2 \sin x = 0$$

has stable focal points at $\dot{x} = 0$, $x = 0$, $\pm 2\pi, \ldots$ with saddle points in between. Sketch the basins of attraction of the unforced underdamped pendulum in the phase plane.

Problem 8-17: Some more basins of attraction
Determine the basins of attraction for

    1. the nested limit cycles in Problem 7-3;

    2. the system

$$\dot{x} = -3x + 4x^2 - \frac{1}{2}xy - x^3, \quad \dot{y} = -2.1y + xy.$$

For this latter system, locate and identify the singular points first.

Problem 8-18: Magnetic basin of attraction

Consider a small metal ball of unit mass attached to a light rigid rod of length l which is allowed to swing above a horizontal plane containing three small identical magnets. Assume that the following conditions prevail:

- $l \gg$ spacing of magnets and the ball moves in the x–y plane a small distance d above the plane containing the three magnets.

- The magnets are tiny and are at the vertices of an equilateral triangle.

- The force between the ball and each magnet obeys the inverse square law.

- Stoke's law of friction applies to the motion of the pendulum.

- The gravitational force is included.

If the three magnets are located at (x_1, y_1), (x_2, y_2), (x_3, y_3) and the ball's instantaneous position is (x, y),

    a. Use Newton's second law to show that the equations of motion describing the ball are of the structure

$$\ddot{x} + b\dot{x} - \sum_{i=1}^{3} \frac{(x_i - x)}{[(x_i - x)^2 + (y_i - y)^2 + d^2]^{3/2}} + cx = 0$$

$$\ddot{y} + b\dot{y} - \sum_{i=1}^{3} \frac{(y_i - y)}{[(x_i - x)^2 + (y_i - y)^2 + d^2]^{3/2}} + cy = 0.$$

    b. In the equations of part (a), take $b = 1.0$, $c = 0.2$, $d = 0.1$, $(x_1 = 0, y_1 = 0)$, $(x_2 = 2, y_2 = 0)$, and $(x_3 = 1, y_3 = \sqrt{3})$. Choosing different initial positions for the ball and taking zero initial velocity, explore the boundaries of the basins of attraction for each magnet. Use symmetry where possible.

8.4 Power Spectrum

Before continuing with the discussion of the period doubling route to chaos, another useful diagnostic tool in nonlinear physics shall be introduced, the "power spectrum". This is a concept used by engineers in digital signal processing [SK89] which can be adapted to studying the frequency content of a solution $x(t)$ to a nonlinear ODE such as the forced Duffing oscillator. We shall keep our discussion of the concept short and simple, referring the reader to, for example, *Numerical Recipes* [PFTV89] for all the gory details that the expert will know that we have omitted. If you are not expert, don't worry. An accompanying Mathematica File, MF32, is provided to allow you to explore forced oscillator systems using the power spectrum.

Suppose that in principle, a nonlinear ODE of physical interest has the time dependent solution $x(t)$, valid for all t $(-\infty < t < \infty)$. If we wish to study the frequency spectrum of the solution, form the Fourier transform

$$X(f) = \int_{-\infty}^{\infty} x(t)e^{-2\pi i f t}dt. \tag{8.31}$$

Here, f is the frequency in Hz, or cycles per second, and is related to ω, the frequency in radians per second, through $\omega = 2\pi f$. Conversely, given $X(f)$ for all f, $x(t)$ can be calculated using the inverse Fourier transform

$$x(t) = \int_{-\infty}^{\infty} X(f)e^{2\pi i f t}df. \tag{8.32}$$

Note that f is allowed to vary over both positive and negative values here. Making use of the Fourier transform and its inverse, a very powerful result called Parseval's theorem

$$\int_{-\infty}^{\infty} |x(t)|^2 dt = \int_{-\infty}^{\infty} |X(f)|^2 df \tag{8.33}$$

can be derived. This theorem is of importance to physical scientists and engineers because of the physical interpretation that can be attached to it. For example, in classical mechanics where $x(t)$ is the instantaneous displacement, the integral on the left is proportional to the total energy. Thus, since the equality holds, $|X(f)|^2$ represents the energy per unit frequency interval. It is customary to refer to $S(f) \equiv |X(f)|^2$ as the power spectrum.[9] The quantity $S(f)$ gives us information on the distribution of energy as a function of frequency. For example, if

$$x(t) = Ae^{-\Gamma t}e^{2\pi i f_0 t} \tag{8.34}$$

for $t \geq 0$, corresponding to a decaying oscillation of frequency f_0, then

$$S(f) = A^2/(\Gamma^2 + 4\pi^2(f - f_0)^2). \tag{8.35}$$

This is a so-called Lorentzian line shape of height A^2/Γ^2 and width proportional to Γ centered at $f = f_0$. As $\Gamma \to 0$, $S(f)$ becomes an infinitely tall "spike" corresponding to having all the power concentrated exactly at the frequency f_0.

For a nonlinear forced oscillator, $x(t)$ cannot generally be determined analytically. Usually one must resort to numerical means to evaluate x at discrete time steps. Instead of having a continuous analytic form for $x(t)$, a sequence of x values is obtained for some finite time domain. For sufficiently small time steps, the sequence will approximate a continuous function. One might imagine that we could use the sequence to numerically evaluate the Fourier transform of $x(t)$ and then $S(f)$. If all of the values of x in the sequence were used, the computing time might grow rather long. A better approach is to sample a relatively small number of x points to give us, we hope, an accurate power spectrum. Such an approach is now described.

---

[9] A suitable normalization is also often introduced in defining $S(f)$.

Assume that $x(t)$ is sampled at evenly spaced time intervals T_s over some finite time range or "window". The sequence of recorded x values is, say, $x_n \equiv x(t_n = nT_s)$ with $n = 0, 1, 2, ..., N - 1$. That is to say, N consecutive values of x spaced a time interval T_s apart are recorded. Further, when calculating the power spectrum for the forced oscillator problem, start sampling at a sufficiently large t, say t_0, so as to ensure that all transients have died away.

For any choice of the sampling interval T_s, there is a very special corresponding frequency called the Nyquist frequency

$$f_{\text{Nyquist}} \equiv \frac{1}{2T_s}. \tag{8.36}$$

Note that from this definition, the sampling frequency $F_s = 1/T_s$ is twice the Nyquist frequency. What is so important about the Nyquist frequency? The answer is provided by the sampling theorem due to Nyquist [Nyq28] and Shannon [Sha49]. This theorem states that if a continuous signal $x(t)$, sampled at an interval T_s, is such that its Fourier transform $X(f) = 0$ for all frequencies $|f| > f_{\text{Nyquist}}$, then $x(t)$ is completely determined by the sampled values x_n. In this case, the Nyquist frequency is clearly greater than the maximum frequency f_{max} in the signal's frequency spectrum. Thus, the sampling frequency $F_s > 2f_{\text{max}}$.

In signal processing, engineers ensure that the sampling theorem prevails by using a lowpass analog filter on their signal to select $f_{\text{max}}(< f_{\text{Nyquist}})$, removing all higher frequencies by the process of attenuation. When $X(f)$ is zero outside the range $-f_{\text{Nyquist}}, ..., f_{\text{Nyquist}}$, they refer to $x(t)$ as being "band-width limited" to frequencies smaller in magnitude than f_{Nyquist}.

What happens if $x(t)$ is not band-width limited to this frequency range, i.e., its Fourier transform $X(f)$ does not vanish outside the range $-f_{\text{Nyquist}}$ to f_{Nyquist}? It turns out that the power outside this frequency range gets "folded back" into the range giving an inaccurate power spectrum. This phenomenon is called "aliasing". To avoid aliasing, one should attempt to make $x(t)$ band-width limited by taking the sampling frequency $F_s > 2f_{\text{max}}$. Unfortunately, for our forced oscillator problem we do not always know a priori what the maximum frequency component is in our signal $x(t)$. On the other hand, suppose that the chosen sampling frequency or sampling interval $T_s = 1/F_s$ is such that $X(f)$ is not zero at the Nyquist frequency. Then increase F_s to check for possible aliasing. Increasing F_s pushes the Nyquist frequency up allowing us to see whether there are indeed higher frequency components present.

Keeping these important aspects in mind, let's continue with the formal derivation of the power spectrum from the N sampled values x_n. It follows from elementary mathematics that, given these N values, we can only generate the Fourier transform at N frequencies. Assuming that the $x(t)$ is band-width limited, we shall take these frequencies to be equally spaced between $-f_{\text{Nyquist}}$ and f_{Nyquist}, viz., the frequencies,

$$f_k \equiv \frac{k}{NT_s} \tag{8.37}$$

with $k = -N/2, .., 0, .., N/2$. The extreme k values generate $-f_{\text{Nyquist}}$ and f_{Nyquist}, respectively. It might seem on counting the k values that we have

$N + 1$ of them but, due to periodicity of the Fourier transform, the extreme k values are not independent but, in fact, are equal.

We now approximate the continuous Fourier transform as

$$
\begin{aligned}
X(f_k) &= \int_{-\infty}^{\infty} x(t)e^{-2\pi i f_k t}dt \\
&\simeq \sum_{n=0}^{N-1} x_n e^{-2\pi i f_k t_n} T_s \\
&= T_s \sum_{n=0}^{N-1} x_n e^{-2\pi i k n/N}.
\end{aligned}
\tag{8.38}
$$

The last sum is referred to in the mathematics literature as the discrete Fourier transform and is expressed as

$$
X_k \equiv \sum_{n=0}^{N-1} x_n e^{-2\pi i k n/N}
\tag{8.39}
$$

with $k = -N/2, ..., N/2$. We can change the k range to $k = 0, ..., N - 1$, thus making it the same as the n range, by noting that X_k is periodic in k with period N. With this standard convention for the range of k, $k = 0$ corresponds to zero frequency, $k = 1, 2, ..., N/2 - 1$ to positive frequencies, $k = N/2 + 1, ..., N - 1$ to negative frequencies, and $k = N/2$ to both f_{Nyquist} and $-f_{\text{Nyquist}}$.

In a similar manner, the inverse discrete Fourier transform can be derived:

$$
x_n = \frac{1}{N} \sum_{k=0}^{N-1} X_k e^{2\pi i k n/N}.
\tag{8.40}
$$

It is left as a problem for the student to show from the discrete Fourier transform pair that Parseval's theorem then becomes

$$
\sum_{n=0}^{N-1} |x_n|^2 = \frac{1}{N} \sum_{k=0}^{N-1} |X_k|^2.
\tag{8.41}
$$

Finally, from the discrete form of Parseval's theorem, the power spectrum (also known as the periodogram spectral estimate [SK89]) can be defined as

$$
S_N(k) = \frac{1}{N} |X_k|^2.
\tag{8.42}
$$

In actually calculating the discrete Fourier transform X_k of a list of data, one can employ Mathematica's Fourier command which makes use of the fast Fourier transform (FFT) algorithm. The FFT is based on the idea of splitting the data set in the discrete Fourier transform into even and odd labeled points and using the periodicity of the exponential function to eliminate redundant operations. A detailed discussion of this conversion may be found in standard numerical analysis texts, for example, in Burden and Faires [BF89] and in *Numerical Recipes* [PFTV89]. Why use this routine? As suggested by the process of eliminating redundant calculations, the FFT is faster than the straightforward evaluation of the discrete Fourier transform. How much faster? A lot!

If N is the number of data points, the discrete Fourier transform involves N^2 multiplications while the FFT turns out to involve about $N \log_2 N$ operations. If, for example, N=1024, then $\log_2 1024 = 10$ so that the discrete Fourier transform requires about 10^6 computations compared to 10,000 for the FFT. In this case the FFT is about 100 times faster. Using the FFT enables us to calculate power spectra on a personal computer. For the serious nonlinear physics researcher, who uses much larger N values, the relative savings in time is even more spectacular.

As an example of a power spectrum calculation relevant to the central theme of this chapter, namely forced nonlinear oscillators, let's apply the accompanying file MF32 to the nonharmonic Duffing oscillator with $\alpha = 0$, $\beta = 1$, $\omega = 1$, $F = 0.2$, $\gamma = 0.04$, $x(0) = 0.27$, $\dot{x}(0) = 0$, and a $\cos(\omega t)$ driving term.

Power Spectrum

In this file, the power spectrum is calculated, using the Fourier command, for the forced Duffing oscillator. The FourierParameters option is set to the "signal processing" convention" {1,-1} so that the Fourier transform conforms to the preceeding theoretical discussion. Mathematica commands in file: Cos, NDSolve, MaxSteps->Infinity, Table, Evaluate, Flatten, Abs, Max, Fourier, FourierParameters, Transpose, ListPlot, PlotRange->All, PlotJoined->True, PlotStyle, Hue, AxesLabel, PlotLabel, StyleForm, ImageSize, ReplacePart

---

It has been observed in the previous section that a period-3 ($n = 3$ subharmonic) solution emerges after the transient has died away. To ensure that the steady-state solution is present, start the data sampling in the file MF32 at $t_0 = 50\pi$. The student can try other values. If this example hadn't been studied, we could begin by assuming that the driving frequency probably represents the highest frequency component in the solution ("signal"). Since $\omega = 1$, assume that $f_{max} = 1/2\pi$. So we must have our sampling frequency $F_s > 2f_{max} = 1/\pi$ or equivalently our sampling interval $T_s < \pi$. Let's choose $T_s = 0.9\pi$ and[10] a large value of N, say $N = 1024$. Why do we want N to be large? The frequency interval in the power spectrum is given by $\delta f = f_{k+1} - f_k = 1/NT_s$. For a fixed sampling interval T_s, increasing N leads to a finer frequency resolution.

Running the Mathematica file with the above parameters, the power spectrum [11] shown in Figure 8.16 is obtained. The power spectrum is essentially zero at the Nyquist frequency corresponding to $k = N/2 = 512$ indicating that aliasing is not a problem. In the positive frequency range $k < 512$, two very sharp peaks occur[12] at about $k = 461$ and $k = 154$. In terms of the frequency $\omega_k = 2\pi f_k = 2\pi k/(NT_s)$, these two peaks are (within experimental error) at the driving frequency $\omega = 1$ and the $n = 3$ subharmonic frequency, respectively, as would be expected from our preliminary discussion in Section 8.3 for a period-3

---

[10]In applying Mathematica's Fourier command it is not necessary that the data list have a length which is an integer power of 2.

[11]In MF32, we have not bothered to normalize the ordinates by dividing by N.

[12]To obtain accurate k values, remove the semi-colon on the plotting points in the file.

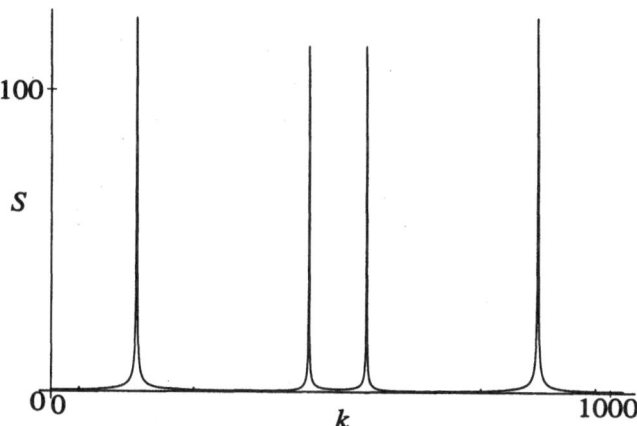

Figure 8.16: Power spectrum for the period-3 solution of the nonharmonic Duffing equation.

solution. In that discussion of the $n = 3$ subharmonic, $x(t)$ was the sum of two cosine terms at the driving and subharmonic frequencies. With the inclusion of damping, finite amplitude peaks appear in the spectrum at the right locations.

Note also how if the negative frequency region corresponding to $k > 512$ were folded over, with the fold at $k = 512$, the negative frequency spectrum is identical to that for positive frequencies. In terms of our preliminary discussion, this isn't too surprising as the cosine terms in that solution are the same if positive ω is replaced with negative ω. Some definitions of the power spectrum make use of this folding feature [PFTV89], adding the folded-over negative frequency region onto the positive region to keep the total power the same and then plotting only positive frequencies. From now on the negative frequency region in our plots is omitted. Only the positive region up to the Nyquist frequency is plotted. This has been done in the second figure generated in MF32. We shall not bother to double the power to account for the missing negative frequency region. In the third figure produced in the file, the power spectrum is plotted against frequency in radians per second. Two spikes are observed at $\omega = 1$ and $\omega = 1/3$ for the present example as expected.

As a second example of applying the power spectrum concept, consider the inverted forced Duffing oscillator with an $F \cos(\omega t + 1)$ driving term and $\alpha = -1$, $\beta = 1$, $\omega = 1$, $\gamma = 0.25$, $x(0) = 0.09$, and $\dot{x}(0) = 0$. For $F = 0.34875$, the Poincaré section approach has indicated that a period-2 solution emerges. What does the power spectrum have to say in this case? Let's again assume that the maximum frequency in the solution spectrum is probably the driving frequency $\omega = 1$ and again take our sampling period to be $T_s = 0.9\pi$. Starting our sampling at $t_0 = 50\pi$ and taking $N = 512$, the plot shown in Figure 8.17 is obtained. The two taller peaks at $k = 230$ and $k = 115$ are easily identified as being in the right locations to correspond to the driving frequency ($\omega = 2\pi \times 230/(512 \times 0.9\pi) \simeq 1$) and the $n = 2$ subharmonic. But what are the other, smaller, peaks in the spectrum? The peak furthest to the left, for example, is at about $k = 51$, so even accounting for experimental error it is in the wrong location to be an $n = 4$ subharmonic. Such a subharmonic is not

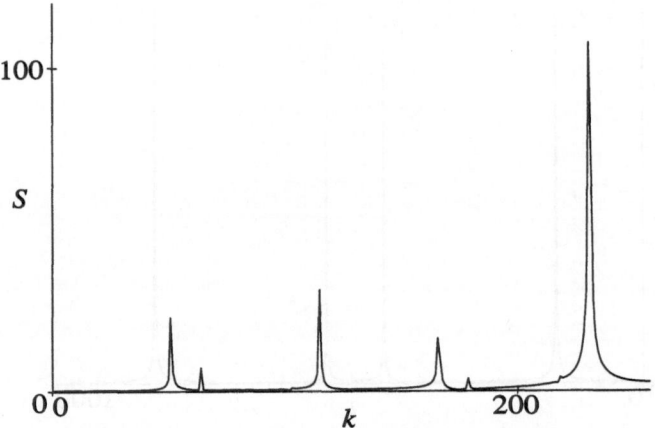

Figure 8.17: An aliased power spectrum for period-2 solution.

expected for period 2, but if it were present it should be at $k = 230/4 = 57.5$, i.e., around $k = 57$ or 58. By now we should begin to suspect that there is aliasing in our spectral calculation and the maximum frequency in the signal is higher than $\omega = 1$. So let's reduce the sampling interval to $T_s = 0.5\pi$ which corresponds to pushing the Nyquist frequency to $\omega_{\text{Nyquist}} = 2$. For $N = 512$, the driving frequency will then correspond to $k = 128$. Running the Mathematica file again we obtain Figure 8.18. Here we have chosen to plot the frequency in radians per second on the horizontal axis, rather than the k values. The tallest

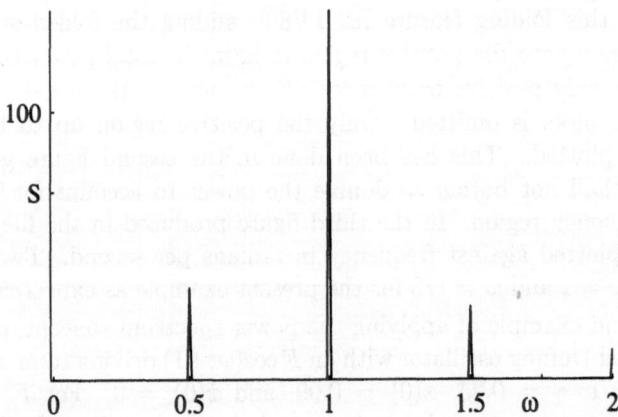

Figure 8.18: Power spectrum for period-2 solution.

peak is located at the driving frequency $\omega = 1$ and only the $n = 2$ subharmonic at $\omega = 0.5$ occurs to the left of it. Notice there is a third peak located above the driving frequency at $\omega = 1.5$. Since this frequency corresponds to a value three times that of the $n = 2$ subharmonic, we can identify it as an $m = 3$, $n = 2$ ultrasubharmonic. Why doesn't it show up in the Poincaré section, giving us three dots instead of two? A simple partial hand-waving argument provides the explanation. In the discussion of the forced linear oscillator, steady-state solutions were derived for the subharmonics and ultrasubharmonics in Section 8.3. If one adds the $n = 2$ subharmonic to the $m = 3$, $n = 2$ ultrasubharmonic, one obtains a resulting solution which is of period 2. If this seems obvious, fine. If not, do the relevant problem in the problem section. Inclusion of the nonlinearity apparently doesn't qualitatively change the argument. Using the same forced linear oscillator results, one can qualitatively account for the difference in amplitudes between the subharmonic and the ultrasubharmonic.

Is the ultrasubharmonic the highest frequency in the spectrum? One can see what appears to be part of a spike at $\omega = 2$ on the far right of the figure, the location corresponding to a second harmonic of the driving frequency. To see if this peak is real or if there are any additional harmonics, decrease T_s to 0.25π. What do you observe? How would you interpret your observation? As you can perhaps begin to appreciate, identifying, say, a period-8 solution through its power spectrum can be a tricky business, and one should try to confirm one's results through other means such as the Poincaré section. Compounding the problem is the fact that the peaks corresponding to higher n subharmonics tend to have very small amplitudes. Semi-log plots are often used in such cases.

The reader who is interested in the measurement of the power spectrum for real data should try the following experimental activity.

Power Spectrum

This activity employs the same experimental setup as in Experimental Activity 24. A fast Fourier transform is used to produce a power spectrum from the measured data.

EXP 27

PROBLEMS

Problem 8-19: Figure 8.16 and other power spectra

Determine the power spectrum for the Duffing oscillator with $\alpha = 0$, $\beta = 2$, $\omega = 1$, $F = 0.2$, $\gamma = 0.04$, $x(0) = 0.27$, $\dot{x}(0) = 0$, and a $\cos(\omega t)$ driving term. Use MF32 with $t_0 = 50\pi$, $T_s = 0.5\pi$, and $N = 1024$. Experiment with changing the β parameter, holding all other parameters the same. Interpret all results.

Problem 8-20: Figures 8.17 and 8.18

Reproduce the spectra shown in Figures 8.17 and 8.18 generated for the inverted Duffing oscillator and confirm the text discussion.

  a. For Fig. 8.17, use MF32 with a $F\cos(\omega t + 1)$ term and $\alpha = -1$, $\beta = 1$, $\omega = 1$, $\gamma = 0.25$, $x(0) = 0.09$, $\dot{x}(0) = 0$, $F = 0.34875$, $t_0 = 50\pi$, $T_s = 0.9\pi$, and $N = 512$.

  b. For Fig. 8.18, take $T_s = 0.5\pi$ and all other parameters as in part (a).

Problem 8-21: Period-2 solution formed by adding an n=2 subharmonic to an m=3, n=2 ultrasubharmonic

Go back to the discussion of subharmonic and ultrasubharmonic solutions for the forced linear oscillator in Section 8.3, and plot the analytic solutions given there for (a) $n = 2$ subharmonic with $A = 4/3$, $F = \omega = 1$, (b) $m = 3$, $n = 2$ ultrasubharmonic with $A = F = \omega = 1$. Add the two solutions and verify by measuring the period that the sum is a period-2 solution and thus would exhibit two points in its Poincaré section. Also account for the observed fact that the amplitude of the subharmonic in the power spectrum is greater than for the ultrasubharmonic.

Problem 8-22: Discrete Parseval's theorem
Making use of the discrete Fourier transform pair, prove the discrete form of Parseval's theorem.

Problem 8-23: Basins of attraction
Making use of the power spectrum approach, verify the basins of attraction picture of the previous section for the nonharmonic Duffing oscillator with $\alpha = 0$, $\beta = 1$, $\omega = 1$, $F = 0.2$, and $\gamma = 0.04$. The $x(0)$ values varied from 0.20 to 0.30 and the $\dot{x}(0)$ from 0.00 to 0.10.

Problem 8-24: Periods 4 and 8
Confirm the period-4 and period-8 solutions whose Poincaré sections were presented in Figure 8.15 using the power spectrum approach.

Problem 8-25: Nonharmonic Duffing oscillator
Using the power spectrum approach, determine the nature of the steady-state solution of the forced oscillator equation

$$\ddot{x} + 0.7\dot{x} + x^3 = 0.75 \cos t$$

subject to the initial conditions $x(0) = \dot{x}(0) = 0$. Confirm your answer by also calculating the Poincaré section.

Problem 8-26: Driven pendulum
Consider the damped, driven, pendulum equation (3.19) with $\gamma = 0.25$, $\omega_0 = 1$, $\omega = 2/3$, $\theta(0) = 0.09$, $\dot{\theta}(0) = 0$, and (i) $F = 1.00$, (ii) $F = 1.07$. Modify MF32 and determine the periodicity of the solution for the two F values.

8.5   Chaotic Oscillations

Before we got waylaid by our side excursion into the realm of power spectra, you may recall that we were on the period doubling route to chaos. Whether

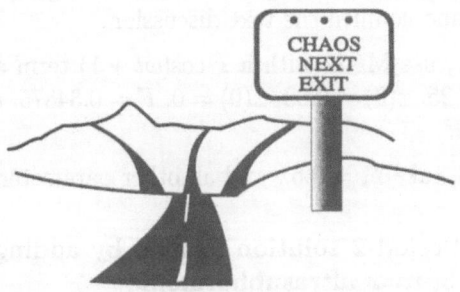

The Route to Chaos

reaching the land of chaos will have been worth the intellectual effort will be up to you to decide. However, at this point in time it is still a somewhat mysterious land with many, as of yet, incompletely explored regions. Even the most intrepid nonlinear scientist, armed with the best of analytical and computational tools, must grope uncertainly on the research frontiers of this strange kingdom which is known to be the home of those entities called "strange attractors".

If these strange attractors are to be observed, we had better hop back into our exploration vehicle which has gotten us to this point, the forced Duffing oscillator, and rejoin the route to chaos. Recall that the inverted Duffing case was being considered with $\alpha = -1$, $\beta = 1$, $\omega = 1$, $\gamma = 0.25$, $x(0) = 0.09$, and $\dot{x}(0) = 0$. The driving term was of the form $F\cos(\omega t + 1)$ and for $F = 0.35797$ a period-8 solution was revealed by observing its Poincaré section. Trying to find the bifurcation points in F at which period 16, 32, 64, ... pop out is not an easy task on a PC because the corresponding F regions for each period get narrower and narrower and the time for steady-state to be achieved generally gets longer and longer. Searching for, say, a period-32 solution, one would like to have hundreds or even thousands of points. This would ensure that, having removed the transient points, each of the 32 Poincaré points was revisited at least several times. Further, if we are going to get a good idea of what strange attractors really look like and wish to explore their internal structure, a large number of points is also needed. In the following Mathematica file, we look at the Poincaré section generated with 10,000 plotting points for the above inverted Duffing oscillator when the driving force amplitude is increased to $F = 0.40$.

Poincaré Sections for a Strange Attractor
In this file, Poincaré sections are generated for a strange attractor. A large number of points are plotted so that the internal structure of the attractor can be observed by magnifying subregions. The file can also be applied to looking at high periodicity solutions, as well as chaotic attractors. Mathematica commands in file: `NDSolve`, `Flatten`, `Evaluate`, `MaxSteps->Infinity`, `Block`, `$DisplayFunction=Identity`, `ListPlot`, `PlotStyle`, `RGBColor`, `Epilog`, `PointSize`, `Graphics`, `Hue`, `Rectangle`, `Show`, `Frame->True`, `PlotRange`, `FrameTicks`, `PlotLabel`, `StyleForm`, `TextStyle`, `ImageSize`

For $F = 0.40$, we do not have any apparent periodicity in $x(t)$ (not shown) and the Poincaré section looks like that shown on the left of Figure 8.19. Is

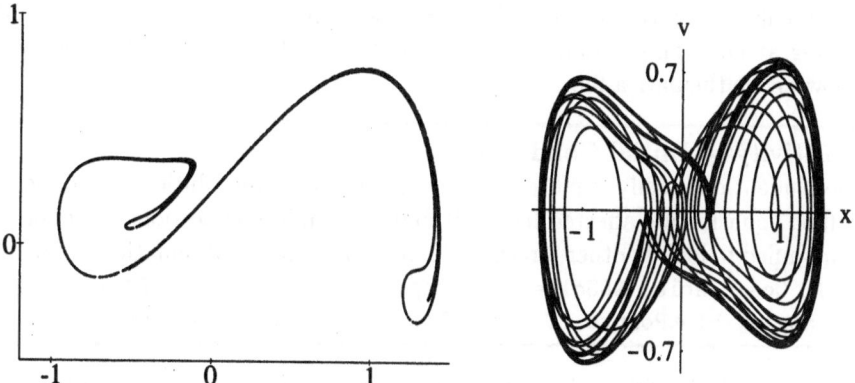

Figure 8.19: Left: Poincaré section for $F = 0.40$. 10,000 points are present! Right: Phase plane portrait for $F = 0.40$.

this strange-looking object one of the strange attractors for which we have been searching? Are we in the land of chaos? Mathematicians do not agree on what the precise characteristics of a strange attractor are. However, loosely speaking, a strange attractor is characterized by having an attractive region in phase space different (i.e., strange looking) from the types of attractors like stable nodal points or limit cycles that we have encountered before. An example of a strange attractor is the butterfly wings attractor of Figure 3.30. Unlike these other attractors, strange attractors have fractal (non-integer) dimensions. A limit cycle, for example, has a dimension which is an integer, namely one, while a nodal point has dimension zero. Strange attractors, because they are chaotic solutions, are also extremely sensitive to initial conditions. Further, unlike the periodic case where one has sharp spikes with nothing in between, the power spectrum in the chaotic regime tends to look like broadband noise with some occasional spikes sticking up. This is because in the chaotic regime, $x(t)$ is nonperiodic so no particular frequencies tend to get singled out. Let's see if, for example, the $F = 0.40$ case whose phase plane portrait is shown on the right of Figure 8.19 has the properties expected of a strange attractor. It looks strange, as does its Poincaré section, when compared to simple attractors. How about the other criterion that we outlined? First, the power spectrum is calculated, the result being shown in Figure 8.20. Although a tall spike is located at the driving

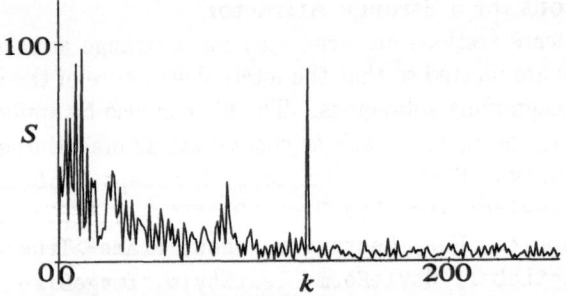

Figure 8.20: Power spectrum for $F = 0.40$ ($N = 512$, $T_s = 0.5\pi$, $t_0 = 50\pi$).

frequency, the spectrum is quite noisy, a feature which would tend to suggest that a chaotic solution emerges for $F = 0.40$. This conclusion is reinforced by looking at how the solution depends on initial conditions. This is done in the following Mathematica file.

Sensitivity to Initial Conditions
The student can easily explore in the same plot how the solutions to the forced Duffing equation are affected by small changes in initial conditions. The file can be modified to look at other forced oscillator equations. Mathematica commands in file: Cos, Table, NDSolve, MaxSteps, Plot, Evaluate, PlotLabel, TextStyle, PlotPoints, Ticks, PlotStyle, Hue, ImageSize

---

Using MF34, Figure 8.21 (top) shows the solutions which would emerge for $F = 0.325$ (recall that this was period 1) for two different values of $x(0)$, namely $x(0) = 0.09001$ and 0.09002, all other conditions being the same. We

did say solutions, plural. There are two solutions in the plot but they would
be indistinguishable even if a much larger scale were used. In the non-chaotic
regime the solution is not sensitive to very tiny changes in initial conditions,
except at the boundaries of basins of attraction. In Figure 8.21 (bottom), which
corresponds to taking $F = 0.40$ with the same two initial conditions, the

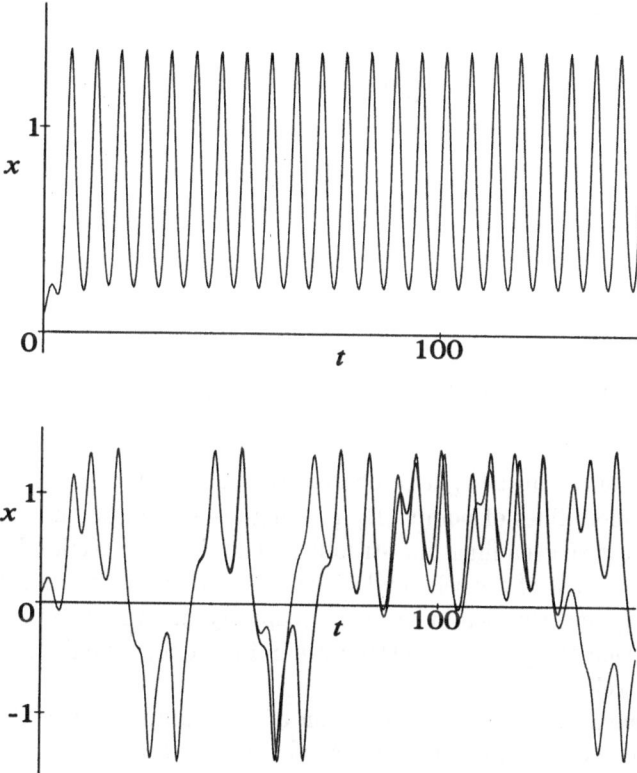

Figure 8.21: Solutions to inverted Duffing equation for two different initial x
values, $x(0) = 0.09001$ and 0.09002: (top) $F = 0.325$; (bottom) $F = 0.40$.

story is very different. Although the two solutions initially follow each other,
they eventually diverge and trace out their own separate paths. This extreme
sensitivity to initial conditions, which is characteristic of chaotic solutions, has
profound implications in the real world. For real physical systems there is always
intrinsic noise so that the initial conditions are never precisely the same. If the
parameters are such that one is in the chaotic regime, one doesn't know what
chaotic solution will emerge. Although a given chaotic solution is deterministic,
the outcome is unpredictable because of the uncertainty in initial conditions.
Now let's examine the Poincaré section for Figure 8.19, and remember that
10,000 points were used to generate the plot. In the small text figure this is not
evident as the points have been blended together.

To see that the points are indeed there, we have magnified two small regions
of the attractor in the file. A small segment, viz., $x = 0.26...0.34$, $y = 0.42...0.50$,
of the straight line region is shown in Figure 8.22. We see at first glance that
the original line is actually composed of three parallel lines of points. Looking

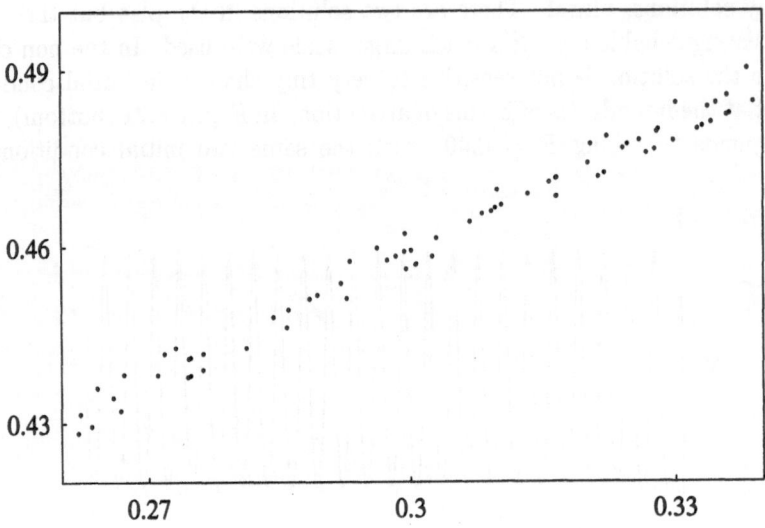

Figure 8.22: Fine structure of Poincaré section.

very closely, one can just begin to observe that the center line is made up in turn of three more lines of points. By plotting even more points and zooming in one could resolve even finer line structure. The Poincaré section has internal structure reminiscent of the self-similar Cantor set that was discussed in Chapter 3. Recalling that the Cantor set had a (capacity) fractal dimension, it is not unreasonable that the geometrical structure in Figure 8.19 should have a fractal dimension as well.

How to calculate the fractal dimension of the Cantor set was discussed in Chapter 3. How would we conceptually go about calculating the capacity fractal dimension D_C of what appears to be clearly a strange attractor? Going back to Equation (3.3), D_C is defined in the limit as $\epsilon \to 0$ through the relation

$$\ln N(\epsilon) = D_C \ln(1/\epsilon) + c \tag{8.43}$$

where c is a constant. If a $\ln N(\epsilon)$ versus $\ln(1/\epsilon)$ plot is generated, then D_C is just the slope of the straight line. This is done by using a box counting approach which is a direct generalization of what we did for the Cantor set. Consider the 3-dimensional phase space for the Duffing oscillator which is the space spanned by the state variables $x(t)$, $y(t) \equiv \dot{x}(t)$, and $\phi(t)$. First locate the region of state space where the attractor "resides" after having carried out the numerical run for a given set of parameters. Then enclose the attractor with a rectangular box which we subdivide into small cubes of side ϵ_0. Then count the number of cubes which have a numerical data point inside it, this giving us $N(\epsilon_0)$. Now halve the size of the cubes so that their sides are of length $\epsilon_1 = \epsilon_0/2$ and count the number $N(\epsilon_1)$. Repeat this process a number of times and plot $\ln N(\epsilon)$ against $\ln(\delta)$ where $\delta = 1/\epsilon$. The slope will then give D_C.

As a simple example, we now estimate the fractal dimension of Barnsley's fern (recall Example 3-1).

Example 8-4: Box Counting Estimate of Fractal Dimension

By generating a random number r between 0 and 1, taking the starting point to be $x_0 = y_0 = 0$, and iterating the 2-dimensional piecewise map

$$(x_{n+1}, y_{n+1}) = \begin{cases} (0, 0.16y_n), & 0.00 < r < 0.01 \\ (0.2x_n - 0.26y_n, 0.23x_n + 0.22y_n + 0.2, & 0.01 < r < 0.08 \\ (-0.15x_n + 0.28y_n, 0.26x_n + 0.24y_n + 0.2), & 0.08 < r < 0.15 \\ (0.85x_n + 0.04y_n, -0.04x_n + 0.85y_n + 0.2), & 0.15 < r < 1.00 \end{cases}$$

from $n = 0$ to $n = N = 40000$, use the box counting approach to estimate the fractal dimension of Barnsley's fern.

Solution: The first part of this code is similar to that in Example 3-1. Here, the 2-dimensional piecewise map will be expressed in the form,

$$x_n = a_i x_{n-1} + b_i y_{n-1} + e_i, \quad y_n = c_i x_{n-1} + d_i y_{n-1} + f_i,$$

where the first branch corresponds to $i = 1$ and is selected if the random number $r < p_1 = 0.01$, the second branch corresponds to $i = 2$, and so on.

```
Clear["Global`*"];
```

Making use of square brackets to index the entries, the coefficient values for each branch of the piecewise relation are given as are the boundaries for the random number r.

```
a[1] =0; a[2] =0.2; a[3] =-0.15; a[4] =0.85;
b[1] =0; b[2] =-0.26; b[3] =0.28; b[4] =0.04;
c[1] =0; c[2] =0.23; c[3] =0.26; c[4] =-0.04;
d[1] =0.16; d[2] =0.22; d[3] =0.24; d[4] =0.85;
e[1] =0; e[2] =0; e[3] =0; e[4] =0;
f[1] =0; f[2] =0.2; f[3] =0.2; f[4] =0.2;
p[1] =0.01; p[2] =0.08; p[3] =0.15;
```

The total number of iterations and the starting coordinates are specified.

```
total =40000; x[0] =0; y[0] =0;
```

The following command line produces a random real decimal number r in the range 0 to 1. A different random number will be generated on each successive iteration of the governing 2-dimensional map.

```
r: =Random[]
```

The maps's piecewise relations are entered as Mathematica functions,

```
x[i_, n_]: =x[n] =a[i] x[n-1] +b[i] y[n-1] +e[i]
y[i_, n_]: =y[n] =c[i] x[n-1] +d[i] y[n-1] +f[i]
```

and the points generated with the Which and Table commands.

```
pts = Table[Which[r < p[1], {x[1, n], y[1, n]},
      r < p[2], {x[2, n], y[2, n]},  r < p[3], {x[3, n], y[3, n]},
      r < 1, {x[4, n], y[4, n]}], {n, 1, total}];
```

The points are now graphed using the ListPlot command. A grid of square boxes is to be laid on top of the fractal fern graph. The total number of boxes along each side of the picture is taken to be, say, six. Thus, $6 \times 6 = 36$ boxes are created. The PlotRange is chosen to be from -0.75 to 0.75 horizontally, and 0 to 1.5 vertically. So the square fractal picture will be 1.5 units in length along each side. Therefore, the size of each box edge is $\epsilon = 1.5/6 = 1/4$, and its reciprocal $\delta = 4$. Using the GridLines option, a grid is placed on the picture with a spacing of 0.25.

```
ListPlot[pts, AspectRatio -> 1, Axes -> False, Frame -> True,
PlotRange -> {{-.75, .75}, {0, 1.5}}, FrameTicks ->
{{-.5, {0, "x"}, .5}, {{.001, "0"}, .5, {.75, "y"}, 1, 1.5}, {}, {}},
PlotStyle -> {RGBColor[.1, 1, .1], PointSize[.007]},
TextStyle -> {FontFamily -> "Times", FontSize -> 16},
GridLines -> {{-.5, -.25, 0, .25, .5}, {.25, .5, .75, 1, 1.25}},
ImageSize -> {400, 400}];
```

Barnsley's fern is displayed in Figure 8.23 with the grid superimposed. For $\epsilon = 1/4$, or $\delta = 1/\epsilon = 4$, examination of the figure reveals that 15 boxes contain one or more points. How many subdivisions and therefore different size

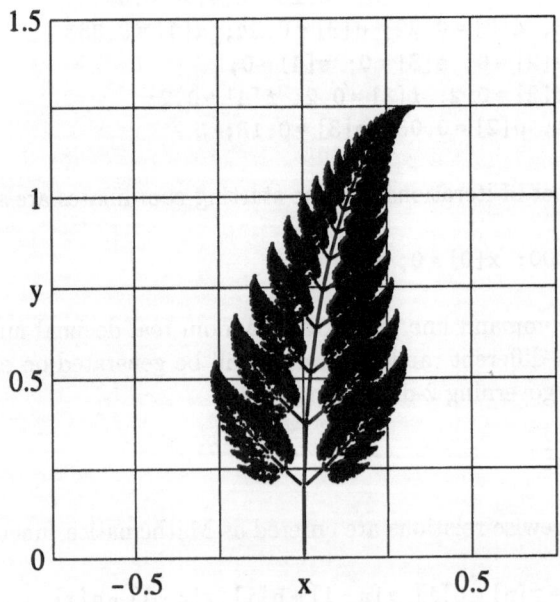

Figure 8.23: Barnsley's fractal fern with square grid superimposed.

grids used will depend on the reader's computer capability. Remember that as the boxes are made smaller, the total number of points should be increased. Running the code with 9, 12, and 15 boxes along an edge, corresponding to $\delta = 9/1.5 = 6$, $\delta = 8$, and $\delta = 10$, we find that the number of occupied boxes is 27, 39, and 58. The logarithms of the delta values and the number of occupied boxes are entered. Decimal zero is added to all the input numbers, so that the logs are explicitly evaluated.

```
logdelta = Log[{4.0, 6.0, 8.0, 10.0}]
```

$\{1.38629, 1.79176, 2.07944, 2.30259\}$

```
lognumber = Log[{15.0, 27.0, 39.0, 58.0}]
```

$\{2.70805, 3.29584, 3.66356, 4.06044\}$

The two preceding lists are formed into a list of lists of plotting points using the Transpose command.

```
plotpoints = Transpose[{logdelta, lognumber}];
```

A linear least squares fit is made to the points,

```
eq = Fit[plotpoints, {1, x}, x]
```

$$1.45093\,x + 0.689695$$

yielding the straight line equation $y = 1.45093\,x + 0.689695$. A graph of this best-fit equation is formed, along with a graph of the data points.

```
gr1 = Plot[eq, {x,1,2.5}, DisplayFunction -> Identity,
    PlotStyle -> {Hue[.6]}];
```

```
gr2 = ListPlot[plotpoints, PlotStyle -> {Hue[0], PointSize[.015]},
    DisplayFunction -> Identity];
```

The two graphs are superimposed,

```
Show[gr1,gr2,DisplayFunction -> $DisplayFunction,PlotRange ->
{{1,2.5},{2.5,4.5}}, Frame->True,FrameTicks->{{1.5,2,2.5},{3,4},
{ },{ }}, Epilog -> {Text["lognumber vs. logdelta",{1.5,4.25}],
Text["Slope of best fit straight line gives Dc",{1.6,4}]},
TextStyle -> {FontFamily -> "Times",FontSize -> 16},
ImageSize -> {600,400}];
```

to produce Figure 8.24. Despite not strictly following the step-halving procedure outlined earlier, it appears that the best-fitting straight line does a reasonably good job of fitting the data points. This gives us some confidence in our procedure.

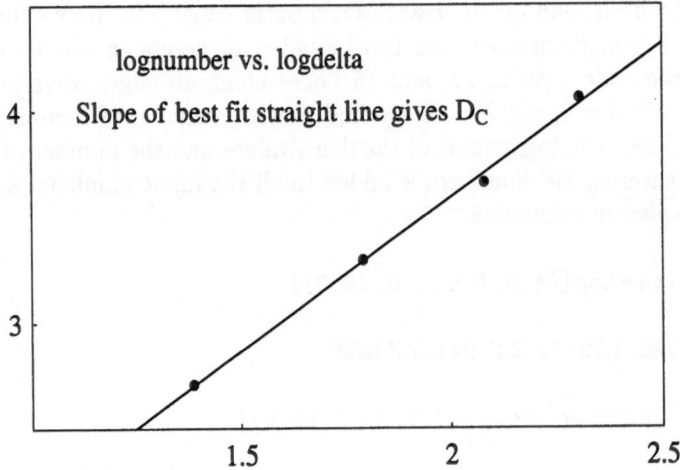

Figure 8.24: Slope of least squares straight line yields fractal dimension.

The slope of the straight line, which is equal to the coefficient of x in **eq**,

   D$_C$ = Coefficient [eq, x]

   1.45093

yields the fractal dimension $D_C \approx 1.45$ for Barnsley's fern. Does this answer make qualitative sense to you?

| End Example 8-4 |

   In practice, there are a number of problems with the direct box counting technique. The procedure doesn't take into account that there often is more than one data point inside a cube or that some cube is empty because the numerical run wasn't carried out for a sufficiently long time for the phase point to pass through that cube. Thus, the $N(\epsilon)$ are too low and, consequently, the estimate of D_C is too low. This wouldn't be a problem except that sometimes D_C differs only slightly from an integer value. For example, for the Lorenz attractor (the butterfly wings attractor of Figure 3.30), Lorenz [Lor84] obtained $D_C =$ 2.06 ± 0.01, indicating that the attractor has non-integer capacity dimension and is thus a strange attractor. A slight underestimation of D_C might leave one with doubts as to whether one had a strange attractor or not. Just looking strange isn't enough to earn the title of strange attractor. A corollary of all this discussion is that one needs a lot of numerical data and very small cubes, which generally becomes prohibitive with a personal computer. Even if more computing power is available, more clever approaches than direct box counting are used as discussed in [PC89]. No attempt will be made here to calculate the fractal dimension of the strange attractor associated with Figure 8.19.

   The capacity dimension is only one of several different fractal dimensions used to characterize strange attractors. One of these others is the Lyapunov dimension. The latter depends on the calculation of so-called Lyapunov exponents which shall be introduced in the next chapter which deals with nonlinear maps. Many of the features seen in forced oscillator systems can be more easily analyzed with the finite difference equations used in mappings rather than solving differential equations. Lyapunov exponents λ are used to characterize the very sensitive dependence on initial conditions that we have noted is an important feature of chaotic behavior. The Lyapunov exponent concept provides us with another tool to distinguish between periodic and chaotic behavior.

Having seen what a strange attractor is, let's continue increasing F for our inverted Duffing oscillator with the $F\cos(\omega t + 1)$ driving term. The strange attractor persists for a range of F (e.g., it's still there at $F = 0.42$) but eventually the Poincaré section begins to break up into "islands" of points and periodicity once again emerges. For example, increasing the force amplitude to $F = 0.459$ we find that a period-5 solution occurs as shown by the Poincaré section on the left of Figure 8.25. Here, five points are seen even though 1000 plotting points were used. The first 50 plotting points were omitted to remove the transient. The periodicity is confirmed by calculating the power spectrum (taking

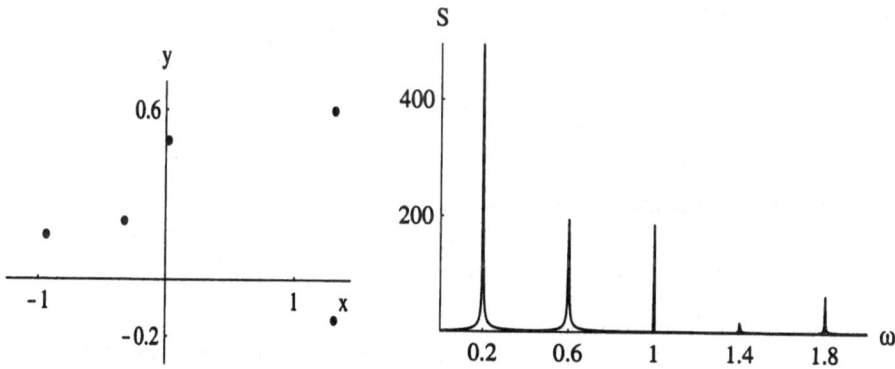

Figure 8.25: Period-5 solution for $F = 0.459$. Poincaré section on the left, power spectrum on the right.

$N = 1024$) which is shown on the right of Figure 8.25. The frequency ω is in radians per second. The power spectrum has isolated spikes indicating a periodic response. To the left of the center spike at $\omega = 1$, which corresponds to the driving frequency, one has the $n = 5$ subharmonic on the far left at $\omega = 0.2$ and an ultrasubharmonic at three times the frequency of the subharmonic. How would you interpret the remaining spikes? By increasing F we have passed right through the land of chaos back into the world of periodicity. Are there more chaotic regions ahead as F is further increased? You will have to continue this journey on your own. Good luck in your explorations! Don't get lost and fail to return as many more exciting nonlinear concepts lie ahead in the following chapters.

PROBLEMS
Problem 8-27: Figures 8.25
Confirm the Poincaré section and power spectrum plots in Figure 8.25. Take $T_S = 0.5\pi$ and $t_0 = 50\pi$ for the power spectrum and look up the remaining parameter values in the text.

Problem 8-28: Cantor-like structure of a strange attractor

Consider the inverted Duffing oscillator with $\gamma = 0.125$, $F = 0.3$, $\alpha = -1$, $\beta = 1$, $x(0) = 1$, $y \equiv \dot{x}(0) = 0.1$, and $\phi(0) = 0$. Calculate the Poincaré section using at least $10,000$ points. Take your viewing range to be $x = -1.5..1.5$, $y = -0.5..1$. Magnify the region $x = 0.6..1$, $y = 0.3..0.6$ and discuss your observations. Then magnify the even smaller region $x = 0.82..0.86$, $y = 0.48..0.51$ and discuss your observations in terms of the self-similar fractal concept.

Problem 8-29: Forced glycolytic oscillator

The equations describing forced oscillations of the glycolytic oscillator are

$$\dot{x} = -x + \alpha y + x^2 y, \qquad \dot{y} = \beta - \alpha y - x^2 y + A + F \cos(\omega t).$$

Taking $\alpha = \beta = 0$, $A = 0.999$, $F = 0.42$, $x(0) = 2$, and $y(0) = 1$, determine the nature of the solutions using the Poincaré section approach for (a) $\omega = 2$, (b) $\omega = 1.75$. Explore the frequency range in between and identify any interesting solutions.

Problem 8-30: The Butterfly Effect

Discuss the statement attributed to Lorenz that even the beating of a single butterfly's wings could affect the accuracy of long range weather forecasting. Even for perfect models of the atmosphere, Lorenz [Lor63] has shown that the weather cannot be predicted beyond about two weeks.

Problem 8-31: Fractal dimension of Koch curve

Use the box counting approach to estimate the fractal dimension of the Koch curve introduced in Chapter 3 and compare your answer with the exact result.

Problem 8-32: Fishbone fern

Estimate the capacity fractal dimension of the "fishbone" fern generated by replacing the parameters in Example 8-4 with the following values:
$a[1] = 0, a[2] = 0.95, a[3] = 0.035, a[4] = -0.04$;
$b[1] = 0, b[2] = 0.002, b[3] = -0.11, b[4] = 0.11$;
$c[1] = 0, c[2] = -0.002, c[3] = 0.27, c[4] = 0.27$;
$d[1] = 0.25, d[2] = 0.93, d[3] = 0.01, d[4] = 0.01$;
$e[1] = 0, e[2] = -0.002, e[3] = -0.05, e[4] = 0.047$;
$f[1] = -0.4, f[2] = 0.5, f[3] = 0.005, f[4] = 0.06$;
$p[1] = 0.02, p[2] = 0.86, p[3] = 0.93$.
Take $N = 10000$ and $x[0] = y[0] = 0$.

Problem 8-33: Cyclosorus fern

Estimate the capacity fractal dimension of the Cyclosorus fern generated by replacing the parameters in Example 8-4 with the following values:
$a[1] = 0, a[2] = 0.95, a[3] = 0.035, a[4] = -0.04$;
$b[1] = 0, b[2] = 0.005, b[3] = -0.2, b[4] = 0.2$;
$c[1] = 0, c[2] = -0.005, c[3] = 0.16, c[4] = 0.16$;
$d[1] = 0.25, d[2] = 0.93, d[3] = 0.04, d[4] = 0.04$;
$e[1] = 0, e[2] = -0.002, e[3] = -0.09, e[4] = 0.083$;
$f[1] = -0.4, f[2] = 0.5, f[3] = 0.02, f[4] = 0.12$;
$p[1] = 0.02, p[2] = 0.86, p[3] = 0.93$.
Take $x[0] = y[0] = 0$ and as large an N value as you can.

Problem 8-34: Fractal dimension of a coastline
Go to an atlas and find an island (e.g., Iceland, Greenland, Great Britain, ...) with a reasonably indented coastline. Use the box counting approach to estimate the fractal dimension of your island's coastline.

Problem 8-35: Dissecting the fern
For Barnsley's fern generated in Example 8-4, determine what each branch of the piecewise algorithm contributes to the overall fractal picture.

Problem 8-36: Fractal tree
Estimate the fractal dimension of the tree produced by the piecewise algorithm given in Problem 3-7 (page 88). Adjust the value of N to what is possible on your computer.

Problem 8-37: Dissecting the tree
For the fractal tree Problem 3-7 (page 88), determine what each branch of the piecewise algorithm contributes to the overall fractal tree picture.

Problem 8-38: Cellular automata
Use the box counting approach to estimate the fractal dimension of the cellular automata generated figure in Example 3-2.

8.6 Entrainment and Quasiperiodicity

To conclude our discussion of forced oscillators, we shall briefly look at two closely related interesting phenomena that occur in the forced Van der Pol oscillator system.

8.6.1 Entrainment

As the student well knows by now, the Van der Pol equation is a self-exciting nonlinear system that naturally evolves towards a periodic solution, the limit cycle, in the absence of an applied periodic force. Now consider the forced Van der Pol equation given by

$$\ddot{x} - \epsilon(1 - x^2)\dot{x} + x = F\cos\omega t. \tag{8.44}$$

When a sufficiently large periodic force (i.e., large F) with a frequency ω close to the natural frequency of the limit cycle is applied, it is possible for the Van der Pol oscillator to give up its natural oscillation and vibrate at the frequency of the applied force. This phenomenon is called entrainment.

Examples of entrainment abound in the real world. The phenomenon was reported some three centuries ago by the Dutch physicist Huygens. Two clocks on the same wall were observed to become synchronized, the coupling being through the wall. A more modern and practical example of entrainment is in the use of an electronic periodic pacemaker to control the heart rhythm in an individual whose heart exhibits irregular oscillations.

By replacing the forced Duffing equation in file MF09 with the forced Van der Pol equation, the phenomenon of entrainment is easily confirmed. For example, take $\epsilon = 0.25$, $\omega = 1.2$, $F = 3$, $x(0) = 2.09$, and $\dot{x}(0) = 0$. The natural period of the unforced Van der Pol oscillator for the given ϵ value is about $2\pi \simeq 6.28$, while the period of the forcing function is $2\pi/1.2 \simeq 5.24$. Carrying out the numerical run yields the phase plane picture in Figure 8.26 which looks like the usual evolution of the Van der Pol equation onto its limit cycle.

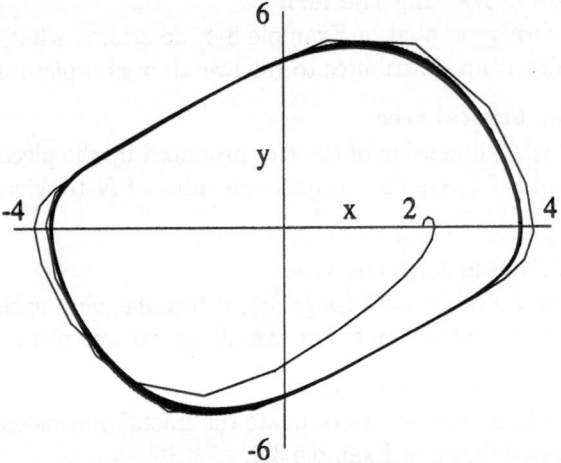

Figure 8.26: Entrainment of the forced Van der Pol oscillator.

However, if $x(t)$ is plotted, as in Figure 8.27, and the number of oscillations in a given time interval is counted, the period is found to be that of the driving force, not the natural period of the Van der Pol equation.

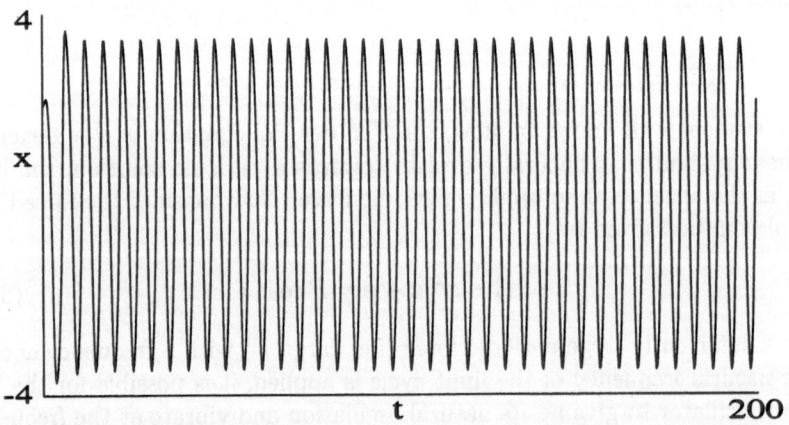

Figure 8.27: $x(t)$ for the forced Van der Pol oscillator in Fig. 8.26.

Entrainment of the Van der Pol oscillator has occurred. For small ϵ and F, an analytic expression for the minimum value of F needed for entrainment to occur can be derived [Jac90]. An alternate approach is to proceed numerically. The latter is left as a problem.

The following experiment illustrates both entrainment and quasiperiodicity, the latter topic being the subject of the next subsection.

Entrainment and Quasiperiodicity
This experiment uses two linked diode oscillators, each diode wired in series
with an inductor. The signals produced by the individual diodes influence each
other to produce entrainment and quasiperiodicity.

EXP 28

PROBLEMS
Problem 8-39: Figures 8.26 and 8.27
Replace the forced Duffing equation in MF09 with the forced VdP equation (8.44)
taking $\epsilon = 0.25$, $\omega = 1.2$, $F = 3$, $x(0) = 2.09$, and $\dot{x}(0) = 0$. Confirm the plots
shown in Figures 8.26 and 8.27 and the text's conclusion about entrainment.

Problem 8-40: Entrainment
Consider the forced Van der Pol equation with $\epsilon = 0.25$, $\omega = 1.2$, $x(0) = 2.09$,
and $\dot{x}(0) = 0$. Numerically determine the approximate minimum value of F at
which entrainment is observed.

Problem 8-41: Other examples of entrainment
By carrying out a literature search, suggest some other examples of entrainment,
providing a careful explanation of the entrainment process.

8.6.2   Quasiperiodicity

When entrainment took place for the forced Van der Pol equation, the driving
frequency "won out" over the natural frequency of the limit cycle. It is possible,
under the right conditions, that neither frequency wins out and so-called quasi-
periodic behavior occurs. Such is the case for the forced Van der Pol oscillator
when we choose $\epsilon = 1$, $\omega = 1.1$, $F = 0.5$, $x(0) = 1.9$, and $\dot{x}(0) = 0$. The phase
plane portrait is shown in Figure 8.28. The phase plane trajectory is confined
to an annular region and is uniformly distributed. The solution $x(t)$ is both
amplitude and frequency modulated.

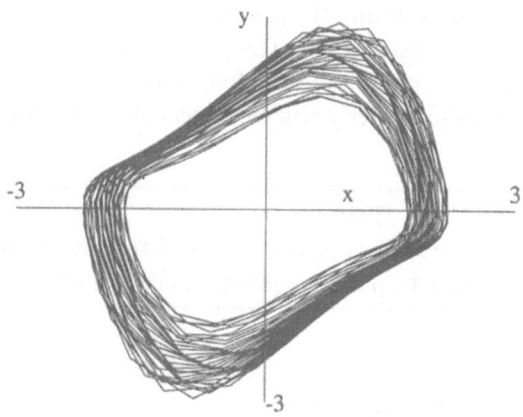

Figure 8.28: Quasi-periodic phase plane picture.

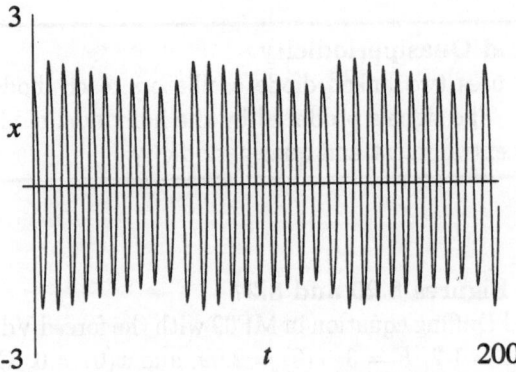

Figure 8.29: Behavior of $x(t)$ for quasi-periodic situation.

The amplitude modulation is clearly seen in Figure 8.29, the frequency modulation being less obvious. If we calculate the power spectrum by modifying file MF32, Figure 8.30 results. The tallest peak is at the natural frequency ($\omega = 1$)

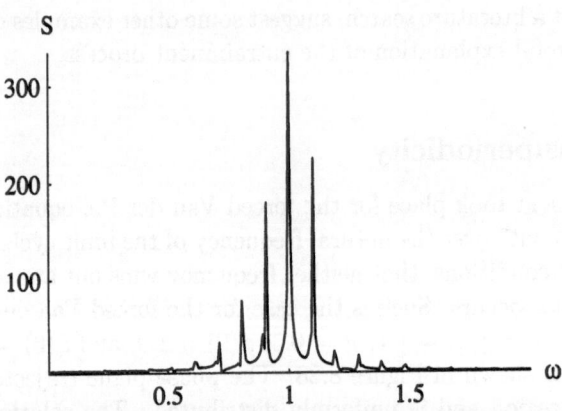

Figure 8.30: Power spectrum for quasi-periodic solution. ($N = 512$, $T = 0.5\pi$)

of the Van der Pol limit cycle and the peak immediately to the right corresponds to the driving frequency ($\omega = 1.1$). There are also smaller side peaks spaced about $\Delta\omega = 0.1$ apart, i.e., at the beat frequency between these two competing frequencies. Quasiperiodicity can be observed in the following neon bulb experiment.

Quasiperiodicity: Neon Bulb
This experiment uses two linked neon bulb oscillator circuits to produce signals that influence each other to produce quasiperiodicity and entrainment.

PROBLEMS
Problem 8-42: Quasiperiodicity
Consider the forced VdP equation (8.44) with $\epsilon = 1$, $\omega = 1.1$, $F = 0.5$, $x(0) = 1.9$, and $\dot{x}(0) = 0$.

a. Confirm the quasiperiodic phase plane picture shown in Figure 8.28.

b. Confirm the behavior of $x(t)$ displayed in Figure 8.29.

c. Confirm the power spectrum shown in Figure 8.30.

8.7 The Rössler and Lorenz Systems

Although they do not represent forced oscillator systems, we would be remiss if some time was not spent on the Rössler and Lorenz systems which are also 3-dimensional in nature and display periodic and chaotic motions, period doubling, and strange attractors. Some relevant aspects of the Lorenz system have already been discussed in Sections 4.7 and 7.6. The Rössler system was assigned for investigation in Problem 4-35.

8.7.1 The Rössler Attractor

We begin with the Rössler system

$$\dot{x} = -(y + z), \quad \dot{y} = x + ay, \quad \dot{z} = b + z(x - c) \tag{8.45}$$

with the "standard" values $a = b = 0.2$ and variable $c > 0$. The Rössler system is simpler than the Lorenz system in that it only involves one nonlinear term. Nevertheless it can display rich behavior as c is increased. Before reading any further, if you have not already done so, you should create a new Mathematica file for the Rössler system by modifying the Lorenz file MF10. As is then easily verified, the Rössler system undergoes a series of period doublings between

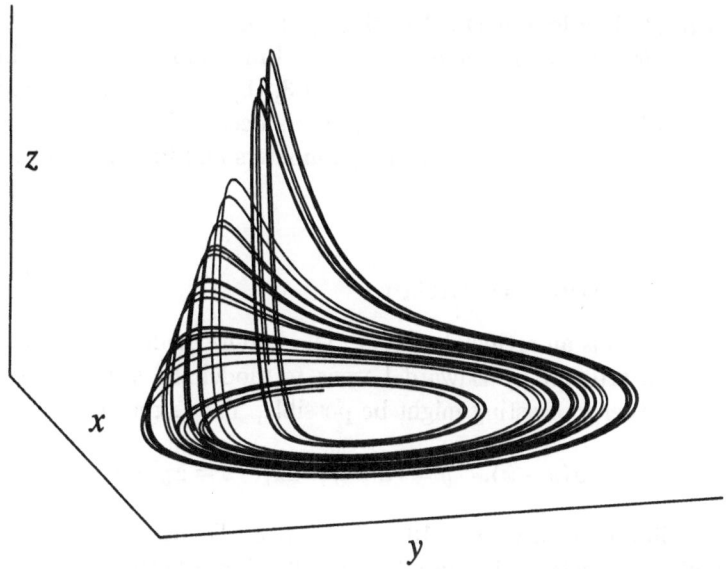

Figure 8.31: Rössler attractor: Initial cond. $x(0) = 4$, $y(0) = z(0) = 0$, t=200.

$c = 2.5$ and $c = 5.0$. For example, at $c = 2.5$ period 1 prevails as seen from a plot of $x(t)$ (not shown), at $c = 3.5$ we have period 2, and at $c = 4.0$ period 4. Increasing c further to 5.0 produces the chaotic Rössler attractor shown in Figure 8.31. In Figure 8.32, we show the behavior of $x(t)$ and $z(t)$. The

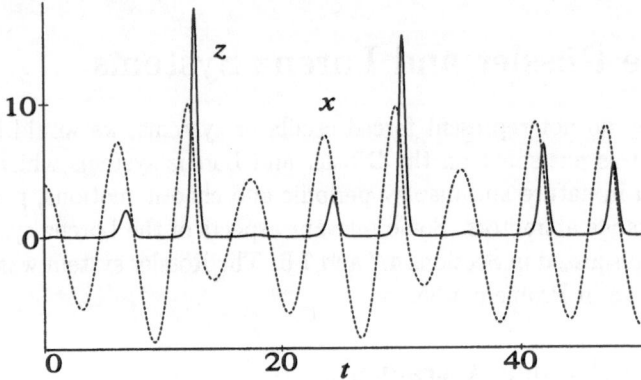

Figure 8.32: Behavior of $x(t)$ and $z(t)$ for the Rössler strange attractor.

trajectory spends most of its time in the x-y plane spiraling out from the origin until at some critical radius it jumps away from the plane, moving in the z direction until some maximum z value is attained, and then dropping back into the x-y plane to undertake a new spiraling out motion, and so on. If one changes the initial conditions, a different transient growth occurs but a similar strange attractor evolves. That the attractor is strange has been confirmed [PJS92] by measuring its fractal dimension, which turns out to be very slightly above 2. By increasing c even further, new period doublings and chaotic behavior may be found.

PROBLEMS
Problem 8-43: Exploring the Rössler system
Explore the Rössler system for $5 < c < 7$. Find values of c for which the following periodic solutions occur: (a) period 3, (b) period 6, (c) period 4, (d) period 8. In what ranges do strange attractors occur? What is the nature of the solution for $c = 8.0$? Take the same parameters and initial conditions as in the text.

8.7.2   The Lorenz Attractor

The Rössler model is an artificial system. The Lorenz model on the other hand arose out of an attempt by Edward Lorenz to model the atmosphere so that long range weather forecasting might be possible. The Lorenz equations are

$$\dot{x} = \sigma(y - x), \quad \dot{y} = rx - y - xz, \quad \dot{z} = xy - bz \qquad (8.46)$$

with the traditional choices $\sigma = 10$ and $b = 8/3$. For the choice $r = 28$, the butterfly wings trajectory of Figure 3.30 results. For a better colorized picture of the trajectory the student should go to file MF10. The trajectory spirals out

The Butterfly Effect

from the center in one of the wings and at some critical radius gets ejected from the wing and is attracted to the other wing where it again spirals outward to be ejected back into the first wing region, and so on. The trajectory stays in the attractive region, continually tracing out new orbits.

As with the Rössler system, period doubling to chaos can be observed for the right range of the parameter r. For example, the period doubling scenario occurs in the range $99.524 < r < 100.795$.

Another famous example of a chaotic butterfly attractor can be found by running Chua's electrical circuit, the subject of the next experiment.

Chua's Butterfly
Chua's circuit is used to produce a "double scroll" trajectory similar to the Lorenz attractor. The op-amp circuit produces a piece-wise linear negative resistance which allows for chaotic behavior.

PROBLEMS
Problem 8-44: Period doubling
Explore the range $99.524 < r < 100.795$ for the Lorenz attractor with the parameters as in the text. Verify that period 2 occurs for $r = 99.98$, period 4 for $r = 99.629$, and period 8 for $r = 99.547$. (Period 16 is supposed to occur for $r = 99.529$ and period 32 for $r = 99.5255$.)

Problem 8-45: Transient Chaos
For $\sigma = 10$, $b = 8/3$, $x(0) = 2$, $y(0) = z(0) = 5$ and $r = 22.4$, show by plotting $x(t)$ that the trajectory in the Lorenz system only appears to be chaotic for an initial transient time and then approaches an attractive rest point.

8.8   Hamiltonian Chaos

To this point in the chapter, we have looked at the dynamical behavior of nonlinear ODE systems that involve three coupled first-order equations and therefore can be described by trajectories in a 3-dimensional phase space. In their ever-widening search for nonlinear systems that display the onset of chaotic behavior when one or more parameters are adjusted, mathematicians and theoretical physicists are currently exploring ODE systems involving phase spaces of dimension higher than three. We shall end this chapter by looking at one particular example involving a four-dimensional phase space, the mathematical treatment of which makes use of the Hamiltonian formalism of classical mechanics.

8.8.1   Hamiltonian Formulation of Classical Mechanics

Consider a system consisting of a single particle[13] with three degrees of freedom, i.e., three independent coordinates are required to locate the particle. If the Lagrangian $L = L(q_i, \dot{q}_i, t)$, with q_i the coordinates, $\dot{q}_i$ the "generalized" velocities, and t the time, a Hamiltonian function $H(q_i, p_i, t)$ can be defined quite generally through the so-called "Legendre transformation" as

$$H = \sum_{i=1}^{3} p_i \dot{q}_i - L \tag{8.47}$$

where the p_i are generalized momenta defined by $p_i = \partial L/\partial \dot{q}_i = p_i(q_1, ..., \dot{q}_1, ...)$. If the system is conservative (the potential energy is velocity independent) and does not contain t explicitly in the Hamiltonian or in its coordinates, it is shown in standard classical mechanics texts [MT95] that the Hamiltonian is a constant of the motion ($dH/dt = 0$) and equal to the total energy (kinetic plus potential) of the system, i.e,

$$H = T + V. \tag{8.48}$$

Further, the equations of motion of the system are given by Hamilton's equations

$$\dot{q}_i = \frac{\partial H}{\partial p_i}, \qquad \dot{p}_i = -\frac{\partial H}{\partial q_i}. \tag{8.49}$$

For example, consider the unforced simple pendulum, which has only one degree of freedom, with no damping present. From Chapter 2, the Lagrangian is given by

$$L = T - V = \frac{1}{2}m(\ell\dot{\theta})^2 - mg\ell(1 - \cos\theta), \tag{8.50}$$

so the generalized momentum is $p = \partial L/\partial\dot{\theta} = m\ell^2\dot{\theta}$. The Hamiltonian $H(p, \theta)$ for the simple pendulum is

$$H = T + V = \frac{p^2}{2m\ell^2} + mg\ell(1 - \cos\theta). \tag{8.51}$$

Hamilton's equations (8.49) yield

$$\dot{\theta} = \frac{\partial H}{\partial p} = \frac{p}{m\ell^2}, \qquad \dot{p} = -\frac{\partial H}{\partial \theta} = -mg\ell\sin\theta \tag{8.52}$$

---

[13]The extension to a system of N particles, with $N > 1$, is straightforward

which are easily combined to yield the familiar simple pendulum equation

$$\ddot{\theta} + \omega^2 \sin\theta = 0 \tag{8.53}$$

on setting $\omega = \sqrt{\frac{g}{\ell}}$. As we are already fully aware from Chapter 4, for the simple pendulum the phase space is 2-dimensional, i.e., a phase plane, when p is plotted against θ.

In this example, there was no advantage to using the Hamiltonian formulation over the Newtonian or Lagrangian treatments. However, the Hamiltonian approach is important in the conceptual development of quantum mechanics and in treating conservative nonlinear dynamical systems whose trajectories move in a four or higher dimensional phase space. We illustrate the latter by introducing the Hénon–Heiles Hamiltonian describing a single particle free to move in a 2-dimensional potential well, so there are two degrees of freedom. If the spatial coordinates are x and y, then four coupled first order Hamiltonian equations will result for the time evolution of x, y, p_x, and p_y. The phase space is then 4-dimensional. However, since the total energy is fixed, only three of the variables are independent and the temporal evolution of the system can be effectively represented by a trajectory in a 3-dimensional phase space.

PROBLEMS
Problem 8-46: The spherical pendulum
Using the Hamiltonian method, find the equations of motion for a spherical pendulum of length ℓ and mass m. If you have forgotten what a spherical pendulum is, see Problem 2-7, page 44.

Problem 8-47: The double pendulum
Instead of employing the Lagrangian approach, redo the double pendulum Problem 2-9 (page 45) using the Hamiltonian method.

Problem 8-48: Motion of a particle on a cylinder
A particle of mass m is constrained to move on the surface of a cylinder, defined by $x^2 + y^2 = R^2$. The particle is subjected to a force directed toward the origin and proportional to the distance it is from the origin. Determine the Hamiltonian function for this system and then derive the equations of motion.

8.8.2   The Hénon–Heiles Hamiltonian

Originally motivated to model the motion of a star inside a galaxy, Hénon and Heiles [HH64] introduced a simple Hamiltonian describing the motion of a particle of unit mass in the two-dimensional potential

$$V(x,y) = \frac{1}{2}x^2 + \frac{1}{2}y^2 + x^2 y - \frac{1}{3}y^3. \tag{8.54}$$

The kinetic energy for the two-dimensional particle motion is

$$T = \frac{1}{2}p_x^2 + \frac{1}{2}p_y^2, \tag{8.55}$$

where p_x and p_y are the x and y components of momentum, respectively, so the Hénon–Heiles Hamiltonian is

$$H = T + V = \frac{1}{2}p_x^2 + \frac{1}{2}p_y^2 + \frac{1}{2}x^2 + \frac{1}{2}y^2 + x^2y - \frac{1}{3}y^3. \qquad (8.56)$$

The geometrical nature of the Hénon–Heiles potential, $V(x,y)$, is examined in the next example.

Example 8-5: Hénon–Heiles Potential

a. Construct a 3-dimensional plot of the Hénon–Heiles potential (8.54).

b. Locate the stationary points of $V(x,y)$ and determine their nature.

c. Determine the value of the potential energy at each stationary point.

d. What is the maximum total energy for a bounded orbit to occur?

e. Construct a 2-D contour plot with the stationary points included.

Solution: The Graphics and Calculus`VectorAnalysis` packages are loaded, the latter being needed in order to use Mathematica's gradient operator, Grad.

```
Clear["Global`*"]; << Graphics`
```

```
<< Calculus`VectorAnalysis`
```

The Hénon–Heiles potential is entered,

```
V= x^2/2+y^2/2+x^2 y-y^3/3
```

$$\frac{x^2}{2} + x^2 y + \frac{y^2}{2} - \frac{y^3}{3}$$

and a 3-dimensional colored picture produced with the Plot3D command. The argument Hue[10V] is used to generate a color variation in the V direction. A better picture results if the mesh is removed with Mesh->False. The remaining plot options should be quite familiar to the reader by now.

```
Plot3D[{V, Hue[10V]}, {x,-1.1,1.1}, {y,-1.1,1.1}, BoxRatios ->
{1,1,1}, PlotRange -> {{-1.1,1.1}, {-1.1,1.1}, {0,.22}},
ViewPoint -> {3,2,2.5}, Mesh -> False, PlotPoints -> 250, Ticks->
{{-1,0,{.5,"x"},1},{-1,0,{.5,"y"},1},{.1,{.15,"V"},.2}},
TextStyle -> {FontFamily -> "Times",FontSize -> 16},
ImageSize -> {400,400}];
```

The resulting 3-dimensional plot may be observed on the computer screen by executing the code, but is not displayed here. The potential V exhibits a symmetric central valley with a single minimum and three symmetrically located saddles surrounding the valley at a higher elevation. If the cubic terms were completely neglected in $V(x,y)$, the potential energy would be of the form $V = \frac{1}{2}x^2 + \frac{1}{2}y^2$, i.e., a paraboloid of revolution. As the reader may easily verify, the equations of motion in the x and y directions would then decouple into two simple harmonic oscillator equations which are easily solved. The presence of the cubic terms alters the parabolic potential into the shape displayed on the computer screen.

To locate the stationary points, we make use of the fact that the force $\vec{F} = -\nabla V$ must vanish at these points. In the next command line, the force (labeled f) is calculated in Cartesian coordinates using Mathematica's gradient operator, Grad.

```
f = -Grad[V,Cartesian[x,y,z]]
```

$$\{-2y\,x - x, -x^2 + y^2 - y, 0\}$$

The output of f yields the x, y, and z components of the force expressed as a list. Of course, the z-component is zero, since the potential has no z dependence here. To locate the stationary points of $V(x, y)$, the x and y force components are set equal to zero and the Solve command applied.

```
sol = Solve[{f[[1]] == 0, f[[2]] == 0}, {x,y}]
```

$$\{\{x \to 0, y \to 0\}, \{x \to 0, y \to 1\}, \{x \to -\frac{\sqrt{3}}{2}, y \to -\frac{1}{2}\}, \{x \to \frac{\sqrt{3}}{2}, y \to -\frac{1}{2}\}\}$$

The x, y coordinates of the four stationary points are given by the above output. The corresponding potential energies at these points are now calculated and displayed as a list.

```
pe = {V/.sol[[1]], V/.sol[[2]], V/.sol[[3]], V/.sol[[4]]}
```

$$\{0, \frac{1}{6}, \frac{1}{6}, \frac{1}{6}\}$$

Therefore, the potential energy at $x = y = 0$ is zero. This is the minimum of the central valley in the 3-dimensional plot. Since there is no energy loss, the origin behaves like a vortex point at very small energies. The potential energy at each of the three saddle points is 1/6. To have a bounded orbit, the particle must have a total energy $H = E$ below that at the saddles, otherwise the particle could escape to infinity.

Finally, a 2-dimensional colored contour plot with the saddles indicated by points is produced. Here, we have chosen to display 70 contours. The three saddle points are colored red and the minimum point colored green using the RGBColor option. The entire plot is given a color variation with the ColorFunction option.

```
ContourPlot[V,{x,-2,2},{y,-2,2},PlotPoints -> 100, Contours -> 70,
ColorFunction->Hue,FrameTicks->{{-2,0,{1,"x"},2},{-2,0,{1,"y"},
2},{ },{ }}, TextStyle -> {FontFamily -> "Times",FontSize -> 16},
Epilog -> {PointSize[.03],{RGBColor[0,1,0],Point[{x,y}
/.sol[[1]]]},{RGBColor[1,0,0],Point[{x,y}/.sol[[2]]],
Point[{x,y}/.sol[[3]]],Point[{x,y}/.sol[[4]]]}},
ImageSize -> {400,400}];
```

A black and white rendition of the resulting contour plot is shown in Figure 8.33. It does not do full justice to the colorful figure observed on the computer screen.

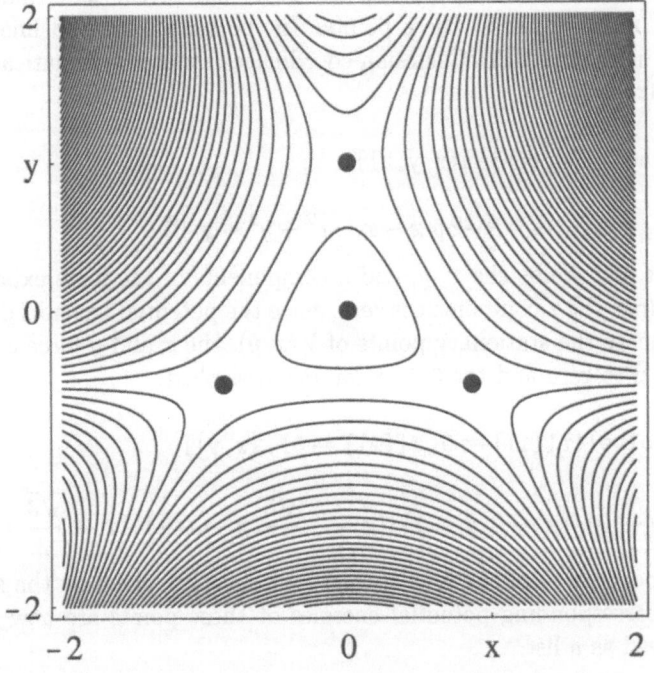

Figure 8.33: 2-dimensional contour plot of the Hénon–Heiles potential.

End Example 8-5

Hamilton's equations for the Hénon–Heiles system are easily calculated:

$$\dot{x} = \frac{\partial H}{\partial p_x} = p_x$$
$$\dot{y} = \frac{\partial H}{\partial p_y} = p_y$$
$$\dot{p}_x = -\frac{\partial H}{\partial x} = -x - 2xy \qquad (8.57)$$
$$\dot{p}_y = -\frac{\partial H}{\partial y} = -y - x^2 + y^2$$

The nonlinear system resides in a 4-dimensional phase space. If desired, the system of four 1st-order ODEs is easily reduced to two coupled 2nd-order ODEs,

$$\ddot{x} = -x - 2xy$$
$$\ddot{y} = -y - x^2 + y^2. \qquad (8.58)$$

In order to numerically integrate the equations, we shall choose to work with the first order system. The system does not have an analytic solution, so its time evolution must be found by numerically integrating forward in time, once the initial values have been specified. One begins by specifying a value $H = E$ for the total energy, with $E < \frac{1}{6}$, as well as $x(0)$, $y(0)$, and $p_y(0)$. The initial value $p_x(0)$ need not be given because its value follows from energy conservation, viz., at $t = 0$,

$$H = E = \frac{1}{2}p_x(0)^2 + \frac{1}{2}p_y(0)^2 + \frac{1}{2}x(0)^2 + \frac{1}{2}y(0)^2 + x(0)^2 y(0) - \frac{1}{3}y(0)^3. \quad (8.59)$$

Thus, $p_x(0)$ is given by

$$p_x(0) = \sqrt{2E - p_y(0)^2 - x(0)^2 - y(0)^2 - 2x(0)^2 y(0) + \frac{2}{3}y(0)^3}, \quad (8.60)$$

on selecting, say, the positive square root. But the energy conservation statement (8.59) holds for all times t, not just at $t = 0$. Now the reader knows that, for example, in geometry the constraint $x^2 + y^2 + z^2 = r^2$ defines a spherical surface (which is 2-dimensional) of radius r in the 3-dimensional x-y-z space. Similarly, here the energy constraint $H = E$ defines a "hypersurface" in our 4-dimensional phase space. The trajectory must live in a 3-dimensional volume in this 4-dimensional space. Thus, the region occupied by the trajectories can be visualized by making a plot of the numerical solution in, say, the 3-dimensional x vs. y vs. p_y space. A Poincaré section can then be formed by selecting a plane which slices through a cross-section of the volume. Traditionally, the Poincaré section is taken for the Hénon–Heiles problem to be the y vs. p_y plane sliced at $x = 0$. Each time a trajectory passes through this plane at $x = 0$ a point will be generated in the plane.

The following Mathematica example illustrates how the above procedure may be carried out.

Example 8-6: Hénon–Heiles Poincaré Sections

---

a. Taking $E = 0.06$, $x(0) = -0.1$, $y(0) = -0.2$, and $p_y(0) = -0.05$, determine $p_x(0)$.

b. Plot the system's trajectory in the 3-dimensional x vs. y vs. p_y space.

c. Form a Poincaré section by viewing the y–p_y plane sliced at $x = 0$.

d. Interpret the Poincaré section picture.

Solution: The Graphics package is loaded.

```
Clear["Global`*"]; << Graphics`
```

Using p1 and p2 to label the x and y momenta, respectively, the kinetic and potential energies are entered and the Hamiltonian formed.

```
T = p1^2/2 + p2^2/2
```

$$\frac{p1^2}{2} + \frac{p2^2}{2}$$

```
V = x^2/2 + y^2/2 + x^2 y - y^3/3
```

$$\frac{x^2}{2} + x^2 y + \frac{y^2}{2} - \frac{y^3}{3}$$

```
H = T + V
```

$$\frac{p1^2}{2} + \frac{p2^2}{2} + \frac{x^2}{2} + x^2 y + \frac{y^2}{2} - \frac{y^3}{3}$$

Hamilton's four equations of motion are calculated.

```
eq1 = x'[t] == D[H, p1] /. p1 -> p1[t]
```

$$x'(t) == p1(t)$$

```
eq2 = y'[t] == D[H, p2] /. p2 -> p2[t]
```

$$y'(t) == p2(t)$$

```
eq3 = p1'[t] == -D[H, x] /. {x -> x[t], y -> y[t]}
```

$$p1'(t) == -x(t) - 2 x(t) y(t)$$

```
eq4 = p2'[t] == -D[H, y] /. {x -> x[t], y -> y[t]}
```

$$p2'(t) == -x(t)^2 + y(t)^2 - y(t)$$

Aside from notation, the above four equations are identical with (8.57). The total energy expression $H = E$ is entered with $E = 0.06$ and the initial values $x = -0.1$, $y = -0.2$, and $p2 = -0.05$.

```
energy = (H /. {x -> -.1, y -> -.2, p2 -> -.05}) == .06
```

$$\frac{p1^2}{2} + 0.0269167 == 0.06$$

The total energy expression is then solved for the initial x-component of the momentum, $p1$.

```
sol = Solve[energy, p1]
```

$$\{\{p1- > -0.257229\}, \{p1- > 0.257229\}\}$$

The positive square root answer is selected in sol and labeled p10.

```
p10 = p1 /. sol[[2]]
```

$$0.257229$$

Hamilton's four equations of motion are now solved numerically over the time interval $t = 0$ to 400, subject to the four initial conditions.

```
sol2 = NDSolve[{eq1, eq2, eq3, eq4, x[0] == -.1, y[0] == -.2,
     p1[0] == p10, p2[0] == -.05}, {x[t], y[t], p1[t], p2[t]},
     {t, 0, 400}, MaxSteps -> 5000];
```

The trajectory is plotted in the x vs. y vs. $p2$ space,

```
gr1 = ParametricPlot3D[Evaluate[{x[t],y[t],p2[t],Hue[x[t]]}/.sol2],
     {t,0,400},Ticks -> {{-.3,0,{.15,"x"},.3},{-.3,0,{.15,"y"},
     .3},{{-.1,"p2"}}},PlotPoints -> 4000, BoxRatios -> {1,1,1},
     ViewPoint -> {1,0,-5}, TextStyle -> {FontFamily->"Times",
     FontSize->16}, DisplayFunction -> Identity];
```

and the planar surface corresponding to $x = 0$ is also created.

```
gr2 = Graphics3D[Polygon[{{0,-.31,-.31},{0,-.35,.35},{0,.35,.35},
     {0,.35,-.31}}], DisplayFunction -> Identity];
```

The two graphs are superimposed to produce Figure 8.34. In this figure the trajectory executes quasiperiodic motion on the surface of a torus, referred to as a KAM (Kolmogorov–Arnold–Moser) torus.

```
Show[gr1,gr2,DisplayFunction- >$DisplayFunction,
     ImageSize -> {500,500}];
```

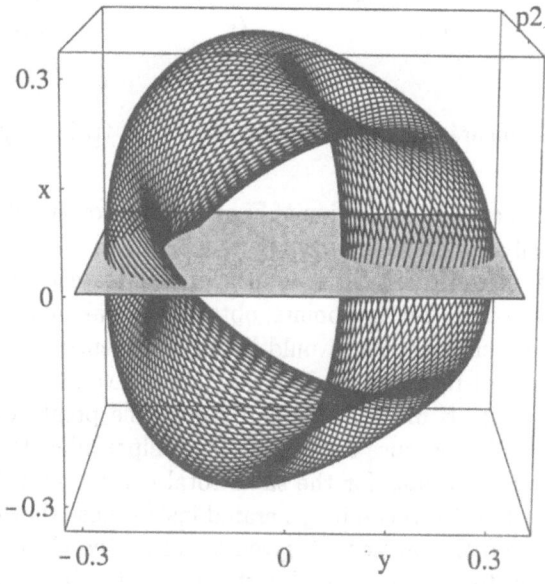

Figure 8.34: A quasi-periodic trajectory on the surface of a KAM torus.

By choosing different initial conditions for the same total energy, other KAM tori can by generated. The $x = 0$ plane which is to be used for obtaining the Poincaré section is also shown. The Poincaré section can be obtained by taking the thin slice $x = 0...0.002$ in the PlotRange option of the ParametricPlot3D command. The resulting Poincaré section is shown in Figure 8.35.

```
ParametricPlot3D[Evaluate[{x[t],y[t],p2[t]}/. sol2],{t,0,400},
PlotPoints->4000, PlotRange->{{0,0.002},{-.4,.4},{-.4,.4}},
ViewPoint->{1,0,0}, AxesLabel->{" ", "y", "p2"},
Ticks->{{ },{-.3, 0, .3}, {-.3, 0, .3}}, TextStyle->
{FontFamily->"Times", FontSize->16}, ImageSize->{500, 500}];
```

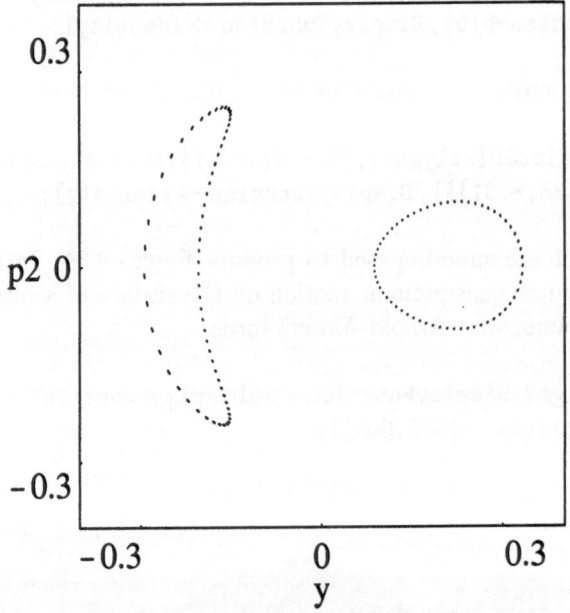

Figure 8.35: Poincaré section for the quasiperiodic trajectory of Fig. 8.34.

The points in the Poincaré section of Figure 8.35 lie on two distorted ellipses. This result is easily understood from Figure 8.34. If the torus were a donut of uniform elliptical crossection, and the quasiperiodic trajectory was confined to the donut's surface, the Poincaré points, obtained by taking a plane perpendicular to the donut's cross-section, would lie on two identical ellipses. Here the torus is twisted and distorted so the Poincaré points lie on two distorted ellipses. If the time of the run is made larger and larger, the points would eventually fill out the two ellipses because the motion is quasiperiodic. By choosing other appropriate initial conditions for the same total energy, nested ellipses corresponding to other KAM tori can be generated inside those shown in Figure 8.35 as well as ellipses on the outside. A given set of nested ellipses resembles the circulation of trajectories seen around vortex points in phase-plane portraits.

End Example 8-6

Now let's look at what happens to the trajectory when the total energy is increased. Running the code in the previous example with $E = 0.16$ and a longer time interval, $t = 0...500$, produces Figures 8.36 and 8.37. To express it in unscientific language, the resulting orbit is now a "mess".

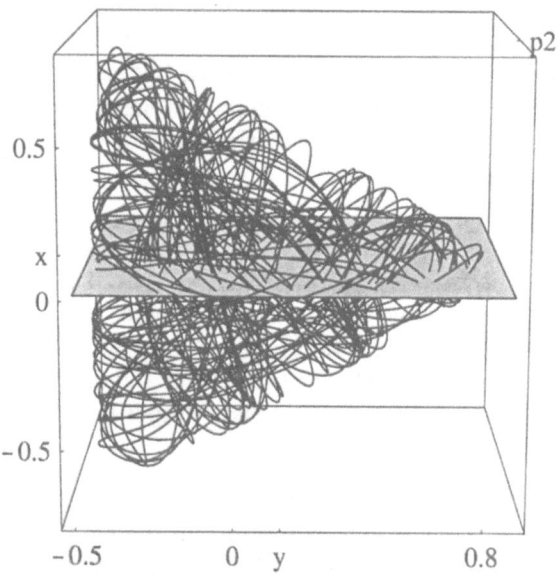

Figure 8.36: Hénon–Heiles chaotic trajectory for $E = 0.16$.

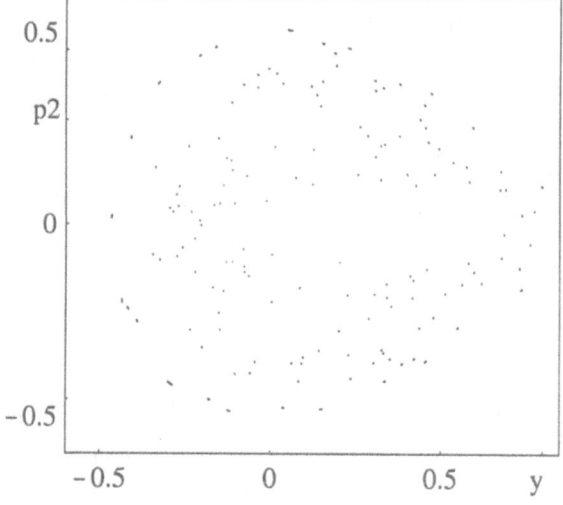

Figure 8.37: Poincaré section for $E = 0.16$.

The trajectory is becoming chaotic in nature and instead of lying on a toroidal surface now tries to fill a 3-dimensional volume. The chaotic behavior is confirmed by the irregular appearance of the Poincaré section. A great deal of mathematical research has gone into understanding the onset of chaos in the Hénon–Heiles system and the "dissolving" of the various KAM tori into chaotic trajectories. A precise mathematical statement of what takes place, known as the KAM theorem, may be found in more advanced nonlinear dynamics texts such as Jackson [Jac90] and Hilborn [Hil94]. In this section we have been content to give the interested reader the computer tools to explore aspects of the KAM tori phenomena and the onset of Hamiltonian chaos discussed in these and other references.

PROBLEMS

Problem 8-49: Neglecting the first cubic term
Neglect the $x^2 y$ cubic term in $V(x, y)$ and run the file given in Example 8-6, keeping all parameter values unchanged. Interpret and discuss the 3-dimensional picture and the Poincaré section produced. Experiment with different total energies and discuss the results.

Problem 8-50: Neglecting the second cubic term
Neglect the $-y^3/3$ cubic term in $V(x, y)$ and run the file given in Example 8-6, keeping all parameter values unchanged. Interpret and discuss the 3-dimensional picture and the Poincaré section produced. Experiment with different total energies and discuss the results.

Problem 8-51: Neglecting both cubic terms
Neglect both cubic terms in $V(x, y)$ and run the file given in Example 8-6. Interpret and discuss the 3-dimensional picture and the Poincaré section produced.

Problem 8-52: Nesting of ellipses
By choosing other initial conditions for $E = 0.06$, produce a Poincaré section plot which demonstrates the nesting of distorted ellipses mentioned in the text.

Problem 8-53: Different initial condition
Keeping all other parameters the same, run the code in Example 8-6 for the input conditions $p2(0) = 0$, $y(0) = 0.2$, and $x(0) = 0$, thus producing a different KAM torus and Poincaré section.

Problem 8-54: A different potential
Paralleling Example 8-5,

   a. Construct a 3-dimensional colored plot of the potential

$$V(x, y) = \frac{1}{2}x^2 + \frac{1}{2}y^2 + x^4 y - \frac{1}{4}y^3.$$

   b. Locate the stationary points of $V(x, y)$ and determine their nature.

   c. Determine the value of the potential energy at each stationary point.

d. What is the maximum value of the total energy for a bounded orbit to occur?

e. Construct a 2-dimensional contour plot with the stationary points included.

Problem 8-55: Trajectories and Poincaré sections

For the potential of the previous problem, carry out the following steps:

a. Generate the Hamiltonian equations.

b. Taking $E = 0.06$, $p_y(0) = -0.05$, $y(0) = -0.2$, and $x(0) = -0.1$, determine $p_x(0)$.

c. Plot the system's trajectory in the 3-dimensional x vs. y vs. p_y space.

d. Form a Poincaré section by viewing the y–p_y plane sliced at $x = 0$.

e. Interpret the Poincaré section picture.

f. Repeat steps (b) to (d) for several different E values, e.g., $E = 0.2$.

Problem 8-56: Toda Hamiltonian

In suitably normalized units the Hamiltonian for a 3-particle molecule, with the forces between particles governed by the Toda potential, can be written ([Jac90]) as $H = T + V$ with

$$T = \frac{p_1^2}{2} + \frac{p_2^2}{2}, \quad V = \frac{1}{3}\left[e^{(x_2 + \sqrt{3}x_1)} + e^{(x_2 - \sqrt{3}x_1)} + e^{(-2x_2)}\right] - 1.$$

Paralleling Examples 8-5 and 8-6,

a. Create a 3-dimensional plot of the potential energy $V(x_1, x_2)$. You will have to play with the plot range to get a suitable picture.

b. Locate and identify the nature of the stationary points of $V(x_1, x_2)$.

c. Generate the Hamiltonian equations.

d. For $E = 1.0$, $p_2(0) = -0.05$, $x_2(0) = -0.2$, $x_1(0) = -0.2$, determine $p_1(0)$.

e. Make a plot of the system's trajectory in the 3-dimensional x_1 vs. x_2 vs. p_2 space. Adjust your viewing box to include the whole trajectory. Is this another example of a KAM torus? Explain.

f. Form a Poincaré section by viewing the x_2–p_2 plane sliced at $x_1 = 0$.

g. Interpret the Poincaré section picture.

h. Explore the Toda Hamiltonian problem for other energy values.

Problem 8-57: KAM Theorem

By consulting the texts by Jackson[Jac90] and Hilborn[Hil94], or any other sources that you can find, discuss the KAM theorem in some detail. Illustrate the theorem with suitable plots of trajectories and Poincaré sections for an example of your choice.

Chapter 9

Nonlinear Maps

In all chaos there is a cosmos, in all disorder a secret order.
Carl Jung (1875-1961), Swiss psychiatrist

9.1   Introductory Remarks

In the study of forced oscillator phenomena, we have avoided plunging into heavy analysis because generally the details can be gory and are probably soon forgotten by the student. The virtue of maps, and the logistic map in particular, is that they are amenable to relatively simple, easily understandable, analysis because they are governed by finite difference equations rather than nonlinear differential equations. Despite their relative simplicity, nonlinear maps can guide us along the road to understanding many of the features that are seen in forced nonlinear oscillator systems such as the period doubling route to chaos, the stretching and folding of strange attractors, and so on. The emphasis will be on understanding rather than trying to establish the direct connection of a given map with a particular nonlinear ODE which is a nontrivial task beyond the scope of this text. Some new concepts like bifurcation diagrams and Lyapunov exponents, which will be encountered in this chapter, could have been introduced in the last chapter but are more easily dealt with in the framework of nonlinear maps.

Trying to understand nonlinear ODE phenomena is not the only reason for studying maps. In mathematical biology, economics, and certain other scientific areas, time is often regarded as discrete rather than continuous. As mentioned in Chapter 3, the biologist, for example, measures animal and insect populations in successive non-overlapping generations. Stock market statistics are recorded daily, and so on. Even in the world of physics, the experimentalist may sample at regular time intervals instead of continuously.

The chapter will begin with a study of one of the most well-known one-dimensional maps, the logistic map, whose physical interpretation was briefly discussed in Section 3.3.4. If the student has forgotten what was said there, please go back and quickly reread this subsection. Much of our attention will be focused on the logistic map because of its simplicity and richness of behavior. For example, the period doubling observed in the logistic map can be related to the period doubling which has been observed experimentally. A second one-dimensional map, the circle map, will also be briefly examined. Then, the important issue of distinguishing between random noise and deterministic chaos will receive our attention. Finally, the chapter finishes with short sections on big topics, namely 2 and 3-dimensional maps, and how to control chaos.

9.2   The Logistic Map

9.2.1   Introduction

As a simple model in mathematical biology which displays very complex behavior, Robert May [May76] championed the introduction of the now famous logistic map into elementary mathematics courses. This nonlinear map has the following form:

$$X_{n+1} = aX_n(1 - X_n) \tag{9.1}$$

with $0 \leq a \leq 4$. This range of the parameter a is chosen to ensure that the mapping produces a value of X in the convenient range 0 to 1. If, for example, $a = 6$ and $X_0 = 0.25$, we would obtain $X_1 = 6 \times 0.25 \times 0.75 = 1.125 > 1$. Using the approach of Example 3-4, the numerical solution of Equation (9.1) can be easily carried out and the result plotted with Mathematica. A slightly different numerical approach is to note that the logistic equation resembles the Euler algorithm for solving nonlinear ODEs where the rhs was evaluated at the previous time step. Indeed, the subscript n in the biological context plays the role of a time step, since it refers to the nth generation of whatever biological species at which "time" the normalized population number of that species is X_n. Unlike the Euler algorithm, however, where one had to take a large number of extremely small time steps in order to get accurate results, the "time step" here is unity and with comparatively few n values one can obtain accurate steady-state results. So, referring to Example 6-2, the logistic map equation may be solved as in the following example.

Example 9-1: The Logistic Map 2

Mimicking the Euler algorithm, numerically solve and plot the logistic equation for $X_0 = 0.2$, $n = 0...$total $= 60$, and $a = 2.8$, 3.3, and 3.8. Use a line style.

Solution: The given input parameters (here, $a = 2.8$) are entered,

```
Clear["Global`*"]

t[0] = 0;  x[0] = 0.2;  a = 2.8;  total = 60;
```

and the functions for iterating the logistic map and the time are formed.

```
x[n_] := x[n] = a x[n-1](1-x[n-1])
t[n_] := t[n] = t[n-1] +1
```

The map is iterated and the plotting points formed with the Table command.

```
pts = Table[{t[n], x[n]}, {n, 0, total}];
```

The ListPlot command is used to produce Figure 9.1(a) for $a = 2.8$. The other two plots, (b) and (c), are obtained for $a = 3.3$ and $a = 3.8$.

```
ListPlot[pts, PlotRange -> {{0,60},{0,1}}, PlotJoined -> True,
PlotStyle -> {RGBColor[0,0,1]}, Ticks -> {{15,30,{45,"n"},60},
{{.001,"0"},.5,{.75,"X"},1}}, TextStyle -> {FontFamily->"Times",
FontSize -> 16}, ImageSize -> {600,400}];
```

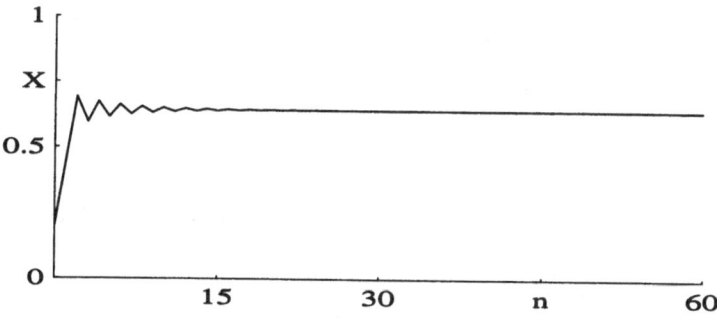

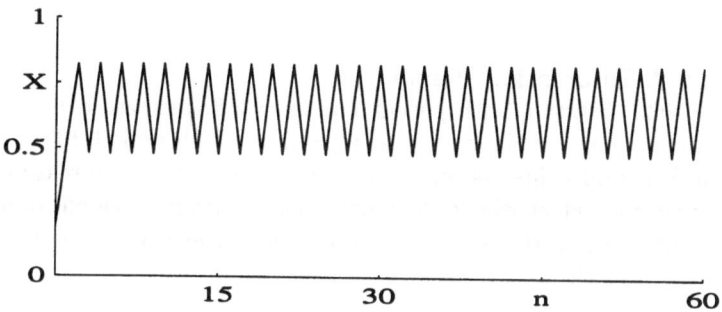

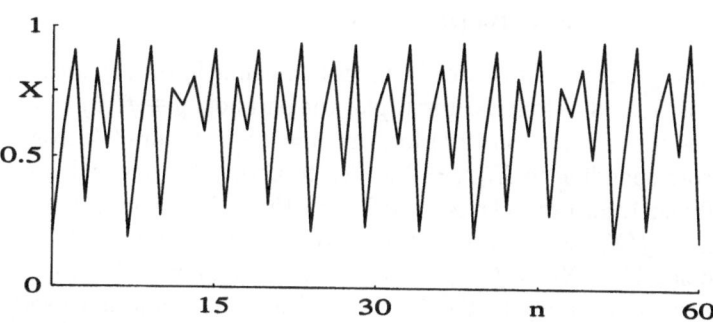

Figure 9.1: Solutions X_n of the logistic map for $X_0 = 0.2$ and (a) (top) $a = 2.8$, (b) (middle) $a = 3.3$, (c) (bottom) $a = 3.8$.

Although the line style was used in Figure 9.1 for easy visualization, it may sometimes cause confusion (e.g., for $a = 3.83$ in Problem 9-1). In such cases, either revert to a point style by removing the `PlotJoined -> True` option, or look directly at the output of the `pts` command line by removing the semi-colon.

End Example 9-1

In Figure 9.1(a), after a short oscillatory transient interval, X_n approaches the constant value 0.642857, this asymptotic number being obtained from the `pts` output. On each successive "time step", the same X_n is produced so that the solution is identified as period 1.

In (b), after a very short transient period, a steady-state periodic solution emerges, oscillating between the two values $X_n = 0.479427$ and 0.823603. Since the repeat time interval for either one of these values is 2, this is a period-2 solution. In the biological context, the population numbers are the same in every second generation.

Finally, in (c), a chaotic solution occurs with no discernible pattern evident, even for very much larger n values than shown here. Our goal is to understand the underlying structure of the logistic equation which gives rise to periodic, chaotic, and other related behaviors that we shall encounter. To this end, it is useful to look at the above results from a geometrical point of view.

PROBLEMS

Problem 9-1: Periodic solutions of the logistic map

Taking $X_0 = 0.2$, identify the periodicity for each of the following a values: a) 3.5, b) 3.56, c) 3.83, d) 3.84.

Problem 9-2: Bifurcation Points

Take $X_0 = 0.2$. Starting with period 1 at $a = 2.8$, increase a and find the approximate a values (to three decimal places) at which period 2 first appears and at which period 4 first occurs. These are examples of bifurcation points for the logistic map which divide the a parameter space into regions of different behavior. Comment on the duration of the transient as each bifurcation point is approached very closely from below.

9.2.2   Geometrical Representation

The so-called geometrical picture of what is happening in the logistic map, as a is increased, is created as follows. First, plot the parabola $y \equiv f(X) = aX(1-X)$, which is the rhs of Equation (9.1) for, say, $a = 2.8$ in Figure 9.2. Next, draw a $45°$ line corresponding to $X_{n+1} = X_n$, i.e., to period 1. Now pick an X_0 value between 0 and 1, for example, $X_0 = 0.2$. Substitute X_0 into (9.1), which yields $X_1 = f(X_0)$. $f(X_0)$ is the intersection of the vertical line from $X_0 = 0.2$ with the parabola. Now $X_1 = f(X_0)$ becomes the new input value to the logistic map so move horizontally to the $45°$ line. Using X_1, again move vertically to the parabola at which point $X_2 = f(X_1) = f(f(X_0)) \equiv f^{(2)}(X_0)$. Repeating this iterative procedure, one eventually approaches an equilibrium or fixed point where the $45°$ line intersects the parabola. At the fixed point, $X_{n+1} = X_n \equiv \tilde{X}_1$

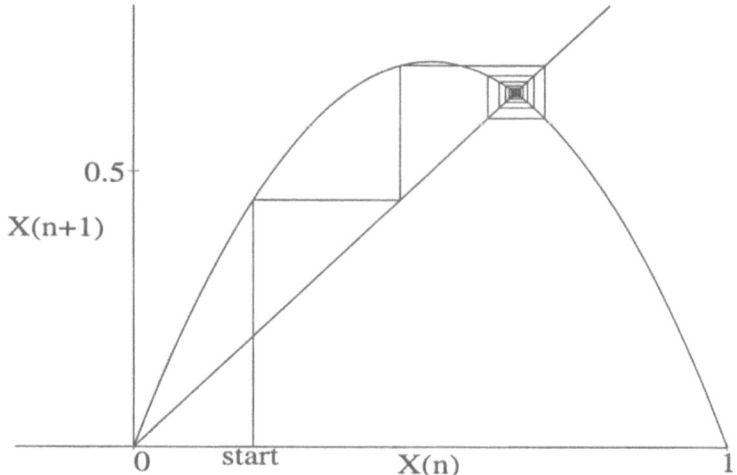

Figure 9.2: Geometric representation of logistic map for $a = 2.8$, $X_0 = 0.2$. The trajectory approaches a fixed point where the $45\,°$ line intersects the parabola.

and period 1 has occurred in agreement with the numerical run. The fixed point here is, of course $\tilde{X}_1 = 0.642857$.

The geometrical construction that has just been outlined, which is sometimes referred to as the "cobweb" diagram, is easily done with Mathematica.

Geometric Representation of The Logistic Map

In this file, the geometric representation of the nonlinear logistic map is generated for any a value between 0 and 4 and X_0 between 0 and 1. Use is made of NestList, which is related to the Nest command. Given the logistic function f[x_]:=a x(1-x), the command Nest[f, x, 3] is equivalent to the operation $f(f(f(x)))$. NestList[f, x, 3] gives a list of the results of applying the function f to x zero through 3 times, i.e., a list of the structure $\{x, f(x), f(f(x)), f(f(f(x)))\}$. Mathematica commands in file: Nest, Flatten, NestList, Partition, Drop, Block, $DisplayFunction=Identity, Table, ListPlot, PlotJoined->True, PlotStyle, Hue, Plot, Show, ImageSize, AspectRatio, Epilog, PlotLabel, PlotRange->All, Ticks, TextStyle

The Mathematica File MF35 was employed to generate Figure 9.2 and the pictures for $a = 3.3$ and 3.8 which follow. Twenty iterations of $f(X)$ were used to produce Figure 9.2.

The corresponding geometric picture for $a = 3.3$ is shown in Figure 9.3 (a). In (a) the "trajectory" is quickly trapped on the rectangle. We know that in this case there are two values between which the steady-state solution oscillates, which will be called the fixed points for period 2. They obviously don't occur at the single nonzero intersection of the $45\,°$ line with the parabola. This is because this geometric picture is suitable for period 1. An appropriate picture is needed for period 2. This is presented in (b).

In (b), the $45\,°$ line ($y = x$) is now interpreted as corresponding to period 2, i.e., $X_{n+2} = X_n$. In addition to plotting the curve $f(X)$, the second-iterate

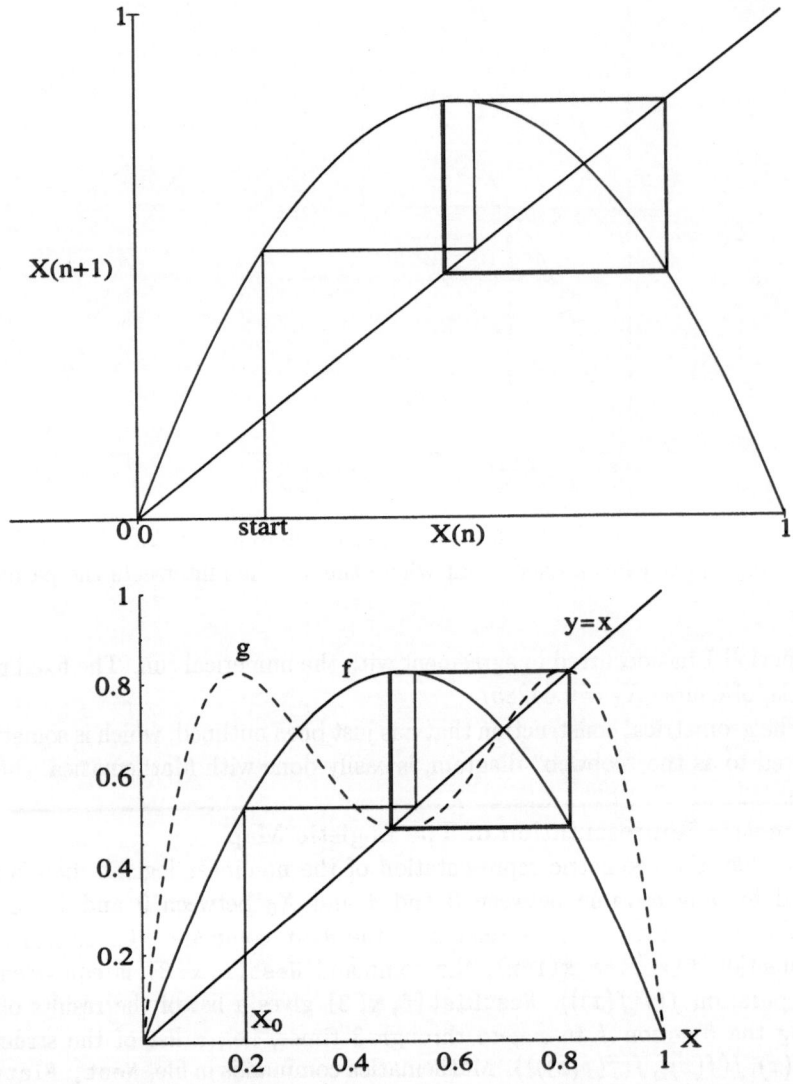

Figure 9.3: Geometric pictures for $X_0 = 0.2$ and $a = 3.3$ (a) (top) using the map $f(X)$, (b) (bottom) using the second-iterate map (dashed curve) $g = f^{(2)}(X)$ to identify the two stable fixed points for period 2.

map $g \equiv f^{(2)}(X) = f(f(X))$, viz., the double-humped dashed curve, has been added. The intersection of the 45° line with g will now yield four fixed points if the origin is included. In steady state, the trajectory cycles around the same rectangle as in (a) which passes through two of the fixed points. These two fixed points are readily identified as being the same two X values that the system oscillated between for period 2 in Figure 9.1 (b). They are referred to as the "stable" fixed points for period 2. The other two fixed points through which the trajectory does not pass are referred to as "unstable fixed points". Stability will be discussed in some depth in the next section.

Example 9-2: Geometric Picture Including 2nd Iterate Map

Verify the geometric plot shown in Figure 9.3(b).

Solution: The values of a, $b \equiv X_0$, and number of iterations are specified.

```
Clear["Global`*"]
```

```
a=3.3; b=0.2; total=60;
```

The functional form of the logistic map is entered, and $f(x)$ calculated.

```
f[x_]:=a x (1-x); f[x]
```

$$3.3\,(1-x)\,x$$

The second iterate map $g(x)$ is created with the **Nest** command.

```
g=Nest[f, x, 2]
```

$$10.89\,(1-x)\,x\,(1-3.3\,(1-x)\,x)$$

To generate the cobweb "trajectory", the **NestList** command is used to apply f repeatedly to the four arguments b. On flattening, a **pts** list results with each entry repeated four times. The first 3 values in this list are then dropped, and the remaining entries partitioned into successive groups of two. The points in **pts2** will produce the cobweb trajectory, when they are joined by straight lines.

```
pts=Flatten[NestList[f, {b, b, b, b}, total]];
```

```
pts2=Partition[Drop[pts, 3], 2];
```

A series of five graphs is now generated. **gr[0]** produces a blue vertical line from $(b, 0)$ to $(b, f(b))$. **gr[1]** joins the points in **pts2** with straight blue lines. **gr2**, **gr[3]**, and **gr[4]** plot $y = x$ (the 45 degree line), $f(x)$, and $g(x)$, respectively. The latter curve is dashed with the **Dashing** option.

```
Block[{$DisplayFunction = Identity},
    gr[0] = ListPlot[{{b,0}, {b,f[b]}}, PlotJoined -> True,
        PlotStyle->{Hue[.6]}];
    gr[1] = ListPlot[pts2, PlotJoined -> True,
        PlotStyle->{Hue[.6]}];
    gr[2] = Plot[x, {x,0,1}, PlotStyle->{Hue[1]}];
    gr[3] = Plot[f[x], {x,0,1}, PlotStyle->{Hue[.3]}];
    gr[4] = Plot[g, {x,0,1}, PlotStyle -> {Dashing[{.02}]}]];
```

The graphs are superimposed with the **Show** command, producing Figure 9.3(b).

```
Show[Table[gr[i],{i,0,4}],AxesLabel -> {"X"," "},PlotRange->
{{0,1},{0,1}},AspectRatio->1,Epilog->{Text["g",{.2,.85},{0,-1}],
```

```
Text["f",{.4,.8},{0,-1}], Text["y=x",{.86,.9},{0,-1}],
Text["X0",{b+.03,.01},{0,-1}], TextStyle->{FontFamily->"Times",
FontSize->16}, ImageSize->{500,500}];
```

$\boxed{\text{End Example 9-2}}$

Finally, to finish off this preliminary discussion, the geometric representation for the chaotic case $a = 3.8$ using the map $f(X)$ is shown in Figure 9.4. The

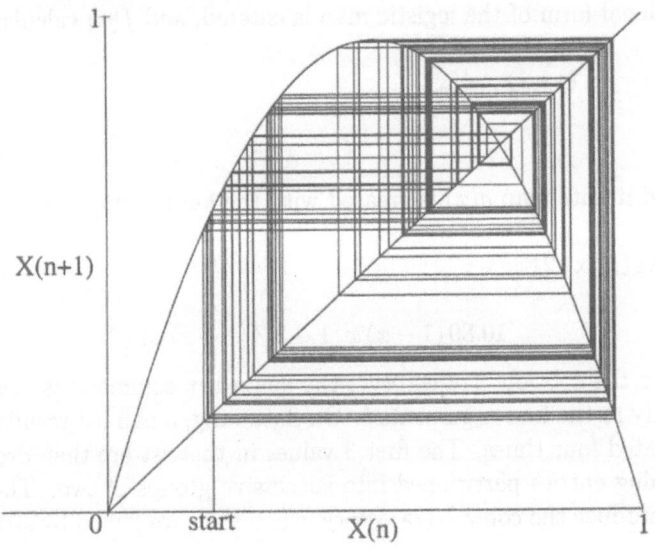

Figure 9.4: Geometric picture for $X_0 = 0.2$ and $a = 3.8$.

trajectory never settles down, continually intersecting the parabola at different locations and gradually filling the enclosed space with new trajectories as the number of iterations is increased.

PROBLEMS

Problem 9-3: Attractive fixed point
Consider $a = 0.8$. Taking $X_0 = 0.1, 0.2, 0.3, ..., 0.9$ and using the geometric representation determine the fixed point to which each trajectory is attracted in each case. Repeat the procedure for $X_0 = 0.2$ and $a = 0.1, 0.2, 0.3, ..., 1.0$. What can you probably conclude?

Problem 9-4: Filling in of chaotic trajectories
Taking $X = 0.2$, $a = 3.8$ and as many iterations as you can, verify that as the number of iterations is increased a progressive filling in of a rectangular region of the first order map $f(X)$ with trajectories takes place.

Problem 9-5: Period four
For $a = 3.53$ and $X_0 = 0.2$ use 100 iterations to locate the steady-state trajectory in the geometric representation. Show that this is period 4 by using the fourth-iterate map. Confirm the four fixed points by directly calculating X_n.

Problem 9-6: Nesting

Create functions for evaluating $\sin(x)$, $\cos(x)$, and then use the `Nest` command to evaluate $\sin(\sin(\sin(\sin(\cos(0.5)))))$. Verify the result by direct calculation.

Problem 9-7: Three iterations

Given $f(x) = ax^3(1-x^2)$, calculate the third iteration of f, i.e., the third-iterate map, by using the `Nest` command. Plot the map for $a = 3$.

Problem 9-8: Cobweb diagram

Taking $a = 1.8$, $X_0 = 0.1$, and 100 iterations, create a cobweb diagram for

$$X_{n+1} = a + X_n - X_n^2.$$

Identify whether the picture represents a periodic or chaotic regime.

Problem 9-9: Cubic map

Taking $a = 2.1$, $X_0 = 0.1$, and 100 iterations, create a cobweb diagram for

$$X_{n+1} = aX_n - X_n^3.$$

By plotting the appropriate iterate map on the same graph, identify the periodicity of the solution.

9.3   Fixed Points and Stability

Fixed point

The ideas that were introduced in the previous section will now be expanded. Two examples of fixed points of the logistic map, namely for period 1 and for period 2, have been encountered. Fixed points are to maps what equilibrium or stationary points are to ODEs. Just as with the latter, the fixed points for maps can be characterized as stable and unstable.

Given a map

$$X_{n+1} = f(X_n) \tag{9.2}$$

the fixed points $\tilde{X}_k$ for period k, with $k = 1, 2, 3, ...$, are obtained by forcing the kth iteration of the map (the kth-iterate map) to return the current value, viz.

$$X_{n+k} = X_n \equiv \tilde{X}_k = f^{(k)}(\tilde{X}_k). \tag{9.3}$$

For the logistic map with $k = 1$, i.e., period 1, we have

$$X_{n+1} = X_n \equiv \tilde{X}_1 = f^{(1)}(\tilde{X}_1) \equiv f(\tilde{X}_1) = a\tilde{X}_1(1 - \tilde{X}_1). \tag{9.4}$$

For the biological problem, this is the condition for zero population growth. Solving for $\tilde{X}_1$ yields $\tilde{X}_1 = 0$ and $\tilde{X}_1 = 1 - 1/a$. Note that the latter fixed point is negative for $a < 1$, zero at $a = 1$ and only becomes positive for $a > 1$. For, say, $a = 2.8$, the non-trivial value of $\tilde{X}_1$ is 0.642857.... The two values of $\tilde{X}_1$ correspond to the two intersections of the 45° line with the parabola in Figure 9.2. With two available fixed points, how does the nonlinear system know which fixed point to evolve toward? It knows about stability, of course!

How does one establish the stability of a fixed point $\tilde{X}_1$? Consider a nearby initial point $X_0 = \tilde{X}_1 + \epsilon$ with ϵ small. A single iteration gives us, on Taylor expanding and keeping first order in ϵ,

$$X_1 = f(X_0) = f(\tilde{X}_1) + \epsilon(df/dX)|_{\tilde{X}_1} = \tilde{X}_1 + (X_0 - \tilde{X}_1)(df/dX)|_{\tilde{X}_1} \quad (9.5)$$

so that

$$|X_1 - \tilde{X}_1| = |X_0 - \tilde{X}_1||(df/dX)|_{\tilde{X}_1}|. \quad (9.6)$$

If $|(df/dX)|_{\tilde{X}_1}| < 1$, then $|X_1 - \tilde{X}_1| < |X_0 - \tilde{X}_1|$. In this case, the iteration has reduced the distance from the fixed point. Since the argument can be repeated, $\tilde{X}_1$ is a stable fixed point if the slope condition $|(df/dX)|_{\tilde{X}_1}| < 1$ holds. On the other hand, if the magnitude of the slope of f is greater than one at the fixed point, the distance to the fixed point increases and the fixed point is unstable. If the magnitude of the slope of f at the fixed point is 1, the fixed point has neutral stability.

The slope condition clearly also applies to the fixed points of higher-iterate maps. To summarize:

A fixed point $\tilde{X}_k$ is stable if the slope $S(\tilde{X}_k)$ of $f^{(k)}(\tilde{X}_k)$ at the fixed point lies between -1 and $+1$, i.e., for an angle between $-45°$ and $+45°$.

Returning to the period 1 case ($k = 1$), the slope at $\tilde{X}_1 = 0$ is $S(0) = a$ and at $\tilde{X}_1 = 1 - 1/a$, $S(1 - 1/a) = 2 - a$. For $a < 1$, there is only one non-negative fixed point located at the origin and it is stable, attracting all trajectories in the interval $0 < X_0 < 1$. Figure 9.5 shows an example of this behavior for $a = 0.9$

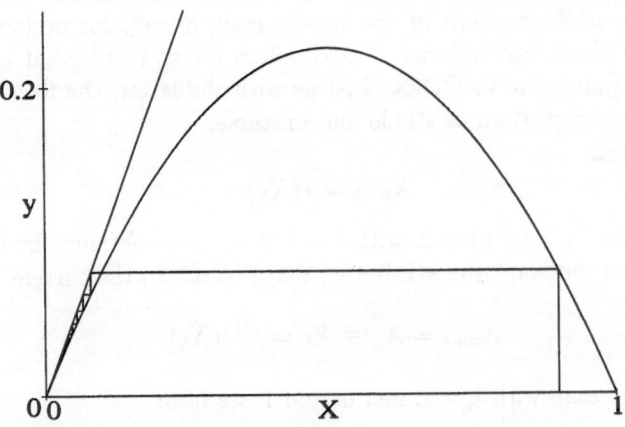

Figure 9.5: Trajectory attracted to fixed point at origin for $a = 0.9 < 1$.

and $X_0 = 0.9$. For $a > 1$, the slope $S(0) > 1$ so the fixed point at the origin is unstable. What about the nontrivial fixed point? For $1 < a < 3$, $S(1 - 1/a) < 1$, so it is a stable fixed point attracting all trajectories in the interval $0 < X_0 < 1$. We have already seen an example for $a = 2.8$ and $X_0 = 0.2$. The student can check that all other X_0 produce trajectories which are attracted to the stable fixed point $1 - 1/2.8 = 0.642857....$ As a passed through 1, the two fixed points exchanged stability. Thus, recalling the discussion of bifurcation values or points in Chapter 4, $a = 1$ is a transcritical bifurcation point. Schematically, using a solid line to denote a stable branch and a dashed line to label an unstable branch of the solution X, a transcritical bifurcation behaves as in Figure 9.6 as a passes through the bifurcation point. The next bifurcation point

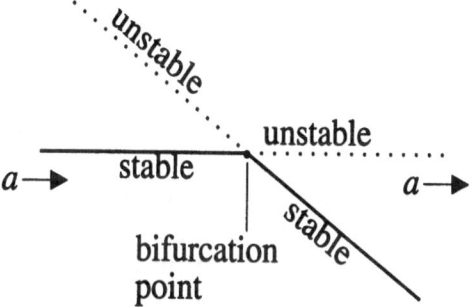

Figure 9.6: Exchange of stability at a transcritical bifurcation point.

occurs at $a = 3$. For $a > 3$, the slope at the nontrivial fixed point has a slope whose magnitude exceeds 1 and thus is unstable. So, both period 1 fixed points are then unstable, so it's not too surprising that period-2 solutions appear as a passes through 3. Mathematicians refer to $a = 3$ as a "flip" bifurcation point. Schematically, a flip bifurcation looks like Figure 9.7. The single stable solution below the bifurcation point loses its stability at the bifurcation point with the birth of a period doubled solution. In the period-2 regime, one has the following relation between the two steady-state values p and q, viz., $q = ap(1 - p)$ and $p = aq(1 - q)$ and hence the origin of the name flip bifurcation. Because of its shape in Figure 9.7, the flip bifurcation is also often referred to as a pitchfork bifurcation. Pitchfork bifurcations were discussed in Chapter 4.

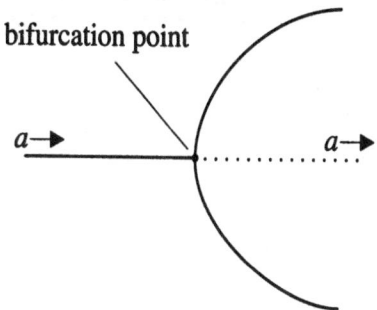

Figure 9.7: Behavior of solution as bifurcation parameter a passes through a flip (pitchfork) bifurcation point. Solid line: stable. Dashed line: unstable.

Returning to Figure 9.3(b), which shows the second-iterate map for $a = 3.3$, the behavior of the period-2 trajectory can be understood. At two of the four fixed points, the magnitude of the slope is less than 1 (i.e., less than the 45° line) so they are stable while at the other two it is greater than 1 so those fixed points are unstable. In steady-state, the period-2 trajectory "cycles" between the two stable fixed points. The two unstable fixed points are simply those that occurred for period 1, viz. at 0 and, for $a = 3.3$, at $1 - 1/a = 1 - 1/3.3 = 0.696969....$

PROBLEMS
Problem 9-10: 3rd and 4th-iterate maps
By forming 2nd-, 3rd- and 4th-iterate maps in the geometric representation, show in terms of slopes that as a is increased from 3.4 that period 4 is "born" from period 2, not period 3.

9.4   The Period-Doubling Cascade to Chaos

As a is further increased from 3.3, the two stable points in the second-iterate map go unstable at the flip bifurcation point $a = 3.449490$ and simultaneously four stable points appear in the fourth-iterate map. To see the latter, modify the `Nest` command line in Example 9-2 to read `g=Nest[f,x,4]`. Figure 9.8 shows the approach of a trajectory starting at $X_0 = 0.2$ to the steady-state period-4 solution for $a = 3.46$. After a few transient legs, the trajectory winds onto the heavy rectangles. The four stable fixed points are easy to spot, corresponding to the four intersections of the fourth-iterate (dashed) curve with the 45° line where the slope of the former is less than 45°. In steady-state, $X_{n+4} = X_n$.

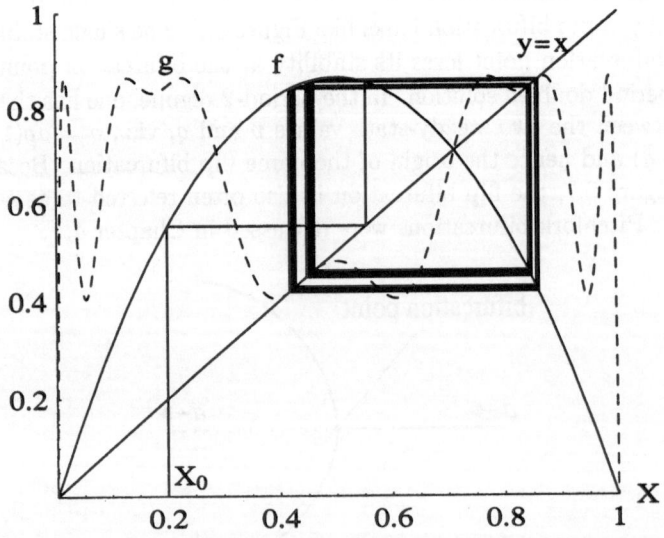

Figure 9.8: The curve g is the 4th-iterate map for $a = 3.46$. The cobweb trajectory achieves steady-state ($X_{n+4} = X_n$) "cycling" through the four stable fixed points which lie at the intersection of the 45° line and curve g.

As a is further increased, an infinite sequence of flip bifurcations takes place with period 2^k solutions (2^k cycles, $k = 1, 2, ...$) being born. The first eight flip bifurcation points a_k are $a_1 = 3$, $a_2 = 3.449490$, $a_3 = 3.544090$, $a_4 = 3.564407$, $a_5 = 3.568759$, $a_6 = 3.569692$, $a_7 = 3.569891$, and $a_8 = 3.569934$. This sequence converges to a limit at $a_\infty = 3.569946...$ according to the formula

$$a_k \approx a_\infty - C/\delta^k \qquad (9.7)$$

where $C = 2.6327...$ and $\delta = 4.669201609.....$ δ is called the Feigenbaum constant or number. Feigenbaum discovered that the constant δ is a universal property of the period doubling route to chaos for maps that are "unimodal" (smooth, concave downward, and having a single maximum), i.e., the number is independent of the specific form of the map. At a_∞ chaos sets in. As discussed in detail in the next section, the period doubling scenario has been confirmed experimentally and the bifurcation points used to determine the Feigenbaum constant through the easily derived relation

$$\delta = \lim_{k \to \infty} \frac{a_k - a_{k-1}}{a_{k+1} - a_k}. \qquad (9.8)$$

The experimental values of δ show a surprising degree of agreement with the value obtained from the logistic map [Cvi84].

A bifurcation diagram can be used to illustrate the period doubling route to chaos. The next Mathematica File, MF36, is used to generate the bifurcation diagram shown in Figure 9.9 for the logistic map. The bifurcation diagram shows clearly the period doubling route to chaos for the logistic equation. For the range between a_∞ and 4 there are small "windows" of periodicity interspersing the chaos. One such window is a period-3 window stretching from $a = 3.828427$ to 3.841499. Can you spot it?

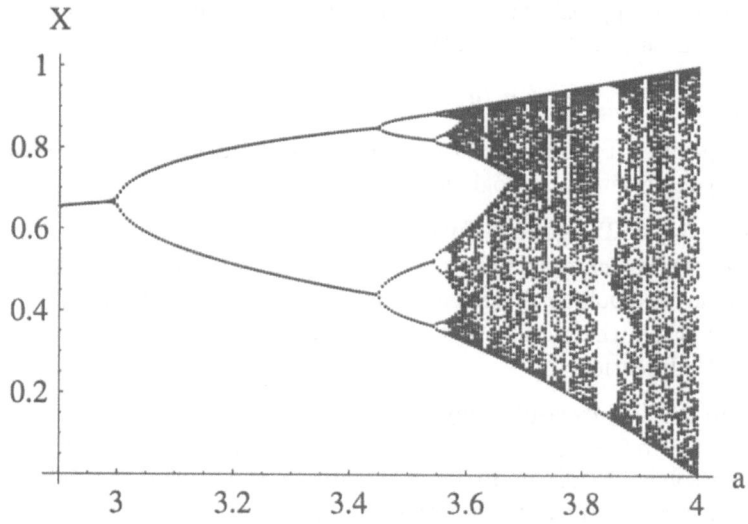

Figure 9.9: Bifurcation diagram for logistic model for $a = 2.9, ..., 4.0$.

The bifurcation point at which period 3 sets in is an example of a tangent bifurcation, the origin of the name being left as part of a problem.

Bifurcation Diagram for the Logistic Map
This file plots the steady-state values (fixed points) X for $X_0 = 0.2$ and $a = 2.9, ..., 4.0$ for the logistic map. The a range is subdivided into 200 equally spaced intervals. For each a value the first 250 transient points are dropped and 150 points plotted. If the reader wishes to look at the fine structure in the map, the a range can be reduced and the program rerun with the new smaller range subdivided into even smaller intervals. Mathematica commands in file: `Compile`, `Map`, `Union`, `Drop`, `NestList`, `Flatten`, `Table`, `ListPlot`, `AxesOrigin`, `PlotStyle`, `AxesLabel`, `AbsolutePointSize`, `TextStyle`, `Hue`, `PlotRange->All`, `ImageSize`

The period doubling route to chaos may be explored in the following two experimental activities.

Route to Chaos
Period doubling and chaos is observed in the oscillations of a small cylindrical bar magnet suspended in a time varying magnetic field.

Driven Spin Toy
A pendulum-like oscillator is pumped by an oscillating magnetic field. The optimum pumping frequencies are investigated for various angular amplitudes. The period doubling route to chaos can be qualitatively explored.

PROBLEMS
Problem 9-11: The Feigenbaum constant
Show from the approximate formula in the text for a_k that the Feigenbaum constant can be calculated from the ratio

$$\delta = (a_k - a_{k-1})/(a_{k+1} - a_k).$$

Strictly speaking it is defined through this ratio in the limit that $k \rightarrow \infty$. What value do you obtain in the logistic model for δ for $k = 7$?

Problem 9-12: Tangent Bifurcation
Obtain the third-iterate map for $a = 3.829$ and $X_0 = 0.2$ and show that there are three stable fixed points. A tangent bifurcation takes place at this (approximate) value of a. From the picture that you generate, discuss the origin of the name "tangent" bifurcation.

Problem 9-13: Intermittency
Intermittency refers to almost periodic behavior interspersed with bursts of chaos. By solving for X_n up to $n = 300$, with $X_0 = 0.5$, show that intermittency occurs for $a = 3.82812$ which is just below the onset of a period-3 window.

Problem 9-14: Period 3 to chaos
Making use of the bifurcation Mathematica File MF36, examine and discuss

the route to chaos starting with the period-3 solution at $a = 3.835$. Hint: You should change your field of view to concentrate on the middle of the three dotted lines and limit the a range. How many period doublings can you find?

Problem 9-15: Flip bifurcation point
What periodic solution begins at the flip bifurcation point a_7? Confirm your answer by calculating X_n just above a_7 and printing out the numbers.

Problem 9-16: The tent map
Derive the bifurcation diagram for the tent map

$$X_{n+1} = 2aX_n, \quad 0 < X \leq 1/2$$

$$X_{n+1} = 2a(1 - X_n), \quad 1/2 \leq X < 1$$

with $0 < a < 1$. Take $X_0 = 0.2$ and 0.6. Summarize the periodic solutions that you see in each case. Are there any differences in the bifurcation diagrams for the two inputs?

Problem 9-17: The sine map
Derive the bifurcation diagram for the sine map

$$X_{n+1} = a\sin(\pi X_n)$$

with $0 \leq a \leq 1$ and $0 \leq X \leq 1$. Is the sine map unimodal? How does the bifurcation diagram qualitatively compare with that for the logistic map if only the range $a = 0.7$ to 1 is plotted?

Problem 9-18: Some miscellaneous maps
Derive the bifurcation diagrams for the following maps over the ranges of a indicated, taking $X_0 = 0.1$ and an appropriate number of a steps:

  a. $X_{n+1} = a + X_n - X_n^2$, $a = 0.5...1.8$,

  b. $X_{n+1} = -(1 + a)X_n + X_n^3$, $a = 0.5...1.5$,

  c. $X_{n+1} = aX_n - X_n^3$, $a = 1.5..2.5$.

Discuss the graphs obtained, particularly determining the ranges over which different periodicities occur and locating the a value at which chaos sets in.

9.5 Period Doubling in the Real World

For the logistic map, we have been able to understand the period doubling route to chaos. What has period doubling in the logistic map to do with the period doubling that was seen in the last chapter for the forced Duffing oscillator or for the Rössler system, much less period doubling that has been observed experimentally? The forced Duffing equation and the Rössler system, which are 3-dimensional in nature and continuous in time, do not appear to have much in common with the 1-dimensional logistic map. Even more amazing, as the following Table 9.1 demonstrates, period doubling has been observed in

Experiment [Reference]	Period Doublings Observed	δ
Hydrodynamic: Water [GMP81] Mercury [LLF82]	4 4	4.3 ± 0.8 4.4 ± 0.1
Electronic: Diode [Lin81] Transistor [AL82] Josephson [YK82]	4 4 4	4.5 ± 0.6 4.7 ± 0.3 4.5 ± 0.3
Laser feedback [Cvi84]	3	4.3 ± 0.3
Acoustic: Helium [Cvi84]	3	4.8 ± 0.6

Table 9.1: Observed period doublings and Feigenbaum constant δ.

real experiments and the Feigenbaum number which has been extracted from the experimental bifurcation points shows quite reasonable agreement with the value of $\delta = 4.669...$ obtained from the logistic map. How can this be so?

Before answering this question, a few words should be said about one of the experiments in the table, because it ties in nicely with the Rayleigh–Bénard convection that was discussed in Section 3.1.3. In the experiment by Libchaber [LLF82] and his co-workers, a box of liquid mercury was heated from below and a temperature gradient established through the thickness of mercury. A dimensionless measure of the temperature gradient is the Rayleigh number R_a which is defined in standard texts on fluid mechanics. For R_a less than a critical value, $R_{critical}$, the fluid remained motionless and the heat flow up through the mercury was via conduction. For $R_a > R_{critical}$, convection occurred and cylindrical rolls formed as schematically illustrated in Figure 3.7. The rolls were stabilized by the application of an external DC magnetic field, the rolls tending to align along the magnetic field direction. For R_a slightly above $R_{critical}$, the temperature at a fixed point on a roll was observed to be constant. As R_a was further increased, another instability occurred with a wave propagating along the roll leading to a temperature variation at the point being monitored. As the temperature gradient (i.e., R_a) was further increased the recorded temperature variations as a function of time indicated a series of period doublings from which the Feigenbaum constant was extracted. In this and the other experiments cited in the table, it was difficult to obtain accurate data for more than about four period doublings. Given that the Feigenbaum constant is strictly defined in the limit of an infinite number of period doublings, the agreement of the experimentally obtained values with that obtained from the logistic map is rather amazing.

As stated earlier, why should there be any agreement at all? A rigorous mathematical answer has been given by Feigenbaum [Fei79], [Fei80] involving the concept of renormalization which is discussed in advanced courses in statistical mechanics. Rather than present this more advanced treatment, we shall opt for an intuitively plausible explanation based on an approach due to Lorenz. Let us consider the Rössler system of the last chapter and try to understand how the period doubling sequence and the Feigenbaum constant for this continuous time 3-dimensional system could possibly be the same as for the discrete time 1-dimensional logistic map. For the Rössler attractor which occurred at $c = 5.0$, the behavior of the variable $x(t)$ was shown in Figure 8.32. Suppose that the maxima of the nth and $n + 1$st oscillations are labeled as x_n and x_{n+1}, respectively, and x_{n+1} is plotted versus x_n for a large number of consecutive n values. Using the following file MF37 this calculation has been carried out and the results presented in Figure 9.10. The data points lie on a 1-dimensional

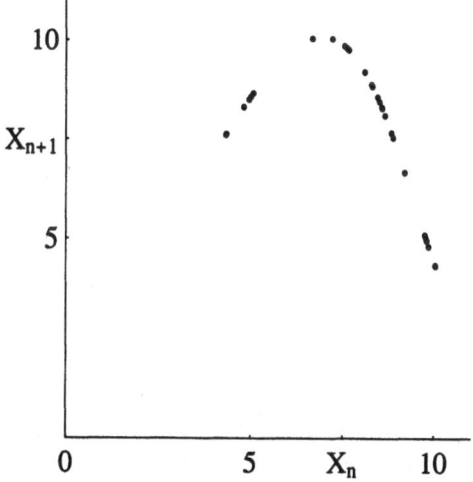

Figure 9.10: Lorenz map for the Rössler attractor.

parabolic-appearing curve if the very small "thickness" in the graph is neglected. So there exists, at least approximately, a functional relationship between x_{n+1} and x_n, viz. $x_{n+1} = f(x_n)$ where the function f is similar in structure to that for the logistic map. A 1-dimensional unimodal map has been extracted! A map constructed in the way just outlined is called a "Lorenz map".

The Lorenz Map

This file demonstrates how to obtain a Lorenz map from a system of three coupled nonlinear ODEs. The Rössler system is taken as the illustrative example. Mathematica commands in file: NDSolve, Method->RungeKutta, MaxSteps, Plot, Evaluate, Ticks, AxesLabel, PlotPoints, PlotStyle, ListPlot, ImageSize, Sign, Partition, Flatten, Table, Evaluate, Dimensions, TextStyle, Cases, AspectRatio, PlotRange, AbsolutePointSize, Hue

Because the Lorenz map for the chaotic Rössler attractor is unimodal, it then follows that the period doubling sequence leading to the attractor and the Feigenbaum constant are the same as for the logistic map. Without going into the details of the experiments, the experiments cited in the table were such that the governing differential equations could be, at least approximately, reduced to 1-dimensional unimodal maps.

PROBLEMS
Problem 9-19: The Lorenz attractor
Derive the Lorenz map for the Lorenz attractor which occurs at $\sigma = 10$, $b = 8/3$, and $r = 28$. Make use of the maxima in the $z(t)$ oscillations as was first done by Lorenz and show that the Lorenz map resembles the tent map (Problem 9.16). Is the Lorenz map unimodal?

Problem 9-20: The universal sequence
According to a theorem due to Metropolis, Stein and Stein [MSS73], for all unimodal maps of the form $X_{n+1} = af(X_n)$, with $f(0) = f(1) = 0$ and a a positive parameter, the order in which stable periodic solutions appear is independent of the detailed structure of the map. The "universal" sequence will consist of period 1, followed by period 2, period 4, period 6 (omitting the higher order period 8, 16, ...), period 5, period 3, ... Check this sequence for the logistic and sine maps by using the relevant bifurcation diagrams. Start with an a value well below the first chaotic region and work up through several of the periodic windows.

Problem 9-21: Literature search
Go to your college or university library and look up one of the experiments listed in Table 9.1. Describe the experiment in your own words and, if possible, discuss qualitatively how the underlying physical equations are reducible to a 1-dimensional unimodal map.

Problem 9-22: Intermittency in the Lorenz system
Intermittency (almost-periodic behavior interspersed with bursts of chaos) is known to occur for the logistic map so it should not prove too surprising after the discussion of this section that it might also occur in the Rössler and, perhaps, the Lorenz systems. Taking the standard values for σ, b, show that periodic behavior occurs in the Lorenz system for $r = 166.0$ and intermittency for $r = 166.2$.

9.6   The Lyapunov Exponent

The Lyapunov exponent[1] λ, of a map is used to obtain a measure of the very sensitive dependence upon initial conditions that is characteristic of chaotic behavior. Consider a general 1-dimensional map

$$X_{n+1} = f(X_n). \tag{9.9}$$

Let X_0 and Y_0 be two nearby initial points in the phase space and consider n iterations with the map to form

$$X_n = f^{(n)}(X_0)$$
$$\tag{9.10}$$
$$Y_n = f^{(n)}(Y_0).$$

---

[1]Named after the Russian mathematician, Alexander M. Lyapunov (1857–1918).

Lyapunov Exponents

For a chaotic situation, nearby initial points will rapidly separate, while for a periodic solution the opposite will occur. Therefore assume, for large n, an approximately exponential dependence on n of the separation distance, viz.,

$$|X_n - Y_n| = |X_0 - Y_0| e^{\lambda n} \tag{9.11}$$

with $\lambda > 0$ for the chaotic situation and $\lambda < 0$ for the periodic case. Taking n large (limit as $n \to \infty$), λ can be extracted from (9.11)

$$\lambda = \lim_{n \to \infty} \frac{1}{n} \ln \left| \frac{X_n - Y_n}{X_0 - Y_0} \right|. \tag{9.12}$$

However, for trajectories confined to a bounded region such as in our logistic map, such exponential separation for the chaotic case cannot occur for very large n, unless the initial points X_0 and Y_0 are very close. Therefore, the limit $|X_0 - Y_0| \to 0$ must also be taken.

Modifying (9.12), we have

$$\begin{aligned} \lambda &= \lim_{n \to \infty} \frac{1}{n} \lim_{|X_0-Y_0| \to 0} \ln \left| \frac{X_n - Y_n}{X_0 - Y_0} \right| \\ &= \lim_{n \to \infty} \frac{1}{n} \lim_{|X_0-Y_0| \to 0} \ln \left| \frac{f^{(n)}(X_0) - f^{(n)}(Y_0)}{X_0 - Y_0} \right| \end{aligned} \tag{9.13}$$

or

$$\lambda = \lim_{n \to \infty} \frac{1}{n} \ln \left| \frac{df^{(n)}(X_0)}{dX_0} \right|. \tag{9.14}$$

Now, $f(X_0) = X_1$, $f(X_1) = X_2$ or $f^{(2)}(X_0) = X_2$, so that for example

$$\frac{df^{(2)}(X_0)}{dX_0} = \frac{df(X_1)}{dX_1} \frac{dX_1}{dX_0} = \frac{df(X_1)}{dX_1} \frac{df(X_0)}{dX_0}. \tag{9.15}$$

Generalizing (9.15), we have

$$\frac{df^{(n)}(X_0)}{dX_0} = \prod_{k=0}^{n-1} \frac{df(X_k)}{dX_k} \tag{9.16}$$

and the Lyapunov exponent λ is given by

$$\lambda = \lim_{n\to\infty} \frac{1}{n} \sum_{k=0}^{n-1} \ln \left| \frac{df(X_k)}{dX_k} \right|. \tag{9.17}$$

For periodic solutions, which starting point X_0 is chosen doesn't matter, but for chaotic trajectories, the precise value of λ will depend on X_0, i.e., in general $\lambda = \lambda(X_0)$. One can, if desired, define an average λ, averaged over all starting points. Whether this is done or not, $\lambda > 0$ should correspond to chaos, $\lambda < 0$ to periodic behavior. Figure 9.11 shows λ (vertical axis) as a function of a for the logistic map for the starting point $X_0 = 0.2$. The figure was generated using the next Mathematica File, MF38.

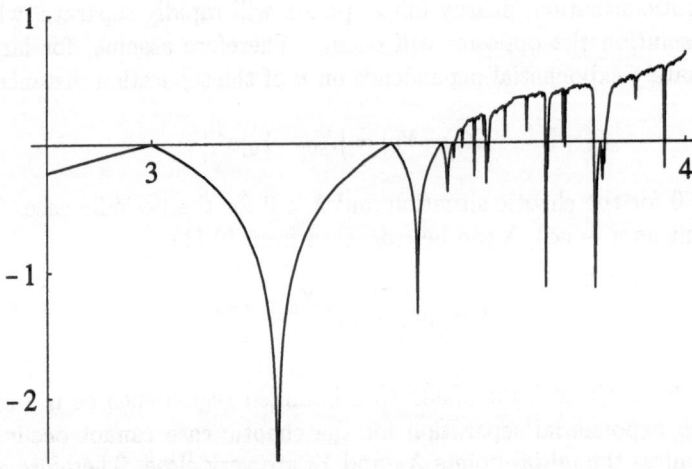

Figure 9.11: Lyapunov exponent λ (vertical axis) as a function of the parameter a (horizontal axis) for the initial point $X_0 = 0.2$.

Lyapunov Exponent for the Logistic Map
The Lyapunov exponent $\lambda(X_0)$ is calculated as a function of a for a specified starting point X_0. Mathematica commands in file: `Compile`, `_Integer`, `With`, `NestList`, `#`, `Apply`, `Plus`, `Log`, `Abs`, `Plot`, `PlotRange`, `AxesOrigin`, `Ticks`, `PlotLabel`, `StyleForm`, `PlotStyle`, `Hue`, `TextStyle`, `ImageSize`

Examining the Lyapunov exponent figure, the reader will find that the regions where $\lambda < 0$ do indeed correspond to the periodic regions in our earlier bifurcation diagram and that $\lambda > 0$ indicates the occurrence of chaos. The Lyapunov exponent concept is readily applied to other maps and can be extended to continuous time nonlinear dynamical systems like those in the last chapter.

PROBLEMS

Problem 9-23: Comparison of Lyapunov exponent and bifurcation diagrams

Compare Figure 9.11 with the bifurcation diagram (Figure 9.9) and show that there is good agreement between the two plots. Particularly check the narrow periodic windows interspersed in the chaotic region of a above a_∞.

Problem 9-24: Lyapunov exponent for the tent map

Analytically prove that the Lyapunov exponent λ for the tent map

$$X_{n+1} = 2aX_n, \quad 0 < X \leq 1/2$$

$$X_{n+1} = 2a(1 - X_n), \quad 1/2 \leq X < 1$$

with $0 < a < 1$ is $\lambda = \ln(2a)$. Determine the regions of a for which periodic and chaotic solutions occur.

Problem 9-25: Lyapunov exponent for the sine map

Determine the Lyapunov exponent as a function of the parameter a $(0 \leq a \leq 1)$ for the sine map $X_{n+1} = a \sin(\pi X_n)$ with $0 \leq X \leq 1$. Discuss your result.

Problem 9-26: Some miscellaneous maps

Derive the Lyapunov exponent as a function of a for the following maps over the ranges of a indicated, taking $X_0 = 0.1$:

a. $X_{n+1} = a + X_n - X_n^2, \quad a = 0.5...1.8,$

b. $X_{n+1} = -(1 + a)X_n + X_n^3, \quad a = 0.5...1.5,$

c. $X_{n+1} = aX_n - X_n^3, \quad a = 1.5..2.5.$

Discuss the graphs obtained, particularly determining the ranges over which different periodicities occur and locating the approximate a value at which chaos sets in.

9.7 Stretching and Folding

For the 1-dimensional logistic map, the 1-dimensional phase space is limited to between 0 and 1. For two neighboring inputs, there will be an exponential divergence of trajectories when chaos prevails. Yet, the trajectories are known to remain in the phase space for $0 < a \leq 4$. This is accomplished by the process of "stretching and folding". This may be easily seen for $a = 4$ where chaos is known to prevail. In this case the logistic map becomes

$$X_{n+1} = 4X_n(1 - X_n). \tag{9.18}$$

The allowable X_n values can be represented by a straight line from 0 to 1 (Figure 9.12). Now calculate X_{n+1} using Equation 9.18. The values $0 \leftrightarrow 1/2$ map into $0 \leftrightarrow 1$. This is the process of (nonuniform) stretching. The values between $1/2 \leftrightarrow 1$ map into $1 \leftrightarrow 0$, i.e., onto the first stretched interval in

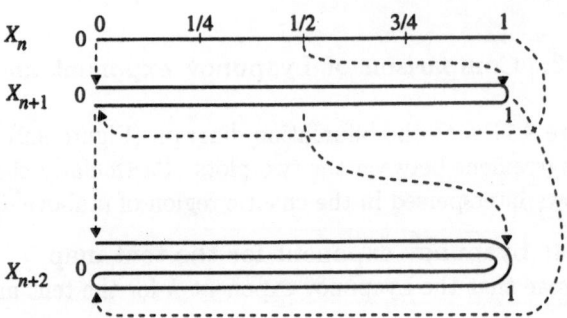

Figure 9.12: Illustration of stretching and folding for logistic map for $a = 4.0$.

reverse order. This is the folding. Repeating the process to get X_{n+2} yields the next sequence in the figure, and so on. The process of stretching and folding is reminiscent of the kneading of dough when making bread. In the case of the 1-dimensional logistic map the kneaded "dough" has no "thickness" since the stretching and folding is along a line. One needs a transverse dimension to acquire thickness and this can be attained through the use of a 2-dimensional map.

As an example of a simple map which includes this feature, the French astronomer Michel Hénon has suggested [Hé76] the following map, which bears his name:

$$X_{n+1} = Y_n + 1 - aX_n^2$$
$$Y_{n+1} = bX_n$$

(9.19)

with $n = 0, 1, 2, \ldots$ and a and b real, positive, constants. The Mathematica code given earlier for the logistic map is easily adapted to handle the Hénon map

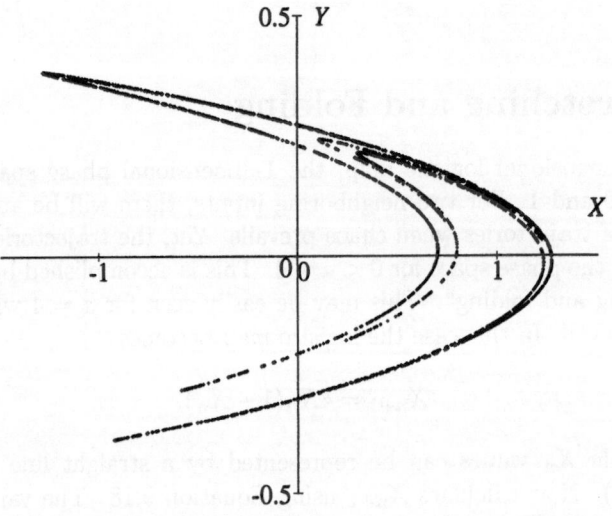

Figure 9.13: The Hénon strange attractor illustrating folding.

and is left as a problem. Alternately, one can modify Example 3-5. Choosing, say, $a = 1.4$, $b = 0.3$, and $X_0 = Y_0 = 0$, then 3000 iterations gives the strange attractor[2], known as the Hénon attractor, shown in Figure 9.13. The first 100 iterations have not been plotted. The folding can be clearly seen in the figure.

To see stretching as well as folding, one can generalize what was done conceptually above for the logistic map where successive applications of the map to the points along a line were looked at. Instead of choosing a single starting point in the 2-dimensional space, consider the square whose boundary is shown on the left of Figure 9.14. One iteration of the Hénon map produces the distorted boundary in the middle figure (note the scale change), while two iterations produces the dramatically stretched and folded figure on the far right.

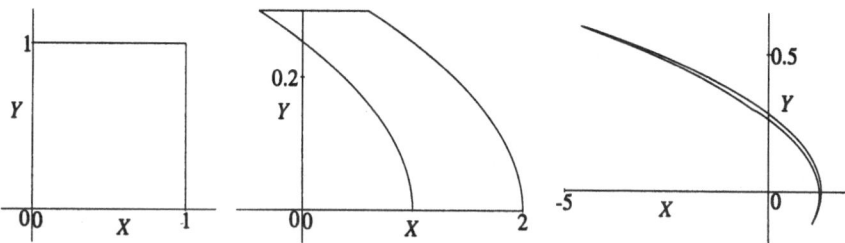

Figure 9.14: Stretching and folding of a square boundary by two successive applications of the Hénon map.

Stretching and folding are also characteristic of chaotic attractors seen in forced oscillator systems. For example, strikingly similar structure to that in

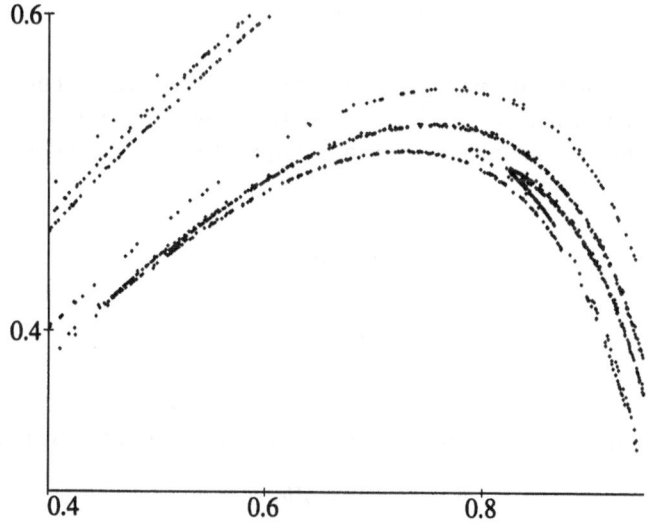

Figure 9.15: Portion of a Duffing strange attractor showing stretching and folding similar to that for the Hénon attractor.

---

[2]Using the box counting technique, it has been established [PJS92] that the Hénon attractor has a fractal dimension about 1.28 and thus is indeed a strange attractor.

Figure 9.13 is seen in Figure 9.15 for the forced inverted Duffing oscillator with $\gamma = 0.125$, $\alpha = -1$, $\beta = 1$, $F = 0.3$, $\omega = 1$, $x(0) = 1$, $y(0) \equiv \dot{x}(0) = 0.1$, and $\phi = 0$. Figure 9.15, with y plotted vertically and x horizontally, shows only a portion of the complete strange attractor which the student was asked to investigate in Problem 8-28 (page 334) dealing with Cantor-like structure. If you didn't do that particular problem, it really is worth going back and solving it to see the entire strange attractor.

PROBLEMS
Problem 9-27: Generating the Hénon map
Adapting the Mathematica code given in the text for the logistic map or modifying Example 3-5, generate the Hénon attractor shown in Figure 9.13.

Problem 9-28: Stretching and folding of a square
Verify Figure 9.14 and show that further applications of the Hénon map yield the Hénon strange attractor. Take the sides of the square to be less than 1.

9.8   The Circle Map

In our introduction to the Poincaré section concept in Chapter 3, it was noted that for second order forced ODEs such as the simple pendulum or Duffing's equation, there must exist some relation between the phase plane coordinates at the end of the $(n + 1)$st driving period and the coordinates at the end of the previous (nth) period, i.e.,

$$X_{n+1} = f_1(X_n, Y_n)$$
$$Y_{n+1} = f_2(X_n, Y_n) \tag{9.20}$$

where the f_i are nonlinear functions whose analytic forms are difficult or impossible to find. Sometimes, approximate analytic forms can be found valid over some limited range of parameters. As an example of how this is done, let's look at the simple pendulum with $\omega = \dot{\theta}$,

$$\theta_{n+1} = f_1(\theta_n, \omega_n)$$
$$\omega_{n+1} = f_2(\theta_n, \omega_n). \tag{9.21}$$

This 2-dimensional Poincaré map reduces to a 1-dimensional map if, after the transients have died away, $\omega_n = g(\theta_n)$, i.e., some function of θ_n. Then,

$$\theta_{n+1} = f(\theta_n, g(\theta_n)). \tag{9.22}$$

So what should one choose for the functional form of the rhs? Jensen, Bak, and Bohr [JBB84] were able to explain certain features observed for the forced pendulum by considering the two parameter "circle" map

$$\phi_{n+1} = \phi_n + \Phi - K \sin(\phi_n) \tag{9.23}$$

or, on setting $\phi_n = 2\pi\theta_n$ and $\Phi = 2\pi\Omega$,

$$\theta_{n+1} = \theta_n + \Omega - \frac{K}{2\pi}\sin(2\pi\theta_n). \tag{9.24}$$

Here Ω and K are real, positive, parameters and periodic boundary conditions are assumed in plotting the map with the angular coordinate θ_n restricted to the range 0 to 1. That is, when θ_n exceeds unity, 1 is subtracted from θ_n to keep it in the range 0 to 1.

One of the most interesting real world applications of the circle map has been the modeling of the heartbeat by Glass and coworkers [GP82], [GSB86].

To see what type of phenomena the circle map allows, use is made of the circle map Mathematica File MF39 which permits us to plot a geometrical picture similar to the cobweb picture for the logistic map.

The Circle Map: Geometric Representation

Although this file is similar to that for the logistic map, it introduces the Mathematica command `FractionalPart[x]` which is used to keep the trajectory in the range 0 to 1 by subtracting the integer part from the generated numbers. For example, `FractionalPart[2.163]` yields 0.163 as its output. In the mathematics literature, the fractional part of x is often written as x mod 1. Mathematica commands in file: `N`, `Sin`, `Flatten`, `NestList`, `Partition`, `Drop`, `FractionalPart`, `Block`, `$DisplayFunction=Identity`, `ListPlot`, `Table`, `PlotJoined->True`, `PlotStyle`, `Hue`, `Plot`, `Show`, `PlotLabel`, `StyleForm`, `AspectRatio`, `PlotRange`, `Ticks`, `TextStyle`, `ImageSize`

To start out, switch off the nonlinearity by setting the parameter $K = 0$ and choosing $\Omega = 0.4$. Take the initial value $\theta_0 = 0.2675$ and consider five iterations. The resulting trajectory is shown in Figure 9.16 produced from the file. The 45° line through the origin corresponds, of course, to $\theta_{n+1} = \theta_n$. The other two straight lines correspond to the rhs of the circle map for $K = 0$, the bottom line being obtained from the top line by subtracting 1. For five

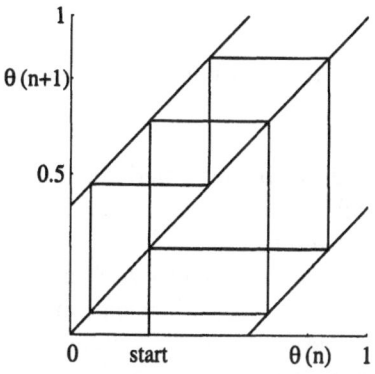

Figure 9.16: Five iterations of the linearized ($K = 0$) circle map. The horizontal axis is θ_n, the vertical axis θ_{n+1}.

iterations, exactly two complete revolutions have been completed with θ back at 0.2675. The "winding number" W for the circle map is defined as

$$W = \lim_{n \to \infty} \frac{(\theta_n - \theta_0)}{n} \tag{9.25}$$

where θ_n refers to the actual values without the integers having been subtracted. These values of θ_n may be extracted from the file by removing the semicolon from the pts command line. For our example we obtain, starting with $n = 1, 2, 3, ...,$ $(0.6675 - 0.2675)/1 = 0.40,\ (1.0675 - 0.2675)/2 = 0.40,\ (1.4675 - 0.2675)/3 = 0.40,....$ The winding number is clearly $W = 0.40$. In this case the winding number is a rational number, since $0.40 = 2/5$. Further, the solution is periodic. Rational winding numbers correspond to periodic solutions. For the linear case, the winding number is equal to the value of the parameter Ω. For the nonlinear $(K \neq 0)$ situation, this is not the case. For our example, the origin of the name winding number is easily seen. There are two complete revolutions of the trajectory in five iterations, and the ratio of these two numbers is 0.4, which is just the value of W.

If, following the choice of Baker and Gollub [BG90], the irrational number $\Omega = 0.4040040004...$ is chosen in the linear case, then the winding number is also irrational. The trajectory does not quite close on successive revolutions as shown in Figure 9.17 generated from the file MF39 with 100 iterations. The

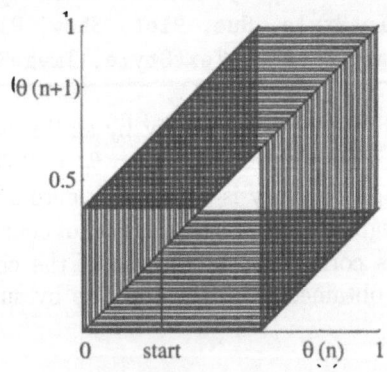

Figure 9.17: Quasi-periodic trajectory for an irrational winding number in the linearized circle map.

trajectory in this case is said to be quasi-periodic. As the number of iterations is increased the diagram fills in and the trajectory comes arbitrarily close to any specified value in the interval 0 to 1.

Now, return to the periodic case and switch on the nonlinearity. In Figure 9.18 we have $\theta_0 = 0.2675$, $\Omega = 0.40$, $K = 0.95$ and have taken 13 iterations. The motion is again quasi-periodic and the winding number is irrational. Note how the nonlinearity has curved the previously straight outer lines between which the trajectory cycles. Periodicity, characterized by the rational winding number $W = 0.40$ can be regained here by slightly adjusting Ω to the irrational number 0.4040040004.... Figure 9.19 demonstrates this, 13 iterations having

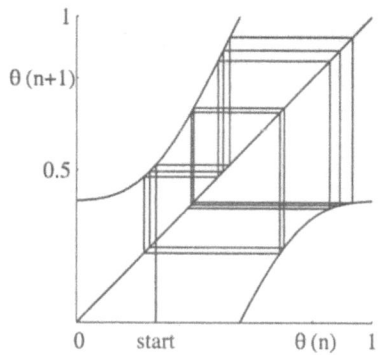

Figure 9.18: Quasi-periodic trajectory for the nonlinear circle map.

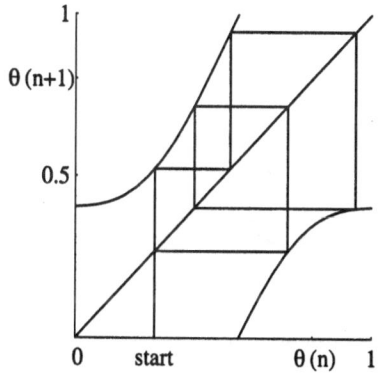

Figure 9.19: Periodic motion with a rational winding number for the nonlinear circle map.

been carried out. For a given nonzero K value, a plot of the winding number W can be generated using the file MF40 as a function of the parameter Ω.

The Devil's Staircase
By plotting the winding number W at a given K value (e.g., $K = 1$) as a function of the parameter Ω, the so-called Devil's staircase is generated. Mathematica commands in file: Compile, _Integer, Map, Drop, #, NestList, Sin, Table, ListPlot, AxesOrigin, PlotLabel, PlotStyle, PlotRange->All, Flatten, Ticks, TextStyle, ImageSize

Figure 9.20 shows the result for $K = 1$. A series of steps corresponding to rational winding numbers and therefore periodic solutions can be seen over certain ranges of Ω. This figure is usually referred to as the "Devil's staircase". If one magnifies the staircase, other rational winding numbers ($W = p/q$ with p and $q > p$ integers) can be observed as small steps. Further magnification would reveal even smaller steps, and so on. Not surprisingly, for Ω confined to the line 0 to 1, the above procedure generates a fractal.

As one decreases K from $K = 1$ through $K = 0.5$ (Figure 9.21) to $K = 0.01$ (Figure 9.22) the widths of the steps decrease and to an overactive imagination

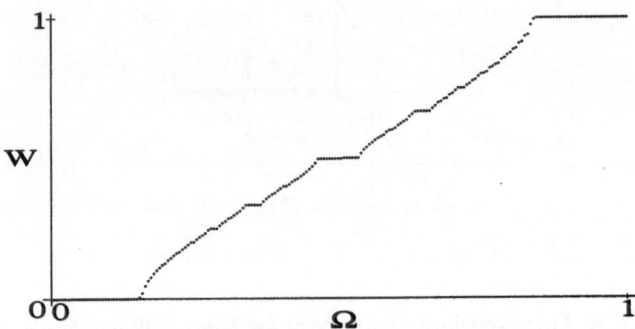

Figure 9.20: The Devil's staircase for $K = 1$.

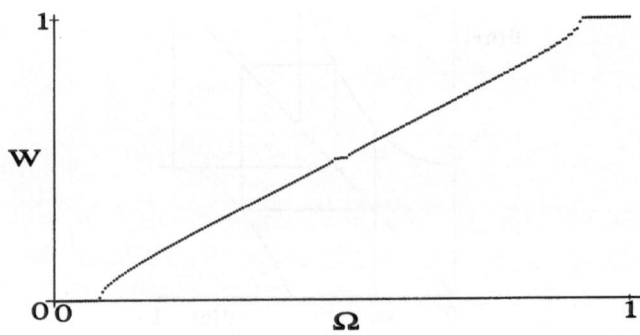

Figure 9.21: The Devil's staircase for $K = 0.5$.

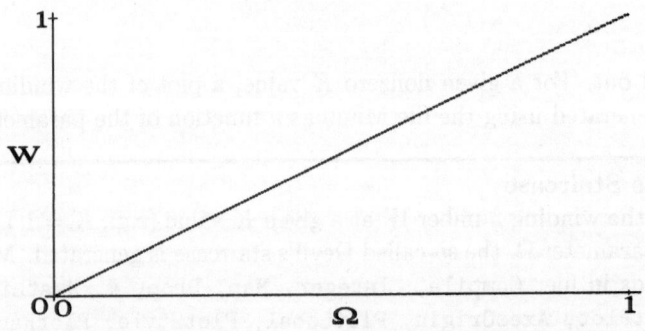

Figure 9.22: The Devil's staircase has vanished for $K = 0.01$.

resemble downward protruding serpent's tongues when the widths are plotted horizontally as a function of K with $K = 1$ at the top of the plot and $K = 0.01$ at the bottom. These tongues, which correspond to rational winding numbers and periodic regions, are referred to as "Arnold's tongues" after the Russian mathematician who discovered them. The phenomena of "phase locking" observed in forced oscillations of the simple pendulum can be related to these tongues. Phase locking refers to the situation when the ratio of the frequency of the pendulum becomes "locked" at p/q with p and q positive integers. For $K > 1$, the tongues overlap and several different periodic solutions can result for given Ω and K depending on the choice of θ_0.

Chaos can also occur for $K > 1$. The route to chaos depends on which region of parameter space for $K < 1$ that one starts in. If, for example, one starts in one of the tongues and increases K at fixed Ω, a period doubling route to chaos occurs. This is not too surprising, as the circle map becomes similar in appearance for $K > 1$ to the logistic map with a quadratic maximum. In mathematical language, the map has become "non-invertible". This has to do with the fact that there are now two values of θ_n for each value of θ_{n+1}. A single unique value is generated in going from θ_n to θ_{n+1}, but not vice versa. It turns out that *a necessary condition for chaos to occur in a 1-dimensional map is that it be non-invertible.*

PROBLEMS
Problem 9-29: $K > 1$
Show that the circle map develops a quadratic maximum similar to the logistic map for $K > 1$.

Problem 9-30: Invertability of the Hénon map
By explicitly solving for X_n, Y_n, show that the Hénon map is invertible provided that $b \neq 0$.

9.9 Chaos versus Noise

Chaos vs Noise: What is the difference?

An extremely important issue in physics and other areas of science is how to distinguish between random noise and deterministic chaos. Typically, the experimentalist has acquired data sampled at regular time intervals and would like to know if there is some underlying chaotic attractor which would allow the data to be interpreted and perhaps future predictions made, or is one dealing with noise from which nothing can be deduced or predicted.

One approach is to argue that if there is some underlying chaotic attractor perhaps it is possible to recover its geometric shape from the sampled time series. Once the shape has been recovered, other analytic tools can then be brought into play.

To illustrate how a chaotic attractor can be reconstructed, if one is present, from a time series, we present three different time series in the following Figure 9.23. The data points are connected by lines for better visualization. At this stage, all that you will be told is that one time series was generated with Mathematica's random number generator while the other two correspond to known chaotic attractors. Can you tell which one is the random sequence?

Let us call the sampling time interval t_s and label the data points as $z_0 =$

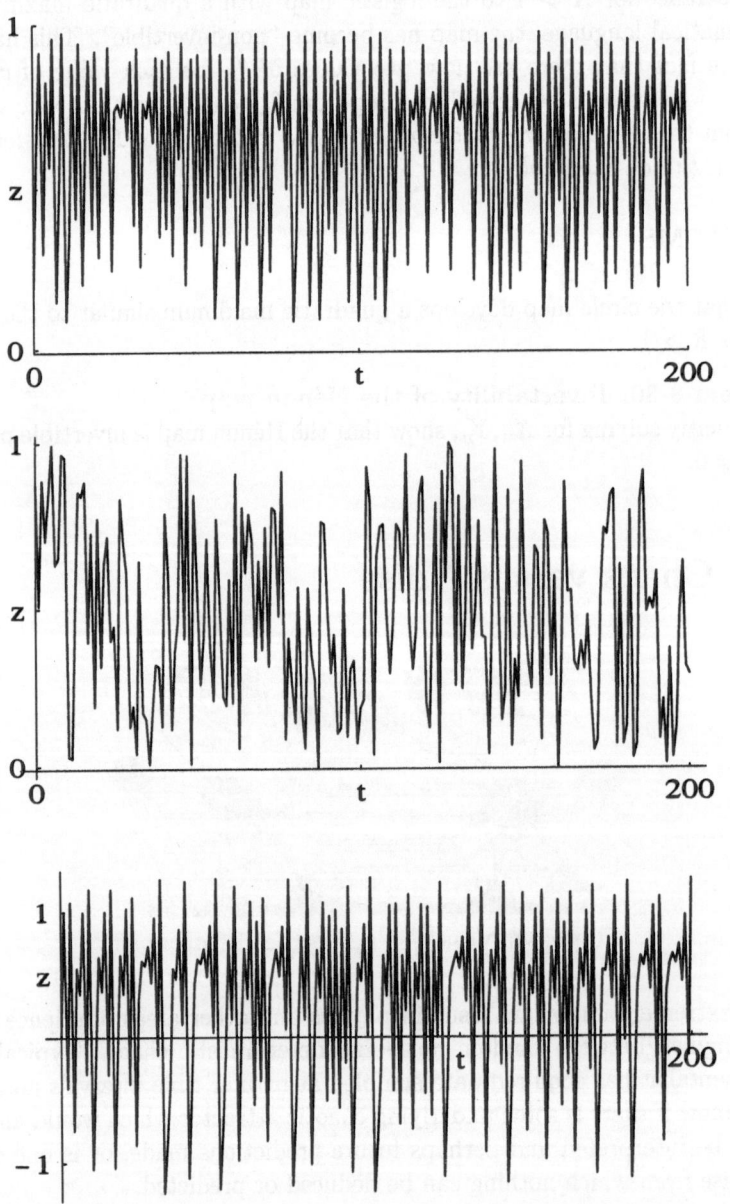

Figure 9.23: Three time series, one of which was produced with a random number generator and the other two corresponding to known chaotic attractors.

$z(0)$, $z_1 = z(t_s)$, $z_2 = z(2t_s)$,...,$z_n = z(nt_s)$,z_{n+1},.... If there is some deterministic relation between the points, then it is expected that z_{n+1} will depend somehow on z_n, z_{n-1}, and so on. In the simplest situation, z_{n+1} will depend only on the previous value z_n, i.e., $z_{n+1} = f(z_n)$. Assuming this is the case, let's plot z_{n+1} versus z_n for each of the time series. This will produce a 2-dimensional space. Mathematica File MF41 carries out this procedure, the results being presented in Figure 9.24 for each of the time series. The order of the figures is the same as in Figure 9.23.

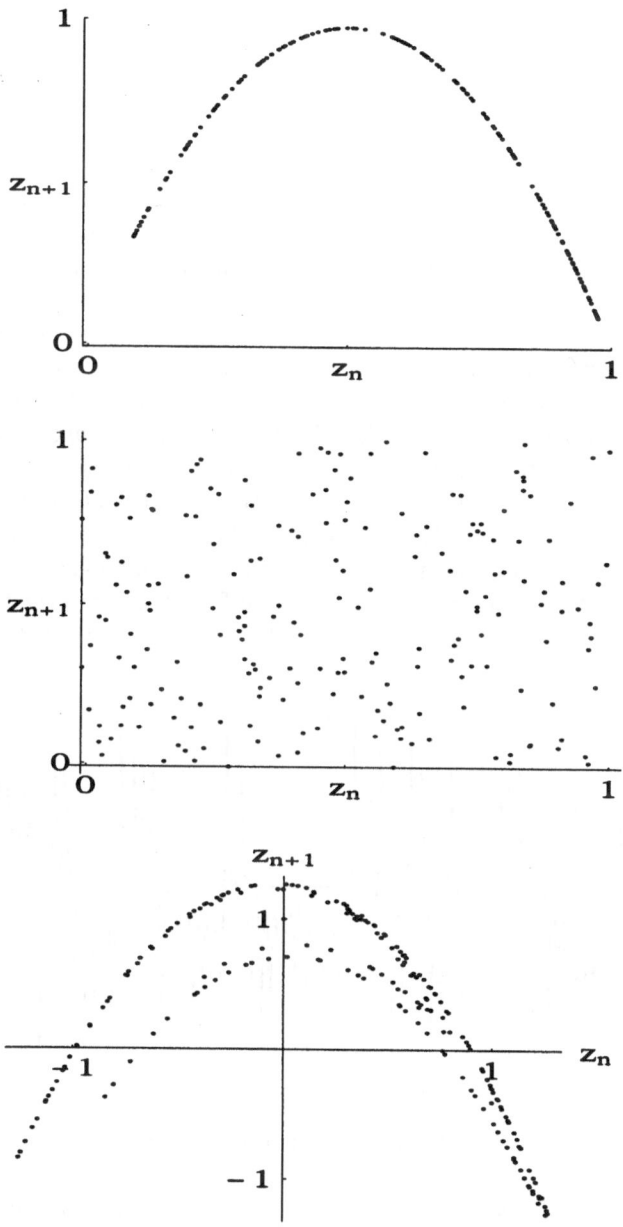

Figure 9.24: Plots of z_{n+1} versus z_n for the three time series.

Now you should be able to tell which figure is the noisy one produced by the random number generator. The other two figures resemble strange attractors that have been previously encountered. Can you identify them?

Noise versus Chaos
With this file, the reader can import data to produce the 3 time series of Fig. 9.23 and then generate the 3 plots in Fig. 9.24. Mathematica commands in file:
`Import, Dimensions, ListPlot, PlotStyle, Hue, PlotJoined->True,`
`ImageSize, AxesLabel, TextStyle, Table, Transpose, AspectRatio`

The above approach, pioneered by Takens [Tak81] and others, can be extended to reconstruct the geometrical structure of chaotic attractors in higher dimensions. To uncover the underlying attractor, the space chosen must be large enough for the trajectory's dimensionality to be revealed. A 2-dimensional space was needed to see the underlying 2-dimensional attractor in the third plot in Figure 9.24. To completely reveal a 3-dimensional attractor, it is necessary to have triplets of numbers to plot. The procedure used above can be generalized. Again, suppose that one has the time series with sampling time t_s. For the first point in 3-dimensional space, (z_0, z_1, z_2) could be chosen, for the second (z_1, z_2, z_3) chosen, and so on. But why choose three consecutive points a single time unit apart rather than, say, a triplet made up of points two time units apart, or three time units, or whatever. More generally, let us call the "time delay" between points going into the triplets $T = mt_s$ with m a positive integer. $m = 1$ corresponds to what was done earlier in two dimensions. Then, the first 3-dimensional plotting point will have coordinates $(z(0), z(T), z(2T))$, the second $(z(t_s), z(t_s + T), z(t_s + 2T))$, the third $(z(2t_s), z(2t_s + T), z(2t_s + 2T))$, etc. What is the optimum choice for T? To answer this, let's look at a concrete example, namely the reconstruction of the butterfly attractor in the Lorenz model from the time series for x shown in Figure 9.25. Here $t_s = 0.02$. Using the next

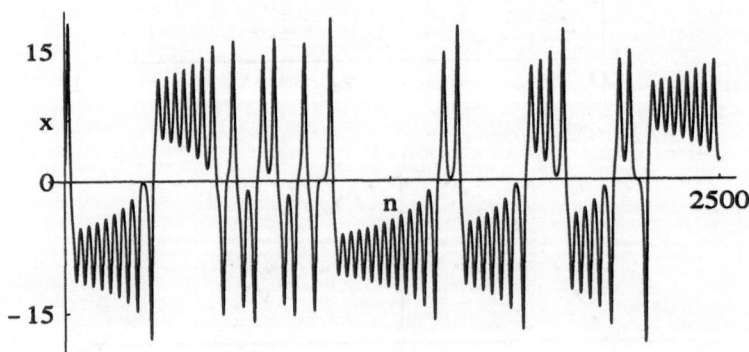

Figure 9.25: Time series for x for the butterfly attractor in the Lorenz model.

Mathematica File MF42, the result of choosing $T = 3t_s$ is shown in Figure 9.26. The points have been joined by lines. It is clear that the procedure outlined above does a reasonably good job in reconstructing the butterfly attractor. In

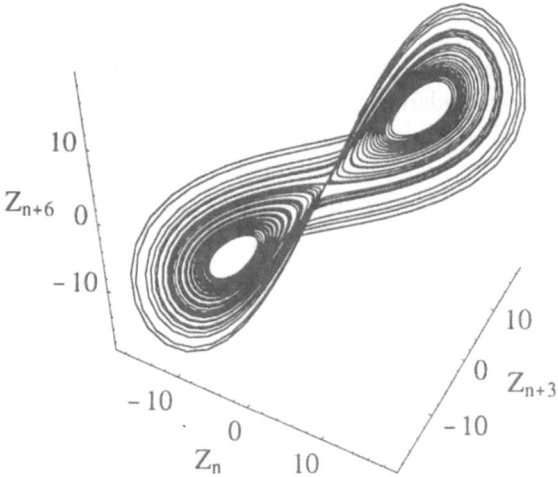

Figure 9.26: Reconstructing Lorenz butterfly attractor from the x time series.

the file MF42, the student can experiment with different values of T for a given t_s and decide which is the optimum value for best reconstructing the attractor.

Reconstructing the Lorenz Attractor
In this file, the Lorenz butterfly attractor is reconstructed from the $x(t)$ time series. Mathematica commands in file: `NDSolve`, `Flatten`, `Table`, `Evaluate`, `ListPlot`, `ScatterPlot3D`, `Boxed->False`, `BoxRatios`, `AxesEdge`

To this point, the reconstruction concept has been tested on known examples. Does the method work for real experimental time series data, particularly when the answer is not known? Well, the method has been successfully applied to the Belousov–Zhabotinskii chemical oscillator reaction (Section 2.4.1) [RSS83], [SWS82], to Taylor–Couette flow in hydrodynamics [AGS84], to ultrasonic cavitation in liquids [LH91], and to measles data for the cities of Baltimore and New York [SK86]. A few words will be said about the last item in this list.

Prior to the introduction of a vaccine which effectively eradicated the disease, major outbreaks of measles occurred every second or third year in New York and less frequently in Baltimore. Monthly data records for the years 1928 to 1963 revealed that there was a seasonal peak each winter plus a tremendous variation from year to year. Using a value for T equal to 3 months, Schaffer and Kot [SK86], were able to reconstruct a 3-dimensional Rössler-like attractor with some noise present. They were then able to calculate Poincaré sections. By plotting successive points on the Poincaré section against each other, they concluded that the measles epidemics were governed by a 1-dimensional unimodal map which allows a period doubling route to chaos. The authors state that the New York results were particularly clean, the Baltimore data showing more scatter.

The following activity shows how a map can be created for a commercially available toy.

Mapping

A triple pendulum toy is modified by removing the two smaller side pendulums and attaching a counterbalancing mass to the central pendulum. As the large sphere on the central pendulum swings over the base of the toy, it receives a magnetic repelling force. The time interval between successive passes of the sphere over the base is measured and a map is generated.

PROBLEMS

Problem 9-31: Butterfly attractor

Try reconstructing the butterfly attractor from the $z(t)$ time series by choosing appropriate values for t_s and T. Discuss why you do not obtain two "wings".

A Butterfly Attractor

Problem 9-32: The Rössler attractor

Using the time series for $x(t)$ for the Rössler system with $a = 0.2$, $b = 0.2$, $c = 5.0$, $x(0) = 4$, $y(0) = z(0) = 0$ reconstruct the Rössler attractor. Try different values for the total elapsed time, t_s, and T and determine which choice gives the best reconstruction, i.e., best resembles Figure 8.31.

9.10   2-Dimensional Maps

9.10.1   Introductory Remarks

An example of a 2-dimensional map, namely the Hénon map, has already been encountered. This map is of the abstract variety, as it was invented by Hénon to illustrate certain features such as the stretching and folding accompanying the occurrence of a strange attractor. Many of the 2-dimensional maps which are currently in fashion are of the abstract variety, although the history of mathematics and physics is full of examples where apparently abstract mathematics turns out to be relevant at some later date to explaining observed physical phenomena. One of the extremely fashionable 2-dimensional maps is the one due to Mandelbrot [Man83]. Mandelbrot's map may be viewed as an extension of the 1-dimensional logistic map to the complex plane. By completing the square on the rhs, the logistic map may be rewritten as

$$Z_{n+1} = Z_n^2 + C \tag{9.26}$$

where $Z = a(1/2 - X)$ and $C = a(2 - a)/4$. The Mandelbrot map results on setting $Z = X + iY$ and $C = p + iq$, i.e., extending the logistic map to complex numbers. Separating into real and imaginary parts yields the 2-dimensional map

$$X_{n+1} = X_n^2 - Y_n^2 + p$$
$$Y_{n+1} = 2X_nY_n + q. \tag{9.27}$$

In contrast to the "abstract" maps, there are some 2-dimensional maps which have a known physical origin or interpretation. Another variation on the logistic map illustrates this, namely the delayed logistic map

$$X_{n+1} = aX_n(1 - X_{n-1}). \tag{9.28}$$

In this model, the term which reflects the tendency for the population to decrease due to negative influences (overcrowding, overeating, environmental toxicity,...) depends not on the previous generation but on the population number two generations ago. That the model is really a 2-dimensional map can be seen by rewriting it as

$$X_{n+1} = aX_n(1 - Y_n), \quad Y_{n+1} = X_n. \tag{9.29}$$

As a second example of a 2-dimensional map which has a physical origin, one should mention the standard map

$$X_{n+1} = X_n + a\sin(Y_n)$$
$$Y_{n+1} = Y_n + bX_{n+1} \tag{9.30}$$

where the variables X, Y are to be evaluated mod 2π. The standard map arises (see Problem 3-29, page 123) when a perfectly elastic ball bounces vertically on a horizontal vibrating plate and the vertical displacement of the plate is neglected relative to the flight of the ball. According to Jackson [Jac90], it also arises for the relativistic motion of an electron in a microtron accelerator as well as in stellerator setups used in plasma fusion experiments.

The exploration of 2-dimensional maps, whether of the abstract or "real" variety, would take us too far afield, so our discussion shall be limited to a few important underlying concepts and examples. Because it is directly parallel to the phase plane analysis in Chapter 4, we begin with the classification of fixed points for 2-dimensional maps. The identification of fixed points can often help with interpreting observed behavior.

9.10.2   Classification of Fixed Points

Consider the general 2-dimensional nonlinear map

$$X_{n+1} = P(X_n, Y_n), \quad Y_{n+1} = Q(X_n, Y_n). \tag{9.31}$$

The fixed points correspond to $X_{n+1} = X_n \equiv \tilde{X}$, $Y_{n+1} = Y_n \equiv \tilde{Y}$ so that they are the solutions of

$$\tilde{X} = P(\tilde{X}, \tilde{Y}), \quad \tilde{Y} = Q(\tilde{X}, \tilde{Y}). \tag{9.32}$$

For other points near a fixed point, write

$$X_n = \tilde{X} + U_n, \quad Y_n = \tilde{Y} + V_n \tag{9.33}$$

with U_n, V_n small and linearize the difference equations so that

$$U_{n+1} = aU_n + bV_n$$
$$\tag{9.34}$$
$$V_{n+1} = cU_n + dV_n$$

with $a \equiv (P_X)_{\tilde{X}, \tilde{Y}}$, $b \equiv (P_Y)_{\tilde{X}, \tilde{Y}}$, $c \equiv (Q_X)_{\tilde{X}, \tilde{Y}}$, and $d \equiv (Q_Y)_{\tilde{X}, \tilde{Y}}$. Eliminating V, we obtain

$$U_{n+2} + pU_{n+1} + qU_n = 0 \tag{9.35}$$

with $p = -(a + d)$ and $q = ad - bc$. To solve this difference equation, assume that $U_n \sim e^{rn} \equiv \lambda^n$ with $\lambda = e^r$. This yields a familiar quadratic equation (see Equation (4.21)) for λ, viz.,

$$\lambda^2 + p\lambda + q = 0. \tag{9.36}$$

Since this equation in general has two solutions, λ_1 and λ_2, the general solution for U_n is of the form $U_n = C_1(\lambda_1)^n + C_2(\lambda_2)^n$ where C_1, C_2 are arbitrary constants. If $|\lambda_1| < 1$ and $|\lambda_2| < 1$, then all orbits are attracted to the fixed point and it is stable. If at least one of the λ has a magnitude greater than one, the fixed point is unstable.

Not surprisingly, much of the phase plane analysis in Chapter 4 can be borrowed, remembering that the present λ is not the same as the λ encountered previously. The "old" λ is equivalent to r here. The following list summarizes the main results for the 2-dimensional mapping case, one special case being left as a problem (see Problem 9-33):

- If λ_1 and λ_2 are real and $0 < \lambda_1 < 1$, $0 < \lambda_2 < 1$, the fixed point is a stable (attracting) node. If $\lambda_1 > 1$, $\lambda_2 > 1$, the fixed point is an unstable (repelling) node.

- If $0 < \lambda_1 < 1$ and $\lambda_2 > 1$, the fixed point is a saddle point.

- If at least one of the λs is negative, it follows from the general solution for U_n that as n increases successive points of a nearby trajectory hop back and forth between two distinct branches.

- If λ_1 and λ_2 are complex conjugate, and their magnitude is not equal to 1, the fixed point is a focal point, stable if the magnitude is less than 1 and unstable if it is greater than 1.

- If λ_1 and λ_2 are complex conjugate and their magnitude is 1, the fixed point is a vortex or center.

PROBLEMS
Problem 9-33: Equal real roots
If one has equal real roots, $\lambda_1 = \lambda_2$, what is the nature of the fixed point? Derive the solution for U_n and relate your discussion to the degenerate root case in Chapter 4.

9.10.3 Delayed Logistic Map

As a first example, we consider the delayed logistic map (9.29) mentioned in the introduction to this section. For $a < 1$ there is a fixed point at the origin,

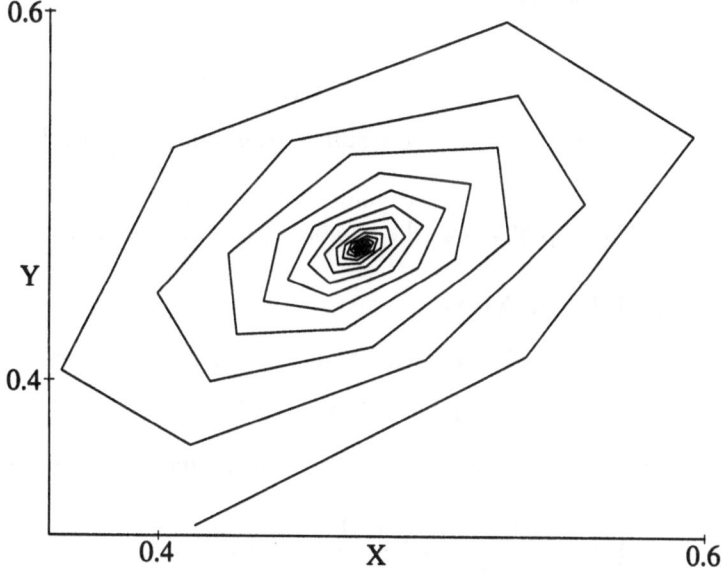

Figure 9.27: Confirmation of stable focal point for delayed logistic map.

while for $a > 1$ a second fixed point occurs at $\tilde{X} = \tilde{Y} = 1 - 1/a$. Further, $P_X = a - a\tilde{Y}$, $P_Y = -a\tilde{X}$, $Q_X = 1$, and $Q_Y = 0$.

For the fixed point at the origin, $p = -a$, $q = 0$, so that $\lambda = 0$ and a. This fixed point is a nodal point for $0 < a \leq 1$ and becomes a saddle point for $a > 1$.

For the second fixed point, $P_X = 1$, $P_Y = 1 - a$, $Q_X = 1$, $Q_Y = 0$ so that $p = -1$ and $q = a - 1$. Thus, $\lambda = (1 \pm \sqrt{(5 - 4a)})/2$. For $1 < a < 2$, $|\lambda| < 1$

and the second fixed point is stable. Below $a = 5/4$, the λs are real and the
fixed point is a stable nodal point, while above this value the λs are complex
conjugate and it is a stable focal point. An illustration of the behavior in the
latter case is given in Figure 9.27 for $a = 1.9$. Joining the points with lines for
viewing convenience, we see that the trajectory winds onto a stable focal point
at $\tilde{X} = \tilde{Y} = 0.47368$.

For $a > 2$, $|\lambda| > 1$ and the λs are complex conjugate, so the second fixed
point is an unstable focal point. Figure 9.28 shows a trajectory for $a = 2.1$

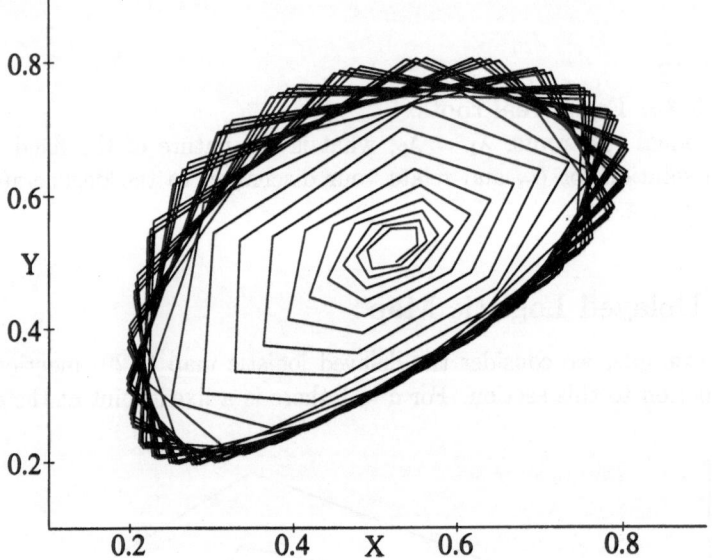

Figure 9.28: Confirmation of unstable focal point for $a = 2.1$.

unwinding in the vicinity of the fixed point at $\tilde{X} = \tilde{Y} = 0.5238$ onto a stable
"braided" loop. The loop fills in as the number of iterations is increased.

9.10.4 Mandelbrot Map

The fixed points, $\tilde{X}$, $\tilde{Y}$, of the Mandelbrot map are determined from

$$\tilde{X} = \tilde{X}^2 - \tilde{Y}^2 + p, \quad \tilde{Y} = 2\tilde{X}\tilde{Y} + q. \tag{9.37}$$

Solving for $\tilde{X}$ in the second equation and substituting into the first yields the
real roots

$$\tilde{Y} = \pm[-\frac{1}{2}(\frac{1}{4} - p) + \frac{1}{2}[(\frac{1}{4} - p)^2 + q^2]^{1/2}]^{1/2}$$

$$\tilde{X} = \frac{1}{2} - \frac{q}{2\tilde{Y}}. \tag{9.38}$$

Next, let's determine the value of λ at the fixed points. One easily finds that
$a \equiv (P_X)_{\tilde{X}, \tilde{Y}} = 2\tilde{X}$, $b = -2\tilde{Y}$, $c = 2\tilde{Y}$, and $d = 2\tilde{X}$. Thus, $-(a + d) = -4\tilde{X}$
and $ad - bc = 4(\tilde{X}^2 + \tilde{Y}^2)$. Then, solving the quadratic equation for λ yields
the complex conjugate roots $\lambda = 2\tilde{X} \pm 2i\tilde{Y}$. For $|\lambda| = 2\sqrt{\tilde{X}^2 + \tilde{Y}^2} < 1$, the
fixed point is a stable focal point, while for $|\lambda| > 1$ it is an unstable focal point.

As a specific example, consider $p = -0.5$, $q = -0.5$. Using the analytic formulae derived above, the singular points are $\tilde{Y} = 0.2751$, $\tilde{X} = 1.4087$ and $\tilde{Y} = -0.2751$, $\tilde{X} = -0.4087$. For the former singular point $|\lambda| = 2.8706 > 1$, so that point is an unstable focal point. For the latter one $|\lambda| = 0.9853 < 1$, so it is a stable focal point. The stable focal point behavior is confirmed in Figure 9.29, where the Mandelbrot map has been iterated 100 times for the

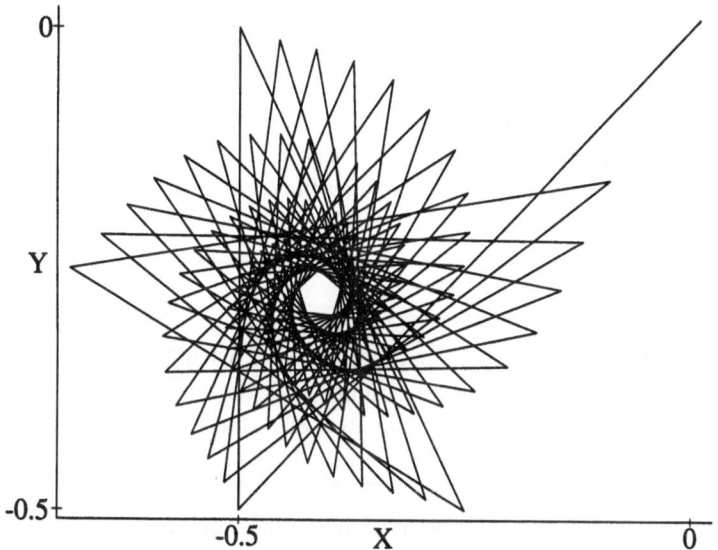

Figure 9.29: Stable focal point behavior for Mandelbrot map.

initial point $X_0 = Y_0 = 0.01$. The "pinwheel" trajectory is winding onto the stable focal point at $\tilde{Y} = -0.2751$, $\tilde{X} = -0.4087$.

In this illustrative example, the value of $|\lambda|$ for the stable focal point is quite close to 1. As q is made more negative for fixed p, $|\lambda|$ exceeds 1, and both singular points are unstable focal points, the trajectory then diverging to infinity (not shown).

PROBLEMS

Problem 9-34: A predator–prey map
As a model of predator–prey interaction, consider the 2-dimensional map

$$X_{n+1} = aX_n(1 - X_n - Y_n), \quad Y_{n+1} = aX_nY_n,$$

with $2 < a \leq 4$. Find and classify the fixed points for $a = 2.40$, 3.0, 3.43 and 3.90. Taking $X_0 = Y_0 = 0.1$, $N = 4000$ iterations, and using both point and line styles, plot the trajectories in each case. Discuss the observed behavior in terms of the fixed points.

Problem 9-35: Confirmation of text pictures
Confirm the following text pictures: (a) the braided loop trajectory in Figure 9.28, (b) the pinwheel trajectory in Figure 9.29.

9.11   Mandelbrot and Julia Sets

A famous picture, referred to as the Mandelbrot set, can be generated by systematically varying p and q in the Mandelbrot map with X_0, Y_0 held fixed. The reader has already seen a beautiful 2-dimensional Mathematica rendition of the Mandelbrot set in the file MF12. Recall that there we chose $X_0 = Y_0 = 0$ and selected the plotting range $p = -1.5...1$, $q = -1..1$. The complex Mandelbrot map was iterated n times, the iterated values either diverging to infinity or being attracted to a finite value of X_n, Y_n. To decide which occurs the iteration was continued while the absolute value of $Z < 2$ and $n < 20$. If $n = 20$, convergence was presumed to occur, while $n < 20$ indicated divergence. The region in the $p - q$ plane corresponding to convergence is the Mandelbrot set, the boundary of this set of points having a complicated fractal structure. As an alternate way of viewing the Mandelbrot set, the following file creates a 3-dimensional picture using the iteration number as the third axis.

Mandelbrot Set

In this file, the code of MF12 is extended to produce a 3-dimensional version of the Mandelbrot set. The `Plot3D` command is used, with the three axes being p and q horizontally and the iteration number n as the vertical axis. Mathematica commands in file: `Module, While, Plot3D, Mesh->False, PlotPoints, Abs, BoxRatios,Ticks, TextStyle, ImageSize`

On the other hand, for p and q fixed in the Mandelbrot map, one can systematically scan a range of starting points, X_0, Y_0. In honor of Gaston Julia, a French mathematician who studied the structure of the complicated boundaries generated between regions of convergence and divergence, the sets of points lying on such boundaries for fixed p, q, are now called "Julia sets". In the following example, the Julia set, known as Douady's rabbit, is generated.

Example 9-3: Douady's Rabbit

For the Mandelbrot map, take $N = 25$ iterations and plot the Julia set corresponding to $p = -0.12$ and $q = -0.74$.

Solution: The values of p and q are entered,

```
Clear["Global`*"]
p=-.12;  q=-.74;
```

and a Julia function of x and y created with a `Module` construct.

```
julia[x_, y_] := Module[{n = 0, X, Y, copyX}, X = x;  Y = y;
         While[X^2 + Y^2 <= 4 && n < 25, copyX = X;
         X = X^2 - Y^2 + p;  Y = 2*copyX*Y + q;  n++];
         If[X^2 + Y^2 > 4, n = 1, n = 0];  n];
```

The initial value of n and the local variables X, Y, and a copy of X, labeled copyX, have been entered as a list. X and Y are obtained from the input values

x and y. If the radius squared, i.e., $X^2 + Y^2$, exceeds 4, it is assumed that the values of X, Y are going to diverge. Therefore, a `While` statement is introduced which allows iterations of the 2-dimensional map as long as the radius squared is less than or equal to 4 and $n < 25$. The copy of X is carried out first and used in the evaluation of Y. To obtain a black and white picture, an `If` statement is inserted to assign the value 1 to regions of divergence (i.e., when $X^2 + Y^2 > 4$) and 0 to regions of convergence. The `Module` ends with the value of n being recorded. If specific x and y values are now given, then the Julia function is evaluated. For example,

```
{julia[0,0], julia[1,0]}
```

```
{0, 1}
```

so that the input point ($x = 0$, $y = 0$) converges while the point $(1,0)$ diverges. The Julia function is now plotted with the `DensityPlot` command for the range $x = -1.4...1.4$, $y = -1.4...1.4$, the result being shown in Figure 9.30.

```
DensityPlot[julia[x,y], {x,-1.4,1.4}, {y,-1.4,1.4}, Mesh->False,
PlotPoints->200, FrameTicks->{{-1,{0,"x"},1},{-1,{0,"y"},1},
{ },{ }}, TextStyle->{FontFamily->"Times",FontSize->16},
ImageSize->{400,400}];
```

The boundary between the two regions can be clearly seen in the figure. The boundary points form the Julia set for the Mandelbrot map. If you mentally

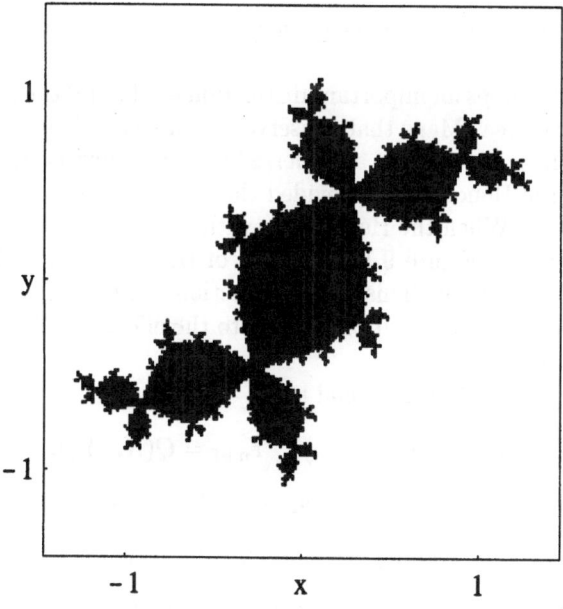

Figure 9.30: Douady's rabbit.

rotate the Julia set slightly, and have a good imagination,[3] you should be able to see Douady's "rabbit".[4] The complicated boundary formed by the Julia set is another example of a fractal structure.

If you wish to see a colored density plot, rather than the black and white version shown here in the text, the If statement should be dropped and the ColorFunction->Hue option added to the DensityPlot command. For the reader's convenience this has been done in Example 09-3 on the CD-ROM.

Other geometrically interesting Julia sets may be generated for appropriate values of p and q. Some particularly good choices of these parameters are given in the Julia set problems which follow.

| End Example 9-3 |

PROBLEMS

Problem 9-36: The San Marco attractor
Generate the so-called San Marco attractor which results from taking $p = -0.75$, $q = 0$ in the Julia function. Take $x = -1.6...1.6$, $y = -1...1$.

Problem 9-37: The octopus
Generate the stylized "octopus" which results on taking $p = 0.27334$, $q = 0.00742$ in the Julia function. Take $N = 100$ and the range $-1.1...1.1$ for both x and y.

Problem 9-38: Other Julia sets
Generate the Julia sets corresponding to

a. $p = -1$, $q = 0$, **b.** $p = 0.32$, $q = 0.043$.

9.12 Nonconservative versus Conservative Maps

For 2-dimensional maps an important distinction is whether the mapping does or does not preserve area. Maps that conserve area are called "conservative maps" and obviously those that do not are referred to as nonconservative. An example of the latter is the Hénon map, provided that the magnitude of the parameter b is not equal to 1. When the Hénon map with $a = 1.4$ and $b = 0.3$ was applied earlier to a square in Figure 9.14, the area of the square shrank as successive iterations were carried out. This area contraction can be confirmed analytically by applying a well-known result of calculus to the effect of a 2-dimensional map on an infinitesimal area.

Consider a general 2-dimensional map

$$X_{n+1} = P(X_n, Y_n), \quad Y_{n+1} = Q(X_n, Y_n). \tag{9.39}$$

Under such a mapping an infinitesimal area $dXdY$ maps into the new area

$$dX'dY' = |\det J(X, Y)| dXdY \tag{9.40}$$

---

[3]The type of imagination needed to see animal shapes in the clouds and in the inkblots of the Rorschach tests administered by psychologists.

[4]Adrien Douady is a French professor of mathematics who does research in the area of holomorphic dynamics.

where det refers to the determinant and $J(X, Y)$ is the Jacobian matrix

$$\begin{pmatrix} P_X & P_Y \\ Q_X & Q_Y \end{pmatrix}.$$

For the Hénon map, $|\det J| = |b|$, so area reduction takes place if $|b| < 1$. In Figure 9.14, we had $b = 0.3$, and thus the area of the square was reduced in addition to being stretched and folded. Because area reduction in the phase plane is a characteristic of dissipative ODE systems,[5] such area contracting maps are also called "dissipative maps".

An example of a conservative map is the standard map for $b = 1$. In this case, the map becomes

$$X_{n+1} = X_n + a\sin(Y_n)$$
$$Y_{n+1} = X_n + Y_n + a\sin(Y_n). \tag{9.41}$$

Then, $|\det J| = |P_X Q_Y - P_Y Q_X| = (1)(1 + a\cos Y) - (a\cos Y)(1) = 1$.

The distinction between conservative and nonconservative maps is important because conservative maps cannot have attractors. The presence of any attractor (focal point, strange attractor, ...) leads to area reduction as all starting points within the attractor's basin of attraction will collapse onto the attractor. So if you hope to encounter a strange attractor in the world of 2-dimensional maps, it's necessary to have a nonconservative map. Despite not allowing the possibility of having strange attractors, conservative maps display many interesting features, the discussion of which is left to more advanced texts [Jac90].

PROBLEMS

Problem 9-39: The Mandelbrot map
Prove that the Mandelbrot map is nonconservative.

Problem 9-40: Hénon's quadratic map
Consider Hénon's quadratic map [Hé69]

$$X_{n+1} = X_n \cos\alpha + (X_n^2 - Y_n)\sin\alpha$$
$$Y_{n+1} = X_n \sin\alpha - (X_n^2 - Y_n)\cos\alpha$$

with $0 \le \alpha \le \pi$.

a. Show that this is a conservative map.

b. Locate and classify the fixed points of this map.

c. For $\cos\alpha = 0.24$ and $X(0) = 0.5$, $Y(0) = 0.3$, $N = 1000$ show that five "islands" are present. How many islands are there when $\cos\alpha = -0.24$?

---

[5]If one considers a square surrounding the origin for the damped pendulum problem, all starting points inside the square lead to trajectories that collapse to the attractor (focal or nodal point) at the origin.

d. Explore the entire range of α and discuss your results.

Problem 9-41: Standard map
In the standard map, set $X_n = 2\pi x_n$, $Y_n = 2\pi y_n$ so that the new variables x, y are to be evaluated mod1. Take $a = b = 1$, $y_0 = 0.1$, $N = 2000$ iterations, and explore the behavior in the x-y plane as x_0 is varied between 0 and 1.

Problem 9-42: Islands and other structures
Consider the map

$$X_{n+1} = Y_n, \quad Y_{n+1} = -X_n + 2CY_n + 4(1 - C)Y_n^2/(1 + Y_n^2)$$

with $0 < C < 1$.

a. Is this map conservative or nonconservative?

b. Locate and classify the fixed points of this map.

c. Take $X_0 = Y_0 = 0.2$ and catalogue the behavior of the map for $N = 3000$ iterations as the parameter C is varied from 0.90 down to 0.001. Find as many different behaviors as you can and plot them. For example, interesting and different behavior occurs for $C = 0.20, 0.15, 0.001$.

Problem 9-43: Lozi map
Consider the Lozi map

$$X_{n+1} = 1 + Y_n - a|X_n|, \quad Y_{n+1} = bX_n$$

with a real and $-1 \le b \le 1$.

a. Is the Lozi map conservative or nonconservative?

b. Locate and classify the fixed points of this map.

c. Taking $X_0 = Y_0 = 0.1$ and $N = 3000$ iterations, explore this map for different values of a and b and see what interesting patterns you can generate. For particularly striking results, attach an appropriate name. For example, try the combinations $a = 1.7, b = 0.5$ ("broken arrow"), $a = 0.43, b = 1.0$ (the "eyeglasses"), $a = -1, b = -1$ (hexagon with "beads"), and $a = 1, b = -1$ (triangle with "beads").

9.13   Controlling Chaos

Although the exploration of chaotic maps is fun and intellectually stimulating to the mathematician, to the engineer the presence of chaos in mechanical and electrical systems is usually not desirable. In recent years, there has been a considerable amount of effort expended in developing ways to "control chaos". In particular, methods have been proposed to suppress chaotic behavior and in some cases render the response of the system to be periodic. One such approach[6] involving proportional feedback has been suggested by Cathal Flynn and Niall

---

[6]A more general method has been given by Ott, Grebogi, and Yorke[OGY90].

Wilson [FW98] and applied to the Hénon attractor as an illustrative example. We shall now describe their method.

Recall that the Hénon map is given by

$$X_{n+1} = Y_n + 1 - aX_n^2, \quad Y_{n+1} = bX_n, \quad (9.42)$$

with a and b real, positive, constants. To render the map into the form used by Flynn and Wilson, we set $X_n = x_n/a$ and $Y_n = (b/a)y_n$, so the Hénon map becomes

$$x_{n+1} = by_n + a - x_n^2, \quad y_{n+1} = x_n. \quad (9.43)$$

First, let's find the fixed points by setting $x_{n+1} = x_n = \tilde{x}$, $y_{n+1} = y_n = \tilde{y}$. The fixed points are then given by $\tilde{y} = \tilde{x}$ where $\tilde{x}$ is the solution of the equation

$$\tilde{x}^2 + (1 - b)\tilde{x} - a = 0. \quad (9.44)$$

As a concrete example, let's take $a = 1.4$ and $b = 0.3$ as in Figure 9.13.

Example 9-4: Fixed Points of Hénon Map

For $a = 1.4$, $b = 0.3$, locate the fixed points of the Hénon map, establish their stability and nature, and plot them on the same graph as the Hénon strange attractor.

Solution: The a and b values are entered,

```
Clear["Global`*"]
a=1.4; b=.3;
```

as well as the quadratic equation (9.44) for the x-coordinates of the fixed points.

```
eq=x^2+(1-b) x-a==0
```

$$x^2 + 0.7\,x - 1.4 == 0$$

The quadratic equation, eq, is numerically solved for the two roots,

```
sol=NSolve[eq, x]
```

$$\{\{x \to -1.5839\}, \{x \to 0.883896\}\}$$

so the coordinates of the first and second fixed points are now known.

```
x1=x/. sol[[1]]; y1=x1
```

$$-1.5839$$

```
x2=x/. sol[[2]]; y2=x2
```

$$0.883896$$

Next, the stability of the fixed points must be established. For the Hénon equations (9.43), we have $P_x = -2x$, $P_y = b$, $Q_x = 1$, $Q_y = 0$, so that $p =$

$-(P_x + Q_y)_{\tilde{x},\tilde{y}} = 2\tilde{x}$ and $q = (P_x Q_y - P_y Q_x)_{\tilde{x},\tilde{y}} = -b$. Thus, the λ values are found by solving the quadratic equation

$$\lambda^2 + p\lambda + q = \lambda^2 + 2\tilde{x}\lambda - b = 0, \qquad (9.45)$$

which is now entered.

```
eq2 = λ^2 + 2 x λ - b == 0;
```

In the following command line, the x-coordinate of the first fixed point is inserted in (9.45) and the λ equation numerically solved.

```
NSolve [eq2 /. x -> x1, λ]
```

$$\{\{\lambda \to -0.0920296\}, \{\lambda \to 3.25982\}\}$$

The first fixed point is unstable because one of the λ values has a magnitude greater than 1. Also, note that one of the λ values is negative. Now, the x-coordinate of the second fixed point is inserted in (9.45), and once again the λ equation numerically solved.

```
NSolve [eq2 /. x -> x2, λ]
```

$$\{\{\lambda \to -1.92374\}, \{\lambda \to 0.155946\}\}$$

The second fixed point is also unstable, with one of the λ values negative. With the stability established, we will iterate the Hénon map 3000 times for the input values $x_0 = y_0 = 0$.

```
x[0] = 0;  y[0] = 0;  total = 3000;
```

The two functions relevant to the Hénon map are formed,

```
x[n_] := x[n] = b y[n-1] + a - (x[n-1])^2
```

```
y[n_] := y[n] = x[n-1]
```

and the map is iterated and plotting points formed with the Table command.

```
pts = Table [{x[n], y[n]}, {n, 0, total}];
```

The Hénon strange attractor is now plotted, the fixed points being included and represented by colored, labeled, suitably-sized, points.

```
ListPlot [pts,AspectRatio->1,PlotStyle->{Hue[.6],PointSize[.005]},
Ticks -> {{-1.5,{.5,"x"},1.5},{-1.5,{.5,"y"},1.5}},Epilog->
{PointSize[.03],{RGBColor[0,1,0], Point[{x1,y1}]},Text["(x_1,y_1)",
{-1.6,-1.5},{0,-1}]},{RGBColor[1,0,0],Point[{x2,y2}],
Text["(x_2,y_2)",{1,1},{0,-1}]}},TextStyle -> {FontFamily->"Times",
FontSize -> 16},ImageSize -> {400,400}];
```

The resulting plot is shown in Figure 9.31.

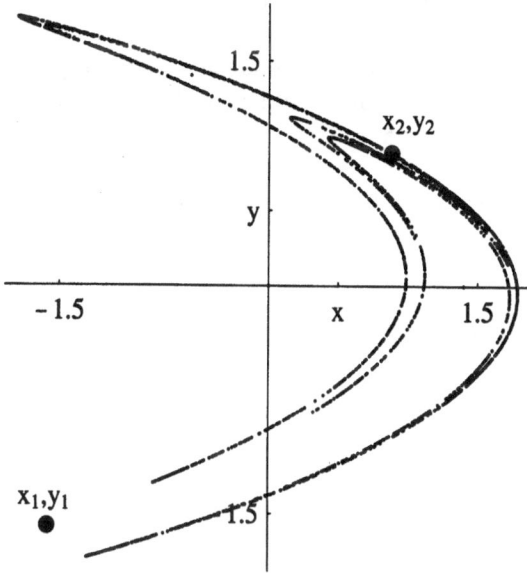

Figure 9.31: Hénon strange attractor with two fixed points shown.

Because each fixed point is unstable with one of the λ values negative, we can qualitatively understand that the two "branches" of the parabolic loops arise from the system hopping back and forth as the number of iterations is increased.

$\boxed{\textbf{End Example 9-4}}$

Next, we outline the chaos control algorithm of Flynn and Wilson. The b parameter will be held fixed at the value 0.3, while the other control parameter a will be allowed to vary slightly around the input value $a0 = 1.4$. The following steps are used to achieve control of the chaos. (The tolerances on the absolute differences in steps 2, 3, and 4 may be adjusted.)

1. Iterate the Hénon map (9.43) to find the next value of x and y.

2. If $x = y$, to within an absolute difference of 0.01, then this point is a fixed point $\tilde{x} = \tilde{y}$ for some particular value of a. Calculate a from the relation (rearranging (9.44)) $a = \tilde{x}^2 + (1 - b)\tilde{x}$ and label it as aa.

3. If the absolute difference between aa and $a0$ is less than 0.1, let $a = aa$.

4. If the absolute difference between aa and $a0$ is greater than 0.2, let $a = a0$. This condition prevents the system from blowing up.

5. Loop back to step (1).

The control algorithm is implemented in the next example for the Hénon map.

Example 9-5: Chaos Control Algorithm

Implement the Flynn–Wilson control algorithm and show that the system can be made to exit from a chaotic regime and lock onto a periodic solution.

Solution: The `Graphics3D` package is loaded, and the parameter values are entered. 5000 iterations of the map will be considered, starting at $x_0 = y_0 = 0$. Initially, we have a=a0 and aa=0.

```
Clear["Global`*"]; << Graphics`Graphics3D`

a0 = 1.4; b = .3; a = a0; aa = 0; x[0] = 0; y[0] = 0; total = 5000;
```

The relevant Hénon map functions are entered,

```
x[n_] := x[n] = a - (x[n - 1])^2 + b y[n - 1]

y[n_] := y[n] = x[n - 1]
```

and iterated subject to the Flynn–Wilson control conditions.

```
pts = Table[If[Abs[x[n] - y[n]] < .01, aa = (x[n])^2 + (1 - b) x[n]];
            If[Abs[a0 - aa] < .1, a = aa];
            If[Abs[aa - a0] > .2, a = a0];
            {n, x[n], y[n]}, {n, 0, total}];
```

A 3-dimensional plot of the points is produced with the `ScatterPlot3D` command, the result being shown in Figure 9.32. The `ViewPoint` has been chosen to show the strange attractor with a single spike emerging after a certain number of iterations.

```
ScatterPlot3D[pts, BoxRatios -> {1,1,1}, ViewPoint -> {6,1.5,1},
    PlotStyle -> {Hue[.6], PointSize[.007]}, Ticks -> {{0,{2500,"n"},
    5000}, {-1,{0,"x"},1}, {-1,{0,"y"},1}}, AxesEdge -> {{1,-1}, {1,-1},
    {1,-1}}, TextStyle->{FontFamily->"Times", FontSize->16},
    ImageSize->{400,400}];
```

Choosing a different viewpoint in the plot command, we can show the behavior of x_n versus n in Figure 9.33. In this figure, the strange attractor is being viewed "edge on", showing up as a chaotic band. The Hénon system locks onto a period-1 solution, the steady-state value of x being 0.852268.

```
ScatterPlot3D[pts, BoxRatios->{2,1,0}, ViewPoint->{0,0,2},
    PlotRange->{{0,1000},{-1.9,1.9},{-2,2}}, PlotStyle->{Hue[1],
    PointSize[.007]}, Ticks->{{0,{600,"n"},1000}, {-1,{0,"x"},1}, { }},
    AxesEdge->{{-1,1},{-1,1},{-1,1}}, TextStyle->{FontFamily->"Times",
    FontSize->16}, ImageSize->{600,400}];
```

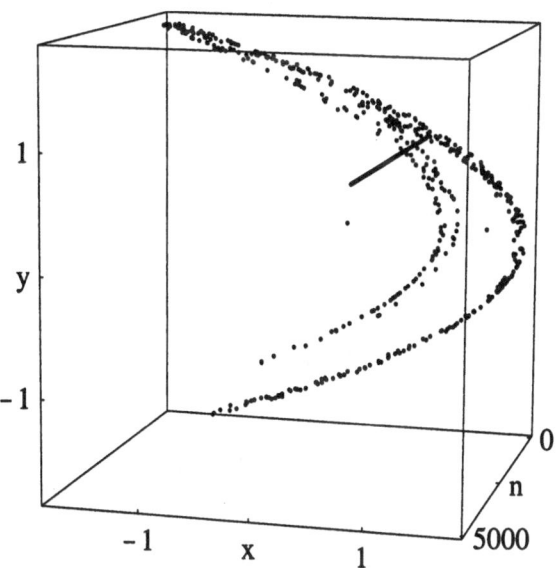

Figure 9.32: Period one spike emerging from the strange attractor.

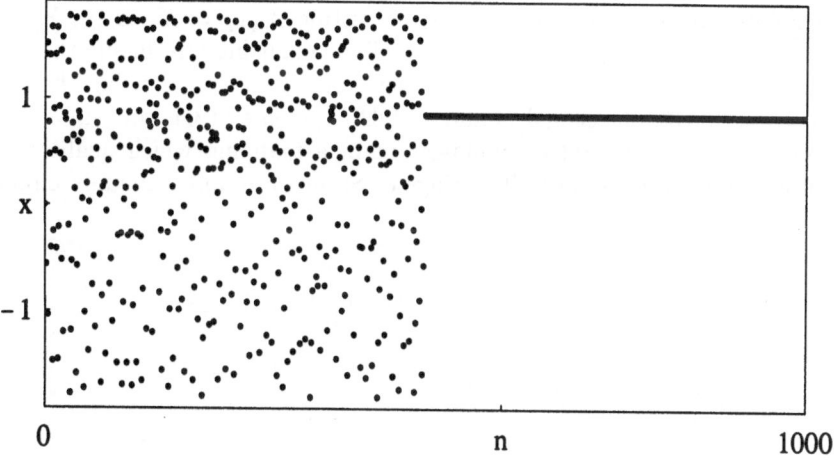

Figure 9.33: Edge-on view of the attractor showing x_n vs. n.

The final calculated values of a and aa are given, as well as the input $a0$ value.

{a0, a, aa}

$$\{1.4,\ 1.32295,\ 1.32295\}$$

The algorithm shifts the value of a slightly from 1.4 to 1.32295.

| End Example 9-5 |

PROBLEMS
Problem 9-44: Tolerances
Experiment with the various tolerances on the absolute differences in the chaos control code to see what effect they have on the outcome.

Problem 9-45: Chaos control for the logistic map
Using the same approach as in the text, control the chaos for the logistic map, taking $a0 = 3.82$ and $X_0 = 0.2$. Use the same tolerances. At what approximate n value does the system leave the chaotic regime and lock onto a period-1 solution? What is the period-1 value of X?

9.14   3-Dimensional Maps: Saturn's Rings

A general 3-dimensional nonlinear map is of the structure

$$X_{n+1} = P(X_n, Y_n, Z_n), \quad Y_{n+1} = Q(X_n, Y_n, Z_n), \quad Z_{n+1} = R(X_n, Y_n, Z_n),$$

with P, Q, and R arbitrary nonlinear functions. Although the location of the fixed points can be readily determined by setting $X_{n+1} = X_n \equiv \tilde{X}$, $Y_{n+1} = Y_n \equiv \tilde{Y}$, and $Z_{n+1} = Z_n \equiv \tilde{Z}$, no attempt will be made here to classify the types of fixed points which can occur in 3 dimensions. Instead, following the same approach as for 3-dimensional nonlinear ODE systems in Chapter 4, we shall consider a specific illustrative nonlinear map, a mapping which qualitatively produces the rings of Saturn. The rings of Saturn are approximately circular

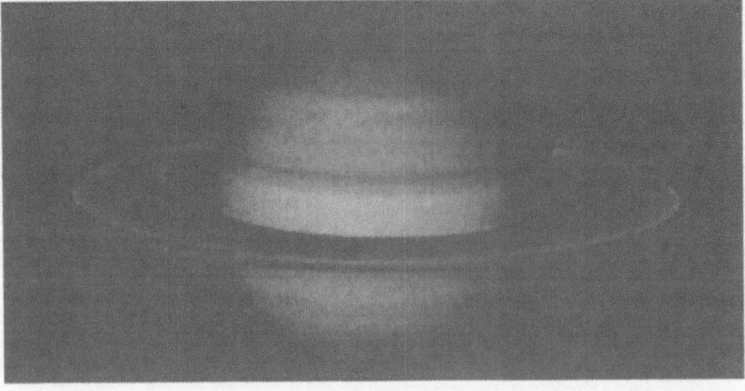

Figure 9.34: Saturn with its rings.

and nearly planar in nature, being 250,000 km across but no more than 1.5 km thick. A NASA photograph of the rings, which are composed primarily of water ice ranging in size from cm to several meters, is reproduced in Figure 9.34. Ice-coated boulders are also probably present, with a few km size rocks likely. The standard classification and vital statistics of the more prominent rings is given in Table 9.2. The large gap between the A and B rings is called the Cassini division in honor of the astronomer Cassini who observed the rings and

Ring	Distance (10^3 km)	Width (10^3 km)	Mass (kg)
D	67.0	7.5	?
C	74.5	17.5	1.1×10^{18}
B	92.0	25.5	2.8×10^{19}
Cassini division			
A	122.2	14.6	6.2×10^{18}
F	140.2	0.5	?
G	165.8	8.0	1×10^7 ?
E	180.0	300.0	?

Table 9.2: Classification and parameters of Saturn's rings. Distances are from Saturn's center to a ring's inner edge.

discovered several of Saturn's moons in the late 1600's. It should be noted that the gaps are not entirely empty and there are further variations within the rings.

In addition to possessing a spectacular ring structure, Saturn has 18 named moons, more than any other planet. Table 9.3 lists the inner seven and outer four of Saturn's moons. Probably because of its substantially larger mass compared to the other inner moons, Mimas[7] plays an important role in the organization of

Moon	Distance (10^3 km)	Radius (km)	Mass (10^{17} kg)	Discoverer (date)
Pan	134	10	?	Showalter (1990)
Atlas	138	14	?	Terrile (1980)
Promethus	139	46	2.70	Collins (1980)
Pandora	142	46	2.20	Collins (1980)
Epimetheus	151	57	5.60	Walker (1980)
Janus	151	89	20.1	Dollfus (1966)
Mimas	186	196	380	Herschel (1789)
....	.....	.....	.....	.....
Titan	1222	2575	1.35×10^6	Huygens (1655)
Hyperion	1481	143	177	Bond (1848)
Iapetus	3561	730	1.88×10^4	Cassini (1671)
Phoebe	12952	110	40	Pickering (1898)

Table 9.3: Some of Saturn's 18 moons.

---

[7]In Greek mythology, Mimas was one of the Titans slain by Hercules.

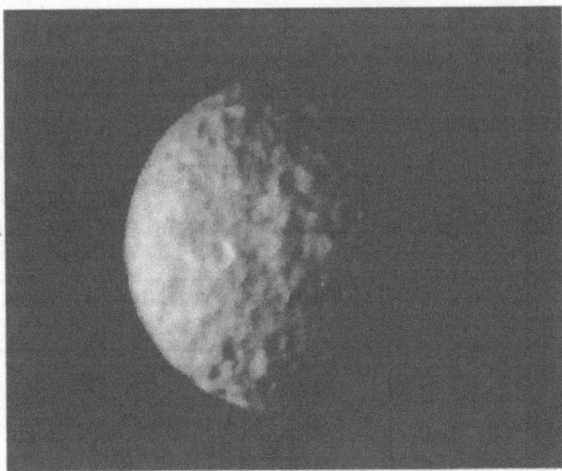

Figure 9.35: Mimas, with its surface dominated by the Herschel impact crater.

Saturn's inner rings. Mimas is also an interesting moon because of its physical appearance. Figure 9.35 shows a Nasa photograph of Mimas which is dominated by the Herschel impact crater 130 km across, which is about $\frac{1}{3}$ of Mimas's diameter. From the length of the shadow cast by the central peak inside the crater, astronomers have deduced that the crater walls are about 5 km high and the central peak rises 6 km from the crater floor, parts of which are 10 km deep. From the data given in Table 9.3, Mimas's density can be calculated to be about 1.2×10^3 kg m^{-3}, which indicates that it is composed mainly of ice with only a small amount of rock present.

There are theoretical reasons, first suggested by the French scientist Edouard Roche in 1848, for thinking that for a given planet there is an inner limiting radius inside of which moons cannot exist. Roche argued that within a critical distance from a planet's center, now called the Roche limit, any orbiting moon would break up because the tidal force on the moon due to the planet would be larger than the gravitational force holding the moon together. The moon would be shredded into smaller particles such as those found in the inner rings of Saturn. The Roche limit for a satellite object of density ρ_s orbiting about a planet of mass M_p, radius R_p, and density ρ_p is given by the formula

$$r = \left[16\frac{\rho_p}{\rho_s}\right]^{1/3} R_p. \tag{9.46}$$

Since Saturn has a radius $R_p = 60.4$ thousand km and a density 0.7 that of water, the Roche limit for a satellite of density comparable to that of Mimas is about 127 thousand km. From Table 9.3, we see that the innermost known moon, Pan, is orbiting at 134 thousand km from Saturn's center, just outside the estimated Roche limit.

A nonlinear mapping which produces a ring pattern qualitatively resembling the rings of Saturn has been developed by Fröyland [Fr92] and discussed by Gould and Tobochnik [GT96]. Letting σ be the radial distance of Mimas from

Saturn's center, r_n the radial distance of a ring particle from Saturn's center after the nth revolution, and θ_n the angular position of a ring particle with respect to Mimas after n revolutions, the Fröyland model equations are of the form

$$\theta_{n+1} = \theta_n + 2\pi \left(\frac{\sigma}{r_n}\right)^{\frac{3}{2}}, \tag{9.47}$$

$$r_{n+1} = 2r_n - r_{n-1} - a\frac{\cos\theta_n}{(r_n - \sigma)^2}, \tag{9.48}$$

with a a positive parameter. Since Equations (9.47) and (9.48) can be rewritten as the finite difference system

$$\theta_{n+1} = \theta_n + 2\pi \left(\frac{\sigma}{r_n}\right)^{\frac{3}{2}},$$

$$r_{n+1} = 2r_n - z_n - a\frac{\cos\theta_n}{(r_n - \sigma)^2}, \tag{9.49}$$

$$z_{n+1} = r_n,$$

we actually have a 3-dimensional nonlinear map. Generalizing the Jacobian matrix of Section 9.12 to three dimensions, it is easy to show that the determinant of the Jacobian matrix is unity here, so the nonlinear mapping is volume-preserving and therefore conservative.

To understand the structure of the model equations (9.47) and (9.48), their physical origin is now briefly discussed. Our presentation is a combination of first principles ideas and some "hand waving" arguments. In the model there are two important forces acting on the ring particles, the dominant effect of Saturn's attractive gravitational force and the perturbing influence of Mimas.

The effect of Saturn is included as follows. Each time Mimas completes an orbit of radius σ with a period T_σ, it undergoes an angular change of 2π radians. If T_n is the period for any other satellite object on its nth revolution, the angle θ that the object makes with respect to Mimas on the $n + 1$st revolution will be given by[8]

$$\theta_{n+1} = \theta_n + 2\pi \left(\frac{T_\sigma}{T_n}\right). \tag{9.50}$$

But Kepler's 3rd law for planetary orbits states that the period T of an object orbiting a planet of mass M_p in a circular[9] orbit of radius r is given by

$$T^2 = \frac{4\pi^2}{GM_p}r^3, \tag{9.51}$$

where $G = 6.67 \times 10^{-11}$ N·m^2/kg^2 is the gravitational constant. Letting r_n be the distance of a ring particle from Saturn's center after n revolutions, then

$$\frac{T_\sigma}{T_n} = \left(\frac{\sigma}{r_n}\right)^{\frac{3}{2}} \tag{9.52}$$

---

[8]To within a term of the structure $2\pi n$ which can be omitted in the model without affecting the results.

[9]For an elliptical orbit, the radius is replaced in the 3rd law with the semi-major axis.

and the first model equation (9.47) immediately follows on substituting (9.52) into (9.50) .

The effect of Mimas is to perturb the radial distance r of a ring particle, causing the distance to change from one orbit to the next. By Newton's 2nd law, a particle's radial acceleration will given by

$$\frac{d^2r}{dt^2} = \frac{F_r}{m},$$

(9.53)

where F_r is the radial component of the gravitational force between Mimas and the particle of mass m. To convert the ODE into a finite difference equation, the second derivative on the lhs of Equation (9.53) is replaced by the standard finite difference[10] approximation $(r_{n+1} - 2r_n + r_{n-1})/(\Delta t)^2$ and the rhs evaluated at time step n. Then, averaging over a complete period T_σ of Mimas, setting $\Delta t = T_\sigma$, and letting n refer to the nth revolution, Equation (9.53) is replaced with

$$r_{n+1} = 2r_n - r_{n-1} + f(r_n, \theta_n)$$

(9.54)

with the form of the function

$$f(r_n, \theta_n) \equiv \frac{(T_\sigma)^2 F_r(r_n, \theta_n)}{m}$$

(9.55)

still to be established. Applying Newton's law of gravitation to the interaction between Mimas (mass M_σ) and a particle, f will be of the structure

$$f = -a\frac{g(\theta_n)}{(r_n - \sigma)^2}$$

(9.56)

with

$$a \equiv GM_\sigma(T_\sigma)^2 = 4\pi^2\sigma^3\frac{M_\sigma}{M_s},$$

M_s being the mass of Saturn, and the angular dependence $g(\theta_n)$ taking on a very complicated form. By symmetry, however, the function g should be an even function of θ_n. The derivation of the precise structure of g is beyond the scope of this text and therefore we use the form suggested in Gould and Tobochnik .

In this reference it is suggested that for modeling purposes, rather than complete realism, one may choose $g(\theta_n) = \cos(\theta_n)$. Noting that $M_s = 5.68 \times 10^{26}$ kg and expressing radial distances r_n in thousands of km, the parameter $a \simeq 17$. However, since the effects of the other moons of Saturn have been completely neglected and the angular dependence is not precise, there is considerable latitude in the choice of the value for the parameter a. For example, Gould and Tobochnik suggest trying $a = 200$ and $a = 2000$. Although the detailed ring structure depends on the value of a, the existence of some sort of ring structure in the model does not. In the following example, the ring structure is generated for $a = 15$, i.e., a value close to our above estimate.

Example 9-6: The Rings of Saturn

Taking $a = 15$, generate a plot of Saturn's ring pattern.

---

[10]See Chapter 6.

Solution: The orbital radius ($\sigma = 185.7$ thousand km) of Mimas, the proportionality constant ($a = 15$), and Saturn's radius (rs=60.4 thousand km), are entered. In the simulation, tot2 =15 input radii are considered and 4000 iterations of the finite difference equations will be performed for each input radius. The smallest initial radius is taken to be ri=65 thousand km and the radii are incremented in steps of $\Delta = 5$ thousand km, so the outermost input radius in the simulation is $65 + 5 \times 15 = 140$ thousand km.

```
Clear["Global`*"]
```

```
σ = 185.7; a = 15; rs = 60.4; ri = 65; Δ=5; tot = 4000; tot2 = 15;
```

Functions are created for producing the input radii and the finite difference equations (9.47) and (9.48).

```
r0[j_]:=ri+Δ j
```

```
tt[n_]:=θ[n+1] =θ[n] +2 N[Pi] (σ/r[n])^3/2
```

```
rr[n_]:=r[n+1] =2r[n] -r[n-1] -a Cos[θ[n]]/(r[n] -σ)^2
```

Since the radial equation is second order, both the values r_0 and r_1 must be inputted. We set r_0=r0[j] and take $r_1 = r_0$. The initial angle is $\theta_0 = 0$ and we set θ_1=tt[0].

```
r[0] = r0[j]; r[1] = r[0]; θ[0] = 0; θ[1] = tt[0];
```

The following command line iterates the governing equations 4000 times for each of the 15 input radii, producing a colored graph for each input radius. Note that, for plotting purposes, Cartesian coordinates are formed for the points.

```
Do[r[0]; pts=Table[{rr[n] Cos[tt[n]],rr[n] Sin[tt[n]]},
{n,1,tot}]; gr[j] = ListPlot[pts,PlotStyle->{Hue[.06j]},
DisplayFunction->Identity],{j,1,tot2}];
```

The Disk command is used to create a filled-in disk, centered at the origin and of radius rs, representing Saturn.

```
grsat = Disk[{0,0},rs];
```

The graphs are superimposed, with Saturn being given a light blue color and the word SATURN placed on the planetary disk.

```
Show[{Table[gr[j],{j,1,tot2}],Graphics[{RGBColor[0,1,1],grsat}]},
DisplayFunction->$DisplayFunction,Axes->False,AspectRatio->1,
Epilog->Text["SATURN",{-2,0}],TextStyle->{FontFamily->"Times",
FontSize->16},ImageSize->{400,400}];
```

For $a = 15$, the model calculation produces the banded ring structure shown in Figure 9.36. Gaps, where no particles are present in the output, can be clearly seen. One can also observe distinct variations in shading in the bands. The shading is much more pronounced in the original color version generated on the computer screen.

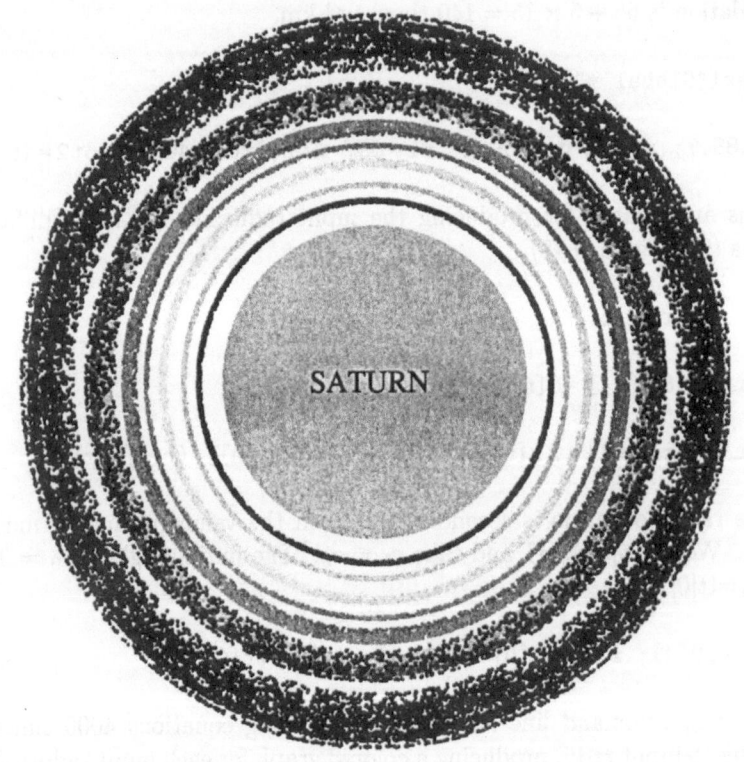

Figure 9.36: Model simulation of the rings of Saturn.

Of course other input radii, initial angles, and values of a could be selected. For certain input radii, the "particles" will either escape towards infinity or will wander inwards and be "captured" by Saturn. These do not contribute to the stable ring structure, so the corresponding graphs should be removed. Similar remarks apply to the situation when varying the parameter a.

| End Example 9-6 |

Let us emphasize that the model calculation presented in this section is intended to show how the ring particles *could* be organized into a ring pattern with gaps, rather than being an accurate predictor of the actual detailed ring pattern which is observed for Saturn. The interested reader might do a literature search to see if a more rigorous and accurate model has been created.

PROBLEMS

Problem 9-46: Different input radii
Plot the banded ring structure for the rings of Saturn for different initial input radii. Remove those graphs which do not contribute to a stable ring structure.

Problem 9-47: Different a values
Plot the banded ring structure for the rings of Saturn for different a values. Remove those graphs which do not contribute to a stable ring structure.

Problem 9-48: Angular dependence in model
Discuss the physical reasonableness (or lack thereof) of using the angular dependendence $g(\theta_n) = \cos(\theta_n)$ in the Saturn rings model.

Problem 9-49: Other angular dependency
Consider other possible angular forms, e.g., $g = (\cos(\theta_n))^3$, $g = (\cos(\theta_n))^2$, $g = \sin(\theta_n)$, etc., for the function $g(\theta_n)$ and discuss the rings patterns, or absence of rings, that occur.

Problem 9-50: Roche limit for the earth
Calculate the Roche limit for a satellite object orbiting the earth. The satellite's density is twice that of the earth, the earth's mass is $M_e = 5.98 \times 10^{24}$ kg, and the earth's average radius about 6.4 thousand km.

Problem 9-51: Artificial satellites
Explain why artificial satellites are not shredded when their orbits lie inside the Roche limit?

Problem 9-52: Jacobian matrix
For the Saturn's rings model, confirm that the 3×3 Jacobian matrix is equal to 1 and thus the map is conservative.

Problem 9-53: Tri-polar model of the arms race
In an article appearing in the text "Chaos Theory in the Social Sciences" (edited by L. D. Kiel and E. Eliot, University of Michigan Press, 1996), Alvin Saperstein has presented a finite-difference arms race model for three competing nations. The 3-dimensional nonlinear mapping is of the form

$$X_{n+1} = 4aY_n(1 - Y_n) + 4\epsilon Z_n(1 - Z_n),$$

$$Y_{n+1} = 4bX_n(1 - X_n) + 4\epsilon c Z_n(1 - Z_n),$$

$$Z_{n+1} = 4\epsilon X_n(1 - X_n) + 4\epsilon c Y_n(1 - Y_n),$$

with all parameters positive. Here X_n, Y_n, and Z_n are the "devotions" of each nation to arms spending. Devotion is defined as the ratio of arms procurement expenditures to the gross national product, and therefore ranges from 0 to 1. The index n labels a budget cycle.

   a. Set $\epsilon = 0$, so that the equations reduces to a bi-polar model. Discuss the structure of the bi-polar equations.

b. Taking $X_0 = 0.01$, $Y_0 = 0.05$, $a = 0.8$, $b = 0.86$, and $N = 150$ iterations, show that a periodic solution results if X_n is plotted vs. n. What is the periodicity?

c. Consider the tri-polar model with the parameters as in part (b) and $\epsilon = 0.2$, $c = 0.2$, and $Z_0 = 0.02$. What effect does including the third nation in the model have on the periodicity of the solution?

d. Saperstein argued that the onset of chaos in his nonlinear models was a precursor to war. He concluded that generally a tri-polar world is more dangerous than a bi-polar one. Relate, if you can, Saperstein's argument to the real world, past and contemporary.

Chapter 10

Nonlinear PDE Phenomena

Humor is emotional chaos remembered in tranquility.
James Thurber (1894–1961), American humorist

10.1   Introductory Remarks

In Section 3.2.1 the reader encountered the Korteweg–deVries (KdV) equation which has been successfully used to describe the propagation of solitons in various physical contexts, the most historically famous being for shallow water waves in a rectangular canal. Using subscripts to denote partial derivatives with respect to the distance x traveled and the time t elapsed, the KdV equation for the (normalized) transverse displacement ψ of the water waves is

$$\psi_t + \alpha\psi\psi_x + \psi_{xxx} = 0. \tag{10.1}$$

Soliton solutions of such a nonlinear PDE are stable (stable against collisions) solitary waves. Recall that a solitary wave is a localized pulse which travels at constant speed without change of shape despite the "competition" between the nonlinearity and other (e.g., dispersive) terms. Not all nonlinear PDEs support solitary waves.

How does one go about finding solitary wave solutions to a given nonlinear PDE or search for other possible physically important solutions? In this chapter, a few of the basic analytic methods for studying nonlinear PDE phenomena will be outlined. The survey will be far from complete, our goal being to study some of the more important phenomena. For simplicity, we shall further confine our attention to PDEs that involve only one spatial dimension and time. In presenting the nonlinear PDE equations, it is also assumed that the student has already skirmished with the diffusion, wave, and Schrödinger equations which form the mathematical backbone for understanding linear diffusive and wave phenomena. The word skirmish is deliberately used, because tremendous expertise at solving these equations is not essential to understanding the ideas to be presented in this chapter.

In Chapter 5, the reader saw that some simple nonlinear ODEs, such as the Bernoulli and Riccati equations, could be reduced to linear ODEs through appropriate transformations. The linear ODEs could then be solved analytically. Further, in the case of the Bernoulli and Riccati equations, Mathematica's analytic ODE solver was able to solve the nonlinear equations directly. On the other hand, for the simple pendulum equation, although Mathematica can solve it with some help, it was not able to solve the pendulum equation directly (recall Mathematica File 01). This chapter begins by looking at a nonlinear PDE example, Burgers' equation which can be analytically solved by introducing a suitable transformation but cannot be directly solved with Mathematica's DSolve command.

10.2   Burgers' Equation

Burgers' equation, which models the coupling between convection[1] and diffusion in fluid mechanics, is of the form

$$U_t + UU_x = \sigma U_{xx} \tag{10.2}$$

with σ a positive parameter.

Example 10-1: Attempt to Analytically Solve Burgers' Equation

Attempt to analytically solve Burgers' equation directly using Mathematica's PDE solver. Then generate a numerical solution and create 3-dimensional plots for the initial profile $u(x, 0) = x(1 - x)$ and boundary conditions $u(0, t) = 0$ and $u(1, t) = 0$. Consider $\sigma = 0.1$ and $\sigma = 0.01$ and discuss the results.

Solution: We enter Burgers' nonlinear PDE (10.2),

    Clear["Global`*"]

    eq = D[u[x, t], t] + u[x, t] D[u[x, t], x] == σ D[u[x, t], x, x]

$$u^{(0,1)}(x, t) + u(x, t)\, u^{(1,0)}(x, t) == \sigma\, u^{(2,0)}(x, t)$$

and seek an analytic solution using the DSolve command.

    DSolve[eq, u[x,t], {x,t}]

$$\text{DSolve}(u^{(0,1)}(x, t) + u(x, t)\, u^{(1,0)}(x, t) == \sigma\, u^{(2,0)}(x, t), u(x, t), \{x, t\})$$

The attempt is unsuccessful. For most nonlinear PDEs general solutions cannot be obtained, but for some PDEs a so-called "complete integral" can be calculated by loading the calculus package

    << Calculus`DSolveIntegrals`

---

[1]If U is effectively a velocity, as will turn out to be the case, the operator on the lhs of Eq. (10.2) is simply the 1-dimensional convective derivative $\frac{d}{dt} = \frac{\partial}{\partial t} + U\frac{\partial}{\partial x}$.

and applying the `CompleteIntegral` command. This function attempts to find a representative family of particular solutions to eq. The complete integral plays a role similar to that of the Green's function for linear second-order PDEs.

```
CompleteIntegral[eq, u[x,t], {x,t}]
```

$$\text{CompleteIntegral}(u^{(0,1)}(x,t) + u(x,t)\,u^{(1,0)}(x,t) = \sigma\,u^{(2,0)}(x,t), u(x,t), \{x,t\},$$

$$\text{IntegralConstants} \rightarrow B)$$

No useful functional form is returned, so let's opt for a numerical solution, entering the given initial and boundary conditions and seeking a numerical solution for $\sigma = 0.1$ and $\sigma = 0.01$.

```
ic=u[x,0] == x (1-x); bc1=u[0,t] == 0; bc2=u[1,t] == 0;

sol[1] = NDSolve[{eq/.σ->.1,ic,bc1,bc2},u[x,t],{x,0,1},{t,0,5}];

sol[2] = NDSolve[{eq/.σ->.01,ic,bc1,bc2},u[x,t],{x,0,1},{t,0,5}];
```

Three-dimensional plots of the two numerical solutions are created,

```
Block[{$DisplayFunction = Identity},
Do[gr[i] = Plot3D[Evaluate[u[x,t] /. sol[i][[1]]],{t,0,5},{x,0,1},
ViewPoint->{3,1,1},PlotRange->All,PlotPoints->{20,20},
Ticks->{{0,{2.5,"t"},5},{0,{.5,"x"},1},{{.1,"u"},.2}},
TextStyle->{FontFamily->"Times",FontSize->16}],{i,1,2}];];
```

and displayed side by side in Figure 10.1 using the `GraphicsArray` command.

```
Show[GraphicsArray[{gr[1],gr[2]}],ImageSize->{800,300}];
```

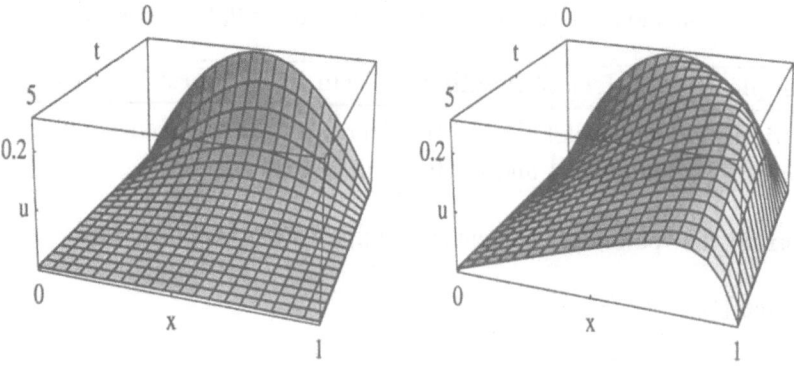

Figure 10.1: Solution of Burgers' eq. for $\sigma = 0.1$ (left), $\sigma = 0.01$ (right).

For $\sigma = 0.1$, the numerical solution on the left displays typical diffusive behavior, σ being large enough for the linear diffusion term (σU_{xx}) to dominate over the nonlinear term $(U U_x)$ in Burgers' equation. For $\sigma = 0.01$, on the other hand, a steepening trend is seen near $x = 1$, the diffusion being sufficiently slow for the nonlinear term to make an appreciable contribution.

$\boxed{\text{End Example 10-1}}$

As was seen in Example 10-1, for Burgers' equation the temporal evolution of an initial profile $U(x, 0)$ depends on the outcome of the competition between the nonlinear and linear diffusive terms. To more fully appreciate what this competition involves, it is instructive to look at the effect of each of the competing terms separately. Let us begin by keeping only the nonlinear term so that

$$U_t + U U_x = 0. \tag{10.3}$$

The general solution of this equation is $U = f(x - Ut)$, where f is an arbitrary (presumably, well-behaved) function. This general solution is easily verified. Setting $z = x - Ut$, $df/dz \equiv f'$, and differentiating the solution with respect to t and x yields

$$U_t = (-tU_t - U)f'$$
$$U_x = (1 - tU_x)f'. \tag{10.4}$$

Then, on multiplying the second of these relations by U and adding and rearranging, we obtain

$$(U_t + U U_x)(1 + tf') = 0 \tag{10.5}$$

so (10.3) follows. What happens when $1 + tf' = 0$ will be discussed shortly.

Let's now examine the general solution of Equation (10.3). For a specified form of $f(z)$, one will generally have an implicit solution for $U(x, t)$. Note also that since $z = x - Ut$ and $U = f(z)$, U plays both the role of a velocity and an amplitude, with larger amplitudes propagating faster to the right than smaller amplitudes. The consequence of this can be easily seen by choosing an illustrative form for $f(z)$. For example, we can take $f(z) = e^{-az^2}$ with $a = 0.1$, and use the `ContourPlot` command with $t = 3n$, $n = 0, 1, ...,$ to extract $U(x, t)$.

$\boxed{\textbf{Example 10-2: Effect of Nonlinear Term in Burgers' Equation}}$

Numerically solve and plot the solution $U(x, t)$ of the transcendental equation $U = e^{-a(x - Ut)^2}$ for $a = 0.1$ and $t = 0$, 3 and 6.

Solution: The parameter a is entered and, setting $t = 3n$,

```
Clear["Global`*"]; << Graphics`

a = .1;
```

the function $F(n) = U - e^{-a(x - 3nU)^2}$ formed.

F[n_] := U - Exp[-a (x - (3 n) U)^2]

The transcendental equation corresponds to setting $F(n) = 0$. This may be accomplished by choosing Contours->{0} in the following ContourPlot command. To see the contour lines, the shading is turned off with the option ContourShading->False. The Do loop is used to create graphs for $n = 0, 1, 2$, i.e., $t = 0, 3, 6$.

```
Do[gr[i] = ContourPlot[F[i],{x,-10,10},{U,0,1.3},Contours -> {0},
      ContourShading -> False,PlotPoints -> 100,
      DisplayFunction -> Identity], {i,0,2}];
```

The three graphs created above are superimposed to produce Figure 10.2.

```
Show[Table[gr[i],{i,0,2}],DisplayFunction -> $DisplayFunction,
PlotRange -> {{-11,11},{0,1.2}},FrameTicks -> {{-10,-5,{0.01,"0"},
{2.5,"x"},5,10},{{.001,"0"},.6,{.8,"U"},1.2},{ },{ }},Epilog->
{Text["t=0",{-.5,1.05}],Text["t=3",{2.75,1.05}],Text["t=6",
{6,1.05}]},TextStyle->{FontFamily->"Times",FontSize->16},
AspectRatio->2/3,ImageSize->{600,400}];
```

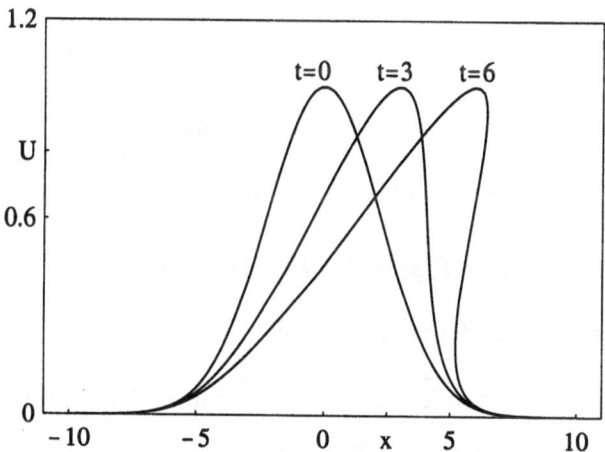

Figure 10.2: Time evolution of the shock structure.

Examination of Figure 10.2 reveals that the input pulse moves to the right with the "top" traveling faster than the "bottom". This is an example of a "shock" structure, such a structure having a point of infinite slope develop after a finite time. This may be seen in the $t = 6$ curve. We can now better understand how the nonlinear term in Burgers' equation leads to the steepening that was observed in Figure 10.1 for $\sigma = 0.01$. In this case the steepening was muted by the presence of small, but nonzero, diffusion.

End Example 10-2

Continuing with our analysis of the shock structure, it follows from the second line in Equation (10.4) that the slope of the U curve is given by

$$U_x = \frac{f'}{1 + f't}. \tag{10.6}$$

Infinite slope occurs when the denominator vanishes, the minimum time to develop being $t_{min} = 1/(-f')_{max}$. The minimum time depends on the maximum negative slope of the initial profile $U(x,0) = f(z)$.

Since, on neglecting the linear 3rd derivative term, the KdV equation has exactly the same structure[2] as Equation (10.3), its nonlinear term will tend to produce the same shock structure in the absence of the competing dispersive 3rd derivative term.

Now, keep only the linear term in Burgers' equation so that

$$U_t = \sigma U_{xx}. \tag{10.7}$$

The student should recognize this as the linear diffusion equation which is used to describe:

- heat flow in solids (U is the temperature and σ the thermal conductivity divided by the product of the density and specific heat);

- diffusion of, e.g., ink in water (U is the concentration of ink molecules and σ the diffusivity);

- diffusion of thermal neutrons in a nuclear reactor;

to mention only a few examples. Given some initial profile $U(x,0) = f(x)$, the linear diffusion equation may be readily solved using a variety of standard mathematical techniques. For example, we can make use of the Laplace transform

$$F(s) = \int_0^\infty f(x,t)\, e^{-st}\, dt \tag{10.8}$$

and its inverse, as in the following example.

Example 10-3: Laplace Transform Solution of Diffusion Equation

Use the Laplace transform approach to analytically solve the diffusion equation (10.7) for the initial profile $U(x,0) = \sin(\pi x)$ and boundary conditions $U(0,t) = U(1,t) = 0$. Create a 3-dimensional plot of the solution for $\sigma = 0.1$.

Solution: The diffusion equation is entered,

```
Clear["Global`*"]; << Calculus`

eq = D[u[x, t], t] - σ D[u[x, t], x, x] == 0
```

$$u^{(0,1)}(x,t) - \sigma\, u^{(2,0)}(x,t) == 0$$

---
[2]Noting that the α factor can be absorbed into the time t.

and the Laplace transform applied to `eq`.

```
lap1 = LaplaceTransform[eq, t, s]
```

$$s\,\mathrm{LaplaceTransform}(u(x,t),t,s) - \sigma\,\mathrm{LaplaceTransform}(u^{(2,0)}(x,t),t,s)$$
$$-u(x,0) == 0$$

The notation is simplified in the `lap1` output and the initial profile substituted.

```
lap2 = lap1 /. {LaplaceTransform[u[x, t], t, s] -> U[x],
    LaplaceTransform[D[u[x, t], x, x], t, s] -> U"[x],
    u[x, 0] -> Sin[Pi x]}
```

$$-\sin(\pi x) + s\,U(x) - \sigma\,U''(x) == 0$$

The ODE `lap2` is solved subject to the two boundary conditions and simplified.

```
lap2sol = DSolve[{lap2, U[0] == 0, U[1] == 0}, U[x], x] // Simplify
```

$$\{\{U(x) \to \frac{\sin(\pi x)}{s + \pi^2\,\sigma}\}\}$$

The desired solution follows on performing the inverse Laplace transform.

```
sol = InverseLaplaceTransform[U[x] /. lap2sol[[1]], s, t]
```

$$e^{-\pi^2 t\,\sigma}\sin(\pi x)$$

The analytic solution is now plotted in 3 dimensions for $\sigma = 0.1$.

```
Plot3D[Evaluate[sol /. σ->.1], {x,0,1}, {t,0,2}, ViewPoint->{3,2,1},
PlotPoints -> {20,20}, Ticks -> {{0,{.5,"x"},1},{0,{1,"t"},2},
{{.5,"u"},1}}, TextStyle->{FontFamily->"Times",FontSize->16},
ImageSize->{600,400}];
```

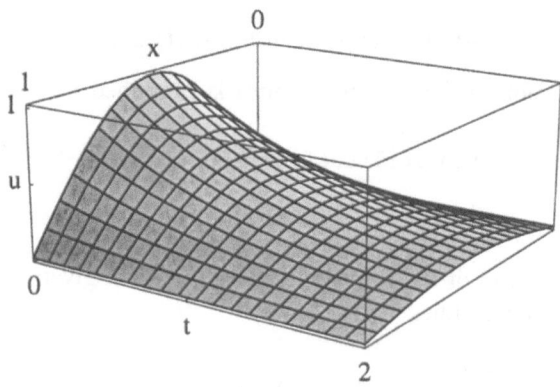

Figure 10.3: Analytic solution of diffusion equation.

In using the Laplace transform approach, it should be noted that Mathematica cannot always find the inverse transform, in particular when the answer involves an infinite sum.

End Example 10-3

When the domain is infinite in extent, the Fourier transform pair

$$f(x,t) = \frac{1}{\sqrt{2\pi}} \int_{-\infty}^{\infty} dk\, F(k,t)\, e^{ikx}, \quad F(k,t) = \frac{1}{\sqrt{2\pi}} \int_{-\infty}^{\infty} dx\, f(x,t)\, e^{-ikx}, \quad (10.9)$$

can be used to solve the diffusion equation for a specified initial profile. For example, suppose that we have a physical situation where the initial concentration of diffusing molecules is in a very tiny region which is approximated as a point. The concentration (number of molecules per unit volume (length in one dimension)) is very large (infinite) at the point and zero outside. If the point is at the origin ($x = 0$), then the initial concentration can be described by the Dirac delta (δ) function[3], viz., $f(x) = \delta(x)$. The following example solves the linear diffusion equation for this profile.

Example 10-4: Fourier Transform Solution of Diffusion Equation
---

Given the initial concentration $f(x) = \delta(x)$, solve the diffusion equation for the time-dependent concentration and plot the analytic solution for $\sigma = 0.5$.

Solution: The linear diffusion term is entered,

```
Clear["Global`*"]; << Calculus`

term = D[u[x, t], x, x]
```

$$u^{(2,0)}(x,t)$$

and the Fourier transform taken.

```
four1 = FourierTransform[term, x, k]
```

$$-k^2 \, \text{FourierTransform}(u(x,t), x, k)$$

The default notation in the output of four1 is simplified.

```
four2 = four1 /. {FourierTransform[u[x, t], x, k] -> U[t]}
```

$$-k^2 \, U(t)$$

The Fourier transform of the diffusion equation is then given by the output of the following command line.

---

[3]Dirac's delta function is defined as $\delta(x - a) = 0$ for $x \neq a$ and equal to infinity for $x = a$. Its area is normalized to unity, $\int_{-\infty}^{\infty} \delta(x - a)dx = 1$. Provided that the function $g(x)$ is not another delta function, the "sifting property" $\int_{-\infty}^{\infty} g(x)\delta(x - a)dx = g(a)$ prevails.

```
eq = D[U[t], t] - σ four2 == 0
```

$$\sigma\, U(t)\, k^2 + U'(t) == 0$$

To solve the above ODE, the Fourier transform of the initial profile is calculated,

```
fouric = FourierTransform[DiracDelta[x], x, k]
```

$$\frac{1}{\sqrt{2\pi}}$$

and the ODE eq solved with fouric as the initial condition.

```
foursol = DSolve[{eq, U[0] == fouric}, U[t], t]
```

$$\{\{U(t) \to \frac{e^{-k^2 t \sigma}}{\sqrt{2\pi}}\}\}$$

The Fourier transformed solution is given by the output of foursol2.

```
foursol2 = U[t] /. foursol[[1]]
```

$$\frac{e^{-k^2 t \sigma}}{\sqrt{2\pi}}$$

The desired analytic solution of the diffusion equation is obtained by taking the inverse Fourier transform of foursol2.

```
sol = InverseFourierTransform[foursol2, k, x]
```

$$\frac{e^{\frac{-x^2}{4 t \sigma}}}{2\sqrt{\pi}\sqrt{t\sigma}}$$

A table of analytic results is created for $\sigma = 0.5$ and $t = 0.25, 0.50, 0.75, 1.0, 1.25$.

```
f = Table[sol /. {σ -> .5, t -> .25 i}, {i, 1, 5}]
```

$$\{0.797885\, e^{-2.\, x^2},\ 0.56419\, e^{-1.\, x^2},\ 0.460659\, e^{-0.666667\, x^2},$$
$$0.398942\, e^{-0.5\, x^2},\ 0.356825\, e^{-0.4\, x^2}\}$$

Using a Do loop, colored plots are produced for each of the above time results,

```
Do[gr[i] = Plot[f[[i]], {x, -4, 4}, PlotRange -> All, PlotStyle ->
      {Hue[.2 i]}, DisplayFunction -> Identity], {i, 1, 5}];
```

and the graphs superimposed in Figure 10.4 using the Show command.

```
Show[Table[gr[i], {i, 1, 5}], DisplayFunction->$DisplayFunction,
Ticks->{{-4,{.001,"0"},{3,"t"},4},{{.2,"U"},.8}},TextStyle->
{FontFamily->"Times",FontSize->16},ImageSize->{600,400}];
```

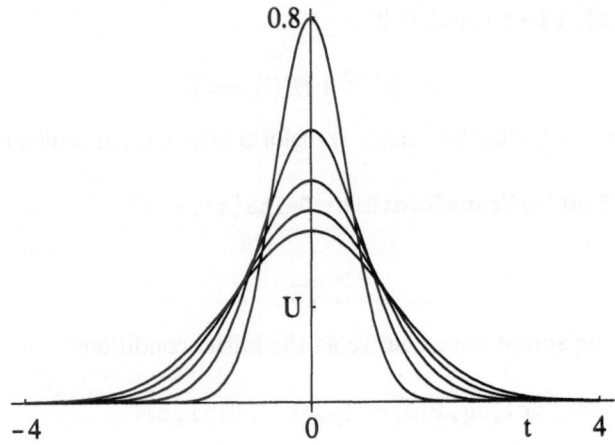

Figure 10.4: Linear diffusion of concentration spike.

A 3-dimensional picture can also be generated with the Plot3D command. This is done in the following command line for the time interval $t = 0.25$ to $t = 2.5$, the result being shown in Figure 10.5.

```
Plot3D[Evaluate[sol/. σ->.5],{x,-4,4},{t,.25,2.5},ViewPoint->
{3,2,1},PlotRange->All,PlotPoints->{40,40},Ticks->{{-4,0,
{2,"x"},4},{.25,{1.25,"t"},2.5},{{.35,"U"},.8}},TextStyle->
{FontFamily->"Times",FontSize->16},ImageSize->{600,400}];
```

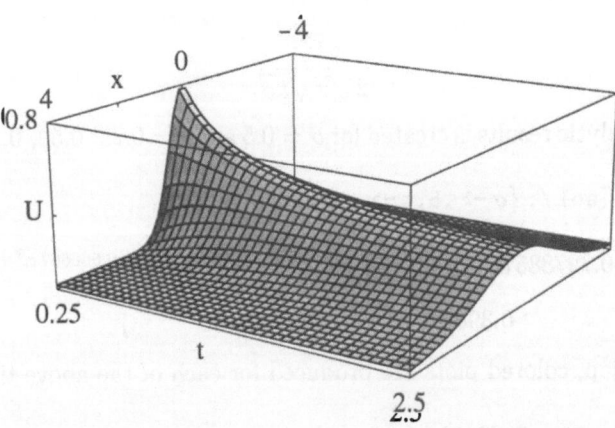

Figure 10.5: 3-dimensional plot of diffusion of concentration spike.

The initial concentration "spike" decreases in height and spreads outwards.

 End Example 10-4

Now that some feeling has been gained about what the competing terms do in Burgers' equation, an analytic solution is sought to the complete equation.

First, set $U = -2\sigma u$ and $\tau = \sigma t$ so that Burgers' equation becomes

$$u_\tau = u_{xx} + 2uu_x. \tag{10.10}$$

In this form, the nonlinear PDE can be linearized by using the Hopf–Cole transformation discovered in 1950–51 by E. Hopf [Hop50] and J. D. Cole [Col51], viz.,

$$u = (\ln V)_x = V_x/V. \tag{10.11}$$

Taking the time (τ) derivative of (10.11) and interchanging the order of the time and spatial derivatives, then

$$u_\tau = (\ln V)_{x\tau} = (V_\tau/V)_x. \tag{10.12}$$

But Burgers' equation (10.10) can be rewritten as

$$u_\tau = (u_x + u^2)_x \tag{10.13}$$

so, on comparing the rhs of the last two equations and using the Hopf–Cole transformation, we have

$$V_\tau = (u_x + u^2)V = u_x V + u(V_x/V)V = (uV)_x = V_{xx}. \tag{10.14}$$

Thus, Burgers' equation has been transformed into the linear diffusion equation

$$V_\tau = V_{xx} \tag{10.15}$$

or

$$V_t = \sigma V_{xx} \tag{10.16}$$

which the reader knows how to solve. Once $V(x, t)$ is obtained, the solution to Burgers' equation in its original form is $U = -2\sigma V_x/V$.

PROBLEMS
Problem 10-1: Minimum time for shock to develop
Neglecting the linear term in Burgers' equation, determine from the analytic formula in the text the minimum time for a point of infinite slope to develop when $f(z) = e^{-az^2}$ with $a = 0.1$. Confirm this answer by running the Mathematica code that was given in Example 10-2.

Problem 10-2: Laplace transform
Using Mathematica, calculate the Laplace transform of the functions
  a. $f(t) = t\,e^{at}$,
  b. $f(t) = t^2 \sin(t)$,
  c. $f(t) = \sin(2t)\sinh(3t)$,
  d. $f(t) = \cos(\sqrt{t})$,
  e. $f(t) = J_0(t)$ (the zeroth order Bessel function).

Problem 10-3: Inverse Laplace transform
Using Mathematica, calculate the inverse Laplace transform of each of the following functions and plot the result over a suitable range of t.

    a. $F(s) = s^2/(s^2+1)^2$,

    b. $F(s) = 1/(\sqrt{s}\,(s^2-4))$,

    c. $F(s) = \arctan(1/s)/s$.

Problem 10-4: Fourier transform

Using Mathematica, calculate the Fourier transform of the functions

    a. $f(x) = e^{-a|x|}$,

    b. $f(x) = \cos(2x)$,

    c. $f(x) = x/(x^2+1)$.

Problem 10-5: Inverse Fourier transform

Using Mathematica, calculate the inverse Fourier transform of each of the following functions and plot the result over the range $x = -3...3$.

    a. $g(k) = \dfrac{1}{1+k^2}$,

    b. $g(k) = \dfrac{\sin k}{k}$,

    c. $g(k) = \dfrac{\sin^2 k}{k^2}$.

Problem 10-6: Example 10-1

In Example 10-1, consider the input profile $u(x,0) = x^n(1-x)$, with $n = 2, 3, 4, ...$, all other conditions being held the same. Discuss what happens as n increases for the two plots.

Problem 10-7: Example 10-2

Rerun the code in Example 10-2, creating a plot for times $t = 3n$, $n = 0, 1, 2, ...10$. Discuss the result. Discuss the similarities and differences of a water wave with respect to the shock behavior.

Problem 10-8: Solving Burgers' equation

In the text, the linear diffusion equation was solved for a Dirac delta function input. If this is the input into the diffusion equation for V, what is the analytic solution to Burgers' equation for $t > 0$. Discuss the result in terms of the two competing terms.

Problem 10-9: Similarity solution of a nonlinear diffusion equation

Consider the following nonlinear diffusion equation for the concentration C

$$C_t = (D(C)C_x)_x$$

with the diffusion constant D replaced with the function $D(C) = C^n$. Buckmaster [Buc77] has taken $n = 3$ to model the spreading of thin liquid films under the action of gravity. Muskat [Mus37] has used $n \geq 1$ to investigate the percolation of gas through porous media. The value $n = 6$ has been used by Larsen and Pomraning [LP80] to study radiative heat transfer by Marshak waves.

Verify by direct substitution that a solution to the above nonlinear diffusion equation is

$$Ct^{1/(n+2)} = (E - \frac{n}{2(n+2)}z^2)^{1/n}$$

where $z \equiv x/t^{1/(n+2)}$ and E is a constant. The solution vanishes (i.e., $C = 0$) outside the z value obtained by setting the bracket on the rhs equal to zero. The variable z is an example of a similarity variable (a new independent variable which is formed from the original two independent variables). Unlike the situation for "normal behaving" diffusion equations, here there is a sharp interface separating regions of nonzero and zero concentration C. Similarity methods for finding similarity solutions such as the one presented here are discussed in Bluman and Cole [BC74].

10.3 Bäcklund Transformations

10.3.1 The Basic Idea

From a known solution of one nonlinear PDE, it is sometimes possible to generate a solution to another nonlinear PDE through a so-called Bäcklund transformation. An auto-Bäcklund transformation, on the other hand, can produce another solution to the same nonlinear PDE.

The difficult part is finding the Bäcklund transformation for the given nonlinear PDE. Various methods are used for different classes of equations and these are discussed in the references given in Daniel Zwillinger's *Handbook of Differential Equations* [Zwi89]. The methods are quite involved, so we shall be content here to give a few illustrative examples of how known Bäcklund and auto-Bäcklund transformations are applied.

10.3.2 Examples

As a first example, once again consider Burgers' equation

$$U_t + UU_x = \sigma U_{xx} \qquad (10.17)$$

for which the solution $U(x, t)$ is known. Let $\phi(x, t)$ be the solution of the linear PDE

$$\phi_t + U(x, t)\phi_x = \sigma \phi_{xx}. \qquad (10.18)$$

Then,

$$V(x, t) = -2\sigma \frac{\phi_x}{\phi} + U(x, t) \qquad (10.19)$$

also satisfies Burgers' equation. The proof is left as a problem.

A simple application of the above transformation is as follows. $U(x, t) = 0$ is obviously a solution of Burgers' equation. Then $\phi(x, t)$ satisfies the linear diffusion equation

$$\phi_t = \sigma \phi_{xx}. \qquad (10.20)$$

From Example 10-4, it is known that one solution of this equation is

$$\phi(x,t) = \frac{e^{-x^2/(4\sigma t)}}{\sqrt{(4\pi\sigma t)}}. \tag{10.21}$$

Thus, applying Equation (10.19), $V(x,t) = x/t$ is obtained as a nontrivial solution of Burgers' equation. This can be confirmed by direct substitution and differentiation.

As a second example, consider the sine-Gordon equation (SGE) which, recall, is a model equation for a Bloch wall between two ferromagnetic domains, viz.,

$$U_{xx} - U_{tt} = (\frac{\partial}{\partial x} - \frac{\partial}{\partial t})(\frac{\partial}{\partial x} + \frac{\partial}{\partial t})U = \sin U. \tag{10.22}$$

Introducing new independent variables $X = \frac{1}{2}(x+t)$, $T = \frac{1}{2}(x-t)$ and using the chain rule of calculus, the SGE can be rewritten in the form

$$U_{XT} = \sin U. \tag{10.23}$$

Example 10-5: Sine-Gordon Transformation

Confirm the independent variable transformation of (10.22) into (10.23).

Solution: The independent variable transformations are entered, with X and T temporarily replaced with the symbols y and z, respectively.

```
Clear["Global`*"]
```

```
y = (x+t)/2;  z = (x-t)/2;
```

Noting that Mathematica knows about the chain rule of calculus, we take the second x derivative of $u(y,z)$ and simplify the result. This is labeled term1.

```
term1 = D[u[y, z], x, x] // Simplify
```

$$\frac{1}{4}\left(u^{(0,2)}\left(\frac{t+x}{2}, \frac{x-t}{2}\right) + 2\,u^{(1,1)}\left(\frac{t+x}{2}, \frac{x-t}{2}\right) + u^{(2,0)}\left(\frac{t+x}{2}, \frac{x-t}{2}\right)\right)$$

Similarly, in term2 the second t derivative of $u(y,z)$ is calculated.

```
term2 = D[u[y, z], t, t] // Simplify
```

$$\frac{1}{4}\left(u^{(0,2)}\left(\frac{t+x}{2}, \frac{x-t}{2}\right) - 2\,u^{(1,1)}\left(\frac{t+x}{2}, \frac{x-t}{2}\right) + u^{(2,0)}\left(\frac{t+x}{2}, \frac{x-t}{2}\right)\right)$$

To evaluate the left-hand side of (10.22), we subtract term2 from term1.

```
lhs = (term1 - term2) // Simplify
```

$$u^{(1,1)}\left(\frac{t+x}{2}, \frac{x-t}{2}\right)$$

Letting $u \to U$, $(t+x)/2 \to X$, and $(x-t)/2 \to T$ yields the lhs of Eq. (10.23).

```
lhs2 = lhs /. {u -> U, (t+x)/2 -> X, (x-t)/2 -> T}
```

$$U^{(1,1)}(X,T)$$

Equating `lhs2` to $\sin(U(X,T)$ yields the complete Equation (10.23).

```
sinegordoneq = lhs2 == Sin[U[X, T]]
```

$$U^{(1,1)}(X,T) == \sin(U(X,T))$$

$\boxed{\textbf{End Example 10-5}}$

An auto-Bäcklund transformation for the sine-Gordon equation (10.23) is given by the pair of nonlinear PDEs

$$
\begin{aligned}
V_X &= U_X + 2a\sin((V+U)/2) \\
V_T &= -U_T + \frac{2}{a}\sin((V-U)/2)
\end{aligned}
\tag{10.24}
$$

where a is an arbitrary parameter. If U is a solution of the SGE and V satisfies the transformation, then V is also a solution of the SGE. The proof, which is quite simple, is again left as a problem.

To see how the transformation works, let's again start with the trivial [4] solution $U(x,t) = 0$. Then, on temporarily setting $\tilde{X} = aX$, $\tilde{T} = t/a$, the pair of transformation equations reduce to

$$V_{\tilde{X}} = 2\sin(V/2), \qquad V_{\tilde{T}} = 2\sin(V/2). \tag{10.25}$$

Subtraction yields

$$V_{\tilde{X}} - V_{\tilde{T}} = 0 \tag{10.26}$$

which has the general solution

$$V = f(\tilde{X} + \tilde{T}) \equiv f(y). \tag{10.27}$$

Now $V_{\tilde{X}} = (df/dy)(dy/d\tilde{X}) = df/dy$ and $V_{\tilde{T}} = df/dy$, so that on adding the equations (10.25), we obtain

$$2\frac{df}{dy} = 4\sin\left(\frac{f}{2}\right). \tag{10.28}$$

This equation is readily integrated yielding

$$\ln\tan\left(\frac{f}{4}\right) = y + \text{constant} \tag{10.29}$$

---

[4]In the literature, this is often referred to as the "vacuum" solution.

or, on solving for f and changing back to our original variables,

$$V = 4\arctan(Ae^{a(X+T/a^2)}) = 4\arctan(Ae^{\pm(x-ct)/\sqrt{(1-c^2)}}) \qquad (10.30)$$

with A an arbitrary constant and $c \equiv (1 - a^2)/(1 + a^2)$. The plus/minus sign is determined by the sign[5] of a, i.e., plus is used if a is positive and minus if a is negative. To see what this new solution looks like, set $A = 1$, $c = 0.5$ and, using Mathematica, plot Equation (10.30) as shown in Figure 10.6. In both

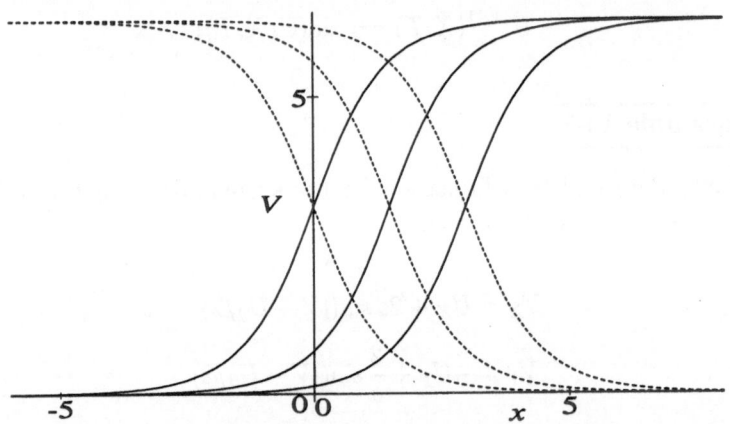

Figure 10.6: Sine-Gordon kink (solid line) and anti-kink (dashed line) solitons obtained by applying auto-Bäcklund transformation to vacuum solution.

cases the solution is propagating to the right (from left to right, $t = 0, 3, 6$) with constant velocity ($c = 0.5$) and unchanging shape. The solutions are the "kink" and "anti-kink" solitons, respectively, which were schematically depicted in Chapter 3 in Figure 3.20. The amplitude of each of these solitons varies from 0 to 2π, so they are often referred to as 2π pulses.

PROBLEMS

Problem 10-10: Bäcklund transformation for Burgers' equation
Prove that if $U(x,t)$ is a solution of Burgers' equation, then $V(x,t)$ given by Equation (10.19) is also a solution of Burgers' equation.

Problem 10-11: Changing the constant A
Confirm the kink and anti-kink soliton profiles in Figure 10.6. What effect does changing the value of the constant A to some other positive value have?

Problem 10-12: Auto-Bäcklund transformation for the SGE
Differentiate the first equation in (10.24) with respect to T and the second with respect to X. Equating the two expressions for V_{XT} and using an appropriate trigonometric identity, show that $U_{XT} = \sin U$. Adding the two expressions, show that $V_{XT} = \sin V$, thus confirming the statements in the text.

---

[5]Since $a = \pm\sqrt{(1 - c)/(1 + c)}$.

10.3.3 Nonlinear Superposition

In the "old" world of linear physics, i.e., the world governed by linear differential equations, the principle of linear superposition of solutions plays an extremely important role. The concepts of Fourier series and Fourier transforms, for example, are based on linear superposition. In the "new" world of nonlinear physics, this basic principle has been unfortunately lost, as was pointed out in Section 1.1. Is it possible to have nonlinear superposition of solutions for nonlinear differential equations? The answer is yes, but one has to work out what form the nonlinear superposition takes on an equation by equation basis. As an illustrative example, a nonlinear superposition formula will be derived for the SGE. Use shall be made of the auto-Bäcklund transformation (10.24).

The basic idea behind the derivation is illustrated schematically in Figure 10.7. Start with a known solution U_0. If the transformation with the

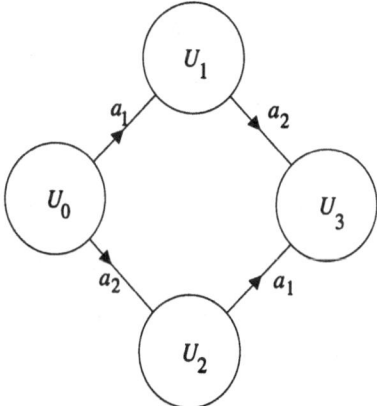

Figure 10.7: Basic approach to deriving a nonlinear superposition formula.

parameter a_1 is employed, the solution U_1 is generated. Choosing a different parameter a_2 produces another solution, U_2. Then applying the transformation to U_1 with the parameter a_2 produces still another solution U_3. But, assuming the order of the operations doesn't matter,[6] applying the transformation to U_2 with the parameter a_1 will generate the same solution U_3.

To ease the burden of writing the transformations, the notation $U_{ij}^{\pm} \equiv (U_i \pm U_j)/2$ shall be introduced. Then, referring to Figure 10.7 and Equation (10.24), the transformation equations are

$$(U_{10}^-)_X = a_1 \sin U_{10}^+ \; ; \; (U_{10}^+)_T = \frac{1}{a_1} \sin U_{10}^-;$$

$$(U_{20}^-)_X = a_2 \sin U_{20}^+ \; ; \; (U_{20}^+)_T = \frac{1}{a_2} \sin U_{20}^-;$$

$$(U_{31}^-)_X = a_2 \sin U_{31}^+ \; ; \; (U_{31}^+)_T = \frac{1}{a_2} \sin U_{31}^-; \qquad (10.31)$$

$$(U_{32}^-)_X = a_1 \sin U_{32}^+ \; ; \; (U_{32}^+)_T = \frac{1}{a_1} \sin U_{32}^-.$$

---

[6]More formally, this is Bianchi's theorem of permutability.

But clearly,

$$U_{32}^- = U_{31}^- + U_{10}^- - U_{20}^- \tag{10.32}$$

so by comparing equations differentiated with respect to X, then

$$a_1(\sin U_{32}^+ - \sin U_{10}^+) = a_2(\sin U_{31}^+ - \sin U_{20}^+). \tag{10.33}$$

Using the trigonometric identity (apply `Simplify` to the rhs to confirm)

$$\sin x - \sin y = 2\sin((x-y)/2)\cos((x+y)/2) \tag{10.34}$$

and noting that $U_{32}^+ + U_{10}^+ = U_{31}^+ + U_{20}^+$, this last equation reduces to

$$a_1\sin((U_{32}^+ - U_{10}^+)/2) = a_2\sin((U_{31}^+ - U_{20}^+)/2) \tag{10.35}$$

or, equivalently,

$$a_1\sin((U_{30}^- - U_{12}^-)/2) = a_2\sin((U_{30}^- + U_{12}^-)/2). \tag{10.36}$$

Applying the trigonometric identity (use `TrigExpand` on the lhs to confirm)

$$\sin(x \pm y) = \sin x \cos y \pm \cos x \sin y \tag{10.37}$$

to this result and dividing through by $\cos(U_{30}^-/2)\cos(U_{12}^-/2)$ yields

$$a_1(\tan(U_{30}^-/2) - \tan(U_{12}^-/2)) = a_2(\tan(U_{30}^-/2) + \tan(U_{12}^-/2)). \tag{10.38}$$

Rearranging and going back to our original notation, the following nonlinear superposition equation is finally obtained:

$$\tan\left(\frac{U_3 - U_0}{4}\right) = \frac{a_1 + a_2}{a_1 - a_2}\tan\left(\frac{U_1 - U_2}{4}\right). \tag{10.39}$$

This superposition result relates four different solutions of the SGE. Although more restrictive than linear superposition, nonlinear superposition can still prove quite useful.

For example, starting with the trivial or vacuum solution $U_0(x,t) = 0$, the auto-Bäcklund transformation for two different positive values of a, a_1 and a_2, can be used to generate the two kink solutions

$$U_1 = 4\arctan(e^{(x-c_1 t)/(\sqrt{(1-c_1^2)})})$$
$$U_2 = 4\arctan(e^{(x-c_2 t)/(\sqrt{(1-c_2^2)})}). \tag{10.40}$$

The nonlinear superposition equation then yields still another solution given by

$$U_3 = 4\arctan\left(\frac{a_1 + a_2}{a_1 - a_2}\tan\left(\frac{U_1 - U_2}{4}\right)\right) \tag{10.41}$$

with [7] $a_1 = \sqrt{((1-c_1)/(1+c_1))}$ and $a_2 = \sqrt{((1-c_2)/(1+c_2))}$.

---

[7] If, for example, a_2 is negative, a minus sign must be inserted before the square root, i.e., $a_2 = -\sqrt{((1-c_2)/(1+c_2))}$, as well as in the exponential in U_2.

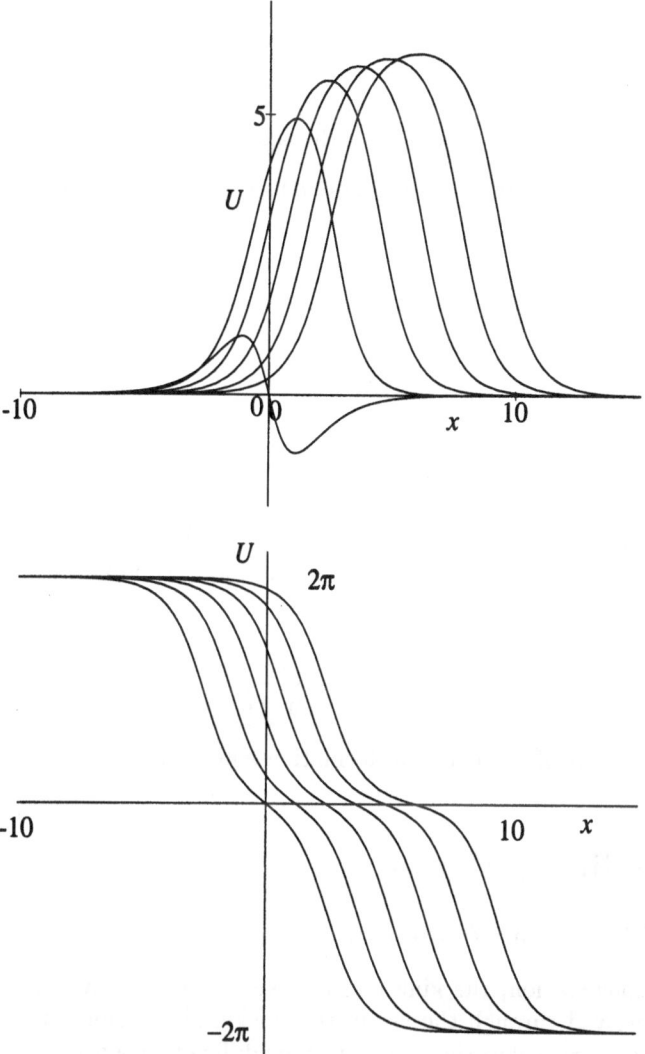

Figure 10.8: Time evolution (left to right) of 0π (top) and 4π (bottom) pulses.

Availing ourselves of Mathematica's plotting capability, the evolution of this new solution is displayed in the top plot of Figure 10.8 for $c_1 = 0.5$, $c_2 = 0.3$. The times shown are for $t = 0, 3, 6, ..., 15$. Do you think that this solution could be a soliton? Clearly not. This solution is referred to as a 0π pulse because its amplitude varies from zero at the "back" ($x \to -\infty$) of the pulse to zero at the "front" ($x \to \infty$).

A 4π pulse, with the amplitude varying from -2π on one side of the pulse to $+2\pi$ on the other, can be created by taking a_1 to be positive and a_2 to be negative. This solution is displayed in the bottom plot of Figure 10.8. Is it a soliton? Look at the distance advanced on each time step for the top and bottom portions of the pulse.

PROBLEMS

Problem 10-13: 4π pulse

Making use of the nonlinear superposition equation, confirm the form and temporal evolution of the 4π pulse shown in the text for $c_1 = 0.5$ and $c_2 = 0.3$. What does the solution U_3 look like if the sign of c_1 is reversed?

Problem 10-14: Nonlinear superposition for the KdV equation

For the KdV equation

$$U_t + 6UU_x + U_{xxx} = 0$$

the auto-Bäcklund transformation is

$$U_x + V_x = a - \frac{1}{2}(U - V)^2$$

$$U_t + V_t = (U - V)(U_{xx} - V_{xx}) - 2(U_x)^2 + U_x V_x - 2(V_x)^2$$

where a is a parameter. If U is a solution of the KdV equation, so is V. Using the notation employed in Figure 10.7, show that the nonlinear superposition equation for the KdV equations is

$$U_3 = U_0 + \frac{2(a_1 - a_2)}{U_1 - U_2}.$$

Hint: Use the first of the two transformation equations.

10.4   Solitary Waves

10.4.1   The Basic Approach

In the previous section, the kink and anti-kink solitary wave solutions to the SGE were derived through the use of the Bäcklund transformation. Since the Bäcklund transformations themselves may be difficult to generate for given nonlinear PDEs, this approach to obtaining solitary wave solutions is not the best way to proceed. A more elementary method is to appeal to the basic definition of a solitary wave as a localized traveling wave whose shape U does not change as it propagates along with a velocity c. That is to say, assume that $U(x,t) = U(z)$ where $z = x - ct$, with $c > 0$ for a wave traveling to the right and $c < 0$ for motion to the left. To localize the solution, further assume that U goes either to zero or to a nonzero constant value, and all derivatives vanish as $|z| \to \infty$. Given a nonlinear PDE of physical interest, we attempt to determine whether a solution of this structure exists. Remember that for solitary waves to be possible, there must exist a "balance" between the nonlinearity and the other terms (e.g., linear diffusive or dispersive terms) in the equation. Not every nonlinear PDE has solitary wave solutions, and even if an equation does have one, it may not be possible to find the exact analytic form of the solitary wave. Note also that by the assumed structure of the solitary wave solution, the number of independent variables has been reduced from two (x, t) to one (z). A consequence of this will be that the nonlinear PDE is reduced to an ODE.

If a solitary wave is found for a given nonlinear PDE, one then can investigate whether or not it is a soliton. Recall that a soliton is a solitary wave which survives a collision with another solitary wave unchanged in shape and traveling with the original velocity.[8] Two of the main approaches to establishing the soliton nature of a solitary wave solution is through numerical simulation and through an analytic method called the "inverse scattering method". The numerical simulation approach looks at collisions between pairs of solitary waves which are initially well separated. The inverse scattering method generates multi-soliton solutions, e.g., the 2-soliton solution which the student encountered in Mathematica File MF08. Both approaches shall be explored in subsequent chapters. Let's now look at a simple method of determining whether solitary wave solutions are possible, and then derive analytic forms for some physically important nonlinear PDEs.

10.4.2 Phase Plane Analysis

The reader might be startled to learn that phase plane analysis, which was applied to nonlinear ODEs in Chapter 4 to obtain a preliminary idea of the possible solutions, can be applied to nonlinear PDEs to determine whether solitary waves are possible. The solitary waves that are being sought come in two kinds of shapes, the "peaked" and the "kink" (or, anti-kink) varieties, as schematically indicated in Figure 10.9. For the former the solution vanishes at $|z| = \infty$ while

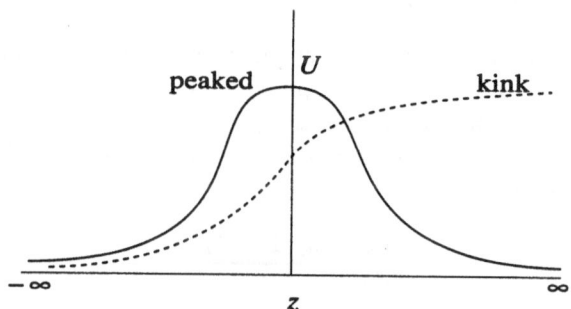

Figure 10.9: Schematic representation of two types of solitary waves.

for the latter the solution goes to two different constant values at $z = -\infty$ and $z = \infty$.

The reason that phase plane analysis is applicable is that the basic assumption that $U(x,t) = U(z = x - ct)$ reduces the nonlinear PDE to a nonlinear ODE. The single solitary wave then will correspond to a separatrix solution as shall be illustrated.

As an example, consider the sine-Gordon equation

$$U_{xx} - U_{tt} = \sin U \tag{10.42}$$

---

[8]In the physics research literature, this strict definition is often relaxed somewhat. Pulses which travel unchanged in shape in the absence of any interaction, but suffer small shape changes as a consequence of a collision, are also referred to as solitons. Strictly speaking, they are not solitons but should be referred to as soliton-like or quasi-solitons.

which the reader already knows has kink and anti-kink solitary wave solutions. Suppose that we had not been able to find these analytic forms, but wanted to establish the possible existence of solitary wave solutions. Assuming that $z = x - ct$, then

$$\frac{\partial}{\partial x} = \frac{d}{dz}\frac{\partial z}{\partial x} = \frac{d}{dz}$$

$$\frac{\partial}{\partial t} = \frac{d}{dz}\frac{\partial z}{\partial t} = -c\frac{d}{dz}. \tag{10.43}$$

Then, calculating the second derivatives in a similar manner, the sine-Gordon equation reduces to the nonlinear ODE

$$(1 - c^2)U_{zz} = \sin U. \tag{10.44}$$

For $c > 1$ this would just be the simple pendulum equation whose phase plane behavior was previously studied. Let's look at the situation[9] for $c < 1$. Setting $Y = U_z$ yields the two coupled first-order equations

$$Y_z = \frac{\sin U}{(1 - c^2)}, \quad U_z = Y, \tag{10.45}$$

or on dividing to eliminate z,

$$\frac{dY}{dU} = \frac{\sin U/(1 - c^2)}{Y}. \tag{10.46}$$

As with the pendulum equation, the singular or stationary points are at $Y_0 = 0, U_0 = n\pi (n = 0, \pm 1, \pm 2, ...)$. We proceed in the same manner as in Section 4.4.1, using the same general notation.

For ordinary points close to $U_0 = Y_0 = 0$, we have on writing $U = U_0 + u, Y = Y_0 + v$ and linearizing in u and v,

$$\frac{dv}{du} = \frac{u/(1 - c^2)}{v}. \tag{10.47}$$

Then[10] identify $a = 0$, $b = 1$, $c = 1/(1 - c^2) > 0$, and $d = 0$. Thus, $p = -(a + d) = 0$ and $q = ad - bc = -1/(1 - c^2) < 0$. From Figure 4.8, the origin can be identified as a saddle point.

For ordinary points near $U_0 = \pi, Y_0 = 0$, the values $a = 0$, $b = 1$, $c = -1/(1 - c^2) < 0$, $d = 0$, $p = 0$, and $q = 1/(1 - c^2) > 0$ result. Thus, the singular point $U_0 = \pi, Y_0 = 0$ is a vortex or focal point. Poincaré's theorem for the vortex establishes that it is a vortex. Similarly, $U_0 = -\pi, Y_0 = 0$ is also a vortex, $U_0 = \pm 2\pi, Y_0 = 0$ saddle points, and so on. Figure 10.10 shows the full phase plane portrait, the heavy curves indicating the separatrixes connecting the saddle points. The arrows indicate the direction of increasing z. A kink solution can be immediately spotted. The separatrix line[11] for positive U and Y connecting the origin and the saddle point at $U_0 = 2\pi, Y_0 = 0$ corresponds

---

[9]What solution corresponds to $c = 1$?

[10]Don't confuse the phase plane label c on the lhs with the speed c on the rhs!

[11]This separatrix line connecting two different saddle (equilibrium) points is an example of what mathematicians call a "heteroclinic orbit".

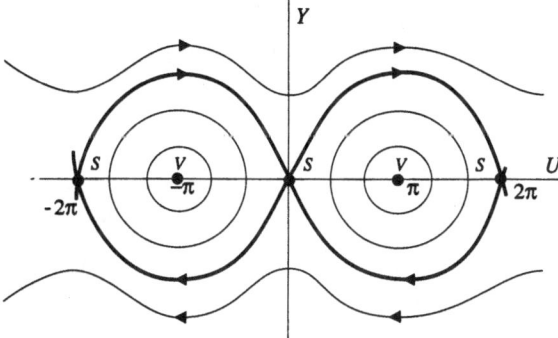

Figure 10.10: Phase plane portrait for the "reduced" sine-Gordon equation, i.e., the nonlinear ODE obtained by assuming $z = x - ct$.

to the kink solution that was obtained previously. Starting at the origin where $z \to -\infty$, as z increases U goes from zero to 2π at the saddle point in the limit that $z \to \infty$. Can you spot the anti-kink solution? Traveling wave solutions can also be seen (e.g., the thinner lines surrounding the vortex at π) where U oscillates up and down as z increases.

As a second example, consider the nonlinear Schrödinger equation (NLSE) which was introduced in Section 3.2.4 as a model equation for the propagation of bright and dark solitons in a transparent optical fiber,

$$iE_z \pm \frac{1}{2}E_{\tau\tau} + |E|^2 E = 0. \tag{10.48}$$

Bright and dark solitons as schematically depicted in Figure 3.23, where the light intensity $|E|^2$ is sketched versus the (normalized) time τ at a fixed distance z along the fiber, are possible for the plus and minus signs respectively. To see that this is so, let's assume a "stationary" solution[12] of the form $E(z,\tau) = X(\tau)e^{i\beta z}$, with X real and β real and positive. The intensity then will be $|E|^2 = (X(\tau))^2$. The plus sign in the NLSE will be examined here, the minus sign case being left as a problem. Substituting the assumed form into the NLSE yields

$$X_{\tau\tau} + 2(X^2 - \beta)X = 0. \tag{10.49}$$

Setting $X_\tau = Y$ gives

$$\frac{dY}{dX} = \frac{2(\beta - X^2)X}{Y}. \tag{10.50}$$

There are three singular points at $X_0 = Y_0 = 0$ and $X_0 = \pm\sqrt{\beta}, Y_0 = 0$. Again writing $X = X_0 + u$, $Y = Y_0 + v$, and linearizing produces the following results:

- $X_0 = Y_0 = 0$: $a = 0$, $b = 1$, $c = 2\beta$, $d = 0$, $p = 0$, $q = -2\beta < 0$, so the origin is a saddle point.

- $X_0 = \pm\sqrt{\beta}, Y_0 = 0$: $a = 0$, $b = 1$, $c = -2\beta$, $d = 0$, $p = 0$, $q = 2\beta > 0$, so these singular points are (using Poincaré's theorem) vortices.

---

[12]In deriving the NLSE from Maxwell's equations, a transformation is made from the laboratory frame to a frame moving with the group velocity. The solitary wave solution derived here is stationary with respect to the latter frame.

A schematic phase plane portrait incorporating these results is given in Figure 10.11, the separatrix lines being shown as thick curves. Again the arrows

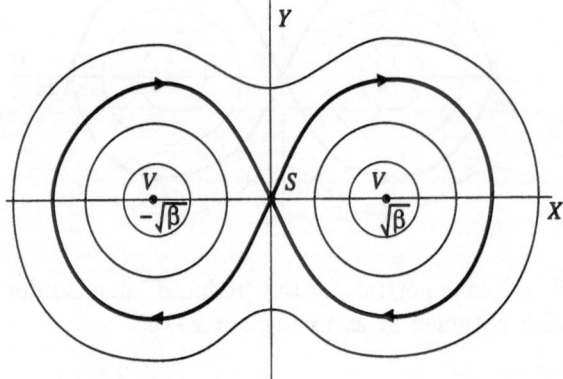

Figure 10.11: Phase plane portrait for "reduced" NLSE.

indicate the direction of increasing (normalized) time τ. The phase plane portrait may be confirmed by adapting the Mathematica code for the tangent field given in Example 4-3 to the current example, the result being shown in Figure 10.12. Referring to Figure 10.11, the separatrix line leaving the origin for

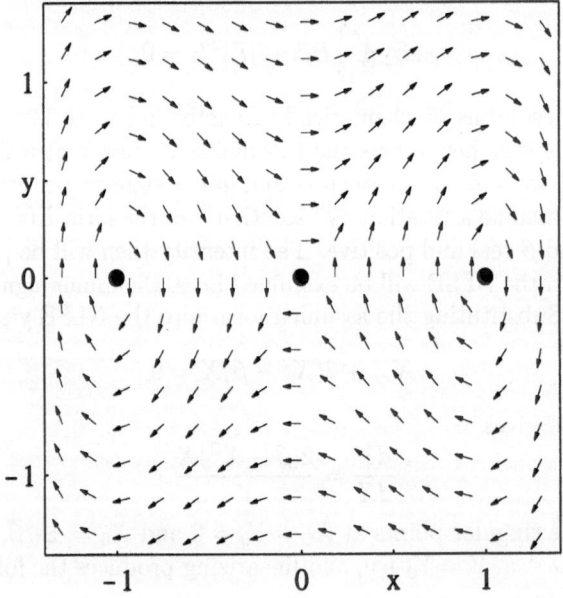

Figure 10.12: Tangent field for the reduced NLSE with $\beta = 1$.

$X > 0$ corresponds to the bright solitary wave. This line[13] tells us that X starts at zero amplitude at $\tau = -\infty$, increases to a maximum positive value at

---

[13]The separatrix line which leaves and returns to the same saddle point is referred to by mathematicians as a "homoclinic orbit".

intermediate τ, and then decreases to zero again as $\tau \to \infty$. Since the intensity involves the square of X, the other separatrix line for $X < 0$ gives the same result. One can also see from the picture that non-solitary wave solutions are possible.

This phase plane approach supplemented by subsequent numerical integration is often used in research calculations where analytic solutions are not possible. The derivation of analytic solitary wave solutions will now be illustrated with several examples.

PROBLEMS
Problem 10-15: Dark solitary wave
Using phase plane analysis confirm that a dark solitary wave exists for the minus sign case in the NLSE. Remember that it is the intensity that is relevant. Confirm your phase plane picture by using the tangent field option.

10.4.3 KdV Equation

What better example is there for producing an analytic solitary wave solution than the historically famous KdV equation

$$\psi_t + \alpha\psi\psi_x + \psi_{xxx} = 0. \tag{10.51}$$

Recall from Section 3.2.1 that the KdV water wave solitons were the first solitons to be noted in nature, when the Scottish engineer John Scott Russell followed them on horseback along a canal in the year 1834.

Assuming $\psi(x,t) = \psi(x - ct) = \psi(z)$, the KdV equation reduces to the third-order nonlinear ODE

$$\psi_{zzz} + (\alpha\psi - c)\psi_z = 0. \tag{10.52}$$

It is straightforward to integrate this equation. The first integration yields

$$\psi_{zz} = A_1 + c\psi - \frac{1}{2}\alpha\psi^2 \tag{10.53}$$

where A_1 is the integration constant. Assuming that the solitary wave has asymptotic boundary conditions $\psi, \psi_{zz} \to 0$ for $|z| \to \infty$, then $A_1 = 0$. The boundary condition corresponds physically to the solution vanishing at plus and minus infinity and the curvature going to zero as well.

Then multiplying this last equation by $2\psi_z dz$ and integrating once again, we obtain

$$(\psi_z)^2 = A_2 + c\psi^2 - \frac{1}{3}\alpha\psi^3 \tag{10.54}$$

with a second integration constant A_2. Assuming that for our solitary wave the slope (ψ_z) also vanishes for $|z| \to \infty$, $A_2 = 0$.

Finally, integrating a third time yields

$$\int^\psi \frac{d\psi}{\psi\sqrt{(1 - \frac{\alpha}{3c}\psi)}} = -2\tanh^{-1}(\sqrt{(1 - \frac{\alpha}{3c}\psi)}) = \sqrt{c}(z + A_3) \tag{10.55}$$

which, on using the trigonometric identity $1 - \tanh^2 \theta = \mathrm{sech}^2 \theta$, can be transformed into

$$\psi = \frac{3c}{\alpha} \mathrm{sech}^2 (\frac{\sqrt{c}}{2}(z + A_3)). \qquad (10.56)$$

Without any loss of generality, set $A_3 = 0$ so that

$$\psi(x, t) = \frac{3c}{\alpha} \mathrm{sech}^2 (\frac{\sqrt{c}}{2}(x - ct)). \qquad (10.57)$$

Of course, solitary wave solutions such as this one can also be derived with Mathematica as illustrated in the next example.

Example 10-6: Derivation of KdV Solitary Wave Solution

Derive the KdV solitary wave solution (10.57) starting with the third-order nonlinear PDE (10.51). Animate the solution and confirm that taller KdV solitary waves travel faster than shorter ones. Take $c = 1$ and $c = 3$ and normalize ψ to remove the factor α.

Solution: A solution of the form $\Psi = U(x - ct)$ is assumed,

```
Clear["Global`*"]; << Graphics`
```

```
Ψ = U[x - c t];
```

and substituted into the KdV equation (10.51).

```
de1 = D[Ψ,t] + α  Ψ  D[Ψ,x] + D[Ψ,x,x,x] == 0
```

$$-cU'(x - ct) + \alpha U(x - ct) U'(x - ct) + U^{(3)}(x - ct) == 0$$

The substitution $x \to z + ct$ is made to simplify the arguments in **de1**.

```
de1 = de1 /. x -> z + c t
```

$$-cU'(z) + \alpha U(z) U'(z) + U^{(3)}(z) == 0$$

The lhs of **de1** is integrated with respect to z and equated to zero. The integration constant is equal to zero for a solitary wave.

```
de2 = Integrate[de1[[1]], z] == 0
```

$$\frac{1}{2}\alpha U(z)^2 - cU(z) + U''(z) == 0$$

To integrate **de2**, let $V = \dfrac{dU}{dz}$ so that

$$\frac{d^2U}{dz^2} = \frac{dV}{dz} = \frac{dV}{dU}\frac{dU}{dz} = V(U)\frac{dV(U)}{dU}.$$

Then the substitutions $U(z) \to U$ and $d^2U/dz^2 \to V(U)\,dV(U)/dU$ are made on the lhs of **de2** and the result set equal to zero.

```
de3 = {de2[[1]] /. {U[z] -> U, D[U[z], z, z] -> V[U] D[V[U], U]}} == 0
```

$$\{\frac{\alpha U^2}{2} - cU + V(U)V'(U)\} == 0$$

As $|z| \to \infty$, both U and $dU/dz = V$ must vanish, so the DSolve command is applied to **de3** subject to the condition $V(U = 0) = 0$,

```
sol = DSolve[{de3, V[0] == 0}, V[U], U]
```

$$\{\{V(U) \to -\sqrt{cU^2 - \frac{U^3\alpha}{3}}\}, \{V(U) \to \sqrt{cU^2 - \frac{U^3\alpha}{3}}\}\}$$

yielding positive and negative square root solutions. The positive square root is selected,

```
sol1 = V[U] /. sol[[2]]
```

$$\sqrt{cU^2 - \frac{U^3\alpha}{3}}$$

and its reciprocal integrated with respect to U to give z as a function of U.

```
eq1 = z == Integrate[1/sol1, U]
```

$$z == -\frac{2U\sqrt{3c - U\alpha}\,\tanh^{-1}\left(\frac{\sqrt{3c - U\alpha}}{\sqrt{3}\sqrt{c}}\right)}{\sqrt{3}\sqrt{c}\sqrt{cU^2 - \frac{U^3\alpha}{3}}}$$

The output of **eq1** is simplified assuming $U > 0$,

```
eq2 = FullSimplify[eq1, U > 0]
```

$$z == -\frac{2\tanh^{-1}\left(\frac{\sqrt{3c - U\alpha}}{\sqrt{3}\sqrt{c}}\right)}{\sqrt{c}}$$

and **eq2** is solved for U.

```
sol2 = Solve[eq2, U]
```

$$\{\{U \to -\frac{3\left(c\tanh^2\left(\frac{\sqrt{c}z}{2}\right) - c\right)}{\alpha}\}\}$$

The output of **sol2** is converted to the form of Equation (10.57) by making use of the identity $\tanh(\theta) = \sqrt{1 - \text{sech}^2(\theta)}$ and recalling that $z = x - ct$.

```
sol3 = Simplify[(sol2 /. Tanh[Sqrt[c] z/2]^2 ->
    1 - Sech[Sqrt[c] z/2]^2) /. z -> x - c t]
```

$$\{\{U \to \frac{3\,c\,\mathrm{sech}^2(\frac{1}{2}\sqrt{c}\,(x-ct))}{\alpha}\}\}$$

This analytic result, whether derived by hand or with computer algebra, is the long sought solitary wave solution of the KdV equation. The height of the pulse is proportional to the speed c and the width inversely proportional to $\sqrt{c}$. This tells us that taller solitary waves should travel faster than shorter ones. This is easily confirmed by animating the solution for two different values of c. First, let's form the normalized solitary wave solution $\phi = (\alpha/3)\,U$.

```
φ = (α/3)(U /. sol3[[1]])
```

$$c\,\mathrm{sech}^2(\frac{1}{2}\sqrt{c}\,(x-ct))$$

Then substitute the speeds $c = 1$ and $c = 3$ into the normalized solution, labeling the resulting expressions as phi1 and phi2, respectively.

```
phi1 = φ /. c -> 1;  phi2 = φ /. c -> 3;
```

On animating phi1 and phi2 and executing the command line, the reader will observe behavior similar to that shown in Figure 10.13. The taller pulse corresponds to $c = 3$, the shorter to $c = 1$.

```
Animate[Plot[{phi1, phi2}, {x,-10,40}, PlotStyle->Hue[.6]], {t,0,10},
Frames -> 20, PlotPoints -> 200, PlotRange -> {{-10,40},{0,3}},
Ticks -> {{-10,{0.01,"0"},10,20,{30,"x"},40},{1,2,{2.5,"φ"},3}},
TextStyle -> {FontFamily -> "Times", FontSize -> 12}]
```

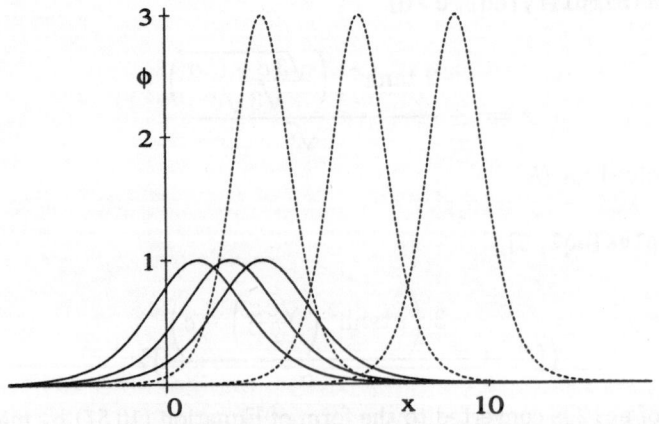

Figure 10.13: Propagation of short and tall KdV solitary waves. Here ϕ is the normalized height and the profiles are for $t = 1, 2, 3$.

End Example 10-6

The fact that taller KdV solitary waves travel faster than shorter ones can be used to test if the solitary waves are solitons. Taking $c > 0$ and the positive x-axis pointing to the right, simply start the taller solitary wave to the left of the shorter one. The taller solitary pulse will overtake the shorter one and it is observed numerically as well as experimentally [BMP83] that both solitary waves survive the collision unchanged in shape and velocity. The reader will learn how to confirm this aspect numerically in the next chapter. An analytic approach, called the inverse scattering method, to demonstrate this collision invariance is the subject of our last chapter. Using this method, one can derive the two soliton solution

$$\psi = \left(\frac{72}{\alpha}\right) \frac{3 + 4\cosh(2x - 8t) + \cosh(4x - 64t)}{[3\cosh(x - 28t) + \cosh(3x - 36t)]^2}, \tag{10.58}$$

encountered in Chapter 3 and the subject of Mathematica File MF08. Most readers will have taken our word that Equation (10.58), which is quite complicated in appearance, satisfies the KdV equation. However, it's possible that a mistake was made in typing the equation into this text. Of course one can verify the solution by substituting it into the lhs of the KdV equation and checking that the lhs reduces to zero. However, this is a very tedious task to do by hand and better accomplished with Mathematica.

Example 10-7: Confirmation of Two Soliton Solution

Confirm that the solution (10.58) satisfies the KdV equation.

Solution: The proposed two soliton solution is entered.

```
Clear["Global`*"]
```

```
Ψ₂ = (72/α) (3 + 4 Cosh[2 x - 8 t] + Cosh[4 x - 64 t])/
    (3 Cosh[x - 28 t] + Cosh[3 x - 36 t])^2
```

$$\frac{72\,(\cosh(64t - 4x) + 4\cosh(8t - 2x) + 3)}{\alpha\,(\cosh(36t - 3x) + 3\cosh(28t - x))^2}$$

On inputting the lhs of the KdV equation (labeled pde), the expression Ψ_2 is automatically substituted into pde and the differentiations performed.

```
pde = D[Ψ₂,t] + α Ψ₂ D[Ψ₂,x] + D[Ψ₂,x,x,x]
```

The lengthy output has been omitted here, but may be viewed on the computer screen on executing the pde command line. Manipulating the output by hand would be a formidable task. On the other hand, applying Simplify to pde,

```
Simplify[pde]
```

$$0$$

quickly yields zero, thus confirming the correctness of the two soliton formula.

End Example 10-7

PROBLEMS

Problem 10-16: Sine-Gordon breather mode

In Chapter 3 it was stated that the SGE permits a moving (velocity v) "breather mode" solution, which is localized in space but oscillatory in time, of the form

$$\psi = 4\arctan\left(\sqrt{\frac{m}{(1-m)}}\frac{\sin((t-vx)\gamma\sqrt{(1-m)})}{\cosh((x-vt)\gamma\sqrt{m})}\right)$$

with $\gamma = 1/\sqrt{1-v^2}$, $-1 < v < 1$, and $0 < m < 1$. Using the `FullSimplify` command, and taking $m = 1/2$ and $\gamma = 1/2$, confirm that the breather solution satisfies the SGE.

Problem 10-17: Traveling wave solution to KdV equation

Show that if the integration constants A_1, A_2 are not set equal to zero, there exists a general traveling wave solution to the KdV equation expressible in terms of an elliptic function and that the solitary wave corresponds to the infinite period limit.

10.4.4   Sine-Gordon Equation

As demonstrated in the following example, the computer algebra approach is quite useful for deriving solitary wave solutions of other nonlinear PDEs. The kink and anti-kink solitary wave solutions were previously derived for the sine-Gordon equation using the Bäcklund transformation. Now, the kink solution will be obtained by assuming that $\psi(x,t) = U(z)$ where $z = x - ct$.

Example 10-8: Sine-Gordon Soliton

Derive the kink solitary wave solution to the sine-Gordon equation and animate the solution for $c = 1/4$. By animating the known kink-kink two soliton solution,

$$\psi = 4\arctan\left(c\frac{\sinh(x/\sqrt{1-c^2})}{\cosh(ct/\sqrt{1-c^2})}\right)$$

confirm that the kink solitary wave is a soliton.

Solution: The assumed form $\Psi = U(x - ct)$ is entered,

```
Clear["Global`*"];  << Graphics`
```

```
Ψ = U[x - c t];
```

and automatically substituted into the following SGE.

```
pde = D[Ψ,x,x] - D[Ψ,t,t] - Sin[Ψ] == 0
```

$$-U''(x-ct)c^2 - \sin(U(x-ct)) + U''(x-ct) == 0$$

The arguments of U and U'' are simplified with the substitution $x \to z + ct$,

```
de1 = pde /. x -> z + c t
```

$$-U''(z)\,c^2 - \sin(U(z)) + U''(z) == 0$$

yielding a second-order nonlinear ODE in $U(z)$. The second derivative terms, $U''(z)$, are collected on the lhs of **de1** and the result equated to zero.

```
de2 = Collect[de1[[1]], U''[z]] == 0
```

$$\left(1 - c^2\right) U''(z) - \sin(U(z)) == 0$$

From the output, we see that nontrivial solutions only occur if c is not equal to 1. As requested, the value $c = 1/4$ will be used. The choice of a specific c value is not necessary but makes the resulting solution simpler in form.

```
c = 1/4;
```

As with the KdV example, we assume that $V = dU/dz$, so that the substitution $d^2 U(z)/dz^2 \to V(U)\,dV(U)/dU$ can be made on the lhs of **de2**. On also letting $U(z) \to U$ and equating to zero, a first-order ODE results which can be analytically solved.

```
de3 = {de2[[1]] /. {U[z] -> U, D[U[z],z,z] -> V[U] D[V[U],U]}} == 0
```

$$\{\frac{15}{16} V(U)\,V'(U) - \sin(U)\} == 0$$

For $z \to -\infty$, we must have $U = 0$ and $V \equiv dU/dz = 0$ for our kink solitary wave. So **de3** is solved subject to $V(U = 0) = 0$,

```
sol = DSolve[{de3, V[0] == 0}, V[U], U]
```

$$\{\{V(U) \to -\sqrt{\frac{2}{15}}\,\sqrt{16 - 16\cos(U)}\}, \{V(U) \to \sqrt{\frac{2}{15}}\,\sqrt{16 - 16\cos(U)}\}\}$$

yielding positive and negative square root solutions. The positive square root will yield the kink solitary wave while the negative square root gives the anti-kink. Let's select the positive square root solution and simplify it.

```
sol1 = Simplify[V[U] /. sol[[2]]]
```

$$4\sqrt{\frac{2}{15}}\,\sqrt{1 - \cos(U)}$$

The solution **sol1** can be further simplified by making the trig substitution $\cos U = 1 - 2\sin^2(U/2)$ and assuming $\sin(U/2) > 0$.

```
sol1 = Simplify[sol1 /. Cos[U] -> 1 - 2 Sin[U/2]^2, Sin[U/2] > 0]
```

$$\frac{8 \sin \left(\dfrac{U}{2} \right)}{\sqrt{15}}$$

Then, $z = x - ct$ must be equal to the integral with respect to U of the reciprocal of sol1.

```
z = x - c  t == Integrate[1/sol1,U]
```

$$x - \frac{t}{4} == \frac{1}{8}\sqrt{15}\left(2 \log\left(\sin\left(\frac{U}{4}\right)\right) - 2 \log\left(\cos\left(\frac{U}{4}\right)\right)\right)$$

The kink solitary wave solution follows on solving the previous output for U,

```
sol2 = Solve[z, U]
```

$$\{\{U \rightarrow -4 \csc^{-1}\left(e^{-\frac{4x}{\sqrt{15}}} \sqrt{e^{\frac{2t}{\sqrt{15}}} + e^{\frac{8x}{\sqrt{15}}}} \right)\},$$

$$\{U \rightarrow 4 \csc^{-1}\left(e^{-\frac{4x}{\sqrt{15}}} \sqrt{e^{\frac{2t}{\sqrt{15}}} + e^{\frac{8x}{\sqrt{15}}}} \right)\}\}$$

and selecting, say, the second answer.

```
U = U /. sol2[[2]]
```

$$4 \csc^{-1}\left(e^{-\frac{4x}{\sqrt{15}}} \sqrt{e^{\frac{2t}{\sqrt{15}}} + e^{\frac{8x}{\sqrt{15}}}} \right)$$

The mathematical structure of U differs from that obtained earlier using the Bäcklund transformation. We could try to force it into the same form as before, but an easier way to see that it really is a kink solution is to animate the solution.

```
Animate[Plot[Chop[U],{x,-10,50}, PlotStyle -> Hue[.6]], {t,0,200},
Frames -> 20,PlotPoints -> 200,PlotRange -> {{-10,50},{0,7}},
Ticks->{{-10,{.01,"0"},10,20,{30,"x"},40,50},{Pi,{4.5,"Ψ"},
2Pi}}, TextStyle -> {FontFamily -> "Times",FontSize -> 12}]
```

On running the animation, the reader will see that we have indeed obtained the kink solution to the sine-Gordon equation. To verify that the solitary wave is a soliton, the two soliton solution is entered, and animated.

```
Φ = 4 ArcTan[c Sinh[x/Sqrt[1-c^2]]/Cosh[c t/Sqrt[1-c^2]]]
```

$$4 \tan^{-1}\left(\frac{1}{4} \operatorname{sech}\left(\frac{t}{\sqrt{15}}\right) \sinh\left(\frac{4x}{\sqrt{15}}\right)\right)$$

```
Animate[Plot[Φ,{x,-50,50},PlotStyle->Hue[.6]],{t,-100,100},
Frames -> 20,PlotPoints -> 200,PlotRange -> {{-50,50},{-7,7}},
Ticks->{{-50,-25,25,{40,"x"},50},{-2Pi,-Pi,Pi,{4.5,"Φ"},2Pi}},
TextStyle -> {FontFamily -> "Times",FontSize -> 12}]
```

The two kink solitary waves "bounce" off each other in the animation and have exactly the same appearance after the collision as before. The kink solitary waves are solitons.

$\boxed{\textbf{End Example 10-8}}$

10.4.5   The Three-Wave Problem

Hitting the wall on the three-wave problem

A problem of interest in nonlinear optics [ER79] is the resonant nonlinear interaction of three collinear plane waves. Two electromagnetic waves with velocities v_1 and v_2 and a sound wave with velocity v_2 ($v_2 \ll v_0, v_1$) interact according to the real amplitude equations

$$(\phi_0)_t + v_0(\phi_0)_x = -\beta_0\phi_1\phi_2$$

$$(\phi_1)_t + v_1(\phi_1)_x = \beta_1\phi_0\phi_2 \qquad (10.59)$$

$$(\phi_2)_t + v_2(\phi_2)_x + \Gamma\phi_2 = -\beta_2\phi_0\phi_1.$$

Here x is the direction of propagation, t is the time, the coupling parameters β_i, $i = 0, 1, 2$, are real and positive, and the sound wave has a real, positive damping coefficient Γ. Although we could use a computer algebra approach, let's tackle this third example by hand.

Setting $z = x - ct$, with the velocity c undetermined, (10.59) reduces to

$$(\phi_0)_z = -\gamma_0\phi_1\phi_2$$

$$(\phi_1)_z = \gamma_1\phi_0\phi_2 \qquad (10.60)$$

$$(\phi_2)_z + \tilde{\Gamma}\phi_2 = -\gamma_2\phi_0\phi_1$$

where $\gamma_i \equiv \beta_i/(v_i - c)$ and $\tilde{\Gamma} \equiv \Gamma/(v_2 - c)$.

Multiplying the first and second equations in (10.60) by ϕ_0/γ_0 and ϕ_1/γ_1 respectively and adding yields

$$\frac{d}{dz}\left(\frac{\phi_0^2}{\gamma_0} + \frac{\phi_1^2}{\gamma_1}\right) = 0, \tag{10.61}$$

or

$$\frac{\phi_0^2}{\gamma_0} + \frac{\phi_1^2}{\gamma_1} = \text{constant} = \alpha^2. \tag{10.62}$$

This latter conservation equation suggests that one might seek solutions of the structure

$$\phi_0 = \alpha\sqrt{\gamma_0}\cos(\psi/2), \qquad \phi_1 = \alpha\sqrt{\gamma_1}\sin(\psi/2), \tag{10.63}$$

where ψ remains to be found. Clearly these forms satisfy the conservation equation. With these assumed forms, the first two equations in (10.60) both reduce to

$$\psi_z = 2\sqrt{(\gamma_0\gamma_1)}\phi_2, \tag{10.64}$$

while (using the identity $\sin\psi = 2\sin(\psi/2)\cos(\psi/2)$) the third becomes

$$(\phi_2)_z + \tilde{\Gamma}\phi_2 = -\frac{1}{2}\alpha^2\sqrt{(\gamma_0\gamma_1)}\gamma_2\sin\psi. \tag{10.65}$$

Combining these last two equations to eliminate ϕ_2 yields a familiar nonlinear ODE for ψ, viz.,

$$\psi_{zz} + \tilde{\Gamma}\psi_z + \kappa^2\sin\psi = 0 \tag{10.66}$$

with $\kappa^2 \equiv \alpha^2\gamma_0\gamma_1\gamma_2$. For κ^2 and $\tilde{\Gamma}$ positive, which is guaranteed if the velocity $c < v_2$, the problem has been reduced to that of the damped simple pendulum. To obtain an analytic result, set $\Gamma = 0$. From the discussion of the undamped pendulum equation in Section 5.2.4, recall that if the transformation

$$\sin(\psi/2) = k\sin\phi \tag{10.67}$$

is introduced, we obtain in the present context

$$\kappa z = \int^\phi \frac{d\phi}{\sqrt{(1 - k^2\sin^2\phi)}} + A, \tag{10.68}$$

where A is the integration constant. For a localized solution, the period must be infinite, so that $k = 1$. Then,

$$\kappa z = \int^\phi \frac{d\phi}{\cos\phi} + A \tag{10.69}$$

or, on integrating,

$$\kappa z = \frac{1}{2}\ln\left(\frac{1 + \sin\phi}{1 - \sin\phi}\right) + A. \tag{10.70}$$

Solving for $\sin \phi$ and using Equation (10.67) with $k = 1$ yields

$$\frac{\psi}{2} = \sin^{-1}(\frac{Ce^{2\kappa z} - 1}{Ce^{2\kappa z} + 1}). \tag{10.71}$$

Particular solutions are obtained by fixing the constant C, which is simply related to the integration constant A. For example, choosing $C = 1$ gives

$$\frac{\psi}{2} = \sin^{-1}(\tanh(\kappa z)) \tag{10.72}$$

and[14]

$$\phi_1 = \alpha\sqrt{\gamma_1} \sin(\psi/2) = \alpha\sqrt{\gamma_1} \tanh(\kappa z) \tag{10.73}$$

$$\phi_0 = \alpha\sqrt{\gamma_0} \cos(\psi/2) = \alpha\sqrt{\gamma_0} \operatorname{sech}(\kappa z). \tag{10.74}$$

To obtain ϕ_2, differentiate Equation (10.72) yielding

$$\psi_z = 2\kappa \operatorname{sech}(\kappa z). \tag{10.75}$$

From Equation (10.64), the following expression for ϕ_2 results:

$$\phi_2 = \alpha\sqrt{\gamma_2} \operatorname{sech}(\kappa z). \tag{10.76}$$

Noting that $\phi_0/(\alpha\sqrt{\gamma_0}) = \phi_2/(\alpha\sqrt{\gamma_2})$, a plot of the $\phi_i/(\alpha\gamma_i)$ is given in Figure 10.14. ϕ_0 and ϕ_2 are qualitatively similar in appearance to the KdV solitary wave while ϕ_1 resembles the sine-Gordon kink solitary wave.

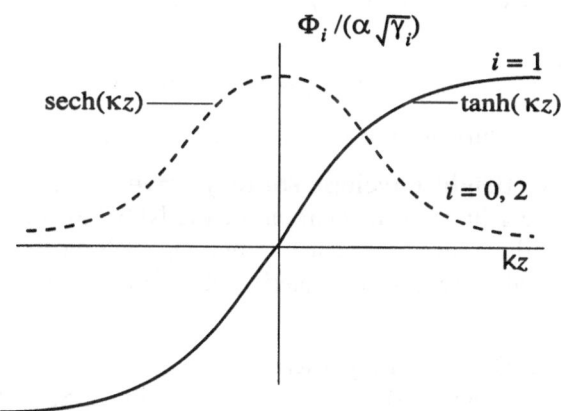

Figure 10.14: Solitary wave forms for the three-wave problem.

From a physical point of view these solitary waves are rather interesting. If the height of ϕ_0 and its velocity c ($c < v_2$) are specified, then the heights and velocities of ϕ_2 and ϕ_1 are automatically fixed. Since typically the sound velocity $v_2 \sim 10^5$ cm/s while the electromagnetic wave velocities $v_0, v_1 \sim 10^{10}$ cm/s, the light pulses will propagate in the positive z direction with a velocity c some five orders of magnitude slower than usual. The soliton nature of these solutions,

---

[14] Using $\sin^{-1} x = \cos^{-1}(\sqrt{(1 - x^2)})$ for ϕ_0.

which have not yet been observed experimentally, has been established by David Kaup and collaborators [Kau76], [KRB79] using the inverse scattering method.

It should be noted that unlike the situation for the KdV equation where dispersive and nonlinear effects balance to yield a solitary wave, in the three-wave problem no such dispersion is present. Instead the nonlinear terms in the three coupled nonlinear PDEs balance. This has interesting consequences which have been discussed by Kaup [Kau76].

PROBLEMS
Problem 10-18: Large damping in the three-wave problem
Analytically determine the solitary wave solutions ϕ_0, ϕ_1, ϕ_2 for the three-wave problem in the limit of very large damping, i.e., when ψ_{zz} is negligible compared to $\tilde{\Gamma}\psi_z$ in the damped pendulum equation. Sketch the solutions indicating their asymptotic values as well as their values at $z = 0$. Discuss the solutions in relation to the phase plane analysis for the damped pendulum.

Problem 10-19: Small sound damping
Making use of the phase plane analysis for the damped pendulum problem, qualitatively discuss how the solitary wave solutions in the three-wave problem are altered when very small sound damping is included.

Problem 10-20: Damping of electromagnetic waves
Show that if damping factors Γ_0 and Γ_1, satisfying the condition $\Gamma_0/(v_0 - c) = \Gamma_1/(v_1 - c) \equiv \bar{\Gamma}$ are inserted into the three-wave equations for ϕ_0 and ϕ_1 respectively, then

$$\phi_0 = \alpha\sqrt{\gamma_0}e^{-\bar{\Gamma}z}\cos(\psi/2), \quad \phi_1 = \alpha\sqrt{\gamma_1}e^{-\bar{\Gamma}z}\sin(\psi/2),$$

where ψ satisfies

$$\psi_{zz} + \tilde{\Gamma}\psi_z + \kappa^2 e^{-2\bar{\Gamma}z}\sin\psi = 0.$$

Can solitary wave solutions exist in this case? Explain.

Problem 10-21: Bright envelope solitary wave
Derive an analytic solitary wave solution for the NLSE for the bright case. In nonlinear optics, this solution is called an envelope soliton as it represents the envelope of an oscillatory electromagnetic pulse. Plot the intensity profile for $\beta = 1$ and $\tau = -4...4$.

Problem 10-22: Dark solitary wave
The dark solitary wave intensity profile for the nonlinear Schrödinger equation is given by $|E|^2 = \rho$ with

$$\rho = \rho_0[1 - a^2\operatorname{sech}^2(\sqrt{\rho_0}a\tau)], \quad a^2 = (\rho_0 - \rho_s)/\rho_0 \leq 1.$$

Plot $|E|^2$ for fixed ρ_0 and several different values of a. Interpret the parameters ρ_0, ρ_s, and a. The phrase "black" solitary wave would correspond to which a value?

Problem 10-23: Toda solitary wave
In Chapter 2, the student was asked to derive the Toda equation of motion

$$\ddot{y}_k(\tau) = 2e^{-y_k} - e^{-y_{k+1}} - e^{-y_{k-1}}$$

describing the vibrations of the 1-dimensional Toda lattice. Show that Toda's equation admits the solitary wave solution

$$e^{-y_k} - 1 = \beta^2 \operatorname{sech}^2[\kappa(k \pm c\tau)]$$

with $\beta \equiv \sinh \kappa$. How is the speed c related to κ?
(Hint: Set $e^{-y_k} - 1 = \dot{S}_k$ and show that

$$\frac{\ddot{S}_k}{1 + \dot{S}_k} = S_{k+1} + S_{k-1} - 2S_k.$$

Further note that S_k is proportional to a tanh function and make use of an identity for $\tanh(u \pm v)$.)

Problem 10-24: Boussinesq solitary wave
Analytically determine the solitary wave solution of the Boussinesq equation

$$\psi_{xx} - \psi_{tt} + 6(\psi^2)_{xx} + \psi_{xxxx} = 0$$

which was discussed in Section 3.2.1. Compare your solitary wave solution with that obtained for the KdV equation.

Problem 10-25: Nerve fiber solitary wave
A simple nerve fiber model [NAY62] is described by the nonlinear diffusion equation

$$\phi_t + \phi(\phi - a)(\phi - 1) = \phi_{xx}.$$

Show by direct substitution that the solitary wave solution is

$$\phi = \frac{1}{1 + e^{(x-ut)/\sqrt{2}}}$$

where the velocity $u = (1 - 2a)/\sqrt{2}$. Is this result easily obtainable by the direct construction procedure of the text? Explain.

Problem 10-26: Burgers' solitary wave
Derive an anti-kink solitary wave solution to Burgers' equation and plot the result. How does the thickness of the anti-kink region depend on amplitude? How does the velocity depend on amplitude?

Problem 10-27: 3-wave solutions
Use a computer algebra approach to derive the soliton solutions to the 3-wave problem, assuming that $\Gamma = 0$. Animate the three profiles in the same plot.

Problem 10-28: Modified KdV equation
Derive a solitary wave solution of the modified KdV equation

$$\psi_t + \alpha \psi^2 \psi_x + \psi_{xxx} = 0,$$

which appears in electric circuit theory [Sco70], the theory [Tor81][Tor86] of double layers in plasmas, and as a model [Ver87] of ion acoustic solitons in a multicomponent plasma.

Problem 10-29: Kadomtsev–Petviashvili solitary wave
In Section 3.2.1 , it was mentioned that the 2-dimensional generalization of the

KdV equation is the Kadomtsev–Petviashvili equation which, for $\alpha = 6$, has the form

$$(U_t + 6UU_x + U_{xxx})_x + 3U_{yy} = 0.$$

Show by direct construction that this equation has the solitary wave solution

$$U = \frac{1}{2}k^2 \operatorname{sech}^2(\frac{1}{2}(kx + ly - vt))$$

where k, l are real constants and $v = k^3 + 3l^2/k$. (You may assume that U is a function of $z = kx + ly - vt$.)

Problem 10-30: Hirota's direct method

A method for obtaining special solutions of nonlinear PDEs, which are independent of any explicit dependence on (x, t), has been developed by R. Hirota [Hir71][Hir76][Hir85b][Hir85a]. In particular, in his 1971 paper Hirota was able to derive an N-soliton solution of the KdV equation. Discuss Hirota's method in detail, illustrating your discussion with a specific example. Comment on the problem of obtaining a Hirota transformation for a given nonlinear PDE. In addition to the cited papers, you might also wish to consult Jackson's "Perspectives of nonlinear dynamics"[Jac90], or Infeld and Rowland's "Nonlinear waves, solitons and chaos"[IR90], or any other reference that you happen to find.

Problem 10-31: Instantons

Go to your college or university library and look up the 1979 *Scientific American* article (Volume 240(2), pages 92-116) on solitons by C. Rebbi. Use this article, and any other journal or book source that you are able to discover, to discuss what is meant by the word "instanton". Give as much detail as you think is necessary to clearly explain the concept.

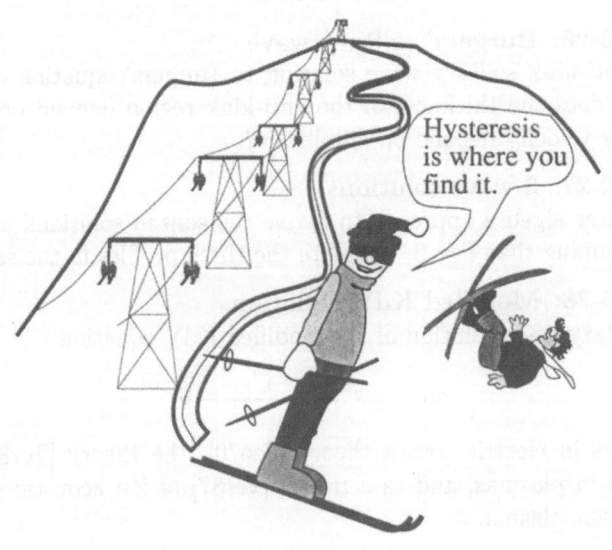

Chapter 11

Numerical Simulation

There is nothing stable in the world; uproar's your only music.
John Keats (1795–1821), English poet

One approach to investigating the collisional stability of solitary waves, as well as the evolution of other input shapes, for the nonlinear PDEs that have been encountered is through the use of numerical simulation. The PDEs can be solved numerically using either explicit or implicit schemes. In either case, the starting point is the representation of the partial derivatives by finite difference approximations. This may be easily accomplished by generalizing the treatment outlined in Chapter 6 for nonlinear ODEs.

11.1   Finite Difference Approximations

Consider a function $\psi = \psi(x, t)$ whose spatial (x) or time (t) partial derivatives are to be calculated. To facilitate the discussion, let's introduce a notation which is fairly standard in the numerical analysis literature. In Figure 11.1, the x–t plane has been subdivided into a rectangular grid or "mesh" with each rectangle having sides of length h and k as shown, where h, k will be taken to be small. The coordinates of a typical intersection or "mesh point" P are

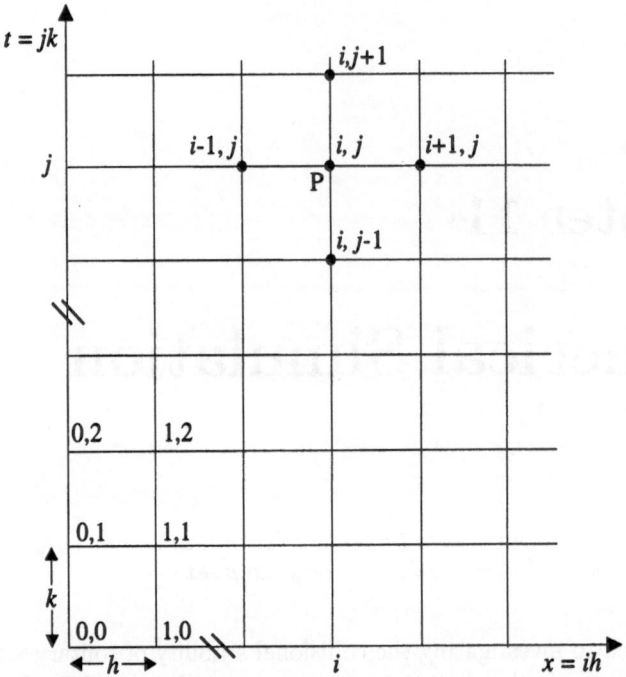

Figure 11.1: Subdividing the x–t plane with a numerical rectangular mesh.

$x = ih$, $t = jk$, with $i, j = 0, 1, 2, ...$ The various mesh points may be labeled by a pair of integers, the point P being denoted by (i, j). The value of ψ at P is written as $\psi_P = \psi(x = ih, y = jk) \equiv \psi_{i,j}$.[1] Expressing the finite-difference approximations in terms of this notation, the various partial derivatives at P may be written down.

For example, borrowing from Chapter 6, in the forward difference approximation the first partial spatial and time derivatives are, respectively,

$$\left(\frac{\partial \psi}{\partial x}\right)_P = \frac{(\psi_{i+1,j} - \psi_{i,j})}{h},$$
$$\left(\frac{\partial \psi}{\partial t}\right)_P = \frac{(\psi_{i,j+1} - \psi_{i,j})}{k}. \tag{11.1}$$

Similarly, the second spatial and time derivatives are given by the central difference approximations,

$$\left(\frac{\partial^2 \psi}{\partial x^2}\right)_P = \frac{(\psi_{i+1,j} - 2\psi_{i,j} + \psi_{i-1,j})}{h^2},$$
$$\left(\frac{\partial^2 \psi}{\partial t^2}\right)_P = \frac{(\psi_{i,j+1} - 2\psi_{i,j} + \psi_{i,j-1})}{k^2}, \tag{11.2}$$

and so on. The accuracy of each approximation is, of course, the same as before. Figure 11.2 shows the mesh points used for the above four partial derivatives and the "weights" (numerical factors) for each term. For high order derivatives

---

[1]To keep the notation compact, we will omit all spaces in the subscripts.

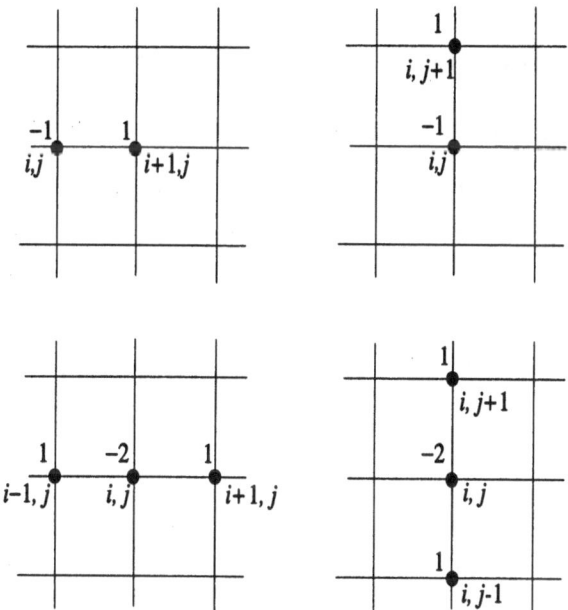

Figure 11.2: Mesh points and weights for four derivatives in (11.1) and (11.2).

of $\psi(x, t)$, the Mathematica **Series** command can prove useful in obtaining or confirming finite difference approximations. If "cross-derivatives" involving differentiation of $\psi(x, t)$ with respect to both x and t are needed, a 2-dimensional Taylor expansion may be used as illustrated in the following example.

Example 11-1: Representation of ψ_{xt}

Use a 2-dimensional Taylor expansion to show that for $k = h$,

$$\left(\frac{\partial^2 \psi}{\partial x \partial t}\right)_P = \frac{1}{4h^2}(\psi_{i+1,j+1} - \psi_{i-1,j+1} + \psi_{i-1,j-1} - \psi_{i+1,j-1}) + O(h^2).$$

Solution: Since ultimately we are going to set $k = h$, clearly a "1-dimensional" expansion in powers of h would suffice to confirm the above representation of ψ_{xt}. However, it is instructive to show how a "2-dimensional" series expansion may be carried out in powers of h and k before specializing to the given situation.

```
Clear["Global`*"]
```

The first term $\psi_{i+1,j+1}$ is entered in the form $\psi(x + h, t + k)$,

```
t1 = ψ[x + h, t + k]
```

$$\psi(h + x, k + t)$$

and a double series expansion in powers of h and k carried out, each expansion keeping terms to fifth order.

```
term1 = Series[t1,{h,0,5},{k,0,5}]; term1 = Normal[term1]
```

$$\psi(x,t) + h\,\psi^{(1,0)}(x,t) + \frac{1}{2}h^2\,\psi^{(2,0)}(x,t) + \frac{1}{6}h^3\,\psi^{(3,0)}(x,t)$$

$$+\frac{1}{24}h^4\,\psi^{(4,0)}(x,t) + \frac{1}{120}h^5\,\psi^{(5,0)}(x,t)$$

$$+k\left(\psi^{(0,1)}(x,t) + h\,\psi^{(1,1)}(x,t) + \frac{1}{2}h^2\,\psi^{(2,1)}(x,t) + \frac{1}{6}h^3\,\psi^{(3,1)}(x,t)\right.$$

$$\left.+\frac{1}{24}h^4\,\psi^{(4,1)}(x,t) + \frac{1}{120}h^5\,\psi^{(5,1)}(x,t)\right)$$

$$+k^2\left(\frac{1}{2}\psi^{(0,2)}(x,t) + \frac{1}{2}h\,\psi^{(1,2)}(x,t) + \frac{1}{4}h^2\,\psi^{(2,2)}(x,t) + \frac{1}{12}h^3\,\psi^{(3,2)}(x,t)\right.$$

$$\left.+\frac{1}{48}h^4\,\psi^{(4,2)}(x,t) + \frac{1}{240}h^5\,\psi^{(5,2)}(x,t)\right)$$

$$+k^3\left(\frac{1}{6}\psi^{(0,3)}(x,t) + \frac{1}{6}h\,\psi^{(1,3)}(x,t) + \frac{1}{12}h^2\,\psi^{(2,3)}(x,t) + \frac{1}{36}h^3\,\psi^{(3,3)}(x,t)\right.$$

$$\left.+\frac{1}{144}h^4\,\psi^{(4,3)}(x,t) + \frac{1}{720}h^5\,\psi^{(5,3)}(x,t)\right)$$

$$+k^4\left(\frac{1}{24}\psi^{(0,4)}(x,t) + \frac{1}{24}h\,\psi^{(1,4)}(x,t) + \frac{1}{48}h^2\,\psi^{(2,4)}(x,t) + \frac{1}{144}h^3\,\psi^{(3,4)}(x,t)\right.$$

$$\left.+\frac{1}{576}h^4\,\psi^{(4,4)}(x,t) + \frac{1}{2880}h^5\,\psi^{(5,4)}(x,t)\right)$$

$$+k^5\left(\frac{1}{120}\psi^{(0,5)}(x,t) + \frac{1}{120}h\,\psi^{(1,5)}(x,t) + \frac{1}{240}h^2\,\psi^{(2,5)}(x,t) + \frac{1}{720}h^3\,\psi^{(3,5)}(x,t)\right.$$

$$\left.+\frac{1}{2880}h^4\,\psi^{(4,5)}(x,t) + \frac{1}{14400}h^5\,\psi^{(5,5)}(x,t)\right)$$

The $O(h^6)$ and $O(k^6)$ terms have been removed from the output with the **Normal** command. The first and second superscripts indicate the number of times partial differentiation with respect to x and t has been undertaken.

The other three terms in the given representation of ψ_{xt} are expanded in a similar manner. Since the lengthy expansions are similar to that for the first term, the outputs are suppressed.

```
t2 = ψ[x-h, t+k]; t3 = ψ[x-h, t-k]; t4 = ψ[x+h, t-k];

term2 = Series[t2,{h,0,5},{k,0,5}]; term2 = Normal[term2];

term3 = Series[t3,{h,0,5},{k,0,5}]; term3 = Normal[term3];

term4 = Series[t4,{h,0,5},{k,0,5}]; term4 = Normal[term4];
```

The sum s=t1-t2+t3-t4 is calculated and the result displayed.

$$s = t1 - t2 + t3 - t4 == \text{Expand}[\text{term1} - \text{term2} + \text{term3} - \text{term4}]$$

$$\psi(x - h, t - k) - \psi(x - h, k + t) - \psi(h + x, t - k) + \psi(h + x, k + t) ==$$

$$4\,h\,k\,\psi^{(1,1)}(x,t) + \frac{2}{3}\,h\,k^3\,\psi^{(1,3)}(x,t) + \frac{1}{30}h\,k^5\,\psi^{(1,5)}(x,t) + \frac{2}{3}\,h^3\,k\,\psi^{(3,1)}(x,t)$$

$$+\frac{1}{9}h^3\,k^3\,\psi^{(3,3)}(x,t) + \frac{1}{180}h^3\,k^5\,\psi^{(3,5)}(x,t) + \frac{1}{30}h^5\,k\,\psi^{(5,1)}(x,t)$$

$$+\frac{1}{180}h^5\,k^3\,\psi^{(5,3)}(x,t) + \frac{1}{3600}h^5\,k^5\,\psi^{(5,5)}(x,t)$$

Since our goal is to confirm the given representation for ψ_{xt}, we now set $k = h$ and again display the sum s.

$$k = h; \quad s$$

$$\psi(x - h, t - h) - \psi(x - h, h + t) - \psi(h + x, t - h) + \psi(h + x, h + t) ==$$

$$4\,h^2\,\psi^{(1,1)}(x,t) + \frac{2}{3}\,h^4\,\psi^{(1,3)}(x,t) + \frac{1}{30}h^6\,\psi^{(1,5)}(x,t) + \frac{2}{3}\,h^4\,\psi^{(3,1)}(x,t)$$

$$+\frac{1}{9}h^6\,\psi^{(3,3)}(x,t) + \frac{1}{180}h^8\,\psi^{(3,5)}(x,t) + \frac{1}{30}h^6\,\psi^{(5,1)}(x,t)$$

$$+\frac{1}{180}h^8\,\psi^{(5,3)}(x,t) + \frac{1}{3600}h^{10}\,\psi^{(5,5)}(x,t)$$

Assuming that $h \ll 1$, we will keep only the largest term of order h^2 on the right-hand side. Since the next largest term is of order h^4, the error in our final result on dividing through by h^2 will be $O(h^4)/O(h^2) = O(h^2)$.

$$s1 = s \,/. \, \{h^4 -> 0, h^6 -> 0, h^8 -> 0, h^{10} -> 0\}$$

$$\psi(x-h, t-h) - \psi(x-h, h+t) - \psi(h+x, t-h) + \psi(h+x, h+t) == 4\,h^2\,\psi^{(1,1)}(x,t)$$

Solving s1 for $\psi^{(1,1)}(x,t)$,

$$s2 = \text{Solve}[s1, D[\psi[x,t], x, t]];$$

and simplifying yields the required form of ψ_{xt}.

$$\psi_{xt} = \text{Simplify}[D[\psi[x,t], x, t] \,/. \, s2[[1]]]$$

$$\frac{\psi(x - h, t - h) - \psi(x - h, h + t) - \psi(h + x, t - h) + \psi(h + x, h + t)}{4\,h^2}$$

End Example 11-1

PROBLEMS

Problem 11-1: Other approximations

Write down the backward and central difference approximations for the first partial spatial and time derivatives in the subscript notation and indicate the mesh points being used and their weights in a sketch of the numerical grid.

Problem 11-2: Another representation of ψ_{xt}

Use a 2-dimensional Taylor expansion to show that to $O(h^2)$ for $k = h$,

$$\left(\frac{\partial^2 \psi}{\partial x \partial t}\right)_P = -\frac{1}{2h^2}(\psi_{i+1,j} - \psi_{i-1,j} + \psi_{i,j+1} + \psi_{i,j-1} - 2\psi_{i,j} - \psi_{i+1,j+1} - \psi_{i-1,j-1}).$$

Indicate the relevant mesh points and weights in a sketch of the numerical grid.

Problem 11-3: Representation of ψ_{xxtt}

Use appropriate Taylor expansions to show that for $k = h$,

$$\left(\frac{\partial^4 \psi}{\partial x^2 \partial t^2}\right)_P = \frac{1}{h^4}(\psi_{i+1,j+1} + \psi_{i-1,j+1} + \psi_{i+1,j-1} + \psi_{i-1,j-1} - 2\psi_{i+1,j}$$

$$-2\psi_{i-1,j} - 2\psi_{i,j+1} - 2\psi_{i,j-1} + 4\psi_{i,j}) + O(h^2).$$

Indicate the relevant mesh points and weights in a sketch of the numerical grid.

Problem 11-4: Representation of ψ_{xxx}

Using Taylor expansions for $\psi(x+h)$, $\psi(x+2h)$, etc., show that a finite difference representation for ψ_{xxx} is given by

$$(\psi_{xxx})_P = \frac{1}{2h^3}(\psi_{i+2,j} - 2\psi_{i+1,j} + 2\psi_{i-1,j} - \psi_{i-2,j}).$$

What is the order in terms of h of the first term that has been neglected in this finite difference approximation? For what nonlinear PDE of physical interest could this approximation prove useful?

Problem 11-5: Representation of Laplacian operator

Using appropriate Taylor expansions, show that a finite difference representation for the Laplacian operator ∇^2 is given by

$$\left(\frac{\partial^2 \psi}{\partial x^2} + \frac{\partial^2 \psi}{\partial y^2}\right)_P = \frac{1}{12h^2}(-60\psi_{i,j} + 16(\psi_{i+1,j} + \psi_{i,j+1} + \psi_{i-1,j} + \psi_{i,j-1})$$

$$-(\psi_{i+2,j} + \psi_{i,j+2} + \psi_{i-2,j} + \psi_{i,j-2})).$$

What is the order in terms of h of the first term which has been neglected in this finite difference approximation? Sketch the numerical grid with the points being used clearly labeled along with their weights. Suggest a physical problem where this approximation could prove useful.

11.2 Explicit Methods

Paralleling the treatment for ODEs, we shall first look at explicit methods for numerically solving PDEs. The approach is illustrated with several examples. Although our goal is to solve nonlinear systems, it is useful to begin with a well-known linear example, the linear diffusion equation.[2] Given some initial input profile, it is desired to determine the profile at some later time subject to specified boundary conditions. This initial value problem is typical of the examples that will be explored in this chapter.

11.2.1 Diffusion Equation

Consider the problem of determining the temporal behavior of the temperature distribution along a thin insulated rod of length L whose ends are held at $T = 0\,°$ Celsius. The temperature distribution $T(z, t)$ is governed by the 1-dimensional linear heat diffusion equation

$$T_t = \sigma T_{zz}. \tag{11.3}$$

Suppose that the initial temperature distribution is

$$T = T_0 \sin(\pi z / L). \tag{11.4}$$

Before devising a numerical scheme, it is useful to make all quantities in the problem dimensionless. This may be accomplished here by defining $x \equiv z/L$, $y \equiv \sigma t/L^2$, and $U \equiv T/T_0$, so that

$$U_y = U_{xx} \tag{11.5}$$

subject to the boundary conditions[3]

$$U(0, y) = U(1, y) = 0, \quad y \geq 0 \tag{11.6}$$

and the initial condition

$$U(x, 0) = \sin(\pi x). \tag{11.7}$$

Using the forward difference approximation for the time (y) derivative and the standard form for the second spatial (x) derivative, Equation (11.5) may be approximated by

$$\frac{(U_{i,j+1} - U_{i,j})}{k} = \frac{(U_{i+1,j} - 2U_{i,j} + U_{i-1,j})}{h^2}, \tag{11.8}$$

or on setting $r \equiv k/h^2$ and solving for $U_{i,j+1}$,

$$U_{i,j+1} = rU_{i-1,j} + (1 - 2r)U_{i,j} + rU_{i+1,j}. \tag{11.9}$$

The various terms in this finite difference formula as well as the given boundary and initial conditions are indicated schematically in Figure 11.3. The unknown

---

[2]Mathematicians classify this equation as a parabolic PDE. The mathematical classification of PDEs will be discussed in a later section.

[3]When the dependent variable is prescribed on the boundaries, Dirichlet boundary conditions are said to prevail.

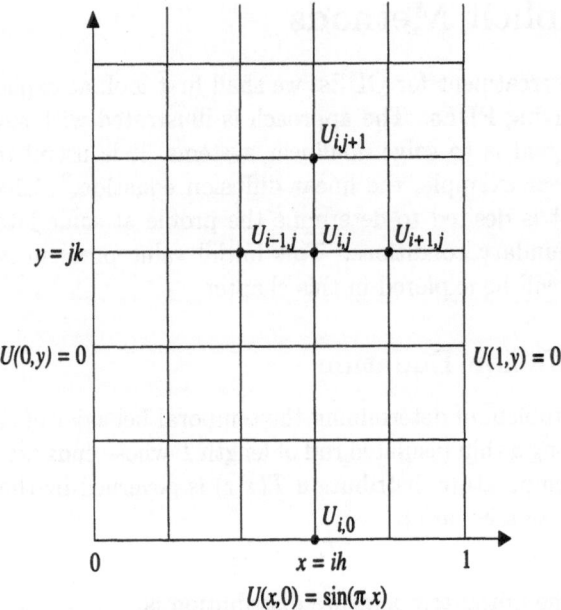

Figure 11.3: Solving the linear diffusion equation with a forward difference explicit scheme subject to the given boundary and initial conditions.

value $U_{i,j+1}$ is to be explicitly determined on the $j + 1$st time step from the three known values $U_{i-1,j}$, $U_{i,j}$, and $U_{i+1,j}$ on the previous jth time step. One starts with the bottom row $(j = 0)$ which corresponds to the initial temperature distribution and calculates U at each internal mesh point of the first $(j = 1)$ time row. With $U_{i,1}$ all known, one proceeds to calculate $U_{i,2}$, and so on.

To carry out this program, the finite difference problem is converted into a matrix representation for calculating U at the internal mesh points. Suppose, for example, that the range 0 to 1 is subdivided into 10 equal intervals, so that the internal x points are $x = 0.1, 0.2, 0.3,...,0.9$. We introduce the "tridiagonal" matrix[4] with $c \equiv 1 - 2r$,

$$A = \begin{bmatrix} c & r & 0 & 0 & 0 & 0 & 0 & 0 & 0 & 0 \\ r & c & r & 0 & 0 & 0 & 0 & 0 & 0 & 0 \\ 0 & r & c & r & 0 & 0 & 0 & 0 & 0 & 0 \\ 0 & 0 & r & c & r & 0 & 0 & 0 & 0 & 0 \\ 0 & 0 & 0 & r & c & r & 0 & 0 & 0 & 0 \\ 0 & 0 & 0 & 0 & r & c & r & 0 & 0 & 0 \\ 0 & 0 & 0 & 0 & 0 & r & c & r & 0 & 0 \\ 0 & 0 & 0 & 0 & 0 & 0 & r & c & r & 0 \\ 0 & 0 & 0 & 0 & 0 & 0 & 0 & r & c & r \\ 0 & 0 & 0 & 0 & 0 & 0 & 0 & 0 & r & c \end{bmatrix} \qquad (11.10)$$

and $U^{(0)} \equiv \{U^{(0)}(x = 0.1), U^{(0)}(0.2),..., U^{(0)}(0.9)\}$ as the initial values of U at the internal mesh points. Then, in going from the zeroth time step to the first

_____
[4]The nonzero elements are located only along the main diagonal and the diagonals just above and below it.

time step, one has

$$U^{(1)} = AU^{(0)} \tag{11.11}$$

and from the zeroth step to the Nth time step,

$$U^{(N)} = A^N U^{(0)}. \tag{11.12}$$

Here A^N means to multiply the input vector by the matrix A a total of N times. Mathematica File MF44 explicitly carries out this procedure. Taking, for example, $k = 0.0005$, $h = 0.1$, so that $r = 0.05$, and increasing N values, a temporal plot similar to that shown in Figure 11.4 is obtained. As time increases the temperature of the rod approaches zero everywhere.

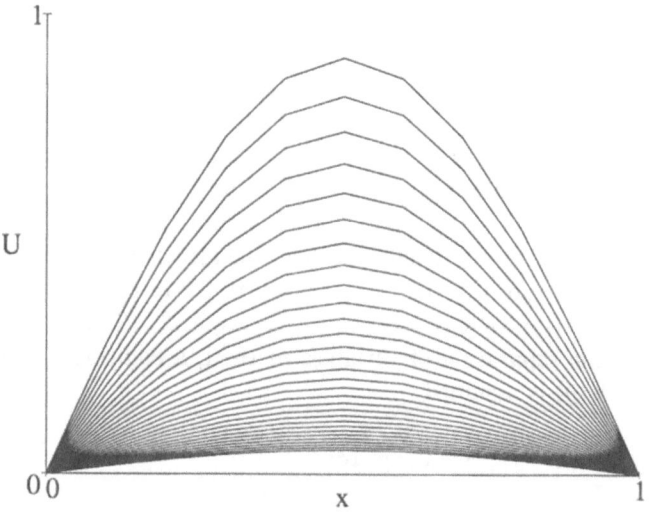

Figure 11.4: Explicit solution (equal time steps) of the linear diffusion equation.

The Explicit Method Applied to the Linear Diffusion Equation
In this file, the explicit scheme outlined in the text is carried out for the linear diffusion equation, and animated and static pictures of the temporal evolution produced. We make use of the Mathematica Dot symbol, ".", which is relevant to products of vectors, matrices, and tensors. The product $a.b$, where a and b are lists with appropriate dimensions, contracts the last index in a with the first index in b. To form the nth matrix power of a matrix a, the command `MatrixPower[a,n]` can be used. Mathematica commands in file: `Sin, Table, MatrixPower, Join, Transpose, ListPlot, Dot, PlotJoined->True, Do, Hue, PlotRange, PlotStyle, AxesLabel, TextStyle, Block, ImageSize, Ticks, Show, $DisplayFunction=Identity`

Of course, the linear diffusion equation is exactly analytically solvable and, as demonstrated earlier in Example 10-3, has a solution (in terms of the present dimensionless variables) of the form

$$U(x, y) = \sin(\pi x)\, e^{-\pi^2 y} \tag{11.13}$$

for the given initial temperature distribution and boundary conditions.

For, say, $N = 1000$ and $k = 0.0005$, the total (normalized) elapsed time is $y = 0.5$. At $x = 0.5$, the analytic formula yields $U = 0.007192$, while numerically $U = 0.007399$, an error of about 3 percent. As with ODEs the accuracy of the explicit scheme can be improved by reducing r, but at the expense of increased computational time. On the other hand, as r is increased the reader may verify by running the file that numerical instability sets in for $r > \frac{1}{2}$.

In arriving at the solution (11.13), the Laplace transform method was used. As pointed out previously, if the input profile is such that the answer involves an infinite series, Mathematica may have trouble performing the inverse Laplace transform. An alternate approach is to use the method of separation of variables. This is illustrated in the following example for a parabolic input profile.

Example 11-2: Separation of Variables

A thin one meter long rod whose lateral surface is insulated to prevent heat flow through that surface has its ends suddenly held at the freezing point, $0°$ Celsius. Taking one end of the rod to be $x = 0$ and the other $x = 1$, the initial temperature distribution is $u(x, y = 0) = x(1 - x)$. Taking the diffusion coefficient to be unity, use the separation of variable method to determine the analytic temperature profile inside the hot rod for an arbitrary time $y > 0$. Animate the solution for the time range $y = 0$ to 0.2 in time steps 0.01.

Solution: In the separation of variables method, a solution of the structure $u(x, y) = X(x)Y(y)$ is assumed.

```
Clear["Global`*"]; << Graphics`
```

```
u[x,y] = X[x] Y[y];
```

Entering the diffusion equation and dividing both sides by $u(x, y)$,

```
pde = D[u[x,y],y]/u[x,y] == D[u[x,y],x,x]/u[x,y]
```

$$\frac{Y'(y)}{Y(y)} == \frac{X''(x)}{X(x)}$$

yields a function of y alone on the left equal to a function of x alone on the right. The above result can only hold if each side of the equation is set equal to a common separation constant, $-a^2$ say.

```
de1 = pde[[1]] == -a^2
```

$$\frac{Y'(y)}{Y(y)} == -a^2$$

```
de2 = pde[[2]] == -a^2
```

$$\frac{X''(x)}{X(x)} == -a^2$$

The individual ODEs, de1 and de2, are then solved for Y and X, respectively.

```
sol1 = DSolve[de1,Y,y]; Y = Y[y] /. sol1[[1]]
```

$$e^{-a^2 y} c_1$$

```
sol2 = DSolve[de2,X,x]; X = X[x] /. sol2[[1]]
```

$$c_1 \cos(a\,x) + c_2 \sin(a\,x)$$

To evaluate the coefficients, the boundary conditions must be applied. We must have $X(0) = X(1) = 0$ for arbitrary (normalized) times y. To satisfy $X(0) = 0$, the coefficient $c_1 = 0$ in X. To satisfy $X(1) = 0$, the coefficient c_2 must be nonzero to have a nontrivial (nonzero) solution and therefore $a = n\pi$, with n a positive integer. Without loss of generality, the coefficient c_2 in X can be set equal to one. These substitutions are made in the following command line. The substitution $a = n\pi$ is also made in Y and for notational convenience the coefficient c_1 in Y is renamed b.

```
X = X /. {C[1] -> 0, C[2] -> 1, a -> n Pi}; Y = Y /. {a -> n Pi, C[1] -> b};
```

The nth term in the general Fourier series solution will be given by the output of the following command line with b a still to be determined function of n.

```
u = X Y
```

$$b\,e^{-n^2 \pi^2 y} \sin(n\,\pi\,x)$$

The complete solution will be a sum over n terms with $n = 1...\infty$. The coefficients b can be evaluated by using the concept of orthogonality. The initial temperature profile at $y = 0$ yields

$$u(x,0) = x\,(1-x) = \sum_{n=1}^{\infty} b_n \sin(n\pi x) \qquad (11.14)$$

where b_n is equivalent to b above. To evaluate b_n, we multiply both sides of (11.14) by $\sin(m\pi x)$ and integrate over the range $x = 0..1$. Orthogonality of the sine functions over this range picks out the term corresponding to $m = n$ in the series. Entering the initial condition,

```
ic = x (1 - x);
```

the above orthogonality procedure is implemented,

```
orthog = Integrate[ic*Sin[n Pi x],{x,0,1}]
        == Integrate[(u /. y -> 0)*Sin[n Pi x],{x,0,1}]
```

$$\frac{2}{n^3 \pi^3} - \frac{2\cos(n\,\pi)}{n^3 \pi^3} - \frac{\sin(n\,\pi)}{n^2 \pi^2} == b\left(\frac{1}{2} - \frac{\sin(2\,n\,\pi)}{4\,n\,\pi}\right)$$

yielding an orthogonality relation which can be solved for b.

```
sol3 = Solve[orthog,b];
```

```
b = b /. sol3[[1]]
```

$$\frac{\dfrac{2\cos(n\pi)}{n^3\pi^3} + \dfrac{\sin(n\pi)}{n^2\pi^2} - \dfrac{2}{n^3\pi^3}}{\dfrac{\sin(2n\pi)}{4n\pi} - \dfrac{1}{2}}$$

To explicitly write out the series solution, a solution function is created.

```
sol[m_] := u /. n -> m
```

Then, using the Sum command, the series to order $m = 7$ is given by the output of the following line.

```
sol = Sum[sol[m], {m,1,7}]
```

$$\frac{8e^{-\pi^2 y}\sin(\pi x)}{\pi^3} + \frac{8e^{-9\pi^2 y}\sin(3\pi x)}{27\pi^3} + \frac{8e^{-25\pi^2 y}\sin(5\pi x)}{125\pi^3}$$

$$+\frac{8e^{-49\pi^2 y}\sin(7\pi x)}{343\pi^3}$$

Examining the exponentials, it is clear that higher-order terms in the series solution become neglibly small for $y > 0$. The series solution is now animated and may be viewed on the computer screen by executing the following command line.

```
Do[Plot[Evaluate[sol /. y->yi],{x,0,1},PlotRange->{{0,1},{0,.25}},
AxesLabel -> {"x","u"},PlotStyle -> {Hue[1]},TextStyle ->
{FontFamily -> "Times",FontSize -> 12}],{yi,0,.2,.01}]
```

End Example 11-2

In Mathematica File MF44, the method of entering the matrix A was one of brute force. If the range from 0 to 1 is to be divided into finer spatial intervals than $h = 0.1$, a more efficient way of entering the matrix should be used. Such an approach is illustrated in the following example.

Example 11-3: A More Efficient Explicit Matrix Method

a. Devise a more efficient algorithm for entering the matrix A in the explicit scheme for solving the linear diffusion equation.

b. Divide the range $x = 0..1$ into 16 equal intervals and explicitly display A.

c. Keeping $r = 0.05$, modify MF44 to handle the larger matrix A and solve the linear diffusion equation for $N = 1400$, creating a time sequence of 70 animated plots.

Solution: The number of plots and the number of internal mesh points is entered. The spatial step size is evaluated,

```
Clear["Global`*"]
```

```
numplots = 70; size = 15; h = 1.0/(size + 1)
```

$$0.0625$$

and determined to be $h = 0.0625$. The matrix A, labeled mat here, is created with the Switch and Table commands. Switch[expression,form$_1$,value$_1$, form$_2$,value$_2$,...] evaluates expression, then compares it with each of the form$_i$, in turn, evaluating and returning the value$_i$ corresponding to the first match found. So, here when $i - j = -1$, the value r is returned; when $i - j = 0$, c is returned; when $i - j = 1$, r is returned; and when $i - j$ equals _ , i.e. Blank, 0 is returned. The desired square matrix of dimensions size by size is then formed with the Table command.

```
mat = Table[Switch[i - j,-1,r,0,c,1,r,_,0],{i,size},{j,size}]
```

$$\begin{bmatrix}
c & r & 0 & 0 & 0 & 0 & 0 & 0 & 0 & 0 & 0 & 0 & 0 & 0 & 0 \\
r & c & r & 0 & 0 & 0 & 0 & 0 & 0 & 0 & 0 & 0 & 0 & 0 & 0 \\
0 & r & c & r & 0 & 0 & 0 & 0 & 0 & 0 & 0 & 0 & 0 & 0 & 0 \\
0 & 0 & r & c & r & 0 & 0 & 0 & 0 & 0 & 0 & 0 & 0 & 0 & 0 \\
0 & 0 & 0 & r & c & r & 0 & 0 & 0 & 0 & 0 & 0 & 0 & 0 & 0 \\
0 & 0 & 0 & 0 & r & c & r & 0 & 0 & 0 & 0 & 0 & 0 & 0 & 0 \\
0 & 0 & 0 & 0 & 0 & r & c & r & 0 & 0 & 0 & 0 & 0 & 0 & 0 \\
0 & 0 & 0 & 0 & 0 & 0 & r & c & r & 0 & 0 & 0 & 0 & 0 & 0 \\
0 & 0 & 0 & 0 & 0 & 0 & 0 & r & c & r & 0 & 0 & 0 & 0 & 0 \\
0 & 0 & 0 & 0 & 0 & 0 & 0 & 0 & r & c & r & 0 & 0 & 0 & 0 \\
0 & 0 & 0 & 0 & 0 & 0 & 0 & 0 & 0 & r & c & r & 0 & 0 & 0 \\
0 & 0 & 0 & 0 & 0 & 0 & 0 & 0 & 0 & 0 & r & c & r & 0 & 0 \\
0 & 0 & 0 & 0 & 0 & 0 & 0 & 0 & 0 & 0 & 0 & r & c & r & 0 \\
0 & 0 & 0 & 0 & 0 & 0 & 0 & 0 & 0 & 0 & 0 & 0 & r & c & r \\
0 & 0 & 0 & 0 & 0 & 0 & 0 & 0 & 0 & 0 & 0 & 0 & 0 & r & c
\end{bmatrix}$$

The matrix mat is clearly of the same form as that shown in Equation (11.10) except that it's larger. The value $r = 0.05$ is now entered and c and k evaluated.

```
r = 0.05; c = 1 - 2 r; k = r h^2
```

$$0.000195313$$

The time step size is $k = 0.000195313$. The function for creating and numerically evaluating the input temperature profile is entered,

```
f[x_] := Sin[Pi x] // N
```

and the list of initial temperature values at each internal mesh point formed.

```
v = Table[f[i/(size + 1)],{i,1,size}];
```

The matrix multiplication is carried out using the `MatrixPower` and `Dot(.)` commands. Since n runs up to `numplots`, $20 \times 70 = 1400$ matrix multiplications are performed. The total time is $1400 \times k \simeq 0.27$.

```
mat2 = Table[{MatrixPower[mat,20 n]}.v,{n,0,numplots}];
```

The x grid points are calculated for plotting purposes.

```
xx = Table[i h,{i,0,size + 1}];
```

A function for listing the time evolved temperature values at each x point is formed. The temperature at each end must be zero because of the boundary conditions. Zeros are added at each end of the list with the `Join` command.

```
yy[i_] := Join[{0},mat2[[i,1]],{0}]
```

Plotting points are formed with the `Transpose` command and graphed with `ListPlot`. The points are joined with blue colored lines.

```
gr[i_] := ListPlot[Transpose[{xx,yy[i]}],PlotJoined -> True,
PlotRange->{{0,1},{0,1}},PlotStyle->Hue[.6],AxesLabel->
{"x","U"},TextStyle->{FontFamily->"Times",FontSize->12}]
```

The `Do` command is used to generate the 70 plots in the time sequence. Double clicking on, say the first plot, produces an animation of the diffusion process.

```
Do[gr[i],{i,1,numplots}]
```

| End Example 11-3 |

PROBLEMS
Problem 11-6: Triangular input shape
Numerically solve the normalized linear diffusion equation subject to the boundary conditions $U(0,y) = U(1,y) = 0$ for $y \geq 0$ and initial condition $U(x,0) = 2x$ for $0 \leq x \leq 1/2$, $2(1 - x)$ for $1/2 \leq x \leq 1$. Consider the three cases: (a) $h = 0.1, k = 0.001$ and 140 time steps, (b) $h = 0.1, k = 0.004$ and 35 time steps, (c) $h = 0.1, k = 0.007$ and 20 time steps. In each case graphically compare the numerical results with the exact analytic solution at $y = 0.140$. Discuss the accuracy of the numerical solution achieved with the different values of $r = k/h^2$.

Problem 11-7: A different input
In Example 11-2, replace the input profile with $u(x,0) = x^2(1 - x)$ and run the code. Try other polynomial input profiles and discuss the results.

Problem 11-8: Comparison
In Example 11-3, replace the input profile with that in Example 11-2 and execute

the code. Experimenting with different r values, compare the numerical results with the exact analytic results of Example 11-2.

Problem 11-9: Another diffusion problem

The temperature at the ends $x = 0$ and $x = 100$ of a rod (insulated on its sides) 100 cm long are held at $0\,°$ and $100\,°$, respectively, until steady-state is achieved. Then at the instant $t = 0$, the temperature of the two ends are interchanged. Determine the resultant temperature distribution in the rod. Animate your solution and discuss the result.

Problem 11-10: Truncation error

Referring to the discussion of truncation error for ODEs in Chapter 6, show that the truncation error at the mesh point $(i, j+1)$ for the forward difference explicit scheme for the normalized linear diffusion equation is $\frac{k}{2}(U_{yy})_{i,j} - \frac{h^2}{12}(U_{xxxx})_{i,j}$.

Problem 11-11: Steady-state temperature distribution

The steady-state temperature distribution in a thin square metal plate 0.5 m on a side satisfies Laplace's equation

$$T_{xx}(x, y) + T_{yy}(x, y) = 0.$$

The boundary conditions on the edges of the plate are $T(0, y) = 0$, $T(x, 0) = 0$, $T(x, 0.5) = 200x$, $T(0.5, y) = 200y$. I.e., the temperature is held at $0\,°$ C on two adjacent edges and allowed to increase linearly from $0\,°$ C to $100\,°$ C on the other two edges.

a. Using the standard central difference approximations for the 2nd derivatives and setting $\Delta x \equiv h$, $\Delta y \equiv k$, $r \equiv h^2/k^2$, show that the numerical algorithm for Laplace's equation is

$$2(1 + r)T_{i,j} - T_{i+1,j} - T_{i-1,j} - rT_{i,j+1} - rT_{i,j-1} = 0.$$

b. Create a mesh which has $3 \times 3 = 9$ equally spaced interior mesh points.

c. Write out the nine mesh equations which have to be solved for the interior temperatures subject to the given boundary conditions.

d. Solve the nine mesh equations for the interior temperatures.

e. Make a plot of the temperature distribution $T(x, y)$ in the plate.

Problem 11-12: Potential distribution

A square inner conductor 3 cm on a side is held at a potential $\Phi = 100$ V. A second square conductor, concentric with the first and 9 cm long on each of its inner sides, is held at 0 V. Take the mesh spacing in both directions to be 1 cm. Φ in the region between the two conductors satisfies Laplace's equation.

a. Make a mesh diagram showing all the interior mesh points for which Φ is to be found.

b. Using central difference approximations for the 2nd derivatives, write out the mesh equations for the interior points. Make use of symmetry to show that only seven interior points need to be used in calculating Φ.

c. Solve the mesh equations and determine Φ at each interior point.

d. Plot Φ in the region stretching from the inner to the outer conductor.

11.2.2   Fisher's Nonlinear Diffusion Equation

As a second example, consider the nonlinear "reaction–diffusion" equation which was originally suggested by Fisher [Fis37] for the spatial spread of a favored gene in a population. Fisher's normalized equation is

$$U_t = U_{xx} + U(1 - U). \tag{11.15}$$

Fisher's equation is simply the logistic equation (see Problem 5-11, page 176) describing population growth with a diffusive term U_{xx} added to account for spatial spreading. Alternately, it can be thought of as the diffusion equation to which a "reaction" term has been added. To achieve some feeling for what effect inclusion of the nonlinear reaction term has on diffusion, let's take the same boundary and initial conditions as in the heat flow example.[5]

To create an explicit scheme connecting $j - 1$ to j, the simplest thing to do is to approximate the nonlinear term by $U_{i,j-1}(1 - U_{i,j-1})$ and include it in the previous derivation. The central diagonal elements in the matrix A will be changed to $(1 - 2r + k - kU_{1,j-1})$, $(1 - 2r + k - kU_{2,j-1})$, $(1 - 2r + k - kU_{3,j-1})$, ... as one moves down the diagonal, with the other diagonals unchanged. Thus, unlike the situation for the linear diffusion case, the new matrix A now changes on each time step, depending on the U values from the previous time step. An algorithm that incorporates these changes is given in the next Mathematica File MF45. Figure 11.5 shows the time evolution of Fisher's equation for $k = 0.0005$, $h = 0.1$, $r = 0.05$, i.e., exactly the same parameters as for the linear diffusion equation. Although the qualitative behavior is the same as for the linear case,

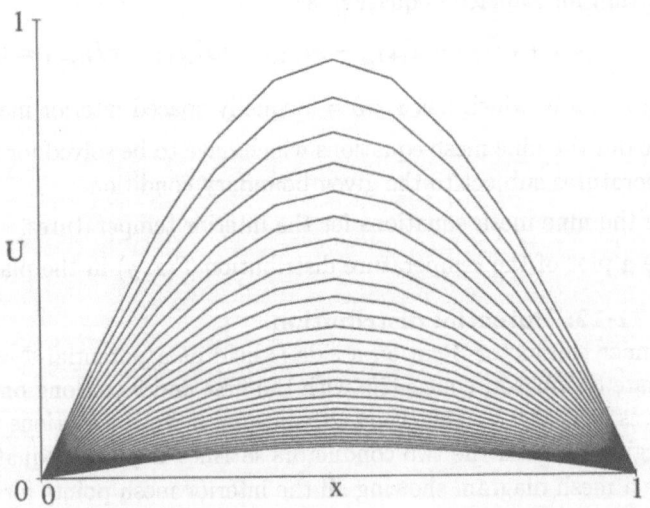

Figure 11.5: Explicit solution (equal time steps) of Fisher's equation.

the reader may verify by running the file that the effect of the nonlinear term is to slow down the diffusion process. At $x = 0.5$, the amplitude U for $N = 1000$ is now about 0.0112, as compared to 0.0074 for the linear problem.

---

[5]In the framework of Fisher's original research, this choice would be rather artificial.

The Explicit Scheme for the Nonlinear Fisher Equation
In this file, the explicit scheme for handling the nonlinear reaction–diffusion (Fisher's) equation is implemented. The same initial and boundary conditions as for the linear diffusion equation are used. The file is animated. Mathematica commands in file: `Sin`, `Table`, `Switch`, `Transpose`, `ListPlot`, `AxesLabel`, `N`, `PlotJoined->True`, `PlotRange`, `PlotStyle`, `Hue`, `TextStyle`, `Block`, `Do`, `Ticks`, `Show`, `ImageSize`, `$DisplayFunction=Identity`

PROBLEMS
Problem 11-13: Onset of numerical instability
Using exactly the same boundary and initial conditions and parameter values as in the text example except for the k value, numerically determine the r ratio at which the explicit scheme becomes numerically unstable for Fisher's equation.

Problem 11-14: Constant source term
Modify the algorithm in MF45 to handle the normalized diffusion equation with a constant source term R, viz.,

$$U_t = U_{xx} + R.$$

With all other parameters as in the file, solve the new equation for $R = 2$ and $R = 20$. Discuss the results.

11.2.3 Klein–Gordon Equation

Both of the previous examples have been of the diffusive type with a single time derivative present. Let's now look at a nonlinear wave equation[6] of the structure

$$\psi_{xx} - \psi_{tt} = a\psi + b\psi^2 \qquad (11.16)$$

with a, b real and subject to the boundary conditions

$$\psi(0, t) = \psi(1, t) = 0, \quad t > 0 \qquad (11.17)$$

and initial conditions for $0 \le x \le 1$

$$\psi(x, 0) = f(x), \quad \psi_t(x, 0) = g(x). \qquad (11.18)$$

Here $f(x)$ and $g(x)$ are real functions which can be specified.

For $a = 1$, $b = 0$, the Klein–Gordon equation of particle physics results with the particle confined to a 1-dimensional "box". Choosing $a = 0$, $b = 0$, and $g(x) = 0$ yields the linear wave equation describing, e.g., the transverse vibrations of an elastic string fixed at both ends which is released from rest. Finally, for a and b nonzero, the equation is a phenomenological nonlinear Klein–Gordon equation.

---

[6]Classified as a hyperbolic PDE by mathematicians.

Only the $a \neq 0, b = 0$ case will be examined here, the other possibilities being left for the reader to explore in the problem section. With $x_i = ih$ $(i = 0, 1, .., m)$ and $t = jk$ $(j = 0, 1, .., N)$, we write

$$\frac{(\psi_{i+1,j} - 2\psi_{i,j} + \psi_{i-1,j})}{h^2} - \frac{(\psi_{i,j+1} - 2\psi_{i,j} + \psi_{i,j-1})}{k^2} = a\psi_{i,j} \qquad (11.19)$$

or, on setting $r \equiv k^2/h^2$ and solving for the $j + 1$st term,

$$\psi_{i,j+1} = (2 - 2r - k^2 a)\psi_{i,j} + r(\psi_{i+1,j} + \psi_{i-1,j}) - \psi_{i,j-1}. \qquad (11.20)$$

This equation holds for the internal mesh points $i = 1, 2, .., (m - 1)$ and for $j = 1, 2, ...$ In this case, the unknown ψ value at time step $j + 1$ depends on ψ values on the previous two time steps, as shown schematically in Figure 11.6. With j starting at the value 1, the second time row is the first row to be

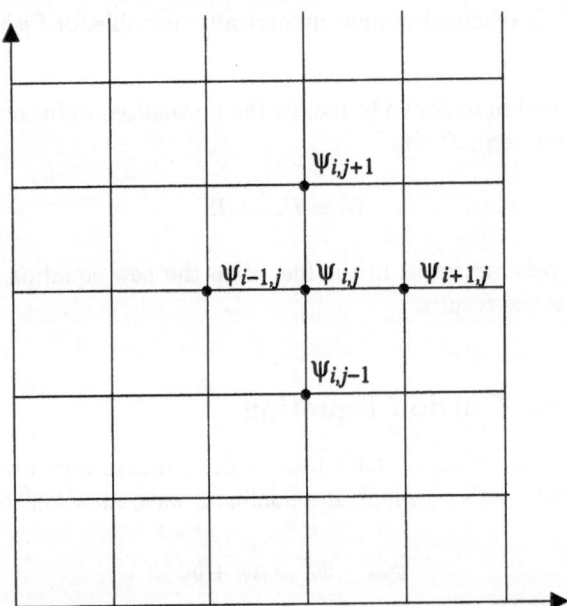

Figure 11.6: Mesh points involved in the explicit scheme for the Klein–Gordon equation.

calculated. Along the bottom (zeroth) row the ψ values are known from the initial condition on $f(x)$, i.e., $\psi_{i,0} = f(x_i)$. The ψ values on the first time row follow from the initial time derivative condition. With a truncation error $O(k)$,

$$\psi_t(x_i, 0) = \frac{(\psi_{i,1} - \psi_{i,0})}{k} = g(x_i) \qquad (11.21)$$

so that

$$\psi_{i,1} = \psi_{i,0} + kg(x_i). \qquad (11.22)$$

Subject to the boundary conditions $\psi_{0,j} = \psi_{1,j} = 0$, the explicit scheme may be expressed in the matrix form

$$\psi^{(j+1)} = A\psi^{(j)} - \psi^{(j-1)} \qquad (11.23)$$

where A is the $(m-1)$ by $(m-1)$ tridiagonal matrix

$$A = \begin{bmatrix} d & r & 0 & 0 & ... & 0 \\ r & d & r & 0 & ... & 0 \\ 0 & r & d & r & ... & 0 \\ ... & ... & ... & ... & ... & ... \\ ... & ... & ... & ... & ... & ... \\ 0 & 0 & ... & r & d & r \\ 0 & 0 & 0 & ... & r & d \end{bmatrix} \qquad (11.24)$$

with $d \equiv 2 - 2r - k^2 a$.

As a specific example, consider the initial conditions $f(x) = \sin(\pi x)$ and $g(x) = 0$ as in the next Mathematica File MF46.

The Klein–Gordon Equation

The explicit finite difference scheme for the Klein–Gordon equation is given in this file with the initial conditions $f(x) = \sin(\pi x)$ and $g(x) = 0$. It may be easily modified to handle different initial conditions, the linear wave equation ($a = b = 0$), and the nonlinear Klein–Gordon ($a \neq 0$, $b \neq 0$). The file is animated. Mathematica commands in file: Table, Switch, Sin, Pi, Table, Dot(.), Join, ListPlot, Transpose, N, PlotJoined->True, PlotRange, Ticks, Hue, AxesLabel, TextStyle, Do, Show, ImageSize

Taking $k = 0.01$, $h = 0.1$ (so that $r = 0.01$), $m = 10$, and $a = 100$, oscillatory behavior similar to that shown in Figure 11.7 is obtained. The motion

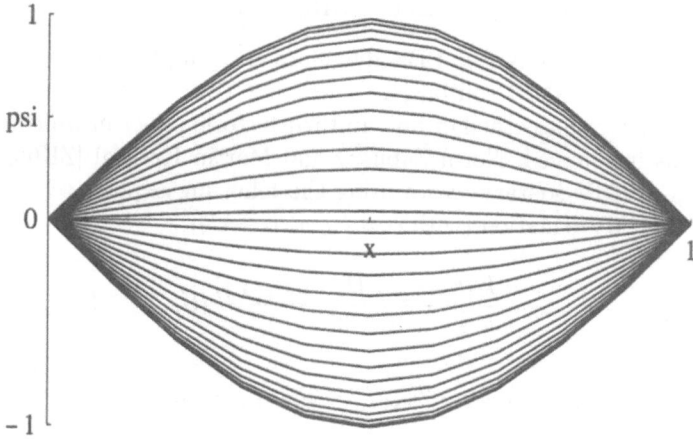

Figure 11.7: Numerical solution of the Klein–Gordon equation.

resembles that for transverse vibrations of an elastic string. This is not too surprising because if a string is embedded in a stretched elastic membrane, the Klein–Gordon equation results. The membrane provides an additional restoring force which in the Hooke's law approximation yields the $a\psi$ term. The student may verify by running the Mathematica file that, compared to the oscillatory behavior for the linear wave equation ($a = 0$), the effect of increasing a is to speed up the oscillations.

PROBLEMS
Problem 11-15: Linear wave equation
Solve the linear wave equation for the same conditions and parameters as in MF46. How do your results differ from the Klein–Gordon equation with $a = 1$? $a = 10$? $a = 100$? For example, compare the number of oscillations for the same total time.

Problem 11-16: Effect of changing the parameter "b"
Solve the nonlinear Klein–Gordon equation for increasing positive values of the parameter b and all other conditions and parameters the same as in MF46. Discuss your results. If you discover interesting behavior, increase the total time. Also investigate negative values of b and discuss.

Problem 11-17: Nonzero initial velocity
Modify the Mathematica file for the Klein–Gordon equation to handle the initial conditions $f(x) = 0$, $g(x) = \sin(\pi x)$ and solve the equation. Compare your results with those in the file.

Problem 11-18: A different initial shape
Modify MF46 to solve the Klein–Gordon equation for the initial conditions $f(x) = x\sin(2\pi x)$, $g(x) = 0$ with all parameters the same as in the file except for the x spacing which should be taken to be $h = 0.05$.

11.2.4   KdV Solitary Wave Collisions

A central problem in nonlinear PDE dynamics is trying to prove that solitary wave solutions are indeed solitons, i.e., whether they survive collisions with other solitary waves unscathed or not. In their historic paper in which the term "soliton" was introduced, Norm Zabusky and Martin Kruskal [ZK65] studied the collision of KdV solitary waves using the following explicit scheme to approximate the KdV equation (with $\alpha = 1$ and $\psi \equiv U$),

$$U_{i,j+1} = U_{i,j-1} - \frac{k}{h}\frac{(U_{i+1,j} + U_{i,j} + U_{i-1,j})}{3}(U_{i+1,j} - U_{i-1,j})$$

$$- \frac{k}{h^3}(U_{i+2,j} - 2U_{i+1,j} + 2U_{i-1,j} - U_{i-2,j}) \qquad (11.25)$$

with $j = 1, 2, \dots$.

Central difference approximations were used for both the first space and first time derivatives to improve the accuracy for given step sizes h and k, respectively. For the third spatial derivative, the finite difference approximation (the last term in (11.25)) used is the same as the student was asked to derive in Problem 11-4 (page 456). Instead of just using the value $U_{i,j}$ for U in the nonlinear term UU_x, an average of the three U contributions from the mesh points $(i+1, j)$, (i, j), and $(i-1, j)$ was employed. Figure 11.8 summarizes all the mesh points used in the Zabusky–Kruskal explicit finite difference scheme.

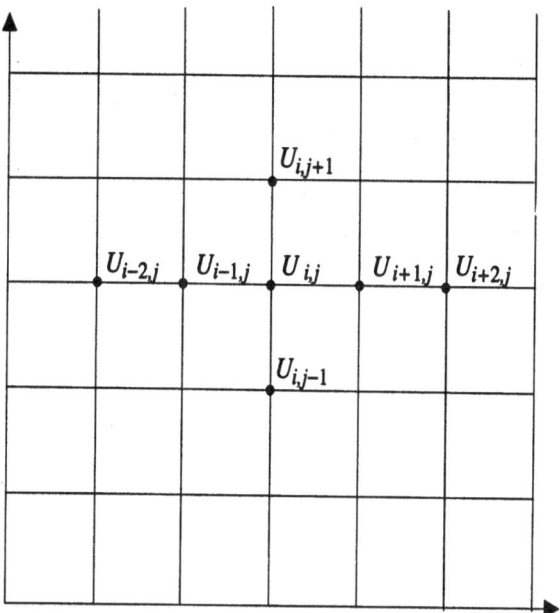

Figure 11.8: The Zabusky–Kruskal explicit scheme to study KdV solitary wave collisions.

As with the Klein–Gordon example, the calculation of the unknown value of U on the $j+1$ time row involves the previous two, j and $j-1$, time rows.

The input solitary wave solutions are specified on the zeroth time row. From Equation (10.57), on setting $\alpha = 1$, a single solitary wave solution is given by

$$U = 3c\,\text{sech}^2(\frac{\sqrt{c}}{2}(x - ct)). \tag{11.26}$$

Assuming that the centroid positions x_1 and x_2 of the two solitary waves are sufficiently far apart at $t = 0$ so that the overlap of their tails is small, the input for the two solitary waves at $(i, 0)$ can be approximated as

$$U_{i,0} = 3c_1\,\text{sech}^2(\frac{\sqrt{c_1}}{2}(x_i - x_1)) + 3c_2\,\text{sech}^2(\frac{\sqrt{c_2}}{2}(x_i - x_2)). \tag{11.27}$$

Since both the solitary waves will travel to the right, we take $x_1 < x_2$ and $c_1 > c_2$ so that the taller solitary wave initially on the left will catch up to the shorter slower solitary wave initially on the right and a collision will take place.

On the first time row, the derivative condition Equation(11.22) gives

$$U_{i,1} = U_{i,0} + k(U_t)_{i,0}$$

$$= 3c_1\,\text{sech}^2 X_i[1 + kc_1\sqrt{c_1}\tanh X_i] + 3c_2\,\text{sech}^2 Y_i[1 + kc_2\sqrt{c_2}\tanh Y_i] \tag{11.28}$$

with $X_i \equiv \frac{\sqrt{c_1}}{2}(x_i - x_1)$, $Y_i \equiv \frac{\sqrt{c_2}}{2}(x_i - x_2)$.

Unlike the Klein–Gordon example, where boundary conditions were imposed, here the numerical spatial domain is taken to be larger than the region containing the two colliding solitary waves to avoid problems at the boundaries.

Because the scheme is explicit, the condition $k/h^3 < 0.3849$ is required for numerical stability. The proof of this is left as a problem in the next section.

What is truly an amazing testimonial to the advances in computing technology is that this numerical procedure which was applied in 1965 by Zabusky and Kruskal on a research computer can now be easily done on a personal computer using Mathematica. This is demonstrated in the next Mathematica File MF47.

KdV Solitary Wave Collisions

The Zabusky–Kruskal finite difference scheme is used to approximate the KdV equation for the study of the collision of two solitary waves. The collision process is animated. The solitary wave solutions pass through each other (virtually) unchanged, indicating that they are (probably) solitons. Qualitatively, the collision process looks like that in Figure 3.18. Mathematica commands in file: Sech, Sqrt, D, Plot, PlotRange, AxesLabel, PlotStyle, Hue, Thickness, TextStyle, ImageSize, $RecursionLimit=Infinity, Table, Chop, Transpose, Do, ListPlot, PlotJoined->True

---

Since a static picture does not do justice to the numerical simulation, we refrain from presenting a picture of the collision process here. You must go the file and run the animation!

In a research context, numerical collision simulations are often used to test for the possible soliton nature of solitary wave solutions. By their very nature, since some numerical error is always present, numerical simulations cannot be 100 percent definitive in proving that the solitary waves are solitons. This fact keeps mathematicians and mathematical physicists occupied trying to analytically nail down their existence. One analytical approach, which works for some classes of nonlinear PDEs, is the inverse scattering method, which is the subject of the last chapter. The KdV two-soliton formula which was the basis of the animation in file MF08 may be derived using this method.

PROBLEMS

Problem 11-19: A different numerical scheme
In the Zabusky–Kruskal finite difference scheme for the KdV equation, U in the nonlinear term UU_x was approximated by the average of three U terms at the grid points $(i+1, j)$, (i, j), and $(i-1, j)$. Compare the results obtained in the Mathematica file with those you would obtain if U was approximated by $U_{i,j}$ alone. Discuss your result.

Problem 11-20: Solitary wave amplification
Modify MF47 by changing the factor $3c_2$ in the input profile (see Equation (11.27)) to $9c_2$. This corresponds to amplifying the smaller solitary wave by a factor of 3. Run the file and interpret the outcome.

Problem 11-21: Three soliton collision
Modify MF47 to include three initially fairly well-separated, solitary waves with the tallest wave to the left of the intermediate height wave which is to the left

of the shortest wave. Choose the locations and c values so that a collision of the three waves is clearly observed. Describe the collision process seen in the run. Is the run consistent with what should happen in a three soliton collision?

Problem 11-22: Radiative ripples
Modify MF47 by changing both sech^2 terms in the input profile (11.27) to sech^4 terms. Describe what happens in the run. Do both or either of the input pulses survive the collision? Be careful in your interpretation.

Problem 11-23: Nonlinear Schrödinger explicit scheme
To study the collision of two initially widely separated solitary wave solutions of the nonlinear Schrödinger equation,

$$iE_z + \frac{1}{2}E_{\tau\tau} + |E|^2 E = 0,$$

a variation on the explicit scheme used in the text may be used. Take $E = U + iV$, where U and V are real functions, to produce two coupled real nonlinear equations. The forward difference approximation may be used to connect the zeroth and first z steps. Write out the explicit scheme for the coupled equations which describes this. The forward difference approximation for the zth derivative can lead to numerical instability even for small r values. For subsequent z steps, this instability can be avoided by then using a central difference approximation for the zth derivative on the jth z step, connecting the $j+1$ and $j-1$ steps. Write out an explicit scheme which describes this. (Note: Although this explicit scheme works well enough, it is slow. In the research literature, the split-step Fourier method (also known as the beam propagation method) is usually used [Agr89].)

11.3   Von Neumann Stability Analysis

In the last section, it was noted that the explicit scheme for the linear diffusion equation becomes numerically unstable for $r \equiv k/h^2 > 1/2$. This may be confirmed analytically by using the Von Neumann stability analysis. Assume that $U = U^0 + u$ where U^0 is the exact solution of the proposed difference scheme and u represents a small error due to roundoff, etc. One then investigates whether u grows or decays with increasing time. If it grows, the scheme is unstable and if it decays, it's stable. The Von Neumann stability analysis is now illustrated with a couple of examples.

11.3.1   Linear Diffusion Equation

For the linear diffusion equation, the explicit scheme that was used is

$$U_{m,n+1} = U_{m,n} + r(U_{m+1,n} - 2U_{m,n} + U_{m-1,n}). \tag{11.29}$$

Because the difference scheme is linear, assuming $U_{m,n} = U^0_{m,n} + u_{m,n}$ yields a finite difference scheme for u identical to (11.29). Making use of Fourier analysis, consider the representative Fourier term

$$u_{m,n} = e^{im\theta}e^{in\lambda}. \tag{11.30}$$

Substituting this into (11.29) yields

$$e^{i\lambda} = 1 + 2r(\cos\theta - 1) = 1 - 4r\sin^2(\theta/2). \qquad (11.31)$$

Setting $\lambda = \alpha + i\beta$, with α and β real, we have

$$|e^{i\lambda}| = e^{-\beta} = |1 - 4r\sin^2(\theta/2)|. \qquad (11.32)$$

For stability $\beta \geq 0$ is needed, because if $\beta < 0$, then $u_{m,n} \sim e^{|\beta|n}$ diverges as n increases (i.e., as time increases). For $\beta \geq 0$, $e^{-\beta} \leq 1$, so $|1 - 4r\sin^2(\theta/2)| \leq 1$. This inequality is valid for $0 \leq r \leq 1/2$. For $r > 1/2$, there exist ranges of θ for which the inequality is violated. The Von Neumann stability analysis confirms our earlier conclusion about the range of r for which the explicit scheme is stable.

11.3.2   Burgers' Equation

As a second example, look at Burgers' nonlinear equation

$$U_t + UU_x = \sigma U_{xx}. \qquad (11.33)$$

The proposed explicit scheme is

$$\frac{(U_{m,n+1} - U_{m,n})}{k} + U_{m,n}\frac{(U_{m+1,n} - U_{m-1,n})}{2h} = \sigma\frac{(U_{m+1,n} - 2U_{m,n} + U_{m-1,n})}{h^2}.$$
$$(11.34)$$

The central difference approximation has been used for U_x because it is more accurate (error $O(h^2)$) than the other difference approximations. In the nonlinear term, U has been approximated by the value at the grid point (m, n). Assuming $U_{m,n} = U_{m,n}^0 + u_{m,n}$ and linearizing in u yields

$$\frac{(u_{m,n+1} - u_{m,n})}{k} + U_{m,n}^0\frac{(u_{m+1,n} - u_{m-1,n})}{2h} = \sigma\frac{(u_{m+1,n} - 2u_{m,n} + u_{m-1,n})}{h^2}.$$
$$(11.35)$$

Note that the very small term involving $(U_{m+1,n}^0 - U_{m-1,n}^0)u_{m,n}$ has been dropped. Then, on setting $r \equiv \sigma k/h^2$, substituting $u_{m,n} = e^{im\theta}e^{in\lambda}$, and using $\cos\theta = 1 - 2\sin^2(\theta/2)$, we have

$$e^{i\lambda} = 1 - 4r\sin^2(\theta/2) - i(kU_{m,n}^0/h)\sin\theta. \qquad (11.36)$$

With $\lambda = \alpha + i\beta$, stability requires that $|e^{i\lambda}| = e^{-\beta} \leq 1$ so that

$$(1 - 4r\sin^2(\theta/2))^2 + (kU_{m,n}^0/h)^2\sin^2\theta \leq 1. \qquad (11.37)$$

Suppose that $\sin\theta = 0$. Then, $\sin(\theta/2) = 0$ or ± 1. For the ± 1 choice, (11.37) yields

$$(1 - 4r)^2 \leq 1 \qquad (11.38)$$

which is satisfied for $0 \leq r \leq 1/2$. On the other hand, the choice of $\sin(\theta/2) = 0$ produces 1 on the lhs of (11.37), so the equation is satisfied.

Now, suppose that $\sin\theta \neq 0$. Expanding (11.37) and using the trigonometric identity $\sin\theta = 2\sin(\theta/2)\cos(\theta/2)$,

$$4r\sin^2(\theta/2)[-2 + 4r\sin^2(\theta/2) + \frac{k(U^0_{m,n})^2}{\sigma}\cos^2(\theta/2)] \leq 0. \tag{11.39}$$

For $\sin(\theta/2) = 0$, the lhs of Equation (11.39) is zero and the equation is satisfied. For $\sin(\theta/2) \neq 0$, we must have from (11.39)

$$-2 + 4r\sin^2(\theta/2) + \frac{k(U^0_{m,n})^2}{\sigma}\cos^2(\theta/2) \leq 0. \tag{11.40}$$

But the largest allowable value of r is $r = 1/2$, so

$$-2 + 2\sin^2(\theta/2) + \frac{k(U^0_{m,n})^2}{\sigma}\cos^2(\theta/2) \leq 0 \tag{11.41}$$

or

$$\sin^2(\theta/2) + \frac{k(U^0_{m,n})^2}{2\sigma}\cos^2(\theta/2) \leq 1. \tag{11.42}$$

Since the trigonometric identity $\sin^2(\theta/2) + \cos^2(\theta/2) = 1$ holds, we have

$$\sin^2(\theta/2) + \frac{k(U^0_{m,n})^2}{2\sigma}\cos^2(\theta/2) \leq \sin^2(\theta/2) + \cos^2(\theta/2). \tag{11.43}$$

Thus, stability requires the second condition

$$\frac{k(U^0_{m,n})^2}{2\sigma} \leq 1 \tag{11.44}$$

in addition to

$$0 \leq r \leq \frac{1}{2}. \tag{11.45}$$

One condition comes from the linear part of the original Burgers' equation, the other, which depends on the "size" of the solution, from the nonlinear part.

PROBLEMS
Problem 11-24: Nonlinear diffusion equation
For the nonlinear diffusion equation

$$U_t = (U^3)_{xx}$$

write down the explicit scheme which results on taking the forward difference approximation for the time derivative and the standard approximation for the second spatial derivative. By applying the Von Neumann stability analysis with $r \equiv k/h^2$, show that stability of the scheme demands that $(U^0_{m,n})^2 r \leq 1/6$.

Problem 11-25: Fisher's equation
Write down an explicit scheme for Fisher's equation and use Von Neumann's stability analysis to determine the upper bound on $r \equiv k/h^2$ for stability of the scheme.

Problem 11-26: Courant stability condition

Using the standard finite difference approximations for second derivatives, show
that the corresponding explicit scheme for the linear wave equation

$$U_{xx} = \frac{1}{v^2} U_{tt}$$

is stable if $r \equiv |v|k/h \leq 1$. This is called the Courant stability condition. Hint:
You have to solve a quadratic equation in $e^{i\lambda}$.

Problem 11-27: KdV equation

In their pioneering work on KdV solitons, Zabusky and Kruskal [ZK65] used
the following finite difference scheme for the KdV equation (with $\alpha = 1$):

$$U_{m,n+1} = U_{m,n-1} - \frac{k}{h} \frac{(U_{m+1,n} + U_{m,n} + U_{m-1,n})}{3}(U_{m+1,n} - U_{m-1,n})$$

$$-\frac{k}{h^3}(U_{m+2,n} - 2U_{m+1,n} + 2U_{m-1,n} - U_{m-2,n}).$$

It is claimed that numerical stability of this scheme requires that $k/h^3 <
2/(3\sqrt{3}) = 0.3849$. Confirm analytically and numerically that this is the case.

11.4   Implicit Methods

In the explicit schemes, use was made of the forward difference approximation
to advance forward in time with the condition that r not exceed some particular
value for numerical stability to prevail. The limitation on r can be waived if a
switch is made to the backward difference approximation.

As an example, consider the normalized linear diffusion equation with the
same boundary and initial conditions as in the explicit scheme treatment. Using the backward difference approximation for the time derivative, the finite
difference scheme is

$$\frac{(U_{i,j} - U_{i,j-1})}{k} = \frac{(U_{i+1,j} - 2U_{i,j} + U_{i-1,j})}{h^2} \tag{11.46}$$

or, setting $r = k/h^2$,

$$-rU_{i-1,j} + (1 + 2r)U_{i,j} - rU_{i+1,j} = U_{i,j-1} \tag{11.47}$$

with $j = 1, 2, 3, \ldots$. Figure 11.9 shows the mesh points connected by the algorithm. We now have an implicit scheme, having to solve a set of simultaneous
equations to obtain the unknown $U_{i,j}$. Again, a matrix approach may be used.
If the range 0 to 1 is divided into 10 intervals as before, and the same notation
is used as in the explicit scheme, then

$$BU^{(j)} = U^{(j-1)} \tag{11.48}$$

where the matrix B is obtained, on comparing (11.47) with (11.9), from the
previous matrix A by replacing r with $-r$. Then, multiplying Eq. (11.48) from
the left by the inverse matrix B^{-1}, we have

$$U^{(j)} = B^{-1}U^{(j-1)}. \tag{11.49}$$

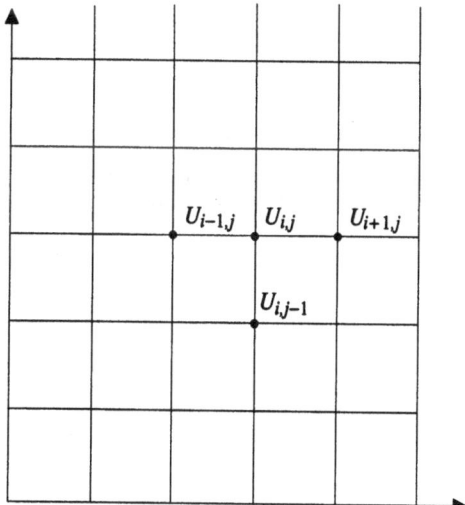

Figure 11.9: Mesh points involved in the backward difference implicit scheme.

Comparing this last line with Equation (11.11), we see that the implicit scheme can be solved for the same initial and boundary conditions as earlier by using file MF43 with the matrix A replaced with B^{-1}. This is left as a problem.

To prove that there is no restriction on the size of r in the backward difference implicit scheme for stability to prevail, the Von Neumann stability analysis is applied, yielding

$$e^{-i\lambda} = 1 + 4r \sin^2(\theta/2). \tag{11.50}$$

Setting $\lambda = \alpha + i\beta$, with α and β positive, we have

$$e^{\beta} = |1 + 4r \sin^2(\theta/2)|. \tag{11.51}$$

For stability, again $\beta \geq 0$ is demanded. This is assured for any real positive r value.

The backward and forward difference approximations have truncation errors of order k in the time derivative. The Crank–Nicolson method, which has a truncation error of order k^2 in the time derivative, results on averaging the forward and backward difference schemes. Again consider the normalized linear diffusion equation. The forward difference and backward schemes connecting the jth and $j+1$st time steps, give respectively,

$$\frac{(U_{i,j+1} - U_{i,j})}{k} = \frac{(U_{i+1,j} - 2U_{i,j} + U_{i-1,j})}{h^2}$$
$$\frac{(U_{i,j+1} - U_{i,j})}{k} = \frac{(U_{i+1,j+1} - 2U_{i,j+1} + U_{i-1,j+1})}{h^2}. \tag{11.52}$$

Forming the average of the two schemes yields the Crank–Nicolson method,

$$\frac{(U_{i,j+1} - U_{i,j})}{k} = \frac{(U_{i+1,j} - 2U_{i,j} + U_{i-1,j})}{2h^2} + \frac{(U_{i+1,j+1} - 2U_{i,j+1} + U_{i-1,j+1})}{2h^2} \tag{11.53}$$

or

$$-rU_{i-1,j+1}+2(1+r)U_{i,j+1}-rU_{i+1,j+1} = rU_{i-1,j}+2(1-r)U_{i,j}+rU_{i+1,j}. \quad (11.54)$$

This finite difference formula relates three unknown values of U on the $j+1$st time row to three known values on the jth time row as shown in Figure 11.10. It is left to the reader to express this formula in matrix language.

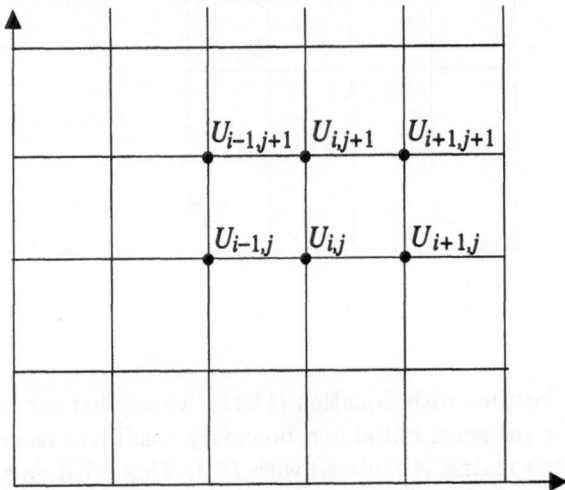

Figure 11.10: Mesh points used in the Crank–Nicolson method.

Applying implicit methods to linear PDEs results in solving sets of simultaneous linear equations. For nonlinear PDEs, sets of nonlinear equations have to be solved. As was the case for nonlinear ODEs, e.g., the fox rabies example in Section 2.2.2, the usual procedure is to linearize the set. This issue will not be pursued here as it is conceptually similar to what was encountered in the ODE case.

PROBLEMS
Problem 11-28: Inverse matrix
Use Mathematica to calculate the inverse of the matrix

$$A = \begin{bmatrix} 1 & 0 & 3 \\ 4 & 5 & 6 \\ 7 & 8 & 9 \end{bmatrix}.$$

What is the inverse if the element 0 is replaced with 2? Explain.

Problem 11-29: Backward difference method
Using the implicit backward difference method, solve the linear diffusion equation for the same conditions and parameters as for the forward difference approximation, except consider $k/h^2 = 0.6$. Contrast the stability with that for the explicit scheme for the same ratio of k/h^2.

Problem 11-30: Truncation error for the Crank–Nicolson method
Show that the truncation error for the Crank–Nicolson method is $O(k^2)+O(h^2)$.

(Hint: Show that the truncation error for the time derivative cancels in the kth order when the forward and backward difference approximations are combined in the Crank–Nicolson method.)

Problem 11-31: Matrix representation of Crank–Nicolson method
Explicitly write out the matrix representation for the Crank–Nicolson method applied to the linear diffusion equation. Then, solve the diffusion equation subject to $U(0, y) = U(1, y) = 0$ for $y \geq 0$ and $U(x, 0) = 2x$ for $0 \leq x \leq 1/2$, $2(1 - x)$ for $1/2 \leq x \leq 1$. Take $h = 0.1$, $k = 0.007$, and calculate U up to the 20th time row. Graphically compare the numerical results with the exact solution on the 20th time row. The explicit scheme displays wild oscillations (Problem 11-6, page 464) for this problem with the step size given.

Problem 11-32: Neumann boundary conditions
Discuss in detail how you would handle the derivative boundary condition $\partial U/\partial x = 0$ at $x = 0$ and 1 (no heat flow across the ends of the rod) in applying the explicit and Crank–Nicolson methods to the linear diffusion equation, $U(x, 0)$ being specified. Mathematicians call such a derivative boundary condition a Neumann boundary condition.

11.5 Method of Characteristics

In applying explicit and implicit schemes to problems involving one spatial coordinate and time, the mesh was rectangular and we moved vertically in the direction of increasing time. For PDEs characterized by wave equation structures, characteristic directions can be introduced which correspond to intersecting diagonal lines whose slopes have values $\pm v$ where v is the speed of the waves. Along these characteristic directions, the integration of the PDEs reduces to an integration of equations involving only total differentials. The method of characteristics will be illustrated with two nonlinear examples.

11.5.1 Colliding Laser Beams

The transient behavior of two intense laser pulses (labeled L and S) as they pass through each other in opposite directions in a resonant absorbing nonlinear medium (absorption coefficient α) of length l can be described [RE76] by the following coupled nonlinear PDEs for the intensities (I),

$$(I_{\text{L}})_z + (1/v)(I_{\text{L}})_t = -gI_{\text{L}}I_{\text{s}} - \alpha I_{\text{L}}$$

$$(I_{\text{s}})_z - (1/v)(I_{\text{s}})_t = -gI_{\text{L}}I_{\text{s}} + \alpha I_{\text{s}}.$$

(11.55)

Here v is the velocity of light in the medium and g a positive constant which depends on the medium parameters and the frequency difference between the L and S beams. The behavior for $\alpha = 0$ was shown in Figure 3.16 for the case of rectangular input pulses. The L beam travels to the right and the S beam to the left. For $g > 0$, the S beam grows at the expense of the L beam. We would now like to show how Figure 3.16 can be numerically generated using a simple version of the method of characteristics.

Let us once again render all quantities into a dimensionless form. Set $x \equiv z/l$, $y \equiv vt/l$, $U \equiv I_L/I_L^0$, and $V \equiv I_s/I_s^l$, where I_L^0 and I_s^l are defined by the input shapes at $z = 0$ and l, respectively. Then the coupled equations reduce to

$$U_x + U_y = e_1, \qquad V_x - V_y = e_2, \tag{11.56}$$

with $e_1 \equiv -g_1 UV - \beta U$, $e_2 \equiv -g_2 UV + \beta V$, $g_1 \equiv gI_s^l l$, $g_2 \equiv gI_L^0 l$, and $\beta \equiv \alpha l$.

To find the "characteristic" directions for this pair of equations, proceed as follows. Since $U = U(x, y)$, $V = V(x, y)$, then

$$dU = U_x dx + U_y dy, \quad dV = V_x dx + V_y dy, \tag{11.57}$$

which may be used to write the coupled PDEs in the form

$$\begin{aligned}
\frac{dU}{dx} + \left(1 - \frac{dy}{dx}\right)\frac{\partial U}{\partial y} &= e_1 \\
\frac{dV}{dx} - \left(1 + \frac{dy}{dx}\right)\frac{\partial V}{\partial y} &= e_2.
\end{aligned} \tag{11.58}$$

Choosing to work along a direction whose slope is $dy/dx = 1$, then $dU/dx = e_1(U, V)$ in this direction, while $dV/dx = e_2(U, V)$ along a line of slope $dy/dx = -1$. The characteristic directions $dy/dx = \pm 1$ form a diamond-shaped grid as

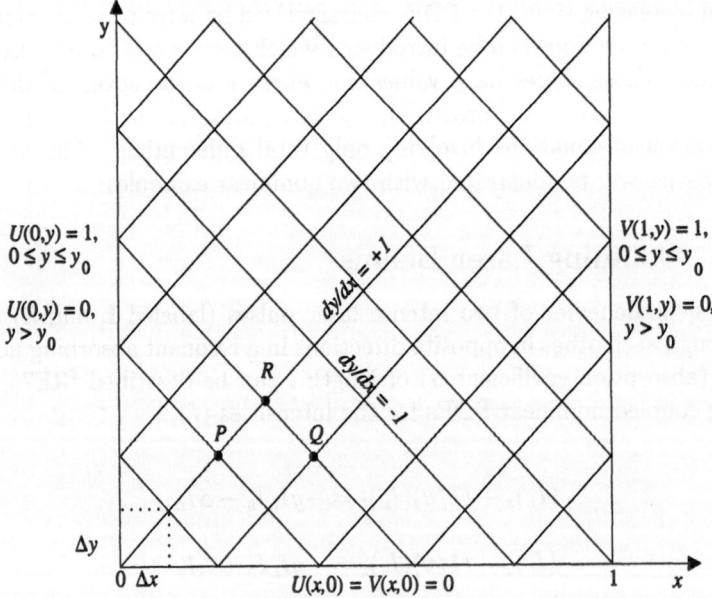

Figure 11.11: Method of characteristics applied to laser beam collision problem.

shown in Figure 11.11 with the grid spacing $\Delta y = \Delta x$. The slopes are just the normalized velocities. To solve for $U(x, y)$ and $V(x, y)$, the following initial and boundary conditions will be imposed:

- $U(x,0) = V(x,0) = 0$ for $0 < x < 1$ (no pulses inside medium initially),

- $U(0,y) = V(1,y) = 1$ for $0 \le y \le y_0$ and zero for $y > y_0$ (rectangular pulses of finite duration fed in at $x = 0$ and 1).

Starting on the bottom time row, one moves along the characteristic direction $dy/dx = 1$ (e.g., from P to R) in calculating the change in U. The value of U at R is approximated by

$$U_R = U_P + (e_1)_P(x_R - x_P) = U_P + (e_1)_P(\Delta x). \qquad (11.59)$$

Similarly, one moves along the characteristic direction $dy/dx = -1$ (e.g., from Q to R) in calculating the change in V. The value of V at R is approximated by

$$V_R = V_Q + (e_2)_Q(x_R - x_Q) = V_Q - (e_2)_Q(\Delta x). \qquad (11.60)$$

A Mathematica File MF48 has been created which solves these equations along the characteristic directions. The parameter values are $g_1 = 0.4$, $g_2 = 20$, $\Delta x = 1/120$ and $y_0 = 0.5$. In Figure 11.12, the behavior of the two beams,

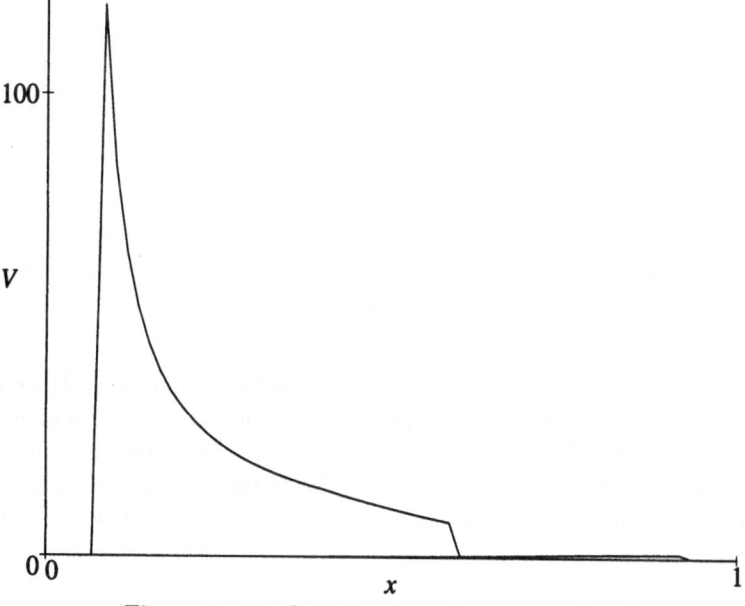

Figure 11.12: Colliding laser beams for $\beta = 0$.

with the S beam traveling to the left and the L beam to the right, is shown for zero absorption ($\beta = 0$). The effect of nonzero absorption may be investigated in the Mathematica file. For zero absorption the S beam (on the left) has grown exponentially on its leading edge, having sucked energy from the L beam (barely visible on the right). The exponential growth can be understood from the basic equation for V on noting that the leading edge of the S beam sees a constant amplitude L beam (i.e., $U = 1$) as long as the two laser beams interact. Because the two beams are traveling in opposite directions, the effective interaction length here is $x_{\text{int}} = 0.25$. Theoretically, then, the leading edge of the S beam should grow exponentially according to the relation $v = e^{g_2 x_{\text{int}}} = e^{20(0.25)} = e^5 = 148.4$. Keeping in mind that the "true" (i.e., theoretical) leading edge of the signal beam lies somewhere between the grid point where the signal beam is numerically largest and the next grid point to the left where the signal beam is zero, the numerical simulation does a reasonable job in confirming the theoretical prediction.

Colliding Laser Beams
The method of characteristics is applied to the coupled nonlinear equations describing colliding intense laser beams. Again, the collision process is animated. The effect of changing the parameter values is easily studied. Mathematica commands in file: `Table`, `Transpose`, `ListPlot`, `PlotStyle`, `Hue`, `Thickness`, `PlotRange`, `AxesLabel`, `TextStyle`, `ImageSize`, `Do`, `Block`, `Show`, `Ticks`, `PlotJoined->True`, `$DisplayFunction=Identity`

PROBLEMS
Problem 11-33: Linear ramp laser pulses
Modify the file MF48 to study the collision of two symmetric linear ramp pulses which grow linearly from zero on their leading edges to one on their trailing edges, each pulse having a (normalized) length of one-half. As in the file, let the leading edges of the pulses enter the medium at $t = 0$. Consider (a) zero absorption ($\beta = 0$), (b) $\beta = 1$. Discuss your results and compare them with the rectangular pulse case.

Problem 11-34: Half-cycle sine wave laser pulses
Repeat the previous problem, but let the input pulses each be a positive half-cycle of a sine wave, each pulse being of unit amplitude and a normalized width of one-half.

Problem 11-35: Pulse train collision
Modify MF48 to study the collision of two symmetric trains of laser pulses traveling in opposite directions, each pulse train consisting of four positive half-cycles of a sine wave. Take each pulse in each train to be of unit amplitude and each train to be of (normalized) width 1/2. The leading edges of the trains enter the medium at time $t = 0$. Consider (a) zero absorption, (b) $\beta = 1$. Discuss your results.

11.5.2   General Equation
The method of characteristics can also be applied to the sine-Gordon equation

$$\frac{\partial^2 \psi}{\partial x^2} - \frac{\partial^2 \psi}{\partial t^2} = \sin \psi. \qquad (11.61)$$

It is a special case of the general equation

$$a\frac{\partial^2 u}{\partial x^2} + b\frac{\partial^2 u}{\partial x \partial y} + c\frac{\partial^2 u}{\partial y^2} + e = 0 \qquad (11.62)$$

resulting when $u \equiv \psi$, $a = 1$, $b = 0$, $c = -1$, $y \equiv t$, and $e \equiv -\sin\psi$. The general equation (11.62) can be solved by the method of characteristics if a, b, c, e are functions of u, $\partial u/\partial x$, $\partial u/\partial y$, but not of higher derivatives.

The procedure is similar in spirit to that in the previous example. Let us set $p \equiv \partial u/\partial x$ and $q \equiv \partial u/\partial y$ so that the general equation becomes

$$a\frac{\partial p}{\partial x} + b\frac{\partial p}{\partial y} + c\frac{\partial q}{\partial y} + e = 0. \tag{11.63}$$

Since $p = p(x, y)$ and $q = q(x, y)$, then $dp = p_x dx + p_y dy$, $dq = q_x dx + q_y dy$, which may be used to rewrite Equation (11.63) in the form

$$a\left[\frac{dp}{dx} - \frac{\partial p}{\partial y}\frac{dy}{dx}\right] + b\frac{\partial p}{\partial y} + c\left[\frac{dq}{dy} - \frac{\partial q}{\partial x}\frac{dx}{dy}\right] + e = 0 \tag{11.64}$$

or, on multiplying through by dy/dx, noting that $\partial q/\partial x = \partial p/\partial y$, and rearranging,

$$\frac{\partial p}{\partial y}\left[a\left(\frac{dy}{dx}\right)^2 - b\left(\frac{dy}{dx}\right) + c\right] - \left[a\frac{dp}{dx}\frac{dy}{dx} + c\frac{dq}{dx} + e\frac{dy}{dx}\right] = 0. \tag{11.65}$$

Then choose characteristic directions whose slopes are given by

$$a\left(\frac{dy}{dx}\right)^2 - b\left(\frac{dy}{dx}\right) + c = 0 \tag{11.66}$$

which, in general, has two roots r given by $r = (b/2) \pm (\sqrt{b^2 - 4ac})/2$. For $b^2 - 4ac$ greater than, equal to, or less than zero, the roots will be, respectively, real and distinct, equal, or complex. Corresponding to these three possibilities, mathematicians refer to the general equation (11.62) as hyperbolic, parabolic, and elliptic. Thus, for example, the sine-Gordon equation ($b^2 - 4ac = 0 - 4(1)(-1) = 4 > 0$) is hyperbolic, while Fisher's equation ($b^2 - 4ac = 0 - 4(1)(0) = 0$) is parabolic. An example of an elliptic equation is Laplace's equation $S_{xx} + S_{yy} = 0$ of electromagnetic theory for which $b^2 - 4ac = 0 - 4(1)(1) = -4 < 0$. The method of characteristics applies to hyperbolic equations.

Along the characteristic directions, from Equation (11.65),

$$a\frac{dp}{dx}\frac{dy}{dx} + c\frac{dq}{dx} + e\frac{dy}{dx} = 0 \tag{11.67}$$

or, in terms of differentials,

$$a\left(\frac{dy}{dx}\right)dp + cdq + edy = 0. \tag{11.68}$$

Corresponding to the two roots $dy/dx = r_1, r_2$ two equations for p and q will result which when solved will enable us to calculate u through the relation (since $u = u(x, y)$)

$$du = \frac{\partial u}{\partial x}dx + \frac{\partial u}{\partial y}dy = pdx + qdy. \tag{11.69}$$

To proceed numerically, (11.68) and (11.69) can be replaced by finite difference approximations. This is demonstrated with the sine-Gordon equation.

11.5.3 Sine-Gordon Equation

For the sine-Gordon equation, we have already identified $u \equiv \psi$, $a = 1$, $b = 0$, $c = -1$, and $e = -\sin u$ so that the two values of dy/dx are $r_1 = 1$ and $r_2 = -1$. The characteristic directions are again straight lines of slope ± 1 as in Figure 11.13.

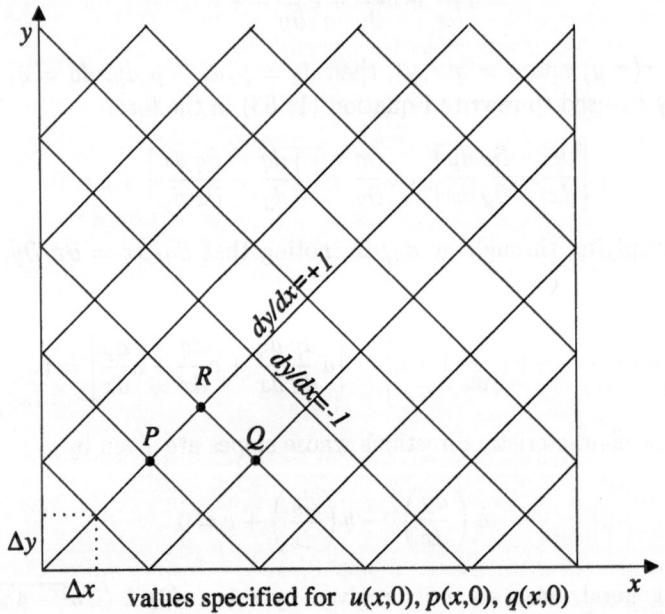

Figure 11.13: Characteristic directions and labeling for the numerical solution of the sine-Gordon equation.

At the grid point R, the quantities p_R, q_R and u_R are to be calculated from the known values of these quantities at the grid points P and Q. Taking $dy/dx = \pm 1$, (11.68) can be approximated along the two characteristic directions by the pair of finite difference equations

$$(p_R - p_P) - (q_R - q_P) = -e_P(y_R - y_P),$$

$$-(p_R - p_Q) - (q_R - q_Q) = -e_Q(y_R - y_Q),$$
(11.70)

which can be solved for p_R and q_R, yielding

$$p_R = \frac{1}{2}(p_Q + p_P) + \frac{1}{2}(q_Q - q_P) + \frac{1}{2}(e_Q - e_P)\Delta y,$$

$$q_R = \frac{1}{2}(p_Q - p_P) + \frac{1}{2}(q_Q + q_P) + \frac{1}{2}(e_Q + e_P)\Delta y,$$
(11.71)

where $\Delta y = y_R - y_P = y_R - y_Q = \Delta x \equiv h$.

To obtain u_R, choose either characteristic and use (11.69). For $dy/dx = 1$,

then

$$u_R = u_P + \frac{1}{2}(p_P + p_R)(x_R - x_P) + \frac{1}{2}(q_P + q_R)(y_R - y_P) \quad (11.72)$$

or

$$u_R = u_P + \frac{1}{2}h(p_P + p_R + q_P + q_R). \quad (11.73)$$

For improved accuracy, the "old" and "new" values of p and q have been averaged. Along $dy/dx = -1$, on the other hand,

$$\begin{aligned} u_R &= u_Q + \frac{1}{2}(p_Q + p_R)(x_R - x_Q) + \frac{1}{2}(q_Q + q_R)(y_R - y_Q) \\ &= u_Q + \frac{1}{2}h(-p_Q - p_R + q_Q + q_R). \end{aligned} \quad (11.74)$$

In general, for a nonlinear hyperbolic PDE the value of u_R calculated along the two different characteristics may not turn out to be numerically the same. Therefore, an average of the two values of u_R is formed,

$$u_R^{AVE} = \frac{1}{2}(u_P + u_Q) + \frac{1}{4}h(p_P - p_Q + q_P + q_Q + 2q_R). \quad (11.75)$$

As an example of this procedure, we shall consider the collision of a sine-Gordon kink solitary wave with an anti-kink solitary wave to see if they are solitons. Making use of Equation (10.30), let's take as input

$$u = 4\arctan(e^{(x-x_1-c_1y)/\sqrt{(1-c_1^2)}}) + 4\arctan(e^{-(x-x_2-c_2y)/\sqrt{(1-c_2^2)}}) \quad (11.76)$$

with $x_1 < 0$, $x_2 > 0$, $c_1 > 0$, $c_2 < 0$, and $y = 0$. The first term by itself is the kink solution, the second the anti-kink solitary wave. Nonzero x_1 and x_2 have been inserted to separate the solitary waves initially. The choice of signs puts the kink to the left of the anti-kink. The signs of the velocities are chosen to ensure a collision. In addition to specifying the input profile, $p(x,0) = u_x|_{y=0}$ and $q(x,0) = u_y|_{y=0}$ are also given. Using the method of characteristics, the temporal and spatial evolution of the input solution has been calculated in file MF49. By running the file, the reader can be the judge of how stable the solitary waves are.

Kink–Antikink Collision

The method of characteristics is applied to the collision of kink and anti-kink solitary wave solutions of the sine-Gordon equation to test if they are solitons. In the file, we take $x_1 = -5$, $x_2 = 5$ and $c_1 = 0.8$, $c_2 = -0.8$. To avoid edge effects, the boundaries of the numerical scheme are taken to be far from where the collision process takes place. In interpreting the profile that results after the collision, remember that if the solitary waves emerge from the collision unchanged except for a possible phase shift, they are solitons. What is the phase shift here? Mathematica commands in file: Sqrt, ArcTan, Exp, D, Table, Transpose, ListPlot, PlotJoined->True, PlotRange, PlotStyle, Hue, Thickness, PlotLabel, Ticks, TextStyle, ImageSize, Sin, Do

MF49

PROBLEMS
Problem 11-36: Kink–kink collision
Using the method of characteristics, verify that two sine-Gordon kink solitary waves can survive a collision with each other unchanged in shape. This is easily done by changing the anti-kink solution in MF49 to a kink.

Problem 11-37: Kink–breather collision
Using the method of characteristics, consider a collision of a sine-Gordon kink solitary wave traveling to the right (velocity $c = 0.8$) with a sine-Gordon breather moving ($c = -0.8$) to the left. The analytic form of the input breather is given in Problem 10-16 (page 442). Take the parameter m in the breather solution to be $m = 0.5$. Take the same initial separation as in file MF49. Does your numerical simulation probably confirm the soliton nature of both the kink and breather solutions?

Problem 11-38: Amplified kink–antikink input
In MF49 double the amplitudes of the input kink and anti-kink solitary wave profiles. Run the file with the amplified input and interpret the outcome.

Problem 11-39: Amplified kink–kink input
Rerun the kink–kink collision problem with the input solitary wave profiles amplified by a factor of four. Interpret the outcome.

Problem 11-40: Curved characteristics and iteration
Consider the nonlinear wave equation

$$U_{xx} - U^2 U_{yy} = 0$$

subject to the following initial conditions along $y = 0$: $U = 0.2 + 5x^2$, $U_y = 3x$.
a) Show that the characteristic directions are given by $dy/dx = \pm U$, so that they depend on the solution itself.
b) Using the notation of the text consider the initial points P $(x = 0.2, y = 0)$ and $Q(x = 0.3, y = 0)$. Find the grid point R between P and Q and with $y > 0$ by approximating the arcs PR and QR as straight lines of slopes r_P and r_Q respectively. With the location of the grid point R determined, find the first approximations to p_R, q_R and then calculate $U_R = U_P + \frac{1}{2}(p_P + p_R)(x_R - x_P) + \frac{1}{2}(q_P + q_R)(y_R - y_P)$.
c) Averaging the "old" slope at P with the "new" slope at R and similarly for the other arc QR, find the next approximation to the location of the grid point R. Then calculate the next approximations to p_R, q_R by replacing e_P and e_Q with the averages $\frac{1}{2}(e_P + e_R)$ and $\frac{1}{2}(e_Q + e_R)$, respectively. Finally, calculate U_R as above. Repeat this process until U_R does not change in its first four significant figures. How many iterations did it take?

11.6   Higher Dimensions

The methods that have been introduced can be extended to handling nonlinear dynamical systems in two and three spatial dimensions. An illustration of an explicit method applied to a 2-dimensional problem is now given.

11.6.1   2-Dimensional Reaction–Diffusion Equations

If we have two or more molecular species which can diffuse and react with each other, then the system can be described by coupled reaction–diffusion equations. Such systems are important in chemical mixing, biological pattern formation, and so on. Cross and Hohenberg [CH93] have explored the following important class of reaction diffusion equations for a two species system:

$$U_t = \sigma_1 \nabla^2 U + f(U^2 + V^2)\,U - g(U^2 + V^2)\,V,$$
$$V_t = \sigma_2 \nabla^2 V + f(U^2 + V^2)\,V + g(U^2 + V^2)\,U. \tag{11.77}$$

Here U and V are the concentrations of the two species, σ_1 and σ_2 are the respective diffusion coefficients, $\nabla^2 = \partial^2/\partial x^2 + \partial^2/\partial y^2$ is the 2-dimensional Laplacian operator, and f and g are smooth functions. Following Gass [Gas98], we shall numerically solve Equations (11.77) in the following Mathematica File MF50 assuming that f and g are sine and cosine functions, respectively, and that periodic boundary conditions prevail. For calculation purposes, each boundary is considered to be adjacent to the boundary on the opposite side. An explicit method is used, the time derivative being replaced with the forward difference approximation and the Laplacian operator with the standard central difference approximations for each of the x and y second derivatives. The initial concentrations are taken to be the Gaussian profiles shown in Figure 11.14.

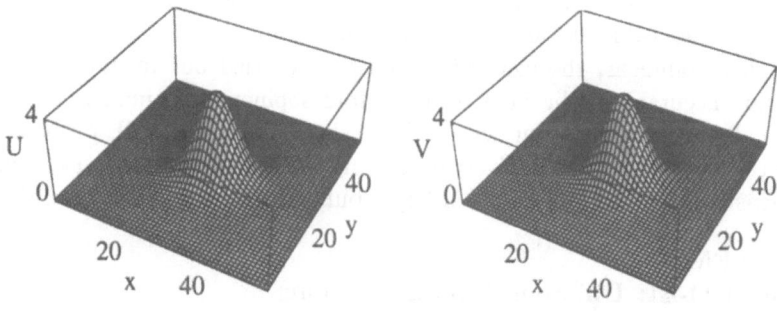

Figure 11.14: Input concentration profiles for the reaction–diffusion equations.

Coupled 2-Dimensional Reaction–Diffusion Equations

In this file, an explicit scheme is carried out for Equations (11.77) for nominal values of the parameters. To input the central difference approximations for the 2-dimensional Laplacian operator, we make use of Mathematica's `RotateRight` and `RotateLeft` commands. These cyclic commands automatically impose periodic boundary conditions at the edges of the computational mesh. The output profiles are animated. Mathematica commands in file: `RotateRight`, `RotateLeft`, `Table`, `Length`, `First`, `ListPlot3D`, `GraphicsArray`, `Show`, `IntegerPart`, `ListPlot3D`, `Block`, `Table`, `Sin`, `Cos`, `Exp`, `N`, `Do`

On execution of Mathematica File MF50, the animated temporal and spatial evolution of the input Gaussian profiles results. The shapes of the two concentration profiles at a later instant in time is illustrated in Figure 11.15.

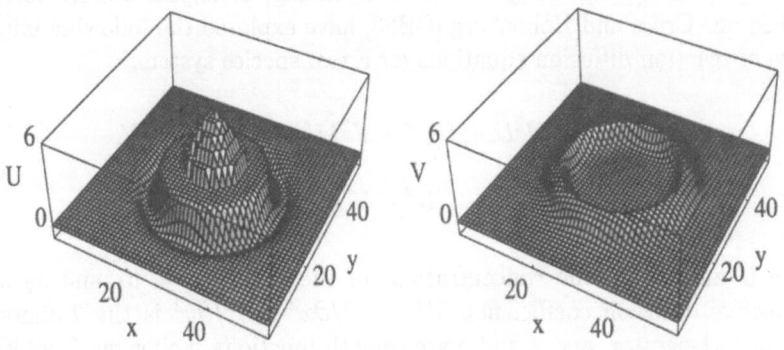

Figure 11.15: Time-evolved profiles for the reaction–diffusion equations.

In this numerical simulation, a modest sized mesh was used and a limited number of time steps calculated. The straight-forward explicit methods that we have discussed generally are much too slow when larger or finer meshes are desired and longer times are involved. The situation becomes even more acute in nonlinear initial value problems in three dimensions. For example, for a mesh 1000 × 1000 × 1000, the computing time is at least one million times as long as the 1-dimensional mesh with 1000 grid points. If the interactions are highly nonlinear, the calculation must be carried out in double-precision to ensure accuracy of the final result. More sophisticated numerical schemes are usually needed in dealing with 3-dimensional nonlinear PDE problems. We finish this chapter by qualitatively describing the approach used in studying the 3-dimensional collisions of so-called "light bullets" mentioned earlier in the text.

PROBLEMS
Problem 11-41: Different f and g functions
Experiment with different f and g functions in MF50, altering the plot ranges as necessary. Discuss your results.

Problem 11-42: Different parameter values
With the functions f and g as in the file, experiment with different parameter values in MF50. Discuss your results.

11.6.2   3-Dimensional Light Bullet Collisions

In Chapter 3, the reader saw examples of initial value problems in three dimensions, namely the collisions of spherical optical solitary waves, or light bullets. In their numerical investigation of light bullet behavior, two of the central goals of Edmundson and Enns [EE95] were to (a) demonstrate the soliton-like character of the colliding solitary waves, and (b) to determine the force law, or interaction potential, operating between the solitons.

This was accomplished by creating a suitable 3-dimensional numerical scheme and, for the repulsive interaction case, performing the analogue of the famous Rutherford scattering experiment, which was instrumental in unraveling the structure of the atom. Light bullets of varying sizes, initial velocities, initial separations, and initial impact parameters[7], were fired at each other and the scattering angle through which the bullets were deflected was measured. A representative collision is illustrated in Figure 11.16, the arrows indicating the

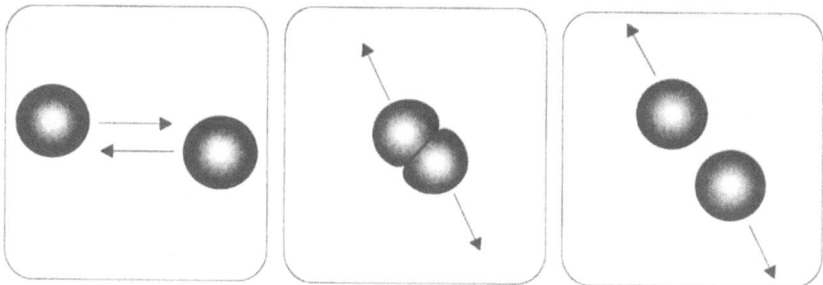

Figure 11.16: Colliding spherical optical solitary waves.

velocity direction for each light bullet. The scattering angle data was then analyzed and fitted by a simple interaction potential which could be qualitatively understood from the structure of the governing model equation, which is of the form

$$iE_z + (E_{xx} + E_{yy} + E_{\tau\tau})/2 + \left(|E|^2/(1 + a|E|^2)\right) E = 0. \qquad (11.78)$$

In suitably normalized units, E is the electric field, z the spatial coordinate in the direction of propagation, x and y the transverse spatial coordinates, τ the time, and a the real, positive, saturation parameter. As already pointed out in Chapter 3, three-dimensional solitary waves solutions are possible because the nonlinear term can balance not only the dispersion in the propagation direction but also the diffraction or spreading in the transverse directions. Further, because of the saturable nature of the nonlinearity, "bistable" solitary waves are possible. That is to say, spherical solitary waves of radically different sizes can exist at the same energy.

As already mentioned, numerically solving highly nonlinear initial value problems in 3 spatial dimensions is not a trivial task, and not one that can be readily carried out on a personal computer. In the light bullet case, Edmundson and Enns used a much more sophisticated numerical technique and carried out their calculations on a supercomputer. Without getting into the messy details, the numerical scheme worked as follows. The nonlinear PDE (11.78) was split into linear (first and second terms) and nonlinear (first and third together) parts and each part solved separately. Although the linear part could be solved by Fourier transform methods, a tremendous saving in time was achieved by replacing the ordinary 3-dimensional Fourier transform with the 3-dimensional

---

[7]The velocity arrows in the left plot of Fig. 11.16 have been drawn so their extensions pass through the centers of the light bullets. The perpendicular distance between the arrows is the impact parameter.

fast Fourier transform (FFT). The nonlinear part was easily integrated numerically. Finally, the two separate results were "spliced" back together to give the numerical solution to the full equation. The method that we have qualitatively outlined is wellknown in the nonlinear optics literature and is referred to as the "split-step Fourier method" or the "beam propagation method". Operator splitting methods similar to this one are discussed in the text *Numerical Recipes* [PFTV89].

Even using a supercomputer, the 3-dimensional computational box could not be too large or the spatial mesh too fine. To handle what happens at the edges of the box, periodic boundary conditions were imposed. As a consequence, for long enough time runs the light bullets were observed to "disappear" through one "wall" of the box and "reappear" through the opposite wall, leading to further collisions.

So what conclusions did Edmundson and Enns reach from their light bullet scattering simulations? The light bullets survived essentially unchanged, indicating that they are indeed solitons, or at least soliton-like. The governing interaction potential turned out to be of the form

$$\frac{V(r)}{V_0} = \frac{e^{-\alpha r}}{r},$$
(11.79)

where r is the separation of the centers of the light bullets and V_0 and α are parameters. The reader might recognize Equation (11.79) as a Yukawa potential. Hideki Yukawa is the 1949 Nobel prize winning physicist who some years earlier formulated a theory of nuclear forces which led to the prediction of the existence of mesons.

It should be noted that depending on the phase relations and other governing parameters, the light bullets can display some very novel behavior. For example, as already pointed out, the model equation allows for bistable behavior, two solitons of dramatically different sizes existing for the same energy. Edmundson exploited this idea by numerically firing a small spherical soliton directly at the center of a much larger spherical soliton. Rather than bounce off the larger light bullet as in the scattering simulations, the small light bullet tunneled right through the larger soliton, emerging unscathed and leaving the big soliton intact as well. Still other interesting phenomena were observed in the numerical investigations which formed the basis of Darran Edmundson's Ph.D. thesis.[8]

---

[8]D. E. Edmundson, *A Dynamical Study of 3-Dimensional Optical Envelope Solitons*, Simon Fraser University, 1996.

Chapter 12

Inverse Scattering Method

The inverse scattering method (ISM) is important because it uses linear techniques to solve the initial value problem for a wide variety of nonlinear wave equations of physical interest and to obtain N-soliton ($N = 1, 2, 3, ...$) solutions. The KdV two-soliton solution was the subject of Mathematica File MF08 where it was animated. The ISM was first discovered and developed by Gardner, Greene, Kruskal and Miura [GGKM67] for the KdV equation. A general formulation of the method by Peter Lax [Lax68] soon followed. This nontrivial formulation is the subject of the next few sections. It is presented to give the reader the flavor of a more advanced topic in nonlinear physics. As you will see, the inverse scattering method derives its name from its close mathematical connection for the KdV case to the quantum mechanical scattering of a particle by a localized potential or tunneling through a barrier.

12.1   Lax's Formulation

Consider a general nonlinear PDE for the function $\phi(x,t)$

$$\phi_t = K(\phi) \tag{12.1}$$

where t is the time variable and K is a nonlinear spatial operator. For the KdV equation with $\alpha = -6$,

$$K = 6\phi\frac{\partial}{\partial x} - \frac{\partial^3}{\partial x^3}. \tag{12.2}$$

Our goal is to solve the nonlinear PDE for $\phi(x,t)$, given some initial profile $\phi(x,0)$. To achieve this goal, we shall avail ourselves initially of a few apparently abstract results. The process begins by stating these results which will at first convey little to the reader. As their meaning and implications are explored and we begin to apply them to the KdV equation, the student shall see the emergence of an intellectually beautiful structure, the inverse scattering method.

Following Lax, suppose that it is possible to find two linear operators $L(\phi)$ and $B(\phi)$ which depend on a solution ϕ of Equation (12.1) and satisfy the operator equation (operating on an arbitrary function $f(x,t)$)

$$iL_t = [B, L] \equiv BL - LB. \tag{12.3}$$

The square bracket, defined by the rhs of this last equation, is called the commutator bracket and occurs as a natural mathematical feature in formal treatments of quantum mechanics. For the KdV equation the L and B operators are

$$L = -\frac{\partial^2}{\partial x^2} + \phi(x,t)$$
$$B = -4i\frac{\partial^3}{\partial x^3} + 3i(\phi\frac{\partial}{\partial x} + \frac{\partial}{\partial x}\phi). \tag{12.4}$$

Noting that L_t is interpreted as being equal to ϕ_t, the following example confirms that L and B satisfy the operator Equation (12.3).

Example 12-1: Lax Formulation of the KdV Equation

Confirm that the operators L and B satisfy Equation (12.3) by showing that this operator equation yields the KdV equation for $\phi(x,t)$.

Solution: Functions are created for the L and B operators.

```
Clear["Global`*"]

L[g_]:=-D[g,x,x]+ϕ[x,t] g

B[h_]:=-4 I D[h,{x,3}]+3 I ϕ[x,t] D[h,x]+D[(ϕ[x,t] h),x])
```

The commutator bracket on the right-hand side of (12.3) is applied to an arbitrary function $f(x,t)$ and the result simplified.

```
rhs = Simplify[(B[L[f[x,t]]] - L[B[f[x,t]]])]
```

$$I f(x,t) \left(6 \phi(x,t) \phi^{(1,0)}(x,t) - \phi^{(3,0)}(x,t) \right)$$

The left-hand side of the operator equation is entered.

```
lhs = I D[φ[x,t],t] f[x,t]
```

$$I f(x,t) \phi^{(0,1)}(x,t)$$

Equating the lhs and the rhs and dividing through by $I f(x,t)$ yields the KdV Equation (12.1) with K given by (12.2).

```
kdVeq = lhs/(I f[x,t]) == rhs/(I f[x,t])
```

$$\phi^{(0,1)}(x,t) == 6 \phi(x,t) \phi^{(1,0)}(x,t) - \phi^{(3,0)}(x,t)$$

Thus, the validity of the operator Equation (12.3), is confirmed. It should be noted that although the proof was trivial to carry out using Mathematica, the manipulations involved in evaluating the commutator bracket are laborious to do by hand.

End Example 12-1

Where did the operators L and B come from? Originally, L, for example, was pulled out of "thin air". Subsequently, more systematic approaches were used to find the operators. The approach of Ablowitz, Kaup, Newell and Segur [AKNS74] will be examined in a later section. Note that the operator L is time dependent through its dependence on $\phi(x,t)$. The notation $L(t)$ will be used when necessary to keep track of the time dependence.

With the time-dependent operator $L(t)$, the eigenvalue problem

$$L\psi = \lambda\psi \tag{12.5}$$

can be formed. If the operator B is self-adjoint,[1] the operator equation (12.3) automatically implies that the eigenvalues λ in the eigenvalue problem (12.5) are independent of time even though L is a time-dependent operator. Further, the eigenfunctions ψ evolve in time according to the equation

$$i\psi_t = B\psi. \tag{12.6}$$

These key results will be proved later. By now you are undoubtedly saying, "So what!" What good are they? Remember that our goal was to find $\phi(x, t > 0)$, given the initial shape $\phi(x,0)$. Well, referring to Figure 12.1, the nonlinear equation (12.1) can be solved by carrying out three linear steps.

---

[1]The adjoint of an operator A, written as $A^\dagger$, is defined here by the integral relation $\int_{-\infty}^{\infty} (AU)^* V dx = \int_{-\infty}^{\infty} U^* A^\dagger V dx$ for arbitrary functions U, V satisfying the boundary conditions that they, and all their derivatives, vanish at $|x| = \infty$. A is said to be self-adjoint if $A^\dagger = A$. Note that the asterisk in the definition of the adjoint refers to the complex conjugate and shouldn't be confused with Mathematica's multiplication symbol.

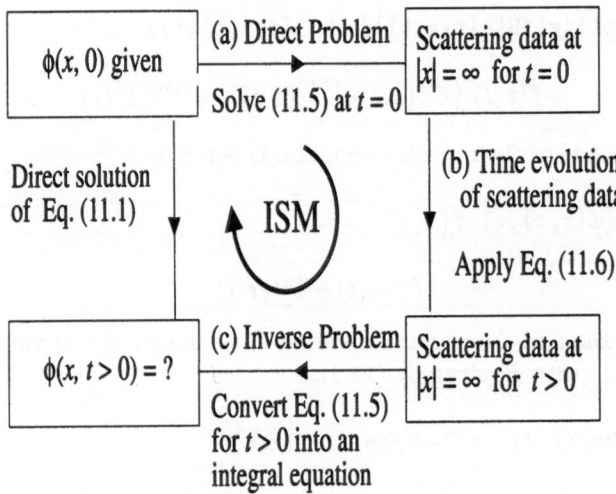

Figure 12.1: Schematic representation of the ISM.

The three steps consist of the so-called "direct problem", the "time evolution of the scattering data", and the "inverse problem".

A) The Direct Problem

Given $\phi(x, 0)$, solve the eigenvalue equation (12.5) at time $t = 0$ for the "scattering data" (e.g., reflection and transmission coefficients) at $|x| = \infty$. For example, for the KdV equation, the eigenvalue equation at $t = 0$ is

$$L(0)\psi = -\frac{\partial^2 \psi}{\partial x^2} + \phi(x, 0)\psi = \lambda\psi. \qquad (12.7)$$

This is just the linear Schrödinger equation of quantum mechanics with the input shape playing the role of the potential function and λ the energy. If the input shape is rectangular, then one has the "square well" (for $\phi(x, 0) < 0$) or the "square barrier" (for $\phi(x, 0) > 0$) problem which the physics student solves at some time in his or her undergraduate career for the "bound state" eigenvalues (energy levels), transmission coefficient, etc.

B) Time Evolution of the Scattering Data

Using Equation (12.6) together with the asymptotic form of the operator B at $|x| = \infty$, the time evolution of the scattering data is calculated next. That is to say, the reflection coefficient, etc., are determined for $t > 0$.

C) The Inverse Problem

From a knowledge of the scattering data of L at time $t > 0$, the solution $\phi(x, t > 0)$ is constructed. This is accomplished by solving an integral equation, derived from (12.5), which depends on the time-evolved data. The name inverse scattering method arises from this last step. Figure 12.1 summarizes the inverse scattering method. The idea is to avoid solving the nonlinear equation (12.1) directly but instead carry out the linear steps (a), (b), and (c).

PROBLEMS
Problem 12-1: Adjoint relations
Prove the following adjoint relations:
(a) $(i\frac{\partial}{\partial x})^\dagger = i\frac{\partial}{\partial x}$,
(b) $(\alpha A)^\dagger = \alpha^* A^\dagger$, α a scalar,
(c) $(AB)^\dagger = B^\dagger A^\dagger$, with A, B operators.

Problem 12-2: Self-adjointness of B
Prove that the operator B for the KdV case is self-adjoint. You may assume that all functions vanish at $|x| = \infty$.

12.2 Application to KdV Equation

To flesh out the ideas that have been introduced, the ISM will be applied to a concrete example, viz., the KdV equation. The direct problem is discussed first.

12.2.1 Direct Problem

Setting the initial shape $\phi(x, 0) \equiv \phi_0(x)$, the eigenvalue problem at $t = 0$ is

$$L(0)\psi(x, 0) = \lambda\psi(x, 0) \tag{12.8}$$

or

$$\psi_{xx}(x, 0) + (\lambda - \phi_0(x))\psi(x, 0) = 0. \tag{12.9}$$

As mentioned earlier, this is the Schrödinger equation with the input shape as the potential and λ the energy. Suppose that $\phi_0(x) = -U_0$ for $|x| < a$ and zero outside this range. Figure 12.2 shows the potential which is recognizable as the square well problem of quantum mechanics. The problem is divided into three regions as shown and the energy λ into two cases, $\lambda > 0$ and $\lambda < 0$. Let's look at $\lambda > 0$ first, setting $\lambda = k^2$ with k real. In region 1, $\phi_0 = 0$, so that the

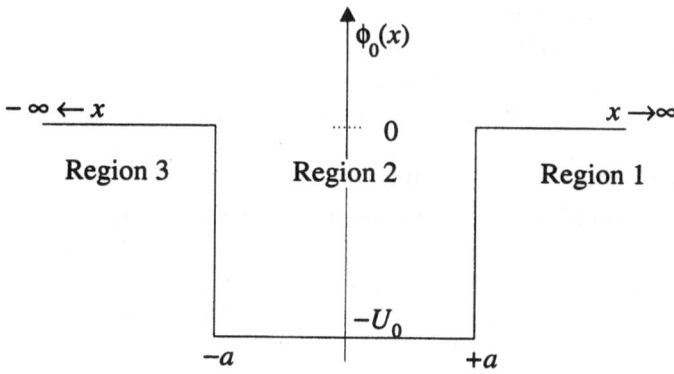

Figure 12.2: Rectangular potential well input for direct problem.

eigenvalue problem is

$$\psi_{xx} + k^2\psi = 0 \tag{12.10}$$

which has the general solution

$$\psi_1 = Ae^{-ikx} + Be^{ikx}. \tag{12.11}$$

To create a scattering problem, consider a "particle" coming from $+\infty$ as shown[2] in Figure 12.3. Without loss of generality, set $A = 1$ and let $B \equiv R$, where R is

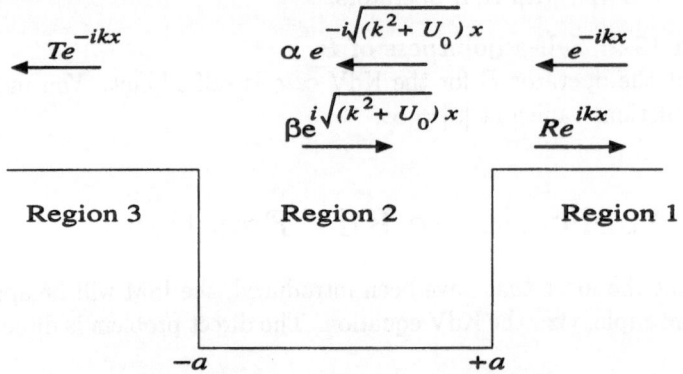

Figure 12.3: The scattering problem for $\lambda > 0$.

the reflection coefficient. Then

$$\psi_1 = e^{-ikx} + Re^{ikx}. \tag{12.12}$$

In region 3, $\phi_0 = 0$, and there is only a transmitted wave traveling to the left described by

$$\psi_3 = Te^{-ikx} \tag{12.13}$$

with T the transmission coefficient.

In region 2, $\phi_0 = -U_0$, and the solution is

$$\psi_2 = \alpha e^{-i\sqrt{(k^2+U_0)}x} + \beta e^{i\sqrt{(k^2+U_0)}x} \tag{12.14}$$

where α and β are constants.

The boundary conditions are $\psi_1(a) = \psi_2(a)$, $(\psi_1(a))_x = (\psi_2(a))_x$, $\psi_3(-a) = \psi_2(-a)$, and $(\psi_3(-a))_x = (\psi_2(-a))_x$. We will not grind through the results here. Mathematica can be very helpful as demonstrated in the file MF51.

Quantum Mechanical Tunneling

In this file, the problem of quantum mechanical tunneling through a rectangular barrier ($\phi_0 = +U_0$) is solved. The physics convention is followed with the incident particle coming from $x = -\infty$ and physical units are used. Mathematica commands in file: DSolve, Solve, Simplify, Abs, Sinh, Sqrt, Pi

---

[2]This convention used in the relevant mathematics literature is opposite to the usual physics convention where the particle approaches the well or barrier from the left.

The important point is that there are four boundary conditions to determine the four unknown coefficients R, T, α, and β. For the ISM, the coefficients R and T, which are dependent on k, are of particular importance. There are no restrictions imposed on k or therefore λ, i.e., they are continuous. So for $\lambda > 0$, one has the "continuous spectrum". Let's now look at $\lambda < 0$, setting $\lambda \equiv -\kappa^2$ with κ real. In region 1,

$$\psi_{xx} - \kappa^2 \psi = 0 \tag{12.15}$$

which has the solutions $\psi \sim e^{\pm \kappa x}$. For ψ to remain finite as $x \to \infty$, the solution $e^{\kappa x}$ must be rejected. Without loss of generality, take $\psi_1 = e^{-\kappa x}$. In region 3 the solutions are the same as in region 1, but now the solution $e^{-\kappa x}$ must be rejected so that $\psi_3 = Ce^{\kappa x}$.

In region 2,

$$\psi_{xx} + (U_0 - \kappa^2)\psi = 0 \tag{12.16}$$

which has the solution $\psi_2 = \alpha e^{-i\sqrt{(U_0 - \kappa^2)}x} + \beta e^{i\sqrt{(U_0 - \kappa^2)}x}$.

Now there are only three coefficients C, α, β, but still four boundary conditions. In this case, a transcendental equation results for κ, the solutions of which lead to discrete values κ_n. The spectrum is now discrete. In quantum mechanics, the discrete κ values are the energy levels in the potential well. The solutions are referred to as the "bound state solutions", while those for $\lambda > 0$ are called the "scattering solutions".

More generally, the input shape would not correspond to a rectangular well, but given a localized ϕ_0, one would still have to find the "scattering parameters" $R(k, t = 0)$, $T(k, 0)$, C_n, κ_n and the number N of bound states to proceed in the ISM. The direct problem is only completed when all of these quantities are known. If the potential only vanishes as $|x| \to \infty$, the scattering data must be evaluated at these limits.

PROBLEMS
Problem 12-3: Bound-state eigenvalues
Using Mathematica, determine the transcendental equation for the bound-state eigenvalues κ_n. Taking $a = 1$, determine the minimum depth of the potential well for one bound state to exist.

12.2.2 Time Evolution of the Scattering Data

The next step is to determine how the scattering data evolves with time. A word of warning before proceeding. This section is easy if you did the problems at the end of the previous section involving the properties of adjoint operators. If you didn't, you would be advised to do so before tackling this subsection.

The second operator B plays the key role here. It is used to define a "time evolution operator"[3] $U(t)$ through the equation

$$iU_t = BU \tag{12.17}$$

with the initial condition $U(0) = I$, i.e., equal to the identity operator (not the Mathematica command I for the square root of minus one).

---

[3]The origin of the name will become quite apparent shortly.

Since B is self-adjoint, i.e., $B^\dagger = B$, $U(t)$ is a unitary operator. That is to say, it satisfies the relation $U^\dagger U = UU^\dagger = I$. The proof that $U(t)$ is a unitary operator goes as follows. Multiply (12.17) from the left by $U^\dagger$ so that

$$iU^\dagger U_t = U^\dagger BU. \tag{12.18}$$

Next take the adjoint of (12.17) yielding

$$(iU_t)^\dagger = -iU_t^\dagger = (BU)^\dagger = U^\dagger B^\dagger = U^\dagger B. \tag{12.19}$$

Multiplying this last equation from the right by U and then subtracting the result from the previous equation yields

$$i(U^\dagger U)_t = 0, \tag{12.20}$$

so that $U^\dagger(t)U(t) = \text{constant} = U^\dagger(0)U(0) = I$.

With this unitary condition proved, it can be shown that the eigenvalues, including the bound state energies ($\lambda_n = -\kappa_n^2$), are independent of time. First it is necessary to demonstrate that the L operator at time $t > 0$, $L(t)$, may be written in terms of $L(0)$, the L operator at $t = 0$, as follows:

$$L(t) = U(t)L(0)U^\dagger(t). \tag{12.21}$$

This result is proved by showing that $L(t)$ given by (12.21) satisfies the operator equation

$$iL_t = [B, L]. \tag{12.22}$$

Taking the time derivative of (12.21),

$$iL_t = iU_t(t)L(0)U^\dagger(t) + iU(t)L(0)U_t^\dagger(t). \tag{12.23}$$

Then using (12.17) in the first term on the rhs and its adjoint equation for the second term, we obtain

$$iL_t = BU(t)L(0)U^\dagger(t) + U(t)L(0)(-U^\dagger(t)B) \tag{12.24}$$

or

$$iL_t = [B, U(t)L(0)U^\dagger(t)] = [B, L]. \tag{12.25}$$

Next consider the eigenvalue problems for $t = 0$ (the direct problem) and for $t > 0$:

$$L(0)\psi(0) = \lambda(0)\psi(0)$$

$$L(t)\psi(t) = \lambda(t)\psi(t). \tag{12.26}$$

We want to show that $\lambda(t) = \lambda(0)$, i.e., that the eigenvalues are independent of time. To do this let's operate on the $t = 0$ equation from the left with $U(t)$ so that

$$U(t)L(0)\psi(0) = \lambda(0)U(t)\psi(0). \tag{12.27}$$

But on the lhs, the term $U^\dagger U = I$ can be inserted, so that with appropriate grouping

$$lhs = U(t)L(0)(U^\dagger(t)U(t))\psi(0) = (U(t)L(0)U^\dagger(t))U(t)\psi(0) = L(t)U(t)\psi(0)$$

(12.28)

where in the last step Equation (12.21) has been used. Then, employing this result and grouping terms on both side of the equation,

$$L(t)(U(t)\psi(0)) = \lambda(0)(U(t)\psi(0)).$$

(12.29)

Comparing this expression with the earlier eigenvalue expression for $t > 0$, one can identify $\psi(t) = U(t)\psi(0)$ and $\lambda(t) = \lambda(0)$, thus proving that the eigenvalues are independent of time. It is also now apparent why $U(t)$ is called the time evolution operator as its action on $\psi(0)$ produces $\psi(t)$.

It also follows immediately that

$$i\psi_t = B\psi,$$

(12.30)

which is proved by differentiating $\psi(t) = U(t)\psi(0)$ with respect to time. Then, on using (12.17),

$$i\psi_t(t) = iU_t(t)\psi(0) = (BU(t))\psi(0) = B(U(t)\psi(0)) = B\psi(t).$$

(12.31)

Since the bound state eigenvalues κ_n are independent of time, it follows that the total number N of bound states does not change with t, remaining at the same value as determined in the direct problem at $t = 0$. The number of bound states depends on the input shape. What about the time evolution of the remaining relevant $t = 0$ scattering data, $R(k, 0)$, $T(k, 0)$, and $C_n(0)$? To find $R(k, t)$, $T(k, t)$, and $C_n(t)$, the detailed form of B must be used. Recall that for the KdV problem

$$B = -4i\frac{\partial^3}{\partial x^3} + 3i(\phi\frac{\partial}{\partial x} + \frac{\partial}{\partial x}\phi).$$

(12.32)

Since the remaining scattering data are evaluated in general at $|x| = \infty$ where it is assumed that $\phi, \phi_x \to 0$, only the asymptotic form of B, $B_\infty = -4i\frac{\partial^3}{\partial x^3}$, need be considered.

First, let's find $C_n(t)$. For the bound states in the limit $x \to -\infty$, $\psi_n(x, t) = C_n(t)e^{\kappa_n x}$. This follows from the fact that the structure of the eigenvalue problem is the same for $t > 0$ as for $t = 0$, but the eigenvalues remain unchanged. Using (12.6) gives

$$i\frac{\partial\psi_n(x, t)}{\partial t} = B_\infty\psi_n(x, t) = -4i\frac{\partial^3}{\partial x^3}\psi_n(x, t)$$

(12.33)

or

$$\frac{\partial C_n(t)}{\partial t} = -4\kappa_n^3 C_n(t).$$

(12.34)

This is easily integrated to yield

$$C_n(t) = C_n(0)e^{-4\kappa_n^3 t}.$$

(12.35)

Next $R(k,t)$ is calculated. For $x \to \infty$,

$$\psi(x,t) = A(t)e^{-ikx} + B(t)e^{ikx}. \tag{12.36}$$

Again, using (12.6) with B_∞ yields

$$\frac{\partial}{\partial t}A(t)e^{-ikx} + \frac{\partial}{\partial t}B(t)e^{ikx} = -4(-ik)^3 Ae^{-ikx} - 4(ik)^3 Be^{ikx}. \tag{12.37}$$

Since the exponentials are linearly independent, their coefficients may be equated and on integrating yield

$$A(t) = A(0)e^{-4ik^3 t}$$

$$B(t) = B(0)e^{4ik^3 t}. \tag{12.38}$$

Comparing with the direct problem,

$$R(k,t) \equiv \frac{B(t)}{A(t)} = \frac{B(0)}{A(0)}e^{8ik^3 t} \equiv R(k,0)e^{8ik^3 t}. \tag{12.39}$$

The time-dependent transmission coefficient can be similarly calculated. From energy conservation, it must satisfy the relation $|R(k,t)|^2 + |T(k,t)|^2 = 1$.

To summarize, the time evolution of the scattering data for the KdV problem has been determined.

12.2.3   The Inverse Problem

Having found the scattering data at t, the data will be used to find the time evolved "potential" $\phi(x,t)$. This inverse problem can be solved by making use of the Gel'fand–Levitan (G-L) (also known as the Marchenko) linear integral equation which is derived from the time-dependent eigenvalue equation. The derivation of the G–L equation is beyond the level of this text. It may be found, for example, in the review paper by Scott, Chu and McLaughlin [SCM73].

The G–L integral equation for the unknown function $\hat{g}_1(x,y;t)$ is

$$\hat{g}_1(x,y;t) + K(x+y;t) + \int_x^\infty K(y+y';t)\hat{g}_1(x,y';t)dy' = 0 \tag{12.40}$$

derived for $y \geq x$. The function K is called the "kernel" of the integral equation and is given here by

$$K(x+y;t) \equiv \hat{R}(x+y;t) + \sum_{n=1}^N m_n(t)e^{-\kappa_n(x+y)} \tag{12.41}$$

where $\hat{R}$ is the Fourier transform of R, i.e.,

$$\hat{R}(x+y;t) \equiv \frac{1}{2\pi}\int_{-\infty}^\infty R(k,t)e^{ik(x+y)}dk \tag{12.42}$$

and

$$m_n(t) \equiv m_n(0)e^{8\kappa_n^3 t}, \tag{12.43}$$

where $m_n(0)$ is a normalization constant determined by the initial shape ϕ_0. It may be evaluated [SCM73] from the expression

$$m_n(0) = \frac{1}{\int_{-\infty}^{\infty} f_1^2(x', i\kappa_n, 0)dx'} \tag{12.44}$$

with $f_1(x, k, 0)$ the solution of the integral equation

$$f_1(x, k, 0) = e^{ikx} - \int_x^{\infty} \frac{\sin(k(x - x'))}{k} \phi_0(x') f_1(x', k, 0)dx'. \tag{12.45}$$

The first term in K is the continuous spectrum contribution while the second term is associated with the discrete spectrum.

On solving the G–L integral equation, $\phi(x, t)$ may be determined by calculating

$$\phi(x, t) = -2\frac{d}{dx}\hat{g}_1(x, x; t). \tag{12.46}$$

It should be noted that the transmission coefficient $T(k)$ doesn't appear in the G–L equation, having been eliminated in its derivation. Knowledge of the structure of $T(k)$ can still be useful, as it can be shown that the bound state eigenvalues κ_n show up as "simple poles" of $T(k)$ on the positive imaginary k axis, i.e., $T(k) \sim 1/(k - i\kappa_n)$. Thus the bound state eigenvalues can be read off from the denominator of the transmission coefficient.

12.3 Multi-Soliton Solutions

To find "pure" soliton solutions, we look for solutions which have no continuous spectrum contribution. To achieve this, set the reflection coefficient $R(k, 0) = 0$, i.e., seek a "reflectionless" potential. Let's check that the ISM works by regaining the one-soliton formula which was obtained by elementary means earlier.

Setting $N = 1$, $R(k, 0) = 0$, and $m_1(0) \equiv m_1$, the G–L integral equation becomes

$$\hat{g}_1(x, y; t) + m_1 e^{8\kappa_1^3 t} e^{-\kappa_1(x+y)} + m_1 e^{8\kappa_1^3 t} \int_x^{\infty} e^{-\kappa_1(y+y')}\hat{g}_1(x, y'; t)dy' = 0. \tag{12.47}$$

The kernel $e^{-\kappa_1(y+y')}$ of the integral equation is an example of a so-called degenerate kernel[4], i.e., it can be factored as a product $e^{-\kappa_1 y}e^{-\kappa_1 y'}$. Integral equations with degenerate kernels can be solved exactly analytically. To accomplish this, look for a solution of the form

$$\hat{g}_1(x, y; t) = e^{-\kappa_1 y}h(x, t). \tag{12.48}$$

Then (12.47) reduces to

$$h(x, t) + m_1 e^{8\kappa_1^3 t} e^{-\kappa_1 x} + m_1 e^{8\kappa_1^3 t} h(x, t) \int_x^{\infty} e^{-2\kappa_1 y'} dy' = 0, \tag{12.49}$$

---

[4]This does not refer to a debauched military individual!

which is easily solved for $h(x, t)$ and thus $\hat{g}_1(x, y; t)$,

$$\hat{g}_1(x, y; t) = \frac{-m_1 e^{8\kappa_1^3 t} e^{-\kappa_1(x+y)}}{1 + \frac{m_1}{2\kappa_1} e^{8\kappa_1^3 t} e^{-2\kappa_1 x}} \tag{12.50}$$

with $y \geq x$. Finally, using (12.46) and simplifying, the one-soliton solution

$$\phi(x, t) = -2\kappa_1^2 \operatorname{sech}^2(\kappa_1(x - 4\kappa_1^2 t) - \delta_1) \tag{12.51}$$

with $\delta_1 = (1/2) \ln(m_1/(2\kappa_1))$ results. Setting the arbitrary phase factor δ_1 equal to zero and recalling that $\alpha = -6$ in our present treatment of the KdV equation, Equation (12.51) is of exactly the same form as the solitary wave solution (10.57) with the identification $\kappa_1 = \sqrt{c}/2$. So the bound state eigenvalue determines the velocity of the soliton. The phase factor δ_1 is just the integration constant A_3 that was set equal to zero in the elementary derivation.

The reader might, quite rightly, argue that this has been a lot of work and a very roundabout way of obtaining the one-soliton solution which was obtained so easily previously. If we were only interested in $N = 1$, this would certainly be a valid criticism. The elementary approach doesn't allow us, for example, to find the two-soliton solution ($N = 2$) given by Equation (3.13) whose animated behavior was seen in Mathematica File 08. The two-soliton solution can be obtained from the G–L integral equation by taking $N = 2$, $R(k, 0) = 0$, assuming a solution of the form

$$\hat{g}_1(x, y; t) = e^{-\kappa_1 y} h_1(x, t) + e^{-\kappa_2 y} h_2(x, t) \tag{12.52}$$

and noting that for $\kappa_1 \neq \kappa_2$, the two exponentials are independent. Perversely, the detailed derivation of the two-soliton solution is left as a problem for the reader. To alleviate the lengthy algebra involved, the reader should make use of Mathematica where possible.

PROBLEMS
Problem 12-4: Two-soliton solution
Using the ISM, derive the two-soliton solution for the KdV equation for $\kappa_1 = \sqrt{2}$, $\kappa_2 = 1$, and $m_1 = m_2 = 1$. Simplify your solution as much as possible and animate it.

Problem 12-5: Fredholm integral equations
Analytically solve the following Fredholm (constant limits present) integral equations for $f(x)$ by exploiting the degenerate nature of the kernel:

  a. $f(x) = x^2 + \int_0^1 xy\, f(y)\, dy$

  b. $f(x) = x + \int_0^1 (xy^2 + x^2 y)\, f(y)\, dy$

Problem 12-6: A Volterra integral equation
Analytically solve the Volterra (variable limit) integral equation

$$f(x) = x + \int_0^x xy\, f(y)\, dy.$$

Problem 12-7: A nonlinear integral equation
Solve the nonlinear integral equation

$$f(x) = \int_0^1 (x+y)f^2(y)dy$$

for $f(x)$ by using the fact that the kernel is degenerate.

Problem 12-8: Conservation laws
Conservation laws are well known in the physical sciences and may be expressed in one dimension in the differential form

$$\frac{\partial N}{\partial t} + \frac{\partial J}{\partial x} = 0$$

where N is the density, J the flux, t the time, and x the spatial coordinate. For example, for 1-dimensional current flow, $N \equiv \rho$ where ρ is the electric charge density and J is the electric current density (charge per second per unit area normal to x). The differential form in this case expresses the conservation of electric charge.

The KdV equation has an infinite number of conservation laws (which do not have such simple interpretations as in the above example) connected to the fact that it possesses multi-soliton solutions. The lowest two conservation laws are easily obtained. Multiply the KdV equation

$$U_t - 6UU_x + U_{xxx} = 0$$

by (a) 1, (b) U, and take, respectively, $N \equiv U$ and $N \equiv U^2$. Determine the structure of J in each case which gives the differential form of the conservation law. Higher order conservation laws are not so easily obtained.

12.4   General Input Shapes

For non-pure-soliton input shapes $\phi_0(x)$, one can in principle use the ISM to find $\phi(x, t > 0)$. The time-evolved shape will be built up out of both discrete spectrum (soliton) and continuous spectrum ($R(k, 0) \neq 0$) contributions. As an example, suppose that

$$\phi_0(x) = -2a\delta(x - x_0) \tag{12.53}$$

with the real parameter $a > 0$. The direct problem then is

$$\psi_{xx} + k^2\psi = -2a\delta(x - x_0)\psi \tag{12.54}$$

to be solved subject to the boundary conditions in Figure 12.4. Continuity of the solution at $x = x_0$ gives us

$$Te^{-ikx_0} = e^{-ikx_0} + Re^{ikx_0}. \tag{12.55}$$

Integrating (12.54) from $x_0 - \epsilon$ to $x_0 + \epsilon$, with $\epsilon \to 0$, we obtain

$$\lim_{\epsilon \to 0}(\psi_x|_{x_0+\epsilon} - \psi_x|_{x_0-\epsilon}) = -2a\psi(x_0) \tag{12.56}$$

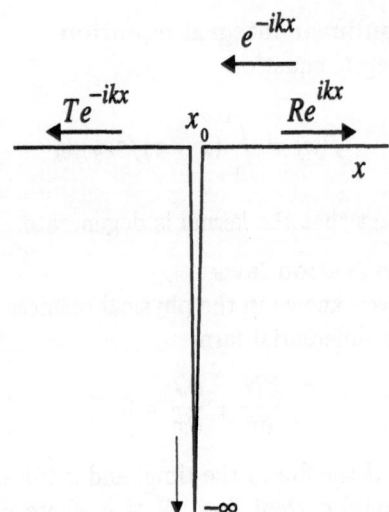

Figure 12.4: Direct problem for a Dirac delta function input shape.

or

$$(-ike^{-ikx_0} + Rike^{ikx_0}) - (-ikTe^{-ikx_0}) = -2aTe^{-ikx_0}. \qquad (12.57)$$

Solving for the transmission and reflection coefficients yields

$$T(k,0) = \frac{1}{(1 - ia/k)} \qquad (12.58)$$

$$R(k,0) = \frac{ia}{k - ia}e^{-2ikx_0}.$$

As a partial check, note that $|T(k,0)|^2 + |R(k,0)|^2 = 1$.

From the denominator of the transmission coefficient, there is one bound state with eigenvalue $\kappa_1 = a$. For the bound state solution, continuity at $x = x_0$ gives us $e^{-ax_0} = C_1(0)e^{ax_0}$ or $C_1(0) = e^{-2ax_0}$.

Finally, $m_1(0)$ must be evaluated. Substituting $\phi_0(x') = -2a\delta(x' - x_0)$ into Equation (12.45) yields

$$f_1(x,k,0) = e^{ikx}, x > x_0 \qquad (12.59)$$

$$f_1(x,k,0) = (1 - ia/k)e^{ikx} + (ia/k)e^{2ikx_0}e^{-ikx}, x < x_0.$$

Then, using this result, Equation (12.44) yields $m_1(0) = ae^{2ax_0}$. Thus, for the direct problem the relevant scattering data at $t = 0$ are

$$N = 1, \quad \kappa_1 = a, \quad m_1(0) = ae^{2ax_0}, \quad \text{and} \quad R(k,0) = (ia/(k - ia))e^{-2ikx_0}.$$

The time-evolved scattering data then are

$$N = 1, \quad \kappa_1 = a, \quad m_1(t) = e^{8a^3t}m_1(0), \quad \text{and} \quad R(k,t) = e^{8ik^3t}R(k,0).$$

Thus, for the inverse problem the kernel of the G–L integral equation is

$$K(x+y;t) = ae^{8a^3 t} e^{2ax_0} e^{-a(x+y)} + \frac{ia}{2\pi} \int_{-\infty}^{\infty} \frac{dk}{(k-ia)} (e^{ik(x+y-2x_0)} e^{8ik^3 t}).$$

(12.60)

Although the integral in the kernel cannot be carried out analytically, the behavior as $t \to \infty$ is easily found. In this limit, the first term completely dominates so that

$$K(x+y;t) \to ae^{8a^3 t} e^{2ax_0} e^{-a(x+y)}.$$

(12.61)

A one-soliton solution emerges as $t \to \infty$, the form being as in (12.51) with $\kappa_1 = a$ and $\delta_1 = ax_0 - (1/2)\ln 2$. The speed and amplitude of the soliton depend on the "strength" a of the delta function.

PROBLEMS

Problem 12-9: Two delta functions input

Suppose that, for the KdV equation, the input shape is the sum of two delta functions of equal strength at $x = x_0$ and $-x_0$, viz.,

$$\phi_0(x) = -2a\delta(x-x_0) - 2a\delta(x+x_0)$$

with $a, x_0 > 0$. Obtain $T(k,0)$ and $R(k,0)$. Using this scattering information, determine how many solitons will emerge from the initial value problem as $t \to \infty$. Be sure to examine different possible values of a and x_0.

Problem 12-10: Amplified one soliton input

For $\alpha = -6$ in the KdV equation, consider the input corresponding to multiplying the one-soliton input by the positive number A. It can be shown that the number N of solitons which emerge is the integer part of

$$N = \frac{1}{2}(-1 + \sqrt{1+8A}).$$

For example, for $A = 3$, two solitons should emerge while for $A = 0.25$ no solitons should appear, the input solution slowly radiating away. Confirm that this is the case by running file MF47 with the following modifications: Take $\alpha = -6$, $x_1 = 20$, $c_2 = 0$ (to remove the second soliton), and insert (a) $A = 3$ with $c_1 = 0.2$, (b) $A = 0.25$ with $c_1 = 0.7$.

12.5 The Zakharov–Shabat/AKNS Approach

A different approach to that of the Lax formulation has been put forth by Zakharov and Shabat [ZS72] and the "AKNS" group of Ablowitz, Kaup, Newell and Segur [AKNS74]. In Lax's treatment, recall that self-adjointness of the operator B led to the result that the eigenvalues λ of the operator $L(t)$ are independent of time. This was a key step in the success of the inverse scattering method.

In the Z–S/AKNS treatment, this latter important result is built into the formulation in a somewhat different way. Consider the linear eigenvalue problem

$$Lv = \lambda v$$

(12.62)

where

$$L = \begin{pmatrix} i\frac{\partial}{\partial x} & -iq(x,t) \\ ir(x,t) & -i\frac{\partial}{\partial x} \end{pmatrix} \tag{12.63}$$

$$v = \begin{pmatrix} v_1 \\ v_2 \end{pmatrix} \tag{12.64}$$

The quantities $q(x,t)$ and $r(x,t)$ are arbitrary functions for the moment. Writing (12.62) out in detail, we have

$$(v_1)_x + i\lambda v_1 = q(x,t)v_2$$

$$(v_2)_x - i\lambda v_2 = r(x,t)v_1. \tag{12.65}$$

Next, assume that the time dependence of v_1 and v_2 is given by

$$(v_1)_t = A(x,t;\lambda)v_1 + B(x,t;\lambda)v_2$$

$$(v_2)_t = C(x,t;\lambda)v_1 - A(x,t;\lambda)v_2 \tag{12.66}$$

where A, B, C are general functions[5] for the moment.

Now comes the key step. Take the t-derivative of (12.65) and the x-derivative of (12.66) while demanding that the eigenvalues λ be constant. That is to say, λ is independent of t and x. This gives us the two sets of equations

$$(v_1)_{tx} + i\lambda(v_1)_t = q_t v_2 + q(v_2)_t$$

$$(v_2)_{tx} - i\lambda(v_2)_t = r_t v_1 + r(v_1)_t \tag{12.67}$$

and, with the time–space derivatives in the opposite order,

$$(v_1)_{xt} = A_x v_1 + A(v_1)_x + B_x v_2 + B(v_2)_x$$

$$(v_2)_{xt} = C_x v_1 + C(v_1)_x - A_x v_2 - A(v_2)_x. \tag{12.68}$$

Using (12.65) and (12.66) to eliminate $(v_1)_t, (v_1)_x, (v_2)_t, (v_2)_x$, noting that

$$\frac{\partial^2}{\partial t \partial x} = \frac{\partial^2}{\partial x \partial t},$$

and then separately equating the coefficients of v_1 and of v_2, the following three

---

[5]More generally, we could have taken the rhs of the second equation to be $C(x,t;\lambda)v_1 + D(x,t;\lambda)v_2$ but it turns out that $D = -A$.

equations for A, B, and C follow:

$$A_x = qC - rB$$

$$B_x + 2i\lambda B = q_t - 2Aq \qquad (12.69)$$

$$C_x - 2i\lambda C = r_t + 2Ar.$$

Although one could be more general, let's pick a specific form for A, B, and C. For example, choose

$$A = -2i\lambda^2 - irq$$

$$B = b_0 + b_1\lambda + b_2\lambda^2 \qquad (12.70)$$

$$C = c_0 + c_1\lambda + c_2\lambda^2$$

where $b_0 = b_0(x,t)$, $b_1 = b_1(x,t)$, etc. Substitute these forms into Equations (12.69) and equate equal powers of λ, since the results must be true for general λ. To carry this out, you can proceed as in the following example.

Example 12-2: Zakharov–Shabat/AKNS Equations

Derive the Z–S/AKNS equations by substituting the forms (12.70) into (12.69) and equating equal powers of λ.

Solution: The expressions for A, B, and C are entered and labeled **aa**, **bb**, and **cc**, respectively.

```
Clear["Global`*"]

aa = -2 I λ² - I r[x,t] q[x,t]
```

$$-2I\lambda^2 - I q(x,t)\, r(x,t)$$

```
bb = b₀ [x,t] + b₁ [x,t] λ + b₂ [x,t] λ²
```

$$b_0(x,t) + b_1(x,t)\,\lambda + b_2(x,t)\,\lambda^2$$

```
cc = c₀ [x,t] + c₁ [x,t] λ + c₂ [x,t] λ²
```

$$c_0(x,t) + c_1(x,t)\,\lambda + c_2(x,t)\,\lambda^2$$

Equations (12.69) are written in the form $A_x - qC + rB = 0$, etc., and the left-hand sides entered. The lengthy output is suppressed here in the text.

```
eq[1] = D[aa,x] - q[x,t] cc + r[x,t] bb
eq[2] = D[bb,x] + 2 I λ bb - D[q[x,t],t] + 2 aa q[x,t]
eq[3] = D[cc,x] - 2 I λ cc - D[r[x,t],t] - 2 aa r[x,t]
```

In the following Do loop, `CoefficientList` is used to collect powers of λ in each of `eq[i]` and put the corresponding coefficients in a list format. The number of entries `n[i]` in each of the three lists is also recorded.

```
Do[rel[i] = CoefficientList[eq[i],λ];n[i]=Length[rel[i]],{i,1,3}];
```

Since the results must be true for arbitrary λ, each coefficient combination must be set equal to zero. This is accomplished in the following three `Table` command lines. `TableForm` is used to achieve a nice ordering of the output.

```
Table[rel[1][[j]] == 0,{j,1,n[1]}] // TableForm
```

$$r(x,t)\,b_0(x,t) - q(x,t)\,c_0(x,t) - I\,r(x,t)\,q^{(1,0)}(x,t)$$

$$-I\,q(x,t)\,r^{(1,0)}(x,t) == 0$$

$$r(x,t)\,b_1(x,t) - q(x,t)\,c_1(x,t) == 0$$

$$r(x,t)\,b_2(x,t) - q(x,t)\,c_2(x,t) == 0$$

```
Table[rel[2][[j]] == 0,{j,1,n[2]}] // TableForm
```

$$-2\,I\,r(x,t)\,q(x,t)^2 - q^{(0,1)}(x,t) + b_0{}^{(1,0)}(x,t) == 0$$

$$2\,I\,b_0(x,t) + b_1{}^{(1,0)}(x,t) == 0$$

$$-4\,I\,q(x,t) + 2\,I\,b_1(x,t) + b_2{}^{(1,0)}(x,t) == 0$$

$$2\,I\,b_2(x,t) == 0$$

```
Table[rel[3][[j]] == 0,{j,1,n[3]}] // TableForm
```

$$2\,I\,q(x,t)\,r(x,t)^2 - r^{(0,1)}(x,t) + c_0{}^{(1,0)}(x,t) == 0$$

$$-2\,I\,c_0(x,t) + c_1{}^{(1,0)}(x,t) == 0$$

$$4\,I\,r(x,t) - 2\,I\,c_1(x,t) + c_2{}^{(1,0)}(x,t) == 0$$

$$-2\,I\,c_2(x,t) == 0$$

From the three sets of output, we can see that there are a total of 11 coefficient equations to be satisfied. Let us now examine the last two sets. From the last line in each set, we see that $b_2(x,t) = 0$ and $c_2(x,t) = 0$. It then follows from the third lines that $b_1(x,t) = 2q(x,t)$ and $c_1(x,t) = 2r(x,t)$. The second lines then yield $b_0(x,t) = \frac{1}{2}i(b_1)_x = iq_x$ and $c_0(x,t) = -\frac{1}{2}i(c_1)_x = -ir_x$. Finally, the first lines produce

$$iq_{xx} = q_t + 2irq^2, \quad -ir_{xx} = r_t - 2ir^2q. \tag{12.71}$$

What about the output equations from the first **Table** command? All three equations yield the tautology $0 = 0$, and thus produce no additional information. Therefore, in this case,

$$A = -2i\lambda^2 - irq$$

$$B = iq_x + 2q\lambda \tag{12.72}$$

$$C = -ir_x + 2r\lambda.$$

End Example 12-2

To this point, the functions $r(x,t)$ and $q(x,t)$ have been arbitrary. If we choose $r = -q^*$, both equations in (12.71) reduce to the nonlinear Schrödinger equation[6]

$$iq_t + q_{xx} + 2|q|^2q = 0. \tag{12.73}$$

So the ISM can be applied to the NLSE. The direct problem is given by solving (12.65) with $r(x, t = 0) = -q^*(x, 0)$. The scattering problem can be set up with a specified input shape $q(x,0)$. The time evolution of the scattering data can be obtained from (12.66) with A, B, C given by (12.72) and $r = -q^*$. Assuming that $q, q_x \to 0$ as $|x| \to \infty$, one only needs to use $A(|x| \to \infty) \equiv A_\infty = -2i\lambda^2$, $B_\infty = 0$, and $C_\infty = 0$. Finally, (12.65) for $t > 0$ may be converted into a pair of coupled integral equations which can, once the time-evolved scattering data has been determined, be solved for $q(x,t)$. The messy details, which are beyond the level of this text, may be found in the AKNS paper [AKNS74].

You might ask how one was so clever as to come up with the original forms for A, B, C and the relation of $r(x,t)$ to $q(x,t)$. Actually, a more general approach is to write down finite polynomial expansions in either λ or $1/\lambda$ for A, B, C and equate equal powers of λ or $1/\lambda$ in (12.69). As we have seen, with Mathematica this is easy to do. All sorts of nonlinear equations, most not yet of any interest to physicists, follow on choosing r and q.

PROBLEMS
Problem 12-11: KdV and modified KdV equations

a. Show that if

$$A = -4i\lambda^3 - 2iqr\lambda + rq_x - qr_x$$

$$B = b_0 + b_1\lambda + b_2\lambda^2 + b_3\lambda^3$$

$$C = c_0 + c_1\lambda + c_2\lambda^2 + c_3\lambda^3,$$

---

[6]Compare with the bright case in (10.48), making the identification $z \equiv t$, $\tau \equiv x$, and $E = \sqrt{2}q$.

then

$$q_t - 6rqq_x + q_{xxx} = 0$$

$$r_t - 6rqr_x + r_{xxx} = 0.$$

b. Show that the KdV equation results for $r = -1$ and that the modified KdV equation

$$q_t \mp 6q^2 q_x + q_{xxx} = 0$$

is obtained for $r = \pm q$. The modified KdV equation is used in solid state physics to describe acoustic waves in certain anharmonic lattices and in plasma physics to describe Alfvén waves in a collisionless plasma.

c. For the KdV case, show that the Z–S/AKNS eigenvalue problem at $t = 0$ reduces to the Schrödinger equation with $q(x, 0)$ as the potential.

Problem 12-12: The sine-Gordon equation
Choose

$$A = (\frac{i}{4} \cos u(x, t))/\lambda$$

$$B = b(x, t)/\lambda$$

$$C = c(x, t)/\lambda$$

and $r = -q = \frac{1}{2}U_x$. Determine $b(x, t), c(x, t)$ (setting the integration constants equal to zero) and the equation that $u(x, t)$ satisfies. By introducing suitable new coordinates, show that the sine-Gordon equation results.

Part II

EXPERIMENTAL

ACTIVITIES

Introduction to Nonlinear Experiments

The test of all knowledge is experiment.
Experiment is the sole judge of scientific truth.
Richard Feynmann, Nobel Laureate in Physics (1965)

As Richard Feynmann reminds us in his wonderful quote [FLS77], science demands that all theory be checked by experiment. It is because nonlinear physics can often be so profoundly counter-intuitive that these laboratory investigations are, in our opinion, so important. Understanding is enhanced when experiments are used to check theory, so please attempt as many of the activities as you can. As you perform them, we hope that you will be amazed and startled by strange behavior, intrigued and terrorized by new ideas, and be able to amaze your friends as you relate your strange sightings! Remember that imagination is as important as knowledge, so exercise yours whenever possible. But please be careful, as exposure to nonlinear activities can be addictive, can provide fond memories, and can awaken an interest that lasts a lifetime.

Although it has been said that a rose by any other name is still a rose, with due apologies to William Shakespeare,

What's in a name? That which we call a rose
By any other name would smell as sweet.
William Shakespeare, Romeo and Juliet, Act II

we have, in an endeavor to encourage the use of these nonlinear investigations, called them experimental activities rather than simply experiments. More importantly, a number of design innovations have been introduced:

A. Each of the included activities may be approached on three levels:

1. simplest—for theoreticians (no insult intended—one of the authors is a theoretician) and non-physicists—the emphasis is on observing and investigating the features of the nonlinear phenomena with the minimum of data gathering and analysis;

2. moderate—for physics majors and engineers—more emphasis on data gathering and analysis;

3. complex—activities designed for a stand-alone course for experimental physicists—a deeper and more profound analysis is required and modifications are suggested to stimulate ideas for research projects.

These three levels have been provided so as to permit instructors and students the freedom to choose the type of activity that best suits their needs. However, a less rigorous and more relaxed approach than that normally used in upper-year undergraduate experiments is encouraged at all three levels.

B. Thirty-three experimental activities are included to provide students and instructors with

1. a large number of permutations and combinations of exploratory paths;

2. diverse investigations that

   (a) best suits their academic or personal interest,

   (b) have a duration which provides a chance for, and encourages, a successful conclusion of the activity,

   (c) are of a suitable level of complexity;

3. a choice of several simple activities or one complex research project.

Designing experimental activities that reward the students' effort with success and the pleasure of accomplishment is a major objective of this book.

C. The experimental activities are directly related to Part I, THEORY

1. to ensure that they are not perceived by the students as an unconnected, irrelevant, and meaningless time-consuming chore;

2. by placing a picture of a stopwatch (shown here) in the margin of the theory portion of the text at the point the activity should be undertaken;

3. to ensure that they reinforce and provide physical examples of the theoretical concepts;

4. by using the text's Mathematica files to explore and confirm the experimental investigations.

The experimental activities are not assigned just for the sake of learning experimental techniques, but rather are designed to deepen and broaden the reader's understanding of the nonlinear physics concepts covered in the text.

D. The experimental activities are designed around the following principles:

1. they are simple and easy to perform;

2. they use apparatus commonly available in undergraduate laboratories;

3. they employ apparatus that is easy to set up;

4. the equipment used is kept as simple as possible in an effort not to confuse the understanding of the apparatus with the understanding of the physics;

5. some activities should be simple enough that they may be done as take-home projects;

6. they should be enlightening and enjoyable;

7. they contain a list of further investigations that probe more deeply into the nonlinear physics involved;

8. they use Mathematica files, which are provided on the accompanying CD-ROM, to check that theory agrees with the observed physical behavior.

Most of the activities in Part II are of relatively short duration and use apparatus that is relatively simple and easy to construct. Brief and uncomplicated theoretical explanations are provided so that the experimenter can quickly explore the nonlinear physics. **However, the detailed interpretation of the physics may not be so easy or simple.** So be prepared to exercise your intellect and imagination.

Some of these activities are completely original, some were designed from theoretical discussions found in books, books that did not include any explanations of how to perform the activity or how to build the apparatus. Many of the activities are modifications of ideas found in diverse sources. The primary source for the borrowed ideas was the *American Journal of Physics* (AJP). We hope the citations and references give credit where credit is due.

We would like to offer some words of caution. Although these experimental activities have been performed by the authors, and all have been student-tested in the laboratory, there is always room to make them better. Accordingly, we welcome any suggestions for improvement of the activities by

- giving a different interpretation of the physics,

- designing better or different apparatus,

- modifying or using a completely different procedure.

These are open–ended experiments; they are not meant to be cookbook laboratory activities. Please feel free to omit certain steps in the procedures, modify others, and just explore as your desires dictate.

As far as we know there is not a comparable nonlinear physics laboratory manual on the market which is closely integrated with a complete theoretical development (Part I of the text) and a computer algebra (Mathematica) approach. We have done our best to reduce the number of ambiguities, typographical errors, and gaps in the procedures' steps. We hope that the number of egregious errors are few so that the experiments perform as stipulated and are rewarding and enjoyable.

Richard H. Enns and George C. McGuire.

Acknowledgements

We would like to thank the students of the University College of the Fraser Valley for their valuable suggestions and in particular David Cooke who independently performed and critiqued the experiments.

Experimental Activity 1

Magnetic Force

Comment: This easy activity should not take more than one hour to complete.

Reference: Section 1.3

Object: To find the mathematical relationship between the force (F) that two thin magnets exert on each other and their separation distance (r).

Theory: Consider two very long thin bar magnets oriented as shown in Figure 1.1 with like poles (e.g., North here) adjacent to each other and separated

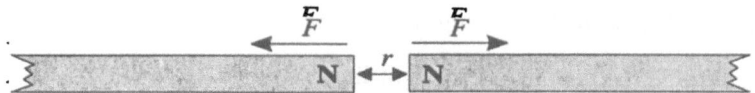

Figure 1.1: Two long bar magnets.

by a distance r. In this situation, the repulsive force will be given approximately by an inverse square law, $F \propto 1/r^2$, a force law characteristic of a monopole–monopole interaction. This is because the North poles behave like isolated monopoles, while the far-distant South poles make very little contribution to the force between the magnets. If the bar magnets are shortened to a point where their lengths are of the same order of magnitude as the radius of a pole face, all four poles will contribute significantly to the overall force. Qualitatively, one would expect that in this case the repulsive force between the two magnets would be reduced, i.e., the exponent of r in the force law to be greater than 2.

In this experimental activity, the mathematical relationship between the force F of repulsion and the separation distance r between two thin cylindrical magnets is investigated. The magnetic force law is assumed to be of the form

$$F = \frac{k}{r^n} \tag{1.1}$$

where k and n are positive constants. The goal is to determine the value of n.

If (1.1) is plotted, an inverse power law graph similar to that shown in Figure 1.2 results. Various ways can be used to determine the value of n, but

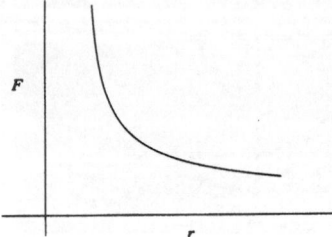

Figure 1.2: An inverse power law between force and separation distance.

one of the simplest is to make a loglog plot of the experimental data and use a least squares fitting procedure. To see how this works, let's take the log of (1.1),

$$\ln\left(F\right) = \ln\left(\frac{k}{r^n}\right) = \ln\left(k\right) - n\ln\left(r\right). \tag{1.2}$$

Setting $y \equiv \ln\left(F\right)$, $b \equiv \ln\left(k\right)$, and $x \equiv \ln\left(r\right)$, Equation (1.2) is just the straight line equation $y = b - nx$ with negative slope of magnitude n.

Thus all one has to do to determine the exponent n is:

1. determine the force F at different separation distances r;

2. make a loglog plot of the data;

3. determine the best-fitting straight line;

4. extract n from the slope of this line.

Although the fitting procedure could be done by hand, a quicker and less tedious approach is to use Mathematica as in the following file.

Magnetic Force Law
This file plots the input magnetic force data as well as making a loglog plot. Making use of Mathematica's `Fit` command, a best-fitting straight line is obtained for the loglog data and the exponent n is extracted. Mathematica commands in file: `ListPlot`, `Fit`, `Plot`, `Show`, `Log`, `Coefficient`, `Exp`, `Part`, `PlotRange`, `AxesLabel`, `PlotStyle`, `Hue`, `PointSize`, `TextStyle`, `ImageSize`, `DisplayFunction->Identity`

X01

Procedure:

1. Using an electronic balance that has a precision of at least 0.10 gram, set up the following apparatus.

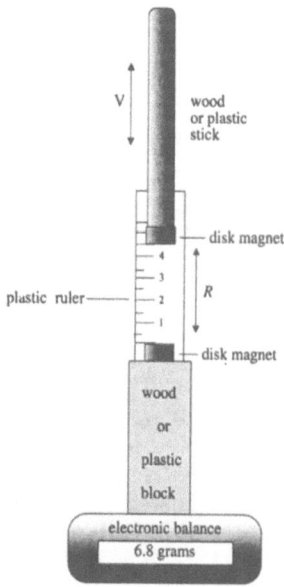

Figure 1.3: Apparatus to find the form of the force law.

2. Make sure you tare (zero) the balance before you bring the second (movable) magnet close to the magnet on the balance.

3. For at least ten different distances—preferably more—measure the separation distance and the resulting force as the movable magnet is brought closer to the magnet on the balance. Place your data in a table.

4. You might wish to repeat the above procedure a few extra times so that you are confident of the accuracy of your data.

5. By inspection determine which exponent value for n would best describe the data.

6. Place your data into the Mathematica file X01 and see what value Mathematica gives for n.

7. How close is your n to an integer? What type of physical interaction does this exponent suggest? Explain.

Things to Investigate

1. Does the force law you found work for an off-axis repulsion?

2. What is the exponent n if a long bar magnet and a thin disk magnet are used? Does your answer make physical sense? Explain.

Experimental Activity 2

Magnetic Tower

Comment: This activity should not take more than one hour to complete.

Reference: Section 1.3

Object: To determine the equilibrium separation distances of thin repelling disk magnets arranged in a vertical tower configuration.

Theory: In this activity a number of identical thin disk magnets are held in a vertical tower position with like poles adjacent. Figure 2.1 shows the arrangement for two disk magnets. The disk magnets have holes through which a vertical wooden dowel passes. A repulsive force exists between the magnets so

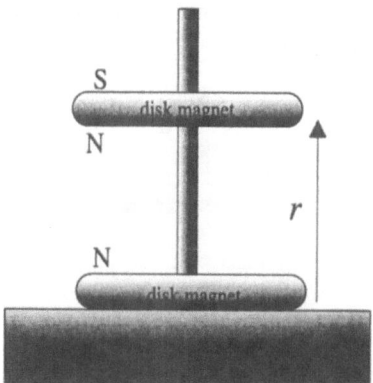

Figure 2.1: Two thin disk magnets supported in a vertical tower configuration.

that a nonzero separation distance results. Assuming that the magnets can be approximated as point dipoles, the interaction force between them can be calculated as follows. The magnetic field $\vec{B}$ at a distance r from a point magnetic dipole with dipole moment $\vec{m}$ is [Gri99]

$$\vec{B}(\vec{r}) = \frac{\mu_0}{4\pi r^3}(3(\vec{m} \cdot \hat{r})\hat{r} - \vec{m}), \tag{2.1}$$

where $\hat{r}$ is the unit vector pointing from the dipole to the observation point at $\vec{r}$ and $\mu_0 = 4\pi \times 10^{-7}$ N/A^2 is the permeability of free space. If we consider the field only along the axis of the magnetic moment $\vec{m}$, then (2.1) reduces to

$$\vec{B} = \frac{\mu_0}{4\pi r^3} 2\vec{m}. \tag{2.2}$$

The magnetic force of repulsion between the two magnets is given by

$$\vec{F} = -\nabla(\vec{m} \cdot \vec{B}), \tag{2.3}$$

so the vertical magnetic force between them is

$$F = \frac{3\mu_0 m^2}{2\pi r^4} \tag{2.4}$$

where r is the separation distance. If the upper disk magnet has a mass M and is in its equilibrium position, then the net force on it is zero so that

$$\frac{3\mu_0 m^2}{2\pi r^4} = Mg, \tag{2.5}$$

where g is the accleration due to gravity. If the magnetic moment m is independently determined and the disk mass M measured, the theoretical separation distance r can be calculated and compared with the observed distance. Conversely, if the magnetic moment is not known, it can be determined from Equation (2.5) by measuring M and r.

To make this activity a little more intriguing and to demonstrate the numerical solving ability of Mathematica, four magnets are used and arranged with the polarities shown in Figure 2.2.

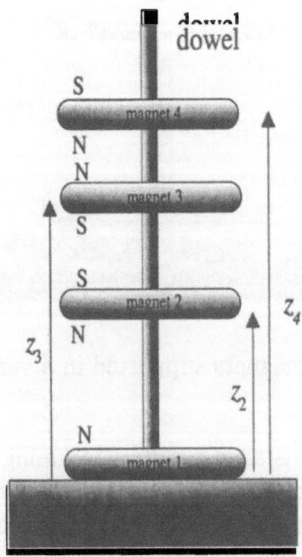

Figure 2.2: Four thin disk magnets in the tower configuration.

The above force analysis for two magnets can be easily extended to the new situation. All four magnets are assumed to be identical, each with the same magnetic moment m and mass M. Frictional effects due to the wooden dowel are assumed to be negligible.

Consider, for example, magnet 2 in Figure 2.2. Magnet 1, resting on the base, exerts a repulsive force of magnitude F_{21} upwards on magnet 2, while magnet 3 exerts a repulsive force of magnitude F_{23} downwards. Magnet 4, on the other hand, exerts an attractive force of magnitude F_{24} upwards. Taking the weight of magnet 2 into account, the net force on it will be

$$F_2 = -Mg + F_{21} - F_{23} + F_{24}. \tag{2.6}$$

Similarly, using the same subscript notation, the net force on magnet 3 due to the other three magnets will be

$$F_3 = -Mg - F_{31} + F_{32} - F_{34}, \tag{2.7}$$

and on magnet 4,

$$F_4 = -Mg + F_{41} - F_{42} + F_{43}. \tag{2.8}$$

By symmetry, note that $F_{42} = F_{24}$, $F_{34} = F_{43}$, etc.

When all four magnets are in their equilibrium positions, the net force on each magnet is zero. Labeling the distances of magnets 2, 3, and 4 from magnet 1 as z_2, z_3, and z_4, respectively, and using the dipole–dipole force law (2.4), Equations (2.6)–(2.8) yield

$$\begin{aligned}
\frac{1}{z_2^4} - \frac{1}{(z_3 - z_2)^4} + \frac{1}{(z_4 - z_2)^4} &= C, \\
-\frac{1}{z_3^4} + \frac{1}{(z_3 - z_2)^4} - \frac{1}{(z_4 - z_3)^4} &= C, \\
\frac{1}{z_4^4} - \frac{1}{(z_4 - z_2)^4} + \frac{1}{(z_4 - z_3)^4} &= C,
\end{aligned} \tag{2.9}$$

where $C = 2\pi Mg/3\mu_0 m^2$. If C is known, the three simultaneous equations can be solved for the three theoretically predicted distances z_2, z_3, and z_4. If C is not known, only the ratios z_2/z_3, z_2/z_4, and z_3/z_4 can be determined. Because the equations are nonlinear, solving them is a nontrivial task, which is better left to Mathematica as in the following file.

Magnetic Tower

This file demonstrates a number of different Mathematica commands. Entering the magnetic dipole field as a list of field components, the magnetic force is calculated using the `DotProduct` and `Grad` commands. Then Equations (2.9) are derived and numerically solved with the `FindRoot` command for the ratios of the separation distances. The theoretically predicted ratios based on the dipole–dipole interaction model can be compared with the experimentally derived ratios which the reader can obtain by carrying out the steps in the following **Procedure**. Mathematica commands in file: `Pi`, `DotProduct`, `Grad`, `FindRoot`, `Print`

X02

Procedure:

1. Find four identical circular thin disk magnets with holes in their centers. Place the magnets on a wooden dowel as shown in Figure 2.2.

2. Measure the separation distances between each of the magnets. A gentle shaking of the dowel will assure that the magnets have settled into their equilibrium positions.

3. Run the Mathematica file X02 to see how the ratios of the measured equilibrium distances compare with the theoretical distances provided by Mathematica.

4. Discuss the accuracy of your results.

Things to Investigate

1. Determine the magnetic moment of one of the disk magnets.

2. Repeat the procedure using five and then six magnets. Discuss the accuracy of your results

Experimental Activity 3

Spin Toy Pendulum

Don't just lie there idly spinning, we aliens must begin our job of making these earthlings understand the importance of nonlinear oscillations.

beep beep

spin toy

Comment: This activity is easy to carry out and should not take more than one hour to complete.

Reference: Section 2.1.1

Object: To investigate a characteristic feature of nonlinear oscillatory motion, namely that the period varies with the amplitude of the oscillations.

Theory: One of the easiest ways of determining if an oscillating object is exhibiting nonlinear motion is to measure the period T for a variety of different initial amplitudes. If T varies with the amplitude, the restoring force is nonlinear. This activity uses a spin toy pendulum which, when oscillating, exhibits an increase in T as the angular amplitude increases, a behavior similar to that for the simple pendulum. The ODE that governs the nonlinear oscillations of a pendulum is

$$\ddot{\theta} + 2\gamma\dot{\theta} + \omega^2 \sin\theta = 0 \tag{3.1}$$

where θ is the angular displacement from equilibrium, γ the linear damping coefficient, and ω the angular frequency.

The inexpensive toy used in this experiment, which is sold under the name

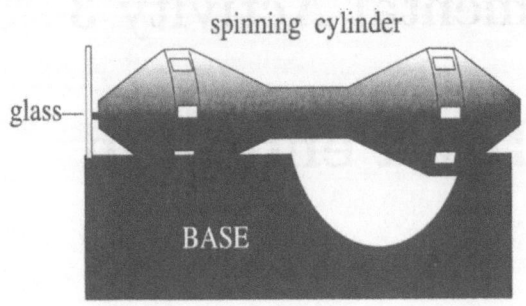

Figure 3.1: The spin toy pendulum.

Revolution: The World's Most Efficient Spinning Device,[1] can be considered as a black box oscillator. The spin toy pendulum is shown in Figure 3.1. The movable cylinder and base of this toy contain magnets that levitate and hold the spinning cylinder against the fixed glass end of the base. If the cylinder is given a sharp twist with the hand, it will continue to spin for a very long time. The toy has a low damping coefficient and therefore a very large quality (Q) factor. The quality factor is a dimensionless parameter that provides a measure of how fast a system dissipates energy. It is defined as $Q = \frac{2\pi E}{\Delta E}$, where E is the total energy and ΔE is the energy lost per cycle. For example, a quartz wrist watch has a $Q \approx 10^4$, whereas a piano string's Q is about ten times lower.

If the cylinder is given a small initial twist, it can be made to oscillate rather than rotate. For sufficiently small oscillations the simple harmonic oscillator equation ((3.1) with $\sin\theta \approx \theta$) prevails and $Q \approx \frac{\omega}{2\gamma}$ if the damping is not too large. In this limit, the period is given by $T = \frac{2\pi}{\omega}$ and is independent of the angular amplitude.

For larger oscillations the period begins to deviate from that predicted by the small oscillation formula and becomes noticeably amplitude-dependent. In this experimental activity, the period of the vibrations is measured against the angular amplitude and the mathematical relationship between these two quantities is investigated.

Procedure:

1. The spinning cylinder used by the authors had white marks placed every 45° around its circumference. These marks can be used to estimate the initial amplitudes. Newer versions of this toy may not contain these white markings. If this is the case, paint them on the cylinder.

---

[1] Arbor Scientific, P.O. Box 2750, Ann Arbor, MI 48106-2750, phone 1-800-367-6695.

2. With a stop watch, and for an initial angular amplitude of less than 10°, measure the period T of the resulting oscillations.

3. Using this small amplitude period, calculate the small amplitude frequency $f = \frac{1}{T}$, and small amplitude angular frequency $(\omega = \frac{2\pi}{T})$.

4. Measure the period for an initial amplitude of 22.5°. Can you detect any difference from the period measured in the previous step?

5. Repeat the above procedure for angles of approximately 45°, 67°, 90°, 112°, 135°, 157°, and for an angle as close to 180° as you can get.

6. Make a graph of the period as a function of the angular amplitude.

7. How do you know the restoring force is nonlinear?

8. For what initial amplitude is the nonlinearity first detectable?

9. Calculate an approximate low period damping coefficient (γ) for this oscillator. This can be done by measuring the length of time (t) it takes a small amplitude to decrease by a factor of two and then use the equation $\gamma = \frac{\ln(2)}{t}$.

10. Calculate the quality factor (Q) for small oscillations.

11. Use the provided Mathematica file X03A to check whether the restoring force is similar to that of a simple pendulum. In the file, substitute your known values for γ and ω into the damped pendulum Equation (3.1). For a given large initial amplitude, the Mathematica file will numerically solve and then make a plot of the angular displacement as a function of time. The file explains how the period may be found from the plot. Compare your experimental data for the period with that given by Mathematica .

Spin Toy Pendulum
This file uses two methods to find the period of oscillation:
1. uses an elliptic integral;
2. numerically solves and plots the solution of the second-order ODE.
Both methods require that you know the period for small oscillations around the equilibrium point. The file will ask you to input this period. Mathematica commands in file: HoldForm, Integrate, Sin, Evaluate, EllipticK, Pi, N, ReleaseHold, NDSolve, PlotPoints, PlotStyle, Hue, AxesLabel, PlotLabel, TextStyle, ImageSize

---

Things to Investigate:

- Tilt the device so that the effect of gravity is reduced. Is the period dependent on the angle that the base is tipped?

- Investigate whether the damping term might be nonlinear.

- Use the Mathematica file X03B to see how accurately a polynomial can be used to approximate the curve shown on your plot of the period vs. angular displacement. Five or six pairs of data from the period vs. displacement graph are needed to run this file. Experiment with changing the order of the polynomial to produce a better fit.

Spin Toy Best Fit

This file produces the best fitting polynomial for the experimental values of period T versus angular amplitude and plots the polynomial on the same graph as the data points. Mathematica commands in file: `Transpose`, `Fit`, `Block`, `ListPlot`, `PlotStyle`, `Hue`, `Plot`, `RGBColor`, `Epilog`, `PointSize`, `Show`, `PlotRange`, `Text`, `$DisplayFunction=Identity`

Experimental Activity 4

Driven Eardrum

Comment: This investigation should not take more than one hour to complete.

References: Section 2.1.2

1. [Hoe80] A good reference for mixer circuits.

2. [MM74] Provided the idea of hearing the modulated frequencies.

Object: To use the nonlinear properties of a diode to model the frequency output of a forced eardrum.

Theory: Although the most common and well-known use for diodes is the rectification of AC currents to DC currents, they can perform other useful functions. For example, the nonlinear properties of the diode can be used to model the frequency output of an eardrum. The asymmetrical loading of an eardrum causes frequencies not originally present in the incoming sound to be produced by the vibrating eardrum and transmitted to the listener.[1] A diode can reproduce this behavior. Electrical engineers and radio buffs may recognize this experiment's circuit as a mixer or modulator.

A diode's nonlinear current–voltage curve is shown in Figure 4.1 and can be described by the equation,

$$i = i_s(e^{\frac{kV}{T}} - 1) \tag{4.1}$$

with V the potential drop across the diode, i_s the reverse saturation current ($i_s \approx 1.0 \times 10^{-8}$ A), k a constant ($k = 11,600/\eta$ where $\eta = 2$ for silicon diodes and $\eta = 1$ for germanium diodes), and T represents the temperature in Kelvin ($T \approx 300$ K). When Equation (4.1) is Taylor expanded about $V = 0$, the first two terms give the approximate nonlinear expression

$$i = a_1 V + a_2 V^2. \tag{4.2}$$

---

[1]On Page 40 of Frank S. Crawford's *WAVES, Berkeley physics course*-volume 3, there is a very nice home experiment that you can perform to test the nonlinearities in your ear. There is also a discussion of the effects of adding a cubic term to the nonlinear expression.

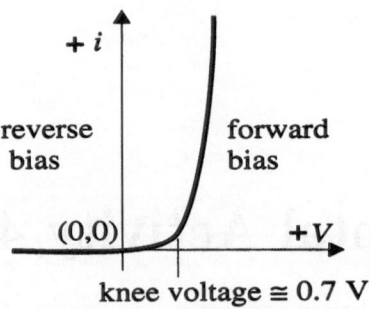

Figure 4.1: A diode's i-V curve.

This activity uses (4.2), which resembles the forcing function for the eardrum, to model the nonlinear properties of the diode.

In the circuit shown in Figure 4.2, three external signals are fed into the

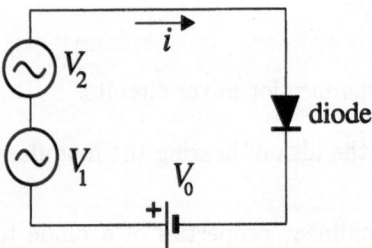

Figure 4.2: A diode circuit for modeling the eardrum

diode. A battery or DC offset (V_0) forward biases the diode and two sinusoidal signals with angular frequencies $\omega_1 = 2\pi f_1$ and $\omega_2 = 2\pi f_2$ are superimposed on the bias signal. The potential drop across the diode is

$$V = V_0 + V_1 \sin(\omega_1 t) + V_2 \sin(\omega_2 t). \tag{4.3}$$

The output signal will contain a variety of new frequencies depending on the values of ω_1 and ω_2. To understand how the diode produces these extra frequencies, substitute (4.3) into (4.2), and simplify the result. Making use of Mathematica file X04, one finds that

$$i = \frac{Is\,k^2 V_1 V_2 \cos(\omega_1 t - \omega_2 t)}{2T^2} - \frac{Is\,k^2 V_1 V_2 \cos(\omega_1 t + \omega_2 t)}{2T^2}$$

$$- \frac{Is\,k^2 V_2{}^2 \cos(2\omega_2 t)}{4T^2} - \frac{Is\,k^2 V_1{}^2 \cos(2\omega_1 t)}{4T^2}$$

$$+ \frac{(Is\,k^2 V_0 V_1 + Is\,kT V_1)\sin(\omega_1 t)}{T^2} + \frac{(Is\,kT V_2 + Is\,k^2 V_0 V_2)\sin(\omega_2 t)}{T^2}$$

$$+ \frac{4\,Is\,kT V_0 + Is\,k^2 V_2{}^2 + 2\,Is\,k^2 V_0{}^2 + Is\,k^2 V_1{}^2}{4T^2}.$$

By examining the individual terms of the above expression, we can see that the first term represents an oscillation at the frequency difference $(\omega_1 - \omega_2)$, the second term represents the frequency sum $(\omega_1 + \omega_2)$, the third and fourth terms represent a doubling of the two initial frequencies, the fifth and sixth term represent the original frequencies, and the final term is a direct current contribution.

Eardrum

This file produces the form of the current–voltage equation which displays sum, difference, etc., terms. An individual term (e.g., the low frequency component) can be selected and its sound heard by using Mathematica's `Play` command. As an option for readers already familiar with the power spectrum concept, the power spectrum of the output signal is calculated. Mathematica commands in file: `Exp`, `Series`, `Normal`, `Block`, `Plot`, `PlotStyle`, `Hue`, `RGBColor`, `Show`, `Epilog`, `TextStyle`, `Hue`, `Sin`, `TrigReduce`, `Expand`, `Part`, `Play`, `Table`, `ListPlot`, `Fourier`, `PlotRange`, `Ticks`, `PlotLabel`, `AxesLabel`, `N`, `Abs`, `PlotJoined->True`

Procedure

1. Build the circuit shown in Figure 4.3 using the diode $1N4001$ or its equivalent. Signal generators with digital readouts make it easier to set the individual frequencies. If the signal generators have a DC offset voltage, it is not necessary to place a battery in the circuit. With the generators turned on but with the amplitude adjustment turned down, set the DC offset voltage at about 0.20 volts. For small amplitude signals, the output signal may need to be amplified before applying it to the speaker.

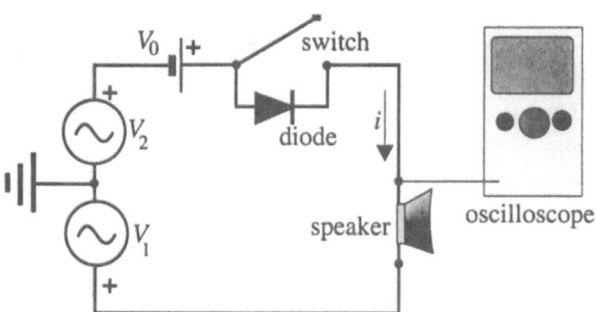

Figure 4.3: Experimental setup

2. With the switch closed (diode not being used), turn on the signal generators and adjust the frequencies and amplitudes to produce frequencies above the range of human hearing $(> 20,000 \text{ Hz})$. Make sure that the individual generators have a frequency difference of about 1000 Hz. Open the switch (the diode is now in the circuit) and you should hear the modulated frequency $\omega_1 - \omega_2$. Repeat for different pairs of high frequencies.

3. Open the switch and drop the individual frequencies down to 1000 Hz and 1400 Hz. Close the switch and see if you can detect both the lower $(\omega_1 - \omega_2)$ and upper $(\omega_1 + \omega_2)$ sideband modulated frequencies.

4. Try a variety of frequency pairs, say 400 and 700 Hz, 800 and 1200 Hz, 2500 and 2900 Hz, etc. For what pair can you detect both the high and low sideband frequencies?

Things to Investigate

1. Using one generator, what is the frequency range of your hearing?

2. With the switch closed (no diode present), how close do the two output frequencies have to be for you to hear an audible beat?

3. Build a low-pass or high-pass filter to extract the frequency that you want from the modulated collection.

4. With a signal analyzer make a fast Fourier transform (FFT) of the output signal. How do the frequencies present in the output signal compare with those predicted by the theory?

5. Use a SPICE program to construct and check if the circuit functions as predicted. Use the SPICE's FFT command to investigate the frequencies in the modulated signal. Are all the predicted frequencies present?

6. For very low frequencies use a paper chart recorder to plot the output. Make an FFT of the output.

7. The experimenter might wish to look for references on diode mixers or radio heterodyning for additional material on similar experiments.

Experimental Activity 5

Nonlinear Damping

Comment: This experiment is not difficult to do, but it will take about three hours to complete including the setup time.

Reference: Section 2.1.3

Object: To investigate the dissipative force created by air drag.

Theory: The magnitude of the dissipative or drag force exerted on a body moving in a straight line through a viscous homogeneous medium depends upon a variety of parameters: the nature and density of the resisting medium as well as the size, shape, and speed of the moving object. To develop an expression for the drag force which correctly relates all these parameters can be quite difficult, especially over a large range in speed. This experiment provides a method of investigating nonlinear damping for low speeds.

Since the speed is the only continuously changing variable, it is reasonable to assume the magnitude of the dissipative force on a body moving in a straight line through air will be a function of the instantaneous speed, i.e., $F(v)$. Even this simplification to one variable can still produce a complicated expression for the force. In an effort to simplify, aid understanding, and provide a workable approach to the problem, we assume that the expression for the dissipative force function is "well-behaved" over a low range of speeds and therefore can be expanded as a Maclaurin power series

$$F(v) = k_0 v^0 + k_1 v^1 + k_2 v^2 + k_3 v^3 + \cdots \tag{5.1}$$

where the coefficients k_0, $k_1, \ldots$ depend on the medium and physical parameters previously mentioned.

When the expression for the force is expressed as a power series it becomes possible to identify, pick out, and then use only one of the terms to model a particular type of drag. For example, in the case of sliding friction, the drag force does not depend on the speed, but only on the material making up the sliding surfaces and the normal force acting on the moving body. When the force produced by the sliding friction is much larger than the force produced by the speed dependent terms, the speed dependent terms may be ignored. Accordingly, only the first term of the above series is needed for sliding friction, so

$$F = k_0 v^0 = k_0. \tag{5.2}$$

If the sliding object with a mass m is on a horizontal surface, the above term may be set equal to the standard expression for the force of sliding friction $(F = -\mu m g)$ so

$$k_0 = -\mu m g. \tag{5.3}$$

Here μ is the coefficient of sliding friction and g is the acceleration due to gravity.

In situations such as a glider moving on a nearly frictionless airtrack, the coefficient of sliding friction is very small and the speed dependent terms dominate. If the airtrack glider is moving slowly through the air and if it produces very little air turbulence, the magnitude of the dissipative force usually varies as a function of the first power of the instantaneous speed (v) so

$$F = k_1 v^1 = k_1 v. \tag{5.4}$$

This assumes that all the other speed-dependent terms are of less importance. This type of dissipative force is known as a Stokes or a laminar damping force.

If the moving airtrack glider produces a lot of turbulence, e.g., due to a higher speed or to its shape, the magnitude of the dissipative force may be more accurately modeled by the term containing the square of the instantaneous speed

$$F = k_2 v^2. \tag{5.5}$$

This type of dissipative force is known as Newtonian damping.

In this activity the instantaneous speeds and dissipative forces are measured for a variety of speeds and then a $\ln(F)$ versus $\ln(v)$ graph is drawn. Assuming that one term of the form $k v^n$ dominates, the values for n and k are found by graphical analysis. After the values for the exponent n and coefficient k are known, it might be possible to identify which of the power series terms best fits the experimental results.

In this experiment, a moving air track glider passes between two photocell timers as shown in Figure 5.1. The net force on the glider can be calculated by first measuring the times (t_1, t_2) it takes the glider to pass by the two photocells. Then the speeds (v_1, v_2) at these two locations are calculated by measuring the

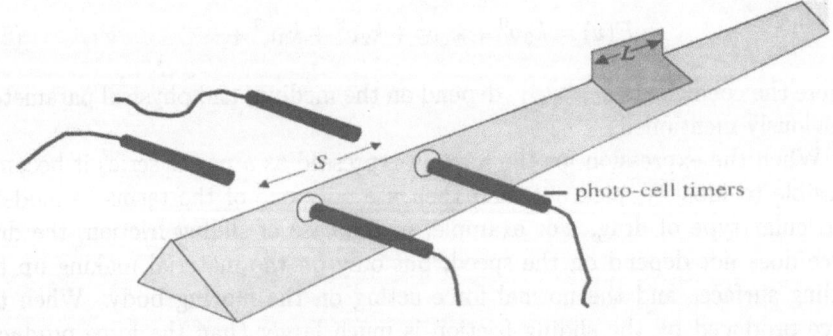

Figure 5.1: An airtrack glider moving between two photocell timers.

length of the glider (L), and dividing by the time the glider takes to pass by each timer, and thus,

$$v_1 = \frac{L}{t_1}, \quad v_2 = \frac{L}{t_2}. \tag{5.6}$$

The dissipative force (F) acting on the glider is found by equating the glider's change in kinetic energy to the work W done on the glider as it moved the distance s between the timers:

$$W = Fs. \tag{5.7}$$

Since the change in the glider's kinetic energy is equal to the work

$$Fs = \frac{mv_2^2}{2} - \frac{mv_1^2}{2}, \tag{5.8}$$

the magnitude of the dissipative force is equal to

$$F = \left| \frac{mv_2^2 - mv_1^2}{2s} \right|. \tag{5.9}$$

If the distance (s) between the timers is kept small, then the two measured speeds (v_1, v_2) will be approximately equal. The speed (v) which creates the dissipative force is close to the average of the two calculated speeds. The average speed ($\bar{v}$) is calculated by using

$$\bar{v} = \frac{v_2 + v_1}{2}. \tag{5.10}$$

Once a number of different forces and their related average speeds have been calculated, the mathematical relationship that unites them is determined.

Procedure:

  1. Set up the apparatus as shown in Figure 5.2.

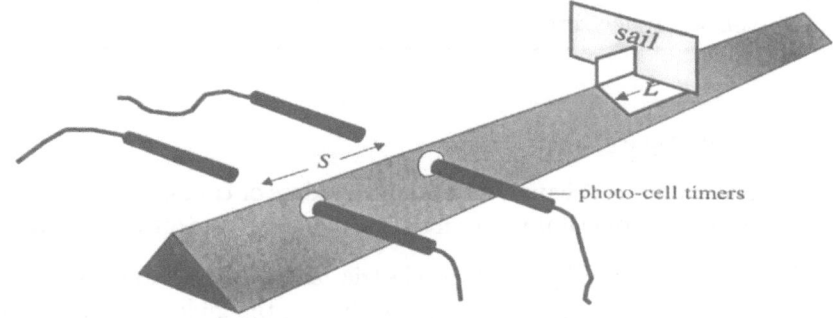

Figure 5.2: Apparatus to investigate force as a function of speed.

  2. Measure the mass m of the glider.

3. Adjust the distance s to be about 20 cm.

4. Make sure that the air track is level.

5. Use a glider that has a length (L) of about 15 cm and a cardboard sail that has an area between 100 and 200 cm^2. The sail creates the majority of the air drag.

6. Launch the glider and record the times (t_1, t_2) it takes to pass by the timers. Using Equation (5.6), calculate the speeds (v_1, v_2).

7. Calculate the dissipative force using Equation (5.9).

8. Calculate the average velocity ($\bar{v}$) using Equation (5.10).

9. Repeat steps 6,7, and 8 for at least six widely different speeds.

10. Draw a graph of $\ln(F)$ versus $\ln(\bar{v})$.

11. Analyze the graph to find the mathematical relationship between F and $\bar{v}$. Determine the values for the exponent n and the constant k.

12. The experimental value for n may not be an integer. If you were to model your experimental result by using only a single term from the Maclaurin series (5.1), what integer would you use for n? Give reasons why the experimental value for n is different from an integer value.

13. Place the experimental data for v and F into Mathematica file X05 to see if the Mathematica file can produce a power series ($y \equiv F$, $x \equiv v$) of the form $y = ax^2 + bx$ or even $y = ax^3 + bx^2 + cx$. Which of the power series best represents your data? With these new equations, use the text's Mathematica file MF02 to numerically solve the equations and then compare the results with the experimental data.

Nonlinear Damping
This is another best fit file for finding the polynomial function $F(v)$ which best describes the experimental data. Mathematica commands in file:
`Transpose, ListPlot, Plot, PointSize, AxesLabel, PlotRange, Fit, Show, ImageSize`

Things to Investigate:

- Use sails of different shapes, for example, wedges or cones, to see how they modify the experimental values for the exponent n and/or coefficient k.

- Modify this activity by using an electric fan to blow across the sail to create the drag force and a spring attached to the glider to measure the force on the stationary glider. Compare the values for n and k with the values found in the above procedure.

- This activity might be modified to measure the drag on larger objects, such as a person coasting on a bicycle.

Experimental Activity 6

Anharmonic Potential

Comment: This activity can be done in under three hours if the apparatus has been previously set up.

Reference: Section 2.1.4

Object: To produce an anharmonic potential, determine its analytic form, and investigate the period of oscillations governed by this potential.

Theory: Anharmonic oscillators have nonlinear restoring forces so, unlike linear oscillators, their periods depend on the magnitude of their amplitudes. The motion of the anharmonic oscillator in this experimental activity is controlled by a potential energy function $(U(x))$ that consists of two terms. One term is the magnetic potential energy $(U_m(x) = \frac{b}{x^n})$ and the other is the gravitational potential energy $(U_g(x) = cx)$, so the total potential energy is

$$U(x) = \frac{b}{x^n} + cx, \qquad (6.1)$$

where b, c, and n are all positive constants. A representative plot of the above function is illustrated in Fig 6.1. Examination of the plot shows that oscillations can occur around the equilibrium point $(x = a)$. For small oscillations about

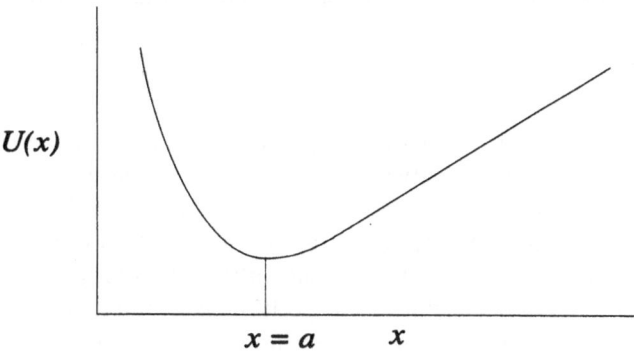

Figure 6.1: Potential energy (U) vs. position (x).

the equilibrium point, Equation (6.1) can be Taylor expanded to yield

$$U(x) = U(a) + U'(a)(x-a) + \frac{1}{2!}U''(a)(x-a)^2 + \frac{1}{3!}U'''(a)(x-a)^3 + \cdots . \quad (6.2)$$

Since the slope of the potential energy curve is equal to zero at $x = a$

$$U'(a) = -\frac{nb}{a^{n+1}} + c = 0, \quad (6.3)$$

the location of the equilibrium point is

$$a = \left(\frac{nb}{c}\right)^{\frac{1}{(n+1)}}. \quad (6.4)$$

Ignoring the terms beyond the quadratic we have

$$U(x) = U(a) + \frac{1}{2!}U''(a)(x-a)^2, \quad (6.5)$$

and since the force $F = -\frac{dU}{dx}$,

$$F = -k(x-a), \quad (6.6)$$

where

$$k = U''(a) = \frac{n(n+1)b}{a^{(n+2)}}. \quad (6.7)$$

This approximation gives the (simple) harmonic limit. In this case, the restoring force is linear in the displacement with k playing the role of the spring constant. For larger displacements the higher order terms in the Taylor expansion must be considered. The restoring force is no longer linear and the potential is referred to as anharmonic.

Assuming that no dissipative forces are acting, the net force on a mass in the anharmonic potential given by Equation (6.1) is found using $F = -\frac{dU}{dx}$, so

$$F = \frac{nb}{x^{(n+1)}} - c. \quad (6.8)$$

From Newton's second law, $F = m\ddot{x}$, the ODE describing the motion is

$$m\ddot{x} - \frac{nb}{x^{(n+1)}} + c = 0. \quad (6.9)$$

In general, an analytical solution of (6.9) is not possible, so numerical methods are usually applied. Mathematica file X06 is used to check the experimental behavior of the oscillator governed by the anharmonic potential.

In this experimental activity, the values for b, c, and n are determined, and then substituted into Equation (6.9). If the initial values for position and speed are also given, a numerical solution for position as a function of time is easily found using Mathematica. The period for large amplitude oscillations is determined from this numerical solution.

Anharmonic Potential

If the experimental values for m, b, n, and c are placed in this file, plots of the potential energy (U) vs. extension (x) and the force (F) vs. extension (x) are produced. The file sets up and numerically solves the ODE that governs the motion and plots the glider's amplitude as a function of time. Mathematica commands in file: `D`, `FindRoot`, `Plot`, `PlotRange`, `PlotStyle`, `RGBColor`, `AxesLabel`, `PlotLabel`, `ImageSize`, `NDSolve`, `TextStyle`, `Hue`

Procedure:

1. Set up the apparatus as shown in Figure 6.2 using a 1.2 m airtrack. Make sure the air track is level.

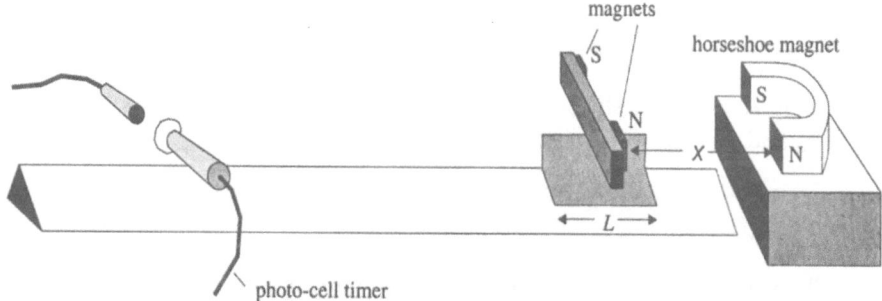

Figure 6.2: Apparatus to determine U_m as a function of x.

2. Mount the gated photocell timer approximately 0.60 m from the securely fastened powerful horseshoe magnet. If a horseshoe magnet is not available, two strong neodymium disk magnets fastened to a steel bar should work as well.

3. Attach two small neodymium disk magnets to the light horizontal plastic bar which is fastened to the top of the glider. The poles of the small magnets should be arranged so that a repulsive force exists between the horseshoe magnet and the glider's small magnets.

4. Measure the mass of the glider (magnets attached).

5. Place the glider near the large repelling magnet. For the first run, the separation between the repelling magnets should be about 0.050 m.

6. Release the glider, and record the time (t) it takes to pass by the photocell at the other end of the track.

7. Calculate the glider's speed ($v = L/t$) as it passes by the timer. (In theory the glider should keep picking up speed as it moves further away

from the repelling magnet, but after it is about 0.50 m from the magnet, the repelling force becomes negligible, and the glider maintains a constant speed.)

8. Calculate the final kinetic energy $(K = \frac{mv^2}{2})$ of the glider. The final kinetic energy is actually equal to the change in kinetic energy because the glider starts from rest. The change in kinetic energy is equal to the negative change in magnetic potential energy. The magnetic potential energy (U_m) is assumed to be zero when the distance between glider and the magnet is large, so the initial magnetic potential energy just equals the final kinetic energy of the glider.

9. Record the initial position (x) of the glider and its magnetic potential energy (U_m).

10. Repeat the above procedure for at least six different initial values of x from 0.05 to 0.20 m.

11. Record the data for x and U_m in an appropriate table.

12. Using the data in the table, find the mathematical relationship between U_m and x. The equation should have the form

$$U_m(x) = \frac{b}{x^n}.$$

The values for b and n can be found by plotting and analyzing a $\ln(U_m)$ vs. $\ln(x)$ graph. (A calculator with a linear regression program may be used to avoid making the plot.)

13. Now incline the air track as shown Figure 6.3. The wood block should have a height of 2 to 3 cm for a 1.2 m long air track. Note: The height of the wood block was chosen to make the maximum gravitational potential roughly match the magnetic potential energy produced by the authors' magnet at $x = 0.10$ m.

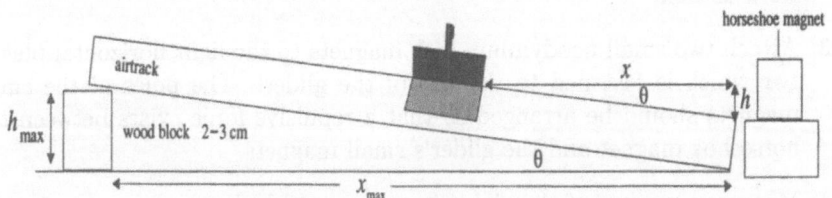

Figure 6.3: Inclined airtrack.

14. Calculate the angle of elevation $(\tan \theta = h_{\max}/x_{\max})$.

15. The gravitational potential energy of the glider at any position on the air track is $U_g = mgh$ where $h = x \tan \theta$ so $U_g = mgx \tan \theta$.

16. The value for c in Equation (6.1) is equal to $mg \tan \theta$. The values b, c, and n are now all known.

17. Use Equation (6.4) to calculate where the theoretical equilibrium point $(x = a)$ is located.

18. Place the glider back on the inclined track and locate the experimental equilibrium point.

19. Compare the theoretical and experimental values for the equilibrium point.

20. Displace the glider a small distance (≈ 0.01 m) from the equilibrium point, release the glider and measure the small oscillation period of the resulting vibrations.

21. Calculate the value for the spring constant using Equation (6.7). After the value for k is known, calculate the small oscillation period by using

$$T = 2\pi \sqrt{\frac{m}{k}}.$$

22. Displace the glider a large distance from the equilibrium point. Record the distance and release the glider. Measure the period of the resulting oscillations.

23. Repeat the above procedure for a different large amplitude.

24. Use Mathematica file X06 to check if the measured values for the large amplitude period agree with the calculated (numerical) periods.

Things to Investigate:

- Check if the measured values for the small oscillation periods can be used to modify the values of the constants b and n so that the periods predicted by the equation agree more accurately with the measured values.

- In the provided Mathematica file X06, change the values of b or n to make the theoretical large amplitude periods conform more accurately to the measured values for the periods.

- Change the model (equation) to account for dissipative forces (sliding or air friction). Alter the Mathematica file so that the predicted amplitudes of the damped vibrations agree with the experimental values.

- Place the whole apparatus on a wheeled platform that can be vibrated back and forth by a strong motor. This motion can be modeled by introducing a sinusoidal driving force which should permit an investigation of possible chaotic motion.

Experimental Activity 7

Iron Core Inductor

Comment: This activity should not take more than two hours to complete.

Reference: Section 2.3.1

Object: To produce current oscillations in a nonlinear tank circuit and to determine the period of these oscillations as a function of the current amplitude.

Theory: One of the most easily identified characteristics of all nonlinear oscillators is that the period varies with the amplitude of the oscillations. This activity attempts to illustrate and verify this behavior. The nonlinear oscillations are created using the nonlinear tank circuit shown in Figure 7.1. The circuit consists of a high voltage capacitor (C) connected to an iron core induc-

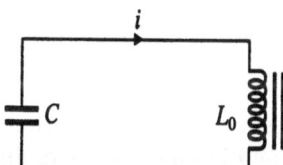

Figure 7.1: A nonlinear tank circuit.

tor. It is the inductor's iron core that makes the circuit behave in a nonlinear manner. To understand how this nonlinearity arises, assume that the inductor is connected to a current source. As the source current increases, the current through the inductor increases, and at first the strength of the magnetic flux keeps pace. However, as the current increases even further, the linearity between current and flux no longer holds. The reason for this loss of linearity is that the magnetic field in the inductor's iron core begins to approach its magnetic saturation limit. Further increases in current produce very small increases in the flux. A plot of the inductor's current i vs. the inductor's flux ϕ will look something like that shown in Figure 7.2. The equation that models the current–flux relationship for this circuit is given by

$$i = \frac{N\phi}{L_0} + A\phi^3,\tag{7.1}$$

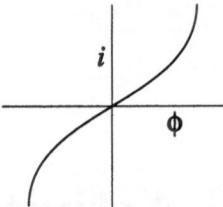

Figure 7.2: Current (i) vs. flux (ϕ) for an iron core inductor.

where L_0 represents the self-inductance with the core present, N is the number of turns in the coil, and A is a constant dependent on the composition and construction of the core.

Using the circuit shown in Figure 7.1, assume that at $t = 0$ seconds the capacitor (C) is fully charged and no current is flowing. As the capacitor begins to discharge, a time varying current i flows in the tank circuit. The potential drops must sum to zero, so

$$\frac{q}{C} + N\frac{d\phi}{dt} = 0. \tag{7.2}$$

Differentiating Equation (7.2) to find the current gives

$$i = -NC\frac{d^2\phi}{dt^2}, \tag{7.3}$$

and if Equations (7.3) and (7.1) are combined, they produce

$$\ddot{\phi} + \alpha\phi + \beta\phi^3 = 0, \tag{7.4}$$

where $\alpha = 1/(L_0 C)$ and $\beta = A/(NC)$. Equation (7.4) is an undamped "hard" spring nonlinear ODE. It has an analytical solution which can be expressed in terms of an elliptic function and has a period of oscillation which decreases with increasing amplitude.[1]

Since the flux ϕ is difficult to measure experimentally, this activity checks the validity of the theory by measuring the current vs. time rather than flux vs. time. The flux is related to the current by Equation (7.1).

Procedure:
Be careful in this activity as high voltages are present which can produce severe electrical shocks. Do not touch the charged capacitor with your fingers and always make sure the capacitor is completely discharged before adjusting the circuit.

1. Wire the circuit shown in Figure 7.3. Use a demonstration transformer for the iron core inductor. Demonstration transformers come with a variety of interchangeable low resistance coils and have a removable or adjustable section of the core. This permits a large number of permutations and combinations of field strengths and inductances to be explored. (The

---

[1]See Section 5.2.4.

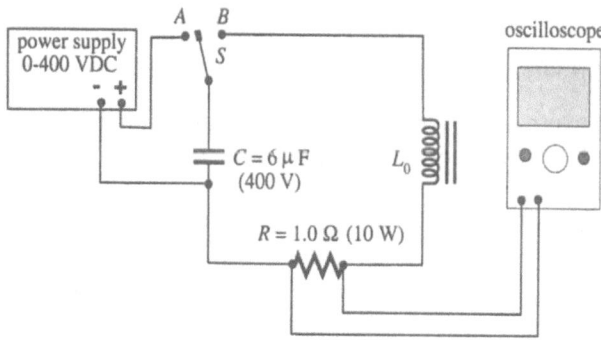

Figure 7.3: The apparatus.

transformer used in this experiment was purchased from CENCO and was listed in their 1995 catalog as the Modular Transformer System, U-core and Yoke #56211T. The coil we used had a stated inductance of 36 mH (no core) and with the core present, $L_0 \approx 1.0$–4.0 H. The inductor had a resistance of 9.5 Ω and consisted of 1000 turns of wire.) The 1.0 Ω resistor is used to sample the signal and since its resistance is 1.0 Ω, the value of the potential drop measured across the resistor equals the value of the current flowing through it. The total resistance of the circuit, due mainly to the inductor, damps the oscillations and permits the period vs. amplitude dependence to be found from a single trace on the cathode ray tube (CRT). Set the movable part of the iron core as shown in Figure 7.4. A storage oscilloscope permits the trace to be retained and studied in a more detailed and leisurely manner.

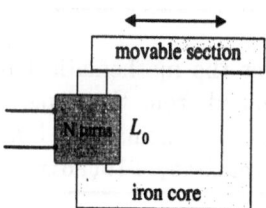

Figure 7.4: The transformer used as the iron core inductor.

2. With the switch at A, charge the capacitor to 5 volts, and then move the switch to position B. Measure the resulting trace's period (T_0) and calculate the angular frequency. Using this angular frequency calculate the value for the circuit's inductance L_0 by using $\omega = \frac{1}{\sqrt{L_0 C}}$.

3. With the power supply voltage set to 50 volts, charge the capacitor and then move the switch from A to B. By observing the amplitude and period of the oscillations, determine whether the decaying oscillations are non-linear. You might wish to use the 10x beam expander on the oscilloscope

to make the trace's pattern more discernible.

4. Repeat for larger and larger voltages. 50 volt steps is suggested.

5. At what voltage do you first see identifiable nonlinear behavior? The shape of the current vs. time trace on the CRT should be similar to that shown in Figure 7.5.

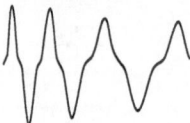

Figure 7.5: Current vs. time trace.

6. If the signal decays too rapidly, try adjusting the movable part of the core to change the size of the overlap region.

7. Examine the CRT trace to confirm that as the amplitude decreases the period increases.

8. Using the oscilloscope trace, record the values for the decreasing amplitude and the lengthening period.

9. Make a graph of the period (T) vs. the current amplitude (i_0). What is the mathematical relationship between the period and amplitude?

10. Mathematica file X07 produces a plot similar to that shown in Figure 7.5. Does the same period–amplitude relationship hold for the Mathematica plot as that found experimentally?

Iron Core Inductor

This file checks whether the mathematical relationship between period and amplitude, determined from the experimental data, is reproduced from similar data given by Mathematica . Mathematica commands in file:
D, NDSolve, Plot, Evaluate, PlotPoints, PlotStyle, TextStyle

11. Try to find the approximate experimental values that can be substituted into this Mathematica file so that they model your results. Do these results confirm or repudiate the nonlinear model given in the theory?

Things to Investigate:

- Repeat the activity using a larger capacitor, e.g., 10 μF. Make sure the voltage rating of the new capacitor is at least 400 V.

- For a set power supply voltage, explore how the CRT signal changes as the core's flux linkage is altered by moving the core. Start with a small core overlap and increase the overlap in small steps. Which position gives the lowest damping?

Experimental Activity 8

Nonlinear LRC Circuit

Comment: This is an easy investigation to perform. It should not take more than two hours to complete.

References: Section 2.3.1

1. [FJB85] This article contains circuits which model both the soft and hard spring oscillators.

Object: To produce a piecewise linear (nonlinear) LRC circuit that models a "soft" spring oscillator and to investigate the relationship between the period and the amplitude of the oscillator.

Theory: A 1-dimensional mechanical soft spring oscillator has a restoring force whose magnitude is given by the expression

$$F = ax - bx^3 \tag{8.1}$$

where a and b are positive constants and x is the displacement of the vibrating object from its equilibrium position. If damping is included ($c \neq 0$), the differential equation describing the object's motion is

$$m\ddot{x} + c\dot{x} + ax - bx^3 = 0 \tag{8.2}$$

or

$$\ddot{x} + 2\gamma\dot{x} + \alpha x - \beta x^3 = 0, \tag{8.3}$$

with $\gamma = \frac{c}{2m}$, $\alpha = \frac{a}{m}$, and $\beta = \frac{b}{m}$.

Oscillations produced by soft springs have periods that increase as the amplitude increases. It is this characteristic that will be used to identify the presence of soft spring oscillations in the electrical circuit used in this activity.

An electrical circuit that has its oscillations governed by an analogous equation to that of (8.3) is shown in Figure 8.1. To understand how the circuit functions, we use Kirchhoff's voltage rule that says that the algebraic sum of the potential drops around the closed loop must sum to zero. Therefore

$$V_L + V_R + V_C = 0 \tag{8.4}$$

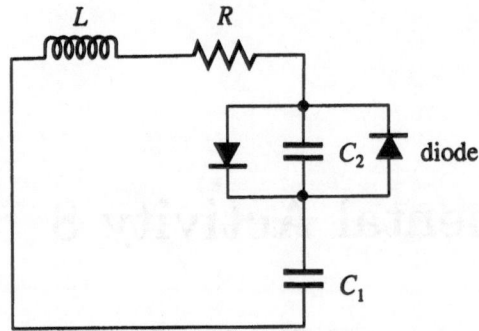

Figure 8.1: Circuit used to produce a soft spring characteristic.

where V_L is the potential drop across the inductor, V_R is the voltage drop across the resistor, and V_C is the voltage drop across both capacitors. Equation (8.4) is equivalent to

$$L\ddot{q} + R\dot{q} + V_C(q) = 0 \tag{8.5}$$

where q is the electric charge. In the above equation, the value for V_C will vary according to the amount of charge on the capacitors C_1 and C_2. A plot of the voltage across both capacitors as a function of the charge on the capacitors is as shown in Figure 8.2. The value of the slope of each linear region is equal to

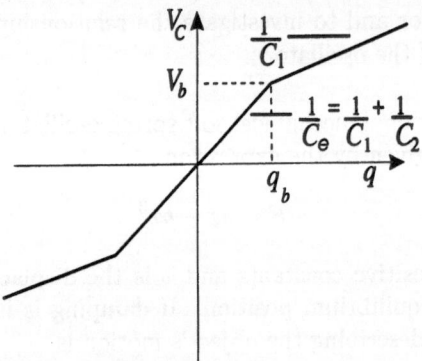

Figure 8.2: Voltage vs. charge across both capacitors.

the reciprocal of the capacitance in that region. The reason for this piecewise linear shape can readily be understood by noting the diodes placed across C_2. When the potential drop across C_2 is lower than a certain voltage—about 0.70 volts for silicon diodes—the diodes do not conduct. In this instance, the diodes act as open switches and so the circuit's equivalent capacitance (C_e) is that of the two capacitors in series, so

$$\frac{1}{C_e} = \frac{1}{C_1} + \frac{1}{C_2}. \tag{8.6}$$

As the charge on the capacitors increases, the potential drop across C_2 will reach 0.7 volts and the diodes begin to conduct. As long as the voltage remains higher than 0.7 volts, the diodes act as closed switches and effectively remove the capacitor C_2 from the circuit. The circuit's equivalent capacitance increases or the slope of the line shown in Figure 8.2 decreases to the value of $\frac{1}{C_1}$. Accordingly, the straight line bends downward at $V_b \approx 0.70$ volts. Remember, the slope is the reciprocal of the capacitance. The shape of the piecewise plot shown in Figure 8.2 has now been explained. In Equation (8.5), the form of $V_C(q)$ is given by the piecewise function

$$V_C = q \begin{cases} C_e^{-1} & |q(t)| < q_b \\ C_1^{-1} & \text{otherwise} \end{cases}. \tag{8.7}$$

To show how the piecewise function in Figure 8.2 might produce behavior similar to that of a "soft" spring, the three straight line segments are approximated by a continuous function of the form

$$V_C = aq - bq^3. \tag{8.8}$$

Plotting Equation (8.8) gives Figure 8.3 which has a shape similar to Figure 8.2

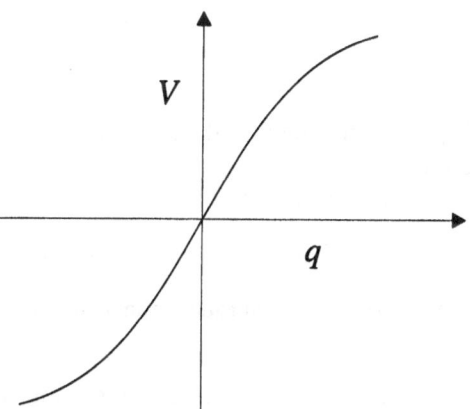

Figure 8.3: Voltage as function of charge.

for a certain limited range of values for the charge q. The reason for this restriction is that if Equation (8.8) is plotted over a large range of q, the curve would turn over at large $|q|$. The piecewise function never does.

Substituting (8.8) into (8.5) gives the ODE for a soft spring,

$$L\ddot{q} + R\dot{q} + aq - bq^3 = 0. \tag{8.9}$$

or

$$\ddot{q} + 2\gamma\dot{q} + \alpha q - \beta q^3 = 0, \tag{8.10}$$

with $\alpha = \frac{a}{L}$ and $\beta = \frac{b}{L}$.

Procedure:

1. Construct the circuit shown in Figure 8.4. The circuit's inductor should have an inductance between 0.80 H, (Berkeley solenoid) and 0.0040 H

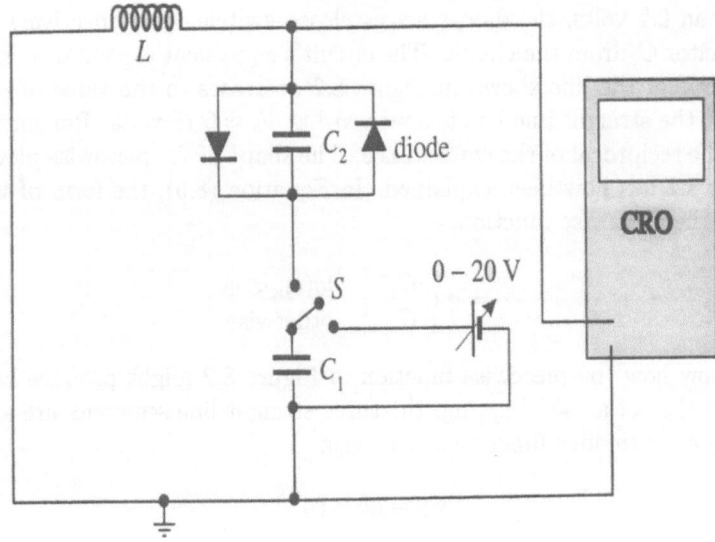

Figure 8.4: Soft spring circuit.

(PSSC solenoid). Perform the initial run with two equal 0.47 μF capacitors. The resistance R shown in Figure 8.1 is provided by the resistance of the wires, diodes, and the solenoid. This resistance damps the oscillations.

2. Set the source voltage at 2 volts and charge the capacitor C_1.

3. Move the switch so that the charged capacitor discharges through the circuit.

4. Observe the trace on the CRT. Can you detect any nonlinearity or the presence of a piecewise function?

5. Repeat the above steps for 4 V, 6 V, 8 V, 10 V,..., 20 V.

6. Select a voltage which produces a CRT trace that gives a clear indication of oscillations occurring in both linear regions. What tells you that the trace originates from a piecewise function rather than a "soft" spring function? Explain the difference.

7. Measure the period and amplitude of the oscillations in both linear regions. Which region has the larger period? What is this period change telling you?

8. Notice that there is a small region where the trace gradually changes from one region to the other. See if you can locate one or two oscillations in the change-over region. What are the amplitudes and periods of these oscillations?

9. Using the given values of the capacitors and the bend voltage, construct a graph similar to that shown in Figure 8.3. Use Mathematica file X08A

to transform the graph to its "soft" spring form. What values does the Mathematica file give for the coefficients a and b? What ratio of C_1 to C_2 would produce a better fit between the soft spring and piecewise linear approximation?

Nonlinear LRC Best Fit
When values for the total electric charge (q) and the potential drop (V) across the capacitors are placed into the two lists, Mathematica gives a best fit equation for the data. Mathematica commands in file: `Transpose`, `ListPlot`, `Fit`, `Plot`, `Show`, `PlotStyle`, `PointSize`, `TextStyle`, `AxesLabel`, `Hue`, `ImageSize`

---

10. Using the values for a and b just determined, calculate the values for α and β defined in Equation (8.10). To compare Mathematica 's approximation with the actual data, a plot of the data points and of the polynomial is constructed on the same graph.

11. After the values for α and β are known, use Mathematica file X08B to solve Equation (8.10) numerically and plot the solution.

Nonlinear LRC Circuit
This file numerically solves and plots the solution of the Duffing equation for the soft spring case. Mathematica commands in file: `NDSolve`, `MaxSteps`, `Plot`, `Evaluate`, `AxesLabel`, `PlotPoints`, `PlotLabel`, `Hue`, `PlotStyle`, `TextStyle`, `ImageSize`, `FontFamily`, `FontSize`, `Ticks`, `Automatic`

---

12. Compare the experimental amplitudes and periods with those produced by the Mathematica file.

13. Use Mathematica file X08C to solve Equation (8.5) numerically with the piecewise expression (8.7) inserted. Plot $q(t)$ vs. t.

LRC Piecewise
This file makes use of Mathematica's `UnitStep` command to correctly solve and plot the solution to the piecewise linear circuit. Mathematica commands in file: `UnitStep`, `Plot`, `Frame->False`, `AxesLabel`, `NDSolve`, `PlotStyle`, `RGBColor`, `FrameTicks`, `PlotLabel`, `TextStyle`, `Ticks`, `ImageSize`, `PlotRange`, `Hue`, `MaxSteps`, `PlotPoints`, `FontFamily`, `Epilog`, `FontSize`

---

14. Repeat all of the above steps for a value of C_1 twice the value of C_2.

15. Repeat the above steps for a value of C_1 five times the value of C_2.

Things to Investigate:

- Use the circuit shown in Figure 8.5 to produce a larger knee or bend voltage. (This method of producing a nonlinearity is also used in Ex-

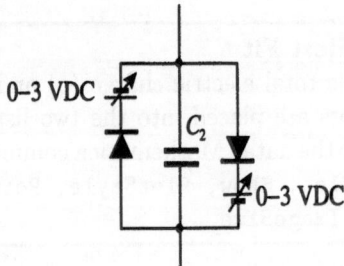

Figure 8.5: Batteries are used to increase the bend voltage.

perimental Activity 30, Chua's Butterfly.) Does this larger bend voltage produce a more easily discernible nonlinearity?

- Repeat the experiment with three or more capacitors. The capacitance of each capacitor should have a different value. All but one of the capacitors should have diodes placed in parallel with them.

- Use three or more identical capacitors but construct different bends by using the set-up shown in Figure 8.5. Each parallel diode circuit should use different potentials (battery values) to create more steps.

Experimental Activity 9

Tunnel Diode Negative Resistance Curve

Comment: An easy activity taking about one hour to complete.

References: Section 2.3.2

1. [BN92] Covers the basic ideas of electronics & solid state circuit devices.

Object: To determine the characteristic I–V curve of a tunnel diode.

Theory: Semiconductor devices, although they are usually made to operate in a linear region, are inherently nonlinear. This activity explores the performance of one of the strangest of all the nonlinear semiconductor devices, the tunnel diode. The reason that tunnel diodes are so strange is that their I–V curves have a negative resistance region. In order to help the electronics neophyte understand the construction and operation of diodes, a brief introduction to normal and tunnel diodes is presented below. If the reader is familiar with the theory and operation of ordinary/tunnel diodes, you may skip to the procedure.

A. Ordinary Diodes Semiconductors are made from elements, e.g., silicon or germanium, that have four electrons in their outermost (valence) shell. To make intrinsic (pure) semiconductors conduct better, small amounts of valence 3 or valence 5 elements are added to their crystal lattice. When these elements are appropriately added, it is known as doping the semiconductor. The rest of this theory explains how these two different doping elements can be used to produce two different types (p, n) of conduction (materials) from intrinsic semiconductors and how diodes are constructed from them.

Figure 9.1 represents an atomic model of a pure silicon (Si^{4+}) semiconductor. Silicon atoms prefer to have eight valence electrons rather than the normal four in their outer shells, so each atom bonds covalently by sharing one of its electrons with its closest neighbor. In Figure 9.1 each straight line represents a bond and a shared electron.

If the silicon has a small percentage of its crystalline atoms replaced (doped) by valence 3 atoms, such as boron (B^{3+}), empty bonding sites (holes) are created. When the semiconductor is doped to produce holes, it is known as a p-type

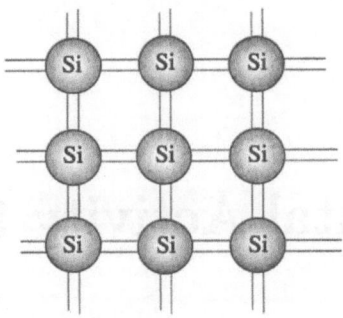

Figure 9.1: The straight lines joining the atoms represent shared electrons.

semiconductor. Figure 9.2 shows a schematic model of a *p*-type semiconductor. The holes can accept stray electrons that drift or wander in from the neighbor-

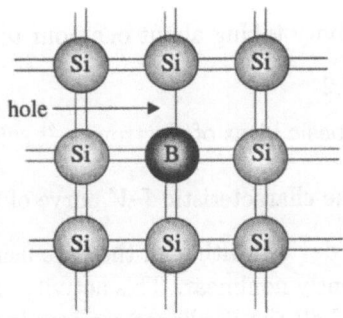

Figure 9.2: Boron, a hole increasing dopant, has been added to the silicon.

ing atoms. An electron leaving one bond to migrate to an empty hole makes it appear as if a hole is moving in the lattice in the opposite direction. The movement of a hole is quantum mechanically equivalent to the movement of a positive charge. As an everyday analogy of the movement of a hole, consider a line of cars stopped at a red light. Assume that between one of the cars and the next stopped car is a large gap. When the light changes to green and the cars (electrons) start to move forward, the gap (hole) appears to move backward. In *p*-type semiconductors it is more conventional to think of the holes moving rather than the electrons.

If a valence 5 atom such as phosphorus (P^{5+}) is used to dope the silicon atoms, free electrons are created when the bonds accept only four electrons per atom and reject one of the donor's five electrons. Figure 9.3 shows an example of this kind of doping. These free electrons, like the positive holes in *p*-type material, are available to drift through the semiconductor. Semiconductors with free electrons are known as *n*-type semiconductors.

A diode is created when a piece of *p*-type material is joined to a piece of *n*-type material. When this joining is properly done, the free electrons in the *n*-type material drift across the *p–n* junction and fall into the holes that exist

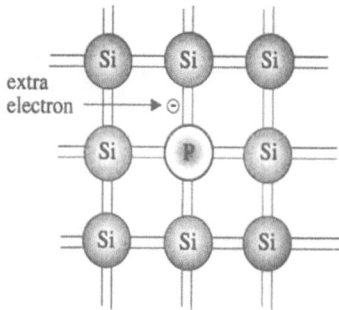

Figure 9.3: Phosphorus, an electron increasing dopant, added to the silicon.

near the junction. This depletes the number of holes near the junction and reduces the number of free electrons available for further depletion. The electron migration causes the net charge of the p-type material to become negatively charged near the junction and at the same time causes the n-type material to become positively charged near the junction, thus producing an electric field $\vec{E}$ as shown in Figure 9.4. Because the region on both sides of the junction

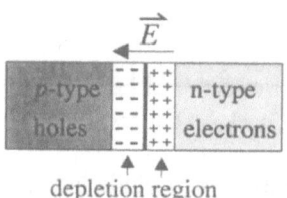

Figure 9.4: A junction p–n diode showing the depletion region.

is now depleted of its majority carriers, it is known as the depletion region. Examining Figure 9.4 shows that the direction of this created electric field acts to hinder further movement of electrons or holes across the junction. As the depletion region and net charge in the region continue to grow, the strength of the electric field and the junction potential increases. The strength of the electric field increases until it is strong enough to stop the migration of electrons across the junction. For silicon, a junction potential of around 0.70 volts[1] (Ge, 0.35 volts) is produced when equilibrium is established.

The circuit symbols for a diode and a tunnel diode along with the shapes of the common enclosing cases are shown in Figure 9.5. The diode's arrow represents the p-type (the anode) material and the vertical bar the n-type material (the cathode). Another way of remembering this convention is that the conventional current flows more easily in the direction of the arrow. Cylindrical diodes have their cathode end marked by a line which encircles the cylinder.

If a normal diode is connected to an external voltage source with the polarities as shown in Figure 9.6, the electric field across the diode's junction is

---

[1]This 0.70 volts is the diode switch voltage mentioned in the last activity.

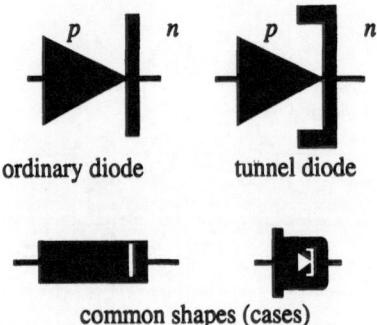

Figure 9.5: Diode and tunnel diode circuit symbols and common cases.

increased. This acts to widen the diode's normal depletion region and makes it even harder for current to cross the junction. A diode wired in this way is said to be in reverse bias and in this mode no current flows in the circuit.

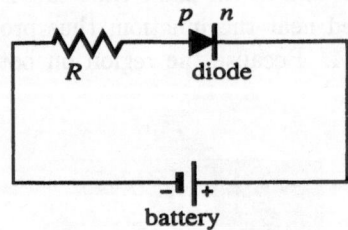

Figure 9.6: A *p–n* diode in reverse bias.

If a normal diode is connected to an external battery with the polarities as shown in Figure 9.7 it is said to be in the forward bias mode. In the forward bias

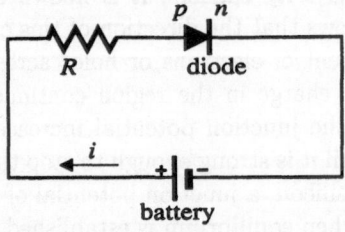

Figure 9.7: A *p–n* diode in forward bias.

the external voltage source, if large enough (> 0.7 V for Si, > 0.35 V for Ge), overcomes the internal junction potential. When this occurs, large numbers of electrons and holes can continuously move across the junction. Sustained electron and hole flow across the junction will produce a relatively large current in the external circuit. For the circuit shown in Figure 9.7, if the battery is replaced by an alternating voltage source, the current will be rectified (becomes a pulsating DC current). The current is large when the AC polarity produces a forward bias, and there is no current when the AC polarity produces a reverse bias. This signal rectification is an important use of an ordinary diode.

B. Tunnel Diodes Tunnel diodes, although constructed similarly to ordinary diodes, function quite differently and are usually made of germanium instead of silicon. Leo Esaki[2] introduced the tunnel diode in 1958. Ordinary diodes have doping concentrations of about 10^{20} m^{-3}, while tunnel diodes have doping concentrations 10,000 times larger which creates many more holes and free electrons in the respective n and p semiconductors and a much smaller depletion region (10^{-8} m instead of 10^{-6} m). With a smaller depletion region, application of the correct forward bias propels the electrons toward the energy barrier created by the depletion region and the electrons are able to quantum mechanically tunnel through, rather than jump over, the barrier. Because this tunneling takes place near the speed of light, a tunnel diode has the capability of being used as a very fast switching device. However, this high speed switching can cause problems for circuit designers because at high frequencies the stray capacitances and inductances contained in the wires and contact points become more important and may introduce unwanted signals into the circuit.

The solid line shown in Figure 9.8 is a sketch of the current (I_d) vs. the potential difference (V_d) for a tunnel diode. The segment OPV represents the

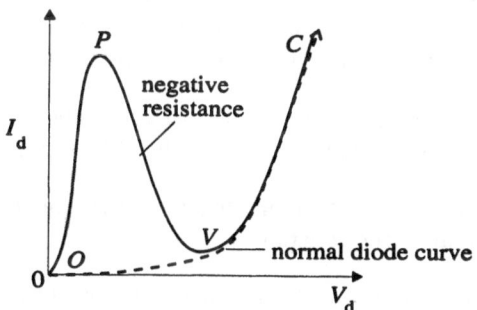

Figure 9.8: Current vs. potential for a tunnel diode.

tunneling portion of the curve while the negative slope segment PV is the negative resistance region. The dashed line represents the current–voltage curve for an ordinary germanium diode. Different tunnel diodes will have different operating characteristics but the shapes of the curves are similar. The main purpose of this experiment is to generate a plot similar to that shown in Figure 9.8. Obtaining this curve makes Experimental Activity 10 easier to perform.

Procedure:

1. Wire the circuit shown in Figure 9.9. The circuit uses the tunnel diodes[3] IN3718 or IN3719, but other tunnel diodes could be used.

---

[2] Esaki, Giaever, and Josephson shared the 1973 Physics Nobel prize for their work on quantum mechanical tunneling in semiconductors.

[3] Tunnel diodes may be purchased from Germanium Power Devices Corporation, 300 Brickstone Sq., Andover, MA 01810 (508-475-1512).

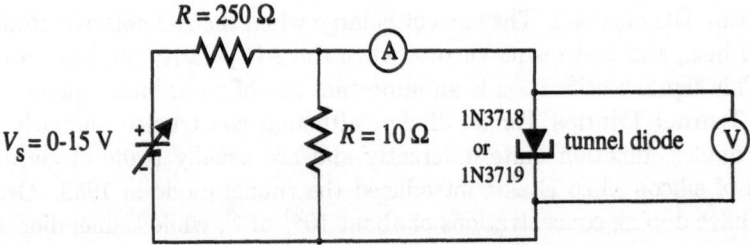

Figure 9.9: Circuit used to produce tunnel diode response curves.

2. Slowly increase the supply voltage (V_s), recording the potential drop across, and the current through, the diode. Tunnel diodes are easily damaged so be careful not to exceed the manufacturer's recommended maximum current (50 mA for the diodes 1N3718 or 1N3719).

3. As the potential drop across the diode is slowly increased, don't be alarmed when a further increase in voltage produces a decrease in current. You have entered the region of negative resistance.

4. Keep increasing the source voltage until the diode is back in its positive resistance range. Continue to record the potential drop across, and the current through, the diode. Do not exceed the diode's maximum current.

5. Plot the values for V_d along the abscissa and the values for I_d on the ordinate. Your curve should look somewhat like Figure 9.8.

6. What is the average value of the negative resistance? The reciprocal of the slope is equal to the resistance.

7. What is the difference between your curve and the one shown in Figure 9.8?

Things to Investigate:

- Replace the 10 Ω resistor with a 50 Ω resistor and rerun the experiment. Explain what happens when the peak voltage is exceeded. After this voltage is reached, slowly decrease the source voltage and watch for any sudden jumps in the measured values. What replacement resistance would stop this hysteresis?

- Assuming the tunnel diode curve is $I_D = A_1 V^3 + A_2 V^2 + A_3 V$, determine the values for the constants. If you have used the previous best-fit Mathematica files, you might wish to use one of these files to help with this task.

- Assuming the reference point is moved to the midpoint of the negative resistance portion of the curve, the equation can be approximated by $i = -av + bv^3$. What are the values for the constants a and b?

Experimental Activity 10

Tunnel Diode Self-Excited Oscillator

Comment: This is an easy experimental activity and should not take more than two hours to complete once the circuit is wired. Doing this experiment will help you understand many of the concepts in the text, e.g., limit cycles, self-excited oscillations, relaxation oscillations, etc.

References: Section 2.3.2

1. Experimental Activity 9: Tunnel Diode Negative Resistance Curve.

2. [BN92] An electronics text with an excellent section on tunnel diodes.

Object: To investigate self-excited oscillations governed by the Van der Pol (VdP) equation with special attention to the changes in shape of the oscillations as one of the control parameters is varied.

Theory: This theory is relatively short because the development of the Van der Pol (VdP) equation governing the oscillations produced by this experimental activity is found on the provided Mathematica file X10. This file also permits a comparison of the experimental results with the theoretical predictions as various control parameters are changed.

Tunnel Diode Oscillator
This file develops and numerically solves the nonlinear ODE governing the tunnel diode circuit of this activity. The numerical solution is plotted. The development is similar to that in Example 2-3. Mathematica commands in file: `D`, `NDSolve`, `Plot`, `Evaluate`, `PlotPoints`, `AxesLabel`, `Ticks`, `ImageSize`, `PlotStyle`, `Thickness`, `TextStyle`, `Expand`, `Collect`, `FontFamily`, `FontSize`, `FontWeight->Bold`

This experimental activity uses the circuit of Example 2-3. The characteristic I–V curve for a tunnel diode is shown in Figure 10.1. Experimental Activity 9 shows how to produce this curve. An approximate equation for the I–V curve

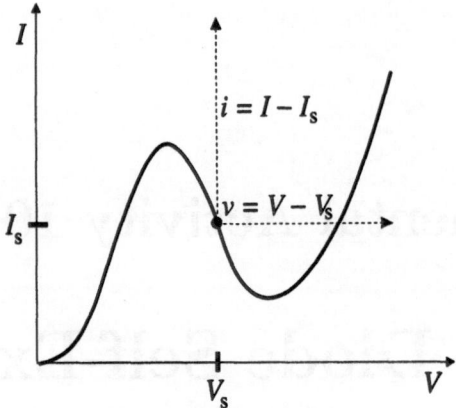

Figure 10.1: Tunnel diode I–V curve.

in the neighborhood of the operating point I_s, V_s is given in the text as

$$i = -av + bv^3 \tag{10.1}$$

where $a = 0.050$ and $b = 1.0$ for the tunnel diode 1N3719. Using Kirchhoff's current and voltage rules produces the unnormalized VdP equation

$$\ddot{v} - (\alpha - \beta v^2)\dot{v} + \omega^2 v = 0 \tag{10.2}$$

where $\alpha = \frac{1}{C}(a - \frac{1}{R})$, $\beta = \frac{3b}{C}$, and $\omega = \frac{1}{\sqrt{LC}}$. In this activity the resistance R is varied to see how this control parameter affects the shape of the oscillations.

Procedure:

1. Wire the circuit as shown in Figure 10.2.

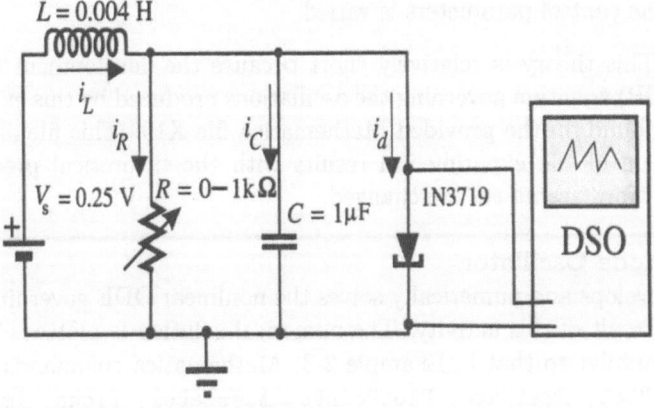

Figure 10.2: The circuit.

When using tunnel diodes, parasitic capacitances and inductances can cause the frequencies to be much higher than expected. So regard this experiment as providing qualitative confirmation of the basic theoretical concepts developed in the Theory part. If a different tunnel diode is used, the values for a and b will have to be calculated. The variable resistor should allow you to change its resistance continously and accurately.

IMPORTANT

- The power supply must be able to deliver a constant voltage of $V_s = 0.25V$. Older power supplies may not be able to maintain a constant output voltage.

- The inductor must have a low resistance so that the circuit does not contain a resistance in the wrong location. The PSSC air core solenoid which has an inductance of 0.004 H and a resistance of less than 1.0 Ω works fine.

2. Turn on the circuit and adjust the power supply to the correct voltage. Set the variable resistance R to about 1 kΩ.

3. Adjust the digital storage oscilloscope (DSO) until a trace showing an oscillation appears on the CRT.

4. Slowly decrease R to see what affect it has on the shape of the CRT trace. As R decreases, does the CRT trace better approximate a sine wave?

5. Make a sketch of the wave shapes for four widely separated R values.

6. What is the critical value for R that makes the trace disappear?

7. Use Mathematica file X10 to compare the CRT trace with that predicted by theory.

Things to Investigate:

- Why does the circuit continue to oscillate when the capacitor is removed from the circuit? What happens to the frequency? What does this tell you about the internal construction of the tunnel diode?

- Repeat the above experiment with different inductors.

- Examine Figure 7.4 in the accompanying text to see what a relaxation oscillation looks like. Does this experiment produce this shape, and if so for what values of R?

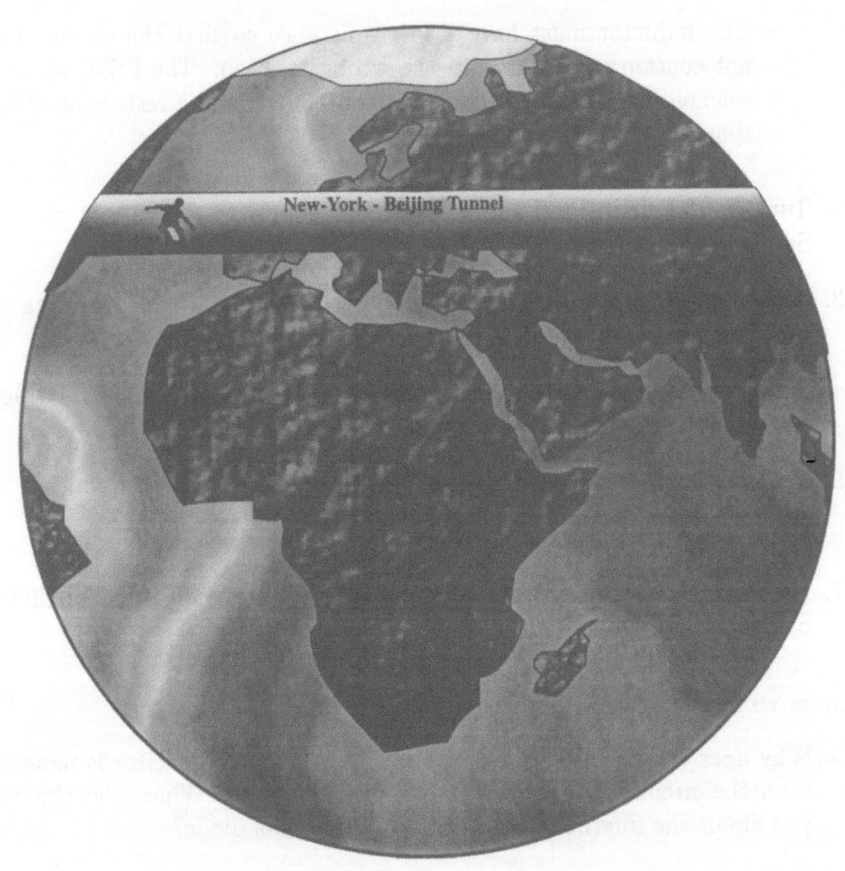

The Ultimate Tunnel Oscillator

Experimental Activity 11

Forced Duffing Equation

Comment: This activity should not take more than two hours to complete.

References: Section 3.3.1

1. [Bri87] This article contains a similar experiment.

2. [DFGJ91] This article contains a discussion of an inverted pendulum and how it can be used to model the Duffing equation.

3. [Pip87] This excellent book has a detailed analysis of the inverted pendulum and as a bonus contains a large number of ideas for additional experiments.

Object: To use an inverted pendulum to investigate the motion, especially the chaotic motion, of an object governed by the forced Duffing equation.

Theory: The importance of the Duffing equation in nonlinear physics cannot be overstated. For example, the accompanying Theory part uses the Duffing equation in a variety of ways and for a variety of reasons:

1. The simplicity of the Duffing equation is deceptive in that the physical behavior it models is remarkably complex;

2. The Duffing equation is a generalization of the linear differential equation that describes damped and forced harmonic motion. This allows many analogies and comparisons to be drawn between the physical behavior modeled by these linear and nonlinear equations;

3. The Duffing equation appears in many disguises and aliases, e.g., forced, unforced, hard spring, soft spring, negative stiffener, and nonharmonic oscillator;

4. The Duffing equation can be used to demonstrate nonlinear behavior such as

   - the variation of the oscillation period with amplitude;
   - jump phenomena and hysteresis;

- bifurcation phenomena, including the period doubling route to chaos;
- the generation of very simple but instructive Poincaré sections.

The negative stiffener Duffing equation is studied in this activity. The negative stiffener is created by using the inverted pendulum as shown in Figure 11.1. The inverted pendulum is made from steel tape similar to that used to make

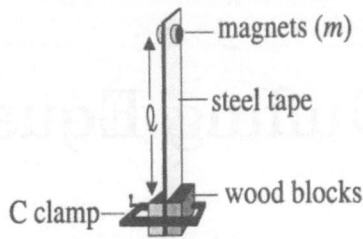

Figure 11.1: The inverted pendulum (Euler strut).

airtrack bumpers or to wrap shipping cartons. Two or more strong (neodymium) magnets are used to provide the mass (m) at the top of the pendulum. The magnets permit the effective length (ℓ) of the pendulum to be easily changed. Using Figure 11.2 as a reference, an approximate expression for the potential

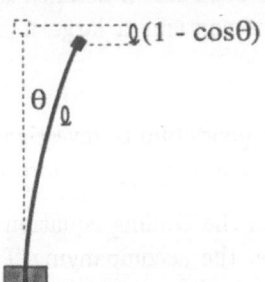

Figure 11.2: Inverted pendulum at angle θ.

energy of the pendulum is

$$V = \frac{k\theta^2}{2} - mgy, \tag{11.1}$$

where $y = \ell(1 - \cos\theta)$ and k is the stiffness constant of the steel spring. The potential energy is assumed to be zero ($V = 0$) at the top of the trajectory. If, in Equation (11.1), the $\cos\theta$ term is written as a power series

$$\cos\theta = 1 - \frac{\theta^2}{2!} + \frac{\theta^4}{4!} + \cdots, \tag{11.2}$$

the potential energy is approximately equal to

$$V = \frac{1}{2}(k - mg\ell)\theta^2 + \frac{1}{24}(mg\ell)\theta^4. \tag{11.3}$$

The terms with exponents greater than four are ignored because it is assumed that θ is small. Notice that in Equation (11.3) the coefficient of the θ^2 term can be positive or negative. A plot of the potential energy (V) vs. (θ) for a varying ℓ is shown in Figure 11.3. The double well curves represent values for ℓ that give

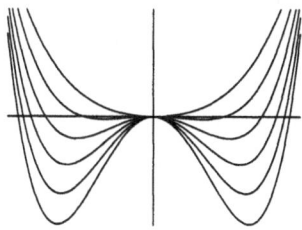

Figure 11.3: Potential (V) vs. angle (θ).

negative coefficients for the θ^2 term, i.e., for $mg\ell > k$. As ℓ increases, the depth of the double wells increases. The double well curves have an unstable saddle point at $\theta = 0$. Although this activity deals with motion under the control of a double well potential, if $mg\ell < k$, a single well potential characteristic of the hard spring Duffing equation is created.

Now that the approximate expression for the inverted pendulum's potential energy has been derived, the Lagrangian $L = T - V$ can be used to produce the differential equation that describes the motion. Neglecting damping and forcing,

$$L = \frac{m\ell^2\dot{\theta}^2}{2} - \frac{1}{2}(k - mg\ell)\theta^2 - \frac{mg\ell}{24}\theta^4, \tag{11.4}$$

and applying the Lagrange equation

$$\frac{d}{dt}\left(\frac{\partial L}{\partial \dot{\theta}}\right) - \frac{\partial L}{\partial \theta} = 0, \tag{11.5}$$

an approximate differential equation that describes the motion of an inverted pendulum is produced:

$$m\ell^2\ddot{\theta} + (k - mg\ell)\theta + \frac{1}{6}(mg\ell)\theta^3 = 0. \tag{11.6}$$

On adding a damping term $b\dot{\theta}$ and a forcing term $\tau_0 \cos(\omega_d t)$, where τ_0 is the maximum torque and ω_d is the driving frequency, Equation (11.6) becomes

$$m\ell^2\ddot{\theta} + b\dot{\theta} + (k - mg\ell)\theta + \frac{1}{6}(mg\ell)\theta^3 = \tau_0 \cos(\omega_d t). \tag{11.7}$$

Expressing Equation (11.7) in a more conventional form gives

$$\ddot{\theta} + 2\gamma\dot{\theta} + \left(\frac{k}{m\ell^2} - \frac{g}{\ell}\right)\theta + \frac{1}{6}\frac{g}{\ell}\theta^3 = \alpha_0 \cos(\omega_d t). \tag{11.8}$$

Equation (11.8) is the equation used to describe the motion in this activity.

This Activity has two procedures. Procedure A is simpler and does not require as many measurements as Procedure B.

Procedure A: Simpler of the Two Procedures

1. Construct the apparatus shown in Figure 11.4. If you do not have this equipment, there is an alternate way of forcing the pendulum. Figure 11.5 shows how a Helmholtz coil can be used to power the pendulum. The inverted pendulum can be made from the steel tape used for airtrack bumpers (recommended), from a hacksaw blade, or from the steel tape

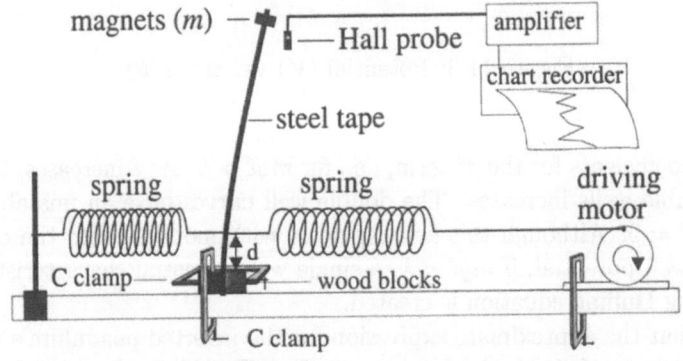

Figure 11.4: Apparatus for the inverted pendulum.

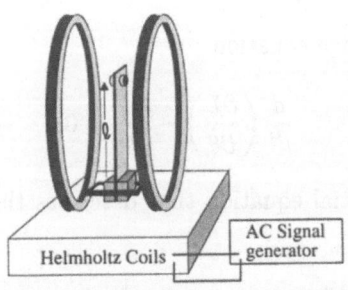

Figure 11.5: An alternate driving method.

used to wrap shipping cartons. A length between 0.20 m and 0.40 m is satisfactory.

2. Connect the two coiled springs to the inverted pendulum. A file can be used to make small v grooves along the edges of the tape. These grooves can be used to hold the springs at a fixed spot. Springs that have a spring constant of about 1.0 N/m are recommended. The 10cm long coiled springs that come with the airtrack are appropriate.

3. When the motor is not running, adjust the position of the magnets until the steel tape is just into its two-well mode. The two-well mode can be recognized by having the magnets stay, when placed, in either one or the other of the two equilibrium positions found on each side of the vertical center position. The apparatus shows two wood blocks clamping the pendulum at its base. The steel tape can be moved up and down between these blocks if adjustments to its length are required. Later, when the motor is running, small adjustments in the springs' tension can be made by sliding the motor back and forth along the table as needed.

4. Place the Hall probe near the top of the inverted pendulum's trajectory. This can be done by taping the Hall probe to a wooden meter stick and then moving the meter stick to the desired position.

5. The output from the Hall probe should be connected to a storage oscilloscope or to a paper chart recorder.

6. Start the motor running. Adjust the frequency or driving amplitude of the motor. A very low frequency ($\nu \approx 0.1 \leftrightarrow 1.0$ Hz) is probably required. After the motor is running you might wish to adjust its horizontal position to make the motion of the pendulum behave more symmetrically. Try to produce an amplitude that remains small enough to keep the Hall probe measuring a signal strength over the whole of the oscillation.

7. The pleasure gained in watching this weird and wonderful motion is, in itself, a worthwhile procedure, so if pressed for time just observe the motion. Attempt to identify period doubling, quadrupling, etc., and confirm your observations by examining the chart recorder or CRT trace.

8. Make a number of different chart recordings. Vary the driving frequency and driving amplitude.

9. It is easy to adjust the motor's frequency or amplitude to produce chaotic motion.

Procedure B: A More Quantitative Procedure

1. Use the same apparatus as in Procedure A.

2. Measure the mass of the small powerful magnets which are attached to the top of the steel tape.

3. Before connecting the coiled springs to the inverted pendulum, measure their stiffness constants (k_s). Strong springs $k_s \approx 10$ N/m should be used if a hacksaw blade is used for the steel strip. Weaker springs can be used if steel bumper tape is used to make the inverted pendulum.

4. Connect the coiled springs to the inverted pendulum.

5. Locate the critical point of the inverted pendulum. Measure d (the distance from the pivot to where the coiled springs are attached to the pendulum) and ℓ.

6. Calculate the linear damping coefficient (γ). An approximate value for the damping coefficient γ can be found by determining the time $t_{1/2}$ it takes the amplitude to halve. Then use the equation $\gamma = \frac{\ln(2)}{t_{1/2}}$ to calculate γ.

7. Calculate the value for the stiffness constant of the inverted pendulum by finding the critical point where the pendulum is moving from a single well to a two-well potential. At this point $k = mg\ell$.

8. Calculate the maximum force $F = k_s x_0$ where x_0 is the maximum amplitude of the driving motor and k_s is the coiled spring constant.

9. Calculate the maximum torque, $\tau_0 = Fd$, where d is the distance from the pivot to the point where the springs are attached to the inverted pendulum; see Figure 11.4.

10. Connect the Hall probe to the chart recorder or a digital storage oscilloscope.

11. With the inverted pendulum at rest in one of its equilibrium positions, start the motor. Try to produce small oscillations so that the Hall probe can detect the pendulum's magnetic field over the full range of its oscillation.

12. Use Mathematica file X11 to compare the chart recorder's plot with the plot produced by the file.

Forced Duffing Equation
Using the experimentally determined parameters, this file can be used to numerically solve and plot the solution of the forced Duffing equation. Mathematica commands in file: `NDSolve`, `MaxSteps`, `Plot`, `Evaluate`, `AxesLabel`, `PlotPoints`, `PlotLabel`, `Ticks`, `ImageSize`, `PlotStyle`, `TextStyle`, `FontFamily`, `FontSize`

Things to Investigate:

- Investigate the motion in the two different potentials (single well and double well), but without using the driving motor to force the motion. Can you detect any nonlinear behavior? What kind of behavior should you be looking for?

- Use different steel tapes to make the inverted pendulum.

- When the pendulum is in its single well mode (hard spring), does it exhibit hysteresis?

- Limit the amplitude of the oscillation by adding an additional nonlinear repelling potential. Fasten two small repelling magnets near the end points of the oscillations. Can you still detect nonlinear behavior?

- How is this experiment changed if the inverted pendulum is inverted? Is this equivalent to negative gravity?

Experimental Activity 12

Focal Point Instability

Comment: This activity should not take more than two hours to complete.

References: Section 4.2

1. [For87] Page 308 gives the circuit used in this activity.

2. [Pip87] This book contains a similar circuit to the one used in this activity. This is an excellent book, easy to read, and it contains many thought-provoking ideas.

Object: To investigate a focal point instability produced by an electrical circuit.

Theory: A damped undriven harmonic oscillator can be described by the following linear second-order homogeneous differential equation:

$$\ddot{q} + 2\gamma\dot{q} + \omega_0^2 q = 0, \tag{12.1}$$

where γ is the damping coefficient. For $0 < \gamma < \omega_0$, the solution is

$$q = Ae^{-(\gamma t)}\cos(\omega t + B), \tag{12.2}$$

where A and B are the integration constants determined by the initial conditions, and $\omega = \sqrt{\omega_0^2 - \gamma^2}$. A plot of Equation (12.2) is shown in Figure 12.1. The amplitude (q) of the oscillations decays exponentially as the time (t) in-

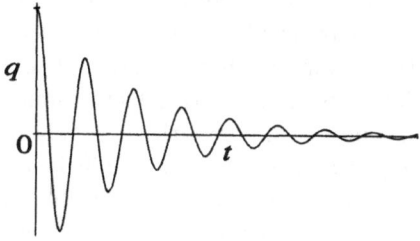

Figure 12.1: Positively damped ($\gamma > 0$) harmonic oscillator.

creases. For a positive damping coefficient, energy is continuously dissipated by the system.

For the less frequently encountered and more unlikely case involving a negative damping coefficient ($\gamma < 0$), the solution to Equation (12.1) for $|\gamma| < \omega_0$ is

$$q = Ce^{(|\gamma|t)}\cos(\omega t + D) \tag{12.3}$$

where C and D are integration constants. A plot of Equation (12.3) is shown in Figure 12.2. The amplitude (q) of the oscillations grows exponentially from some very small initial value as the time (t) increases. Energy is continuously absorbed by the system.

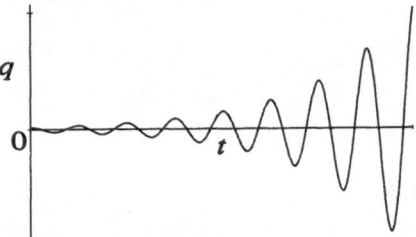

Figure 12.2: Negatively damped ($\gamma < 0$) harmonic oscillator.

Phase plane plots for both of these solutions are shown in Figure 12.3. For Equation (12.2), the trajectory winds onto a stable focal point (F) as shown in

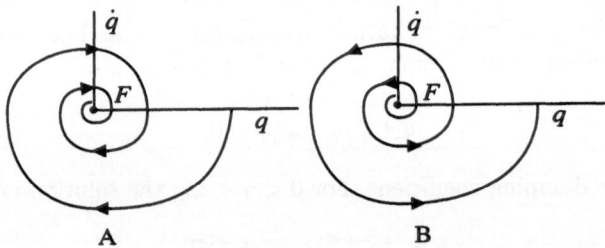

Figure 12.3: Phase plane trajectories for positive (A) and negative (B) damping.

plot A. Plot B is the phase plane plot for Equation (12.3) and illustrates that the oscillations unwind from an unstable focal point. The negative damping coefficient means that any movement, however tiny, acts as a seed to start the oscillations. Once the oscillations have started, their amplitude increases until the oscillator self-destructs or some other factor limits its growth. Real or actual examples which produce such a growth are rarely encountered in linear physics. Situations which give negative damping coefficients are hard to imagine. The presence of a negative damping coefficient indicates the oscillator absorbs energy from its surroundings and when encountered is very surprising, witness the collapse of the Tacoma Narrows Bridge. (Try to think of other physical examples of where γ is negative.) This activity provides an opportunity to investigate the phenomenon of a negative damping coefficient and its associated focal point instability.

In this activity an electric circuit is used to produce oscillations that behave as if a negative damping coefficient is present. The schematic of the circuit is shown in Figure 12.4. The circuit contains a positive resistance (R_e), a negative

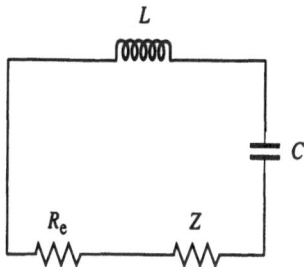

Figure 12.4: Circuit to produce an unstable focal point.

resistance ($Z = -r$), an inductor (L) and a capacitor (C). (Don't worry if you do not know how to purchase a negative resistance, all will be made clear.) Kirchhoff's voltage law says that the sum of the potential drops around the loop must sum to zero, so

$$V_L + V_{R_t} + V_C = 0, \tag{12.4}$$

where $V_L = L\frac{d^2q}{dt^2}$, $V_R = R_t\frac{dq}{dt}$, $V_C = \frac{q}{C}$, and $R_t = R_e + Z = R_e - r$ is the total resistance. Thus, the differential equation relating electrical charge (q) to time (t) is

$$\ddot{q} + \frac{(R_e - r)}{L}\dot{q} + \frac{1}{LC}q = 0. \tag{12.5}$$

Comparing Equations (12.1) and (12.5) gives the relationship $2\gamma = \frac{(R_e-r)}{L}$ and $\omega_0^2 = \frac{1}{LC}$. Now if r is greater than the positive resistance (R_e) the circuit is negatively damped and the oscillations, once started, will grow exponentially. If the initial conditions for the circuit are $q = 0$ and $\dot{q} = 0$, the circuit is stable and quiescent. However, if the initial conditions have even the smallest nonzero values, the charge begins to oscillate and the amplitude of the oscillations exponentially increases. The term "unstable focal point" is well chosen. In this experiment, circuit noise will provide the tiny signal needed to seed, or initiate, the oscillations.

The negative resistance $Z = -r$ in this activity is produced by an operational amplifier (op-amp) wired as shown in Figure 12.5. The triangular circuit symbol is the op-amp. To understand how an ideal op-amp functions, consider it to be a black box that operates under the control of a few simple rules:

1. The negative sign ($-$) on the inverting terminal of the op-amp does not indicate a polarity. It indicates that the output signal, if connected directly, is 180 degrees out of phase with the input signal;

2. The positive sign ($+$) on the noninverting input of the op-amp does not indicate a polarity. It indicates that the output signal, if connected directly, is in phase with the input signal;

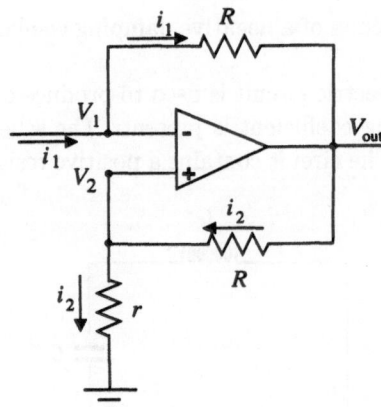

Figure 12.5: Op-amp circuit to produce negative resistance.

3. No current flows into the inverting $(-)$ or noninverting $(+)$ inputs. The ideal op-amp is considered to have an infinite input impedance;

4. The voltages at the inverting $(-)$ and noninverting $(+)$ inputs are equal. This is a consequence of the above rule;

5. An op-amp has no (or very low) output impedance.

The op-amp's power supplies are not shown in Figure 12.5.

To show how the op-amp acts as a negative resistance, consider Figure 12.5 and the above enumerated rules. The input impedance (Z) of the circuit is defined by the equation

$$Z = \frac{V_1}{i_1}. \tag{12.6}$$

The value for Z is to be found in terms of r. Since the op-amp draws no current, the current i_1 does not enter the op-amp but must travel through the top resistor (R) so

$$V_1 - V_{\text{out}} = i_1 R. \tag{12.7}$$

Solving Equation (12.7) for i_1 and substituting into Equation (12.6) gives

$$Z = \frac{V_1 R}{V_1 - V_{\text{out}}}. \tag{12.8}$$

The output voltage is given by

$$V_{\text{out}} = i_2 r + i_2 R \tag{12.9}$$

and since $V_1 = V_2$ (Rule 4), $i_2 = \frac{V_1}{r}$ and then Equation (12.9) becomes

$$V_1 - V_{\text{out}} = -\frac{V_1 R}{r}. \tag{12.10}$$

Substituting Equation (12.10) into (12.8) gives

$$Z = -r, \tag{12.11}$$

which shows that an op-amp wired in this configuration acts as a negative resistance with a magnitude r. It should also be noted that if r is replaced with a capacitor (inductor), the negative sign changes the impedance from a capacitance to an inductance or vice versa.

When the sub-circuit shown in Figure 12.5 is connected to a capacitor (C),

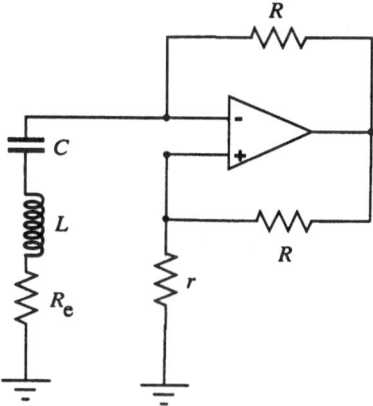

Figure 12.6: Equivalent circuit to Figure 12.4.

inductor (L), and resistance (R_e) as shown in Figure 12.6, the circuit is equivalent to the circuit shown in Figure 12.4. This equivalent circuit permits the total resistance to be easily and slowly changed from negative to positive. The speed with which the resulting oscillations exponentially grow or decay can be controlled by changing the value (positive or negative) of the total resistance.

Procedure:

1. Wire the circuit shown in Figure 12.7. Any dual inline 8-pin mini dip op-amp should work but μA741 is a nice choice. It is much easier if the circuit is mounted on a breadboard that comes with the +15 and −15 volt power supply. The values of the inductor and capacitor are not critical, but the solenoid should not have a ferromagnetic core. A digital storage oscilloscope makes it easier to study the trace.

2. Slowly adjust the variable resistor (r) until its value is approximately equal to the resistance of the external circuit. A dial resistance box, if graduated in small enough divisions, can be used for the variable resistor.

3. When the resistor (r) is larger than R_e the trace should explode onto the CRT. What is this critical value for r? The signal should reach a maximum value of around 28 volts.

4. If the resistance of r is just a bit larger than R_e, you will see a trace with a slow but exponential increase in the amplitude of the oscillation. With proper adjustment of r you should be able to make this growth take a relatively long time ($\approx$ 30 s or more) to reach its saturation voltage.

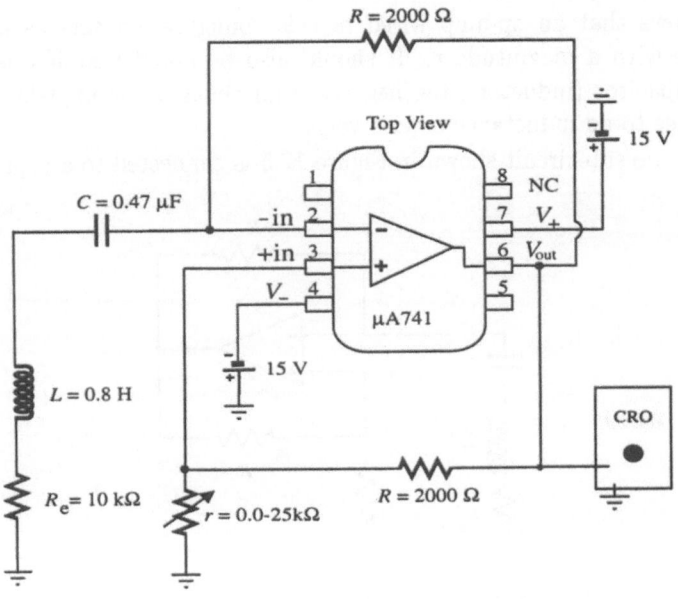

Figure 12.7: Circuit to produce unstable focal point.

5. With the trace in its saturated state, reduce r by a small amount and watch the exponential decay of the trace.

Things to Investigate:

- Have some fun with this very unusual circuit. Adjust the values any way you wish but keep the resistors marked R above 1500 Ω. This keeps the output current from exceeding the op-amp's specification of 25 mA.

- Try using an inductor with a ferromagnetic core. Can you detect any nonlinear effects when you do this?

- Connect a signal generator in series with the external circuit. Using a very low power signal and a negative resistance below the total external resistance, look for a signal trace that shows the transient and steady state solutions of a forced oscillator. (Make sure you include the output impedance of the signal generator in your total external resistance.)

- Explain why the signal generator should not be used to input a signal when the negative resistance is larger than the external resistance of the circuit.

- If you have access to a SPICE program such as MICRO-CAP try simulating the operation of the circuit shown in Figure 12.7 on the computer.

Experimental Activity 13

Compound Pendulum

Comment: This activity should take less than one hour to complete.

References: Section 5.2.4

1. [Cro95] This article provides many ideas for additional and more complex investigations of rigid rod oscillations.

Object: To measure the period of a pendulum that has a large initial amplitude.

Theory: In this experiment, the period of a pendulum swinging with a large amplitude is measured and then compared with the theoretical value predicted by Mathematica file X13.

Compound Pendulum Period

This file calculates the period of the compound pendulum (meter stick) shown in Figure 13.1 for a large initial angular amplitude and compares the result with that given by the small angle formula. Mathematica commands in file: HoldForm, Integrate, Sqrt, Sin, Pi, Part

To ensure that the pendulum swings in a circular arc, a compound pendulum (a meter stick) is substituted for the more traditional string and bob pendulum. Figure 13.1 is a sketch of the pendulum. If the air resistance and sliding friction

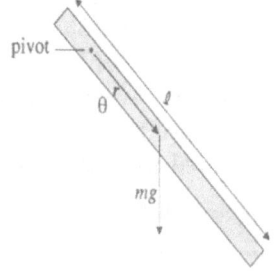

Figure 13.1: The compound pendulum.

at the pivot are ignored, the ODE for the torque on the pendulum is

$$I\ddot{\theta} = -rmg\sin\theta \tag{13.1}$$

where I is the moment of inertia, m the mass of the pendulum, g the gravitational field strength, and r the distance from the pivot to the center of mass. Using the parallel axis theorem, $I = \frac{m\ell^2}{12} + mr^2$, where ℓ is the length of the stick. The above ODE can be written in the standard simple pendulum form

$$\ddot{\theta} + \omega_0^2 \sin\theta = 0, \tag{13.2}$$

with $\omega_0 = \sqrt{\frac{rg}{r^2 + \frac{\ell^2}{12}}}$. The small angle frequency ω_0 should be calculated before doing the experiment. The period for small oscillations can be calculated using $T = \frac{2\pi}{\omega_0}$, while for large oscillations $T = \frac{4}{\omega_0}K(k)$, where $K(k)$ is the complete elliptical integral of the first kind, $k = \sin(\frac{\theta_{max}}{2})$, and θ_{max} is the maximum angle of oscillation.

Procedure:

1. Set up the apparatus as shown in Fig. 13.2. Suspend the meter stick so that the distance from its center of mass to the fulcrum is about 40 cm.

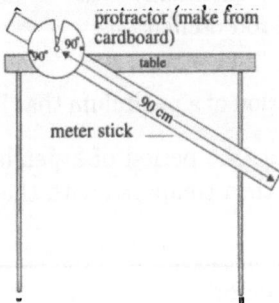

Figure 13.2: Apparatus for measuring the period of a compound pendulum.

2. Calculate the theoretical period for small oscillations.

3. Measure the small oscillation period. Compare the measured and theoretical values.

4. Start the pendulum with a large initial amplitude of sixty degrees. With a stop watch, measure the period for one complete swing. Note the maximum return angle. If friction makes the return amplitude much smaller than the starting angle, use an average value for the amplitude. Make several measurements to reduce the error. Mathematica file X13 may be used to compare the measured and calculated values for the period.

5. Repeat Step 4 using angles of 80°, 100°, 120°, 140°,....

Things to Investigate:

At what minimum angle can measurement discern a difference between the small oscillation period and the actual period?

Experimental Activity 14

Damped Simple Pendulum

Comment: This activity will take about two hours to complete.

References: Section 5.2.4 and AJP article by Crawford [Cra75]

Object: To determine the mathematical structure of the drag force exerted by air on a large simple pendulum bob.

Theory: Consider a simple pendulum whose bob consists of a large, relatively light, sphere of mass m on the end of a long pendulum arm of length r. In the absence of damping, the pendulum equation takes the familiar form

$$\ddot{\theta} + \omega^2 \sin\theta = 0, \tag{14.1}$$

where θ is the angle with the vertical and the frequency $\omega = \sqrt{g/r}$, g being the acceleration due to gravity. Letting x be the horizontal displacement of the bob from equilibrium and limiting the motion to small θ, then $\sin\theta \approx \theta = x/r$. So Equation (14.1) reduces to the undamped simple harmonic oscillator equation,

$$\ddot{x} + \omega^2 x = 0. \tag{14.2}$$

Now let's include air resistance due to the large bob, assuming the drag force (per unit mass) to be of the form $F_{\text{drag}} = -b\,|\dot{x}|^{n-1}\,\dot{x}$, where $\dot{x}$ is the velocity and b and n are positive constants. Stokes' and Newton's laws of resistance correspond to $n = 1$ (linear damping) and $n = 2$ (nonlinear damping), respectively. Including the drag force, the equation of motion then becomes

$$\ddot{x} + b\,|\dot{x}|^{n-1}\,\dot{x} + \omega^2 x = 0. \tag{14.3}$$

Multiplying Equation (14.3) by $\dot{x}$ and integrating with respect to t yields the energy conservation statement,

$$\frac{1}{2}\dot{x}^2 + b\int |\dot{x}|^{n-1}\dot{x}^2\,dt + \frac{1}{2}\omega^2 x^2 = C, \tag{14.4}$$

where the integration constant C corresponds to the total energy.

Let us now consider one complete oscillation of the pendulum, starting from rest ($\dot{x} = 0$) with initial maximum displacement $x = a_i$. At the end of the oscillation, the velocity $\dot{x}$ is again equal to zero and the final maximum displacement

$a_f < a_i$. Substituting the initial and final conditions into (14.4) and subtracting the two resulting equations yields

$$\frac{1}{2}\omega^2(a_i^2 - a_f^2) = b \int_0^T |\dot{x}|^{n-1}\dot{x}^2 \, dt, \tag{14.5}$$

where T is the time for one complete oscillation.

If the energy loss per cycle is not too large, the integral in (14.5) can be approximately evaluated by assuming that a linear solution $x = a\cos(\omega t)$ prevails with $a = (a_i + a_f)/2$. Then the velocity is $\dot{x} = -a\omega \sin(\omega t)$. Setting $\tau = \omega t$, Equation (14.5) becomes

$$a_i^2 - a_f^2 = 2\,b\,a^{n+1}\,\omega^{n-2} \int_0^{2\pi} |\sin(\tau)|^{n-1}\sin^2(\tau)\,d\tau. \tag{14.6}$$

But the left-hand side of (14.6) can be factored, viz.,

$$a_i^2 - a_f^2 = (a_i + a_f)(a_i - a_f) = 2\,a\,(a_i - a_f) \equiv 2\,a\,\Delta a,$$

so that on using Mathematica file X14 to evaluate the integral, (14.6) yields

$$\Delta a = \frac{2\sqrt{\pi}\,b\,a^n\,\omega^{n-2}\,\Gamma(1+n/2)}{\Gamma(3/2+n/2)}. \tag{14.7}$$

The result is expressed in terms of Gamma functions. Note that only for $n = 2$ is Δa independent of the length r (contained in ω) of the pendulum. Since Δa is proportional to a^n, the slope of a log-log plot of Δa vs. a will produce the damping exponent n. Once n is known, the constant b can be determined.

Damped Simple Pendulum

This file evaluates the integral in the expression for Δa and plots the ratio $\Delta a/a$ versus a for different n values. Mathematica commands in file: HoldForm, ReleaseHold, Integrate, Abs, Sin, Plot, Evaluate, Table

Procedure:

1. Set up the pendulum, taking the bob to be a large Styrofoam sphere (radius $\approx$ 3 cm) and the length (r) of the pendulum to be over 2.0 m.

2. Displace the bob a measured horizontal distance a_i from its equilibrium position and release it from rest. Measure the maximum horizontal displacement a_f of the bob after one complete oscillation.

3. Repeat the above procedure for a number of different initial displacements but do not go above a maximum angle of 20 degrees.

4. Create a table of $a = (a_i + a_f)/2$ and $\Delta a = a_i - a_f$.

5. Use the file X01 to determine the value of n which best fits the data.

6. Then use Equation (14.7) to calculate b.

7. Discuss your results.

Experimental Activity 15

Stable Limit Cycle

Comment: After the circuit has been constructed, this investigation should not take more than two hours to complete.

References: Section 7.1

1. [HCL76] The Wien bridge equations may be found in this article.

2. [HN84] An article that discusses a similar experiment.

3. [Mal93] Contains an excellent explanation of the Wien bridge oscillator.

Object: To use a Wien bridge circuit to investigate a signal that has its origin in an unstable focal point and ends in a stable limit cycle. The signal will be studied in its transient and steady state regimes.

Theory: This activity uses a Wien bridge oscillator to produce a stable limit cycle. Elementary electronic courses study the Wien bridge circuit because it can produce very precise and stable sine waves. In these courses the nonlinear properties of the Wien bridge oscillator are rarely mentioned or explored. The oscillations produced by a Wien bridge are autonomous, i.e., the forcing function is not periodic or time dependent. The final periodic motions that result from autonomous systems are known as limit cycles. This activity should be of special interest to electrical engineers who might want to look at both the linear and nonlinear aspects of this important circuit.

The Wien bridge oscillator uses a counterbalancing negative and positive feedback to produce a very stable sine wave signal. When the circuit is first turned on, the positive feedback is larger than the nonlinear negative feedback. Any white noise that is present in the circuit provides the seed that causes the circuit to leave its unstable focal point. As the amplitude of the oscillations increases, the current increases through a nonlinear resistance which usually is an incandescent bulb or other thermistor. As the current through the nonlinear resistance increases, the negative feedback controlled by the resistance of this resistor grows until it balances the positive feedback. When this balance is attained the circuit has reached its limit cycle. The Wien bridge circuit is usually operated near its threshold and in this case the voltage vs. time limit cycle is a near perfect sine wave. The phase plane limit cycle is a circle (ellipse).

A schematic of the Wien bridge oscillator is shown in Figure 15.1. The

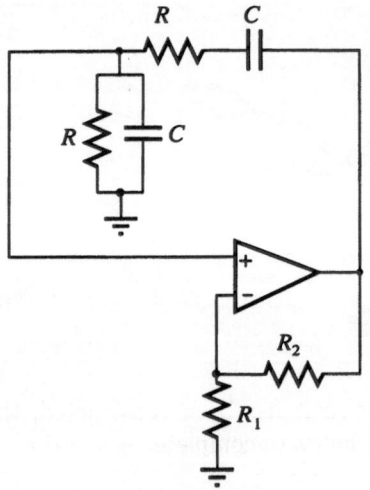

Figure 15.1: Wien bridge oscillator.

triangular circuit symbol represents an operational amplifier (op-amp). In an effort to understand how the complete circuit in Figure 15.1 functions, the circuit is first broken into its positive and negative feedback sections and then these sections are analyzed separately and in combination with each other.

The positive feedback section of the Wien bridge oscillator shown in Figure 15.1 is reconstructed in Figure 15.2. The equivalent circuit is also shown with the rectangular boxes representing the complex impedances Z_1 and Z_2. The positive feedback circuit is analyzed to see how much of the AC signal present at the op-amp's output (V_2) is sent back to the input (V_1). All signal

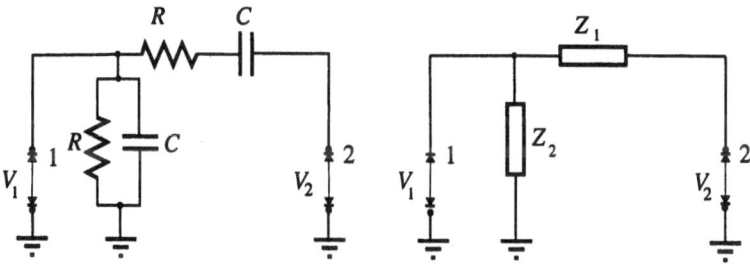

Figure 15.2: Positive feedback circuit and its equivalent.

strengths are measured relative to a ground level of 0.0 volts. An AC output signal at terminal 2 (V_2), causes a current (i) to flow through the complex impedances (Z_1, Z_2). The value for the current is

$$i = \frac{V_2}{Z_1 + Z_2},$$ (15.1)

where $Z_1 = R - jX$, $Z_2 = \frac{-jRX}{R-jX}$, and the capacitor's reactance is $X = \frac{1}{2\pi\nu C}$. The symbol j represents the imaginary number $\sqrt{-1}$ and ν represents the frequency of the AC signal. The strength of the AC signal between ground and terminal 1 is $V_1 = iZ_2$ so

$$V_1 = V_2 \frac{Z_2}{Z_1 + Z_2}.$$ (15.2)

In terms of R and X, Equation (15.2) becomes

$$V_1 = V_2 \frac{(-jRX)}{(R^2 - X^2) - j(3RX)}.$$ (15.3)

Equation (15.3) predicts that the maximum positive feedback occurs when $R = X$. Accordingly,

$$V_1 = \frac{V_2}{3}$$ (15.4)

which gives the maximum positive feedback of $\frac{1}{3}$ of the output voltage back to the input terminal.

Figure 15.1 shows how this positive feedback subsection of the circuit is connected to the op-amp. If $R = X$, one-third of the output signal is fed back into the noninverting terminal (+) of the op-amp. A nonzero value for the positive feedback means that if an external signal, no matter how small, is present, the op-amp quickly goes to saturation. For example, assume the initial input signal had a unit value. The positive feedback would first increase it to $1 + 1/3$, then to $1.33 + 1.33/3$, then to $1.77 + 1.77/3$ ... and so on to saturation. Saturation occurs because the circuit cannot keep increasing the signal's amplitude (voltage) past the op-amp's capability.

The phase shift between the output and positive feedback input signals must also be known. If there were a phase shift between the input and output signal, the positive feedback could destructively interfere with the input signal. Multiplying the numerator and denominator of Equation (15.3) by the denominator's complex conjugate, the phase shift is given by

$$\tan \phi = -\frac{(R^2 - X^2)}{3RX}.$$ (15.5)

Equation (15.5) shows that when $X = R$, there is no phase shift between V_2 and V_1, so the phase shift can be ignored. The positive feedback remains at $1/3$ as previously calculated.

The frequency (ν) which makes $R = X = 1/(2\pi\nu C)$ is

$$\nu = \frac{1}{2\pi RC}.$$ (15.6)

All other frequencies produce phase shifts. The circuit's white noise will contain a frequency that corresponds to the resonance frequency. This frequency is quickly amplified until saturation is reached. Anyone who has heard a sound system accidentally amplify its own sound is aware of the effects of positive feedback. Something must be done to limit this electronic catastrophe. The Wien bridge oscillator uses a nonlinear negative feedback to control the linear positive feedback.

The negative feedback section of the Wien bridge circuit shown in Figure 15.1 is shown in Figure 15.3. Before attempting an explanation of how the negative

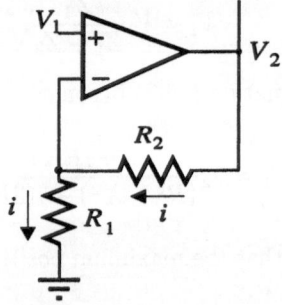

Figure 15.3: Negative feedback section of Wien bridge.

feedback section of the circuit operates, the rules that describe how an ideal op-amp function are reviewed:

1. The negative sign ($-$) on the inverting terminal of the op-amp does not indicate a polarity. It indicates that the output signal, if connected directly, is 180 degrees out of phase with the input signal;

2. The positive sign ($+$) on the noninverting input of the op-amp does not indicate a polarity. It indicates that the output signal, if connected directly, is in phase with the input signal;

3. No current flows into the inverting ($-$) or noninverting ($+$) inputs. The ideal op-amp is considered to have an infinite input impedance;

4. The voltages at the inverting (−) and noninverting (+) inputs are equal. This is a consequence of the above rule;

5. An op-amp has no (or very low) output impedance.

With these rules in mind, the amount of negative feedback is easily calculated. The voltage V_2 is

$$V_2 = i(R_2 + R_1). \tag{15.7}$$

The voltage at the inverting (−) input of the op-amp is

$$V_- = iR_1 \tag{15.8}$$

and since $V_- = V_+ = V_1$, the negative feedback (B) is the ratio of the output voltage to the input voltage and is equal to

$$B = \frac{V_2}{V_1} = \frac{(R_1 + R_2)}{R_1}. \tag{15.9}$$

Normally, op-amps are considered to have a very large gain (100,000 or more) for low frequencies. The negative feedback given by B reduces the total gain to some smaller and more easily managed value.

Now, the total gain G is the product of the constant positive feedback of $1/3$ and the variable negative feedback (B), so

$$G = \frac{B}{3}. \tag{15.10}$$

If in (15.9), $R_2 = 2R_1$ then $B = 3$ and the full circuit has a net gain of one. If $R_1 < 0.5R_2$, then $B > 3$ and the net gain is larger than one. If $R_1 > 0.5R_2$, then $B < 3$ and any signal present at V_2 decays to zero. Therefore the critical value for the negative feedback gain is 3. Wien bridge circuits produce stable sine waves by designing the negative feedback portion of the circuit to maintain a value of 3 automatically. When $B=3$, $G=1$, and stability has been reached.

The Wien bridge is a nonlinear circuit because the R_1 resistor has a nonlinear I–V curve. This resistance is usually provided by an incandescent light bulb. The resistance of this bulb grows as the potential drop across the bulb increases as shown in Fig. 15.4. For small V, the resistance of the bulb is a "well behaved function" of V which can be Taylor expanded to give

$$R_1(V) = R_0 + R'(0)V + \frac{1}{2!}R''(0)V^2 + \cdots. \tag{15.11}$$

Since the slope is zero at $V = 0$ then, for sufficiently small V, the higher order terms can be ignored and

$$R_d \equiv R_1 - R_0 = kV^2, \tag{15.12}$$

where $k \equiv \frac{1}{2!}R''(0)$ and R_d represents the dynamic or incremental resistance of the bulb. The symbol R_0 represents the resistance of the bulb when no current is flowing through it.

The rate of change of the dynamic resistance (R_d) is controlled by two competing mechanisms:

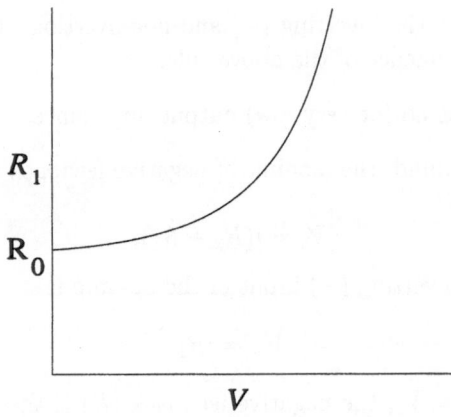

Figure 15.4: Resistance R_1 as a function of V.

- the forcing function due to the voltage drop V across the bulb and

- the tendency of the bulb to cool back to room temperature.

Assuming that the rate of change of R_d with time is proportional to R_d gives

$$\frac{dR_d}{dt} = -\frac{R_d}{\tau} + kV^2. \tag{15.13}$$

The first term on the right-hand side of the equal sign is the cooling contribution (note the minus sign) with the relaxation time τ a measure of the rate of cooling. The relaxation time is the time it takes the resistance to decrease from some value to $1/e$ of this value. The second term is the forcing contribution which follows from Equation (15.12).

For the Wien bridge circuit to "turn on", the value for the initial resistance of the bulb $(R_1 = R_0)$ must have a value less than $0.5R_2$, because Equation (15.9) shows that in this case $B > 3$, which results in a net gain $(G = B/3)$ that is greater than one. As the current through the bulb increases, its resistance (R_1) increases until the value for B decreases to 3. At this point the circuit is then producing a signal of constant amplitude and frequency—a near perfect sine wave. If the current would for some strange reason increase even further, the resistance (R_1) of the bulb would increase and the value for B would decrease below 3 and the gain would drop below 1 and the circuit would correct itself by lowering the value of the output signal until the value for B once again reaches 3. The relaxation time (τ) of the resistor (bulb) is usually much larger than the natural period of the oscillator. If the relaxation time was smaller than the oscillation period, the corrections would distort the signal. Relaxation times for incandescent bulbs are usually in the 0.2–2.0 second range, so the Wien bridge oscillator should have periods at least ten times smaller than these relaxation times. This means that the Wien bridge can produce reliable sine waves with a frequency of 5 Hz and up. This activity is the first of two that investigates the nonlinear properties of a Wien bridge circuit.

Procedure A: Producing the limit cycle.
Do not dismantle this circuit, it can be used in the next activity.

1. With a digital meter measure the initial resistance (R_0) of the incandescent bulb. A Radio Shack 12 V, 25 mA mini-bulb works very well for R_1. This value for R_0 gives a rough idea of the value needed for R_2 to make the Wien bridge leave its focal point.

2. Wire the circuit shown in Figure 15.5. Any generic op-amp from the 741 family should work, but μA741 is a reasonable and cheap op-amp to use. It is much easier if the circuit is wired on a breadboard that comes with the +15 and −15 volt power supplies. The values for the two resistances (R) or the two capacitors (C) are not critical, but make sure the capacitors (C) and resistors (R) have equal values and produce a frequency above 5 Hz. For example, $R = 1000\ \Omega$ and $C = 0.010\ \mu$F are reasonable values to use.

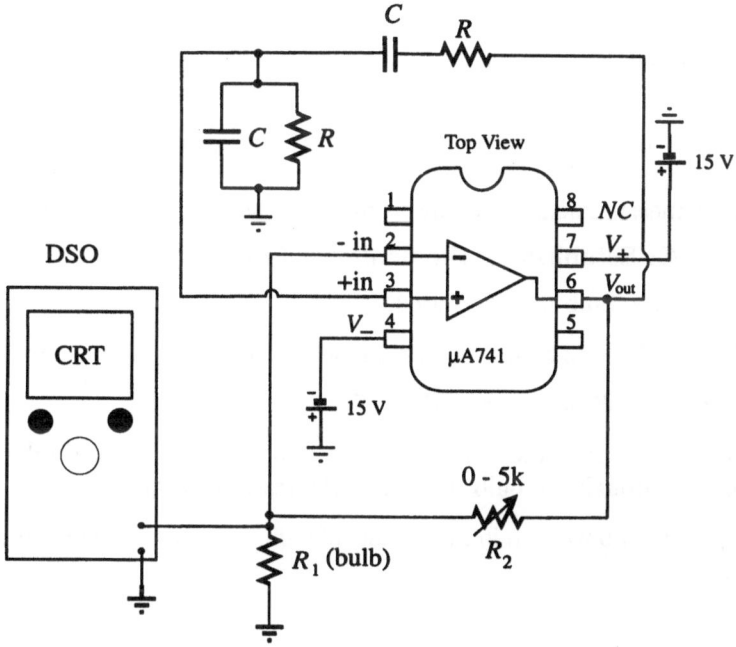

Figure 15.5: Wien bridge circuit.

3. For the resistor marked R_2 use a variable resistance box which permits the resistances to be read directly off the box. A storage oscilloscope makes it easier to study the trace.

4. Use (15.6) to calculate the value for the resonant frequency of your circuit.

5. Turn on your circuit.

6. Adjust R_2 ($R_2 \approx 2R_0$) until a sine wave is detected on the CRT. When the wave occurs you know that $2R_1 = R_2$.

7. What is the critical value for R_2? Calculate the value for R_1. What is the dynamic resistance R_d, where $(R_d \equiv R_1 - R_0)$?

8. Measure the value of the frequency of the limit cycle and compare it with the theoretical value.

9. After the CRT is displaying a sinusoidal trace, turn off the power to the circuit.

10. Turn the circuit back on and watch the signal reach its limit cycle.

11. Examine and try to explain the transient wave pattern.

12. Repeat the above steps for larger values of R_2.

If you wish, you may stop here or if you are interested you might wish to try the following procedures.

Procedure B: Finding the relaxation time τ

1. Turn the circuit back on and observe the transient waves. Measure the modulation period of the transient wave and then calculate the modulation frequency.

2. Measure the relaxation time (τ). This is approximately the time for the modulated wave's amplitude to decrease to $1/e$ or $\approx 1/3$ of its initial value. These measurements can only be made in the transient regime.

Procedure C: Establishing the nonlinearity of R_1

1. Remove the bulb from the circuit.

2. Place the bulb in series with an ammeter and a DC power source.

3. Connect a voltmeter across the bulb.

4. Start with a potential difference of 0.05 volts and record the value of the current through the bulb. Calculate the resistance R_d.

5. Repeat the above step in increments of 0.05 volts up to a maximum of 1.0 volts.

6. Make a plot of R_d vs. V.

7. Assuming that Equation (15.12) holds for lower voltages, what is the value for k? Does this relationship hold for larger voltages?

Things to Investigate:

- Have some fun with this circuit. Change the values for R and C and/or use a different type of incandescent bulb.

- Investigate the behavior of the circuit far from its threshold. This can be done by making $R_2 \gg 2R_0$. Watch for different signal shapes and amplitude changes.

- Place the incandescent bulb in a beaker of cold water. How does this effect the transient portion of the wave?

Experimental Activity 16

Van der Pol Limit Cycle

Comment: This investigation should not be attempted unless Experimental Activity 15 has been completed. However, this activity is easy, useful and should not take more than two hours to complete.

References: Section 7.1

1. [DW83] This article contains numerical values that can be used with Mathematica to produce similar phase plane plots. It contains a nonlinear circuit that was once proposed to control the movement of an elevator, not really a good idea, but the circuit might provide an interesting idea for a self-designed experiment.

2. [HN84] Discusses a similar experiment using the Wien bridge oscillator governed by the VdP equation.

3. [HCL76] The source of the Wien bridge equation used in [HN84].

Object: To produce and investigate the limit cycle oscillations predicted by a Van der Pol-like equation.

Theory: This activity uses the same Wien bridge oscillator that was used in Experimental Activity 15. For convenience the circuit is shown again in Figure 16.1. In the last activity the bridge produced, when operated near threshold, self-excited oscillations that left an unstable focal point and grew into near perfect sine waves. The oscillatory signal was sampled by measuring the RMS voltage (V) across the resistor R_1. It is reasonable to assume the oscillations are governed by a differential equation similar to

$$\ddot{V} + 2\gamma\dot{V} + \omega^2 V = f(t), \tag{16.1}$$

where γ is the damping coefficient and $f(t)$ represents the seed noise to start the circuit running. Remembering that the dimensionless quality factor (Q), for small damping, is given by $Q = \frac{\omega}{2\gamma}$ and for convenience letting $\Gamma = \frac{1}{Q}$ puts the above equation into the form

$$\ddot{V} + \Gamma\omega\dot{V} + \omega^2 V = f(t). \tag{16.2}$$

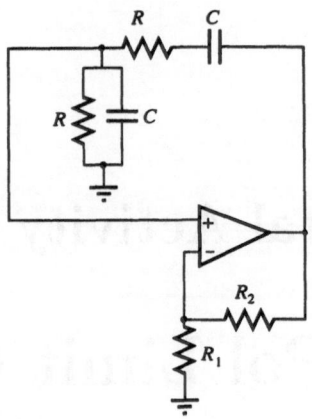

Figure 16.1: Wien bridge oscillator.

For the Wien bridge circuit to reach its limit cycle the coefficient Γ must be able to oscillate between a positive and a negative value. To understand how this might be accomplished, recall that in the last experiment the total gain of the Wien bridge circuit was shown to be

$$G = \frac{B}{3} \tag{16.3}$$

where the negative feedback B is

$$B = \frac{R_1 + R_2}{R_1}. \tag{16.4}$$

Further, recall that the value for B depends on the nonlinear resistor (the incandescent bulb) and that when the circuit is turned on the value for B starts with some value larger than 3 and therefore $G > 1$. When $G > 1$ the oscillations must grow so the coefficient Γ must be negative. The value for B decreases as the current through the resistor R_1 grows. As B decreases, G decreases below the value of 1. When $G < 1$ the oscillations decay so the coefficient Γ must be positive. One way of producing this sign oscillation in Γ is to write it as

$$\Gamma = -\beta(1 - \frac{1}{G}) = -\beta(1 - \frac{3}{B}) \tag{16.5}$$

where β is some positive constant.

Combining Equations (16.2) and (16.5) produces

$$\ddot{V} - \beta\omega(1 - \frac{3}{B})\dot{V} + \omega^2 V = 0 \tag{16.6}$$

where the forcing term has been dropped ($f(t) = 0$) because this forcing term is only needed to start the oscillations, not to maintain them. Equation (16.6) can be shown to be equivalent to the Wien bridge Equation (2)

$$\ddot{V} + 9\omega(\alpha - \alpha_c)\dot{V} + \omega^2 V = 0 \tag{16.7}$$

found in [HN84]. To show the equivalence between Equations (16.6) and (16.7), we note that $\alpha = \frac{1}{B}$, $\alpha_c = \frac{1}{3}$, and $\beta = 3$.

Substituting Equation (16.4) into Equation (16.6) and noting that the resistance R_1 (the incandescent bulb) is given by (from Experimental Activity 15 with $\frac{dR}{dt} \approx 0$ near equilibrium)

$$R_1 = R_0 + \tau k V^2, \tag{16.8}$$

we obtain

$$\ddot{V} - 3\omega \left(1 - \frac{3(R_0 + \tau k V^2)}{R_0 + R_2 + \tau k V^2}\right) \dot{V} + \omega^2 V = 0, \tag{16.9}$$

or

$$\ddot{V} - 3\omega \left(\frac{(R_2 - 2R_0 - 2\tau k V^2)}{(R_0 + R_2 + \tau k V^2)}\right) \dot{V} + \omega^2 V = 0. \tag{16.10}$$

Assuming

$$R_0 + R_2 \gg R_2 - 2R_0$$

and further that

$$R_0 + R_2 \gg \tau k V^2,$$

then Equation (16.10) yields

$$\ddot{V} - 3\omega \left(\frac{(R_2 - 2R_0 - 2\tau k V^2)}{R_0 + R_2}\right) \dot{V} + \omega^2 V = 0. \tag{16.11}$$

Equation (16.11) is an unnormalized Van der Pol equation. The purpose of this activity is to confirm experimentally the limit cycle oscillations predicted by the VdP-like equation. The values for k and τ measured in Experimental Activity 15 are required.

Procedure:

1. The relevant circuit is shown in Figure 16.2. It is the same as that used in Experimental Activity 15 so you should not have to rewire it if you did that experiment. Make sure the same incandescent bulb is used. This saves you the effort of remeasuring the values for R_0, τ, and k.

2. Use $\nu = \frac{1}{2\pi RC}$ to calculate the value for the oscillation frequency of your circuit.

3. Turn on your circuit.

4. Adjust R_2 ($R_2 \approx 2R_0$) until a sinusoidal signal is just detected on the CRT. What is the critical value for R_2? What is the value for R_1? Why is the value for R_1 different than the value for R_0?

5. Adjust R_2 just past the critical value and make sure that a sine wave appears on the CRT. Measure the frequency and compare it with the theoretical value.

6. After the trace has reached its limit cycle, measure or find the values for R_2 and V_{RMS}.

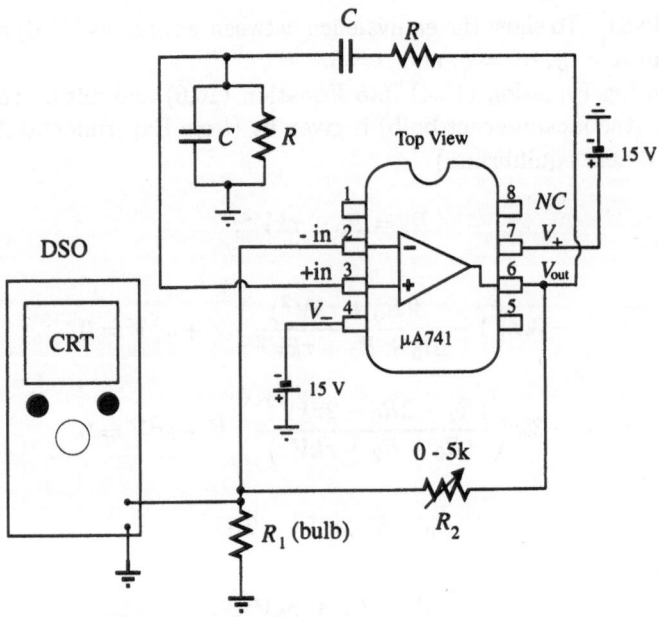

Figure 16.2: Wien bridge oscillator circuit.

7. Increase the value of R_2 and repeat the above steps.

8. Place your experimental values into Mathematica file X16 and check to see if the file reproduces the experimental results.

Van der Pol Equation
This file uses the unnormalized Van der Pol (VdP) equation derived in the text to model the behavior of the Wien bridge oscillator. Mathematica commands in file: `D, Sqrt, NDSolve, MaxSteps, Ticks, Evaluate, AxesLabel, PlotPoints, Hue, PlotLabel, PlotStyle, TextStyle, FontWeight, Plot, ImageSize`

Things to Investigate:

- Will the threshold approximation fail for large values of R_2? What shape of trace would tell you the threshold approximations are becoming less valid?

- Change the values for R and C to alter the frequency ν. Repeat the above procedure and see if the near-threshold approximations are still valid for the same value of R_2.

Experimental Activity 17

Relaxation Oscillations: Neon Bulb

Comment: This activity should not take more than one hour to complete.

Reference: Section 7.2

Object: To investigate the relaxation oscillations produced by a constant voltage source applied to a circuit that contains a neon glow lamp wired in parallel with a capacitor.

Theory: Stable periodic oscillations that are caused by autonomous (time-independent) forcing functions are known as limit cycles. If the dependent variable exhibits fast changes near certain time values, with relatively slowly varying regions between, the oscillator is said to exhibit relaxation oscillations. In this experimental activity the negative resistance characteristic of a neon glow bulb in conjunction with a capacitor and constant voltage power supply is used to produce relaxation oscillations.

Figure 17.1 is a sketch of a neon glow lamp. The small bulb is about 1.0 cm long and contains two vertical metal electrodes which are separated by approximately 2 mm. The electrodes are surrounded by a moderately low pressure neon

Figure 17.1: Neon glow lamp.

gas and when a potential difference is applied across the electrodes an electric field is created. This electric field accelerates the electrons and neon ions inside the bulb. If the electric field is large enough, the electrons reach a speed that is sufficient to start an avalanche of electrons. The avalanche is caused by an electron knocking out of the neon atom a second electron, thereby providing two electrons for the next collision, and four for the next collision, and so on until saturation is reached.

Accordingly, a neon glow lamp is considered to have a very high resistance until the correct potential difference exists across the electrodes. At the critical potential difference—the firing voltage (V_f)—the presence of a seed electron creates an avalanche and the bulb begins to conduct. This conduction reveals itself by making the lamp glow. Once the neon bulb is glowing, its resistance decreases, so a current-limiting resistance is usually placed in series with the bulb. The current-limiting resistance is normally selected to have a value that permits the bulb to stay glowing without burning it out. This happens because as the current through the bulb increases and the potential drop across the bulb decreases, the potential drop across the resistance increases. If the correct resistance is selected, the bulb stays lit. If the current-limiting resistance is very large, the bulb turns off. If the current-limiting resistance is too small, the current can continue to increase until the bulb burns out. Because neon bulbs consume very little power and have a very long life they are often used to indicate if a device such as a power supply is turned on.

The neon bulb in this activity is used a little differently than that given in the description above. The circuit shown in Figure 17.2 produces relaxation

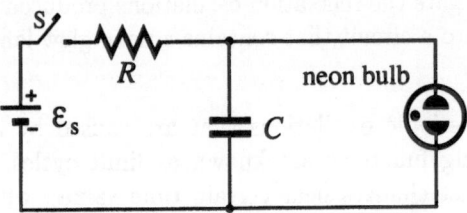

Figure 17.2: Neon tube circuit.

oscillations. When the switch (S) is closed, the capacitor slowly (relatively) charges through the large resistance (R). The potential drop across the capacitor increases until the bulb's firing voltage (V_f) is reached and the bulb turns on. As soon as the bulb turns on, its resistance drops and the capacitor rapidly discharges through the bulb. As the capacitor discharges, the potential drop across the capacitor (bulb) decreases until the extinction voltage is reached and the bulb turns off. When the bulb turns off, its resistance rapidly regains its large initial value (infinity), the capacitor begins to recharge and the limit cycle is in operation.

Figure 17.3 shows the same cycle as a function of the voltage and the current. Be careful not to assume that the lengths of the line segments (DA, AB, BC, CD) are proportional to the time. As the voltage across the bulb increases from 0 to A, no (or very little) current flows through the bulb. Once the firing voltage (V_f) is reached, the bulb's resistance drops dramatically and the bulb begins to conduct. This reduction in the bulb's resistance allows for a rapid increase in the current through it. This is represented by the movement from point A to point B. Now the circuit relaxes as the capacitor discharges and the

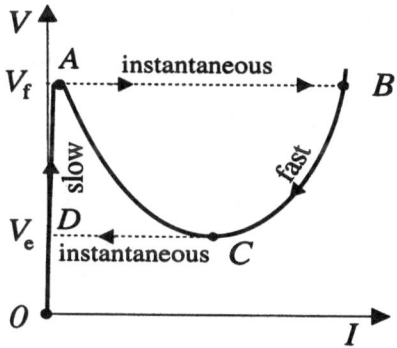

Figure 17.3: Voltage (V) vs. current (I) for a neon tube.

current decreases from B to C. This discharge happens very rapidly because the resistance of the bulb when it is discharging is much smaller than the external resistance R. When the voltage across the bulb reaches the extinction voltage (V_e) at C, the bulb turns off and the current decreases from C to D. The voltage now restarts its slow climb from D to A. The cycle is complete. (This process is dependent on having a value for the external resistance R that is large enough to let the voltage drop across the bulb decrease past the extinction value. If the value for R was smaller than some critical value, it might be possible for the bulb to stay on because the extra current could be supplied by the battery.)

A voltage vs. time graph across the bulb is shown in Figure 17.4. The exter-

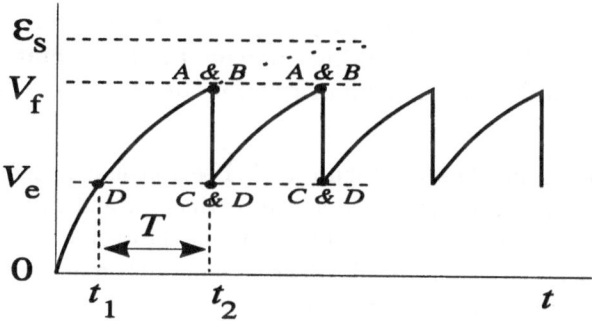

Figure 17.4: Voltage vs. time response curves for a neon tube.

nal resistance (R) is chosen to make the time to charge the capacitor through R much larger than the time to discharge the capacitor through the bulb. Equation (17.1) gives the relationship between the voltage across the charging capacitor as a function of time

$$V = \varepsilon_s \left(1 - e^{-t/(RC)}\right), \tag{17.1}$$

and can be used to calculate the period (T) of the relaxation oscillations. In Equation (17.1), ε_s represents source voltage which would be the maximum voltage across the capacitor if the bulb was removed from the circuit. The period $T = t_2 - t_1$ is calculated by solving (17.1) for the time t_1 by first rearranging

$$e^{-t_1/(RC)} = \frac{(\varepsilon_s - V_e)}{\varepsilon_s} \tag{17.2}$$

and then solving for t_1 by taking the logarithm of each side of (17.2) to give

$$t_1 = -RC \ln \left| \frac{\varepsilon_s - V_e}{\varepsilon_s} \right|. \tag{17.3}$$

Similarly,

$$t_2 = -RC \ln \left| \frac{\varepsilon_s - V_f}{\varepsilon_s} \right|. \tag{17.4}$$

The period of the oscillation is given by $T = t_2 - t_1$, so

$$T = RC \ln \left| \frac{\varepsilon_s - V_e}{\varepsilon_s - V_f} \right|. \tag{17.5}$$

The main purpose of this activity is to see the shape of the relaxation oscillations shown in Figure 17.4 and determine the period T.

Procedure A: Determining V_e and V_f

This procedure is for determining the firing voltage (V_f) and the extinction voltage (V_e). It can also be used to determine the voltage vs. current response curve for a neon bulb.

This activity uses high voltages, so please use caution. High voltage can be lethal. Before making changes to the circuit always turn the circuit off and wait a second or two for the capacitors to discharge.

1. Build the circuit shown in Figure 17.5. Use a neon glow lamp of the class Ne-2. These bulbs have a variety of extinction and firing voltages but the firing voltage is usually between 70 and 110 volts and the extinction voltage between 50 and 70 volts. The variable DC power source (ε_s) should be able to produce a maximum of 200 volts. The 50k resistor should be at least 1 watt. You do not need to use the ammeter if you are only determining the extinction and firing potentials.

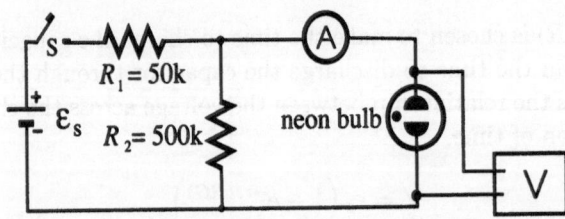

Figure 17.5: Apparatus to determine V_e and V_f.

2. Slowly increase the value of the source voltage and record the potential drop (V_f) across the neon bulb at the instant the bulb first begins to glow.

3. After the bulb lights, slowly decrease the source voltage and record the potential drop (V_e) at which the bulb turns off.

4. Repeat the above steps a couple of times to ensure that your values are reasonably accurate.

Making a plot of the voltage vs. current through a neon bulb.

1. Increase the voltage of the power supply until it makes the bulb turn on. Record the firing voltage.

2. Keep increasing the power supply's emf while at the same time record the current through the bulb and voltage across the bulb. Do not let the power supply's emf increase past 200 volts or the 50k resistor might burn out.

3. Record the current through and potential drop across the bulb as you slowly decrease the power supply's emf. Record the extinction voltage.

4. Make a plot of the voltage vs. the current.

Procedure B: Relaxation Oscillations

1. Build the circuit shown in Figure 17.6. Use values for R of 0.50 $M\Omega$ and C of 0.47 μF (or a variable capacitor) and a variable DC emf (0–200 volts) power supply (ε_s).

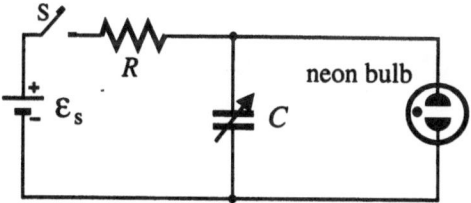

Figure 17.6: Relaxation oscillation circuit.

2. Turn on the power supply and slowly increase the voltage until the light begins to flash. Now increase the voltage approximately 10% – 20% above the firing voltage and record the power supply's emf.

3. With a stop watch, measure the period of the flashing neon bulb. You might wish to adjust the value of C or R if the period is too short to measure easily with the stop watch.

4. Use Equation (17.5) and the known values for V_f, V_e, and ε_s to calculate the period. How do the measured and calculated periods compare?

Procedure C: Observing the Phase Plane of the Relaxation Oscillator

1. Build the phase plane circuit shown in Figure 17.7.

2. Adjust the source emf (voltage) until the bulb is flashing quite rapidly, i.e., $T \approx 0.20$ s.

3. Use the x–y and AC settings on the dual trace oscilloscope and observe the phase plane portrait.

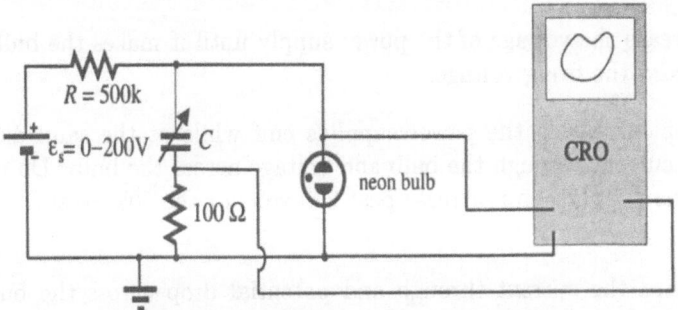

Figure 17.7: Phase plane circuit.

Experimental Activity 18

Relaxation Oscillations: Drinking Bird

Comment: This activity should not take more than one hour to complete.

Reference: Section 7.2

Object: To investigate the relaxation oscillations produced by the toy known as the drinking bird.

Theory: This activity uses an inexpensive toy, sold under the name Drinking Bird,[1] as a relaxation oscillator. Figure 18.1 is a sketch of this well known toy. The toy normally operates by first wetting the cloth wrapped around the

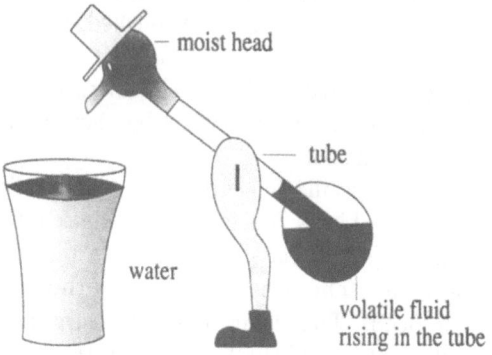

Figure 18.1: The drinking bird.

bird's head with water. The energy needed to evaporate the water, lowers the temperature of the head, and subsequently the air-pressure inside the head. The reduced pressure causes the volatile fluid contained within the tail of the bird to rise through a vertical tube to the head. When the fluid reaches the head, its

---

[1]Arbor Scientific, P.O. Box 2750, Ann Arbor, MI 48106-2750, phone 1-800-367-6695.

weight causes the bird to "drink" from the glass of water. The drinking action as shown in Figure 18.2 performs three functions:

1. it keeps the bird's head wet by dipping it into the glass of water;

2. it lifts the lower end of the tube out of the liquid in the tail, thus equalizing the pressure inside the bird;

3. it provides a way for the fluid to flow back to the tail.

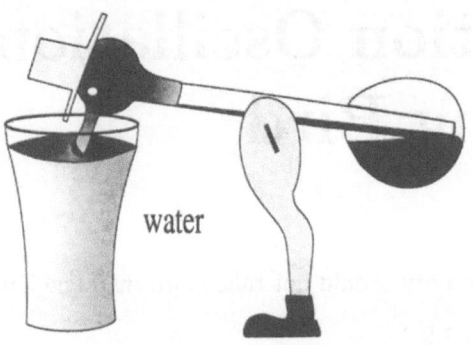

Figure 18.2: The bird drinks.

Accordingly, when the tube opens, the pressure equalizes, and the force of gravity pulls the fluid back to the tail and the bird upright. After the bird reaches its upright position, the fluid covers the bottom of the tube, the water that has soaked into the cloth wrapped around the bird's head evaporates, a pressure difference between the head and the tail is reestablished, and the bird is forced to drink once again. The relaxation oscillations have started. The bird's bobbing (drinking) is a relaxation oscillation because the energy source is not pulsating and because the bird's motion is characterized by short and long durations for different segments of its oscillation.

In this activity, the bird's oscillations are produced in a slightly different way. The glass bulb which is the bird's tail is painted black (the colloidal solution used to make electrostatic pith balls works well as a paint) and a bright light is focused on the black tail. The heat produced by the absorbed light causes the fluid to rise to the head. The rising fluid forces the bird into its drinking position which pulls the tail out of the light. Once the bird is in the drinking position, the fluid flows back to the tail, the bird becomes erect, the tail reenters the light beam, and the oscillations have started.

The strength of the light flux shining on the bird's glass bulb tail can be adjusted to alter the bird's bobbing rate. This is an easy to perform activity because the bobbing rate of the bird is slow enough to permit hand measurements of its period of oscillation. This activity qualitatively investigates the relationship between the strength of the light flux and how it controls the bird's oscillations.

Procedure:

1. Set up the following apparatus. The light source should be a 250 watt

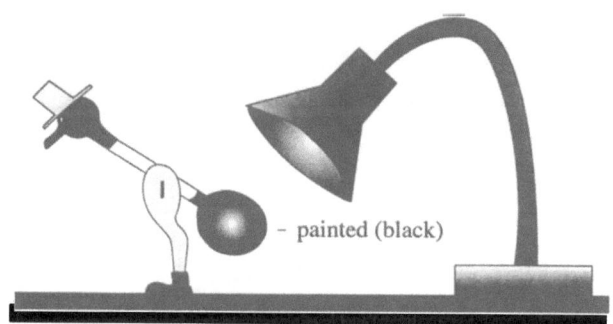

Figure 18.3: Experimental setup.

nonfrosted light bulb. **Make sure the lamp can safely handle this size of bulb (wattage).** The light from an overhead projector also works well. A Fresnel lens may be used to focus the light on the bird's tail or to create a collimated beam of light.

2. Focus the light on the bird's tail so the bird begins to oscillate.

3. Measure the period of the bird's oscillations. Is the time between oscillations constant or does it vary with some pattern?

4. Plot a graph of the time and position of the bird's bobbing head.

5. Repeat the above steps with different light intensities to modify the bird's oscillation rate.

6. Is there a critical flux (or distance from the light source) which makes the bird's oscillations begin or stop?

Things to Investigate:

- Investigate how the combination of the light shining on the bird's tail and the bird's head bobbing into the water affects the rate of the oscillations.

- Use a dripping faucet to let water drip slowly onto the bird's head. Control the rate of the drips and investigate the effects on the bird's bobbing. Is there a critical drip rate?

Experimental Activity 19

Relaxation Oscillations: Tunnel Diode

Comment: This is an easy experimental activity and should not take more than two hours to complete.

References: Section 7.2

1. Experimental Activity 9: Tunnel Diode Negative Resistance Curve

2. [BN92] A useful book for extra information on tunnel diodes.

Object: To investigate self-excited relaxation oscillations produced by an electric circuit that consists of a constant energy source connected to a tunnel diode, inductor and resistor. To produce a tunnel diode's negative resistance curve from the relaxation oscillations.

Theory: Self-excited or auto-catalytic relaxation oscillators are highly nonlinear systems that spontaneously oscillate even though they are driven by a constant energy source. Self-excited relaxation oscillators have the ability to continuously absorb energy, but usually give off their energy in bursts or at specific locations in their phase space. The study of self-exited oscillators has played an important role in deepening the understanding of nonlinear systems. Relaxation oscillators, e.g., the Van der Pol oscillator, have been discussed in some detail in Reference 1. Self-excited oscillations are intimately connected with the formation of limit cycles.

The occurrence of self-excited oscillators in nature are many, one of the most important being your beating heart, and one of the most famous the wind driven torsional oscillations of the Tacoma Narrows bridge. The collapse of this bridge in 1940 was not due to the wind gusting in resonance with the natural frequency of the structure. It was due to the bridge's ability to absorb the wind's energy continuously and then release the energy in resonance with the bridge's natural frequency. Local residents observing the bridge being built, and after its completion, reported oscillations for winds blowing with speeds as low as 5 m/s (10 mph). The bridge was the destination of many Sunday drives, not just to see the marvelous structure, but also to feel its strange vibrations. Rumor has it that many of the local inhabitants predicted the collapse of this the world's largest unintentional swinging bridge long before it happened.

The electric circuit shown in Figure 19.1 is used to explain how self-excited relaxation oscillations might occur. The circuit is designed so that it produces

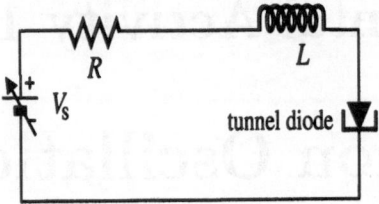

Figure 19.1: Relaxation oscillation circuit.

its operating point (V_0) near the middle of the unstable negative resistance region of the tunnel diode; see Figure 19.2. When the voltage of the source shown

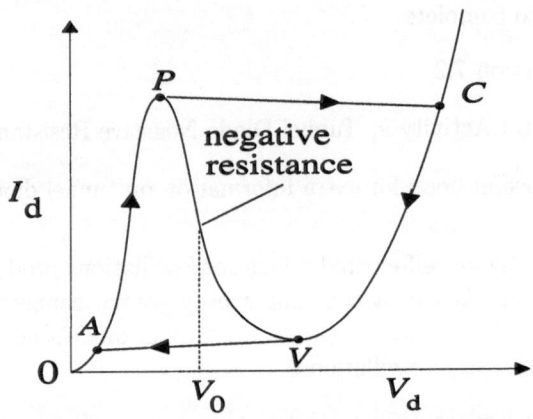

Figure 19.2: Current vs. potential for a tunnel diode.

in Fig. 19.1 is made to increase very slowly toward V_0, the potential drop across the inductor can be ignored. The current continues to increase until the current in the tunnel diode reaches point P. If the voltage of the source is increased still further in an attempt to push the current even higher, the voltage drop across the diode must jump from P to C. The current cannot enter the negative resistance region because in that region the current would be decreasing. The jump from point P to C is very fast. During this jump the inductor quickly changes its polarity in an effort to maintain Kirchhoff's voltage rule. At point C the potential drop across the diode is larger than the source potential, so the reversed potential of the inductor keeps the algebraic sum of the potentials around the circuit equal to zero. The reversed polarity will cause the current to decrease or relax toward V, hence the term *relaxation oscillations*. As the current reaches V the potential is still above the operating point (V_0) of the diode, so it must decrease even further. It cannot enter the negative resistance region because that would mean the current would increase, so the potential must quickly jump from V to A. During the jump the inductor quickly reverses its polarity to keep Kirchhoff's voltage rules intact. The potential of the source and inductor now cause the current to climb to P, and the process is repeated. This qualitative description presupposes that the inductance is sufficiently large to allow these large jumps from P to C and V to A. For small L it is possible for the system to oscillate in the negative resistance region. In this case the oscillations can appear very different. For very small oscillations in the negative resistance region, they are sinusoidal. A voltage time plot for this relaxation oscillation process is shown in Figure 19.3. The shape of the plot is character-

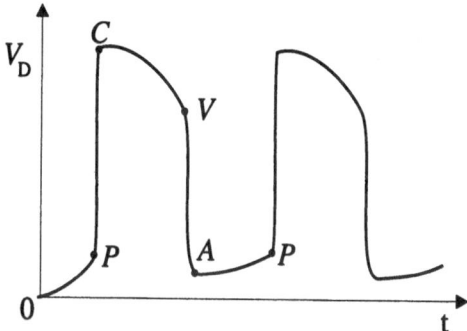

Figure 19.3: Voltage vs. time for the relaxation oscillator.

istic of many, but not all, relaxation oscillators. This activity should reproduce a replica of this plot.

** Tunnel diodes are very difficult circuit devices to get to function as predicted. The reason is that they like to oscillate at very high frequencies which make any small parasitic inductances and capacitances change the circuit's parameters. This is very bothersome to circuit designers. **

Procedure:

1. Wire either the circuit shown in Figure 19.4 or Figure 19.5. If the second circuit is used, the power supply must be able to maintain a set operating potential. Any tunnel diode should work, but if in doubt 1N3718 or 1N3719[1] is recommended.

2. If the circuit in Figure 19.4 is used, adjust the variable resistance (R_1) to find an operating point in the negative resistance region of the tunnel diode's I–V response curve. You will know that you are in this region when the characteristic relaxation oscillation curve (Figure 19.3) appears on the CRT.

3. If the circuit in Figure 19.5 is used, adjust the voltage between 0.05 volts and 0.30 volts to put the operating point somewhere in the negative resistance region.

---

[1]Tunnel diodes may be purchased from Germanium Power Devices Corporation, 300 Brickstone Sq., Andover, MA 01810 (508-475-1512).

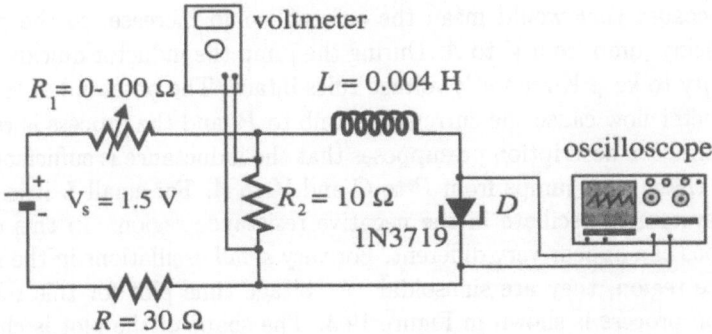

Figure 19.4: A tunnel diode circuit for producing relaxation oscillations.

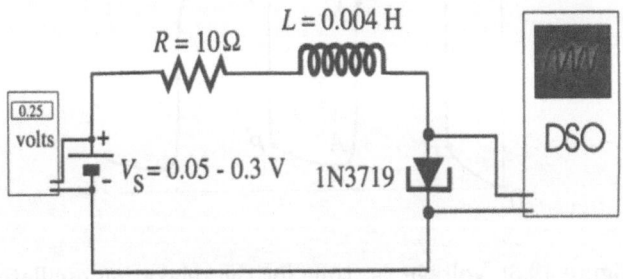

Figure 19.5: An alternate circuit for producing relaxation oscillations.

4. Using the oscilloscope, measure the period of the oscillations.

5. Measure the values for the voltages at points A, P, V, and C. If you completed Experimental Activity 9, compare these values with those obtained in that activity. If you did not complete Experimental Activity 9, sketch a rough plot of the response curve using the values measured here.

Things to Investigate:

- Remove the inductor from the circuit. Do relaxation oscillations still occur?

- Place a variable capacitor in parallel with the tunnel diode. Investigate how the relaxation oscillation is affected by changing the capacitance.

Experimental Activity 20

Hard Spring

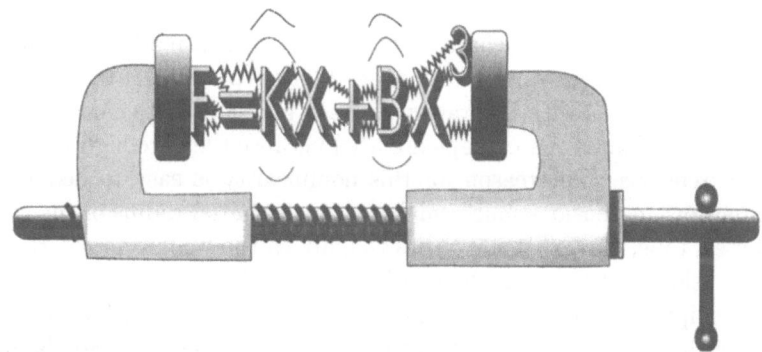

Comment: This activity should take less than one hour to complete.

Reference: Section 8.1

Object: To find the mathematical relationship between a nonlinear spring's extension and the force required to produce that extension. To investigate the period vs. amplitude relationship for a hard spring.

Theory: The study of the motion that occurs under the control of a hard spring has played an important role in increasing the understanding of nonlinear physics. A spring is classified as a nonlinear hard spring when the restoring force increases faster than that given by Hooke's law. A graph of the force (F) as a function of the extension (x) for a typical hard spring is shown in Figure 20.1. A possible equation for this curve is

$$F = ax + bx^3,$$ \hfill (20.1)

where a and b are positive constants. Damped oscillations governed by this nonlinear forcing function can be described using Newton's second law,

$$\ddot{x} + 2\gamma\dot{x} + \alpha x + \beta x^3 = 0$$ \hfill (20.2)

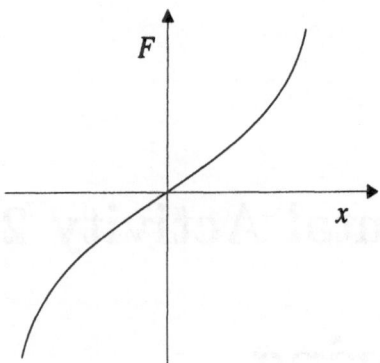

Figure 20.1: Force (F) and extension (x) for a hard spring.

where γ is the damping coefficient, $\alpha = \frac{a}{m}$, $\beta = \frac{b}{m}$, and m is the mass. A hard spring oscillator has a period that becomes smaller as the amplitude increases.

In this activity, a nonlinear hard spring arrangement is made by forming two ellipses from identical lengths of steel tape. (This type of steel tape is used to wrap cartons or comes with laboratory airtrack and is used to construct glider bumpers.) These bands of steel tape are used because they are highly nonlinear for large extensions. The reason for this nonlinearity is easy to deduce. For small extensions, extensions much smaller than the initial radius of the circular band of steel tape, the extension varies roughly as the pulling force. However, as the extension increases, the shape of the band of steel tape becomes more and more elliptical, and very large forces are now required to produce small increases in the extension. The limiting case would be where the circular spring is so deformed that it acts as two parallel steel tapes. Figure 20.2 shows how the two springs will be used in this experiment. Two springs of the same size

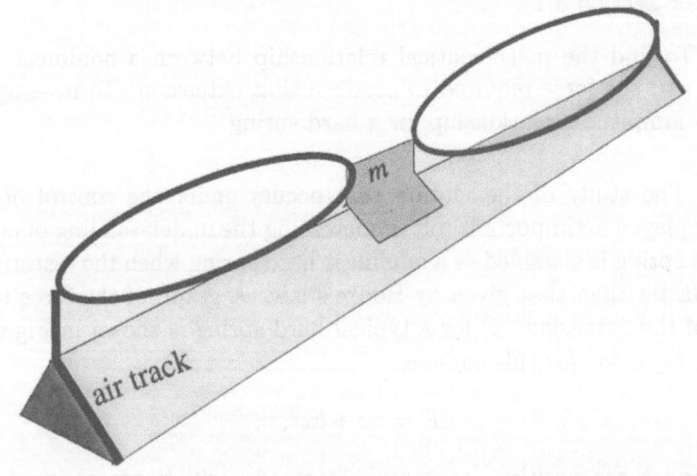

Figure 20.2: The nonlinear springs attached to an airtrack glider.

and shape are needed to ensure that symmetrical restoring forces occur for both positive and negative extensions.

Procedure:

1. Connect the glider to the springs as shown in Figure 20.3. The circular springs can be made from the same steel tape as used to make the airtrack bumpers. Remember that the springs should be the same size and shape. Masking tape is a convenient method of attaching the steel springs to the glider. Connect the springs so that they produce an oval shape.

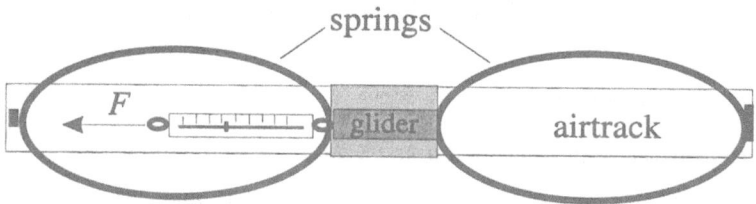

Figure 20.3: Top view: airtrack, airtrack glider and two nonlinear springs.

2. Attach a force meter to the glider. Measure and record the extensions for different forces. Make at least ten measurements on both sides ($+$ and $-$) of the equilibrium point. Average the magnitudes of the plus and minus extensions to eliminate any asymmetry in the springs.

3. Draw a graph of the force (F) as a function of the extension (x).

4. Find the values for the constants a and b in Equation (20.1). You should use Mathematica file X20A to help find the best-fit values for the equation's constants. Just follow the instructions in the file.

Hard Spring Best Fit
When your experimental values for the extension (the x values) and the force (the F values) are substituted into this file's two lists, Mathematica produces the best fit polynomial for the data. To compare the best fit equation with the actual values, a plot of the points and of the polynomial is constructed. Mathematica commands in file: `Transpose`, `ListPlot`, `PlotStyle`, `PointSize`, `PlotLabel`, `StyleForm`, `FontWeight->Bold`, `FontSlant->Italic`, `FontSize`, `Frame->True`, `FrameStyle`, `Plot`, `Hue`, `Fit`, `RGBColor`, `TextStyle`, `Chop`, `ImageSize`, `Show`

---

5. Check the accuracy of the equation by measuring the period for various initial amplitudes. Measure the mass of the glider, and then use Mathematica file X20B to calculate the periods for different amplitudes to see if there is agreement between the measured and theoretical values. (Measuring the period can be difficult because the springs are so heavily damped and therefore the oscillations decay very quickly. Measure the period for

one or two oscillations only and then repeat a number of times to find the average.)

Hard Spring

Using the best fit parameters for the hard spring, Duffing's equation is solved and plotted. The behavior of the period with amplitude can be determined from the graph. Mathematica commands in file: `NDSolve`, `MaxSteps`, `Evaluate`, `Ticks`, `PlotLabel`, `StyleForm`, `FontFamily`, `Plot`, `FontSize`, `FontWeight`, `TextStyle`, `ImageSize`, `AxesStyle`, `FrameTicks`, `FontSlant->Italic`, `Frame->True`, `FrameStyle`

Things to Investigate:

- Repeat the activity using an airtrack glider of different mass.

- Repeat the activity using circular springs with different radii.

- Determine the value for γ by finding the time it takes for the amplitude of small oscillations to decrease by a factor of 2.

Experimental Activity 21

Nonlinear Resonance Curve: Mechanical

Comment: This activity should take less than one hour to complete.

Reference: Section 8.2

Object: To produce a nonlinear resonance response curve for a mechanical oscillator.

Theory: Nonlinear and linear resonance response curves are constructed by plotting the modulus of the steady state amplitude $|A|$ versus the driving frequency (ω). Figure 21.1 shows typical linear and nonlinear resonance response

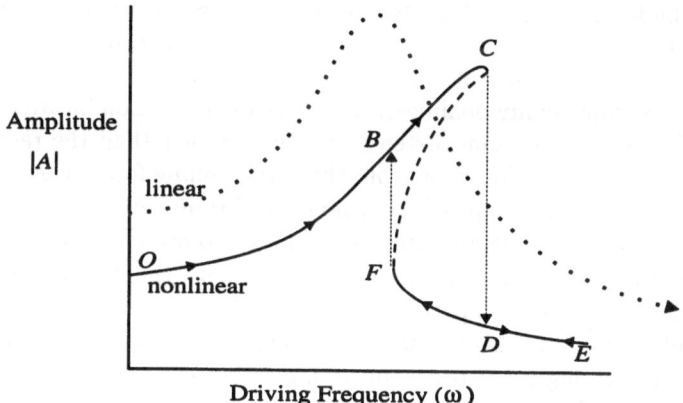

Figure 21.1: Linear and nonlinear resonance response curves.

curves. In the nonlinear case, as the driving frequency is slowly increased from O, a point C is reached where the value for the amplitude undergoes a sudden drop from C to D and then continues to point E and beyond. However, if the frequency is slowly decreased from point E, the amplitude will undergo a quick jump from F to B and then will continue along the curve to O. The frequency at which the jump from F to B occurs is different than that of the drop frequency. The phenomenon of following different system paths when moving in different directions is known as hysteresis.

The nonlinear hard spring is modeled by the equation that relates the pulling force (F) to the spring's extension (x) in the following manner,

$$F = ax + bx^3 \tag{21.1}$$

where a and b are positive constants. A graph of the force as a function of extension is shown in Figure 21.2. In this experiment the nonlinear hard springs

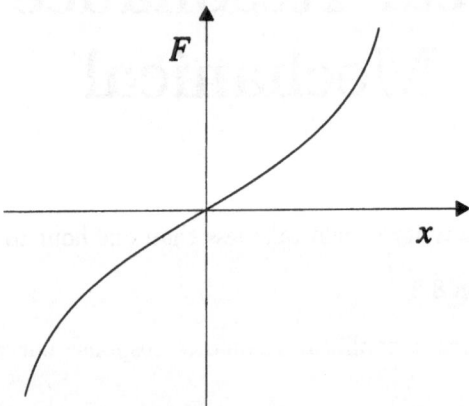

Figure 21.2: Force (F) and extension (x) for a hard spring.

are made by forming circles from lengths of steel tape. (This type of steel tape is used to wrap or secure cartons or a similar kind of tape comes with laboratory airtrack. It is used to construct the airtrack bumpers.) These circular bands are for large extensions highly nonlinear. The reason for this nonlinearity is easy to deduce. For small extensions, extensions much smaller than the radius of the circular band, the extension varies roughly as the pulling force. However, as the extension increases, the shape of the band of steel tape becomes more and more elliptical, and larger and larger forces are required to produce the same increase in the extension. The limiting case would be where the circular spring is so deformed that it acts as two parallel steel tapes. Figure 21.3 shows how the two springs will be used in this experiment. Two springs are needed to ensure that symmetrical restoring forces occur for both positive and negative extensions. A strong sinusoidal driving motor is connected to one of the springs. The motor is used to force the vibrating airtrack glider into its steady state. The motion of the vibrating glider can now be modeled by the forced Duffing equation

$$\ddot{x} + 2\gamma\dot{x} + \alpha x + \beta x^3 = A_0 \sin(\omega t). \tag{21.2}$$

The constants α, β, γ, and A_0 can be determined experimentally but are not needed in this activity.

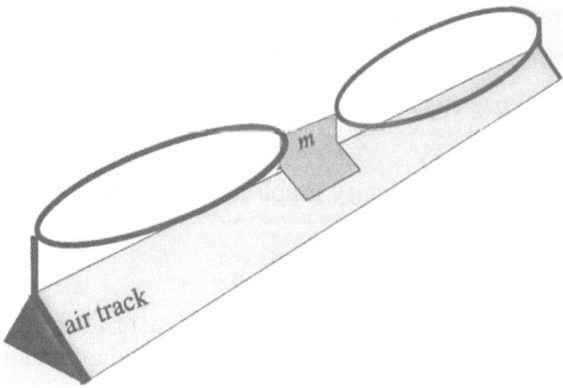

Figure 21.3: The nonlinear springs connected to an airtrack glider.

Procedure:

1. Set up the following apparatus. The circular springs initially should have of about 0.4 m. Connect and stretch each spring to produce an oval shape of approximately 0.5 m by 0.3 m.

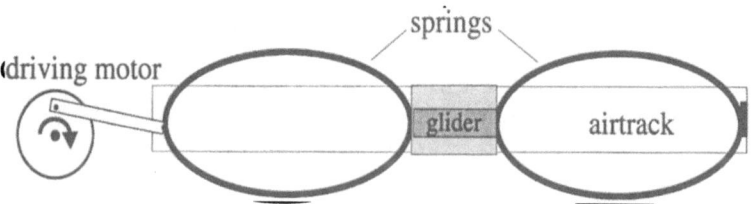

Figure 21.4: Two nonlinear springs connected to a forcing motor.

2. The driving motor should be connected to a variable voltage power supply so that the frequency of the motor can be adjusted.

3. With the motor running at a low frequency, measure the motor's frequency (or voltage) and the steady state amplitude of the air track glider. The circular springs are quite heavily damped (internally), so you should not have to wait too long for the glider to reach its steady state amplitude.

4. Increase the voltage (motor's frequency) a little and repeat the above measurement.

5. Repeat the above step until a sudden drop in the glider's amplitude occurs.

6. With the motor running at a frequency well above the drop frequency, measure the glider's amplitude.

7. Slowly reduce the motor's frequency, and for each reduction measure the glider's steady state amplitude.

8. Repeat the above step until a sudden jump in the glider's amplitude occurs.

9. Draw an amplitude–frequency graph for your data.

Things to Investigate:

- Repeat the activity with small strong magnets attached to the glider. (These magnets produce even stronger damping due to inducing a current in the airtrack. Lenz's law then can be used to explain the magnetic damping.) Does the hysteresis loop still occur at the same frequencies?

- Repeat the activity using the same circular springs but with a different initial elliptical (oval) setting. What is the effect on the data?

- Investigate the behavior of the resonance frequency of the apparatus.

- Look for examples of period doubling.

Experimental Activity 22

Nonlinear Resonance Curve: Electrical

Comment: This activity should not take more than two hours to complete.

References: Section 8.2

1. Experimental Activity 8 explains how the piecewise linear circuit used in this activity functions.

2. [FJB85] This article shows how similar electrical circuits can be used to model the hysteresis effects of "hard spring" and "soft spring" oscillators.

Object: To investigate how the resonance frequency changes as a function of the forcing amplitude (voltage) for a piecewise linear (nonlinear) circuit. The nonlinearity is similar to that of a soft spring oscillator.

Theory: Resonance has played an important role in the studying and classifying of linear and nonlinear systems. One must be careful to understand how the resonance frequency differs for linear and nonlinear oscillations. For example,

1. a 1-dimensional linear system has one resonance frequency while a nonlinear system can have an infinite number;

2. a linear system has one resonance frequency regardless of the pumping force, while nonlinear systems do not.

This activity explores and investigates these two properties.

A modification of the electric circuit in Activity 8 is used here and is shown in Figure 22.1. The nonlinearity (piecewise linear) property of this circuit is created by the two diodes connected in parallel with the capacitor C_2. A plot of the voltage across both capacitors as a function of charge would appear as shown in Figure 22.2 and it should be noted that the reciprocal of the slope in a given region is the capacitance for that region. The diodes act as open switches when the voltage is below some critical level ($V_b \approx 0.7$ volts) and act as electrical shorts to eliminate the capacitor C_2 when the voltage exceeds this critical value.

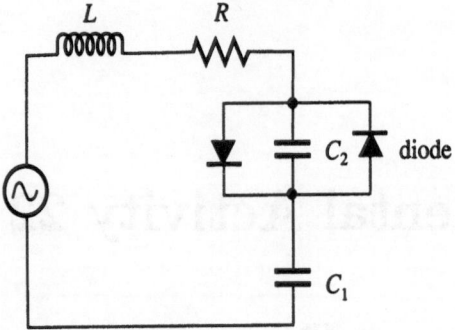

Figure 22.1: Circuit for producing a piecewise linear response.

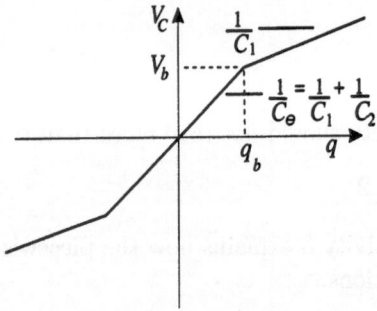

Figure 22.2: Voltage across both capacitors vs. charge.

When the diodes are acting as open switches both capacitors are in operation, so the equivalent capacitance of the two capacitors in series is $C_e = \frac{C_1 C_2}{C_1 + C_2}$ and the corresponding high resonance frequency is given by $\nu_{\text{high}} = \frac{1}{2\pi\sqrt{LC_e}}$. When the diodes are acting as shorts, C_2 is effectively removed from the circuit. Then $C_e = C_1$ and the low resonance frequency is given by $\nu_{\text{low}} = \frac{1}{2\pi\sqrt{LC_1}}$. For a

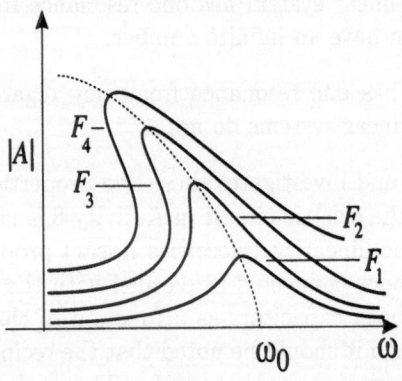

Figure 22.3: Amplitude $|A|$ vs. frequency (ω) for a soft spring oscillator.

soft spring oscillator the magnitude of the amplitude $|A|$ versus the frequency ω is as shown in Figure 22.3. Notice that as the forcing amplitude F increases $(F_4 > F_3 > F_2 > F_1)$ all the resonance frequency curves tilt to the left. This causes the respective resonance frequency (ω_0) curve (the dashed line) to also tilt to the left.

The piecewise linear function produces a similar tilting. The resonance curve for a small forcing amplitude (voltage) starts at ν_{high} and cannot go any lower than ν_{low}. A plot of the forcing amplitude as a function of resonance frequency is shown in Figure 22.4. For comparison, a plot of the forcing amplitude–resonance

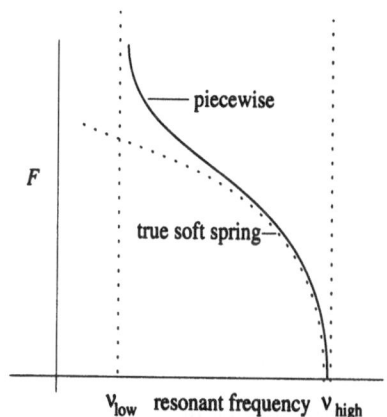

Figure 22.4: Forcing amplitude vs. resonance frequency.

frequency for a true "soft spring" oscillator is also shown on the graph. In this activity a plot similar to the solid curve in Figure 22.4 is produced.

Procedure:

1. Construct the circuit as shown in Figure 22.5.

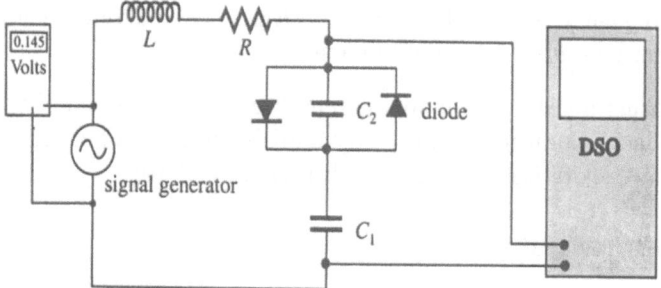

Figure 22.5: Piecewise linear circuit.

2. Use a signal generator with a low output impedance (50 Ω). The value of the inductance is not critical, but generally the larger the values the

better. The 0.80 H, 67 Ω, Berkeley lab solenoid works well. Use capacitors that range from 1 μF to 0.047 μF. Keep C_1 larger than C_2 so that the piecewise bend is pronounced. A reasonable value for R is 50 Ω or lower. You want the circuit to have a large quality factor (Q) because a large Q means a sharp resonant spike. Remember that for small damping, $Q = \frac{\omega}{2\gamma}$. Here $2\gamma = \frac{R}{L}$ and $\omega = \frac{1}{\sqrt{LC_e}}$ and therefore $Q = \frac{\sqrt{L}}{R\sqrt{C_e}}$. To get this large Q, you might wish to leave the external resistance R out of the circuit and rely only on the small intrinsic resistance of the wiring. However, if you do remove the external resistance, it makes it more difficult to keep the forcing amplitude constant as you near resonance. The reason for this is that the impedance of the external circuit gets smaller and smaller as you approach resonance; so if there was no external resistance and if the signal generator had no output impedance, the current in the circuit would be infinite. Signal generators are given an output impedance to prevent this catastrophe.

3. Calculate the two limiting resonant frequencies, ν_{high} and ν_{low}.

4. If the high resonance frequency is below 1000 Hz, you should be able to use a digital multimeter to check the signal generator's output voltage (RMS). If not, use an oscilloscope.

5. With the signal generator producing a low amplitude signal of around 0.10 volts (RMS), locate the circuit's resonant frequency. Do this by slowly increasing or decreasing the signal generator's frequency and watching the CRT trace (or meter) until a maximum amplitude is observed. After the resonant frequency has been located, note the forcing amplitude as it probably has changed a little.

6. Increase the forcing amplitude to 0.20 volts and repeat the above step.

7. Continue to increase the forcing amplitude in steps of 0.10 volts and repeat the above measurements.

8. When the forcing amplitude nears 1.0 volts, you will be approaching the low frequency limit of this circuit. Decide for yourself when and where the last measurement should be made.

9. Construct a graph of the forcing amplitude as a function of the resonant frequency. Confirm the similarity and the difference between your curve and the theoretical curve for a soft spring oscillator?

Things to Investigate:

- Repeat the experiment with capacitors that have different values.

- Did you see any evidence of hysteresis as the frequency was varied? If you wish to modify the circuit to investigate hysteresis, see [FJB85].

- Simulate this circuit's performance with a SPICE program or with a similar program such as MICRO-CAP.

Experimental Activity 23

Nonlinear Resonance Curve: Magnetic

Comment: This activity should take less than two hours to complete.

Reference: Section 8.2

Object: To investigate and construct a nonlinear resonance response curve for the forced hard spring Duffing equation.

Theory: Nonlinear and linear resonance response curves are constructed by plotting the modulus of the steady state amplitude $|A|$ versus the driving frequency (ω). Figure 23.1 shows typical linear and nonlinear resonance response curves. In the nonlinear case as the driving frequency is slowly increased from O, a point C is reached where the value for the amplitude undergoes a sudden drop from C to D and then continues to point E and beyond. However, if the

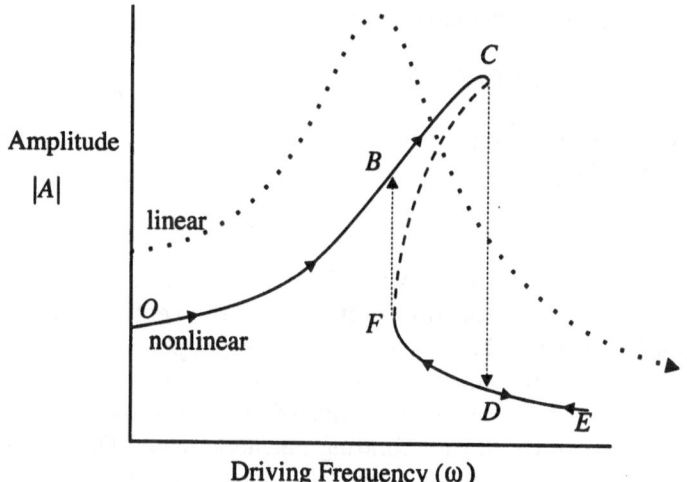

Figure 23.1: Linear and nonlinear resonance response curves.

frequency is slowly decreased from point E, the amplitude will undergo a quick jump from F to B and then will continue along the curve to O. The frequency at which the jump from F to B occurs is different than that of the drop frequency. The phenomenon of following different system paths when moving in different directions is known as hysteresis.

In this experiment a small cylindrical bar magnet will be suspended in an oscillating magnetic field. The torque (τ) between the magnetic moment (μ)

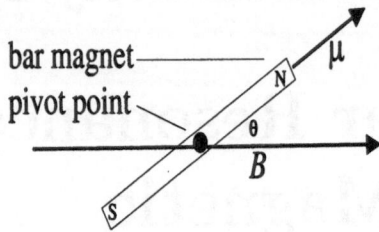

Figure 23.2: Magnet suspended in an external magnetic field.

and the vibrating magnetic field (B) provides the forcing function

$$\tau_f = \mu B \sin \theta. \tag{23.1}$$

The magnet is suspended by using VHS video recording tape, held under tension. The tape acts as a hard spring. It is assumed the tape provides a restoring torque similar in form to the equation

$$\tau_r = -\alpha\theta - \beta\theta^n, \tag{23.2}$$

where n is approximately 3 and α and β are positive constants. The net torque is $\tau = I\ddot{\theta}$ where I is the moment of inertia of the cylindrical magnet. The vibrating magnetic field has the form $B = B_0 \cos(\omega t)$ where B_0 is the maximum value of the magnetic field. Combining the above equations, and taking $n = 3$, produces the nonlinear differential equation

$$I\ddot{\theta} + \alpha\theta + \beta\theta^3 = \mu B_0 \cos(\omega t) \sin \theta. \tag{23.3}$$

As the driving frequency ω is varied, hysteresis may occur. This is the subject of the present activity.

Procedure:

1. Construct the apparatus shown in Figure 23.3. Place the magnet as close as possible to the solenoid or better yet at the center of the solenoid or Helmholtz coils. The magnet's supporting apparatus was built with a wooden Tinker Toy construction set. This supporting apparatus allows the tape's tension to be easily adjusted, and also allows the magnet to be placed correctly in the vibrating magnetic field. The magnet is held between a piece of doubled over VHS video tape. Use masking tape to pinch the VHS tape just below where the magnet is to be supported. Adjust the VHS tape's tension if the resulting oscillations are too small

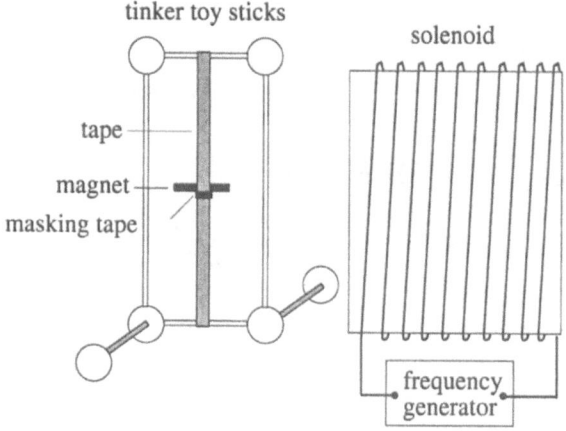

Figure 23.3: Solenoid, suspended magnet, and power supply.

or large. The solenoid that comes with the Berkeley Physics Experiments works very well in this activity, but any large solenoid or Helmholtz coils could be used. A cylindrical magnet about 5 cm long with a diameter of 0.50 cm works well, but any small strong bar magnet should work.

2. With the frequency generator set to produce a large (maximum) amplitude signal, start from a low frequency and slowly increase the driving frequency. Record the frequency where the amplitude undergoes a sudden drop.

3. Starting at a high frequency slowly reduce the driving frequency and record the frequency where the sudden jump occurs. Was hysteresis present?

4. Repeat the steps, but now record the values for the amplitude as the frequency is increased and decreased. Take two or three measurements past the drop and jump frequency. The amplitude can be measured by casting a shadow of the vibrating magnet on a wall. For a more accurate measurement of the amplitude, attach a small mirror to the magnet and reflect a laser beam off the mirror and onto a wall.

5. Construct an amplitude (magnitude) vs. frequency graph.

Things to Investigate:

- Repeat the activity using a different magnet (size or strength).

- Repeat the activity using a different tape tension.

- Design an experiment to find the relationship between the tape's restoring torque and the angular displacement.

- Calculate the magnetic moment (μ) by carrying out the following steps:

    1. Calculate the value of the external uniform magnetic field (B_0);

2. Displace the magnet a small angle, say 10°, from its equilibrium position;

3. Release the magnet and measure the period (T) of its small vibrations;

4. Develop and use the equation

$$\mu = \frac{4\pi^2 I}{T^2 B_0}$$

to calculate the magnetic moment (μ).

Experimental Activity 24

Subharmonic Response: Period Doubling

Comment: This activity should take less than one hour to complete.

Reference: Section 8.3

Object: To investigate period doubling and chaotic behavior.

Theory: One of the main precursors that a oscillating system might become chaotic as a control parameter is changed is that it display period doubling. Figure 24.1 is a sketch of a period-2 signal, the pattern repeating itself after two

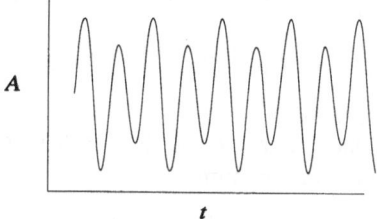

Figure 24.1: A signal showing a period 2 characteristic.

oscillations. In this activity, a small cylindrical bar magnet is suspended in a sinusoidal time-varying magnetic field with a tape which provides a hard spring restoring force. See Experimental Activity 23 for a derivation of the relevant nonlinear ODE.

Procedure:

1. Construct the apparatus in Figure 24.2. Make sure the magnet is as close as possible to the solenoid or, better yet, at the center of the solenoid.

2. The magnet's supporting apparatus can be built with a Tinker Toy construction set. This set allows for easy adjustment of the tape's tension by pushing the sticks in or out of their supporting blocks, and it also allows

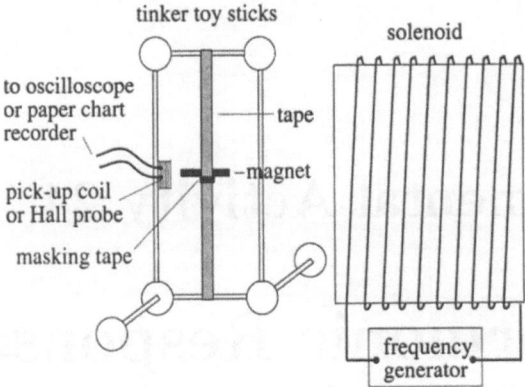

Figure 24.2: Solenoid, magnet, power supply, and pick-up coil.

for building different sizes of supports so that the magnet can be correctly placed in the vibrating magnetic field. The magnet is held between a piece of doubled-over VHS video tape. Use masking tape to pinch the VHS tape just below where the magnet is to be supported. If the magnet's amplitude is too small or too large, adjust the tape's tension. The driving solenoid that comes with the Berkeley physics experiments package works very well, but any large solenoid or Helmholtz coil could be used. A 5 cm long, 0.5 cm diameter, cylindrical magnet was used, but any small strong bar magnet should work. A small pick-up coil (CENCO's #79735-01T) or a Hall probe can be used to monitor the signal.

3. Connect the Hall probe or pick-up coil to a digital storage oscilloscope.

4. With the frequency generator set at suitable (usually maximum) power and at a low frequency, increase the driving frequency until the magnet's amplitude is just at its maximum.

5. Slowly reduce the frequency while watching the CRT trace for period doubling, quadrupling, etc.

6. If you are using a scope, draw sketches of the waveforms.

7. For a period-2 signal, measure the time between two large spikes and between a large spike and a smaller spike. Calculate the driving period. Which of the measured times is equal to the driving period?

8. Try to adjust the driving frequency to produce a period-4 solution. Repeat the previous step.

Things to Investigate:

- Repeat the activity using a different magnet (size or strength) and/or a different tape tension.

- Design an experiment to find the relationship between the tape's restoring torque and the angular displacement.

Experimental Activity 25

Diode: Period Doubling

Comment: This activity should not take more than one hour to complete.

References: Section 8.3

1. [HJ93] Source of the circuit used in this activity.

2. [Bri87] Discussion of how diodes produce period doubling.

Object: To investigate the period doubling route to chaos.

Theory: Although one of the most common uses for a diode is to rectify AC current, here it is used to produce a signal that exhibits period doubling and quadrupling. Figure 25.1 shows the simple circuit used in this experiment. Initially there was some dispute of how a diode produces a period doubling

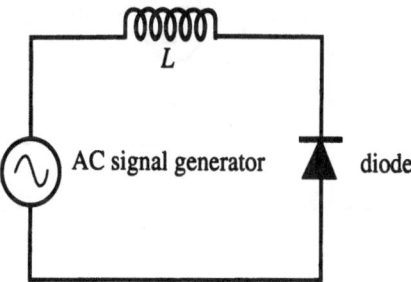

Figure 25.1: Diode circuit.

signal, but more recently a consensus has been reached. An explanation based on [Bri87] and [HJ93] is now presented.

For the circuit of Figure 25.1, period doubling takes place if the circuit frequency is near the diode's resonance frequency. This frequency is determined by the diode's internal capacitance and the external inductance. The diode's resonance frequency does not remain constant because the diode capacitance changes as the voltage across the diode increases (see Figure 25.2.) When the

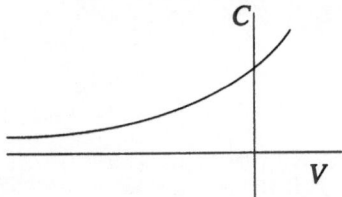

Figure 25.2: Capacitance vs. voltage for a diode

correct frequency set by the AC signal generator is present, the amount of current flowing through the diode can affect the time it takes the diode to recover to its reverse bias equilibrium. The nonlinearity arises from the unrecombined charges that have crossed the diode's p–n junction when the diode is in the forward bias mode. These charges do not instantly recombine when the diode switches to the reverse bias mode, but diffuse back to their "home" region. Initially, these unrecombined charges act like a battery that produces a transient reverse bias current. This transient reverse bias current can be much larger than the reverse bias saturation current and even stranger, it may be larger than the forward bias current. To see how this large reverse bias current occurs, consider a signal generator wired in series with a resistance (R) and a diode. Assume the signal generator is producing a square wave of amplitude V_0. See Figure 25.3.

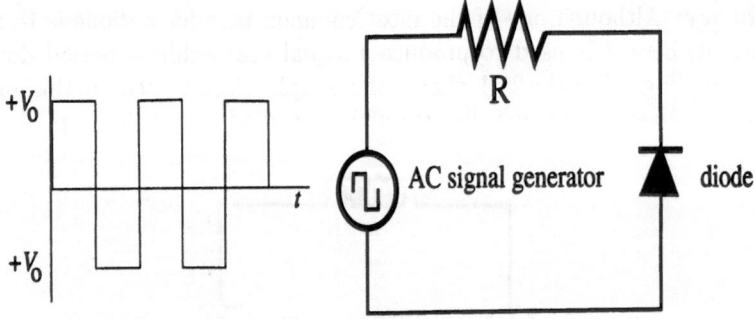

Figure 25.3: Square wave input for diode circuit

When in the forward bias, the value for the current is given by $i = \frac{V_0 - V_d}{R}$. When the square wave instantly switches polarity, the diode is placed in the reverse bias mode. The trapped unrecombined charges maintain the forward diode potential of V_d, so the reverse current is given by $i = \frac{-V_0 - V_d}{R}$ which (momentarily) has a larger magnitude than the forward bias current. In fact, if V_0 is just a little larger than V_d, then this transient reverse bias current can be many times larger than the forward bias current. The trapped unrecombined charges act to delay the recovery of the diode. As the charges flow back across the junction, the depletion region widens and the transient reverse bias current decreases until its value reaches the value of the normal reverse bias current. The larger the forward current, the greater the number of unrecombined charges and the longer it takes for the diode to reach its reverse bias equilibrium. Figure 25.4 shows a set of plots that indicate the behavior of the diode as it switches from forward to reverse bias. The 1st plot in Figure 25.4 shows current vs. time. Here t_s

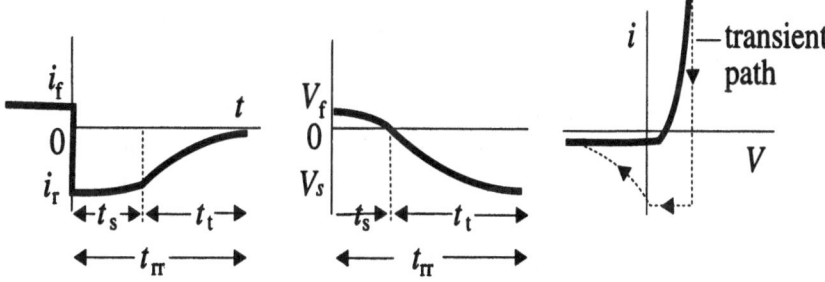

Figure 25.4: Curves for a diode switching from forward bias to reverse bias.

is the storage time (the time that unrecombined charges are still providing a potential), t_t is the transition time (the time required for the large reverse bias current to return to the normal reverse bias current), and t_{rr} is the total reverse recovery time. The 2nd plot shows the diode potential difference vs. time. The 3rd plot shows current vs. voltage for a normal diode with the transient current path overlaid upon it.

For certain frequencies, the circuit can switch back to its forward bias mode before the reverse bias equilibrium has been fully established. When this occurs, the behavior of the diode's next cycle depends on the parameters of the previous cycle. Each subsequent cycle could have a different initial condition. If this were to occur, chaos would result. To produce a period-2, period-4, etc., solution, it is necessary that the initial conditions are appropriately repeated.

Procedure:

1. Set up the apparatus as shown in Figure 25.5. The diode 1N4001 is inexpensive and works well.

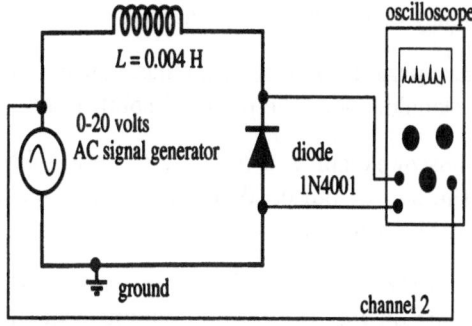

Figure 25.5: Period doubling circuit.

2. The solenoid recommended for the PSSC apparatus works well as the inductor. It has very low resistance with an inductance of about 4 mH.

3. The signal strength and frequency of the generator is read from Channel 2. This signal acts as a reference for the signal displayed on Channel 1. The trace displayed on Channel 1 is the voltage–time signal for the diode. It is this signal which shows the period doubling.

4. Explore generator frequencies between 10 kHz and 90 kHz. For the first measurement, set the generator to a low frequency and a low amplitude signal. With the generator producing a constant frequency slowly increase the strength (amplitude) of the signal. Watch Channel 1 on the oscilloscope for signs of period doubling. (The pattern repeats every two oscillations.) If no period doubling occurs, reduce the generator's amplitude to zero, increase the generator's frequency by 5 kHz and slowly increase the signal's amplitude.

5. Repeat the above procedure until a period-2 signal occurs. Record the frequency and amplitude of the signal.

6. When period doubling is observed, record the generator's frequency and amplitude (signal strength) from the trace shown on Channel 2.

7. Continue increasing the generator's amplitude to see if period 4 or even period 8 can be detected. When first observed, record the generator's amplitude. (When the authors performed this activity, a period-4 signal was located at 50 kHz with an AC signal voltage of about 4 volts.)

8. For what generator frequency range does the period doubling sequence occur?

9. To produce phase plane portraits, switch the oscilloscope to its x–y setting. Sketch the CRT traces for period 2, 4, etc.

Things to Investigate:

• Use a larger inductance and repeat the above steps. Does this lower or raise the frequency at which period 2 starts?

• At a point where period 2 is just occurring, slowly insert a ferromagnetic core into the solenoid. Does the period-2 solution remain?

• When the oscilloscope displays a period-2 or period-4 signal, set the scope to trigger on the largest pulse. Watch to see if the smaller pulse always occur at exactly the same place.

• If the strength of the signal is really the parameter that causes the nonlinearity, speculate on why the period doubling sequence disappears when the signal strength is increased past a certain value.

Experimental Activity 26

Five-Well Magnetic Potential

Comment: A mainly qualitative activity taking about an hour to complete.

References: Section 8.3

1. See the magnetic basin Problem 8-18 on page 316 .

2. [Moo92] Contains explanations of multi-well potentials.

Object: To explore the chaotic motion of a spherical pendulum moving in a five-well magnetic potential.

Theory: This activity uses the commercially available toy shown in Figure 26.1. The light (massless) plastic pendulum rod has a disk magnet attached to its lower end (the bob). A small magnet at the top of the pendulum's plastic rod

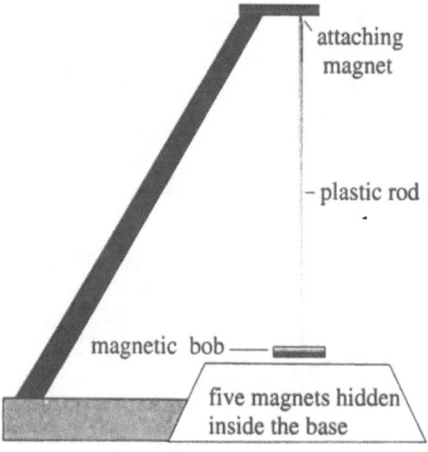

Figure 26.1: Spherical pendulum: A five-well potential toy.

allows it to adhere to a metal plate in the supporting column. This pivoting arrangement allows the pendulum to swing in all directions. The supporting base of the toy contains five permanent disk magnets that are equally spaced around the bob's central, stable equilibrium position. These five magnets exert a repelling force on the pendulum's magnetic bob. Also located in the bottom of the toy's supporting structure is a simple circuit that consists of an iron core solenoid, a transistor, and a nine volt battery. Every time the swinging bob passes over its central equilibrium position, it induces a current in the solenoid. This small induced current is applied to the base terminal of the transistor which momentarily turns it on, thus allowing a much larger current to flow through the solenoid. The electromagnet (iron-core solenoid) produces a repelling magnetic field which gives the pendulum's bob a sharp repelling push every time it passes over its central equilibrium position. In this manner energy is added to the pendulum and it keeps swinging chaotically as long as the circuit is functioning.

A two-dimensional x–y equipotential diagram for a five-well magnetic arrangement is given in Figure 26.2. This diagram was created with Mathematica file X26 assuming a $V \propto \frac{1}{R^3}$ ($F \propto \frac{1}{R^4}$) magnetic potential.

5-Well Magnetic Potential

This file is used to study the motion of the spherical pendulum in a 5-well magnetic potential. Figures 26.2 and 26.3 are generated with this file. Mathematica commands in file: Sum, Do, Cos, Sqrt, Solve, ImplicitPlot, Plot3D, Table, SetCoordinates, Grad, Part, NDSolve, ParametricPlot3D, Cos, ParametricPlot, MaxSteps, UnitStep, BoxRatios, Evaluate

---

Figure 26.2: Equipotential lines for a five-well magnetic potential.

The equipotential lines show the locations of the five unstable equilibrium positions, located above the five repelling magnets, and the central stable equilibrium position of the bob. The five unstable equilibrium positions indicate that there should be many unstable saddle points in the bob's phase space. Subsequently, the pendulum bob has a high probability of exhibiting chaotic motion. To get a better feeling for how the bob's path is controlled by the potential well, it is instructive to construct a three dimensional plot of the above 2D plot. Figure 26.3 shows a plot of the potential V vs. the coordinates x and y, also generated with file X26. Seen in this way, it is easier to imagine how a

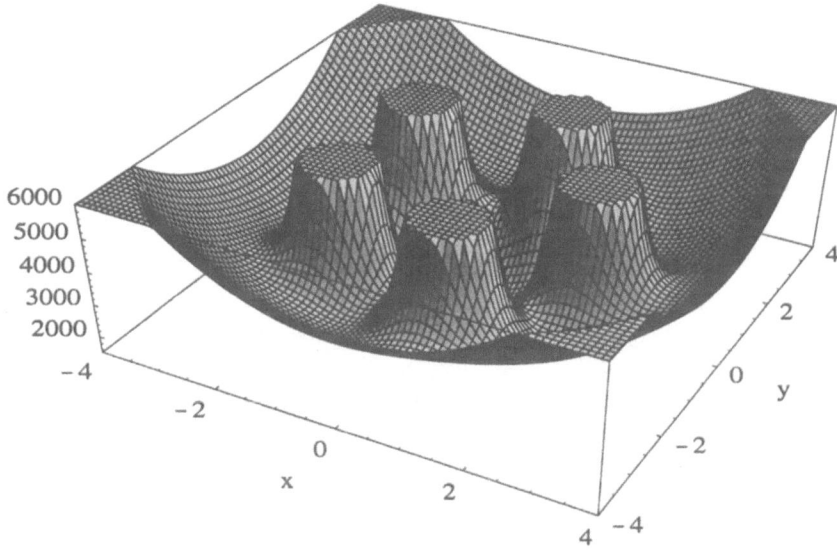

Figure 26.3: A 3D Plot for a five-well magnetic potential.

low friction sliding glider moves in the well for you can easily spot the peaks and valleys of the well. Mathematica file X26 may be used to explore various orientations of the last two plots or to produce your own contour plots.

The total potential energy (V_t) of the bob (mass m) moving in the magnetic and gravitational field is

$$V_t = \sum_{j=1}^{n} \frac{k_j}{R_j^3} + mgz, \qquad (26.1)$$

where n is the number of magnets, k_j are constants, and R_j is the distance from the jth magnet to the x, y, z position of the bob. The differential equations describing the motion are of the structure

$$m\ddot{\mathbf{r}}_i + 2\gamma\dot{\mathbf{r}}_i + \nabla V_t = F(\mathbf{r}_i, \dot{\mathbf{r}}_i, t) \qquad (26.2)$$

where $\mathbf{r}_i$ is the vector position $(i = 1, 2, 3)$ of the bob with respect to its stable equilibrium position, γ is the drag coefficient, and F is a forcing function. The component equations are complicated and tedious to develop and solve by hand so this is relegated to Mathematica file X26.

Procedure A: This procedure (which may be omitted if you have done Experimental Activity 1) checks to see if the magnets repel each other with a force that varies as the inverse fourth power of their separation distance. This was the assumed form in producing the equipotential and 3-dimensional plots.

1. Using two thin strong neodymium magnets oriented so they exert repelling forces, construct the apparatus shown in Figure 26.4.

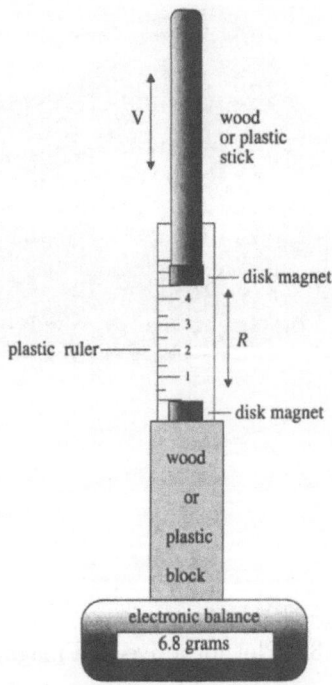

Figure 26.4: Apparatus to find the form of the force law.

2. With the second movable magnet far away from the magnet fastened on the balance, zero the balance.

3. Slowly bring the second magnetic disk towards the secured magnet. Record the values for the separation distances and repelling forces.

4. What is the relationship between the force and the separation distance?

5. What is the theoretical prediction for two dipoles? How does your force law compare with this theoretical prediction?

6. Is the repulsive force always pointing radially away from the pole face of the disk magnet? Does this create a problem in this experiment?

Procedure B: This procedure involves exploring Mathematica file X26 and comparing with what is experimentally observed.

1. If the driving piecewise force in the Mathematica file is made equal to zero (i.e., removed) does the spherical pendulum damp to the center position? If it does not, reduce the strength of the repelling force (potential) until it does. What happens when you remove the battery from the toy?

2. With no damping or piecewise force, see if you can make the pendulum's bob follow one of the equipotential lines shown in Figure 26.2. To do this you will need to apply an initial speed. Which line should be the easiest to follow? Can you do this with the actual toy?

3. Make a hard copy plot of the $x(t)$ vs. t motion. What does this plot (motion) indicate about the motion? Is it chaotic? Can you detect any periodicity? How about for the actual toy?

4. Change the form for the potential from $V \propto 1/R^3$ to other inverse power laws to see how Mathematica handles these changes. Discuss the results.

Procedure C: In this procedure, the location of the bob versus the time the pendulum is at that location is recorded. The data are plotted and the plot is examined for signs of chaos.

1. Use a Hall probe connected to a paper chart recorder or storage oscilloscope to produce a plot of the time the bob was at some specific location.

2. Compare the plot made when the Hall probe is located at some outside position with that when it is located at the central equilibrium position.

3. Do the plots indicate any regular or recurring pattern? Is chaos indicated?

4. You might wish to use more than one Hall probe to record on the same paper, or on two different strips of paper, a record of its motion.

5. If you make hard copy records, you might wish to store them and use the results in a later experimental procedure.

An English student's five-well problem.

Experimental Activity 27

Power Spectrum

Comment: This experiment should take less than two hours to complete.

Reference: Section 8.4

Object: To use a fast Fourier transform (FFT) to construct a power spectrum from experimental data.

Theory: A power spectrum is an important and useful diagnostic tool in trying to ascertain the frequency content of an oscillating time series. Section 8.4 should be reviewed before attempting this activity.

One of the main precursors of the onset of chaotic motion is the appearance of an oscillation that exhibits period doubling, a typical plot being shown in Figure 27.1. Data can be lifted from time series plots so that a corresponding

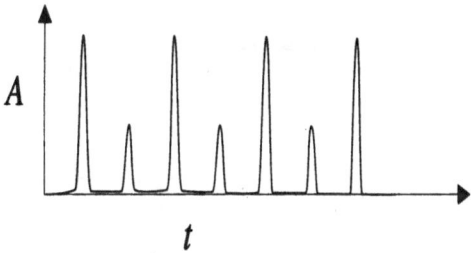

Figure 27.1: A signal showing a period-2 characteristic.

power spectrum can be constructed. For example, if the above graph contained a large number of peaks, and if the sampling data was collected correctly, a plot similar to the one shown in Figure 27.2 would result. The tall peak is at the driving frequency and the smaller peak to the left of the tall peak is a subharmonic at half the driving frequency and thus indicates period doubling. The spike to the right of the center peak is called an ultrasubharmonic and is at three times the frequency of the subharmonic peak. In this experiment a complicated time series graph will be analyzed to produce a power spectrum.

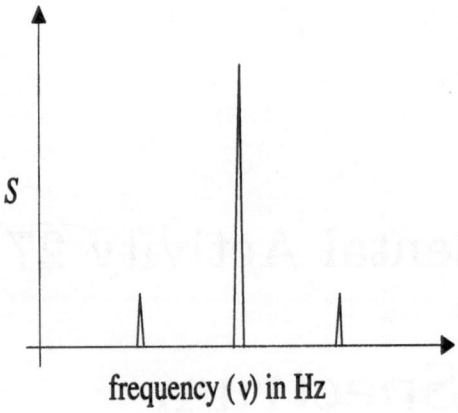

Figure 27.2: Power spectrum S for a period-2 solution.

Procedure:

1. If you have a hard copy of data from any previous activity, analyze that data and omit the following data collection steps.

2. If you do not have hard copy, construct the apparatus shown in Figure 27.3. The details of construction and use may be found in Experimental Activity 24.

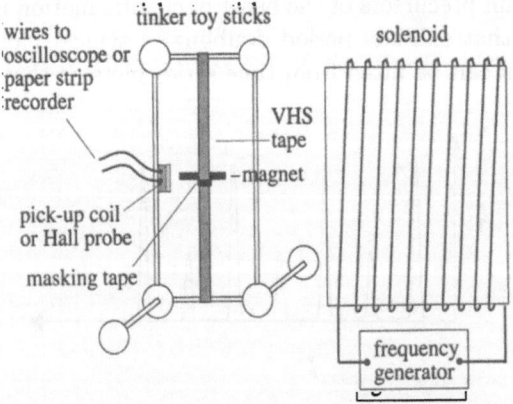

Figure 27.3: Solenoid, suspended magnet, power supply, and pick-up coil.

3. Connect the pick-up coil or a Hall probe to a paper strip recorder.

4. With the frequency generator set at a suitable (usually maximum) power level and at a low frequency, increase the driving frequency until the magnet's amplitude is just at its maximum. The magnet should vibrate, not rotate.

5. Slowly reduce the frequency while watching the strip recorder's plot for examples of period doubling, tripling, etc. When a reproducible pattern is noticed, leave the signal generator at this frequency.

6. After some time and when the trace shows a large number (> 20) of similar oscillations, record the signal generator's driving frequency and the speed (cm/s) of the paper moving through the strip recorder.

7. Use the trace on the strip of paper to calculate the fundamental frequency and compare it with the frequency of the signal generator.

8. Select a suitable portion of the trace and subdivide it into equal time intervals. Each interval is equal to the sampling period (T_s). For each marked point T_s, $2T_s$, $3T_s$, ..., record the amplitude of the wave at that specific spot. Experience has shown that about 64 points are a minimum number to sample, but the more the merrier.

9. Mathematica file X27 calculates the power spectrum. Place the measured amplitudes into this file and run it. The file provides additional instructions to help you do this.

Power Spectrum
This file can be used to calculate the power spectrum from the experimental data. Mathematica commands in file: `ListPlot`, `PlotRange->All`, `PlotJoined->True`, `PlotStyle`, `TextStyle`, `ImageSize`, `Length`, `Fourier`, `Take`, `Chop`, `Ticks`, `Hue`, `PlotLabel`, `AxesLabel`

---

10. Does the power spectrum confirm the periodicity that you originally observed on the trace?

11. Locate the peak that corresponds to the fundamental frequency. Locate peaks to the left of the fundamental frequency which indicate the occurrence of subharmonics. Identify these peaks by their frequencies.

12. Locate the peaks to the right of the fundamental frequency. Identify them in terms of multiples of the fundamental and subharmonic frequencies.

13. Repeat the procedure for twice as many data points.

Things to Investigate:

- Do a power spectrum of a stock market graph showing a stock's fluctuating value as a function of time.

- Do a power spectrum of some other type of data, such as weather or climatic effects, biological data, etc.

Experimental Activity 28

Entrainment and Quasiperiodicity

Comment: This activity should not take more than one hour to complete.

References: Section 8.6.1

1. [HJ93] This article provided the idea for the following activity.

Object: To investigate entrainment and the quasiperiodic route to chaos.

Theory: In this experiment, a diode is wired in series with an inductor and then placed in parallel with another diode and inductor as shown in Figure 28.1. Experimental Activity 25 showed that a single branch of this circuit can produce

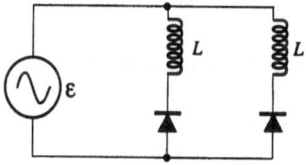

Figure 28.1: Circuit to exhibit entrainment and quasiperiodicity.

period doubling when operating at the correct frequency and signal strength. When in tandem, the parallel branches modulate the behavior of each other. As the signal strength increases and as each branch begins to exhibit period doubling, the frequencies remain close but incommensurate (the frequency ratio is irrational), but as the signal strength is increased even further, the frequencies can phase lock or become entrained. At even higher signal strengths, quasiperiodicity arises as the signal moves back and forth between each inductor's preferred frequency.

Procedure:

1. Set up the apparatus as shown in Figure 28.2. For the diodes, use any general purpose diode, e.g., 1N4001.

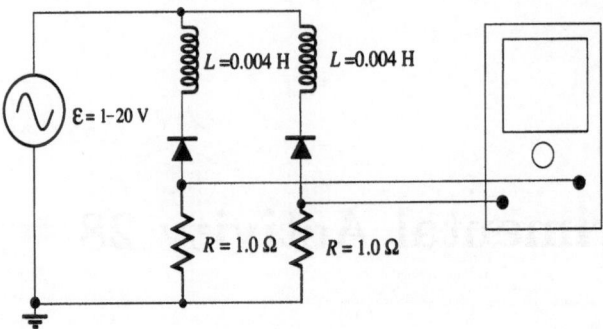

Figure 28.2: Experimental setup.

2. The solenoid recommended for the PSSC apparatus works well as the inductor. It has very low resistance with an inductance of about 4 mH.

3. The two signal strengths are monitored across each of the resistors and fed into channels 1 and 2 on the oscilloscope.

4. If Experimental Activity 25 has been completed, the signal strengths and frequencies needed to produce period 2 are known. If not, frequencies between 10 kHz and 90 kHz should produce the desired effects.

5. Set the generator so that it produces a low amplitude signal. With the generator producing a constant frequency, slowly increase the signal strength. Watch the oscilloscope in its dual channel mode for signs of period doubling in each branch of the circuit. Switch to the x–y setting to observe the coupling of the two signals. Figure 28.3 is a schematic representation of a quasiperiodic shape that can be observed in the x–y mode. Switching the scope back and forth from its x–y setting to its dual-channel mode may help you decipher what the individual and coupled signals are doing.

Figure 28.3: The folded torus known as the Klein bottle condom.

6. Alter the amplitude or frequency and search for evidence of entrainment.

7. Continue increasing the signal amplitude to see if you can detect when chaos begins.

Things to Investigate:

- Use one solenoid with a little smaller inductance than the other and see if you can still make the frequencies entrain.

Experimental Activity 29

Quasiperiodicity

Comment: This experiment should not take more than one hour to complete.

References: Section 8.6.2

1. [Moo92] The idea for the circuit in this activity was found in this book.

2. Experimental Activity 17 dealing with a neon bulb relaxation oscillation.

Object: To investigate the various ways that two neon bulb relaxation oscillators can interact to produce quasiperiodic and chaotic flashes of light.

Theory: Quasiperiodicity occurs when neither frequency of two coupled oscillators wins out and the coupled system jumps from one frequency to the other. This activity investigates this behavior for two coupled neon glow lamps. A glow lamp (see Figure 17.1) consists of a 1 cm long bulb containing two vertical metal electrodes separated by about 2 mm and surrounded by a low pressure neon gas. When a potential difference is applied across the electrodes, it creates an electric field that accelerates charged ions and electrons inside the bulb that exist due to cosmic radiation and natural radioactivity. A sufficiently large electric field can accelerate the electrons to a speed that is sufficient to create an electron avalanche. The negative electrode attracts the positive neon ions and the electrons leaving the electrode recombine with the positive neon ions to produce a soft glow. Accordingly, a neon glow lamp is considered to have a very high resistance until the correct potential difference exists across the electrodes. At the critical potential difference—the firing voltage (V_f)—the presence of a seed electron creates an avalanche of electrons and the bulb begins to conduct and the lamp glows. Once this occurs, the lamp's resistance decreases dramatically, so a current limiting resistance is usually placed in series with the bulb, its value normally selected so that the bulb stays glowing without burning out or turning off. As the current through the bulb increases and the potential drop across the bulb decreases, the potential drop across the resistance increases. Accordingly, three operating conditions are possible. If the current limiting resistance is very large, the bulb turns off; if it is too small, the current continues to increase and the bulb burns out; if it is correctly selected, the bulb remains lit. Because neon bulbs consume very little power and have a very long life, they are used to indicate if a device such as a power supply is turned on. For a more complete description of how the glow lamp functions see Experimental Activity 17.

Procedure: The voltages used in this activity can produce severe or even fatal electrical shocks. Always turn off the circuit and let the capacitors discharge before making changes in the circuit.

1. Build the circuit shown in Figure 29.1. The variable DC power source

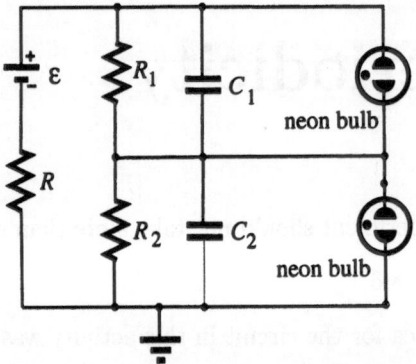

Figure 29.1: Coupled neon bulb circuit.

(ε) should be able to produce values from 0 to 400 V. The resistors R_1 and R_2 should each be 1 MΩ. The capacitors C_1 and C_2 should have identical values around 0.47 μF. The capacitors should be rated at least half the maximum voltage used, so 250 V is a safe value. The glow lamp is of the class Ne-2, such bulbs having a variety of extinction and firing voltages but V_f is usually 70 to 110 V and the extinction voltage 50 to 70 V. Bulbs with lower V_f are preferable, because this permits lower voltage power supplies to be used.

2. Slowly increase the source voltage until the lights just begin to flash.

3. Slowly increase or decrease the source voltage to produce flashes that might indicate that chaotic oscillations could be imminent. Hint: Do the flashes indicate one or more of the following properties: quasiperiodicity, entrainment (mode-locking), period doubling, or intermittency?

4. Try to adjust the source voltage to produce chaos.

Things to Investigate:

- Replace R_1 with a variable resistor that differs from R_2 by 5% to 10%. Repeat the above steps.

- Add a third circuit to make three coupled oscillators.

Experimental Activity 30

Chua's Butterfly

Comment: This activity should not take more than three hours to complete.

References: Section 8.7.2

1. [HJ93] This article is the source of the circuit used in this activity.

Object: To produce a chaos-exhibiting double-scroll strange attractor similar to the Lorenz attractor.

Theory: In this activity, an op-amp circuit is used to produce a negative resistance (r) similar to that encountered in previous activities, but with one main difference. The negative resistance is made piecewise linear so that the current–voltage relation is as shown in Figure 30.1. The negative resistance

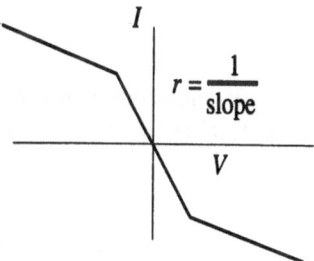

Figure 30.1: Piecewise linear negative resistance.

allows an oscillating signal to occur and its piecewise linear nature allows the system to become chaotic.

An equivalent circuit to the one used in this activity is shown in Figure 30.2. Although the purpose here is mainly to observe the period doubling route to chaos, the governing equations of this circuit are now developed using the state variables method. The state variables method is particularly useful when applied to nonlinear systems because it allows the system to be studied by locating and identifying the stationary points. First the state variables needed to describe the system are determined. For electrical systems the common state

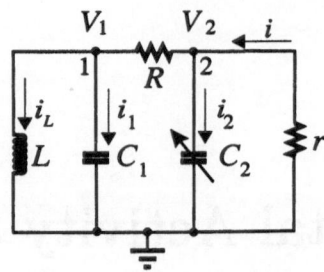

Figure 30.2: Equivalent circuit for state variable analysis.

variables are the voltage drops across the capacitors and the currents flowing through the inductors, but the choice of state variables is not unique. A few simple rules can be used to help set up the equations:

1. Identify the smallest number of variables that are needed to solve the system. The minimum number is equal to the number of degrees of freedom;

2. A rule of thumb is that the number of degrees of freedom is equal to the number of primary loops. This is the same idea as identifying the degrees of freedom in a mechanical system;

3. The number of equations that must be constructed is equal to the number of state variables;

4. The left side of the equations should only contain first-order time derivatives of one of the state variables;

5. No derivatives should appear on the right side of any of the equal signs;

6. The right side of the set of equations contains only the state variables, their coefficients, and the forcing functions if any are present;

7. Kirchhoff's voltage and current rules are used to set up the equations.

As an example of how the state variables method works, consider the circuit of Figure 30.2. The circuit has 3 primary loops so it has 3 degrees of freedom and a minimum of 3 state variables (and, thus, 3 equations) are required. The variables for this circuit are chosen to be V_1 (the potential drop across C_1), V_2 (the drop across C_2), and i_L (the current through L). The sum of the currents into junction 2 must equal the sum of the currents out of junction 2, so

$$i = C_2 \frac{dV_2}{dt} + \frac{V_2 - V_1}{R}. \tag{30.1}$$

The above current i is equal to the voltage drop across the negative resistor divided by the negative resistance, so $i = -V_2/r$ and Equation (30.1) becomes

$$\frac{dV_2}{dt} = \frac{V_1 - V_2}{C_2 R} - \frac{V_2}{r C_2}. \tag{30.2}$$

Current conservation at nodal point 1 gives

$$\frac{V_2 - V_1}{R} = i_L + i_1. \qquad (30.3)$$

Since $q_1 = C_1 V_1$, then $i_1 = C_1 \frac{dV_1}{dt}$ and Equation (30.3) becomes

$$\frac{dV_1}{dt} = \frac{V_2 - V_1}{C_1 R} - \frac{i_L}{C_1}. \qquad (30.4)$$

Finally, by inspection of the circuit, the third circuit equation is

$$\frac{di_L}{dt} = \frac{V_1}{L}. \qquad (30.5)$$

The required system of three equations ((30.2), (30.4), and (30.5)) is complete. Nonlinear theory has shown that when a physical system is modeled by three coupled first-order nonlinear differential equations, chaos is possible. The investigation of chaotic behavior produced by the circuit which is governed by the above equations is the purpose of this activity.

Procedure:

1. Wire the circuit shown in Figure 30.3. Notice the two diodes which create the piecewise function. These diodes act as switches which do not turn

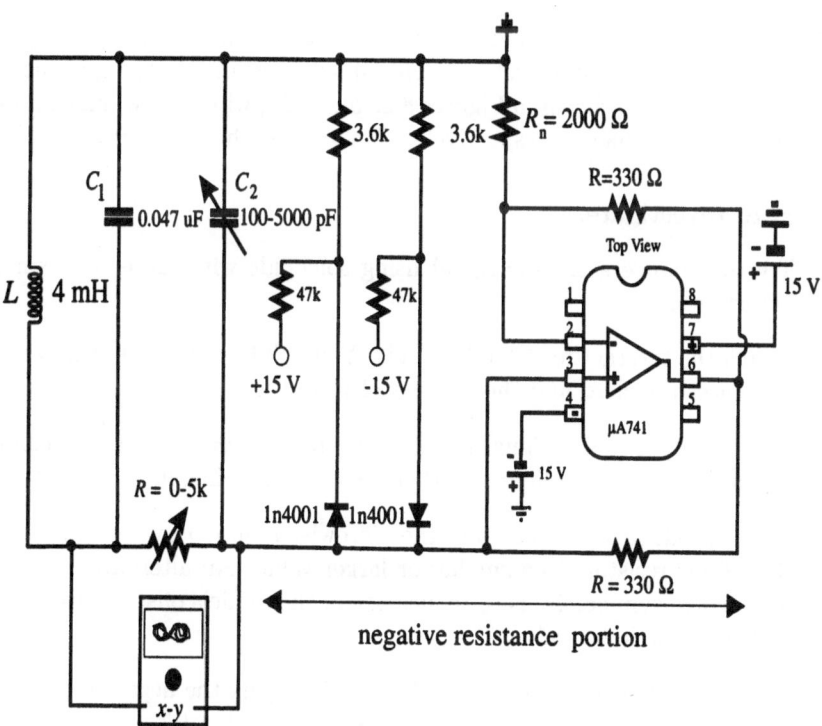

Figure 30.3: Chua's circuit to produce a double scroll strange attractor.

on until a potential drop of approximately 1.7 volts appears across the ground and the bottom wire on the diagram. When the diodes turn on, the 3.6 k resistors become part of the 2000 Ω negative resistance.

2. The PSSC solenoid works well as the inductor. It has very low resistance ($\approx 1\Omega$) with an inductance of about 4 mH.

3. A 0–5.0 kΩ analog variable resistor seemed to work better than a digital dial type.

4. The signal strength is read across the variable external resistor. This resistor is considered to be the positive resistor R in the equivalent circuit Figure 30.2. To see the double-scroll strange attractor, the CRT must be set to its x–y setting.

5. Slowly reduce the resistance of the variable resistor from its maximum value to a point close to the initial negative resistance (2000 Ω in the circuit shown here). When this is done, the CRT trace changes from a single closed loop to one indicating period doubling (the trajectories cross), quadrupling, etc. This circuit is very sensitive to small changes in resistance, so carefully explore the critical region.

6. With finer tuning of the variable resistance, the jump to chaos is easily identified by the appearance of the double scroll. If you have difficulty producing a double scroll, try changing the value of the variable capacitor.

7. When in the double scroll mode, switch the oscilloscope from its x–y setting to its dual channel setting. This should permit you to reproduce the time series plots shown in Figures 8.29 and 9.25, which show quasiperiodic and chaotic behavior, respectively. Can you see the similarities?

Things to Investigate:

- Try to produce a chaotic signal using solenoids with larger and smaller inductances.

- Search the Internet for Chua's circuit. You will be surprised at how many references are cited and the number of ways the circuit is used.

- If you can produce a signal in the audio range (100–15,000 Hz), connect the output to an amplifier and then to a speaker. Listen to the chaos!

- Try to produce asymmetries in the piecewise function by changing one of the 3.6 kΩ resistors to a smaller or larger value. An alternate method of producing asymmetries is to reduce one of the diode's offset voltages from 15 V to some lower value.

- Write a Mathematica file to plot V_2 vs. V_1, taking the negative resistance to be the piecewise function $r = \begin{cases} -2000 & |V_2(t)| < 1.7 \\ -2015 & \text{otherwise} \end{cases}$, $C_2 = 1.1 \times 10^{-9}$ F, and $R \approx 2012\ \Omega$.

Experimental Activity 31

Route to Chaos

Comment: This activity should not take more than two hours to complete.

Reference: Section 9.4

1. [Bri87] This article discusses a similar experiment.

2. [Chi79] Gives a parameter which can be used to decide if the motion in this activity is chaotic.

Object: To investigate forced nonlinear oscillatory motion.

Theory: A cylindrical bar magnet which is free to rotate horizontally about a vertical axis perpendicular to its horizontal longitudinal axis is placed in an external magnetic field (B) as shown in Figure 31.1. In this configuration the magnet will experience a restoring torque (τ) given by

$$\tau = -\mu B \sin \theta \qquad (31.1)$$

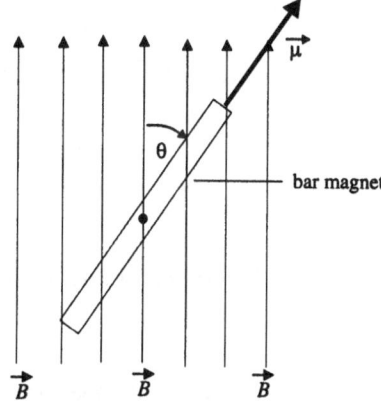

Figure 31.1: Bar magnet in time varying magnetic field.

where μ represents the magnetic moment of the magnet and θ represents the angle between the magnetic field (B) and the magnetic moment. The negative sign in Equation (31.1) indicates the restoring torque is in the opposite direction to the angular displacement (θ). If the magnitude of the magnetic field is a sinusoidal function of the time (t), then B varies according to $B = B_0 \cos(\omega t)$. Ignoring damping and the small restoring force exerted by the earth, Newton's second law $\tau = I\ddot{\theta}$ gives

$$\ddot{\theta} = -\frac{\mu B_0}{I} \cos(\omega t) \sin\theta, \qquad (31.2)$$

where I is the moment of inertia of the magnet. If the length ℓ of the cylindrical bar magnet (mass m) is much larger than the end radius, then the magnet's moment of inertia is approximately equal to

$$I = \frac{1}{12} m\ell^2. \qquad (31.3)$$

By setting $\tau = \omega t$, Equation (31.2) can be put into the form

$$\ddot{\theta}(\tau) = -\frac{s^2}{2} \cos\tau \sin\theta \qquad (31.4)$$

where $s = \sqrt{\frac{2\mu B_0}{I\omega^2}}$ plays the role of a control parameter.

This activity investigates motion of the bar magnet governed by Equation (31.2) or (31.4). The motion exhibits a variety of nonlinear phenomena, including intermittency and chaos.

Procedure:

1. Set up the apparatus as shown in Figure 31.2.

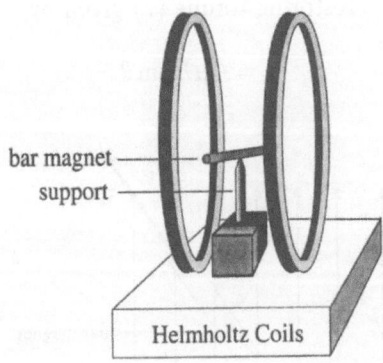

Figure 31.2: Experimental apparatus.

2. If the motion is to be observed without measurements, instead of a bar magnet, use a standard classroom demonstration dipping needle or a demonstration compass. If measurements are to made, then a stronger

magnet must be used and a frictionless method of suspending the magnet must be found. We rely on your ingenuity. One approach is to attach two neodymium magnets to a demonstration compass in a teeter-totter fashion to maintain the horizontal balance.

3. To monitor the signal, use a Hall probe or a small pick-up coil connected to a chart recorder or to a digital storage oscilloscope (DSO).

4. With the signal generator strength set at maximum, alter the driving frequency until the magnet is set into a rotating motion.

5. Reduce the frequency until the motion appears chaotic.

6. What is the value for the critical driving frequency where the motion becomes chaotic?

7. Change the strength of the driving signal and driving frequency. See if you can find or locate regions of intermittency. Intermittency is when the motion remains in one periodic state for some time before leaping to another periodic state. The decreasing length of time spent in any given periodic state is a measure of approaching chaos.

8. Keeping the external magnetic field constant, measure the period T of small oscillations around the equilibrium point.

9. Measure the mass m and length ℓ of the oscillating magnet. Calculate the magnet's moment of inertia I.

10. Calculate the value for the magnetic field B_0. Use the equation given for the Helmholtz coils, usually written on the apparatus. If a solenoid is used, calculate the strength of its magnetic field using standard textbook equations.

11. Calculate the magnetic moment (μ) of the magnet using

$$\mu = \frac{4\pi^2 I}{T^2 B_0}.$$

12. Write a Mathematica file that numerically solves Equation (31.2). For a given frequency ω, compare the $\theta(t)$ predicted by the file with that observed experimentally.

13. Calculate the parameter s using the value found for the critical driving frequency. It has been predicted [Chi79] that chaotic motion occurs when $s > 1$. How does your value of s compare with the theoretical prediction?

14. Compare the frequency needed to efficiently pump the magnet from rest to a large amplitude with the frequency for small oscillations.

Things to Investigate:

- Modify Mathematica File 11 and investigate the Poincaré sections produced for Equation (31.4) for different s values.

- Rewrite Equation (31.2) to include the restoring force of the earth's magnetic field. Write a Mathematica program to investigate the difference this makes in the predicted motion. Is the ignoring of the earth's magnetic field justified?

- Repeat the steps in the procedure with the magnet suspended so it can rotate vertically.

- Suspend the magnet so it can rotate vertically, but has its center of mass initially below the fulcrum. This is now analogous to a compound pendulum. Try pumping this pendulum and study its motion. Write the differential equation that describes this situation and compare its predictions with the observations.

- Analyze the signal using a fast Fourier transform (FFT) to produce a power spectrum. Does the power spectrum indicate possible chaos?

- Add a damping term to Equation (31.2) and then write a Mathematica file to explore and compare the experimental behavior with the behavior predicted by Mathematica .

Experimental Activity 32

Driven Spin Toy

Comment: This activity should not take more than one hour to complete.

Reference: Section 9.4

Object: To investigate forced nonlinear oscillatory motion.

Theory: This activity uses an inexpensive toy, sold under the name Revolution: The World's Most Efficient Spinning Device, as a black box oscillator[1]. Figure 32.1 is a diagram of the device. The movable cylinder and base of this

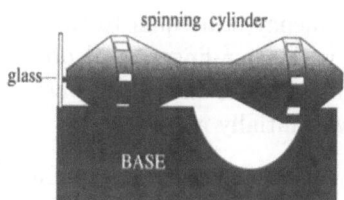

Figure 32.1: The spin toy oscillator.

toy contain magnets. These magnets levitate and hold the cylinder against the fixed glass end of the base. If the cylinder is placed in an oscillating magnetic field, the magnetic field acts as a forcing function. With the addition of an oscillating and pumping external magnetic field, the cylinder can be made to exhibit period doubling and chaotic motion. This investigation is mainly qualitative in nature. The quantitative portion of this activity explores the forcing frequencies and amplitudes required to produce chaotic motion.

Procedure:

1. Set up the following apparatus. The solenoid that comes with the Berkeley Lab course works well. The power supply (function generator) should be able to be adjusted to produce sine waves of 0.1 to 1.0 Hz. The core is a laminated piece of iron that comes with a demonstration transformer set.

---

[1]Arbor Scientific, P.O. Box 2750, Ann Arbor, MI 48106-2750, phone 1-800-367-6695.

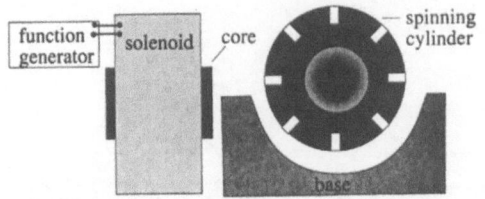

Figure 32.2: The forced oscillator.

2. The spinning cylinder has white marks placed at 45° around its circumference. These marks can be used to measure the angular amplitude of the oscillations.

3. Measure the period T of the oscillating cylinder for a very small initial amplitude. Calculate the natural frequency ($\nu = \frac{1}{T}$) and the natural angular frequency (ω_0) for the spinning cylinder. This gives an approximate value for the pumping frequency.

4. With the cylinder initially at rest, and with the external magnetic field in resonance with the small oscillation frequency, observe the motion of the cylinder. At this pumping frequency, is it possible to make the cylinder complete a full revolution? Explain.

5. With the driving frequency still equal to the small angle frequency, give the cylinder a spin with your fingers and observe the resulting motion. Does the motion finally settle into the same pattern as that produced when the cylinder was initially at rest?

6. With the cylinder initially at rest, find a pumping frequency that makes the cylinder oscillate with a large amplitude. (Hint: A pendulum (a swing) is more efficiently pumped with a frequency nearly double that of its natural frequency.) Observe the motion to see if period doubling occurs. Be careful. Small frequency increases can make the cylinder rotate and chaotic motion might be the result.

7. Give the cylinder an initial spin with your fingers. With the same pumping frequency as that found in the previous step, observe the motion to see if it is qualitatively different from the previous motion.

8. Is the motion chaotic? What tests or measurements would have to be performed to see if the motion is chaotic?

9. When the cylinder is oscillating, change the driving amplitude or the driving frequency. Explore the effect this has on the cylinder.

Experimental Activity 33

Mapping

Comment: This activity should not take more than one hour to complete.

References: Section 9.9

1. [DH91] This article contains a mapping procedure similar to the one in this activity.

Object: To construct a map for a forced nonlinear oscillator.

Theory: This activity uses commercially available toys, two of which are shown in Figure 33.1. In the bottom supporting structure of each toy is a transistor

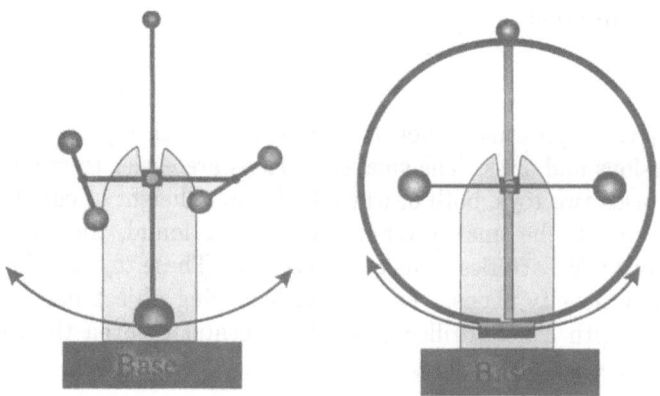

Figure 33.1: Original toys.

circuit that is turned on every time a swinging magnet passes over it. The magnet moving over the top of the supporting base induces a current in a small coil hidden inside the base. This induced current when applied to the base terminal of the transistor momentarily turns the transistor on. At this point a much larger current flows through a larger solenoid which is used to produce a repelling magnetic field. In this manner the swinging magnet is given a sharp push when it passes over the base of the toy. The time interval between successive pushes is measured. A map is created by plotting one time interval against the next time interval.

If the toys are left as pictured in Figure 33.1, the pumping is very regular, but the rotation of the smaller pendulums is chaotic. To make the main rotation less periodic and to permit full rotations, the smaller pendulums are removed. (Sometimes a counterbalancing object is needed to make the toy complete a full rotation.) Figure 33.2. shows the modified toys. If the toys you buy are

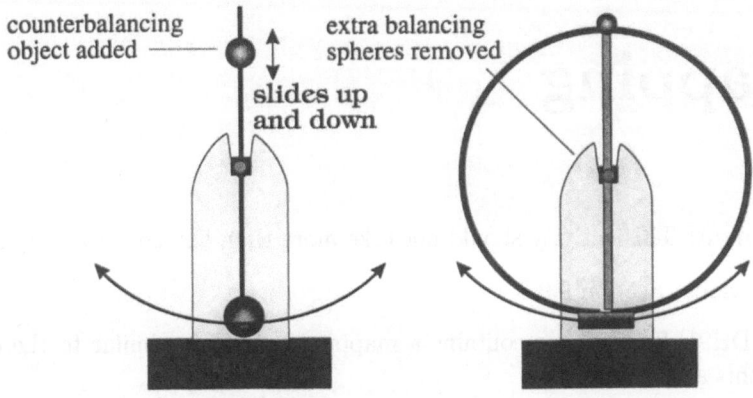

Figure 33.2: Modified toys.

different than those shown, then we rely on your ability to modify them so they will make full rotations.

Procedure:

1. Set up the apparatus as shown in Figure 33.3. This toy comes in a number of versions and sizes. The smaller versions are easier to pump. (The authors used two toys, both of which had a stem height or circular diameter of 15 cm. If the smaller versions cannot be found, then drive the toys with two 9 V batteries connected in series.) These toys can be purchased in many novelty shops. The counterbalancing object may be a rubber stopper with a hole drilled through its center or even the lead weights normally used on fish lines.

2. Remove, but do not disconnect, the 9 volt battery from the base of the toy. Attach two additional wires to the terminals of the battery and connect as shown in the diagram. The capacitor is essential and is used to block the 9 volt DC signal, but lets the pulsed signal pass.

3. Start the toy moving by giving the pendulum a little shove.

4. Adjust the movable counterbalancing object up or down until an interesting nonlinear rotational motion occurs.

5. Turn on the strip recorder and note the speed of the chart paper.

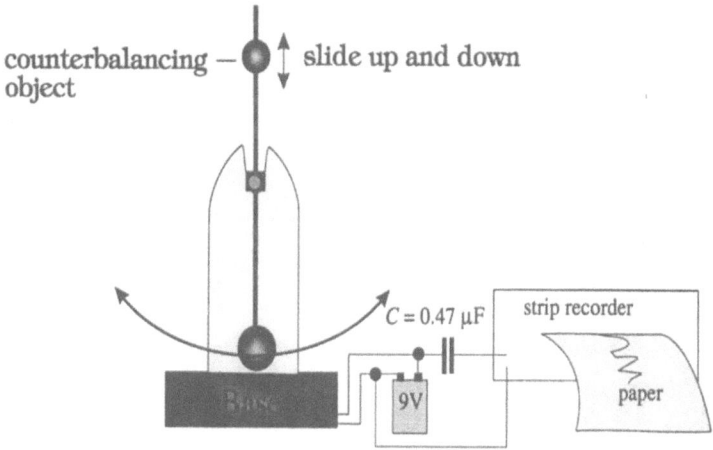

Figure 33.3: One of the toys used as a forced oscillator.

6. From the paper strip, measure the time intervals, Δt_1, Δt_2, Δt_3, ... between pushes by the repelling coil in the base. An alternate method for measuring the pumping times is to use a computer to monitor the times.

7. Create a map by placing on the ordinate Δt_{n+1} and on the abscissa Δt_n. Input the data into the provided Mathematica file X33 to see the map and explore various plotting configurations.

Mapping
This Mathematica notebook is used to create the map of Δt_{n+1} vs. Δt_n. Mathematica commands in file: `Length`, `ListPlot`, `Table`, `PlotRange`, `TextStyle`, `ImageSize`

---

8. What does the plot tell you about the motion? Is there any evidence for a period doubling route to chaos? How can you tell if the motion is chaotic?

Things to Investigate:

- Interesting maps and very strange attractors were produced by collecting a very large number of points. When this was done, the authors used commercially available software to record the data. You might wish to do the same.

- Plot the amplitude of the strip recorder's signal against the time.

- Compare the strange attractors produced by this activity with those produced in the dripping faucet article [DH91].

Bibliography

[Agr89]   G. P. Agrawal. *Nonlinear Fiber Optics*. Academic Press, New York, 1989.

[AGS84]   R. H. Abraham, J. P. Gollub, and H. L. Swinney. Testing nonlinear dynamics. *Physica D*, 11:252, 1984.

[AJMS81]   R. M. Anderson, H. C. Jackson, R. M. May, and A. M. Smith. Population dynamics of fox rabies in Europe. *Nature*, 289:765, 1981.

[AKNS74]   M. J. Ablowitz, D. J. Kaup, A. C. Newell, and H. Segur. The inverse scattering transform—Fourier analysis for nonlinear problems. *Studies in Applied Mathematics*, 53:249, 1974.

[AL82]   F. T. Arecchi and F. Lisi. Hopping mechanism generating $1/f$ noise in nonlinear systems. *Physical Review Letters*, 49:94, 1982.

[ALSB73]   J. Aroesty, T. Lincoln, N. Shapiro, and G. Boccia. Tumor growth and chemotherapy: mathematical methods, computer simulations, and experimental foundations. *Mathematical Bioscience*, 17:243, 1973.

[ANST86]   S. V. Antipov, M. V. Nezlin, E. N. Snezhkin, and A. S. Trubnikov. Rossby autosoliton and stationary model of the Jovian great red spot. *Nature*, 323:238, 1986.

[AS72]   M. Abramowitz and I. A. Stegun. *Handbook of Mathematical Functions (AMS 55)*. National Bureau of Standards, Washington, 1972.

[Bar88]   M. Barnsley. *Fractals Everywhere*. Academic Press, San Diego, 1988.

[BC31]   V. Bush and S. H. Caldwell. Thomas–Fermi equation solution by the differential analyzer. *Physical Review*, 38:1898, 1931.

[BC74]   G. W. Bluman and J. D. Cole. *Similarity Methods for Differential Equations*. Springer-Verlag, New York, 1974.

[Bel58]   B. P. Belousov. Oscillation reaction and its mechanism. In *Collection of Abstracts on Radiation Medicine*. Medgiz, Moscow, 1958.

[BEMS71]   A. Barone, F. Esposito, C. J. Magee, and A. C. Scott. Theory and applications of the sine-Gordon equation. *Rivista Del Nuovo Cimento*, 1:227, 1971.

[BEP71]   I. P. Batra, R. H. Enns, and D. Pohl. Stimulated thermal scattering of light. *Physica Status Solidi*, (b) 48:11, 1971.

[BF89]   R. L. Burden and J. D. Faires. *Numerical Analysis*. PWS-Kent Publishing Company, Boston, 4th edition, 1989.

[BG90]   G. L. Baker and J. P. Gollub. *Chaotic Dynamics*. Cambridge University Press, Cambridge, 1990.

[BH62]   W. A. Bentley and W. J. Humpreys. *Snow Crystals*. Dover, New York, 1962.

[BJGLS84] E. Ben-Jacob, N. Goldenfeld, J. S. Langer, and G. Schön. String models of interfacial pattern formation. *Physica D*, 12:245, 1984.

[BMP83]   A. Bettini, T. A. Minelli, and D. Pascoli. Solitons in the undergraduate laboratory. *American Journal of Physics*, 51:977, 1983.

[BN92]   R. Boylestad and L. Nashelsky. *Electronic Devices & Circuit Theory*. Prentice Hall, New Jersey, 5th edition, 1992. See pages 807–812.

[Bou72]   J. Boussinesq. *Journal of Mathematics. Pure and Applied Series 2*, 17:55, 1872.

[Bri87]   K. Briggs. Simple experiments in chaotic dynamics. *American Journal of Physics*, 55:1083, 1987. The idea for using a steel tape to model the Duffing equation was found in this article.

[Buc77]   J. Buckmaster. Viscous sheets advancing over dry beds. *Journal of Fluid Mechanics*, 81:735, 1977.

[Cam87]   D. Campbell. Nonlinear science—from paradigms to practicalities. *Los Alamos Science*, 15:218, 1987.

[CH80]   D. R. J. Chillingworth and R. J. Holmes. Dynamical systems and models for reversals of the earth's magnetic field. *Mathematical Geology*, 12:41, 1980.

[CH93]   M. C. Cross and P. C. Hohenberg. Pattern formation outside of equilibrium. *Rev. Mod. Physics*, 65:851, 1993.

[CH94]   L. Cruzeiro-Hansson. Two reasons why the Davydov soliton may be thermally stable after all. *Physical Review Letters*, 73:2927, 1994.

[Cha39]   S. Chandresekhar. *An Introduction to the Study of Stellar Structure*. Dover Reprint, Chicago, 1939.

[Chi79]   B.V. Chirikov. A universal instability of many-dimensional oscillator systems. *Phys. Rep.*, **52**:265, 1979.

[Cho64]   W. F. Chow. *Principles of Tunnel Diode Circuits*. Wiley, New York, 1964.

[Col51]   J. D. Cole. *Quart. Appl. Math.*, 9:225, 1951.

[Cra75]   F. Crawford. Damping of a simple pendulum. *American Journal of Physics*, **43**:276, 1975.

[Cro95]   A. Cromer. Many oscillations of a rigid rod. *American Journal of Physics*, **63**:112, 1995. This article is highly recommended. It provides ideas for supporting an oscillating rod, diagrams illustrating different pendulum configurations and the equations for calculating their periods, and suggestions for additional and more complex investigations of rigid rod oscillations.

[Cun64]   W. J. Cunningham. *Introduction to Nonlinear Analysis*. McGraw Hill, New York, 1964.

[Cvi84]   P. Cvitanovic. *Universality in Chaos*. Adam Hilger, Bristol, 1984.

[Dav62]   H. T. Davis. *Introduction to Nonlinear Differential and Integral Equations*. Dover, New York, 1962.

[DFGJ91]   B. Duchesne, C. Fischer, C. Gray, and K. Jeffrey. Chaos in the motion of an inverted pendulum: An undergraduate laboratory experiment. *American Journal of Physics*, **59**:987, 1991. This article delves deeper into the theory of the inverted pendulum than does the Briggs' article.

[DH91]   K. Dreyer and F. Hickey. The route to chaos in a dripping water faucet. *American Journal of Physics*, **59**:619, 1991. This article contains maps produced by dripping faucets.

[DK76]   A. S. Davydov and N. I. Kislukha. Solitons in one-dimensional molecular chains. *Physica Status Solidi*, (b)75:735, 1976.

[Duf18]   G. Duffing. *Erzwungene Schwingungen bei Veränderlicher Eigenfrequenz*. F. Vieweg und Sohn, Braunschweig, 1918.

[DW83]   W.F. Drish and W.J. Wild. Numerical solutions of Van der Pol's equation. *American Journal of Physics*, **51**:439, 1983. This article contains values for the coefficients of the VdP equation that might be used by Mathematica to reproduce the article's plots. The article also contains a discussion of a nonlinear electric circuit that was proposed for use with elevators.

[EE95]   D. E. Edmundson and R. H. Enns. Particle-like nature of colliding 3-dimensional optical solitons. *Physical Review A*, 51:2491, 1995.

[EGSWO81] M. Eigen, W. Gardiner, P. Schuster, and R. Winkler-Oswatitsch. The origin of genetic information. *Scientific American*, April:88, 1981.

[Eig71]   M. Eigen. Self-organization of matter and evolution of biological macromolecules. *Die Naturwissenschaften*, 58:465, 1971.

[Emm77]   G. S. Emmerson. *John Scott Russell—A Great Victorian Engineer and Naval Architect*. John Murray, London, 1977.

[ER79]   R. H. Enns and S. S. Rangnekar. The 3-wave interaction in nonlinear optics. *Physica Status Solidi*, (b) 94:9, 1979.

[ER87a]   R. H. Enns and S. S. Rangnekar. Bistable solitons and optical switching. *IEEE Journal of Quantum Electronics*, QE-23:1199, 1987.

[ER87b]   R. H. Enns and S. S. Rangnekar. Optical switching between bistable soliton states of the highly nonlinear Schrödinger equation. *Optics Letters*, 12:108, 1987.

[FC86]   G. R. Fowles and G. L. Cassiday. *Analytical Mechanics*. Saunders College Publishing, Orlando, 1986.

[Feh70]   E. Fehlberg. Klassische Runge–Kutta Formeln vierter und niedrigerer ordnung mit schrittweiten-kontrolle und ihre andwendung auf wärmeleitungsprobleme. *Computing*, 6:61, 1970.

[Fei79]   M. J. Feigenbaum. The universal metric properities of nonlinear transformations. *Journal of Statistical Physics*, 21:69, 1979.

[Fei80]   M. J. Feigenbaum. Universal behavior in nonlinear systems. *Los Alamos Science*, 1:4, 1980.

[Fer28]   E. Fermi. Statistical method of investigating electrons in atoms. *Zeitschrift für Physik*, 48:73, 1928.

[Fis37]   R. A. Fisher. The waves of advance of advantageous genes. *Annals of Eugenics*, 7:353, 1937.

[FJB85]   E.L.M. Flerackers, H.J. Janssen, and L. Beerden. Piecewise linear anharmonic lrc circuit for demonstrating "soft" and "hard" spring nonlinear resonant behavior. *American Journal of Physics*, **53**:574, 1985.

[FKN72]   R. J. Field, E. Körös, and R. M. Noyes. Oscillations in chemical systems, Part 2. Thorough analysis of temporal oscillations in the bromate–cerium–malonic acid system. *Journal of the American Chemical Society*, 94:8649, 1972.

[FLS77]   R. P. Feynman, R. B. Leighton, and M. Sands. *The Feynman Lectures on Physics*. Addison-Wesley, Reading, Massachusetts, first edition, 1977. Volume 1, Page I-1.

[FN74]   R. J. Field and R. M. Noyes. Oscillations in chemical systems, IV. Limit cycle behavior in a model of a real chemical reaction. *Journal of Chemical Physics*, 60:1877, 1974.

[For87]   L.R. Fortney. *Principles of Electronics: Analog & Digital.* Harcourt Brace Jovanovich, Toronto, first edition, 1987. See page 308.

[FPU55]   E. Fermi, J. R. Pasta, and S. M. Ulam. Studies of nonlinear problems. *Los Alamos Science Laboratory Report*, LA-1940, 1955.

[Fr92]   J. Fröyland. *Introduction to Chaos and Coherence.* Institute of Physics Publishing, 1992.

[Fro34]   M. Frommer. Uber das auftreten von wirbeln und strudeln (geschlossener und spiraliger integralkurven) in der umgebung rationaler unbestimmtheitsstellen. *Math. Annalen*, 109:395, 1934.

[FW98]   Cathal Flynn and Niall Wilson. A simple method for controlling chaos. *American Journal of Physics*, 66:730, 1998.

[Gar70]   M. Gardner. *Scientific American*, 223 (10):120–0, 1970.

[Gar71a]   M. Gardner. *Scientific American*, 224 (2):112, 1971.

[Gar71b]   M. Gardner. *Scientific American*, 224 (4):117, 1971.

[Gas98]   Richard Gass. *Mathematica for Scientists and Engineers.* Prentice Hall, Upper Saddle River, N.J., 1998.

[GGKM67]   C. S. Gardner, J. M. Greene, M. D. Kruskal, and R. M. Miura. Method for solving the Korteweg de Vries equation. *Physical Review Letters*, 19:1095, 1967.

[GMK89]   S. Grossmann and G. Mayer-Kress. Chaos in the international arms race. *Nature*, 337:701, 1989.

[GMM71]   N. S. Goel, S. C. Maitra, and E. W. Montroll. On the Volterra and other nonlinear models of interacting populations. *Reviews of Modern Physics*, 43:231, 1971.

[GMP81]   M. Giglio, S. Musazzi, and V. Perini. Transition to chaotic behavior via a reproducible sequence of period-doubling bifurcations. *Physical Review Letters*, 47:243, 1981.

[GP82]   L. Glass and R. Perez. Fine structure of phase locking. *Physical Review Letters*, 48:1772, 1982.

[Gri71]   J. S. Griffith. *Mathematical Neourobiology.* Academic Press, New York, 1971.

[Gri99]   D. J. Griffiths. *Introduction to Electrodynamics.* Prentice Hall, Upper Saddle River, N.J., 1999.

[GSB86]   L. Glass, A. Schrier, and J. Bélair. Chaotic cardiac rhythms. In A. V. Holden, editor, *Chaos*. Princeton University Press, Princeton, 1986.

[GT96]   H. Gould and J. Tobochnik. *An Introduction to Computer Simulation Methods*. Addison-Wesley, Reading, Mass., 1996.

[Hé69]   M. Hénon. Numerical study of quadratic area-preserving mappings. *Q. Appl. Math.*, 27:219, 1969.

[Hé76]   M. Hénon. A two-dimensional mapping with a strange attractor. *Communications in Mathematical Physics*, 50:69, 1976.

[Has90]   A. Hasegawa. *Optical Solitons in Fibers*. Springer-Verlag, New York, 2nd edition, 1990.

[Hat21]   A. S. Hathaway. Pursuit in a circle. *American Mathematical Monthly*, 27:93, 1921.

[Hay64]   C. Hayashi. *Nonlinear Oscillations in Physical Systems*. McGraw Hill, New York, 1964.

[HCL76]   P. Horn, T. Carruthers, and M. Long. Threshold instabilities in nonlinear self-excited oscillators. *Physical Review A.*, **14**:833, 1976. The original equation for the Wien bridge was found in this article.

[Hel95]   H. Helmholtz. *Sensation of Tone*. Longmans Green, London, 1895.

[HH64]   M. Hénon and C. Heiles. The applicability of the third integral of motion: some numerical experiments. *Astrophys. J.*, 69:73, 1964.

[Hil94]   R. C. Hilborn. *Chaos and Nonlinear Dynamics*. Oxford University Press, Oxford, 1994.

[Hir71]   R. Hirota. Exact solution of the Korteweg-de Vries equation for multiple collisions of solitons. *Physical Review Letters*, 27:1192, 1971.

[Hir76]   R. Hirota. Direct methods of finding exact solution of nonlinear evolution equations. In R. M. Miura, editor, *Bäcklund Transformations*. Springer-Verlag, New York, 1976.

[Hir85a] R. Hirota. Direct methods in soliton theory. In R. K. Bullough and P. J. Caudrey, editors, *Solitons: Topics of Modern Physics*. Springer-Verlag, New York, 1985.

[Hir85b] R. Hirota. Fundamental properties of the binary operators in soliton theory and their generalization. In S. Takeno, editor, *Dynamic Problems in Soliton Systems*. Springer-Verlag, New York, 1985.

[HJ93]   E. Hunt and G. Johnson. Keeping chaos at bay. *IEEE Spectrum*, page 32, 1993. This article contains a number of nonlinear circuits.

[HN84]   Y. Hayashi and T. Nakagawa. Transient behaviors in the Wien bridge oscillator: A laboratory experiment for the undergraduate student. *American Journal of Physics*, **52**:1021, 1984. The equation for the Wien bridge was found in this article.

[Hoe80]   Stuart A. Hoenig. *How to Build and Use Electronic Devices Without Frustration, Panic, Mountains of Money, or an Engineering Degree*. Little Brown and Company, Boston, 2nd edition, 1980.

[Hop50]   E. Hopf. *Communications in Pure and Applied Mathematics*, 3:201, 1950.

[HP95]   A. C. Hindmarsh and L. R. Petzold. Algorithms and software for ordinary differential equations and differential-algebraic equations, part 1: Euler methods and error estimation. *Computers in Physics*, 9:34, 1995.

[Ing73]   A. P. Ingersoll. Jupiter's great red spot: A free atmospheric vortex. *Science*, 182:1346, 1973.

[IR90]   E. Infeld and G. Rowlands. *Nonlinear Waves, Solitons and Chaos*. Cambridge University Press, Cambridge, 1990.

[Jac90]   E. A. Jackson. *Perspectives of Nonlinear Dynamics*. Cambridge University Press, Cambridge, 1990.

[JBB84]   M. H. Jensen, P. Bak, and T. Bohr. Transition to chaos by interaction of resonances in dissipative systems, 1. Circle maps. *Physical Review A*, 30:1960, 1984.

[JER76]   B. L. Jones, R. H. Enns, and S. S. Rangnekar. On the theory of selection of coupled macromolecular systems. *Bulletins in Mathematical Biology*, 38:15, 1976.

[Kau76]   D. J. Kaup. The three–wave interaction: A nondispersive phenomena. *Studies in Applied Mathematics*, 55:9, 1976.

[KB43]   N. Krylov and N. Bogoliubov. *Introduction to Nonlinear Mechanics*. Princeton University Press, Princeton, 1943.

[KdV95]   D. J. Korteweg and G. de Vries. On the change of form of long waves advancing in a rectangular canal, and a new type of long stationary wave. *Philosophical Magazine*, 39:422, 1895.

[KG95]   D. Kaplan and L. Glass. *Understanding Nonlinear Dynamics*. Springer-Verlag, New York, 1995.

[KK58]   L. V. Kantorovich and V. I. Krylov. *Approximate Methods of Higher Analysis*. Interscience Publishers, New York, 1958.

[KM27]   W. O. Kermack and A. G. Mc Kendrick. Contributions to the mathematical theory of epidemics-i. *Proceedings of the Royal Society*, 115A:700, 1927.

[Kos74]   E. L. Koschmieder. *Advances in Chemical Physics*, 26:177, 1974.

[KP70]   B. B. Kadomtsev and V. I. Petviashvili. On the stability of solitary waves in weakly dispersive media. *Soviet Physics Doklady*, 15:539, 1970.

[KRB79]   D. J. Kaup, A. Rieman, and A. Bers. Space-time evolution of nonlinear three-wave interactions. *Reviews of Modern Physics*, 51:275, 1979.

[Kro55]   W. S. Krogdahl. Stellar pulsation as a limit cycle phenomena. *Astrophysical Journal*, 122:43, 1955.

[Kru91]   J. A. Krumhansl. Unity in the science of physics. *Physics Today*, 44, No.3:33, 1991.

[Lau86]   H. A. Lauwerier. Two-dimensional iterative maps. In A. V. Holden, editor, *Chaos*. Princeton University Press, Princeton, 1986.

[Lax68]   P. D. Lax. Integrals of nonlinear equations of evolution and solitary waves. *Communications in Pure and Applied Mathematics*, 21:467, 1968.

[LH91]   W. Lauterborn and J. Holzfuss. Acoustic chaos. *International Journal of Bifurcation and Chaos*, 1:13, 1991.

[Lin81]   P. Linsay. Period doubling and chaotic behavior in a driven anharmonic oscillator. *Physical Review Letters*, 47:1349, 1981.

[LJH78]   D. Ludwig, D. D. Jones, and C. S. Holling. Qualitative analysis of insect outbreak systems: the spruce budworm and forest. *J. Anim. Ecol.*, 47:315, 1978.

[LLF82]   A. Libchaber, C. Laroche, and S. Fauve. Period doubling cascade in mercury, a quantitative measurement. *J. Physique Lett.*, 43:L211, 1982.

[Lor63]   E. N. Lorenz. Deterministic nonperiodic flow. *Journal of Atmospheric Science*, 20:130, 1963.

[Lor84]   E. N. Lorenz. The local structure of a chaotic attractor in four dimensions. *Physica D*, 13:90, 1984.

[LP80]   E. W. Larsen and G. C. Pomraning. Asymptotic analysis of nonlinear Marshak waves. *SIAM Journal of Applied Mathematics*, 39:201, 1980.

[LS71]   L. Lapidus and J. H. Seinfeld. *Numerical Solution of Ordinary Differential Equations*. Academic Press, New York, 1971.

[LS74]   S. Leibovich and A. R. Seebass, editors. *Nonlinear Waves*. Cornell University Press, Ithaca, N.Y., 1974.

[Mal93]   A.P. Malvino. *Electronic Principles*. Macmillan/McGraw-Hill, New York, 5th edition, 1993. pages 794–798.

[Man75]   B. B. Mandelbrot. *Les Objets Fractals*. Flammarion, Paris, 1975.

[Man77]   B. B. Mandelbrot. *Fractals: Form, Chance and Dimension*. W. H. Freeman, San Francisco, 1977.

[Man83]   B. B. Mandelbrot. On the quadratic mapping $z \rightarrow z^2 - \mu$ for complex μ and z. *Physica D*, 7:224, 1983.

[Mar10]   Martienssen. *Phys. Zeit.*, 11, 1910.

[May76]   R. M. May. Simple mathematical models with very complicated dynamics. *Nature*, 261:459, 1976.

[May80]   R. M. May. Nonlinear phenomena in ecology and epidemiology. *Annals of the New York Academy of Science*, 357:267, 1980.

[MER88]   D. M. McAvity, R. H. Enns, and S. S. Rangnekar. Bistable solitons and the route to chaos. *Physical Review A*, 38:4647, 1988.

[MH67]   S. McCall and E. Hahn. Self-induced transparency by pulsed coherent light. *Physical Review Letters*, 18:908, 1967.

[MH69]   S. McCall and E. Hahn. Self-induced transparency. *Physical Review*, 183:457, 1969.

[Min64]   N. Minorksy. Nonlinear problems in physics and engineering. In H. Margenau and G. Murphy, editors, *The Mathematics of Physics and Chemistry, Vol.2*, chapter 6. D. Van Nostrand, New York, 1964.

[MM74]   Richard E. Miers and William D.C. Moebs. Simple demonstration of amplitude modulation and beats. *American Journal of Physics*, **42**:603, 1974.

[Moo92]   Francis C. Moon. *Chaotic and Fractal Dynamics*. Wiley, New York, first edition, 1992. An excellent book, with some very nice ideas for experiments.

[Mor48]   P. M. Morse. *Vibrations and Sound*. McGraw Hill, New York, 1948.

[MP94]   A. B. Murray and C. Paola. A cellular model of braided rivers. *Nature*, 371:54, 1994.

[MSS73]   N. Metropolis, M. L. Stein, and P. R. Stein. On finite limit sets for transformations on the unit interval. *J. Combin. Theor.*, 15:25, 1973.

[MT95]   J. B. Marion and S. T. Thornton. *Classical Dynamics of Particles and Systems*. Saunders College Publishing, Orlando, 1995.

[Mur89]   J. D. Murray. *Mathematical Biology.* Springer-Verlag, Berlin, New
          York, 1989.

[Mus37]   M. Muskat. *The Flow of Homogeneous Fluids Through Porous
          Media.* McGraw Hill, New York, 1937.

[NAY62]   J. Nagumo, S. Arimoto, and S. Yoshizawa. An active pulse trans-
          mission line simulating nerve axon. *Proc. IRE*, 50:2061, 1962.

[NB95]   A. H. Nayfeh and B. Balachandran. *Applied Nonlinear Dynamics.*
          Wiley, New York, 1995.

[New80]   C. M. Newton. Biomathematics in oncology: Modelling of cellular
          systems. *Ann. Rev. Biophys. Bioeng.*, 9:541, 1980.

[NT75]   R. B. Neff and L. Tillman. *Hewlett-Packard Journal*, Nov., 1975.

[Nyq28]   H. Nyquist. Certain topics in telegraph transmission theory. *Trans.
          AIEE*, 47 (April):617, 1928.

[OGY90]   E. Ott, C. Grebogi, and J. A. Yorke. Controlling chaos. *Physical
          Review Letters*, 64:1196, 1990.

[OSS84]   M. Olsen, H. Smith, and A. C. Scott. Solitons in a wave tank.
          American Journal of Physics, 52:826, 1984.

[PC89]   T. S. Parker and L. O. Chua. *Practical Numerical Algorithms for
          Chaotic Systems.* Springer-Verlag, Berlin, New York, 1989.

[PFTV89] W. H. Press, B. P. Flannery, S. A. Teukolsky, and W. T. Vetterling.
          Numerical Recipes. Cambridge University Press, Cambridge, 1989.

[Pip87]   A.P. Pippard. *Response and Stability.* Cambridge University Press,
          Cambridge, England, first edition, 1987. A very interesting book.
          It contains many ideas for other simple experiments.

[PJS92]   H. O. Peitgen, H. Jürgens, and D. Saupe. *Fractals for the Class-
          room, part 2.* Springer-Verlag, Berlin, New York, 1992.

[PL68]   I. Prigogine and R. Lefever. Symmetry breaking instabilities in dis-
          sipative systems, II. *Journal of Chemical Physics*, 48:1695, 1968.

[PLA92]   M. Peastrel, R. Lynch, and A. Armenti. Terminal velocity of shut-
          tlecock in vertical fall. In A. Armenti, editor, *The Physics of
          Sports*, chapter 9. American Institute of Physics, New York, 1992.

[PR86]   H. O. Peitgen and P. H. Richter. *The Beauty of Fractals.* Springer-
          Verlag, Berlin, New York, 1986.

[Rö76]   O. E. Rössler. An equation for continuous chaos. *Physics Letters*,
          57A:397, 1976.

[Rö79]   O. E. Rössler. An equation for hyperchaos. *Physics Letters*,
          71A:155, 1979.

[Rap60]   A. Rapoport. *Fights, Games and Debates.* University of Michigan
            Press, 1960.

[Ray83]   Lord Rayleigh. On maintained vibrations. *Philosophical Magazine*,
            15:229, 1883.

[RE76]   S. S. Rangnekar and R. H. Enns. Numerical solution of the tran-
            sient gain equations for stimulated backward scattering in absorb-
            ing fluids. *Canadian Journal of Physics*, 54:1564, 1976.

[Ric60]   L. F. Richardson. Mathematics of war and foreign politics. In
            N. Rashevsky and E. Trucco, editors, *Arms and Insecurity: A
            Mathematical Study of the Causes and Origins of War.* Boxwood,
            Pittsburgh, 1960.

[Rob79]   K. A. Robbins. Periodic solutions and bifurcation structure at high
            R in the Lorenz model. *Journal of Applied Mathematics*, 36:457,
            1979.

[RSS83]   J. C. Roux, R. H. Simoyi, and H. L. Swinney. Observation of a
            strange attractor. *Physica D*, 8:257, 1983.

[Rus99]   Heikki Ruskeepää. *Mathematica Navigator.* Academic Press, San
            Diego, 1999.

[Rus44]   J.S. Russell. Report on waves. *British Assoc. Adv. Sci.*, 14th
            Meeting, 1844.

[Sal62]   B. Saltzman. Finite amplitude free convection as an initial value
            problem. *International Journal of Atmospheric Science*, 19:329,
            1962.

[Sap84]   A. M. Saperstein. Chaos—a model for the outbreak of war. *Nature*,
            309:303, 1984.

[SC64]   G. Sansone and R. Conti. *Nonlinear Differential Equations.* Perga-
            man Press, New York, 1964.

[SCM73]   A. C. Scott, F. Y. F. Chu, and D. W. McLaughlin. The soliton:
            A new concept in applied science. *Proceedings of the I.E.E.E.*,
            61:1443, 1973.

[Sco70]   A. C. Scott. *Active and Nonlinear Wave Propagation in Electron-
            ics.* Wiley-Interscience, New York, 1970.

[Sco81]   A. C. Scott. Introduction to nonlinear waves. In R. H. Enns,
            B. L. Jones, R. M. Miura, and S. S. Rangnekar, editors, *Nonlinear
            Phenomena in Physics and Biology.* Plenum, New York, 1981.

[Sco87]   D. E. Scott. *An Introduction to Circuit Analysis.* McGraw-Hill,
            New York, 1987.

[Sel68]   E. E. Sel'kov. Self-oscillations in glycolysis. *European Journal of
            Biochemistry*, 4:79, 1968.

[Sha49]   C. E. Shannon. Communication in the presence of noise. *Proceedings of the IRE*, 37:10, 1949.

[SK86]   W. M. Schaffer and M. Kot. Differential systems in ecology and epidemiology. In A. V. Holden, editor, *Chaos*. Princeton University Press, Princeton, New Jersey, 1986.

[SK89]   R. D. Strum and D. E. Kirk. *Discrete Systems and Digital Signal Processing*. Addison Wesley, Reading, Mass., 1989.

[Spa82]   C. Sparrrow. *Bifurcations in the Lorenz Equation*. Springer-Verlag, 1982.

[Str94]   S. H. Strogatz. *Nonlinear Dynamics and Chaos*. Addison-Wesley, Reading, MA, 1994.

[SWS82]   R. H. Simoyi, A. Wolf, and H. L. Swinney. One-dimensional dynamics in a multi-component chemical reaction. *Physical Review Letters*, 49:245, 1982.

[Tak81]   F. Takens. Detecting strange attractors in turbulence. In D. A. Rand and L. S. Young, editors, *Lecture Notes in Mathematics*. Springer-Verlag, New York, 1981.

[Tho27]   L. H. Thomas. The circulation of atomic fields. *Proceedings of the Cambridge Philosophical Society*, 23:542, 1927.

[Tod89]   M. Toda. *Theory of Nonlinear Lattices*. Springer-Verlag, Berlin, New York, 2nd edition, 1989.

[Tor81]   S. Torvén. Modified Korteweg-de Vries equation for propagating double layers in plasmas. *Physical Review Letters*, 47:1053, 1981.

[Tor86]   S. Torvén. Weak double layers in a current carrying plasma. *Physica Scripta*, 33:262, 1986.

[Tys76]   J. J. Tyson. The Belousov–Zhabotinskii reaction. In *Lecture Notes in Biomathematics, Vol. 10*. Springer-Verlag, New York, 1976.

[Ued79]   Y. Ueda. Randomly transitional phenomena in systems governed by Duffing's equation. *Journal of Statistical Physics*, 20:181, 1979.

[Van26]   B. Van der Pol. On relaxation oscillations. *Philosophical Magazine*, 2:978, 1926.

[Ver87]   F. Verheest. Ion acoustic solitons at critical densities in multi-component plasmas with different ionic charges and temperatures. *International Conference on Plasma Physics*, 2:115, 1987.

[Ver90]   F. Verhulst. *Nonlinear Differential Equations and Dynamical Systems*. Springer-Verlag, Berlin, New York, 1990.

[Vol26]   V. Volterra. Variazionie fluttuazioni del numero d'individui in specie animali conviventi. *Mem. Acad. Lincei*, 2:31, 1926.

[VV28]   B. Van der Pol and J. Van der Mark. The heart beat considered
           as a relaxation oscillator and an electrical model of the heart.
           Philosophical Magazine, 6:763, 1928.

[Win72]   A. T. Winfree. Spiral waves of chemical activity. *Science*, 175:634,
           1972.

[Wol86]   S. Wolfram, editor. *Theory and Applications of Cellular Automata*.
           World Scientific, Singapore, 1986.

[Wol99a]   *Wolfram Research, Mathematica 4.0 Standard Add-on Packages*.
           Wolfram Media, Champaign,IL., 1999.

[Wol99b]   Stephen Wolfram. *The Mathematica Book*. Cambridge University
           Press, Cambridge, 1999.

[YK82]   W. J. Yeh and Y. H. Kao. Universal scaling and chaotic behavior
           of a Josephson junction analog. *Physical Review Letters*, 49:1888,
           1982.

[ZK65]   N. J. Zabusky and M. D. Kruskal. Interaction of "solitons" in a
           collisionless plasma and the recurrence of initial states. *Physical
           Review Letters*, 15:240, 1965.

[ZS72]   V. E. Zakharov and A. B. Shabat. Exact theory of two-dimensional
           self-focusing and one-dimensional self-modulation of waves in non-
           linear media. *Soviet Physics JETP*, 34:62, 1972.

[Zwi89]   D. Zwillinger. *Handbook of Differential Equations*. Academic
           Press, New York, 1989.

[ZZ70]   A. N. Zaikin and A. M Zhabotinskii. Concentration wave propa-
           gation in two-dimensional liquid-phase self-organizing system. *Na-
           ture*, 225:535, 1970.

Index

Also by the authors

Nonlinear Physics with MAPLE for Scientists and Engineers, 2nd edition, Birkhäuser Boston, 0-8176-4119-X, 2000

Computer Algebra Recipes: A Gourmet's Guide to the Mathematical Models of Science, Springer-Verlag, New York, 0-3879-5148-2, 2001

Reviewer comments on the Maple edition of
NONLINEAR PHYSICS

An...excellent book...the authors have been able to cover an extraordinary range of topics and hopefully excite a wide audience to investigate nonlinear phenomena...accessible to advanced undergraduates and yet challenging enough for graduate students and working scientists.... The reader is guided through it all with sound advice and humor.... I hope that many will adopt the text.

—American Journal of Physics

Its organization of subject matter, clarity of writing, and smooth integration of analytic and computational techniques put it among the very best... Richard Enns and George McGuire have written an excellent text for introductory nonlinear physics.

—Computers in Physics

...correctly balances a good treatment of nonlinear, but also nonchaotic, behavior of systems with some of the exciting findings about chaotic dynamics...one of the book's strength is the diverse selection of examples from mechanical, chemical, electronic, fluid and many other systems.... Another strength of the book is the diversity of approaches that the student   is encouraged to take...the authors have chosen well, and the trio of text...software, and lab manual gives the newcomer to nonlinear physics quite an effective set of tools.... Basic ideas are explained clearly and illustrated with many examples."

—Physics Today

... the care that the authors have taken to ensure that their text is as comprehensive, versatile, interactive, and student-friendly as possible place this book far above the average.

—Scientific Computing World